D0866178

TO CONVERT FROM	TO	MULTIPLY BY
poise	Pa · s	10^{-1}
microinch	micron	3.937×10
erg	J	10^{-7}
dyne	N	10^{-5}
micron	m	10^{-6}
angstrom	m	10^{-10}
ton (short)	lb	2.000×10^3
ton (long)	lb	2.240×10^3

TEMPERATURE CONVERSIONS

$°C = \frac{5}{9}(°F - 32)$

$°F = \frac{9}{5}°C + 32$

$K = °C + 273$

$°R = °F + 460$

MULTIPLICATION FACTORS

PREFIX	SYMBOL		PREFIX	SYMBOL	
tera	T	10^{12}	centi	c	10^{-2}
giga	G	10^{9}	milli	m	10^{-3}
mega	M	10^{6}	micro	μ	10^{-6}
kilo	k	10^{3}	nano	n	10^{-9}
deci	d	10^{-1}	pico	p	10^{-12}

MANUFACTURING PROCESSES FOR ENGINEERING MATERIALS

Serope Kalpakjian

Illinois Institute of Technology

Manufacturing Processes for Engineering Materials

 ADDISON-WESLEY PUBLISHING COMPANY

Reading, Massachusetts

Menlo Park, California

London

Amsterdam

Don Mills, Ontario

Sydney

Sponsoring Editor: Thomas Robbins
Production Manager: Martha K. Morong
Production Editor: Laura Skinger
Designer: Catherine L. Dorin
Design Coordinator: Patricia O. Williams
Cover Designer: T. A. Philbrook
Art Editors: Susanah H. Michener and Marcia S. Strykowski
Copy Editor: Deborah A. Shukis
Illustrator: Parkway Illustrated Press
Manufacturing Supervisor: Ann E. DeLacey

Library of Congress Cataloging in Publication Data

Kalpakjian, Serope,
    Manufacturing processes for engineering materials.

    Bibliography: p.
    Includes index.
    1. Manufacturing processes.   I. Title.
TS183.K34   1984   670   83-15456
ISBN 0-201-11690-1

Reprinted with corrections, July 1985

IJ—DO—898

To Jean, Claire, and Kent

Preface

Manufacturing is a broad activity comprising many subjects; among these are the behavior and properties of metallic and nonmetallic materials, metallurgy, stress analysis, chemistry, heat transfer, lubrication, industrial engineering, and economics. In view of these diverse topics, the proper method of teaching manufacturing processes at the undergraduate level has been a subject of considerable discussion and evaluation.

Traditionally, this subject has been taught in a descriptive and qualitative manner, covering both the principles of manufacturing processes and the equipment involved. However, this approach has often proved inadequate when judged against the more scientifically oriented engineering courses where the subject matter is analysed in greater detail. As a consequence, this area of study generally has not acquired the stature and significance that it rightly deserves. Professional societies, notably the Society of Manufacturing Engineers, have made great efforts in presenting to students the importance of manufacturing and the challenges that it offers as a professional activity. Regrettably, too many students still think of manufacturing as a dingy old forge shop or a machine shop. This situation has been particularly serious in the United States, unlike in other industrialized countries where manufacturing is indeed a prestigious field of study.

During the past few years, there has been a healthy trend toward a more quantitative approach in teaching manufacturing, without losing sight of the applied aspects of processing and equipment. This book is an attempt to present manufacturing processes with a proper balance of analysis and application. It aims to establish quantitative relationships among material properties and process variables, to describe materials used in engineering applications and the operations and equipment involved with them, and to discuss the economic and competitive aspects of manufacturing.

The text covers all basic manufacturing processes that have been traditionally taught in undergraduate curricula and also includes the latest developments. The analytical aspects are presented to the extent of establishing quantitative relationships, such that the student can recognize the important parameters influencing a particular process. Lengthy derivations of formulas are avoided; these are included in the Solutions Manual for the instructor. The text also includes illustrative problems, examples, and case studies as an aid to the student in understanding the subject matter.

Numerous illustrations and tables are provided to supplement the discussions in the text and to serve as reference material in classroom exercises. The book can also serve as a reference for the practicing engineer. The bibliography at the end of each chapter contains books and periodicals. The list is more comprehensive than is usually found in similar textbooks. A set of problems and questions is included in each chapter. The glossary at the end of the book includes words and terms that are commonly used in materials and manufacturing processes, and it is designed as a quick reference for the benefit of students.

Although introductory, the book is comprehensive in its coverage of manufacturing processes and is designed such that certain sections or chapters can be deleted at the discretion of the instructor. The text is suitable for mechanical, metallurgical, and industrial engineering curricula, depending on where such courses are taught.

It is hoped that, in reading this text, the student will appreciate the importance of manufacturing processes, and that, with the approach taken in this book, he or she will find the subject matter as exciting and challenging as any other engineering discipline.

ACKNOWLEDGMENTS

I am grateful to a great number of people who so kindly helped me on many aspects in the preparation of this book.

It is a pleasure to express my deep gratitude to two eminent authorities who have had a great influence on my professional life. Milton C. Shaw was my first instructor and adviser in materials processing research while I was a graduate student at MIT. This was followed by research in metalworking processes at Cincinnati Milacron Inc., under the guidance of M. Eugene Merchant. It has been a unique privilege to be associated with two pioneers of manufacturing engineering science, who continue to be such helpful colleagues.

I would like to thank my colleagues K. J. Weinmann, Michigan Technological University; S. Malkin, Technion—Israel Institute of Technology; B. F. von Turkovich, University of Vermont; S. Kobayashi, University of California, Berkeley; and S. Ramalingam, University of Minnesota for their support during the early stages of this book.

I am grateful to my colleagues at various institutions for their critical review of individual chapters: B. F. von Turkovich, University of Vermont; R. Komanduri, General Electric Co.; B. S. Levy, Inland Steel Corp.; B. S. Thakkar, Bell Laboratories; S. Rajagopal, IIT Research Institute; K. L. Johnson, FMP Corp.; S. Misra, National Can Co.; E. S. Nachtman, Tower Oil & Technology Co.; K. M. Kulkarni, Cabot Corp.; W. Wilson, A. Finkl & Sons; and E. L. Kotzin, American Foundrymens' Society.

I wish to acknowledge the help of my associates at IIT who read all or parts of this book critically and offered many helpful suggestions: K. P. Meade, S. Sreenivasan, A. Cowie, P. D. L. Rao, A. Bhatnagar, V. Aronov, R. C. Dix, S. Mostovoy, S. Kumar, V. Richards, E. Johnson, and my students H. K. Tseng and K. Carlson.

Many thanks to my colleague S. Mostovoy and to my son, Kent, for their efforts in trying to teach me the mysteries of word processors, to N. N. Breyer for his help in locating various sources of information, to R. C. Daehn (Packer Engineering Associates Inc.) and H. M. Nagib for their help with photography, to P. H. Francis for his encouragement, and to many colleagues and organizations who supplied various figures and gave permission for their reproduction. The figures for chapter openers were courtesy of Ford Motor Co. (Ch. 1), Bundesanstalt für Materialprüfung (3 and 4), General Motors Corp. (5), American Iron and Steel Institute (6), Vollrath Co. (7), Cincinnati Milacron Inc. (8 and 13), American Metal Powder Industries Federation (10), and Chicago Bridge and Iron Co. (12).

It is a pleasure to acknowledge the assistance of the able staff at Addison-Wesley Publishing Company in all aspects of producing this book. Many thanks to my editor Thomas Robbins, and to Laura R. Skinger, Susanah H. Michener, Patricia O. Williams, and Deborah A. Shukis. It was a joy to work with such a knowledgeable and helpful team; no author could have asked for more.

Finally, I would like to thank my family for their help, understanding, and patience during the three years that it took to write and produce this book.

Chicago, Illinois   S.K.
March 1984

Contents

3
Manufacturing Properties of Metals and Alloys   97

4
Surfaces: Measurement, Properties, and Treatment   177

5
Casting Processes   239

6
Bulk Deformation Processes   287

7

Sheet-Metal Forming Processes   389

8

Material-Removal Processes: Cutting   461

9

Material-Removal Processes: Abrasive, Chemical, Electrical, and High-Energy Beams 555

10

Plastics: Properties and Processing Methods 595

11
Processing of Powder Metals and Ceramics   647

12
Fastening and Joining Methods   695

13
Manufacturing Economical Products   753

1 Introduction to Manufacturing Processes

1.1 MANUFACTURING

The word manufacturing is derived from the Latin *manu factus*, meaning made by hand. Manufacturing is making useful products from raw materials by various processes, following a well-organized plan for all aspects involved.

The importance of manufacturing can be best appreciated by observing that with few exceptions, all objects around us had a different shape at one time. Furthermore, every object is made of one or more materials and is shaped into a product by any of a wide variety of processing methods. Therefore it is not surprising that in industrialized countries manufacturing comprises approximately one third of the gross national product (i.e., the value of all goods and services produced) and two thirds of the wealth producing activities.

Manufacturing processes date back to the earliest civilizations with the production of various articles made of wood, ceramics, stone, and metal (Table 1.1). The processes first employed have been developed gradually on different continents, using newer materials and at increasing rates of production. The greatest advances have been made beginning in the 1800s. Today, a wide variety of materials and manufacturing methods are available for products ranging from single-component parts, such as steel balls 0.010 in. (0.25 mm) in diameter, to highly sophisticated products, such as computers and supersonic aircraft.

Manufacturing even a simple product requires the consideration of many factors. For example, a paper clip must be designed to meet the functional requirement of holding pieces of paper together with sufficient clamping force so the papers will not slip away from one another. They are usually made of steel wire, although plastic ones are also available. The required length of wire is cut from a longer piece and is bent a number of times to give it the proper shape. The wire itself is made by a process called wire drawing. This process reduces the cross-sectional area of a long rod through dies and also imparts certain mechanical properties to the material, such as strength and hardness. The rod, in turn, is made by processes such as drawing and extrusion from a round billet.

The manufacturing of a paper clip thus involves design, selecting a suitable material, and selecting manufacturing methods to meet service requirements. These selections are made not only on the basis of technical requirements, but also on economic considerations—to minimize the costs involved so the product will be competitive in the marketplace.

The design process for a part involves a clear understanding of the *functions* and the *performance* expected from that part (Fig. 1.1). The product designer, or the design engineer, specifies the shape and dimensions of the product, its appearance, and the materials to be used. Prototypes of the product are made first. It is possible at this stage that modifications will have to be made, either in the original design or in the materials selected, if difficulties arise in making the product economically. Appropriate manufacturing methods are then chosen by the manufacturing engineer.

Selection of material requires a knowledge of the *function* and *service requirements* of the product, and the materials that are available to fulfill these requirements. The

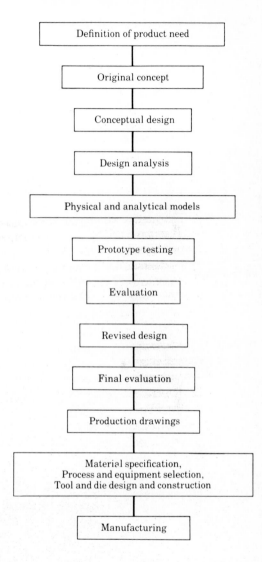

FIGURE 1.1 Chart showing the procedure required in the design and manufacturing of a component. Depending on the complexity of the product, the span of time between the original concept and the marketing of a product may range from a few months to a few years.

material selected for a paper clip should have a certain amount of elasticity and strength. For example, if the modulus of elasticity is too high, a great deal of force will be required to open the clip. If the modulus is too low, the clip will not exert sufficient clamping force. Also, if the yield stress of the wire material is too low, the clip will bend permanently and will be difficult to reuse. These factors also depend on the diameter of the wire and the design of the clip.

The selection of the material also involves consideration of cost, appearance, surface texture, and corrosion resistance, and of the manufacturing processes involved. Many questions must be properly answered. Will the chosen material be able

TABLE 1.1
HISTORICAL DEVELOPMENT OF MATERIALS AND MANUFACTURING PROCESSES (DATES ARE APPROXIMATE). (After J. A. Schey, C. S. Smith, R. F. Tylecote, T. K. Derry, T. I. Williams and S. Kalpakjian)

	PERIOD	METALS AND CASTING	FORMING PROCESSES
Egypt: ~3100 BC to ~300 BC Greece: ~1100 BC to ~146 BC Roman empire: ~500 BC to AD 476 Middle ages: ~476 to 1492 Renaissance: 14th to 16th centuries	Before 4000 BC	Gold, copper, meteoritic iron	Hammering
	4000–3000 BC	Copper casting, stone and metal molds, lost wax process, silver, lead, tin, bronze	Stamping, jewelry
	3000–2000 BC	Bronze casting	Wire by cutting sheet and drawing; gold leaf
	2000–1000 BC	Wrought iron, brass	
	1000–1 BC	Cast iron, cast steel	Stamping of coins
	AD 1–1000	Zinc, steel	Armor, coining, forging, steel swords
	1000–1500	Blast furnace, type metals, casting of bells, pewter	Wire drawing, gold and silver smith work
	1500–1600	Cast iron cannon, tinplate	Water power for metalworking, rolling mill for coinage strips
Industrial revolution: ~1750 to 1850	1600–1700	Permanent mold casting, brass from copper and metallic zinc	Rolling (lead, gold, silver), shape rolling (lead)
	1700–1800	Malleable cast iron, crucible steel	Extrusion (lead pipe), deep drawing, rolling (iron bars and rods)
	1800–1900	Centrifugal casting, Bessemer process, electrolytic aluminum, nickel steels, babbitt, galvanized steel, powder metallurgy, tungsten steel, open-hearth steel	Steam hammer, steel rolling, seamless tube piercing, steel rail rolling, continuous rolling, electroplating
WWI	1900–1920		Tube rolling, hot extrusion
WWII	1920–1940	Die casting	Tungsten wire from powder
Space age	1940–1950	Lost wax for engineering parts	Extrusion (steel), swaging, powder metals for engineering parts
	1950–1960	Ceramic mold, nodular iron, semiconductors, continuous casting	Cold extrusion (steel), explosive forming, thermomechanical treatment
	1960–1970	Squeeze casting, single crystal turbine blades	Hydrostatic extrusion
	1970–1980s	Compacted graphite, vacuum casting, organically bonded sand, automation of molding and pouring, large aluminum castings for aircraft structures, rapid solidification technology	Precision forging, isothermal forging, superplastic forming, die design by analytical methods

TABLE 1.1 *(continued)*

JOINING PROCESSES	TOOLS, TOOL MATERIALS, AND MACHINING	CERAMICS, PLASTICS, ETC.
	Tools of stone, flint, wood, bone, ivory, composite tools	Earthenware, glazing, natural fibers
Soldering (Cu-Au, Cu-Pb, Pb-Sn)	Corundum	
Riveting, brazing	Hoe making, hammered axes, tools for ironmaking and carpentry	Glass beads, potter's wheel, glass vessels
Forge welding of iron and steel, gluing	Improved chisels, saws, files, woodworking lathes	Glass pressing and blowing
	Etching of armor	Venetian glass
	Sandpaper, windmill driven saw	Crystal glass
	Hand lathe (wood)	Cast plate glass, flint glass
	Boring, turning, screw cutting lathe, drill press	Porcelain
	Shaping, milling, copying lathe for gunstocks; turret lathe, universal milling machine, vitrified grinding wheel	Window glass from slit cylinder, vulcanization, rubber processing, polyester, styrene, celluloid, rubber extrusion, molding
Oxyacetylene; arc, electrical resistance, and thermit welding	Geared lathe, automatic screw machine, hobbing, high-speed steel tools, aluminum oxide and silicon carbide (synthetic)	Automatic bottle making, Bakelite
Coated electrodes	Tungsten carbide, mass production, transfer machines	Development of plastics, casting, molding, PVC, cellulose acetate, polyethylene
Submerged arc welding		Acrylics, synthetic, rubber, epoxies
TIG, MIG and electroslag welding, explosive welding	Electrical and chemical machining, automatic control	ABS, silicones, fluorocarbons polyurethane, float glass, tempered glass
Plasma arc and electron beam, adhesive bonding	Titanium carbide, synthetic diamond, numerical control	Acetals, polycarbonates, cold forming of plastics, reinforced plastics, filament winding
Laser beam, diffusion bonding (also combined with superplastic forming)	Cubic boron nitride, coated carbides, computer integrated manufacturing, adaptive control, industrial robots, flexible manufacturing systems, unmanned factory	Adhesives, composite materials, optical fibers, structural ceramics, ceramic components for automotive and aerospace engines

to undergo bending during the forming process without breaking? Can the wire be easily cut from a long piece? Will it cause excessive wear on the tooling? Will the cutting process produce a smooth edge or will it leave a burr on the wire, which is undesirable in a paper clip? Does the material have to be metallic? If it is made of steel, what is the optimal carbon content and which carbon steel, meeting all design and service requirements, is the least expensive?

1.2 INTERDISCIPLINARY NATURE OF MANUFACTURING PROCESSES

Manufacturing involves many disciplines. Among these are the mechanical and physical properties and the behavior of materials, physical and mechanical metallurgy, physics, plasticity, stress analysis, physical and organic chemistry, lubrication, and heat transfer. Other aspects of manufacturing involve product design, machine design and construction, planning for production, automation, computers, microprocessors, quality control, assembly, cost control, safety, productivity, and marketing.

Manufacturing is generally referred to as an *activity* because it involves many areas. It is apparent that cooperation between all parties involved is essential for a manufacturing operation to be successful and profitable. The design engineer, for instance, must be aware of the implications of specifying unnecessarily close tolerances, fine surface finish, or the type, quality, and quantity of the materials selected. Surveys have indicated that many products are overdesigned, presumably because of concern on the part of the designer to produce a safe product or the uncertainty of design calculations. Many designs are based on past experience and intuition, rather than on a truly scientific approach.

The capabilities and limitations of each manufacturing process must be recognized by the product designer. Brittle and hard materials, for instance, cannot be formed or machined easily. Materials with high melting temperatures or parts with very thin cross-sections cannot be cast properly. Alternative materials, currently available forms of materials, and all suitable methods of manufacturing must be investigated, always bearing in mind cost considerations.

1.2.1 EXAMPLES OF PRODUCTS

Examples of some common products and the major considerations involved in the selection of materials and processes in their manufacture are given below.

Bicycle
Consider the design, material selection, and processing methods of the major components of a bicycle (Fig. 1.2). The important factors are safety, strength, corrosion resistance, and appearance. Note that a number of materials have been selected for the various components, although these are mostly plain carbon steels because of their relatively low cost and sufficient strength. There are about a dozen manu-

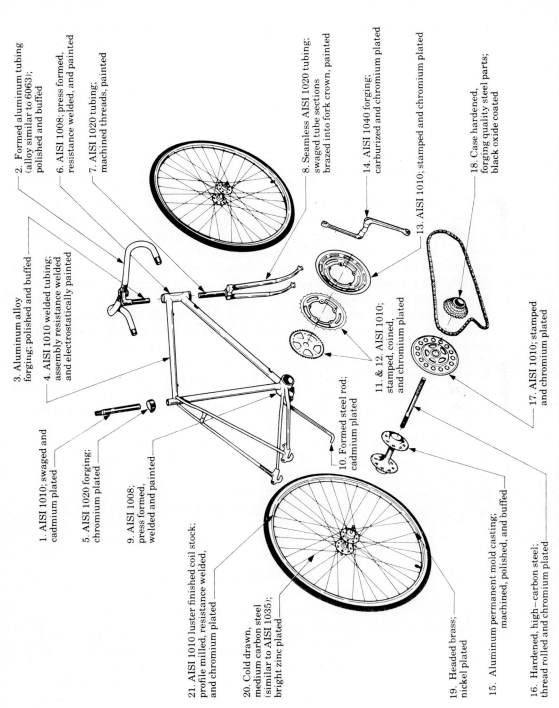

2. Formed aluminum tubing (alloy similar to 6063); polished and buffed

6. AISI 1008; press formed, resistance welded, and painted

7. AISI 1020 tubing; machined threads, painted

8. Seamless AISI 1020 tubing; swaged tube sections brazed into fork crown, painted

14. AISI 1040 forging; carburized and chromium plated

13. AISI 1010; stamped and chromium plated

18. Case hardened, forging quality steel parts; black oxide coated

3. Aluminum alloy forging; polished and buffed

4. AISI 1010 welded tubing; assembly resistance welded and electrostatically painted

1. AISI 1010; swaged and cadmium plated

5. AISI 1020 forging; chromium plated

9. AISI 1008; press formed, welded and painted

10. Formed steel rod; cadmium plated

11. & 12. AISI 1010; stamped, coined, and chromium plated

17. AISI 1010; stamped and chromium plated

21. AISI 1010 luster finished coil stock; profile milled, resistance welded, and chromium plated

20. Cold drawn, medium carbon steel (similar to AISI 1035); bright zinc plated

19. Headed brass; nickel plated

15. Aluminum permanent mold casting; machined, polished, and buffed

16. Hardened, high–carbon steel; thread rolled and chromium plated

FIGURE 1.2 Components of a bicycle (Schwinn Continental) showing materials and processes used in manufacturing this bicycle. Note that there are about a dozen materials and more than a dozen individual manufacturing processes involved. *Source: Metal Progress.* July 1973, p. 64, © American Society for Metals.

facturing processes involved in producing this bicycle, such as forging, machining, casting, welding, plating, and polishing. Although not shown in Fig. 1.2, each part is also specified by the design engineer with a range of dimensions, tolerances, and surface finish.

The handlebar (item 2 in Fig. 1.2) is made of aluminum tubing, which is shaped and then polished and buffed for appearance. There are many types of aluminum; the 6063 alloy was selected by the manufacturer as the optimal material for this application. The stem of the handlebar (item 3) is made of an aluminum forging. Malleable iron and aluminum castings were considered for this part, but these materials were unacceptable because of weight and safety reasons. Handlebars tend to be misused; the bicycle may be dropped on a hard surface or somebody may sit on the handlebar during the ride.

The stem and the front fork (item 8) on a bicycle are regarded by the designers as the most important components of a bicycle from a safety standpoint. These highly stressed parts must be designed and the materials selected carefully while keeping the weight down. For light racing bicycles, just as for aircraft and aerospace applications, strength-to-weight and stiffness-to-weight ratios are important factors in material selection.

The spokes of the wheels of the bicycle are made of steel wire with a specified minimum tensile strength of about 150,000 psi (1000 MPa). They are bright zinc electroplated for corrosion resistance and appearance.

Vacuum Cleaner

The major components of the lower sections of a household vacuum cleaner are shown in Fig. 1.3. The selection of materials (steel, aluminum, plastics, rubber) and processes (stamping, casting, extruding, molding, plating) are based on criteria similar to those concerning the bicycle example described above.

Refrigerator Door Liner

In the selection of materials for a door liner for refrigerators (where eggs, butter, salad dressings, small bottles, etc., are stored), in addition to the temperature range to which the liner is subjected, the following factors are taken into consideration:

Mechanical Requirements: strength, toughness (impact, door slamming, shelving), stiffness, resilience, and resistance to scratching and wear.

Physical Requirements: thermal properties, dimensional stability, and electrical insulation.

Chemical Requirements: resistance to staining, odor, chemical reactions with food and beverages, and cleaning fluids.

Appearance: color, stability of color, surface finish, texture, and feel.

Manufacturing Properties: methods of manufacturing and assembly, effects of processing on material properties and behavior over a period of time, compatibility with other components in the door, and cost of materials and manufacturing.

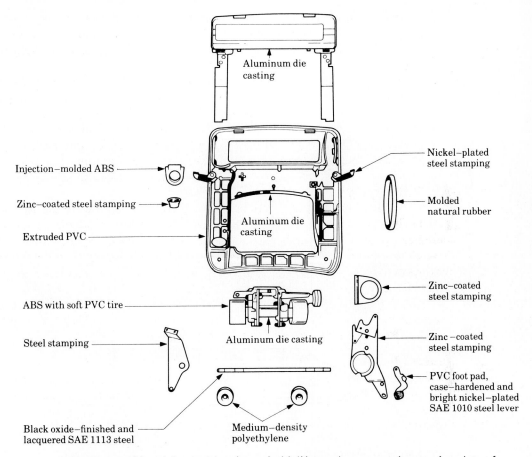

FIGURE 1.3 Materials used in a household (Hoover) vacuum cleaner. A variety of metallic and nonmetallic materials is used in most products today. Materials are selected on the basis of various technical and economic criteria.

In one extensive study concerning this subject, it was shown that, considering all the factors involved, the candidate materials for door liners were two plastics: ABS (acrylonitrile–butadiene–styrene) and HIPS (high-impact polystyrene). One aspect of the study was the effect of vegetable oils (e.g., salad dressing stored in the door shelf)) on the strength of these plastics. It was found that the load-bearing capacity of HIPS was significantly reduced in the presence of vegetable oils. HIPS becomes brittle in the presence of oils (stress cracking), whereas ABS is not affected to any significant extent.

Space Shuttle Orbiter

During ascent and reentry, the space shuttle is subjected to a wide range of temperatures, as shown in Fig. 1.4. To protect the surfaces of the craft against high

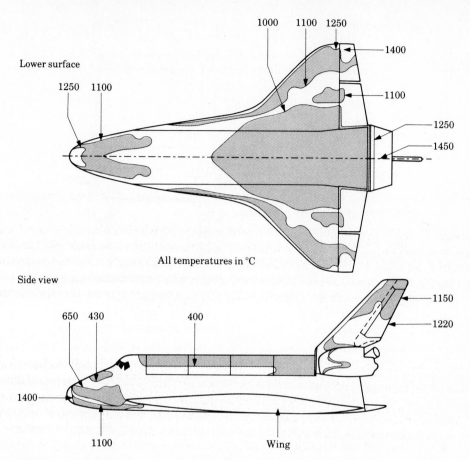

FIGURE 1.4 Approximate temperature distribution on the space shuttle orbiter during ascent and reentry. Each shuttle has over 34,000 ceramic silica tiles, each with a unique geometry and bonded into an individual nesting place on the aluminum structure. The bonding material consists of several layers of silicone-based adhesives.

temperatures, two design approaches were considered. One is the use of ablative materials. Ablation is the process of extracting heat from a surface by allowing the surface to erode by melting and evaporation at high temperatures. However, this means that material is lost from the surface of the craft and that certain parts have to be replaced after each space mission.

The second approach (the one selected) is to use materials that can withstand high temperatures so that parts need not be replaced after each mission. Because most materials undergo severe oxidation at these temperatures, carbon-reinforced composite materials were selected with a silicon carbide–based coating. Silicon carbide is a hard refractory ceramic with a melting point of about 5000°F (2750°C) and is therefore suitable for this application.

Cookware

Cooking pots and pans should conduct heat evenly and rapidly. Copper is an excellent material for this application, but it is more expensive than other suitable materials. Aluminum is also a good conductor of heat. However, it is corroded by acids and develops small pores on its surface where dirt and bacteria can gather, thus affecting the flavor of the food. Carbon steel is another candidate material; however, it tarnishes and corrodes. Stainless steel has good corrosion resistance and is easy to clean because of its smooth surface. Also, it has good forming characteristics. However, stainless steel does not conduct heat as well as the other materials.

One solution, where cost is minimized, is the development of a three-ply construction. A core of carbon steel is sandwiched between two layers of stainless steel. The outside is buffed to a fine finish. The inner cooking surface has a coarser finish so that scratches made during cooking or cleaning are not readily visible.

The handles and knobs of pots and pans are made of thermosetting resins. These plastics are suitable materials because they have sufficient heat resistance and low thermal conductivity, and they can be easily molded into any shape. The handles are attached by studs and brackets that are welded to the utensils. Screws or rivets are usually not used, as they tend to become loose over a period of time and collect dirt or food particles that may be difficult to clean.

Coins

There are five general criteria in the selection of materials for coins.

The first criterion is subjective and involves factors such as the appearance of the coin, its color, weight, and ring (the sound made when striking). Also included in this criterion is the feel of the coin. This term is difficult to describe as it combines many factors. It is similar to the feel of a fine piece of wood, polished stone, or fine leather.

The second is the life of the coin. This involves resistance to corrosion and wear while the coin is in circulation. These two factors basically determine how long the surface imprint of the coin will be visible and also the ability of the coin to retain its original luster.

The third criterion involves the manufacturing of the coin. This includes factors such as the formability of the candidate coin materials, life of the dies used in the coining operation, and the capability of the materials and processes to resist counterfeiting.

The fourth criterion concerns the suitability of the coin for use in coin-operated devices such as vending machines, turnstiles, and pay telephones. These machines are generally equipped with detection devices that first test the coins for proper diameter, thickness, and surface condition and, secondly, for electrical conductivity and density. The coin is rejected if it fails these tests.

The final criterion is the cost of the raw materials, processing costs, and sufficient supply of the coin material.

Electric Drill

The various components of a common type of hand-held electric drill are shown in Fig. 1.5. Based on considerations discussed in the above examples, you may inspect

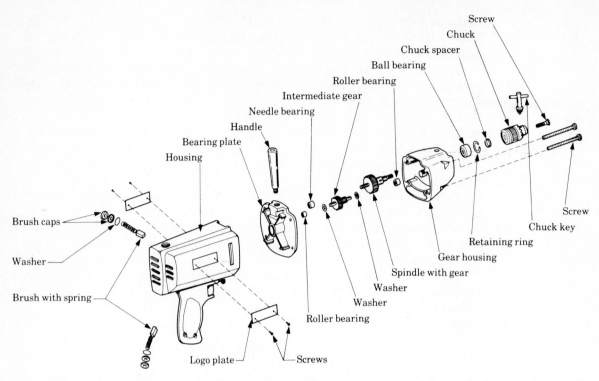

FIGURE 1.5 Components of an electric drill. Note that, in addition to the electric motor and the chuck, there are over 30 individual components, each one requiring an appropriate material and an economic process to manufacture it.

this drawing and make suggestions as to what kind of materials might be suitable for the various parts. In considering each part, it is important to understand clearly the functions of that part and how it relates to the rest of the product.

From the foregoing examples, you can see that manufacturing involves a large number of considerations, each one requiring a different technical discipline.

1.2.2 EXAMPLES OF PART FAILURE

It is possible that, in spite of all the efforts made to select appropriate materials and manufacturing methods, some key element may be left out in the overall consideration. Consider, for example, the following case history.

Bolt Failure in a Missile
A series of bolts on the thrust control valves of a missile failed. They were made of stainless steel. Fracture of the bolts was intergranular, i.e., the fracture path was

along the grain boundaries of the metal. There was no evidence of grinding, burning, or mechanical damage due to the manufacturing processes used, or of any metallurgical defects.

It was observed that the failed bolt was located in the vicinity of an aluminum-alloy flange that had corroded. It became apparent that the presence of aluminum accelerated a failure process, known as stress-corrosion cracking, of the stainless steel bolt. (Aluminum is anodic to stainless steel, hence it can be used as the sacrificial anode in restricting the corrosion of the steel, which is the cathode. Similarly, the zinc coating acts as the sacrificial anode in restricting the corrosion of galvanized steel sheet.) However, stress-corrosion was not the main cause of failure of the bolt. The only possible explanation was hydrogen evolving at the bolt during corrosion. This caused the strength and toughness of the steel bolt to drop significantly, a phenomenon known as hydrogen embrittlement of steel.

The method suggested for eliminating such failure was to improve the toughness and resistance to hydrogen embrittlement of the bolt. The procedure was to heat treat (age) the bolt for an additional one hour at 1000°F (540°C).

Another method suggested was to electrically insulate the bolt from the aluminum alloy flange to stop galvanic cell reaction. When graphite in grease was applied at the stainless steel–aluminum interface, there were no failures. The grease acted as an insulator and eliminated cell reaction and corrosion. Nickel–cadmium plating was also tried, but it gave inadequate protection because the plating was porous and allowed the diffusion of hydrogen into the steel bolt.

Numerous examples can be found in the technical literature describing a wide variety of failures. A component or an assembly of parts is generally considered to have failed when:

a. It stops functioning, e.g., a broken shaft,

b. It does not perform its function properly, e.g., a worn bearing, or

c. It becomes unsafe or unreliable for further use, e.g., a frayed cable in a winch or a surface fatigue crack on a shaft.

Failure of a component or structure may be caused by one or more of the following reasons:

a. Design Deficiencies: improper cross-sections, clearances, location of holes, incorrect estimate of loads, temperatures.

b. Improper Selection of Materials: failure to recognize environmental effects, compatibility with other components in the system, the required mechanical and physical properties.

c. Defects in Materials: flaws, cracks, discontinuities, nonuniformity of properties.

d. Defect Due to Processing of the Material: improper practices in heat treatment, welding, machining, grinding.

e. Improper Assembly of the Components (particularly for moving members): misalignment, improper torque in tightening nuts or bolts, incorrect selection and use of fasteners.

f. Improper Use of the Product: incorrect start-up procedures, excessive forces or impact, inadequate maintenance.

1.3 CRITERIA FOR MATERIAL SELECTION

It is evident that there are many factors that must be considered in material selection for processing and use. These requirements are outlined below.

General Requirements

Mechanical Properties: strength, ductility, hardness, elasticity, toughness, fatigue, creep; also, strength-to-weight and stiffness-to-weight ratios, particularly for aerospace and automotive applications.

Physical Properties: density; thermal expansion and conductivity; melting point; electrical, optical, and magnetic properties.

Chemical Properties: oxidation, corrosion, general degradation, toxicity, flammability.

Appearance: color, feel, surface texture.

Cost and Availability of Raw Materials

Factors to Be Considered

Available Forms of Materials: dimensions, tolerances, surface texture.

Manufacturing Properties of Materials: can they, with relative ease, be formed, machined, cast, welded, surface treated, heat treated, etc.?

Effect of Processing on Final Properties of Materials: what are the beneficial and adverse effects of processing these materials on their properties and service life?

Time- and Service-Dependent Effects: wear, fatigue failure (mechanical and thermal), creep, environmental effects on mechanical and physical properties, corrosion, dimensional stability, compatibility with other materials, consequence of part failure, safety.

Reliability of Supply of the Materials and its Effect on Costs: for instance, U.S. reliance on imported raw materials (Fig. 1.6). Other countries have similar reliance on various raw materials.

Costs of Processing by Different Methods

Recycling and Disposal of the Materials: after their useful service life.

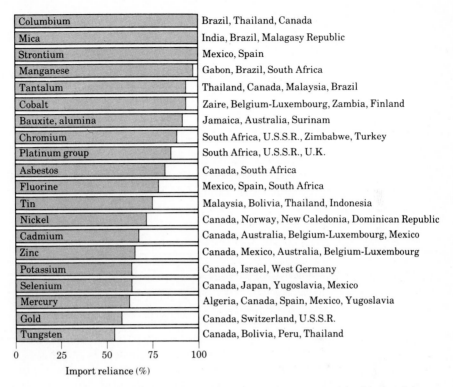

Material	Countries
Columbium	Brazil, Thailand, Canada
Mica	India, Brazil, Malagasy Republic
Strontium	Mexico, Spain
Manganese	Gabon, Brazil, South Africa
Tantalum	Thailand, Canada, Malaysia, Brazil
Cobalt	Zaire, Belgium-Luxembourg, Zambia, Finland
Bauxite, alumina	Jamaica, Australia, Surinam
Chromium	South Africa, U.S.S.R., Zimbabwe, Turkey
Platinum group	South Africa, U.S.S.R., U.K.
Asbestos	Canada, South Africa
Fluorine	Mexico, Spain, South Africa
Tin	Malaysia, Bolivia, Thailand, Indonesia
Nickel	Canada, Norway, New Caledonia, Dominican Republic
Cadmium	Canada, Australia, Belgium-Luxembourg, Mexico
Zinc	Canada, Mexico, Australia, Belgium-Luxembourg
Potassium	Canada, Israel, West Germany
Selenium	Canada, Japan, Yugoslavia, Mexico
Mercury	Algeria, Canada, Spain, Mexico, Yugoslavia
Gold	Canada, Switzerland, U.S.S.R.
Tungsten	Canada, Bolivia, Peru, Thailand

Import reliance (%)

FIGURE 1.6 The United States' reliance on imported raw materials. Many of these materials are essential in the production of tool and die materials, high-strength metals and alloys, and critical components of jet and rocket engines. Each country has a similar list, indicating its reliance on specific imported raw materials. Finding appropriate substitutes, either because of scarcity or economic reasons, is an important aspect of materials development. *Source*: U.S. Bureau of Mines, data adapted by E. Verink.

There exists an ever-growing variety of available materials. These range from traditional ferrous and nonferrous metals and alloys to ceramics, plastics, and composite materials. The following examples indicate the importance of the care required in material selection.

In order to save weight and reduce fuel consumption and thus increase the payload, airplanes are built of materials that have a high strength-to-weight ratio. The structural materials used in a Boeing 747, for example, are 82% aluminum, 13% steel, 4% titanium, and 1% fiberglass. The Boeing 757, a later model, uses similar amounts of metal but 3% composite materials, which includes 0.5% fiberglass.

Composites are generally epoxy resins reinforced with graphite or glass fibers. They have a high strength-to-weight ratio. However, their cost is approximately ten times that for metals. Also, they are made largely by hand so production costs are presently high. Considering fuel costs, overall economics is such that, in spite of these drawbacks, the trend is to use higher percentages of composites in airplanes. In fact, small aircraft are now being made of nonmetallic materials. Graphite fiber–reinforced plastics are also used in other applications, such as helicopter blades, turbojet fans, golf clubs, and tennis rackets.

The selection of appropriate materials in manufacturing is a complex problem. The long-range significance of the selections made may be inadvertently ignored during deliberations. For example, graphite fiber–reinforced materials, when accidentally or purposely incinerated, can release sufficient fibers into the atmosphere to short out electrical equipment. Military authorities have been concerned with the effects on electrical equipment of a cloud of graphite fibers deliberately or accidentally set on fire and released into the atmosphere.

1.4 CRITERIA FOR PROCESS SELECTION

In manufacturing processes it is often found that there is more than one method of manufacturing a part. The selection of a particular process over others depends on a large number of factors.

Consider, for instance, the following example of forming a simple dish-shaped part from sheet metal (Fig. 1.7a). Such a part can be formed by placing a flat piece of sheet metal between a pair of male and female dies and closing the dies by applying a force on the upper one. A large number of parts can be formed at high production rates by this method, known as stamping or pressworking. The dies are activated by a machine, which can be either a mechanical or a hydraulic press.

Assume now that the size of the part is very large, e.g., 80 in. (2 m) in diameter. Further assume that only 50 of these parts are required. In response to these constraints, one has to reconsider the total operation and ask a number of questions. Is it economical to manufacture a set of dies 80 in. in diameter (which are very costly) when the production run (quantity of production) is so low? Are machines available, with sufficient capacity, that can accommodate such large dies? What are the alternative methods of manufacturing this part? Does it have to be made in one piece?

This part can also be made by welding together smaller pieces of metal formed by other methods. (Ships and water tanks are made by this method.) Would a part manufactured by welding be acceptable for its intended purpose? Will it have the required properties and the correct shape after welding, or will it require additional processing?

The same part can be made by a spinning process similar to making ceramic dishes and pottery on a potter's wheel. It can also be made by the explosive forming process shown in Fig. 1.7(b). In this method, the male die is replaced by water and the shock wave from the explosion generates sufficiently high pressure to form the part.

In comparing the two processes in Fig. 1.7, one notes that, in the explosive forming process, the deformation of the material takes place at a very high rate. In selecting

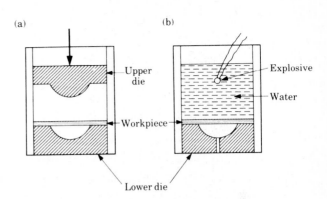

FIGURE 1.7 Two methods of forming a dish-shaped part from sheet metal. (a) Conventional mechanical or hydraulic press, using a pair of male and female dies. (b) Explosive forming method using only one die. In manufacturing operations, choice of the process for a particular application depends on a wide variety of technical and economic considerations.

this process over conventional pressworking, one has to consider various factors. For instance, is the material capable of undergoing deformation at high rates without fracture? Does the high rate have any detrimental effect on the final properties of the formed part? Can tolerances be held within acceptable limits? Is the life of the die sufficiently long under the high transient pressures generated in this process? Can this operation be performed in a manufacturing plant in a city or should it be carried out in open country? Although having only one die is an advantage, is the overall operation economical?

Many questions of this nature must be asked prior to the selection of one process over others. In this particular example, it was noted that there are a number of processes that could be employed to make this part. This is true for almost all manufactured parts and components (Table 1.2). Furthermore, the final product is usually a result of many different processes. For example, the steel wire in the example of the paper clip discussed in Section 1.1 is a product that has been subjected to many different processes, starting with raw materials, as shown in Fig. 1.8.

Process selection generally depends on the following factors:

Type of material and its properties;

Final desired properties with respect to part performance;

Size, shape, thickness and complexity of the part;

Tolerances and surface finish requirements;

Subsequent processing involved;

The design and cost of tooling, the effect of workpiece material on tool and die life;

Scrap generated and its value;

Availability of equipment and operating experience in the plant;

Lead time required to initiate production;

Number of parts required and the desired production rate;

Overall cost of processing.

TABLE 1.2
COMMON METHODS OF MANUFACTURING VARIOUS PRODUCTS

Advertising signs (lighted)	Thermoforming
Aircraft skin panels	Stretch forming
	Peen forming
Beverage cans	Deep drawing, ironing
	Roll forming, welding, soldering or adhesive bonding
Bolts and screws	Thread rolling
	Machining
Bowls (metal)	Deep drawing
	Spinning
Coins	Roll bonding, punching, coining
Crankshafts	Forging, machining, grinding
	Casting, machining, grinding
Engine blocks	Casting, machining
Footballs (plastic)	Rotational molding
Gears, cams	Casting
	Forging
	Powder metals
	Molding
Hose	Extrusion
Laminated sheet metal	Adhesive bonding
	Roll bonding
	Diffusion bonding
	Brazing
Musical strings	Wire drawing
Nails	Wire drawing, heading
Picture frames (metal)	Roll forming
Plastic beverage bottles	Blow molding
Pressure vessels	Welding
	Filament winding
Railroad wheels	Casting
	Forging
Rocket nose cones	Spinning
Trash cans (plastic)	Rotational molding
Trays for candy (plastic)	Thermoforming
Turbine blades and vanes	Forging, machining
	Casting, machining
Vase (ceramic)	Slip casting, drying, firing

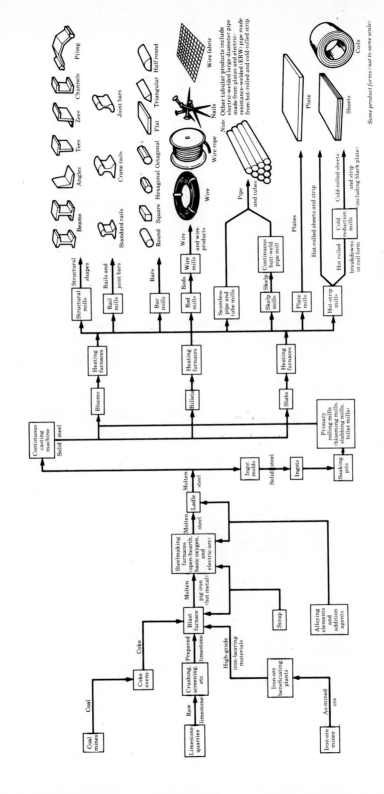

FIGURE 1.8 Chart showing the conversion of raw materials into different shapes of steel. Although iron and steel were first cast during the period of 1000–1 BC., steelmaking did not begin until about the 14th century. *Source: The Making, Shaping, and Treating of Steel*, 9th ed., copyright © 1971 by United States Steel Corporation.

SUMMARY

It is evident that manufacturing a product, whether it is a light bulb, a typewriter, or an automobile, requires a comprehensive approach involving design, materials, and processing. It is also evident that close cooperation among the product designer, materials engineer, and the manufacturing engineer is essential for manufacturing a product that functions well and is competitive in the marketplace. The more complicated the product, the greater the need for coordination and communication among the parties involved. In view of the economic significance and competitive aspects in manufacturing today, the need for thorough assessment of the advantages and limitations of each material and each process is more important than ever.

The rest of this book describes in detail the properties and behavior of materials and principles of all manufacturing methods, including economics of manufacturing.

BIBLIOGRAPHY

General References

Alexander JM, Brewer RC. *Manufacturing Properties of Materials.* London: Van Nostrand, 1963.

Alting L. *Manufacturing Engineering Processes.* New York: Marcel Dekker, 1982.

Amstead BH, *et al. Manufacturing Processes,* 7th ed. New York: Wiley, 1977.

Bolz RW. *Production Processes,* 5th ed. New York: Industrial Press, 1981.

Datsko J. *Material Properties and Manufacturing Processes.* New York: Wiley, 1966.

DeGarmo EP. *Materials and Processes in Manufacturing,* 5th ed. New York: Macmillan, 1979.

Dieter GE. *Engineering Design: A Materials and Processing Approach.* New York: McGraw-Hill, 1983.

Doyle LE. *et al. Manufacturing Processes and Materials for Engineers,* 2d ed. Englewood Cliffs, N.J.: Prentice-Hall, 1961.

Edgar C. *Fundamentals of Manufacturing Processes and Materials.* Reading, Mass.: Addison–Wesley, 1965.

Farag MM. *Materials and Process Selection in Engineering.* London: Applied Science Publishers, 1979.

Hanley DP. *Introduction to the Selection of Engineering Materials.* New York: Van Nostrand Reinhold, 1980.

Lindberg RA. *Processes and Materials of Manufacture,* 3d ed. Boston: Allyn and Bacon, 1983.

Moore HD, Kibbey DR. *Manufacturing: Materials and Processes.* Columbus, Ohio: Grid, 1975.

Myers JG, *et al. Energy Consumption in Manufacturing.* Cambridge, Mass.: Ballinger, 1974.

Niebel BW, Draper AB. *Product Design and Process Engineering.* New York: McGraw-Hill, 1974.

Patton WJ. *Modern Manufacturing Processes and Engineering*. Englewood Cliffs, N.J.: Prentice-Hall, 1970.

Pollack HW. *Manufacturing and Machine Tool Operations*, 2d ed. Englewood Cliffs, N.J.: Prentice-Hall, 1979.

Radford JD, Richardson DB. *Production Engineering Technology*, 2d ed. London: Macmillan, 1974.

Rhine CR. *Machine Tools and Processes for Engineers*. New York: McGraw-Hill, 1971.

Schey JA. *Introduction to Manufacturing Processes*. New York: McGraw-Hill, 1977.

Stokes VL. *Manufacturing Processes*. Columbus, Ohio: Merrill, 1975.

Thomas GG. *Production Technology*. New York: Oxford, 1970.

Trucks HE. *Designing for Economical Production*. Dearborn, Mich.: Society of Manufacturing Engineers, 1974.

Yankee H. *Manufacturing Processes*. Englewood Cliffs, N.J.: Prentice-Hall, 1979.

References

The Encyclopedia of Engineering Materials and Processes. New York: Reinhold, 1963.

Kirk-Othmer. *Encyclopedia of Chemical Technology*, 3d ed. New York: Wiley, 1978.

Machinery's Handbook, 21st ed. New York: Industrial Press, 1980.

Manufacturing Engineers' Manual. New York: McGraw-Hill, 1971.

Standard Handbook for Mechanical Engineers, 8th ed. New York: McGraw-Hill, 1978.

Tool and Manufacturing Engineers Handbook, 3d ed. New York: McGraw-Hill, 1976.

Conference Proceedings

Annals of the International Institution for Production Engineering Research (CIRP), Paris, annual.

Colwell LV, *et al.*, eds. *International Conference: Manufacturing Technology*. Dearborn, Mich.: Society of Manufacturing Engineers, 1967.

Proceedings of the North American Manufacturing Research Conference, annual since 1973. Dearborn, Mich.: Society of Manufacturing Engineers,

Shaw MC, *et al.*, eds. *International Research in Production Engineering*, American Society of Mechanical Engineers, New York, 1963.

Tobias SA, *et al.*, eds. *Advances in Machine Tool Design and Research*, annual since 1960.

Selected Periodicals in Manufacturing and Related Subjects (*See also Bibliography in Each Chapter*)

American Machinist

Cutting Tool Engineering

Foundry Management and Technology

International Journal of Machine Tool Design and Research

International Journal of Production Research

Journal of Applied Metalworking

Journal of Engineering for Industry

Journal of Engineering Materials and Technology

Journal of Manufacturing Systems

Journal of Mechanical Working Technology

Lubrication Technology

Machine and Tool Blue Book

Machinery

Manufacturing Engineering

Manufacturing Engineering Transactions

Materials Engineering

Materials Science and Engineering

Metal Progress

Modern Machine Shop

Modern Plastics

Plastics Engineering

Plastics Technology

Precision Engineering

Production

Production Engineer

SAMPE Quarterly

Sheet Metal Industries

Tribology International

Wear

Welding Design and Fabrication

Wire Journal International

Abstracts

Applied Mechanics Reviews

Applied Science and Technology Index

Engineering Index

Metals Abstracts

Information Centers

Machinability Data Center, Metcut Research Associates, Inc., 3980 Rosslyn Dr., Cincinnati, OH, 45209

Mechanical Properties Data Center, Battelle Columbus Laboratories, Columbus, OH, 43201

Metals and Ceramics Information Center, Battelle Columbus Laboratories, Columbus, OH, 43201

Professional Associations

Abrasive Engineering Society (AES), 1700 Painters Run Rd., Pittsburgh, PA, 15243

The Aluminum Association (AA), 818 Connecticut Ave. N.W., Washington, DC, 20006

American Ceramic Society (ACS), 65 Ceramic Dr., Columbus, OH, 43214

American Die Casting Institute (ADCl), 2340 Des Plaines Ave., Des Plaines, IL, 60018

American Foundrymen's Society (AFS), Golf and Wolf Rds., Des Plaines, IL, 60016

American Gear Manufacturers Association (AGMA), 1901 N. Ft. Myer Dr., Arlington, VA, 22209

American Institute of Mining, Metallurgical and Petroleum Engineers (AIME), 345 E. 47th St., New York, NY, 10017

American Iron and Steel Institute (AISI), 1000 16th St., N.W., Washington, DC, 20036

American National Standards Institute (ANSI), 1430 Broadway, New York, NY, 10018

American Powder Metallurgy Institute (APMI), 105 College Rd. E., Princeton, NJ, 08540

American Society for Metals (ASM), Metals Park, OH, 44073

American Society for Nondestructive Testing (ASNT), 4135 Arlington Plaza, Columbus, OH, 43228

American Society for Quality Control (ASQC), 230 W. Wells St., Milwaukee, WI, 53203

American Society for Testing and Materials (ASTM), 1916 Race St., Philadelphia, PA, 19103

American Society of Lubrication Engineers (ASLE), 838 Busse Highway, Park Ridge, IL, 60068

American Society of Mechanical Engineers (ASME), 345 E. 47th St., New York, NY, 10017

American Welding Society (AWS), 550 N.W. LeJeune Rd., Miami, FL, 33126

Copper Development Association (CDA), 405 Lexington Ave., New York, NY, 10174

Electronic Industries Association (EIA), 2001 Eye St., NW, Washington, DC, 20006

Forging Industry Association (FIA), 1121 Illuminating Bldg., Cleveland, OH, 44113

Grinding Wheel Institute (GWI), 712 Lakewood Center N., Cleveland, OH, 44107

Institute of Industrial Engineers (IIE), 25 Technology Park/Atlanta, Norcross, GA, 30092

International Institution for Production Engineering Research (CIRP), 10, rue Mansart, 75009 Paris, France

Metal Powder Industries Federation (MPIF), 105 College Rd. E., Princeton, NJ, 08540

The Metallurgical Society of AIME, 420 Commonwealth Dr., Warrendale, PA, 15086

National Bureau of Standards (NBS), Washington, DC, 20234

National Machine Tool Builders Association (NMTBA), 7901 Westpark Dr., McLean, VA, 22102

Numerical Control Society (NCS), 519 Zenith Dr., Glenview, IL, 60025

Society for the Advancement of Material and Process Engineering (SAMPE), Box 613, Azusa, CA, 91702

Society of Automotive Engineers (SAE), 400 Commonwealth Dr., Warrendale, Pa, 15086

Society of Die Casting Engineers, 2000 North 5th Ave., River Grove, IL, 60171

Society of Manufacturing Engineers (SME), One SME Dr., Dearborn, MI, 48128

Society of Plastics Engineers (SPE), 14 Fairfield Dr., Brookfield Center, CT, 06805

Steel Founders' Society of America (SFSA), 455 State St., Des Plaines, IL, 60016

2 Fundamentals of the Mechanical Behavior of Materials

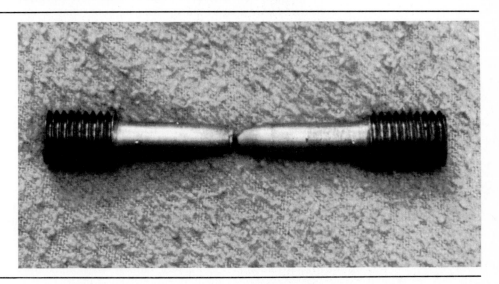

2.1 INTRODUCTION

The manufacturing methods and techniques by which materials can be shaped into useful products were outlined in Chapter 1. In manufacturing, one of the most important groups of processes is *plastic deformation*, namely, shaping materials by applying forces in various ways (also known as deformation processing). It includes bulk deformation (forging, rolling, extrusion, rod and wire drawing) and sheet-forming processes (bending, drawing, spinning, and general pressworking).

This chapter deals with the fundamental aspects of the mechanical behavior of materials during deformation. Individual topics are deformation modes, stresses, forces, effects of rate of deformation and temperature, hardness, residual stresses, yield criteria, and various methods of analyzing metalworking processes in order to determine force and power requirements.

In stretching a piece of metal to make an object such as a fender of an automobile or a length of wire, the material is subjected to tension. A solid cylindrical piece of metal is forged in the making of a turbine disk, thus the material is subjected to compression. Sheet metal undergoes shearing stresses when, for instance, a hole is punched in it. A piece of plastic tubing is expanded by internal pressure to make a bottle, thus subjecting the material to tension in various directions.

In all of these processes, the material is subjected to one or more of the basic modes of deformation shown in Fig. 2.1, namely, tension, compression, and shear. The degree of deformation to which the material is subjected is defined as *strain*. For tension or compression the *engineering strain*, or *nominal strain*, is defined as

$$e = \frac{\ell - \ell_o}{\ell_o}. \tag{2.1}$$

We note that in tension the strain is positive and in compression it is negative. The *shear strain* is defined as

$$\gamma = \frac{a}{b} \tag{2.2}$$

In order to change the geometry of the bodies or elements in Fig. 2.1, forces must be applied to them as shown by the arrows. The determination of these forces as a function of strain is very important in manufacturing processes. The forces have to be known in order to design the proper equipment, to select the tool and die materials for proper strength, and to determine whether or not a specific metalworking operation can be accomplished on certain equipment.

Thus, the relation between a force and the deformation it produces is an essential parameter in manufacturing. In this relationship, strain is an *independent* variable, whereas the force required to cause it is a *dependent* variable. The basic material property tests that are performed to establish these relationships are covered in the sections below.

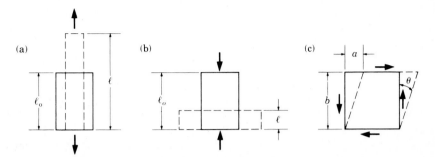

FIGURE 2.1 Types of strain. (a) Tensile. (b) Compressive. (c) Shear. All deformation processes in manufacturing involve strains of these types. Tensile strains are involved in stretching sheet metal to make car bodies, compressive strains in forging metals to make turbine disks, and shear strains in making holes by punching.

2.2 THE TENSION TEST

The tension test, because of its relative simplicity, is the most common test for determining the *strength-deformation characteristics* of materials. It involves the preparation of a test specimen (according to *ASTM* specifications) and testing it under tension on any of a variety of available testing machines.

The specimen (Fig. 2.2a) has an original length ℓ_o and an original cross-sectional area A_o. Although most specimens are solid and round, flat sheet or tubular specimens are also tested under tension. The original length is the distance between *gage marks* on the specimen and is generally 2 in. (50 mm). Longer lengths may be used for larger specimens such as structural members.

Typical results from a tension test are shown in Fig. 2.2(b). The *engineering stress*, or *nominal stress*, is defined as the ratio of the applied load to the original area,

$$\sigma = \frac{P}{A_o} \tag{2.3}$$

and the engineering strain is

$$e = \frac{\ell - \ell_o}{\ell_o}.$$

When the load is applied, the specimen elongates proportionately up to the *yield point Y*. If the load is removed within this range, the specimen returns to its original length. This is the range of *linear elastic behavior*. The *modulus of elasticity*, or *Young's modulus*, E is defined as

$$E = \frac{\sigma}{e}. \tag{2.4}$$

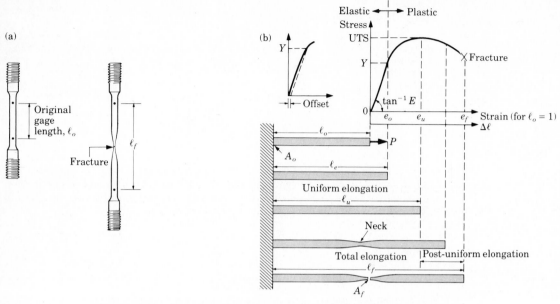

FIGURE 2.2 (a) Original and final shape of a standard tensile test specimen. (b) Outline of a tensile test sequence showing different stages in the elongation of the specimen. See also Fig. 10.6(a). Tensile test data are important in calculating forces and power requirements, and predicting the behavior of materials in manufacturing processes.

This linear relationship between stress and strain is known as *Hooke's law* (after R. Hooke, 1635–1703), the more generalized forms of which are given in Section 2.8.3.

The elongation of the specimen is accompanied by a contraction of its lateral dimensions. The absolute value of the ratio of the lateral strain to longitudinal strain is known as *Poisson's ratio*, v, (after S. D. Poisson, 1781–1840). Some typical values for a variety of materials are given in Table 2.1.

The area under the stress–strain curve up to the yield point Y of a material is known as the *modulus of resilience*,

$$\text{Modulus of resilience} = \frac{Ye_o}{2} = \frac{Y^2}{2E}. \tag{2.5}$$

This area has the units of *energy per unit volume* and indicates the *specific energy* that the material can store elastically (Table 2.2).

With increasing load, the specimen begins to yield, that is, it begins to undergo *plastic (permanent) deformation* and the relationship between stress and strain is no longer linear. Since the rate of change in the slope of the stress–strain curve beyond the yield point is rather small for most materials, the determination of Y may be difficult.

The usual practice is to define the yield stress as the point on the curve that is offset by a strain of (usually) 0.2%, or 0.002 (Fig. 2.2b). Other offset strains may also

TABLE 2.1
MODULUS OF ELASTICITY E AND POISSON'S RATIO v FOR VARIOUS MATERIALS

METALS	E		v
	psi $\times 10^6$	GPa	
Aluminum and its alloys	10–11.5	69–79	0.31–0.34
Cast irons	15–22	105–150	0.21–0.30
Copper and its alloys	15–21.5	105–150	0.33
Ductile iron	22–25	150–170	0.28
Lead and its alloys	2	14	0.43
Magnesium and its alloys	6–6.5	41–45	0.35
Malleable iron	26–27	180–185	0.27
Molybdenum	47	325	0.32
Nickel and its alloys	26–31	180–214	0.31
Steel (plain carbon)	29	200	0.33
Steel (austenitic stainless)	27.5–29	190–200	0.28
Tantalum and its alloys	21.5–27	150–186	0.35
Titanium and its alloys	11.5–19	80–130	0.31–0.34
Tungsten	58	400	0.27
NONMETALLIC			
Acrylics	0.2–0.5	1.4–3.4	0.35–0.40
Epoxies	0.5–2.5	3.5–17	0.34
Nylons	0.2–0.4	1.4–2.8	0.32–0.40
Rubbers	0.0015–0.015	0.01–0.1	0.5
Concrete	3–5	20–35	0.12–0.25
Glass and porcelain	10	70	0.24
Diamond	120–150	820–1050	
Graphite (bulk)	1	7	

TABLE 2.2
MODULUS OF RESILIENCE, MODULUS OF ELASTICITY, AND YIELD STRESS FOR VARIOUS METALS

	MODULUS OF RESILIENCE, in.-lb/in^3	E, psi $\times 10^6$	Y, psi $\times 10^3$
Copper, annealed	3	16	10
Lead	0.25	2	1
Magnesium	8.5	6	10
Medium-carbon steel, annealed	27.5	29	40
Spring steel	385	29	150
Titanium alloy, annealed	570	16	135

be used and should be specified in reporting the yield stress. The terms *elastic limit* and *proportional limit* can also be used to specify the point where the stress and strain are no longer proportional.

As the specimen continues to elongate under increasing load, its cross-sectional area decreases uniformly throughout its length. The load (hence the engineering stress) reaches a maximum and then begins to decrease. The maximum stress is known as the *tensile strength* or *ultimate tensile strength* (*UTS*) of the material.

When the specimen is loaded beyond its UTS, it begins to *neck* (Fig. 2.2) and the elongation is no longer uniform. That is, the change in the cross-sectional area of the specimen is no longer uniform along the length of the specimen but is concentrated locally in a "neck" formed in the specimen (known as *necking*, or *necking down*). As the test progresses, the engineering stress drops further and the specimen finally fractures in the necked region. The final stress level (marked by an × in Fig. 2.2b) is known as *breaking* or *fracture stress.*

2.2.1 DUCTILITY

Ultimate tensile strength is a practical measure of the overall strength of a material. Likewise, the strain at fracture is a measure of its *ductility*, that is, how large a strain the material withstands before failure. Note from Fig. 2.2(b) that until the UTS is reached, elongation is uniform. The strain up to the UTS is known as *uniform strain.* The elongation at fracture is known as the *total elongation.* This is measured between the original gage marks after the two pieces of the broken specimen are placed together, as shown in Fig. 2.2(a).

Two quantities commonly used to define ductility in a tension test are percent elongation and reduction of area. Percent *elongation* is defined as

$$\% \text{ elongation} = \frac{\ell_f - \ell_o}{\ell_o} \times 100 \tag{2.6}$$

and is based on the total elongation.

Necking is a *local* phenomenon. If we put a series of gage marks at different points on the specimen, pull, and break it under tension and then calculate the percent elongation for each pair of gage marks, we find that with decreasing gage length the percent elongation increases (Fig. 2.3). The closest pair of gage marks undergo the largest elongation because they are closest to the necked region. However, the curves do not approach zero elongation with increasing gage length. This is because the specimens have all undergone a finite uniform and permanent elongation before fracture. It is thus important to report gage length in conjunction with elongation data, whereas other tensile properties are generally independent of gage length.

A second measure of ductility is *reduction of area*, defined as

$$\% \text{ reduction of area} = \frac{A_o - A_f}{A_o} \times 100. \tag{2.7}$$

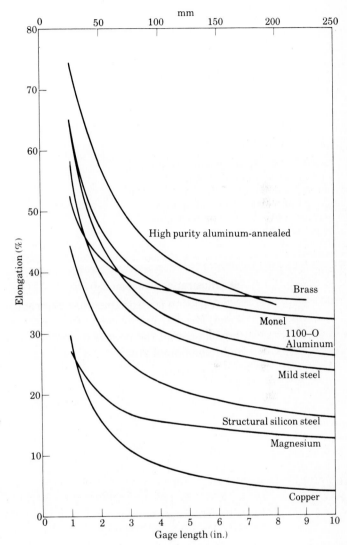

FIGURE 2.3 Total elongation in a tensile test as a function of original gage length for various metals. Because necking is a local phenomenon, elongation decreases with gage length. Standard gage length is usually 2 in. (50 mm), although shorter ones can be used if larger specimens are not available.

Thus, a material that necks down to a point at fracture, such as a glass rod at elevated temperature, has 100% reduction of area.

The elongation and reduction of area are generally related to each other for many common engineering metals and alloys, as shown in Fig. 2.4. Elongation ranges approximately between 10 and 60%, for most materials, while values between 20 and 90% are typical for reduction of area. *Thermoplastics* (Chapter 10) and *superplastic* materials (Section 2.2.7) exhibit much higher ductility. Brittle materials, by definition, have little or no ductility. Examples are glass at room temperature or a piece of chalk.

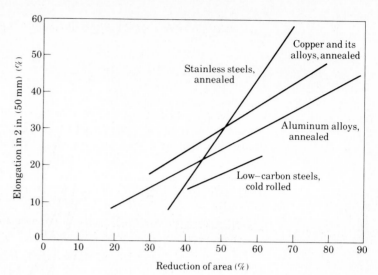

FIGURE 2.4 Approximate relationship between elongation and reduction of area for different groups of metals. Both elongation and reduction of area are measures of the ductility of a material.

● Illustrative Problem 2.1

A tension test specimen has an original gage length of 50 mm and a diameter of 4 mm. The maximum load during the test is 10 kN. The final gage length is 80 mm and the diameter of the necked region is 3 mm. Calculate the UTS, percent elongation, and reduction of area.

SOLUTION.

$$\text{UTS} = \frac{P}{A_o} = \frac{10{,}000}{\pi \dfrac{4^2}{4}} = 796 \text{ N/mm}^2 = 796 \text{ MPa.}$$

$$\text{Elongation} = \frac{80 - 50}{50} \times 100 = 60\%.$$

$$A_o = \pi \frac{4^2}{4} = 12.57 \text{ mm}^2.$$

$$A_f = \pi \frac{3^2}{4} = 7.07 \text{ mm}^2.$$

$$\text{Reduction of area} = \frac{12.57 - 7.07}{12.57} \times 100 = 43.8\%. \quad ●$$

2.2.2 TRUE STRESS AND TRUE STRAIN

It is apparent that, since stress is defined as the ratio of force to area, *true stress* should be defined as

$$\sigma = \frac{P}{A}, \tag{2.8}$$

where A is the actual (hence true) or instantaneous area supporting the load.

TABLE 2.3
COMPARISON OF ENGINEERING AND TRUE STRAINS IN TENSION

e	0.01	0.05	0.1	0.2	0.5	1	2	5	10
ϵ	0.01	0.049	0.095	0.18	0.4	0.69	1.1	1.8	2.4

Likewise, the complete tension test may be regarded as a series of incremental tension tests where, for each succeeding increment, the original specimen is a little longer than the previous one. We can now define *true strain* (or *natural* or *logarithmic strain*) ϵ as

$$\epsilon = \int_{\ell_o}^{\ell} \frac{d\ell}{\ell} = \ln\left(\frac{\ell}{\ell_o}\right). \tag{2.9}$$

Note that, for small values of engineering strain, $e = \epsilon$ since the value of $\ln(1 + e) = \epsilon$. For larger strains, however, the values diverge rapidly, as seen in Table 2.3.

The volume of a metal specimen in the plastic region of the test remains constant (Section 2.8.5). Hence, the true strain within the uniform elongation range can be expressed as

$$\epsilon = \ln\left(\frac{\ell}{\ell_o}\right) = \ln\left(\frac{A_o}{A}\right) = \ln\left(\frac{D_o}{D}\right)^2 = 2\ln\left(\frac{D_o}{D}\right). \tag{2.10}$$

Once necking begins, the true strain at any point in the specimen can be calculated from the change in cross-sectional area at that point. Thus, by definition, the largest strain is at the narrowest region of the neck.

We have seen that, at small strains, the engineering and true strains are very close and hence either one can be used in calculations. However, for the large strains encountered in metalworking, the true strain should be used. This is the true measure of the strain and can be illustrated by the following two examples.

Assume that a tension specimen is elongated to twice its original length. This involves a deformation equivalent to that of compressing a specimen to one half of its original height. Using the subscripts t and c for tension and compression, respectively, it is found that $\epsilon_t = 0.69$ and $\epsilon_c = -0.69$, whereas $e_t = 1$ and $e_c = -0.5$. Thus, true strain is a correct measure of strain.

For the second example, assume that a specimen 10 mm in height is compressed to a final thickness of zero. Thus, $\epsilon_c = -\infty$, whereas $e_c = -1$. The fact is that we have deformed the specimen infinitely, which is exactly what the value of the true strain indicates.

It is obvious from these examples that true strains are consistent with the actual physical phenomenon and that the engineering strains are not.

2.2.3 TRUE STRESS—TRUE STRAIN CURVES

The relation between engineering and true values for stress and strain can now be used to construct true stress–true strain curves from a curve such as that in Fig. 2.2.

The procedure for this construction is given in Illustrative Problem 2.2 below.

A typical true stress–true strain curve is shown in Fig. 2.5. For convenience, such a curve can be approximated by the equation

$$\sigma = K\epsilon^n. \tag{2.11}$$

Note that this equation indicates neither the elastic region nor the yield point Y of the material. These quantities, however, are easily available from the engineering stress–strain curve. (Since the strains at the yield point are very small, the difference between the true and engineering yield stress is negligible for metals. This is because the difference in the cross-sectional areas A_o and A at yielding is very small.)

When the curve in Fig. 2.5 is plotted on a log–log scale, it is composed of two straight lines. One straight line indicates the elastic region and the other straight line, with the slope n, indicates the plastic region, Fig. 2.6. The slope n is known as the *strain-hardening exponent*, and K (the true stress at a true strain of one) is known as the *strength coefficient*. The agreement between the data points and Eq. (2.11) may not always be as good as shown in Fig. 2.6; however, in most cases the curve can be drawn with a reasonably representative value of n.

Values of K and n for a variety of engineering materials are given in Table 2.4. The true stress–true strain curves for some materials are given in Fig. 2.7. Some differences between Table 2.4 and these curves may exist because of different sources of data and test conditions.

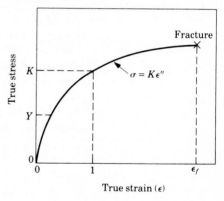

FIGURE 2.5 True stress-true strain curve in tension. Note that, unlike in an engineering stress-strain curve, the slope is always positive, and that the slope decreases with increasing strain. Whereas in the elastic range stress and strain are proportional, in the plastic range the curve can be approximated by the power expression shown in the figure.

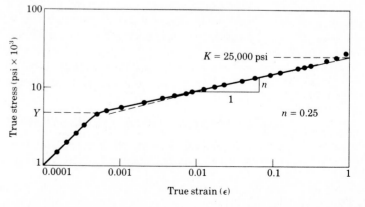

FIGURE 2.6 True stress-true strain curve in tension for 1100–O aluminum plotted on log-log scale. Note the large difference in the slopes in the elastic and plastic ranges. *Source*: After R. M. Caddell and R. Sowerby, *Bull. Mech. Eng. Educ.*, vol. 8, 1969, pp. 31–43.

TABLE 2.4
TYPICAL VALUES FOR *K* AND *n* IN EQ. (2.11) AT ROOM TEMPERATURE

	K		
	psi × 10³	MPa	*n*
Aluminum, 1100-O	26	180	0.20
2024-T4	100	690	0.16
5052-O	30	210	0.13
6061-O	30	205	0.20
6061-T6	60	410	0.05
7075-O	58	400	0.17
Brass, 60-39-1 Pb, annealed	115	800	0.33
70-30, annealed	130	895	0.49
85-15, cold-rolled	84	580	0.34
Bronze (phosphor), annealed	105	720	0.46
Cobalt-base alloy, heat-treated	300	2070	0.50
Copper, annealed	46	315	0.54
Molybdenum, annealed	105	725	0.13
Muntz metal, annealed	115	800	0.50
Steel, low-carbon annealed	77	530	0.26
1045 hot-rolled	140	965	0.14
1112 annealed	110	760	0.19
1112 cold-rolled	110	760	0.08
4135 annealed	147	1015	0.17
4135 cold-rolled	160	1100	0.14
4340 annealed	93	640	0.15
17-4 P-H annealed	175	1200	0.05
52100 annealed	210	1450	0.07
302 stainless, annealed	190	1300	0.30
304 stainless, annealed	185	1275	0.45
410 stainless, annealed	140	960	0.10
Vanadium, annealed	112	770	0.35

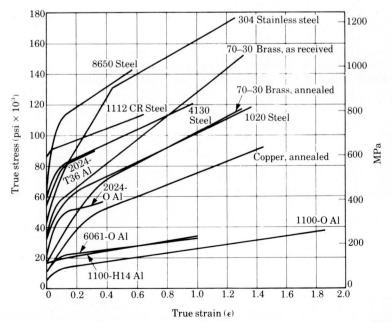

FIGURE 2.7 True stress-true strain curves in tension at room temperature for various metals. The point of intersection of each curve at the ordinate is the yield stress *Y*; thus the elastic portions of the curves are not indicated. Compare with Fig. 2.5. The area under the curve is toughness. See also Fig. 10.8. *Source*: S. Kalpakjian.

The area under the true stress–true strain curve is known as *toughness* and can be expressed as

$$\text{Toughness} = \int_o^{\epsilon_f} \sigma \, d\epsilon, \tag{2.12}$$

where ϵ_f is the true strain at fracture. Toughness is defined as the energy per unit volume (specific energy) that has been dissipated up to fracture.

It is important to remember that this specific energy pertains only to that volume of material at the narrowest region of the neck. Any volume of material away from the neck has undergone less strain and, hence, has dissipated less energy.

It should also be pointed out that here toughness is different from the concept of fracture toughness as treated in textbooks on fracture mechanics. Fracture mechanics (the study of the initiation and propagation of cracks in a solid medium) is beyond the scope of this text. It is of limited relevance to metalworking processes, except in die design and die life.

● **Illustrative Problem 2.2**

The following data are taken from a stainless steel tension test specimen:

$A_o = 0.056 \text{ in}^2$

$A_f = 0.016 \text{ in}^2$

$\ell_o = 2 \text{ in.}$

Elongation in 2 in. = 49 %

LOAD, P, lb	EXTENSION, $\Delta\ell$, in.
1600	0
2500	0.02
3000	0.08
3600	0.20
4200	0.40
4500	0.60
4600 (max.)	0.86
3300 (fracture)	0.98

Draw the true stress–true strain curve for the material.

SOLUTION. The load–extension curve is shown in Fig. P2.1. In order to determine the true stress and true strain, the following relationships are used:

$$\text{True stress } \sigma = \frac{P}{A}$$

$$\text{True strain } \epsilon = \ln\left(\frac{\ell}{\ell_o}\right) \text{ up to necking}$$

$$\epsilon_f = \ln\left(\frac{A_o}{A_f}\right) \text{ at fracture}$$

$$\ell = \ell_o + \Delta\ell \text{ up to necking}$$

Assuming that the volume of the specimen remains constant, we have

$$A_o\ell_o = A\ell.$$

Thus,

$$A = \frac{A_o\ell_o}{\ell} = \frac{(0.056)(2)}{\ell} = \frac{0.112}{\ell} \text{ in}^2.$$

Using these relationships, we obtain the following data:

$\Delta\ell$, in.	ℓ, in.	ϵ	A, in^2	TRUE STRESS, psi
0	2.00	0	0.056	28,600
0.02	2.02	0.01	0.0555	45,000
0.08	2.08	0.039	0.054	55,800
0.20	2.20	0.095	0.051	70,800
0.40	2.40	0.182	0.047	90,000
0.60	2.60	0.262	0.043	104,500
0.86	2.86	0.357	0.039	117,300
0.98	2.98	1.253	0.016	206,000

The true stress and true strain are plotted (solid line) in Fig. P2.2. The point at necking

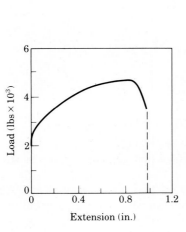

FIGURE P2.1

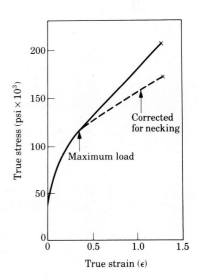

FIGURE P2.2

is connected to the point at fracture by a straight line because there are no data on the instantaneous areas after necking begins. The correction on this curve (broken line) is explained in Section 2.8. ●

2.2.4 INSTABILITY IN THE TENSION TEST

As noted previously the onset of necking in a tension test corresponds to the ultimate tensile strength (UTS) of the material. The slope of the load–elongation curve at this point is zero, and it is here that instability begins. That is, the specimen begins to neck and cannot support the load because the neck is becoming smaller in cross-sectional area.

Using the following relationships

$$\epsilon = \ln\left(\frac{A_o}{A}\right),$$

$$A = A_o e^{-\epsilon},$$

and

$$P = \sigma A = \sigma A_o e^{-\epsilon},$$

one finds that at necking,

$$\frac{dP}{d\epsilon} = A_o\left(\frac{d\sigma}{d\epsilon} e^{-\epsilon} - \sigma e^{-\epsilon}\right) = 0.$$

Hence,

$$\frac{d\sigma}{d\epsilon} = \sigma.$$

However,

$$\sigma = K\epsilon^n,$$

and consequently,

$$nK\epsilon^{n-1} = K\epsilon^n$$

$$\epsilon = n. \tag{2.13}$$

Thus, the true strain at the onset of necking (termination of uniform elongation) is numerically equal to the strain hardening exponent, n. Hence, the higher the value of n, the greater the strain to which a piece of material can be stretched before necking begins.

It can be seen from Table 2.4 that metals such as annealed copper, brass, and stainless steel can be stretched uniformly to a greater extent than the other materials listed. These observations are covered in greater detail in Chapters 7 and 10 because of their relevance to sheet-forming processes.

Instability in a tension test can be viewed as a phenomenon where two competing processes are simultaneously taking place. As the load on the specimen is increased, its cross-sectional area becomes smaller, this being more pronounced in the region where necking begins. On the other hand, with increasing strain, the material is becoming stronger due to strain hardening. Since the load on the specimen is the product of area and strength, instability sets in when the rate of decrease in area is greater than the rate of increase in strength. This is also known as *geometric softening*.

● **Illustrative Problem 2.3**

A material has a true stress–true strain curve given by

$$\sigma = 100{,}000\epsilon^{0.5} \text{ psi}.$$

Calculate the true ultimate tensile strength and the engineering UTS of this material.

SOLUTION. Since the necking strain corresponds to the maximum load and the necking strain for this material is

$$\epsilon = n = 0.5,$$

we have as the true ultimate tensile strength

$$\sigma = Kn^n$$
$$\sigma = 100{,}000 \, (0.5)^{0.5} = 70{,}710 \text{ psi}.$$

The true area at the onset of necking is obtained from

$$\ln\left(\frac{A_o}{A_{neck}}\right) = n = 0.5.$$

Thus,

$$A_{neck} = A_o e^{-0.5}$$

and the maximum load P is

$$P = \sigma A = \sigma A_0 e^{-0.5},$$

where σ is the true ultimate tensile strength. Hence,

$$P = (70{,}710)(0.606)(A_o) = 42{,}850 A_o \text{ lb}.$$

Since UTS $= P/A_o$,

$$\text{UTS} = 42{,}850 \text{ psi}. \quad ●$$

2.2.5 TYPES OF STRESS–STRAIN CURVES

Every material has a differently shaped stress–strain curve. Its shape depends on its composition and many other factors to be treated in detail later. Some of the major types of curves are shown in Fig. 2.8.

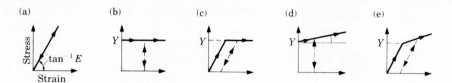

FIGURE 2.8 Schematic illustration of various types of idealized stress-strain curves. (a) Perfectly elastic. (b) Rigid, perfectly plastic. (c) Elastic, perfectly plastic. (d) Rigid, linearly strain-hardening. (e) Elastic, linearly strain hardening. The broken lines and arrows indicate unloading and reloading during the test. Most engineering metals exhibit a behavior similar to curve (e). See also Fig. 10.9.

A perfectly elastic material behaves like a spring with stiffness E. The behavior of brittle materials, such as glass, ceramics, and some cast irons, may be represented by such a curve (Fig. 2.8a). There is a limit to the stress the material can sustain, after which it breaks.

A rigid, perfectly plastic material has, by definition, an infinite value of E. Once the stress reaches the yield stress Y, it continues to undergo deformation at the same stress level. When the load is released, the material has undergone permanent deformation, with no elastic recovery, as seen in Fig. 2.8(b).

An elastic, perfectly plastic material is a combination of the first two; it undergoes elastic recovery when the load is released (Fig. 2.8c).

A rigid, linearly strain-hardening material requires an increasing stress level to undergo further strain. Thus, its *flow stress* (magnitude of the stress required to maintain plastic deformation at a given strain) increases with increasing strain. It has no elastic recovery upon unloading (Fig. 2.8d). See Fig. 2.51 for flow stress.

An elastic, linearly strain-hardening curve (Fig. 2.8e) is an approximation of the behavior of most engineering materials, with the modification that the plastic portion of the curve has a decreasing slope with increasing strain (i.e., Fig. 2.5).

Remember that the slopes of the elastic portions of the models in Fig. 2.8 are highly exaggerated and that the initial slopes are actually very steep. This can be seen by noting the magnitude of the modulus of elasticity of engineering materials. Also note that some of these curves can be expressed by Eq. (2.11) by changing the value of n (Fig. 2.9) or by other equations of a similar nature.

2.2.6 EFFECTS OF TEMPERATURE

In this and subsequent sections, the various factors that have an influence on the shape of stress–strain curves will be discussed. The first factor is temperature. Although somewhat difficult to generalize, increasing temperature usually increases ductility and toughness and lowers the modulus of elasticity, yield stress, and ultimate tensile strength. These effects are shown in Figs. 2.10 and 2.11.

Temperature also affects the strain-hardening exponent n of most metals, in that n decreases with increasing temperature (Fig. 2.12). Depending on the type of material and its composition and level of impurities, elevated temperatures can have other

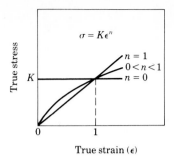

FIGURE 2.9 The effect of strain-hardening exponent n on the shape of true stress-true strain curves. When $n = 1$, the material is elastic, and when $n = 0$ it is rigid, perfectly plastic. See Table 2.4 for n values.

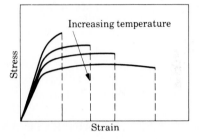

FIGURE 2.10 Typical effects of temperature on engineering stress-strain curves. Temperature affects the modulus of elasticity, yield stress, ultimate tensile strength, and toughness of materials. See also Fig. 10.12.

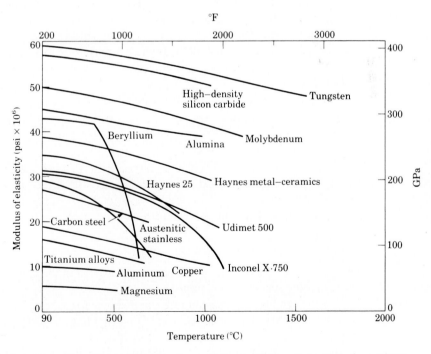

FIGURE 2.11 The effect of temperature on the modulus of elasticity for various materials. Note also the maximum useful temperature range for these materials. See also Figs. 10.13 and 11.14. The properties of these materials are described in Chapter 3.

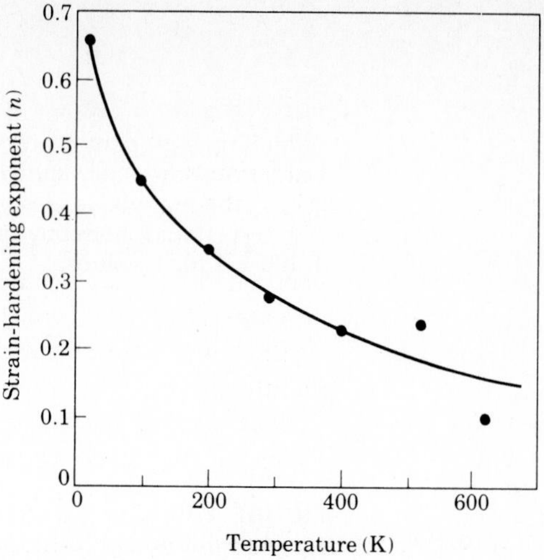

FIGURE 2.12 The effect of temperature on the strain-hardening exponent *n* for pure aluminum. *Source*: After R. P. Carreker and W. R. Hibbard, Jr., *Trans. TMS-AIME*, vol. 209, 1957, pp. 1157–1163.

significant effects, as detailed in Chapter 3. However, the influence of temperature is best discussed in conjunction with strain rate for the reasons explained below.

2.2.7 STRAIN RATE AND ITS EFFECTS

Depending on the particular manufacturing operation and equipment, a piece of material may be formed at low or at high speeds. In performing a tension test, the specimen can be strained at different rates to simulate the actual deformation process.

Whereas the *deformation rate* may be defined as the speed (in feet per minute or meters per second, for instance) at which a tension test is being carried out (e.g., the rate at which a rubber band is stretched), the *strain rate* is a function of the geometry of the specimen.

The *engineering strain rate* $\dot{e}$ is defined as

$$\dot{e} = \frac{de}{dt} = \frac{d\left(\dfrac{\ell - \ell_o}{\ell_o}\right)}{dt} = \frac{1}{\ell_o} \cdot \frac{d\ell}{dt} = \frac{v}{\ell_o} \tag{2.14}$$

and the *true strain rate* $\dot{\epsilon}$ as

$$\dot{\epsilon} = \frac{d\epsilon}{dt} = \frac{d\left(\ln\left(\dfrac{\ell}{\ell_o}\right)\right)}{dt} = \frac{1}{\ell} \cdot \frac{d\ell}{dt} = \frac{v}{\ell}, \tag{2.15}$$

where v is the speed of deformation (e.g., the speed of the jaws of the testing machine in which the specimen is clamped).

Although the deformation rate v and engineering strain rate $\dot{e}$ are equivalent, the true strain rate $\dot{\epsilon}$ is not identical to v. Thus, in a tension test with v constant, the true strain rate decreases as the specimen becomes longer. In order to maintain a constant $\dot{\epsilon}$, the speed must be increased accordingly. However, for small changes in length of the specimen during a test, this difference is not significant.

Typical deformation speeds employed in various metalworking processes, and the strain rates involved, are shown in Table 2.5. Note that there are considerable differences in the magnitudes. Because of this wide range, strain rates are generally quoted in orders of magnitude, such as 10^2 per second, 10^4 per second, etc.

The typical effects of temperature and strain rate on the strength of metals are shown in Figs. 2.13 and 2.14. These figures clearly indicate that increasing strain rate increases strength, and that the sensitivity of the strength to the strain rate increases with temperature. Note, however, that for the materials in Fig. 2.13 this effect is relatively small at room temperature; this may not be true for other metals and alloys. For instance, the yield stress at room temperature in Fig. 2.14 shows considerable sensitivity to strain rate.

It can be seen from Fig. 2.13 that the same strength can be obtained either at low temperature, low strain rate or at high temperature, high strain rate. This is important in estimating the resistance of materials to deformation when processing them at various strain rates and temperatures.

The effect of strain rate on strength also depends on the particular level of strain; this effect increases with strain (Fig. 2.15). The strain rate has also been found to affect the strain hardening exponent n in that it decreases with increasing strain rate (Fig. 2.16).

The effect of strain rate on the strength of materials is generally expressed by

$$\sigma = C\dot{\epsilon}^m, \qquad (2.16)$$

where C is the *strength coefficient*, similar to K in Eq. (2.11), and m is the *strain-rate sensitivity exponent* of the material.

Equation (2.16) can be plotted in a manner similar to Fig. 2.6, with temperature as a parameter, as seen in Fig. 2.13. Note that the value of m increases with increasing temperature (Fig. 2.17).

A general range of values for m is as follows: up to 0.05 for cold working, 0.05 to 0.4 for hot working, and 0.3 to 0.85 for superplastic materials. For a Newtonian fluid,

TABLE 2.5
TYPICAL RANGES OF STRAIN, DEFORMATION SPEED, AND STRAIN RATES IN METALWORKING PROCESSES

PROCESS	TRUE STRAIN	DEFORMATION SPEED, m/s	STRAIN RATE, s^{-1}
Cold working			
Forging, rolling	0.1–0.5	0.1–100	1–10^3
Wire and tube drawing	0.05–0.5	0.1–100	1–10^4
Explosive forming	0.05–0.2	10–100	10–10^5
Hot working and warm working			
Forging, rolling	0.1–0.5	0.1–30	1–10^3
Extrusion	2–5	0.1–1	10^{-1}–10^2
Machining	1–10	0.1–100	10^3–10^6
Sheet-metal forming	0.1–0.5	0.05–2	1–10^2
Superplastic forming	0.2–3	10^{-4}–10^{-2}	10^{-4}–10^{-2}

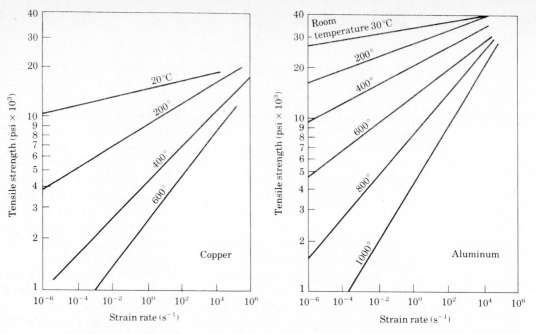

FIGURE 2.13 The effect of strain rate on the ultimate tensile strength of copper and aluminum. Note that as temperature increases, the slope increases. Thus, tensile strength becomes more and more sensitive to strain rate as temperature increases. *Source*: After J. H. Hollomon.

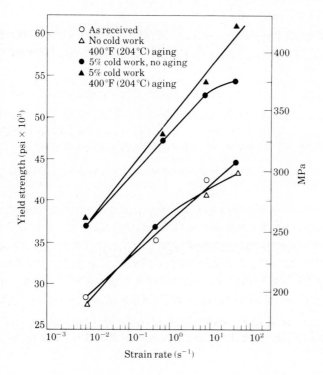

FIGURE 2.14 The effect of strain rate on the yield strength *Y* for cold-rolled aluminum-killed steel (Section 5.5). Note the wide range of strain rates. *Source*: After A. Saxena and D. A. Chatfield, SAE Paper 760209, 1976.

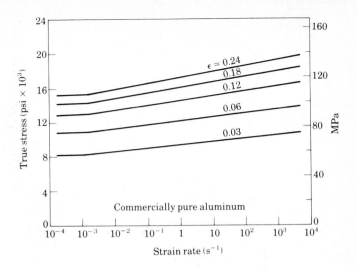

FIGURE 2.15 True stress as a function of strain rate at various strains, for commercially pure aluminum. The true stress required to sustain plastic deformation at a particular strain is known as the flow stress. *Source*: After U. S. Lindholm, *Behavior of Materials Under Dynamic Loading*, ASME, 1965, p. 51.

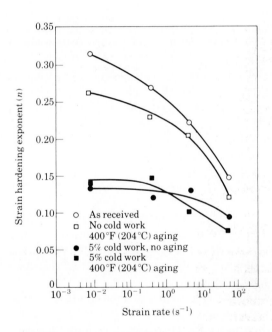

FIGURE 2.16 The effect of strain rate on the strain hardening exponent n for cold-rolled rimmed steel (Section 5.5). *Source*: After A. Saxena and D. A. Chatfield, SAE Paper 760209, 1976.

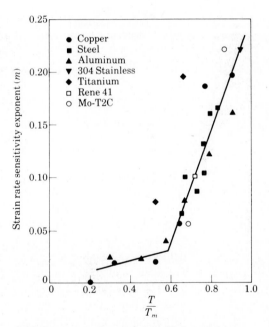

FIGURE 2.17 Dependence of the strain-rate sensitivity exponent m on the homologous temperature T/T_m for various materials. T is the testing temperature and T_m is the melting point of the metal, both on the absolute scale. The transition in the slopes of the curve occurs at about the recrystallization temperature of the metals. See also Fig. 3.17. *Source*: After F. W. Boulger, DMIC Report 226, Battelle Mem. Inst., 1966, pp. 13–37.

TABLE 2.6
APPROXIMATE RANGE OF VALUES FOR *C* **AND** *m* **IN EQ: (2.16) FOR VARIOUS ANNEALED METALS AT TRUE STRAINS RANGING FROM 0.2 TO 1.0.** (After T. Altan and F. W. Boulger, *Jr. Eng. Ind.* 95:1009–1019, 1973.)

MATERIAL	TEMPERATURE, °C	C psi × 10³	C MPa	m
Aluminum	200–500	12–2	82–14	0.07–0.23
Aluminum alloys	200–500	45–5	310–35	0–0.20
Copper	300–900	35–3	240–20	0.06–0.17
Copper alloys (brasses)	200–800	60–2	415–14	0.02–0.3
Lead	100–300	1.6–0.3	11–2	0.1–0.2
Magnesium	200–400	20–2	140–14	0.07–0.43
Steel				
Low-carbon	900–1200	24–7	165–48	0.08–0.22
Medium-carbon	900–1200	23–7	160–48	0.07–0.24
Stainless	600–1200	60–5	415–35	0.02–0.4
Titanium	200–1000	135–2	930–14	0.04–0.3
Titanium alloys	200–1000	130–5	900–35	0.02–0.3
Ti-6Al-4V*	815–930	9.5–1.6	65–11	0.50–0.80
Zirconium	200–1000	120–4	830–27	0.04–0.4

* At a strain rate of 2 × 10⁻⁴/s.

Note: As temperature increases, *C* decreases and *m* increases. As strain increases, *C* increases and *m* may increase or decrease, or it may become negative within certain ranges of temperature and strain.

where the shear stress increases linearly with rate of shear, the value of *m* is 1. Some specific values for *C* and *m* are given in Table 2.6.

The term *superplastic* refers to the capability of some materials to undergo large uniform elongation prior to failure. The elongation may be on the order of a few hundred to as much as 2000%. An extreme example is hot glass. Among other materials exhibiting superplastic behavior are polymers at elevated temperatures, very-fine-grain alloys of zinc-aluminum and titanium alloys.

The magnitude of *m* has a significant effect on necking in a tension test. Experimental observations show that, with high *m* values, the material stretches to a greater length before it fails. This is an indication that necking is delayed with increasing *m*. When necking is about to begin (i.e., the cross-sectional area of the specimen in the neck region becomes smaller) its strength, with respect to the rest of the specimen, increases due to strain hardening. However, the strain rate in the neck region is also higher than the rest of the specimen because the material is elongating faster there. Since the material in the necked region is becoming stronger as it is strained at a higher rate, this region exhibits a greater resistance to necking. This increased resistance to necking thus depends on the magnitude of *m*.

As the test progresses, necking becomes more diffuse and the specimen becomes longer before fracture. Thus, total elongation increases with increasing *m* value (Fig. 2.18). As expected, the elongation after necking (postuniform elongation) also

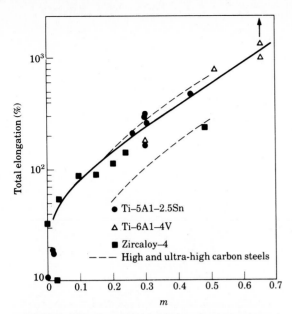

FIGURE 2.18 The effect of strain-rate sensitivity exponent m on the total elongation for various metals. Note that elongation at high values of m approaches 1000%. This phenomenon is utilized in superplastic forming of metals, described in Section 7.13. *Source*: After D. Lee and W. A. Backofen, *Trans. TMS-AIME*, vol. 239, 1967, p. 1034.

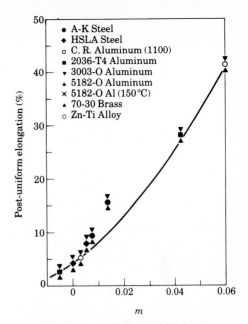

FIGURE 2.19 The effect of strain-rate sensitivity exponent m on the post-uniform (after necking) elongation for various metals. See Fig. 2.2(b). *Source*: After A. K. Ghosh, *Jr. Eng. Mat. Tech.*, vol. 99, 1977, pp. 264–274.

increases with increasing m (Fig. 2.19). It has also been observed that the value of m decreases with metals of increasing strength (Fig. 2.20).

The effect of strain rate on ductility is difficult to generalize, as can be seen from Fig. 2.21. Some of these phenomena are explained in Chapter 3. Since the formability of materials depends largely on their ductility, it is important to recognize the effect of temperature and strain rate on ductility. Generally, higher strain rates have an adverse effect on the ductility of materials. (The increase in ductility due to the strain-rate sensitivity of materials has been exploited in superplastic forming, as will be described in Section 7.13.)

2.2.8 EFFECTS OF HYDROSTATIC PRESSURE

Note that the tests described thus far have been carried out at ambient pressure. Tests have been performed under hydrostatic conditions, with pressures ranging up to 10^5 psi (10^3 MPa). Important observations have been made concerning the effects of high hydrostatic pressure on the behavior of materials:

a. Hydrostatic pressure substantially increases the strain at fracture (Fig. 2.22).

b. It has little or no effect on the shape of the true stress–true strain curve, but only extends it (Fig. 2.23).

c. It has no effect on the strain or the maximum load at which necking begins.

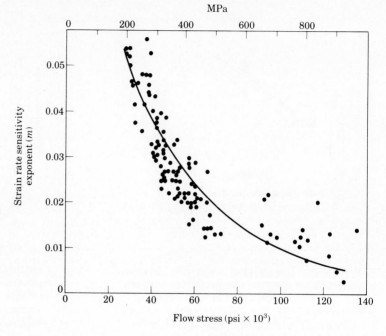

FIGURE 2.20 Relationship between flow stress for hot- and cold-rolled low-carbon steels and their *m* value at room temperature. Stronger steels are less sensitive to strain rate. Flow stress in this figure is the same as the uniaxial yield stress. *Source*: After A. Saxena and D. A. Chatfield, SAE Paper 760209, 1976.

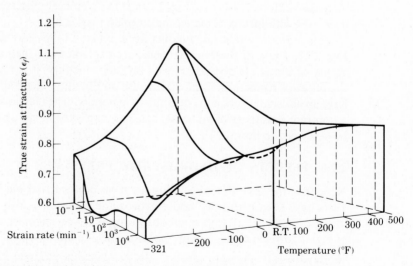

FIGURE 2.21 True strain at fracture for 303 stainless steel as a function of strain rate and temperature. This curve shows that it is difficult to generalize the combined effects of temperature and strain rate on the ductility of metals. *Source*: After G. W. Form and W. M. Baldwin, Jr., *Trans. ASM*, vol. 48, 1956, pp. 474–485.

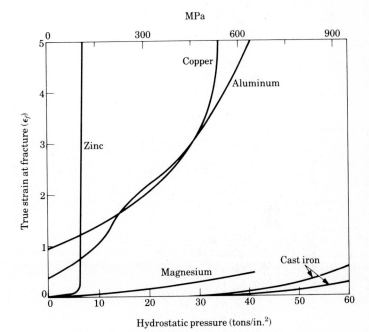

FIGURE 2.22 The effect of hydrostatic pressure on true strain at fracture in tension for various metals. Even cast iron becomes ductile under high pressure. *Source*: After H. Ll. D. Pugh and D. Green, *Proc. Inst. Mech. Eng.*, vol. 179, Part 1, No. 12, 1964–1965.

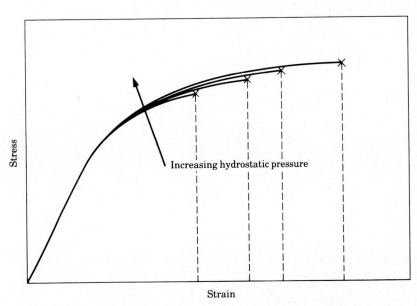

FIGURE 2.23 Schematic illustration of the effect of hydrostatic pressure on true stress-true strain curves. Note that pressure does not change the shape of the curve, it simply extends it. Ductility (true strain at fracture) and toughness of metals increase substantially with hydrostatic pressure, an observation that is significant in metalworking processes. See, for instance, Section 6.16.4.

The increase in ductility due to hydrostatic pressure has also been observed in other tests, such as compression and torsion. This increase in ductility has been observed not only with ductile metals, but also with brittle metals and nonmetallic materials. Materials such as cast iron, marble, and various rocks have been found to acquire some ductility (or increase in ductility), and thus deform plastically, when subjected to hydrostatic pressure. The level of the pressure required to enhance ductility depends on the material.

Experiments have shown that generally the mechanical properties of metals are not altered after being subjected to hydrostatic pressure.

2.2.9 EFFECTS OF RADIATION

In view of the nuclear applications of many metals and alloys, studies have been conducted on the effects of radiation on material properties. Typical changes in the mechanical properties of steels and other metals exposed to high-energy radiation are increased yield stress and tensile strength and hardness, and decreased ductility and toughness (Fig. 2.24). The magnitudes of these changes depend on the material and its condition, temperature, and level of radiation.

2.2.10 FLEXURE (BEND) TESTS

For materials with insufficient ductility or for brittle materials (certain tool and die materials made of ceramics and carbides), it can be difficult to prepare and test tension-test specimens. The test commonly used for such materials is the 3-point or 4-point flexure, or bend, test, as described in Section 3.7.2.

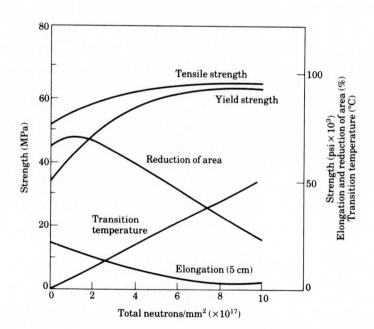

FIGURE 2.24 The effect of radiation on the strength and ductility for a carbon-silicon steel. Such curves are useful in predicting the service behavior of materials for nuclear applications. *Source*: After C. O. Smith, ORSORT, Oak Ridge, TN.

2.3 THE COMPRESSION TEST

Many operations in metalworking, such as forging, rolling, and extrusion, are performed with the workpieces under compressive loads. The compression test, where the specimen is subjected to a compressive load as shown in Fig. 2.1(b), can give useful information for these processes, such as stresses required and the behavior of the material under compression.

However, the deformation shown in Fig. 2.1(b) is ideal. The compression test is usually carried out by compressing a solid cylindrical specimen between two flat platens; the friction between the specimen and the dies is an important factor. Friction causes *barreling* (Fig. 2.25). In other words, friction prevents the top and bottom surfaces from expanding freely.

This situation makes it difficult to obtain relevant data and to construct properly the compressive stress–strain curve for the following reasons:

a. The cross-sectional area of the specimen changes along its height,

b. Friction dissipates energy and this energy is supplied through an increased compressive force (Fig. 2.26). Thus it is difficult to obtain results that are truly indicative of the properties of the material. With effective lubrication or other means (see Chapter 4) it is, of course, possible to minimize friction, and hence barreling, to obtain a reasonably constant cross-sectional area during this test.

FIGURE 2.25 Barreling in compressing a round solid cylindrical specimen (7075–O aluminum) between flat dies. Barreling is due to friction at the die-specimen interfaces, which retard the free flow of the material. See also Figs. 6.1 and 6.2. *Source*: K. M. Kulkarni and S. Kalpakjian.

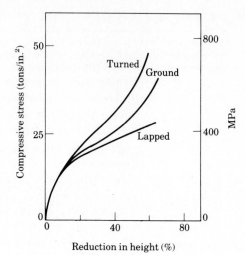

FIGURE 2.26 The effect of surface roughness of dies on the compressive stress for annealed copper. The stress must increase with roughness because additional work has to be provided to overcome friction. *Source*: After M. Cook and E. C. Larke, *Jr. Inst. Met.*, vol. 71, 1945, p. 386.

The engineering strain rate $\dot{e}$ in compression is given by

$$\dot{e} = -\frac{v}{h_o},\tag{2.17}$$

where v is the speed of the die and h_o is the original height of the specimen. The true strain rate $\dot{\epsilon}$ is given by

$$\dot{\epsilon} = -\frac{v}{h},\tag{2.18}$$

where h is the instantaneous height of the specimen. It can be seen that, if v is constant, the true strain rate increases as the test progresses.

In order to conduct this test at a constant true strain rate, a *cam plastometer* has been designed that, through a cam action, reduces v proportionately as h decreases during the test.

The compression test can also be used to determine the ductility of a metal by observing the cracks that form on the barreled cylindrical surfaces of the specimen. These are discussed in Section 6.7 in greater detail. It should be noted here that hydrostatic pressure has a beneficial effect in delaying the formation of these cracks.

It is apparent that, with a sufficiently ductile material and effective lubrication, compression tests can be carried out uniformly to large strains. This is unlike the tension test, where, even for very ductile materials, necking sets in after relatively little elongation.

Plane Strain Compression Test

Another test is the *plane strain compression test* (Fig. 2.27), which simulates processes such as rolling (see Section 6.10). In this test, the die and workpiece geometries are such that the width of the specimen does not undergo any significant change during

FIGURE 2.27 Schematic illustration of the plane-strain compression test. The dimensional relationships shown should be satisfied for this test to be useful and reproducible. This test gives the yield stress of the material in plane strain, Y'. *Source:* After A. Nadai and H. Ford.

$w > 5h$

$w > 5b$

$b > 2$ to $4h$

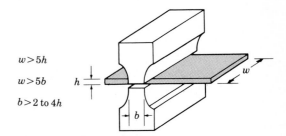

compression; that is, the material under the dies is in the condition of plane strain (see Section 2.8.3).

The yield stress of a material in plane strain Y' can be shown to be

$$Y' = \frac{2}{\sqrt{3}} Y = 1.15Y \qquad (2.19)$$

according to the distortion-energy criterion (see Section 2.8.3).

As can be seen from the recommended geometric relationships given in Fig. 2.27, the test parameters must be chosen properly to make the results meaningful. Furthermore, caution should be exercised in test procedures, such as preparing the die surfaces and aligning the dies, lubricating the surfaces, and accurately measuring the load.

When the results of tension and compression tests on the same material are compared, it is found that, for ductile metals, the true stress–true strain curves for both tests coincide (Fig. 2.28). However, this is not true for brittle materials, particularly in regard to ductility.

● **Illustrative Problem 2.4**

A 40-mm-high cylindrical specimen is being compressed between flat platens at a speed of 0.1 m/s. Calculate the engineering and true strain rates to which the material is being subjected when its height is 10 mm.

FIGURE 2.28 True stress-true strain curve in tension and compression for aluminum. For ductile metals the curves for tension and compression are identical. This is not true for brittle materials. *Source:* After A. H. Cottrell, *The Mechanical Properties of Matter*, New York: Wiley, 1964, p. 289

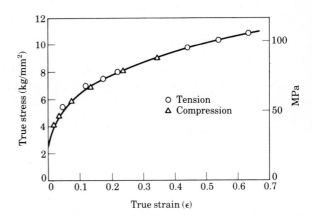

SOLUTION.   The engineering strain rate is

$$\dot{e} = -\frac{v}{h_o} = -\frac{0.1}{0.040} = -2.5 \text{ s}^{-1}.$$

The true strain rate is

$$\dot{\epsilon} = -\frac{v}{h} = -\frac{0.1}{0.010} = -10 \text{ s}^{-1}.$$

Note the large difference in the numbers obtained.   ●

2.3.1 BAUSCHINGER EFFECT

There are situations in deformation processing of materials where a workpiece is first subjected to tension and then to compression, or vice versa. Examples are bending and unbending, roller leveling (see Section 6.12.3), and reverse drawing (see Section 7.9.4). It has been observed that when a metal with a tensile yield stress Y is subjected to tension into the plastic range and then the load is released and applied in compression, the yield stress in compression is lower than that in tension (Fig. 2.29).

This phenomenon is known as the *Bauschinger effect* (after J. Bauschinger, 1886) and is exhibited in varying degrees by all metals and alloys. This effect is also observed when the loading path is reversed, i.e., compression followed by tension. Because of the lowered yield stress in the reverse direction of load application, this phenomenon is also called *strain softening* or *work softening*.

2.3.2 THE DISK TEST

For brittle materials a disk test is available where the disk is subjected to diametral compression. This test is covered in Section 3.7.2.

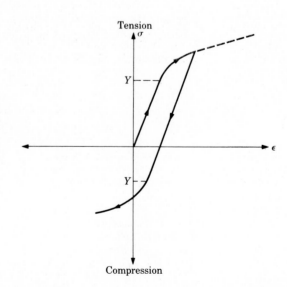

FIGURE 2.29 Schematic illustration of the Bauschinger effect. Arrows show loading and unloading paths. Note the decrease in the yield stress in compression after the specimen has been subjected to tension. The same result is obtained if compression is applied first, followed by tension, whereby the yield stress in tension decreases.

2.4 THE TORSION TEST

Another test method for determination of material properties is the torsion test. In order to obtain an approximately uniform stress and strain distribution along the cross-section, this test is generally carried out on a tubular specimen (Fig. 2.30).

The shear stress τ can be determined from the equation

$$\tau = \frac{T}{2\pi r^2 t},\tag{2.20}$$

where T is the torque, r is the mean radius, and t is the thickness of the reduced section of the tube.

The shear strain γ is determined from the equation

$$\gamma = \frac{r\phi}{\ell},\tag{2.21}$$

where ℓ is the length of the reduced section and ϕ is the angle of twist, in radians. With the shear stress and shear strain thus obtained from torsion tests, one can construct the shear stress–shear strain curve of the material (see Section 2.8.7).

In the elastic range, the ratio of the shear stress to shear strain is known as the *shear modulus* or the *modulus of rigidity, G,*

$$G = \frac{\tau}{\gamma}.\tag{2.22}$$

It can be shown that the shear modulus and the modulus of elasticity E are related by the formula

$$G = \frac{E}{2(1 + v)}.\tag{2.23}$$

Thus, for most metals E is about 2.6 times G.

Note that unlike tension and compression tests, one does not have to be concerned with changes in the cross-sectional area of the specimen in torsion testing. The

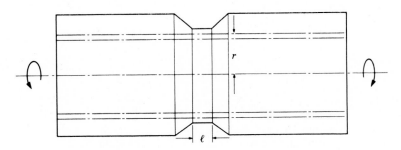

FIGURE 2.30 Notched-tube specimen for torsion testing. Torsion tests do not involve necking (as in tensile testing), or friction (as in compression testing).

shear stress–shear strain curves in torsion increase monotonically, hence they are analogous to true stress–true strain curves.

Torsion tests are performed on solid round bars at elevated temperatures in order to estimate the formability of a metal in forging. The greater the number of twists prior to failure, the greater the forgeability of the material (see Section 6.7).

Tests have also been conducted on round bars that are compressed axially. The maximum shear strain at fracture is measured as a function of the compressive stress. It has been found that the shear strain at fracture increases substantially as the compressive stress increases (Fig. 2.31). This observation again indicates the beneficial effect of a compressive environment on the ductility of materials. Other experiments show that, with tensile normal stresses, the curves in Fig. 2.31 follow a trend downward and to the left, signifying reduced ductility.

The effect of compressive stresses on increasing the maximum shear strain at fracture has also been observed in metal cutting, as treated in Section 8.3.

The normal compressive stress has been found to have no effect on the magnitude of shear stresses required to cause yielding or to continue the deformation, just as hydrostatic pressure has no effect on the general shape of the stress–strain curve (see Fig. 2.23).

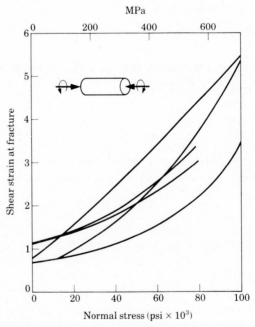

FIGURE 2.31   The effect of axial compressive stress on the shear strain at fracture in torsion for various steels. Note that the effect on ductility is similar to that of hydrostatic pressure, shown in Fig. 2.22. *Source*: Based on data in P. W. Bridgman, *Large Plastic Flow and Fracture*, New York: McGraw-Hill, 1952.

2.5 THE HARDNESS TEST

One of the most common tests for assessment of the mechanical properties of materials is the hardness test. Hardness of a material is generally defined as its resistance to permanent indentation. Less commonly, hardness may also be defined as resistance to scratching or to wear, as treated in Section 4.8.

Various techniques have been developed to measure the hardness of materials using different indenter materials and geometries. However, it has been found that the resistance to indentation depends on the shape of the indenter and the load applied. Thus, hardness is not a fundamental property.

Among the most common standardized hardness tests are the Brinell, Rockwell, Vickers, Knoop, and Scleroscope tests.

2.5.1 BRINELL TEST

In this test, introduced by the Swedish metallurgist J. A. Brinell in 1900, a steel or tungsten carbide ball 10 mm in diameter is pressed against a surface with a load of 500, 1500, or 3000 kg (Fig. 2.32). The Brinell hardness number (HB) is defined as the ratio of the load P to the curved area of indentation,

$$\text{HB} = \frac{2P}{(\pi D)(D - \sqrt{D^2 - d^2})} \text{ kg/mm}^2, \tag{2.24}$$

where D is the diameter of the ball and d is the diameter of the impression in millimeters.

Depending on the condition of the material, different types of impressions are obtained on the surface after performing a Brinell hardness test. Annealed materials generally have a rounded profile, whereas cold-worked (strain-hardened) materials have a sharp profile (Fig. 2.33). The correct method of measuring the indentation diameter d is shown in Figs. 2.33(a) and (b).

Because the indenter (with a finite elastic modulus) also undergoes elastic deformation under the applied load P, hardness measurements may not be as correct as expected. One method of minimizing this effect is to use tungsten carbide balls, which, because of their high modulus of elasticity (see Table 8.7), deform less than steel balls. Also, since harder workpiece materials produce very small impressions, a 1500-kg or 3000-kg load is recommended in order to obtain impressions that are sufficiently large for accurate measurement. Tungsten carbide balls are generally recommended for Brinell hardness numbers higher than 500. In reporting the test results for these high hardnesses the type of ball used should be cited.

Because the impressions made by the same indenter at different loads are not geometrically similar, the Brinell hardness number depends on the load used. Thus the load employed should also be cited with the test results. The Brinell test is generally suitable for materials of low to medium hardness.

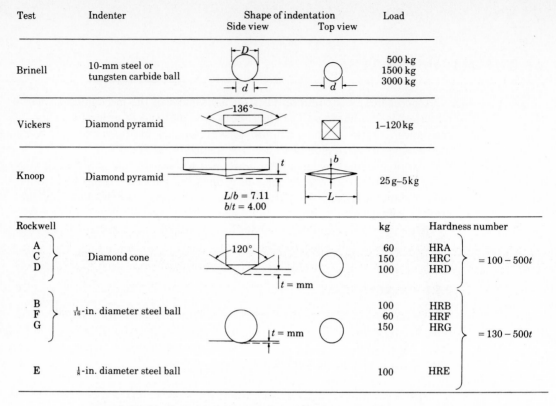

Test	Indenter	Shape of indentation Side view	Shape of indentation Top view	Load
Brinell	10-mm steel or tungsten carbide ball			500 kg 1500 kg 3000 kg
Vickers	Diamond pyramid	136°		1–120 kg
Knoop	Diamond pyramid	$L/b = 7.11$ $b/t = 4.00$		25 g–5 kg

Rockwell				kg	Hardness number
A C D	Diamond cone	120° $t = \text{mm}$		60 150 100	HRA HRC HRD $= 100 - 500t$
B F G	$\frac{1}{16}$-in. diameter steel ball	$t = \text{mm}$		100 60 150	HRB HRF HRG $= 130 - 500t$
E	$\frac{1}{8}$-in. diameter steel ball			100	HRE

FIGURE 2.32 General characteristics of hardness testing methods. The Knoop test is known as a microhardness test because of the light load and small impressions. *Source*: After H. W. Hayden, W. G. Moffatt, and V. Wulff.

Brinelling is a term used to describe permanent indentations on a surface between contacting bodies, for example, a component with a hemispherical protrusion or a ball bearing resting on a flat surface. Under fluctuating loads or vibrations, such as during transportation or vibrating foundations, a permanent indentation may be produced on the flat surface due to dynamic loading.

Meyer Hardness

Whereas the Brinell test is based on the curved area of indentation, the *Meyer test* (after E. Meyer, 1908) uses the projected area of indentation,

$$\text{Meyer hardness} = \frac{4P}{\pi d^2}. \tag{2.25}$$

This equation may be rewritten in the form

$$P = C_1 d^n, \tag{2.26}$$

where C_1 is a constant for the material and n is known as the Meyer hardness exponent, which is a measure of the strain hardening capability of the material. However, the Meyer test is not commonly used.

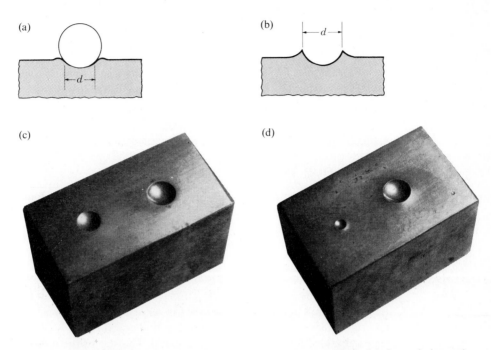

FIGURE 2.33 Indentation geometry for Brinell hardness testing. (a) Annealed metal. (b) Work-hardened metal. (c) Hardness impression on annealed steel. (d) Hardness impression on cold-worked steel. Note the difference of metal flow at the periphery of the impressions.

2.5.2 ROCKWELL TEST

In this test, developed by the metallurgist S. P. Rockwell in 1922, the depth of penetration is measured. The indenter is pressed on the surface, first with a minor load and then a major one. The difference in the depth of penetration is a measure of the hardness.

There are several Rockwell hardness test scales that use different loads and indenter materials and geometries. Some of the more common hardness scales and the indenters used are listed in Fig. 2.32. The Rockwell hardness number, which is read directly from a dial on the testing machine, is expressed as follows: If the hardness number is 55 using the C scale, then it is written as 55 HRC. *Rockwell superficial hardness* tests have also been developed using lighter loads and the same type of indenters.

The Rockwell hardness test is used for a wide range of hardnesses. The test is rapid and can be automated.

2.5.3 VICKERS TEST

The *Vickers* hardness test, also known as the *diamond pyramid hardness* test, uses a pyramid-shaped diamond indenter (see Fig. 2.32) with a load ranging from 1 to

120 kg. The test was developed in England in 1922 by R. Smith and G. Sandlund. The Vickers hardness number (HV) is given by the formula

$$HV = \frac{1.854P}{L^2}.$$ (2.27)

The impressions are typically less than 0.5 mm on the diagonal. The Vickers test gives essentially the same hardness number regardless of the load. It is suitable for testing materials with a wide range of hardness, including very hard steels.

2.5.4 KNOOP TEST

This test uses a diamond indenter in the shape of an elongated pyramid (see Fig. 2.32) with loads ranging generally from 25 g to 5 kg. The *Knoop* hardness number (HK) is given by the formula

$$HK = \frac{14.2P}{L^2}.$$ (2.28)

The Knoop test (after F. Knoop, 1939) is a *microhardness* test because of the light loads it employs and hence is suitable for very small or thin specimens and for brittle materials, such as gem stones, carbides, and glass. This test is also used in measuring the hardness of individual grains in a metal. The size of the indentation is generally in the range of 0.01 to 0.10 mm; thus surface preparation is very important. Because the hardness number obtained depends on the applied load, test results should always cite the load employed.

2.5.5 SCLEROSCOPE

The Scleroscope is an instrument in which a diamond-tipped indenter (hammer), enclosed in a glass tube, is dropped on the specimen from a certain height. The hardness is determined by the rebound of the indenter; the higher the rebound, the harder the specimen. Indentation is slight. Since the instrument is portable it is useful for measuring the hardness of large objects.

2.5.6 MOHS HARDNESS

This test is based on the capability of one material to scratch another. The Mohs hardness (after F. Mohs, 1822) is based on a scale of 10, with 1 for talc and 10 for diamond (hardest substance known). Thus, a material with a higher Mohs hardness can scratch ones with a lower hardness (Fig. 2.34).

The Mohs scale is used generally by mineralogists and geologists. However, some of the materials tested are of interest to manufacturing engineers, as discussed in Chapters 8, 9, and 11. Although the Mohs scale is qualitative, good correlation is obtained with the Knoop hardness. Soft metals have a Mohs scale of 2 to 3, hardened steels about 6, and aluminum oxide (used in grinding wheels) a scale of 9.

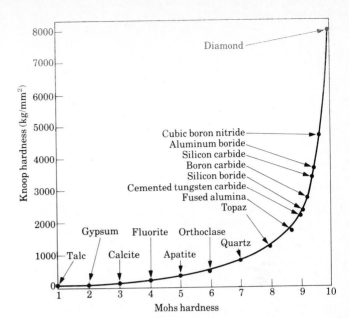

FIGURE 2.34 Comparison of Knoop and Mohs hardness scales for various materials. Diamond is the hardest substance known. Diamonds that are used in manufacturing industries are synthetically made.

2.5.7 HARDNESS AND STRENGTH

Since hardness is the resistance to permanent indentation, it is equivalent to performing a compression test on a small portion of the surface of a material. Thus, one would expect some correlation between hardness and yield stress Y in the form of

$$\text{Hardness} = cY, \tag{2.29}$$

where c is a proportionality constant.

Theoretical studies, based on plane strain slip-line analysis (Section 2.10.2) with a smooth flat punch indenting the surface of a semi-infinite body, have shown that for a perfectly plastic material of yield stress Y the value of c is about 3. This is in reasonably good agreement with experimental data (Fig. 2.35). Note that cold-worked materials (which are close to being perfectly plastic in their behavior) show better agreement than annealed ones. The higher c value for the annealed materials is explained by the fact that, due to strain hardening, the average yield stress they exhibit during indentation is higher than their initial yield stress.

The reason for hardness, as a compression test, giving higher values than the uniaxial yield stress Y of the material can be seen in the following analysis. If we assume that the volume under the indenter is a column of material (Fig. 2.36), then it would exhibit an uniaxial compressive yield stress Y. However, the volume being deformed under the indenter is, in reality, surrounded by a rigid mass (Fig. 2.37). The surrounding mass prevents this volume of material from deforming freely. In fact, this volume is under *triaxial compression*. As will be seen in Section 2.8 on yield criteria, this material requires a normal compressive yield stress that is higher than the uniaxial yield stress of the material.

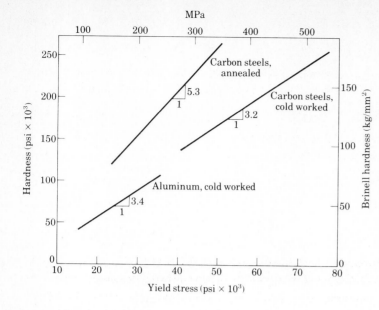

FIGURE 2.35 Relation between Brinell hardness and yield stress for aluminum and steels. For comparison, the Brinell hardness (which is always measured in kg/mm²) is converted to psi units on the left scale.

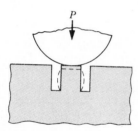

FIGURE 2.36 Simulation of a hardness test on a uniaxial column of metal, showing unconstrained deformation of the metal under the indenter. Because in actual hardness testing the metal is constrained, hardness values are about three times that of the uniaxial yield stress of the metal. *Source*: After E. Orowan.

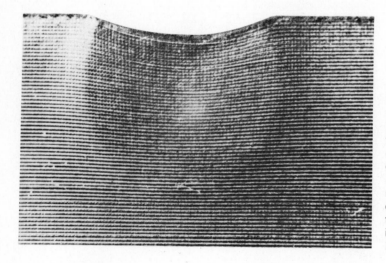

FIGURE 2.37 Bulk deformation in mild steel under a spherical indenter. Note that the depth of the deformed zone is about one order of magnitude larger than the depth of indentation. For a hardness test to be valid, the material should be allowed to fully develop this zone. This is why thinner specimens require smaller impressions. *Source*: Courtesy of M. C. Shaw and C. T. Yang.

More practically, a relationship has also been observed between the ultimate tensile strength (UTS) and Brinell hardness number (HB) for steels, as follows:

$$\text{UTS} = 500(\text{HB}), \tag{2.30}$$

where *UTS* is in psi and HB in kg/mm^2 as measured with a load of 3000 kg. In *SI* units, the relationship is given by

$$\text{UTS} = 3.5(\text{HB}), \tag{2.31}$$

where UTS is in MPa.

2.5.8 HARDNESS TESTING PROCEDURES

For a hardness test to be meaningful and reliable, the *zone of deformation* under the indenter (see Fig. 2.37) must be allowed to develop freely. Consequently, the location of the indenter with respect to the edges of the specimen to be tested and the thickness of the specimen are important considerations. It is generally recommended that the location be at least two diameters (of the ball) from the edge of the specimen, and that the thickness should be at least 10 times the depth of penetration of the indenter. These can be appreciated by reviewing Fig. 2.37.

Moreover, the indentation should be sufficiently large to give a representative hardness value for the bulk material. If hardness variations are to be detected in a small area, or if the hardness of individual constituents in a matrix or an alloy is to be determined, the indentations should be very small, like those in Vickers or Knoop tests obtained under light loads.

While surface preparation is not critical for the Brinell test and somewhat more important for the Rockwell test, it is important for the other hardness tests because of the small size of the indentations. Surfaces may have to be polished in order to measure correctly the dimensions of the impression.

The values obtained from different hardness tests can be correlated and converted to different scales (Fig. 2.38). Charts are available in the technical literature for such conversions. Care should be exercised in using these charts because of the many variables involved with regard to material characteristics and indentation geometry.

Although most hardness testers require relatively small specimens or parts, portable testers have been developed (in addition to the Scleroscope) for soft metals and nonmetallic materials. These testers, which can be used on large parts, usually operate on the principle of a spring-loaded indenter mounted in a suitable frame. New hardness testing methods are also being developed. One is based on the eddy current principle and utilizes the electrical and magnetic properties of the workpiece material. In another method, the deceleration of an impacting indenter is measured.

Hot hardness tests can also be carried out using conventional testers with certain modifications, such as surrounding the specimen and indenter with a small electric furnace. The hot hardness of materials is important in applications where the materials are subjected to elevated temperatures, such as in cutting tools in machining and dies for metalworking (see Fig. 8.52).

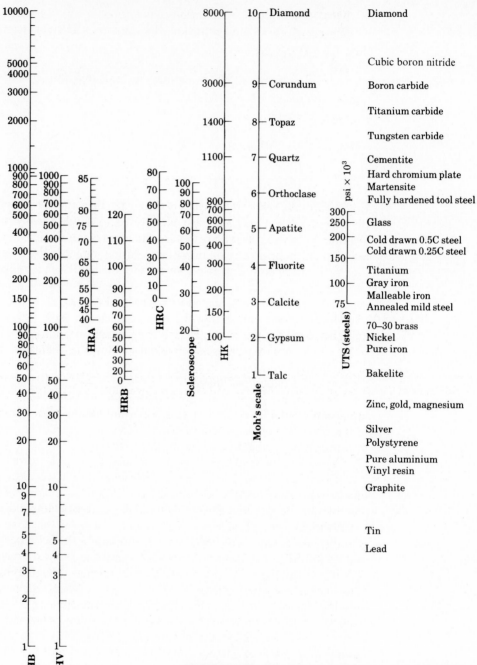

FIGURE 2.38 Chart for converting various hardness scales. Because of the various factors involved, hardness conversions are approximate.

● **Illustrative Problem 2.5**

A piece of steel is highly deformed at room temperature. Its hardness is found to be 300 HB. Estimate the modulus of resilience for this material in lb/in^2.

SOLUTION. Since the steel has been subjected to large strains at room temperature, we may assume that its stress–strain curve has flattened considerably, thus approaching the shape of a perfectly plastic curve. According to Eq. (2.29) and using a value of $c = 3$, we obtain

$$Y = \frac{300}{3} = 100 \text{ kg/mm}^2 = 142,250 \text{ psi.}$$

The modulus of resilience is defined as in Eq. (2.6),

$$\text{Modulus of resilience} = \frac{Y^2}{2E}.$$

For steel, $E = 30 \times 10^6$ psi. Hence,

$$\text{Modulus of resilience} = \frac{(142,250)^2}{2 \times 30 \times 10^6} = 337 \text{ in. } \cdot \text{lb/in}^3. \quad ●$$

2.6 DEFORMATION-ZONE GEOMETRY

The observations made concerning Fig. 2.37 are important in estimating and calculating forces in metalworking operations. As shown above for hardness testing, the compressive stress required for indentation is, ideally, about 3 times the yield stress Y required for uniaxial compression. It was also noted that:

a. The deformation under the indenter is localized, making the overall deformation highly nonuniform,

b. The deformation zone is relatively small compared to the dimensions of the specimen.

On the other hand, in a simple frictionless compression test with flat dies, the top and bottom surfaces of the specimen are always in contact with the dies and the specimen deforms uniformly.

You can visualize different situations covering the wide range between these two extreme examples of specimen-deformation geometry. The *deformation zones* and the pressures required are shown in Fig. 2.39 for the frictionless condition as obtained from slip-line analysis (Section 2.10.2). Note that the ratio h/L is the important parameter in determining the inhomogeneity of deformation. It is also important to emphasize the frictionless nature of these examples since, as will be shown in Section 2.10.1, friction has a significant effect on forces, particularly at small values of h/L.

Deformation-zone geometry depends on the particular metalworking process and such parameters as the die geometry and percent reduction of the material (Fig. 2.40). Details of these processes and the role of the deformation geometry are covered in Chapter 6.

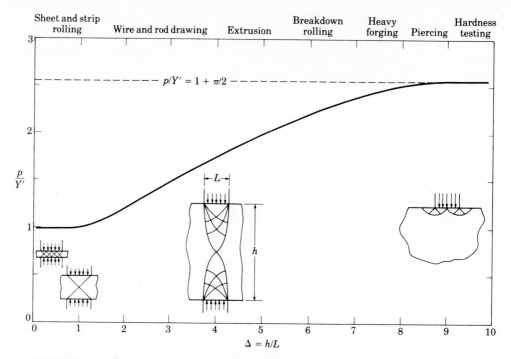

FIGURE 2.39 Die pressures required in frictionless plane-strain conditions for a variety of metalworking operations. The geometric relationship between contact area of the dies and workpiece dimensions is an important factor in predicting forces in plastic deformation of materials. *Source*: After W. A. Backofen, *Deformation Processing*, Reading, Mass.: Addison-Wesley, 1972, p. 135.

2.7 RESIDUAL STRESSES

In this section it will be shown that inhomogeneous deformation, such as that discussed above, leads to *residual stresses*—stresses that remain within a part after it has been deformed and all external forces have been removed.

A typical example of inhomogeneous deformation is the bending of a beam (Fig. 2.41). The bending moment first produces a linear elastic stress distribution. As the moment is increased, the outer fibers begin to yield and, for a typical strain-hardening material, the stress distribution shown in Fig. 2.41(b) is eventually obtained. After the part is bent (permanently, since it has undergone plastic deformation) the moment is removed by unloading. This unloading is equivalent to applying an equal and opposite moment to the beam.

As shown in Fig. 2.8, all recovery is elastic. This means that in Fig. 2.41(c) the moments of the areas *oab* and *oac*, about the neutral axis, must be equal. (For the purposes of this treatment it is assumed that the neutral axis does not shift.)

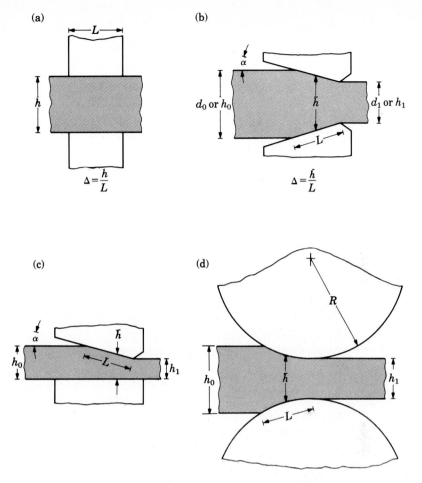

FIGURE 2.40 Examples of plastic deformation processes in plane strain showing the h/L ratio. (a) Indenting with flat dies. This is similar to the cogging operation shown in Fig. 6.18. (b) Drawing or extrusion of strip with a wedge-shaped die, described in Chapter 6. (c) Ironing. See also Fig. 7.46. (d) Rolling, described in Section 6.11. As shown in Fig. 2.39, the larger the h/L ratio, the higher the die pressure. In actual processing, however, the smaller this ratio the greater is the effect of friction at the die-workpiece interfaces. This is because contact area, hence friction, increases with decreasing h/L ratio.

The difference between the two stress distributions produces the residual stress pattern within the beam. Note that there are compressive residual stresses in layers *ad* and *oe*, and tensile in layers *do* and *ef*. Since there are no external forces, the residual stress system in the beam must be in static equilibrium.

Although this example involves stresses in one direction only, in most situations in deformation processing the residual stresses are three-dimensional.

In the example described above, the equilibrium of residual stresses may be disturbed by altering the geometry of the beam, such as by removing a layer of material.

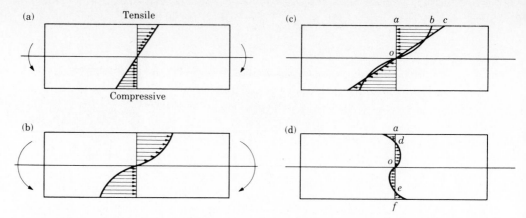

FIGURE 2.41 Residual stresses developed in bending a beam made of an elastic, strain-hardening material. Note that unloading is equivalent to applying an equal and opposite moment to the part, as shown in (b). Because of nonuniform deformation, most parts made by plastic deformation processes contain residual stresses. Note that the forces and moments due to residual stresses must be internally balanced.

The beam will then acquire a new radius of curvature in order to ensure the balance of internal forces. Another example of this effect is the drilling of round holes on surfaces with residual stresses. It may be found that, as a result of removing this material, the equilibrium of the residual stresses is disturbed and the hole becomes oval.

Such disturbances of residual stresses lead to *warping*, some simple examples of which are shown in Fig. 2.42. The equilibrium of residual stresses may also be disturbed by *relaxation* of these stresses over a period of time; this results in instability of the dimensions and shape of the component. These dimensional changes can be an important consideration for precision machinery and measuring equipment.

Residual stresses are also caused by *phase changes* in metals during or after processing due to density differences between phases (e.g., between ferrite and martensite in steels). This causes microscopic volumetric changes and results in residual stresses. This phenomenon is important in warm and hot working and in heat treatment following cold working (e.g., deformation at room temperature).

Residual stresses can also be caused by *temperature gradients* within a body, such

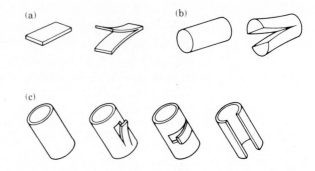

FIGURE 2.42 Distortion of parts with residual stresses after cutting or slitting. (a) Rolled sheet or plate. (b) Drawn rod. (c) Thin-walled tubing. Because of the presence of residual stresses on the surfaces of parts, a round drill may produce an oval-shaped hole because of relaxation of stresses when a portion is removed.

as during the cooling cycle of a casting, applying brakes to a railroad wheel, or in a grinding operation. The expansion and contraction due to temperature gradients are analogous to nonuniform plastic deformation.

2.7.1 EFFECTS OF RESIDUAL STRESSES

Tensile residual stresses on the surface of a part are generally considered to be undesirable because they lower the fatigue life and fracture strength of the part. A surface with tensile residual stresses can sustain lower additional tensile stresses (due to external loading) than a surface that is free of any residual stresses. This is particularly true for relatively brittle materials where fracture takes place with little or no plastic deformation.

Tensile residual stresses in manufactured products can also lead to *stress cracking* or *stress–corrosion cracking* over a period to time (see Section 3.7.1).

On the other hand, compressive residual stresses on a surface are generally desirable. In fact, in order to increase the fatigue life of components, compressive residual stresses are imparted on surfaces by techniques such as *shot peening* and *surface rolling* (see Section 4.13.2).

Residual stresses on a surface can also have an effect on the action of lubricants, as described in Section 4.11.

2.7.2 REDUCTION OF RESIDUAL STRESSES

Residual stresses may be reduced or eliminated either by *stress-relief annealing* (Section 3.5) or by further *plastic deformation*. Given sufficient time, residual stresses may also be diminished at room temperature by *relaxation*. The time required can be greatly reduced by raising the temperature of the component.

Relaxation of residual stresses by stress-relief annealing is generally accompanied by warpage of the part. Hence a "machining allowance" is commonly provided to compensate for dimensional changes during stress relieving.

The mechanism of reduction or elimination of residual stresses by plastic deformation is as follows: Assume that a piece of metal has the residual stresses shown in Fig. 2.43, namely, tensile on the outside and compressive on the inside. The part with these stresses, which are in the elastic range, is in equilibrium. Also assume that the material is elastic–perfectly plastic, as shown by the diagram in the figure.

The level of residual stresses are shown on the stress–strain diagram, both being below the yield stress Y. If a uniformly distributed tension is applied to this specimen, points σ_c and σ_t in the diagram move up on the stress–strain curve, as shown by the arrows. The maximum level that these stresses can reach is the tensile yield stress Y. With sufficiently high loading, the stress distribution becomes uniform throughout the part, as shown in Fig. 2.43(c). If the load is now removed, the stresses recover elastically and the part has no residual stresses.

Note that very little stretching is required to relieve these residual stresses. This is because the elastic portions of the stress–strain curves for metals are very steep, hence the elastic stresses can be raised to the yield stress with very little strain.

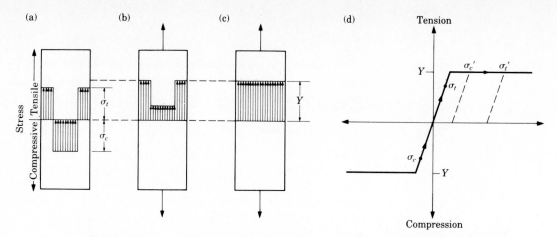

FIGURE 2.43 Elimination of residual stresses by stretching. See Section 2.7.2 for details. Residual stresses can also be reduced or eliminated by thermal treatments, such as stress relieving or annealing.

The technique for reducing or relieving residual stresses by plastic deformation, such as by stretching as described above, requires sufficient straining to establish a uniformly distributed stress in the part. It is therefore apparent that a material such as the elastic, linearly strain-hardening type, Fig. 2.8(e), can never reach this condition since the compressive stress $\sigma_{c'}$ will always lag behind $\sigma_{t'}$. If the slope of the stress–strain curve in the plastic region is small, then the difference between $\sigma_{c'}$ and $\sigma_{t'}$ will be rather small and little residual stresses will be left in the part after unloading.

● **Illustrative Problem 2.6**

In Fig. 2.43 assume that $\sigma_t = 140$ MPa and $\sigma_c = -140$ MPa. The material is aluminum and the length of the specimen is 0.25 m. Calculate the length to which this specimen should be stretched so that, when unloaded, it will be free of any residual stresses. Assume that the yield stress of the material is 150 MPa.

SOLUTION. The stretching should be to the extent whereby σ_c reaches the yield stress in tension, Y. Thus the total strain should be

$$\epsilon_{\text{total}} = \frac{\sigma_c}{E} + \frac{Y}{E}.$$

For aluminum, let $E = 70$ GPa. Thus,

$$\epsilon_{\text{total}} = \frac{140}{70 \times 10^3} + \frac{150}{70 \times 10^3} = 0.00414.$$

Hence the stretched length should be

$$\ln\left(\frac{\ell_f}{0.25}\right) = 0.00414$$

or

$$\ell_f = 0.2510 \text{ m}.$$

Since the strains are very small, we may use engineering strains in these calculations. Thus,

$$\frac{\ell_f - 0.25}{0.25} = 0.00414$$

or

$$\ell_f = 0.2510 \text{ m}. \quad \bullet$$

2.7.3. DETERMINATION OF RESIDUAL STRESSES

Various experimental techniques have been developed for the determination (also called measurement) of residual stresses in manufactured parts. The most common nondestructive technique is by x-ray analysis, where the interatomic spacing of a set of lattice planes is measured and is compared with that of a stress-free specimen.

The most common destructive technique involves removal of layers of material from a part and then measuring the dimensional changes, such as changes in the length or curvature of the part. These dimensional changes are very small and are generally measured with the use of strain gages. For large deflections, other measuring instruments can be used. The residual stresses are then calculated from a set of equations relating stresses to strains. The work involved is somewhat tedious and time consuming, especially in the three-dimensional case.

Another technique for determining residual stresses is to place strain gages on a surface and then drill a small hole near it, usually 0.020 in. (0.5 mm) in diameter. The strain gages measure the strains around the drilled hole. Based on these strains, the residual stresses can then be computed. This is a semidestructive method. In addition to the determination of surface residual stresses by this method, a variation of this technique (called incremental drilling method) can be used to compute residual stresses in the underlying layers of a material.

2.8 YIELD CRITERIA

In most manufacturing operations involving deformation processing, the material, unlike that in a simple tension or compression test specimen, is generally subjected to *triaxial* stresses. For example, in the expansion of a thin-walled spherical shell under internal pressure, an element in the shell is subjected to equal biaxial tensile stresses (Fig. 2.44a). In drawing a rod or wire through a conical die, an element in the deformation zone is subjected to tension in its length direction and to compression on its conical surface (Fig. 2.44b). An element in the flange in deep drawing of sheet metal is subjected to a tensile radial stress and compressive stresses on its surface and in the

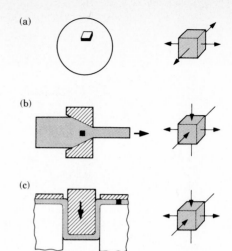

FIGURE 2.44 The state of stress in various metalworking operations. (a) Expansion of a thin-walled spherical shell under internal pressure. (b) Drawing of round rod or wire through a conical die to reduce its diameter. (c) Deep drawing of sheet metal with a punch and die to make a cup.

circumferential direction (Fig. 2.44c). As will be shown in subsequent chapters, many similar examples can be given where the material is subjected to various normal and shear stresses during processing.

In a simple tension or compression test, when the applied stress reaches the uniaxial yield stress Y, the material will deform plastically. However, if the material is subjected to a more complex state of stress there are relationships between these stresses that will predict yielding. These relationships are known as *yield criteria*, the most common ones being the maximum-shear-stress criterion and the distortion-energy criterion.

2.8.1 MAXIMUM-SHEAR-STRESS CRITERION

This criterion (also known as the *Tresca criterion*, after H. Tresca, 1864) states that yielding occurs when the maximum shear stress within an element is equal to or exceeds a critical value. As will be seen in Section 3.2.2, this critical value of the shear stress is a material property called *shear yield stress* (k). Thus, for yielding,

$$\tau_{max} \geq k. \tag{2.32}$$

The most convenient way of determining the stresses on an element is by the use of *Mohr circles* (after O. Mohr, 1914). Some typical examples are shown in Fig. 2.45. Note that in these examples the normal stresses are *principal stresses*; in other words, they act on planes on which there are no shear stresses. (The construction of Mohr circles and the determination of principal stresses will not be discussed here as they are covered in textbooks on strength of materials.)

If the maximum shear stress, as determined from Fig. 2.45 or from appropriate equations, is equal to or exceeds k, then yielding will occur. This can be best visualized by the construction in Fig. 2.46. In order to cause yielding, the largest circle must touch the "ceiling" represented by the shear yield stress k. Note that there are many combinations of stresses (or states of stress) that can give the same maximum shear stress.

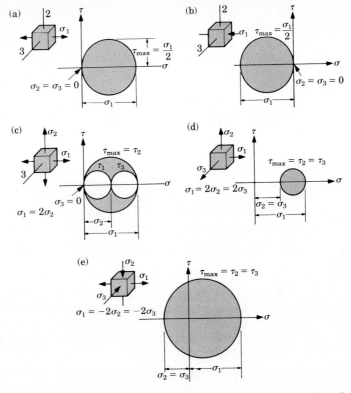

FIGURE 2.45 Mohr's circles for various states of stress. (a) Uniaxial tension. (b) Uniaxial compression. (c) Biaxial tension (plane stress). (d) Triaxial tension. (e) Biaxial compression with tension. These states of stress are encountered in most metalworking processes.

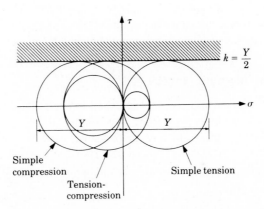

FIGURE 2.46 Three sets of Mohr's circles representing stresses that are large enough to cause yielding. Note that there is an infinite number of states of stress that produce circles of the same maximum diameter, hence yielding. For any state of stress, the radius of the largest circle must be equal to the shear yield stress k to cause yielding.

If, for some reason, we are unable to increase the stresses on the element in order to cause yielding, then the solution is simply to lower the ceiling by raising the temperature of the material. This is the basis and one major reason for hot working of materials (see Section 3.6).

From the simple tension test, we find that

$$k = \frac{Y}{2}. \tag{2.33}$$

It should be pointed out that the material is assumed to be *continuous*, *homogeneous*, and *isotropic*, i.e., it has the same properties in all directions. Also, tensile stresses are positive and compressive stresses negative, and the yield stress in tension and in compression are assumed to be equal. These are important assumptions.

The maximum-shear-stress criterion can now be written as

$$\sigma_{max} - \sigma_{min} = Y = 2k. \tag{2.34}$$

This means that the maximum and minimum normal stresses produce the largest circle and hence the largest shear stress. Consequently, the intermediate stress has no effect on yielding. It should be emphasized that the left-hand side of Eq. (2.34) represents the applied stresses and the right-hand side is a material property.

2.8.2 DISTORTION-ENERGY CRITERION

This criterion (also known as the *von Mises criterion*, after R. von Mises, 1913) states that yielding will occur when the relationship between the principal applied stresses and the uniaxial yield stress Y of the material is as follows:

$$(\sigma_1 - \sigma_2)^2 + (\sigma_2 - \sigma_3)^2 + (\sigma_3 - \sigma_1)^2 = 2Y^2. \tag{2.35}$$

Note that, unlike the maximum-shear-stress criterion, the intermediate principal stress is included in this equation. Here again, the left-hand side of the equation represents the applied stresses and the right-hand side a material property.

● **Illustrative Problem 2.7**

A thin-walled spherical shell is under internal pressure p. The shell is 20 in. in diameter and 0.1 in. thick. It is made of a perfectly plastic material with a yield stress of 20,000 psi. Calculate the pressure required to cause yielding of the shell according to both yield criteria discussed above.

SOLUTION. For this shell under internal pressure, the membrane stresses are given by

$$\sigma_1 = \sigma_2 = \frac{pr}{2t},$$

where $r = 10$ in. and $t = 0.1$ in. The stress in the thickness direction, σ_3, is negligible because of the high r/t ratio of the shell. Thus, according to the maximum-shear-stress criterion,

$$\sigma_{max} - \sigma_{min} = Y$$

or

$$\sigma_1 - 0 = Y$$

and

$$\sigma_2 - 0 = Y.$$

Hence, $\sigma_1 = \sigma_2 = 20,000$ psi.

The pressure required is then

$$p = \frac{2tY}{r} = \frac{(2)(0.1)(20,000)}{10} = 400 \text{ psi.}$$

According to the distortion-energy criterion,

$$(\sigma_1 - \sigma_2)^2 + (\sigma_2 - \sigma_3)^2 + (\sigma_3 - \sigma_1)^2 = 2Y^2$$

or

$$0 \quad + \quad \sigma_2^2 \quad + \quad \sigma_1^2 \quad = 2Y^2.$$

Hence $\sigma_1 = \sigma_2 = Y.$

Thus the answer is the same, or

$$p = 400 \text{ psi. } \bullet$$

● **Illustrative Problem 2.8**

Explain why a correction factor has to be applied in the construction of a true stress–true strain curve, shown in Fig. P2.2, from tensile test data.

SOLUTION. The stress distribution at the neck of a specimen is shown in Fig. P2.3, where we note that in this region there is a triaxial state of stress. This is so because each element in the region has a different cross-sectional area; the smaller the area, the greater the tensile stress on the element. Hence, element 1 will contract more than element 2, and so on. However, element 1 is restrained from contracting freely by element 2, and element 2 is restrained by element 3, and so on. This restraint causes radial and circumferential tensile stresses in the necked region. This situation results in an axial tensile stress distribution as shown in Fig. P2.3.

The true uniaxial stress in tension is σ, whereas the calculated value of true stress at fracture is the average stress. Hence, a correction has to be made. A mathematical analysis by P. W. Bridgman gives the ratio of true to average stress as

$$\frac{\sigma}{\sigma_{av}} = \frac{1}{(1 + 2R/a)[\ln(1 + a/2R)]}, \tag{2.36}$$

where R is the radius of curvature of the neck and a is the radius of the specimen at the neck. Since R is not easy to measure during a test, an empirical relation has

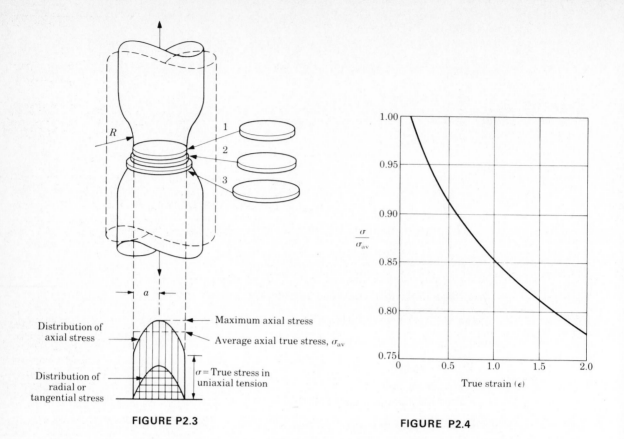

Distribution of
axial stress

Distribution of
radial or
tangential stress

Maximum axial stress

Average axial true stress, σ_{av}

σ = True stress in
uniaxial tension

FIGURE P2.3

FIGURE P2.4

been established between a/R and the true strain at the neck, Fig. P2.4. The corrected true stress–true strain curve is shown in Fig. P2.2. ●

2.8.3 PLANE STRESS AND PLANE STRAIN

These two states of stress are important in the application of yield criteria. *Plane stress* is the state of stress in which one or two of the pairs of faces on an elemental cube are free of any stresses. An example is the torsion of a thin-walled tube. There are no stresses normal to the inside or outside surface of the tube; hence the state of the stress of the tube is one of plane stress. Other examples are shown in Figs. 2.44(a) and 2.47(a).

The state of stress where one of the pairs of faces on an element undergoes zero strain is known as *plane strain*. An example is shown in Fig. 2.47(b) depicting a piece of material being compressed in a die; note that one pair of faces is touching the walls (groove) of the die and cannot expand. Another example is the *plane strain compression test* shown in Fig. 2.27. Here, by proper choice of specimen dimensions, the width of the specimen is kept essentially constant. (Note that an element does not have to be physically constrained on the pair of faces for plane strain conditions to exist.) A third example is the torsion of a thin-walled tube in which the wall thickness remains constant (see Section 2.8.7).

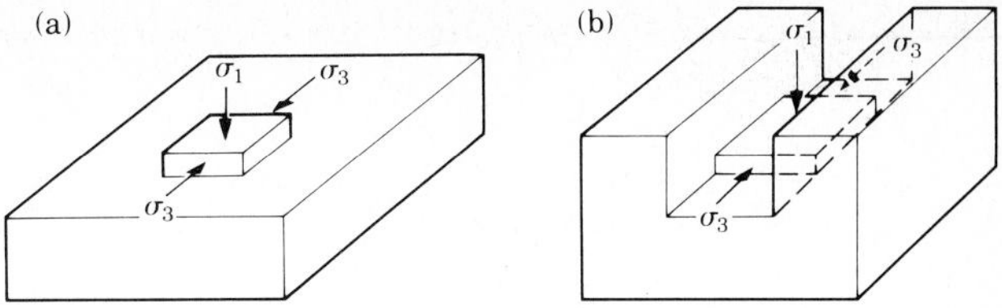

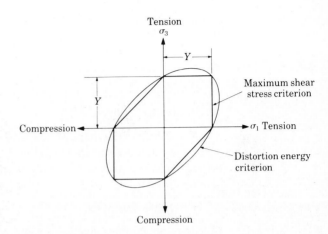

FIGURE 2.47 Two examples of states of stress. (a) Plane stress. (b) Plane strain.

FIGURE 2.48 Plane stress diagrams for maximum shear stress and distortion energy criteria. The stresses applied should fall on or outside these curves to cause yielding. (It is assumed that the yield stress Y in tension and compression are equal, as they are for ductile but not for brittle materials.)

A review of the two yield criteria outlined above indicates that the plane stress condition can be represented by the diagram in Fig. 2.48. It can be seen that the maximum-shear-stress criterion gives an envelope of straight lines. The distortion-energy criterion for plane stress reduces to the equation

$$\sigma_1^2 + \sigma_3^2 - \sigma_1\sigma_3 = Y^2 \tag{2.37}$$

and is shown graphically in the figure. Whenever a point (with its coordinates representing the two principal stresses) falls on these boundaries, the element will yield.

Yielding under plane strain conditions requires the determination of the stress level, if any, on the faces of the element undergoing plane strain (Fig. 2.47b). This is done through the use of *generalized Hooke's law* equations shown below.

$$\epsilon_1 = \frac{1}{E}\left[\sigma_1 - v(\sigma_2 + \sigma_3)\right]$$

$$\epsilon_2 = \frac{1}{E}\left[\sigma_2 - v(\sigma_1 + \sigma_3)\right] \tag{2.38}$$

$$\epsilon_3 = \frac{1}{E}\left[\sigma_3 - v(\sigma_1 + \sigma_2)\right]$$

For plane strain, one of the strains in these equations is zero. This means that there exists a particular relationship between the principal stresses. At yielding, the Poisson's ratio v is 0.5 (since the volume of an element undergoing plastic deformation is found to remain constant) and, therefore, all three stresses can be determined. Thus,

77

for the case in Fig. 2.47(b),

$$\sigma_2 = \frac{\sigma_1 + \sigma_3}{2} \tag{2.39}$$

Note that σ_2 is now an intermediate stress.

For the plane strain compression of Fig. 2.47, the distortion-energy criterion (which includes the intermediate stress) reduces to

$$\sigma_1 - \sigma_3 = \frac{2}{\sqrt{3}} Y = 1.15Y = Y'.$$

● **Illustrative Problem 2.9**

A specimen in the shape of a cube 10 mm on each side is being compressed without friction in a die cavity as shown in Fig. 2.47(b), where the width of the groove is 15 mm. Assume that the material is made of an elastic, linearly strain-hardening material given by

$$\sigma = 100 + 20\epsilon \text{ MPa}.$$

Calculate the compressive force required when the height of the specimen is 3 mm, according to both yield criteria.

SOLUTION. It can be seen that, when the height of this specimen is reduced to 3 mm, the surface area A will be

$$(10)(10)(10) = (3)(A).$$

Hence,

$$A = 333.3 \text{ mm}^2.$$

Since the groove is only 15 mm wide, the specimen will touch the walls of the groove, because $(15)(15) = 225 \text{ mm}^2$, which is smaller than the final surface area required. Thus, this is both a plane strain and also a plane stress problem.

The true strain after deformation is (in absolute value)

$$|\epsilon| = \ln\left(\frac{10}{3}\right) = 1.204.$$

Hence the strength level (flow stress, see Fig. 2.51) that the material will exhibit at this strain is

$$Y_f = 100 + (20)(1.204) = 124.08 \text{ MPa}.$$

According to the maximum-shear-stress criterion, the force required is

$$P = Y_f A = (124.08)(333.3) = 41,350 \text{ N} = 41.35 \text{ kN}.$$

According to the distortion-energy criterion,

$$Y_f = (1.15)(124.08) = 142.69 \text{ MPa}.$$

Hence,

$$P = (142.69)(333.3) = 47,560 \text{ N} = 47.56 \text{ kN} \quad \bullet$$

2.8.4 EXPERIMENTAL VERIFICATION OF YIELD CRITERIA

The yield criteria described above have been tested experimentally. A suitable specimen commonly used is a thin-walled tube under internal pressure and/or torsion. Under such loading, it is possible to generate different states of plane stress. Various experiments, with a variety of ductile materials, have shown that the distortion energy criterion agrees better with the experimental data than does the maximum-shear-stress criterion (Fig. 2.49).

This would suggest that one should use the distortion-energy criterion for the analysis of metalworking processes since the latter generally makes use of ductile materials. On the other hand, the simpler maximum-shear-stress criterion can also be used, particularly by designers, since the difference between the two is negligible for most practical applications.

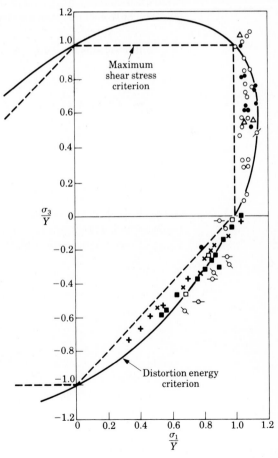

FIGURE 2.49 Verification of yield criteria by experimental data for various materials, superimposed on plane-stress diagrams. Note that the data points fall closer to the distortion-energy criterion than to the maximum-shear-stress criterion.

2.8.5 VOLUME STRAIN

By summing the three equations of the generalized Hooke's law it is seen that

$$\epsilon_1 + \epsilon_2 + \epsilon_3 = \frac{1 - 2v}{E}(\sigma_1 + \sigma_2 + \sigma_3), \tag{2.40}$$

where the left-hand side of the equation can be shown to be the volume strain or *dilatation* Δ. Thus,

$$\Delta = \frac{\text{vol. change}}{\text{orig. vol.}} = \frac{1 - 2v}{E}(\sigma_1 + \sigma_2 + \sigma_3). \tag{2.41}$$

It can now be shown that, in the plastic range, where $v = 0.5$, the volume change is zero. Hence, in plastic working,

$$\epsilon_1 + \epsilon_2 + \epsilon_3 = 0, \tag{2.42}$$

which is a convenient means of determining a third strain if two strains are known.

The *bulk modulus* is defined as

$$\text{Bulk modulus} = \frac{\sigma_{av}}{\Delta} = \frac{E}{3(1 - 2v)}, \tag{2.43}$$

where

$$\sigma_{av} = \tfrac{1}{3}(\sigma_1 + \sigma_2 + \sigma_3). \tag{2.44}$$

From Eq. (2.41) it can be seen that, in the elastic range, where $0 < v < 0.5$, the volume of a tension-test specimen increases and that of a compression specimen decreases.

2.8.6 EFFECTIVE STRESS AND EFFECTIVE STRAIN

A convenient means of expressing the state of stress on an element is the *effective* (*equivalent* or *representative*) *stress* $\bar{\sigma}$ and *effective strain* $\bar{\epsilon}$ given by the equations below. For the maximum-shear-stress criterion,

$$\bar{\sigma} = \sigma_1 - \sigma_3, \tag{2.45}$$

and for the distortion-energy criterion,

$$\bar{\sigma} = \frac{1}{\sqrt{2}}[(\sigma_1 - \sigma_2)^2 + (\sigma_2 - \sigma_3)^2 + (\sigma_3 - \sigma_1)^2]^{1/2}. \tag{2.46}$$

The factor $1/\sqrt{2}$ is chosen so that, for simple tension, the effective stress is equal to the uniaxial yield stress Y.

The strains are likewise related to the effective strain. For the maximum-shear-stress criterion,

$$\bar{\epsilon} = \tfrac{2}{3}(\epsilon_1 - \epsilon_3), \tag{2.47}$$

and for the distortion-energy criterion,

$$\bar{\epsilon} = \frac{\sqrt{2}}{3} [(\epsilon_1 - \epsilon_2)^2 + (\epsilon_2 - \epsilon_3)^2 + (\epsilon_3 - \epsilon_1)^2]^{1/2}. \tag{2.48}$$

Again, the factors 2/3 and $\sqrt{2}/3$ are chosen so that for simple tension the effective strain is equal to the uniaxial tensile strain. It is apparent that stress–strain curves may also be called effective stress–effective strain curves.

2.8.7 COMPARISON OF NORMAL STRESS–STRAIN AND SHEAR STRESS–STRAIN CURVES

Stress–strain curves in tension and torsion for the same material are, of course, comparable. Also, it is possible to construct one curve from the other since the material is the same. The procedure for this conversion is outlined below.

The following observations are made in regard to Fig. 2.50, showing the tension

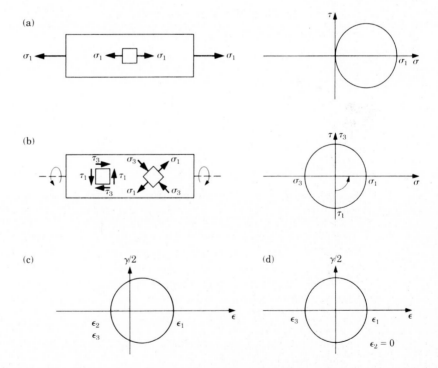

FIGURE 2.50 Mohr's circle diagrams for stress and strain in uniaxial tension, (a) and (c), and torsion, (b) and (d). Guy wires for antennas, or the spokes of a bicycle, are subjected to uniaxial tension (because they are thin and long), whereas the drive shaft of an automobile is subjected to the stresses shown in (b).

and torsional states of stress:

a. In the tension test, the uniaxial stress σ_1 is also the effective stress and the principal stress.

b. In the torsion test, the principal stresses occur on planes whose normals are at 45 degrees to the longitudinal axis; the principal stresses σ_1 and σ_3 are equal in magnitude but opposite in sign.

c. The magnitude of the principal stress in torsion is the same as the maximum shear stress.

We now have the following relationships:

$$\sigma_1 = -\sigma_3,$$

$$\sigma_2 = 0,$$

$$\sigma_1 = \tau_1.$$

Substituting these stresses in Eqs. (2.45) and (2.46) for effective stress, the following relationships are obtained: For the maximum-shear-stress criterion,

$$\bar{\sigma} = \sigma_1 - \sigma_3 = \sigma_1 + \sigma_1 = 2\sigma_1 = 2\tau_1, \tag{2.49}$$

and for the distortion-energy criterion,

$$\bar{\sigma} = \frac{1}{\sqrt{2}} [(\sigma_1 - 0)^2 + (0 + \sigma_1)^2 + (-\sigma_1 - \sigma_1)^2]^{1/2} = \sqrt{3}\sigma_1 = \sqrt{3}\tau_1. \tag{2.50}$$

With regard to strains, the following observations can be drawn from Fig. 2.50:

a. In the tension test, $\epsilon_2 = \epsilon_3 = -\dfrac{\epsilon_1}{2}$.

b. In the torsion test, $\epsilon_1 = -\epsilon_3 = \dfrac{\gamma}{2}$.

c. The strain in the thickness direction of the tube is zero, i.e., $\epsilon_2 = 0$.

Observation (c) is true because the thinning due to the principal tensile stress is countered by the thickening under the principal compressive stress of the same magnitude. Hence, $\epsilon_2 = 0$. Since σ_2 is also zero, a thin-walled tube under torsion is both a plane stress and a plane strain situation.

Substituting these strains in Eqs. (2.47) and (2.48) for effective strain, the following relationships are obtained: For the maximum-shear-stress criterion,

$$\bar{\epsilon} = \tfrac{2}{3}(\epsilon_1 - \epsilon_3) = \tfrac{2}{3}(\epsilon_1 + \epsilon_1) = \tfrac{4}{3}\epsilon_1 = \tfrac{2}{3}\gamma, \tag{2.51}$$

and for the distortion-energy criterion,

$$\bar{\epsilon} = \frac{\sqrt{2}}{3} [(\epsilon_1 - 0)^2 + (0 + \epsilon_1)^2 + (-\epsilon_1 - \epsilon_1)^2]^{1/2} = \frac{2}{\sqrt{3}}\epsilon_1 = \frac{1}{\sqrt{3}}\gamma. \tag{2.52}$$

This set of equations provides a means by which tensile-test data can be converted to torsion-test data, and vice versa.

2.9 WORK OF DEFORMATION

In this section the work required for plastic deformation of materials will be discussed. Since work is defined as the product of force and distance (collinear with force), a quantity equivalent to work per unit volume is the product of stress and strain. Since the relation between stress and strain in the plastic range depends on the particular stress–strain curve, this work is best calculated by referring to Fig. 2.51.

Note that the area under the true stress–true strain curve for any strain ϵ_1 is the *energy per unit volume u (specific energy)* of the material deformed. This is expressed as

$$u = \int_0^{\epsilon_1} \sigma \, d\epsilon. \tag{2.53}$$

As seen in Section 2.2.3, true stress–true strain curves can be represented by the expression

$$\sigma = K\epsilon^n.$$

Hence, Eq. (2.53) can be written as

$$u = K \int_0^{\epsilon_1} \epsilon^n \, d\epsilon$$

or

$$u = \frac{K\epsilon_1^{n+1}}{n+1} = \bar{Y}\epsilon_1, \tag{2.54}$$

where $\bar{Y}$ is the average flow stress of the material.

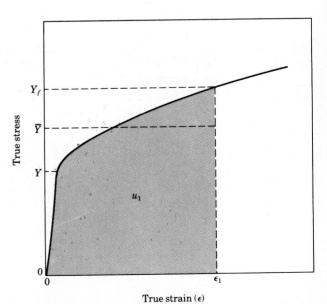

FIGURE 2.51 Schematic illustration of true stress-true strain curve showing yield stress Y, average flow stress $\bar{Y}$, specific energy u_1, and flow stress Y_f. Flow stress is defined as the true stress required to continue plastic deformation at a particular true strain. Thus, for strain-hardening materials there is a flow stress for each strain.

This energy represents the work dissipated in uniaxial deformation. For a more general condition, where the workpiece is subjected to triaxial stresses and strains, the effective stress and effective strains can be used. The energy per unit volume is then

$$u = \int_0^{\bar{\epsilon}} \bar{\sigma} \, d\bar{\epsilon}. \tag{2.55}$$

To obtain the work expended, we multiply u by the volume of the material deformed. Thus,

$$\text{Work} = (u)(\text{volume}). \tag{2.56}$$

The energy represented by Eq. (2.54) is the minimum energy or the *ideal* energy required for uniform (homogeneous) deformation. The energy required for actual deformation involves two additional factors. One is the energy required to overcome friction at the die–workpiece interfaces. The other is the *redundant work* of deformation, which is described as follows:

In Fig. 2.52(a), a block of material is being deformed into shape, such as by forging, extrusion, or drawing through a die as described in Chapter 6. As shown in sketch (b) this deformation is uniform, or homogeneous. In reality, however, the material more often than not deforms as in sketch (c) due to the effects of friction and die geometry. The difference between (b) and (c) is that (c) has undergone additional shearing along horizontal planes.

This shearing requires expenditure of energy since additional plastic work has to be done in subjecting the various layers to shear strains. This is known as *redundant work*; the word redundant is due to the fact that this work does not contribute to the shape change of the material. [Note that (b) and (c) have the same overall shape and dimensions.]

The total specific energy required can now be written as

$$u_{\text{total}} = u_{\text{ideal}} + u_{\text{friction}} + u_{\text{redundant}}. \tag{2.57}$$

The efficiency of a process is defined as

$$\eta = \frac{u_{\text{ideal}}}{u_{\text{total}}}. \tag{2.58}$$

(a)   (b)   (c)

FIGURE 2.52 Deformation of grid patterns in a workpiece. (a) Original pattern. (b) After ideal deformation. (c) After inhomogeneous deformation, requiring redundant work of deformation. Note that (c) is basically (b) with additional shearing, especially at the outer layers. Thus, part (c) requires greater work of deformation than part (b).

The magnitude of this efficiency varies widely, depending on the particular process, frictional conditions, die geometry, and other process parameters. Typical values are estimated to be 30 to 60% for extrusion and 75 to 95% for rolling.

● **Illustrative Problem 2.10**

A thin-walled spherical shell, made of a perfectly plastic material of yield stress Y, original radius r_o and thickness t_o, is being expanded by internal pressure. Calculate the work done in expanding this shell to a radius of r_f. If the diameter expands at a constant rate, what changes take place in the power consumed as the radius increases?

SOLUTION. The membrane stresses are given by

$$\sigma_1 = \sigma_2 = Y$$

(from Illustrative Problem 2.7), where r and t are instantaneous dimensions. The true strains in the membrane are given by

$$\epsilon_1 = \epsilon_2 = \ln\left(\frac{2\pi r_f}{2\pi r_o}\right) = \ln\left(\frac{r_f}{r_o}\right).$$

Because an element in this shell is subjected to equal biaxial stretching, the specific energy is

$$u = \int_0^{\epsilon_1} \sigma_1 \, d\epsilon_1 + \int_0^{\epsilon_2} \sigma_2 \, d\epsilon_2 = 2\sigma_1\epsilon_1 = 2Y \ln\left(\frac{r_f}{r_o}\right).$$

Since the volume of the shell material is $4\pi r_o^2 t_o$, the work done is

$$W = (u)(\text{volume}) = 8\pi Y r_o^2 t_o \ln\left(\frac{r_f}{r_o}\right)$$

The specific energy can also be calculated from the effective stresses and strains. Thus, according to the distortion-energy criterion,

$$\bar{\sigma} = \frac{1}{\sqrt{2}} [(0)^2 + (\sigma_2)^2 + (-\sigma_1)^2]^{1/2} = \sigma_1 = \sigma_2$$

and

$$\bar{\epsilon} = \frac{\sqrt{2}}{3} [(0)^2 + (\epsilon_2 + 2\epsilon_2)^2 + (-2\epsilon_2 - \epsilon_2)^2]^{1/2} = 2\epsilon_2 = 2\epsilon_1.$$

(The thickness strain $\epsilon_3 = -2\epsilon_2 = -2\epsilon_1$ because of volume constancy in plastic deformation, where $\epsilon_1 + \epsilon_2 + \epsilon_3 = 0$.) Hence,

$$u = \int_0^{\bar{\epsilon}} \bar{\sigma} \, d\bar{\epsilon} = \int_0^{2\epsilon_1} \sigma_1 \, d\epsilon_1 = 2\sigma_1\epsilon_1.$$

Thus the answer is the same.

Power is defined as

$$\text{Power} = \frac{dW}{dt}.$$

The expression for work can be written as

$$W = k \ln\left(\frac{r}{r_o}\right) = k(\ln r - \ln r_o),$$

since all other factors in the expression are constant. Thus,

$$\text{Power} = \frac{k}{r}\frac{dr}{dt}.$$

Because the shell is expanding at a constant rate, $dr/dt = $ constant. Hence the power is related to the instantaneous radius r by

$$\text{Power} \propto \frac{1}{r}. \quad \bullet$$

2.9.1 WORK AND HEAT

Almost all the mechanical work of deformation in plastic working is converted into *heat*. This conversion of work to heat is not 100% because a small portion of this energy is stored within the deformed material as elastic energy. This is known as *stored energy* (discussed in Section 3.5). Stored energy is generally 5 to 10% of the total energy input. However, it may be as high as 30% in some alloys.

In a simple frictionless process and assuming that work is completely converted into heat, the temperature rise is given by

$$\Delta T = \frac{u_{\text{total}}}{\rho c}, \tag{2.59}$$

where u_{total} is the specific energy from Eq. (2.57), ρ is the density, and c is the specific heat of the material. It can be seen that higher temperatures are associated with large areas under the stress–strain curve and smaller values of specific heat.

The theoretical temperature rise for a true strain of 1 (such as a 27-mm-high specimen compressed down to 10 mm) has been calculated to be as follows:

Aluminum	165°F	(75°C)
Copper	285	(140)
Low-carbon steel	535	(280)
Titanium	1060	(570)

The temperature rise given by Eq. (2.59) is for an ideal situation, where there is no heat loss. In actual operations heat is lost to the environment, to the tools and dies, and to any lubricants or coolants used. If the process is performed very rapidly, these losses are relatively small.

Under extreme conditions, an adiabatic process is approached, with very high temperature rise, leading to *incipient melting*. This rise in temperature can be calculated provided the stress–strain curve used is at the appropriate strain rate level. On the other hand, if the process is carried out slowly, the actual temperature rise will be a small portion of the calculated value. It should also be pointed out that properties such as specific heat and thermal conductivity depend on temperature and this should be taken into account in the calculations.

● **Illustrative Problem 2.11**

Calculate the total work done for the specimen in Illustrative Problem 2.2. Calculate the specific energy for an element in the necked area and the theoretical temperature rise.

SOLUTION.   The total work done on the specimen can be obtained from the area under the curve in Fig. P2.1. This area can be determined graphically and is found to be

Work = 3950 in. · lb.

The specific energy of an element in the necked area, i.e., at fracture, is the area under the true stress–true strain curve (corrected) in Fig. P2.2, namely,

$$u = \int_0^{1.253} \sigma \, d\epsilon = 155,000 \text{ in.} \cdot \text{lb/in}^3.$$

The theoretical temperature rise, i.e., adiabatic, is obtained from Eq. (2.59), where, for stainless steel, $\rho = 0.29$ lb/in^3 and $c = 0.12$ Btu/lb · °F. Thus,

$$\Delta T = \frac{155,000}{(0.29)(0.12)(778)(12)} = 477°F = 247°C.$$

Note that in these calculations only the work of plastic deformation is considered since no friction or redundant work is involved in a simple tension test. Also, the actual temperature rise will be lower because of the heat loss from the necked zone to the rest of the specimen and to the environment. ●

2.10 METHODS OF ANALYSIS OF METALWORKING PROCESSES

The treatments thus far in this chapter pertain generally to simple situations of tension, compression, and torsion where the state of stress can easily be defined. In actual metalworking processes, however, the material is generally subjected to a complex state of stress along with the additional component of friction at the die–workpiece interfaces. Furthermore, these stresses and friction generally vary along the interfaces.

The accurate calculation or prediction of loads, forces, stresses, and temperatures is essential not only for the proper design of equipment but also for studying the behavior of the material during processing.

In this section, the major methods of analysis used for deformation processing of materials will be reviewed briefly. The advantages and limitations of each technique are described individually.

2.10.1 SLAB METHOD OF ANALYSIS

This is one of the simpler methods of analyzing the stresses and loads in plastic deformation of materials. It requires the selection of an element in the workpiece and identifying all the normal and frictional forces acting on this element.

As an example, let us take the case of a simple compression process with friction (Fig. 2.53a), which is the basic deformation mode in forging. The purpose of this analysis is to determine the die pressure distribution from which one can then calculate the load required.

As the flat dies compress the part, it is reduced in thickness and, since the volume remains constant, the part expands laterally. This relative movement at the die–workpiece interfaces causes frictional forces acting in opposition to the movement of the piece. These frictional forces are shown by the horizontal arrows in Fig. 2.53(a). For simplicity, let us also assume that the deformation is in plane strain, i.e., the workpiece is not free to flow in the direction perpendicular to this page.

Let us now take an element, as shown in the figure, and indicate all the forces acting on it (Fig. 2.53b). Note the correct direction of the frictional forces. Also note the difference in the horizontal forces acting on the sides of the element; this difference is due to the presence of frictional forces on the element. It is assumed that the lateral stress distribution σ_x is uniform along the height h.

The next step in this analysis is to balance the horizontal forces on this element since the element must be in static equilibrium. Thus,

$$(\sigma_x + d\sigma_x)h + 2\mu\sigma_y\, dx - \sigma_x h = 0$$

or

$$d\sigma_x + \frac{2\mu\sigma_y}{h}\, dx = 0.$$

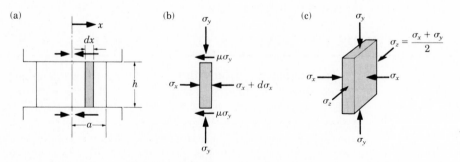

FIGURE 2.53 Stresses on an element in plane-strain compression (forging) between flat dies. The stress σ_x is assumed to be uniformly distributed along the height h of the element. Identifying the stresses on an element (slab) is the first step in the slab method of analysis for metalworking processes, described in Section 2.10.1, and used extensively in Chapter 6.

Note that we have one equation but two unknowns. The second equation is obtained from the yield criteria as follows. As shown in Fig. 2.53(c), this element is subjected to triaxial compression (compare this element with Fig. 2.47b). Using the distortion-energy criterion for plane strain, the following expression is obtained:

$$\sigma_y - \sigma_x = \frac{2}{\sqrt{3}} Y = Y'. \tag{2.60}$$

Thus,

$$d\sigma_y = d\sigma_x.$$

Note that σ_y and σ_x are assumed to be principal stresses. In the strictest sense, σ_y cannot be a principal stress because a shear stress is also acting on the same plane. However, this assumption is acceptable for low values of the coefficient of friction μ and is the standard practice in this method of analysis.

There are now two equations that can be solved by noting that

$$\frac{d\sigma_y}{\sigma_y} = -\frac{2\mu}{h} dx$$

or

$$\sigma_y = Ce^{-2\mu x/h}.$$

The boundary conditions are such that at $x = a$, $\sigma_x = 0$ and thus $\sigma_y = Y'$ at the edges of the specimen. (Since all stresses are compressive, we may ignore negative signs for stresses. This is traditional in such analysis. See also Chapter 6.) Hence the value of C becomes

$$C = Y'e^{2\mu a/h},$$

and thus

$$p = \sigma_y = Y'e^{2\mu(a-x)/h}. \tag{2.61}$$

Also,

$$\sigma_x = \sigma_y - Y' = Y'[e^{2\mu(a-x)/h} - 1]. \tag{2.62}$$

Equation (2.61) is plotted qualitatively in Fig. 2.54 in dimensionless form. Note

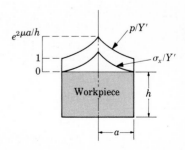

FIGURE 2.54 Distribution of die pressure p in plane-strain compression with sliding friction. Note that the pressure at the left and right boundaries is equal to the yield stress in plane strain, Y'. Sliding friction means that the frictional stress is directly proportional to the normal stress. See Section 4.6.

that the pressure increases exponentially toward the center of the part and also that it increases with the a/h ratio. Further coverage of these formulas is given in Section 6.3.1 on forging.

Because of its simplicity, this method of analysis has been used widely for almost all metalworking processes and is quite useful in estimating forces in processing. Other examples are given in Chapter 6.

2.10.2 SLIP-LINE ANALYSIS

This method has been applied generally to plane-strain conditions. The deforming body is assumed to be rigid, perfectly plastic, and isotropic. The technique consists of the construction of a family of straight or curvilinear lines that intersect each other orthogonally (see also Fig. 2.39). These lines, known as a *slip-line field*, correspond to the directions of yield stress of the material in shear, k.

The network of slip-lines, whose construction depends largely on intuition and experience and are postulated a priori, must satisfy certain conditions. These are: static equilibrium of forces, yield criteria, and boundary conditions. The slip-line fields must be checked for equilibrium conditions so that, for instance, the slip-lines meet at a free surface at a 45-degree angle. Also, they must be compatible with the velocity field. This means that the movement of the material must be such that mass continuity is maintained.

A simple two-dimensional example using this method is given in Fig. 2.55, simulating the hardness test with a flat rectangular indenter. In this model, the material is assumed to deform along shear planes as shown. The movements of the triangular blocks of material should be such that they can slide along each other, but

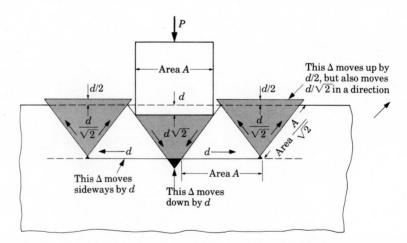

FIGURE 2.55 A simple two-dimensional model of slip-line analysis for indentation of a solid body with a flat rectangular punch. This model is similar to stacking of triangular wooden blocks. *Source*: Adapted from M. A. Ashby and D. R. H. Jones, *Engineering Materials*, Elmsford, N.Y.: Pergamon, 1980, p. 106.

cannot separate from each other. Thus the velocity normal to a plane of shear must be the same for points on either side of the plane. This is a simplified model. Instead of three triangular blocks, the deformation could consist of a large number of smaller blocks.

We can now determine the hardness of this material (P/A) as follows. The work done by the indenter is the product of P and d (assuming that P is not a function of the magnitude of d). This external work must be equal to the internal work done in shearing the blocks against each other. Letting k be the shear yield stress and noting that each triangular block has two sides, the following relation is obtained:

$$Pd = 2k\left(\frac{A}{\sqrt{2}}\,d\sqrt{2} + Ad + 2\,\frac{A}{\sqrt{2}}\,\frac{d}{\sqrt{2}}\right)$$

or

$$P = 6Ak.$$

Hence,

$$\text{Hardness} = \frac{P}{A} = 6k = 3Y. \tag{2.63}$$

Even with this very simplified model, one is able to obtain an expression for hardness that is in good agreement with Eq. (2.29).

Because a number of slip-line fields that satisfy all the conditions can be constructed, this method cannot give a unique solution for a particular deformation process. A solution must be verified by checking it with experimental data. (Another example of slip-line analysis is given in Fig. 8.11 in regard to chip formation in metal cutting.)

The slip-line method has been used successfully in the analysis of a number of metalworking processes (forging, rolling, extrusion) to predict stresses, loads, directions of material flow, and temperature variations within a workpiece. The frictional conditions at the die–workpiece interfaces can also be included in the analysis. Some success has also been obtained in studying axisymmetric cases (compression of a cylinder), strain-hardening effects, and anisotropic materials.

Many details are involved in the application of this technique. Therefore further treatment of this topic is beyond the scope of this text.

2.10.3 UPPER-BOUND TECHNIQUE

In this method, the overall deformation zone is divided into a number of smaller zones within which the velocity of a particle is continuous. The particle velocities in adjacent zones may be different. However, just as in slip-line analysis, at the boundaries between the zones, or between a zone and the die surface, all movement must be such that discontinuities in velocity occur only in the tangential direction.

This technique, which has been used successfully for axisymmetric cases, such as extrusion and rod drawing, is lengthy and gives results similar to those in slip-line analysis. In this analysis, the total power consumed in an operation is the sum of the following:

a. *Ideal* power of deformation,

b. Power consumed in *shearing* the material along velocity discontinuities, and

c. Power required in overcoming *friction* at the die–workpiece interfaces.

In the final analysis, a velocity field that minimizes the total calculated power is taken as the actual one and is subsequently compared with experimental data.

Theorems have been developed to obtain lower- and upper-bound solutions for the required loads. The lower-bound solution underestimates the load and hence is of no practical value. The upper-bound solutions, on the other hand, overestimate the load and are therefore of practical interest in metalworking operations.

2.10.4   VISIOPLASTICITY

This is an experimental technique developed to determine the strain rates and the stress distributions in the deformation zone. It consists of placing a grid pattern on a flat surface (or at an interface, such as the median plane in a solid cylinder) and observing the distortion of the grid after subjecting the specimen to an incremental (small) deformation. The process is then repeated a number of times.

From the distortions, the strain rates are calculated (since the strain and the time of incremental deformation are known) and then, with the use of plasticity equations, the stresses are calculated.

The visioplasticity technique can be used for plane strain and for axisymmetric cases. It is also used to validate predictions of stress distributions obtained from other analytical or numerical methods. Accuracy depends on the precision with which the grid patterns are placed and measured after each incremental deformation. Small errors in measurement can lead to significant errors in the calculation of stresses.

2.10.5   FINITE-ELEMENT METHOD

In this method, the deformation zone in an elastic–plastic body is divided into a number of elements interconnected at a finite number of nodal points. The actual velocity distribution is then approximated for each element. A set of simultaneous equations is then developed representing unknown velocity vectors. From the solution of these equations actual velocity distributions and the stresses are calculated.

This technique can incorporate friction conditions at the die–workpiece interfaces and actual properties of the workpiece material. It has been applied to relatively complex geometries in bulk deformation and in sheet-forming problems. Accuracy is influenced by the number and shape of the finite elements, the deformation increment, and the methods of calculation. To ensure accuracy, complex problems require extensive computations. The finite-element method gives a detailed outline of the actual stresses and strain distributions throughout the workpiece.

SUMMARY

1. The mechanical behavior of materials in manufacturing processes is related to their strength, ductility, elasticity, hardness, and the energy required for plastic deformation.

2. The behavior of materials depends on the particular material and on a number of other variables, such as temperature, strain rate, and state of stress.

3. Tests have been developed to obtain the relation between stress and strain as a function of the above variables. These relationships are important in assessing the behavior of a particular material in a manufacturing process, especially in regard to forces required and the capability of the material to undergo the desired deformation without failure.

4. Two important parameters are the strain-hardening capability (indicated by the strain-hardening exponent n) and the strain-rate sensitivity of the material (indicated by the exponent m).

5. How a material is subjected to a shape change is also important. This requires the study of the deformation-zone geometry. The manner in which a material is subjected to plastic deformation is important in determining the nature and extent of residual stresses in the part after all external forces are removed. These stresses are important in subsequent processing of the part or during its service life.

6. Since in actual metalworking operations the material is generally subjected to three-dimensional stresses, yield criteria have been developed to establish relationships between the uniaxial yield stress of the material (generally obtained from a tension test) and the stresses applied.

7. Various methods of analysis are available to calculate the stresses, forces, and energies required in deformation processing of materials. These analyses are important not only for the selection of appropriate equipment for metalworking, but also in their design.

BIBLIOGRAPHY

General Introductory Texts

Alexander JM, Brewer RC. *Manufacturing Properties of Materials.* London: Van Nostrand, London, 1963.

Ashby MF, Jones DRH. *Engineering Materials: An Introduction to their Properties and Applications.* New York: Pergamon, 1980.

Biggs WD. *The Mechanical Behavior of Engineering Materials.* New York: Pergamon, 1965.

Cook NH. *Manufacturing Analysis.* Reading, Mass.: Addison–Wesley, 1966.

Cottrell AH. *The Mechanical Behavior of Matter.* New York: Wiley, 1964.

Crane FAA. *Mechanical Working of Metals.* New York: Macmillan, 1964.

Davis HE, Troxell GE, Hauck GFW. *The Testing of Engineering Materials*, 4th ed. New York: McGraw-Hill, 1982.

Dieter GE. *Mechanical Metallurgy*, 2d ed. New York: McGraw-Hill, 1976.

Honeycombe RWK. *Plastic Deformation of Metals*. London: Edward Arnold, 1968.

Marin J. *Mechanical Behavior of Engineering Materials*. Englewood Cliffs, N.J.: Prentice-Hall, 1962.

McClintock FA, Argon AS, eds. *Mechanical Behavior of Materials*. Reading, Mass.: Addison–Wesley, 1966.

Moss JB. *Properties of Engineering Materials*. London: Butterworths, 1971.

Mott BW. *Micro-Indentation Hardness Testing*. London: Butterworths, 1956.

O'Neill H. *Hardness Measurement of Metals and Alloys*. London: Chapman and Hall, 1967.

Parkins RN. *Mechanical Treatment of Metals*. London: Allen and Unwin, 1968.

Polakowski NH, Ripling EJ. *Strength and Structure of Engineering Materials*. Englewood Cliffs, N.J.: Prentice-Hall, 1964.

Pugh HLD. *Mechanical Behaviour of Materials under Pressure*. New York: Elsevier, 1971.

Schey JA. *Introduction to Manufacturing Processes*, New York: McGraw-Hill, 1977.

Suh NP, Turner APL. *Elements of the Mechanical Behavior of Solids*. New York: McGraw-Hill, 1975.

Tabor D. *The Hardness of Metals*. New York: Oxford, 1951.

Advanced Texts

Avitzur B. *Handbook of Metal-Forming Processes*. New York: Wiley, 1983.

Avitzur B. *Metal Forming: Processes and Analysis*. New York: McGraw-Hill, 1968.

Backofen WA. *Deformation Processing*. Reading, Mass.: Addison–Wesley, 1972.

Calladine CR. *Engineering Plasticity*. New York: Pergamon, 1969.

Ford H, Alexander JM. *Advanced Mechanics of Materials*, 2d ed. New York: Halsted, 1977.

Hill R. *The Mathematical Theory of Plasticity*. New York: Oxford, 1950.

Hoffman O, Sachs G. *Introduction to the Theory of Plasticity for Engineers*. New York: McGraw-Hill, 1953.

Johnson W, Mellor PB. *Engineering Plasticity*. New York: Van Nostrand, 1973.

Mendelson A. *Plasticity: Theory and Application*, New York: Macmillan, 1968.

Nadai A. *Theory of Flow and Fracture of Solids*, 2d ed. New York: McGraw-Hill, vol. I, 1950, vol. II, 1963.

Prager W, Hodge PG. *Theory of Perfectly Plastic Solids*. New York: Wiley, 1951.

Slater RA. *Engineering Plasticity: Theory and its Application to Metal Forming Processes*, New York: Halsted, 1974.

Thomsen EG, Yang CT, Kobayashi S. *Mechanics of Plastic Deformation in Metal Processing*. New York: Macmillan, 1964.

PROBLEMS

2.1. Explain why an offset method is used to determine the yield stress on a stress–strain curve in tension. Would it be necessary to use this method for a highly strain-hardened metal?

2.2. Using the same scale for stress, it can be seen that the tensile engineering stress–strain curve for a material is lower than its true stress–true strain curve. Is this also true for a compression test on the same material?

2.3. Assume that a material, with a uniaxial yield stress Y, yields under a stress system of principal stresses $\sigma_1, \sigma_2, \sigma_3$, where $\sigma_1 > \sigma_2 > \sigma_3$. Show that the superposition of a hydrostatic stress p on this system (such as placing the specimen in a pressurized chamber) does not affect yielding. In other words, the material will still yield according to the yield criteria.

2.4. A cylindrical specimen of a brittle material 1 in. high and 1 in. in diameter is subjected to compression. It is found that fracture takes place at an angle of 45 degrees under a load of 30,000 lb. Calculate the shear stress and the normal stress on the fracture surface.

2.5. A paper clip is made of wire 1 mm in diameter. If the original material from which the wire is made is a rod 10 mm in diameter, calculate the longitudinal and lateral (diametral) strains that the wire has undergone during processing.

2.6. A thin-walled tube is subjected to tension in the elastic range. Using the generalized Hooke's law equations, show that both the diameter and the thickness of this tube decrease when it is under tension.

2.7. Calculate the work involved in frictionless compression of a solid cylinder 30 mm high and 20 mm in diameter to a reduction of 50% for the following materials in Table 2.4: 6061-T6 aluminum, annealed copper, annealed 4340 steel, annealed 304 stainless steel.

2.8. In Illustrative Problem 2.2, calculate the toughness of the material by the area under the stress-strain curve. Check your answer by first determining the K and n values for this material and then obtaining the toughness by integration.

2.9. Explain how the modulus of resilience of a strain-hardening material changes as it is cold-worked.

2.10. What is the volume of a 50-mm diameter solid steel sphere when subjected to a hydrostatic pressure of 2 GPa?

2.11. A rod is 10 in. long. It is stretched in two steps, first to a length of 12 in. and then to 17 in. Show that the true strains are additive, whereas engineering strains are not.

2.12. If you remove the layer of material *ad* from the specimen in Fig. 2.41, such as by machining or grinding, which way will the specimen curve? (*Hint*: Assume that the bar in sketch (d) is composed of four horizontal springs held at the ends. Thus, from the top down: compression, tension, compression, tension springs.)

2.13. Explain why deformation rate and true strain rate are not equivalent.

2.14. In Fig. 2.7, identify two materials that have the lowest and highest uniform elongation, respectively. Calculate these quantities as percentages.

2.15. A material has a strength coefficient $K = 100,000$ psi and $n = 0.2$. Assuming that a tensile-test specimen made from this material begins to neck at a true strain of 0.2, show that the ultimate tensile strength (engineering) of this material is 59,340 psi.

2.16. Modify the curves in Fig. 2.8 to indicate the effects of temperature on material properties.

2.17. Is it possible to completely remove residual stresses in a piece of material, by the technique described in Fig. 2.43, if the material is elastic, linearly strain hardening?

2.18. Using the generalized Hooke's law equations (Eq. (2.38)), show that a thin-walled tube does not undergo any thickness change when subjected to torsion.

3 Manufacturing Properties of Metals and Alloys

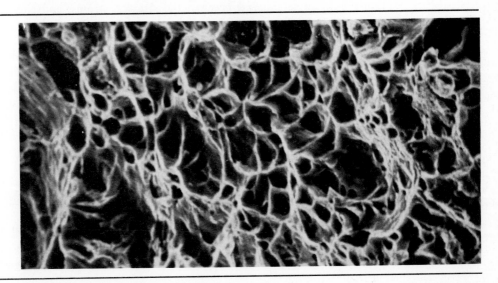

3.1 INTRODUCTION

The workpiece or specimen was treated in Chapter 2 as a geometric entity on which shape changes were induced by applying forces in different directions. Relationships were established between the stresses, strains, and energies involved in deformation and the effects of various parameters on those relationships were studied.

Because of the great variety of available materials and the large number of factors involved, the study of material behavior is an extensive subject. In this chapter, the fundamentals of the behavior of metals in regard to their structure and manufacturing properties will be reviewed. This will allow us to explain the behavior of materials that was observed in Chapter 2, and will serve as a guide in predicting the behavior and performance of materials during their processing history. We will consider the materials of the workpieces and also of the tools and dies that are used to shape a variety of metallic and nonmetallic materials into products.

The properties and behavior of materials are complex functions of many factors, such as composition, impurities, environment, size, and surface condition, as well as such parameters as temperature and strain rate.

We will also review some metallurgical fundamentals but will discuss mainly those aspects that enhance our understanding of manufacturing processes. Materials other than metals and alloys are covered in various chapters throughout this text.

3.2 THE CRYSTAL STRUCTURE OF METALS

The basic structure of metals consists of an orderly arrangement of atoms. When metals solidify, the atoms arrange themselves into various configurations, called *crystals*, in which the smallest group of atoms showing the characteristic lattice structure of that metal is known as a *unit cell*. The three basic structures found in most metals are: *body-centered cubic* (bcc), *face-centered cubic* (fcc), and *hexagonal close-packed* (hcp).

These structures are represented by the models shown in Fig. 3.1. The top and bottom planes in the hcp cell are called *basal planes*. Of these three structures, the fcc and hcp crystals have the most densely packed configurations. The order of magnitude of the distance between the atoms in these crystal structures is 10^{-8} in. (10^{-7} mm).

The crystal structure of certain metals (such as iron, titanium, cobalt, chromium,

(a)   (b)   (c)

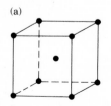 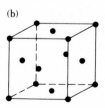 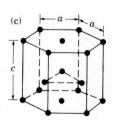

FIGURE 3.1 Crystal structures in metals. (a) Body-centered cubic unit cell, bcc. (b) Face-centered cubic unit cell, fcc. (c) Hexagonal close-packed, hcp. See also Table 3.1.

TABLE 3.1
CRYSTAL STRUCTURE OF SOME METALS

BODY-CENTERED CUBIC	FACE-CENTERED CUBIC	HEXAGONAL CLOSE-PACKED	c/a
Chromium	Aluminum	Beryllium	1.57
Columbium	Cobalt (β)	Cobalt (α)	1.62
Iron (α)	Copper	Magnesium	1.62
Molybdenum	Gold	Titanium (α)	1.59
Tantalum	Iron (γ)	Zinc	1.86
Titanium (β)	Lead	Zirconium (α)	1.59
Vanadium	Nickel		
Zirconium (β)	Silver		

and zirconium) depends on the temperature. The appearance of more than one type of crystal structure is known as *polymorphism* or *allotropism* (Table 3.1).

3.2.1 DEFORMATION OF SINGLE CRYSTALS

There are two basic mechanisms by which permanent (plastic) deformation may take place in crystal structures. One is by *slipping* of one plane of atoms over an adjacent plane (Fig. 3.2). The second mechanism is *twinning* in which there is a mirror image across the plane of twinning. Twins form abruptly and are the cause of the creaking sound (tin cry) when tin or zinc is bent at room temperature.

The deformation of single crystal specimens by slip and twinning is shown in Fig. 3.3. Note that the slip mechanism in the first two examples can be simulated with a deck of cards or blocks sliding along each other.

The illustrations in Fig. 3.2 show the basic mechanisms by which *plastic deformation* takes place—through slipping or shear. Thus, there must be a shear stress of sufficient magnitude within a piece of metal to cause plastic deformation.

Note in Fig. 3.1 that it is possible to pass a number of planes through crystals. Depending on the orientation of a plane, there is a different number of atoms per unit

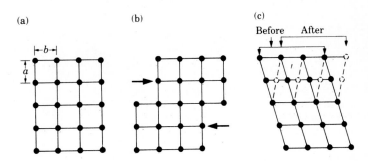

FIGURE 3.2 Deformation of a crystal structure. (a) Before deformation. (b) Deformation by slip. (c) Deformation by twinning.

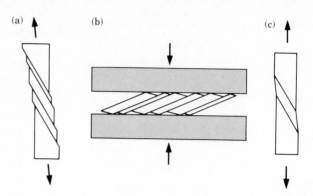

FIGURE 3.3 Deformation of single crystals.
(a) Slip in tension. As deformation
progresses the slip planes tend to align
themselves in the direction of pulling.
(b) Slip in compression. The slip planes
tend to align themselves parallel to the platens.
(c) Twinning in tension.

area in a plane and also a different spacing between these parallel planes. The planes of greatest atomic density are those farthest apart.

The maximum shear stress required to cause slip can be shown to be directly proportional to the ratio b/a in Fig. 3.2(a). Since a is the spacing of the slip planes and b is inversely proportional to the atomic density in the slip plane, it can be seen that low ratios of b/a require less shear stress to cause slip. Therefore it can be stated that slip takes place on planes of maximum atomic density. Likewise, slip takes place in the direction of maximum atomic density.

It is evident from the foregoing that a single crystal is highly *anisotropic*; that is, its properties are different in different directions.

The combination of a slip plane and its direction is known as a *slip system*. In fcc crystals there are 12 possible slip systems and in bcc crystals 48. The hcp crystal has 3 slip systems at low temperatures, but other systems become active at elevated temperatures.

The portions that slip rotate from their original angular position toward the direction of the tensile force (see Fig. 3.3a). Also note that slip takes place on certain planes only. With the use of electron microscopy it has been shown that what appears to be a single slip plane is actually a *slip band* composed of a number of slip planes (Fig. 3.4). These bands are about 10,000 atom diameters apart, i.e., 10^{-4} in. (10^{-3} mm).

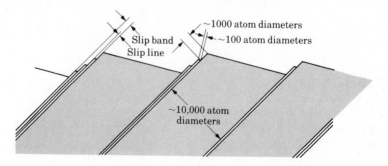

~1000 atom diameters
~100 atom diameters
Slip band
Slip line
~10,000 atom diameters

FIGURE 3.4   Slip lines and slip bands in a single crystal. 10,000 atom diameters is about 10^{-4} in. (10^{-3} mm).

3.2.2 THE STRENGTH OF SINGLE CRYSTALS

It is possible to calculate the maximum theoretical shear stress required to cause slip or permanent deformation in a single crystal. When there are no stresses, the atoms in the crystal structure are in equilibrium. We can move a layer of atoms in the direction of x in Fig. 3.5 by applying a shear stress. Since each atom in the upper row is attracted to the nearest atom in the lower row, there is nonequilibrium at positions 2 and 4. The direction of the shear stress reverses during this movement because the atom past position 3 is now attracted to the middle atom in the lower row. At position 3 the shear stress is zero since this is a symmetric position.

In Fig. 3.5 we assume, as a first approximation, that the shear stress varies sinusoidally across one atomic distance. It can then be shown that the maximum theoretical *shear strength* is given by

$$\tau_{max} \simeq \frac{G}{2\pi}. \tag{3.1}$$

More refined calculations have set the value of this stress between $G/10$ and $G/30$.

With a similar approach, the theoretical or ideal *tensile strength* of a crystal can also be calculated. In the bar shown in Fig. 3.6 the interatomic distance, when no tensile stress is applied, is a. In order to pull the specimen apart, the cohesive force between the atoms has to be overcome, the cohesive force being zero in the unstrained (equilibrium) condition. When the tensile stress reaches σ_{max}, the bonds between neighboring atomic planes break. This stress is the ideal tensile strength.

It can be shown that for most solids this strength is given by,

$$\sigma_{max} \simeq \frac{E}{10}. \tag{3.2}$$

Note that the strength levels obtained from Eqs. (3.1) and (3.2) are much higher than the actual strengths of metals, the difference being one to two orders of magnitude.

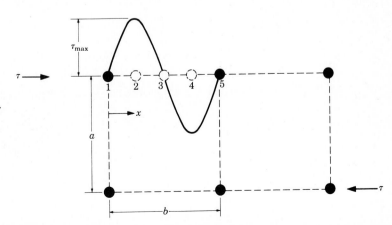

FIGURE 3.5 Variation of shear stress in moving a row of atoms over another. Note that at position 3 the shear stress is zero since an atom at this position is attracted equally to the two lower atoms.

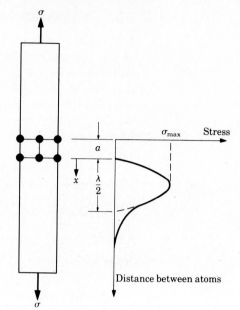

FIGURE 3.6   Variation of cohesive force in tension as a function of distance between atoms. Position a is the equilibrium (stable) position. When the tensile stress reaches its maximum, the bonds between two neighboring atomic planes break. This maximum is the ideal tensile strength of the material.

This discrepancy is explained in Section 3.2.3 in terms of *imperfections* in the crystal structure.

Needlelike crystals of metal, called *whiskers*, are produced by special techniques. These are single crystal filaments a few millimeters long and 1 to 10 μm in cross-section. Whiskers, because of their size (see Section 3.9), are either free of imperfections or their imperfections do not greatly affect their strength. The strength of whiskers approaches the theoretical strength of the material.

Iron whiskers have a tensile strength of almost 2×10^6 psi (14 GPa) as compared with a strength of 10^4 psi (70 MPa) for bulk iron. Graphite whiskers are produced with strengths up to 3.5×10^6 psi (24 GPa). (See also Table 10.4.)

● **Illustrative Problem 3.1**

Calculate the theoretical shear and tensile strengths for aluminum, plain-carbon steel, and tungsten using the data in Table 2.1.

SOLUTION.   The theoretical shear strength is obtained from the expression

$$\tau = \frac{G}{2\pi},$$

where

$$G = \frac{E}{2(1 + v)}.$$

The data are tabulated below:

	E, GPa	ν	G, GPa	THEORETICAL STRENGTH, MPa	
				SHEAR	TENSILE
Aluminum alloy	79	0.34	29	4,600	7,900
Plain-carbon steel	200	0.33	75	12,000	20,000
Tungsten	400	0.27	157	25,000	40,000

When compared with the data in Table 3.3, the theoretical strengths are found to be as much as 50 times higher than actual tensile strengths. ●

3.2.3 IMPERFECTIONS IN CRYSTAL STRUCTURES

The basic types of imperfections in crystals are as follows:

a. Line defects, called dislocations (Fig. 3.7),

b. Point defects, such as a vacancy or an impurity atom in the space of the atom of the pure metal,

c. Volume imperfections, such as inclusions or voids,

d. Planar imperfections, such as grain boundaries.

Dislocations are defects in the orderly arrangement of the atomic structure in the crystal lattice. They are the most significant defects in explaining the discrepancy between the theoretical and actual strength of metals. The two types of dislocations are *edge* and *screw* dislocations (Fig. 3.7).

A plane of slip containing a dislocation will require less shear stress to slip than a perfect lattice. This is because some of the atoms do not have as strong a bond as others due to their offset positions (Fig. 3.8).

Analogies have been proposed to describe the movement of dislocations and the lower stresses associated with them. One example is the earthworm, which moves

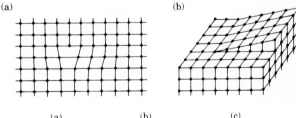

FIGURE 3.7 Two types of dislocations in a single crystal. (a) Edge dislocation. (b) Screw dislocation.

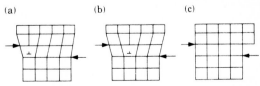

FIGURE 3.8 Movement of a dislocation across the crystal lattice under shear stress. Dislocations explain why the actual strength of metals is much lower than that predicted by theory, based on Fig. 3.5.

forward by means of a hump that starts at its tail and moves toward the head. A second example is the moving of a large carpet by forming a hump at one end and moving the hump to the other end.

Although the presence of a dislocation reduces the shear strength, dislocations can become entangled and interfere with each other, thus increasing the shear stress required for slip. The density of dislocations (total length of dislocation line per unit volume) increases with increasing plastic deformation by as much as 10^6 at room temperature. The dislocation density for some conditions are shown below:

Very pure single crystals, 0 to 10^3

Annealed single crystals, 10^5 to 10^6

Annealed polycrystals, 10^7 to 10^8

Highly cold-worked metals, 10^{11} to 10^{12}

The increase in the shear stress, and hence the overall strength of the material, is known as *strain hardening*. Thus, the greater the strain, the greater the density of dislocations and the more the entanglements, leading to an increase in strength. The movement of dislocations can be impeded not only by other dislocations, but also by barriers such as *impurities* and *grain boundaries*.

Note in Fig. 2.5 that the *rate* of strain hardening decreases with increasing plastic strain. This is because the incremental increase in dislocation density becomes relatively less significant with increasing strain.

The physical properties of metals such as melting point, specific heat, coefficient of thermal expansion, and elastic constants are not sensitive to lattice defects. On the other hand, mechanical properties such as yield and fracture strength, as well as electrical conductivity, are affected by these defects.

3.3 GRAINS AND GRAIN BOUNDARIES

When a metal solidifies, many independent crystal nuclei begin to form with random orientations at various points within the liquid. During solidification, these nuclei grow into *grains*, which impinge upon one another and form *grain boundaries* (Fig. 3.9). Each grain consists of a single crystal (for pure metals) or a polycrystalline aggregate (for alloys).

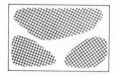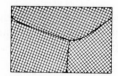

FIGURE 3.9 Schematic illustration of the solidification of a liquid mass and formation of grain boundaries. Note the difference in crystal orientation of each grain. Grain boundaries have a major influence on the behavior of metals.

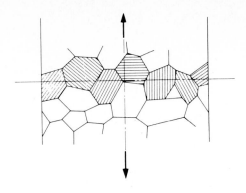

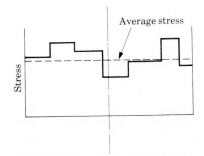

FIGURE 3.10 Variation of normal stress across a plane of poly-crystalline metal specimen under tension. The strength exhibited by each grain depends on its orientation with respect to the direction of pulling. The average stress in a tensile test is the ratio of load to the cross-sectional area of the specimen. *Source:* After L. H. Van Vlack.

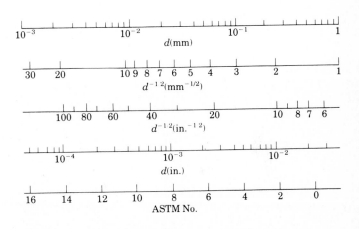

FIGURE 3.11 ASTM grain sizes and corresponding grain diameters. Note that the larger the ASTM number, the smaller the grain size. Most sheet metals used in manufacturing various products have a grain size number 7 or higher. See also Table 3.2.

Because of the usually random crystallographic orientation of its grains, the behavior of a *macroscopic* piece of metal is thus *isotropic* (Fig. 3.10). Note that the crystallographic orientation changes abruptly from one grain to the next across the grain boundaries.

The *grain size* can be measured with a microscope by counting the number of grains in a given area, or the number of grains that intersect the length of line drawn randomly on an enlarged photograph of the grains. The ASTM grain size number n is related to the number of grains N per square inch at a magnification of $100 \times$ (equal to 0.0645 mm^2 of actual area) by

$$N = 2^{n-1}. \tag{3.3}$$

Figure 3.11 shows the standard ASTM grain sizes and the equivalent grain diameters.

TABLE 3.2
GRAIN SIZES

ASTM NO.	GRAINS/mm²	GRAINS/mm³
−3	1	0.7
−2	2	2
−1	4	5.6
0	8	16
1	16	45
2	32	128
3	64	360
4	128	1,020
5	256	2,900
6	512	8,200
7	1,024	23,000
8	2,048	65,000
9	4,096	185,000
10	8,200	520,000
11	16,400	1,500,000
12	32,800	4,200,000

Grains are generally extremely small and hence there are many grains in a unit volume of metal (Table 3.2). For instance, a grain size of 7 is generally acceptable for sheet metals. This means that this material has 23,000 grains per mm³. Grain sizes between 5 and 8 are generally considered fine grains.

Grain size has a significant influence on mechanical properties. At room temperature, a large grain size is generally associated with a low yield and tensile strength, low hardness, and high ductility. The yield strength Y is the most sensitive property and is related to grain size by the empirical formula

$$Y = Y_i + kd^{-1/2}, \tag{3.4}$$

where Y_i is a basic yield stress that can be regarded as the stress opposing the motion of dislocations, k is a constant indicating the extent to which dislocations are piled up at barriers, such as grain boundaries, and d is the grain diameter. Equation (3.4) has been found to be valid below the recrystallization temperature of the material (see Section 3.5).

● **Illustrative Problem 3.2**

   a) Calculate the number of grains in the spherical head of a pin 1 mm in diameter if the ASTM grain size is 8.
   b) Estimate the number of grains in a paper clip if the grain size is 5.

SOLUTIONS.

a) The volume of the head of the pin is

$$V = \tfrac{4}{3}\pi r^3 = \tfrac{4}{3}\pi (0.5)^3 = 0.52 \text{ mm}^3.$$

ASTM grain size 8, according to Table 3.2, contains 65,000 grains/mm^3. Hence

Total number of grains = (0.52)(65,000) = 33,800.

b) Assume that the wire from which the paper clip is made is 0.8 mm in diameter and that it is 100 mm long. The volume of the material is

$$V = \frac{\pi}{4}(0.8)^2(100) = 50.27 \text{ mm}^3.$$

ASTM grain size 5 contains 2900 grains/mm^3. Hence

Total number of grains = (50.27)(2900) = 145,780. ●

3.3.1 ROLE OF GRAIN BOUNDARIES

Grain boundaries play an important role in the strength and ductility of metals and alloys. The effects depend on parameters such as temperature, strain rate, and the type and amount of impurities in the grain boundary.

At elevated temperatures and in materials with high strain-rate-sensitivity index m, plastic deformation also takes place by grain boundary sliding. The *creep* (elongation under tensile stress over a period of time, usually at elevated temperatures) mechanism is also through grain boundary sliding.

Embrittlement

It has been observed that a normally ductile and strong metal, when brought into intimate (atomic) contact with certain low-melting-point metals, can crack at a very low stress level when subjected to tension.

Embrittlement of grain boundaries can occur at room temperature or at elevated temperatures. Examples are copper wetted with lead or bismuth (Fig. 3.12) and aluminum with a mercury–zinc amalgam or liquid gallium. These elements weaken the grain boundaries by reducing their surface energy. Thus they act in a manner similar to stress raisers, such as notches.

Although the term *liquid metal embrittlement* has been used for such phenomena (because the embrittling element is in a liquid state), it has been demonstrated that this can also occur at temperatures well below the melting point of the embrittling element. This is known as *solid metal embrittlement*. Liquid metals usually produce one crack, whereas solid metals generally produce multiple cracks with crack propagation speeds that are much lower than those induced by liquid metals.

Another example of grain boundary embrittlement is *hot shortness*. This is due to local melting of a constituent or an impurity in the grain boundary at a

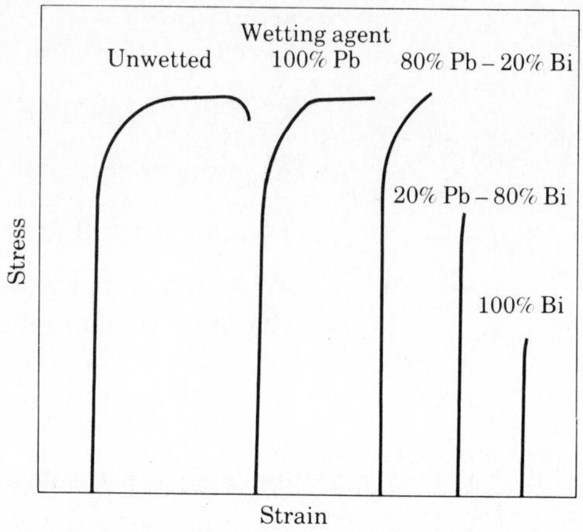

FIGURE 3.12   Embrittlement of copper by lead and bismuth at 660°F (350°C). Embrittlement has important effects on the strength, ductility, and toughness of materials. *Source:* After W. Rostoker, *et al., Embrittlement by Liquid Metals*, Reinhold, 1960, p. 22.

temperature below the melting point of the metal itself. When such a material is subjected to plastic deformation, it crumbles and disintegrates. Examples are leaded steels and brass, resulfurized steels (see Fig. 6.32), and antimony in copper. To avoid hot shortness, the material is usually deformed at a lower temperature to prevent melting at the grain boundaries.

Another form of embrittlement, *temper embrittlement* in alloy steels, is caused by segregation of impurities in the grain boundaries at temperatures between 900 and 1100°F (480 to 590°C).

3.4 DEFORMATION OF POLYCRYSTALLINE METALS

If a piece of metal, with uniform *equiaxed grains* (spherical, as shown in Fig. 3.13), is subjected to plastic deformation at room temperature (*cold working*), either by compressing it as shown or by subjecting it to tension in the horizontal direction, the grains become elongated. When tested, this deformed metal, if it is strain hardening, exhibits greater strength and lower ductility than before (Fig. 3.14). As described in

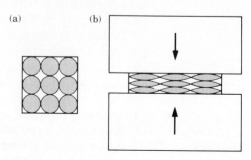

FIGURE 3.13   Deformation in compression of idealized equiaxed grains in a workpiece, such as is done in forging or rolling of metals, described in Chapter 6. (a) Before, and (b) after plastic deformation.

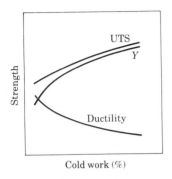

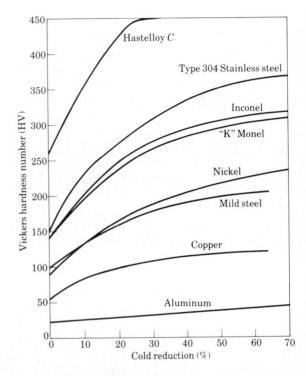

FIGURE 3.14 Typical effects of cold working (deformation at room temperature) on the strength and ductility of metals. Note that with increasing deformation, the difference between the yield stress (Y) and ultimate tensile strength (UTS) decreases. This is one way of knowing whether or not a piece of metal has been cold worked. See Section 3.6.

FIGURE 3.15 The effect of cold working on the Vickers hardness for various metals. Note that the effect of cold working on hardness is similar to the effect on strength, shown in Fig. 3.14. *Source:* International Nickel Co.

Section 2.5.7, increasing strength also means increasing hardness. The amount of increase depends on the strain to which the metal is subjected (Fig. 3.15).

As a result of cold working, changes in the physical properties of the metal may also take place. Generally, there is a decrease in electrical conductivity and corrosion resistance, a slight decrease in density, and an increase in the coefficient of thermal expansion.

3.4.1 ANISOTROPY (TEXTURE)

It can be seen in Fig. 3.13 that, as a result of deformation, the grains have become elongated in one direction and contracted in the other. As a consequence, the material has become anisotropic. In other words, the properties in the vertical direction are different from those in the horizontal direction. There are two general types of anisotropy: *crystallographic anisotropy* and *mechanical fibering*.

As shown in Fig. 3.3(a), when a metal crystal is subjected to tension the sliding blocks rotate toward the direction of pulling. Thus, slip planes (or bands) tend to align with the direction of deformation. Likewise, for a polycrystalline aggregate (Fig. 3.10), with grains in various orientations, all slip directions tend to align themselves with the direction of pulling. Under compression, the slip planes tend to align themselves in a direction perpendicular to the direction of compression (Figs. 3.3b and 3.13). In deformation processes more complex than simple tension or compression, slip planes tend to align themselves along the general direction of deformation.

This type of alignment is called *preferred orientation*. It results in anisotropy in the properties of the material. When tested in tension, the piece in Fig. 3.13 is stronger in the horizontal direction and weaker in the vertical. The ductility variation depends on many factors and cannot be generalized.

Mechanical fibering is due to the alignment of impurities and inclusions (stringers), voids, etc., during deformation. Note, for instance, that if the spherical grains in Fig. 3.13 were coated with impurities, these impurities would align themselves generally in a horizontal direction after deformation. Since the impurities weaken the grain boundaries, it follows that the piece would be weak and less ductile in the vertical direction.

Anisotropy influences mechanical and physical properties. For example, transformer sheet steel is rolled in such a way that the resulting preferred orientation imparts anisotropic magnetic properties, thus reducing hysteresis losses.

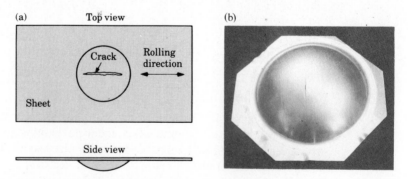

FIGURE 3.16   (a) Schematic illustration of a crack in a bulged sheet metal. Note the orientation of the crack with respect to the rolling direction. (b) Aluminum sheet with a crack developed in a bulge test. Note the longitudinal marks on the sheet. These marks are parallel to the rolling direction (and to the crack) and are variously known as roping, ridging, or looper lines. They are observed in aluminum alloys and ferritic stainless steels, and have been attributed to the segregation of alloying elements in the ingot from which the sheet is made. *Source:* Courtesy of J. S. Kallend, Illinois Institute of Technology.

As will be shown in subsequent chapters, many products develop anisotropy of mechanical properties during processing. This fact must be taken into account to ensure proper orientation of the material during subsequent processing or use. Note, for instance, that the transverse ductility of a rolled sheet is lower than in its longitudinal direction (Fig. 3.16a and b). This observation would be important in further processing of this sheet such as in bending it or making a closed-end, thin-walled pressurized vessel. In the latter example, the hoop stress will be twice the axial stress; hence the sheet must be cut and processed accordingly.

Another example of anisotropy is the unidirectional solidification of turbine blades to obtain properties in the optimal direction. This is discussed in Section 5.10.

3.5 RECOVERY, RECRYSTALLIZATION, AND GRAIN GROWTH

It was noted that plastic deformation at room temperature results in the deformation of grains and grain boundaries, a general increase in strength and lowering of ductility, and causes anisotropy. These effects can be reversed and the properties brought back to their original levels by heating the deformed piece of material to a specific temperature range and holding it there for a time. The temperature and length of time depend on the material and other factors to be discussed below.

Three thermally activated events take place during this heating process:

a. Recovery: during recovery (in a certain temperature range), the stresses in the highly strained regions are relieved. Subgrain boundaries begin to form with no appreciable change in mechanical properties. This is known as *polygonization.*

b. Recrystallization: this is the process where, in a certain temperature range, new equiaxed and strain-free grains are developed, replacing the older grains. The temperature for recrystallization is approximately $0.4 \, T_m$, where T_m is the melting point of the metal on the absolute scale (Fig. 3.17). Recrystallization results in a decrease in the density of dislocations, lower strength and higher ductility of the material (Fig. 3.18). (It can be seen from Fig. 3.17 that metals such as lead, tin, cadmium, and zinc recrystallize at about room temperature. Thus, when these metals are deformed at room temperature, they do not strain harden.)

c. Grain Growth: if the temperature is raised further, the grains begin to grow and may eventually exceed the original grain size. This is known as *grain growth* and it affects mechanical properties to a small extent. However, large grains produce a rather rough surface appearance on sheet metals when subsequently stretched to form a part (Fig. 3.19), or when bulk metal is subjected to plastic deformation, such as in compression and forging (Fig. 3.20). This phenomenon is known as *orange peel.* If this is objectionable, care must be exercised during the recrystallization process to prevent grain growth.

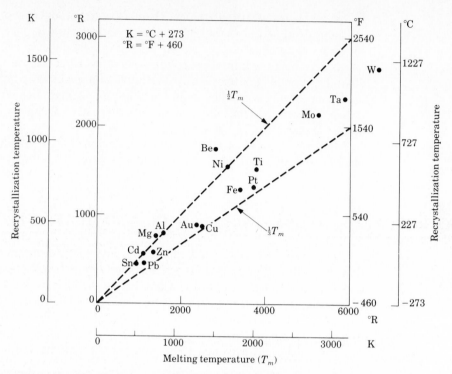

FIGURE 3.17   Recrystallization temperature for various metals. This temperature generally lies between one-half and one-third of the melting point of the metals, on the absolute scale. Molybdenum and tantalum are known as refractory metals because of their high melting point. *Source:* After L. H. Van Vlack.

Note from Fig. 3.18 that most of the change occurs during recrystallization. The energy required to cause the changes during the recrystallization process is supplied by the *stored energy* of cold work. This energy is due to the fact that a dislocation has elastic energy because it distorts and strains the crystal lattice. Since cold work increases the density of dislocations by a factor of as much as 10^6 we see that the strain energy, and hence the stored energy, increases with increasing cold work. This energy is generally on the order of 5 to 10% of the total energy of deformation and depends on the material, degree of deformation, temperature of deformation, and grain size. However, it has been shown that it can be as high as 30% with some alloys.

From the foregoing it follows that recrystallization depends on the degree of prior cold work and that the higher the cold work, the lower the temperature required for recrystallization. It is also apparent that recrystallization is a function of time, since it involves diffusion of atoms across grain boundaries.

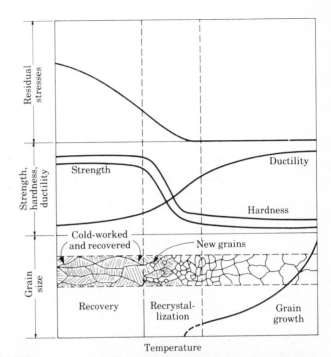

FIGURE 3.18   Schematic illustration of the effects of recovery and recrystallization on grain structure. Note the formation of small strain-free grains during recrystallization. *Source:* After G. Sachs.

FIGURE 3.19   Orange peel in a pressworked steel sheet. The coarser the grain, the rougher is the surface appearance. Thus grain size control is important. *Source:* Courtesy of P. Noe, Inland Steel Corp.

FIGURE 3.20   Orange peel on the barreled surface of an aluminum specimen in compression. Note the similarity of surface roughness to the part shown in Fig. 3.19. *Source:* A. Mulc and S. Kalpakjian.

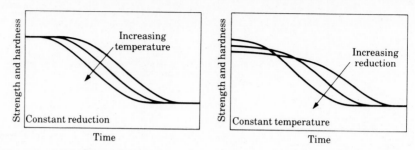

FIGURE 3.21   Variation of strength and hardness with recrystallization temperature, time, and prior cold work. Note that the more a metal is cold worked, the less time it takes to recrystallize. This is because of the higher stored energy from cold working due to increased dislocation density.

The effects on recrystallization of temperature, time, and reduction in thickness by cold working are shown in Figs. 3.21 and 3.22. These effects may be summarized as follows:

a. The greater the prior cold work, the lower the temperature required for recrystallization.

b. For a constant degree of cold work, the time required for recrystallization decreases with increasing temperature, and vice versa.

c. The higher the degree of cold work, the smaller the grain size after recrystallization.

FIGURE 3.22   Tensile strength of pure copper (75% cold worked, by drawing) as a function of annealing temperature and time. Note that the annealing time required increases rapidly with decreasing temperature.

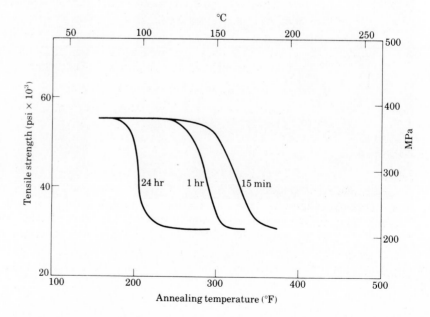

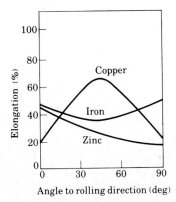
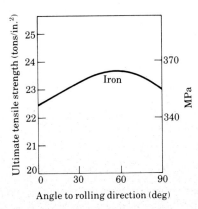
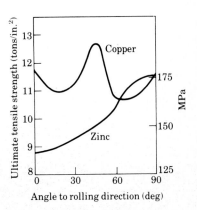

FIGURE 3.23 Anisotropy in recrystallized sheets. Iron: cold-rolled to 69% reduction and annealed for two hours at 1330°F (720°C). Copper: cold-rolled to 96% reduction and annealed for 1/2 hour at 930°F (500°C). Zinc: cold-rolled to 95% reduction and tested after six days at room temperature. *Source:* After W. Boas, *Physics of Metals and Alloys*, New York: Wiley, 1947, p. 117.

d. As a first approximation based on experimental observations, the recrystallized grain size may be assumed to be independent of the recrystallization temperature.

e. Anisotropy due to preferred orientation might persist after recrystallization (Fig. 3.23). To restore isotropy, a temperature higher than that for recrystallization may be required.

Annealing

The general term used to describe the return of a cold-worked metal to its original properties is *annealing*. It consists of heating a piece to a specific temperature range, holding it at that temperature for a time, and then cooling it at a slow rate, such as 20°F (10°C) per hour in a furnace. This is called *full annealing*.

Typical annealing temperatures are given below:

Aluminum-base alloys	560–800°F (290–430°C)
Carbon steels	1450–1650 (790–900)
Copper-base alloys	660–1560 (350–850)
Nickel-base alloys	1470–2190 (820–1200)

To avoid excessive softness in annealing, the cooling cycle may be done completely in still air. For steels, this is called *normalizing*, and it results in somewhat higher strength and lower ductility than full annealing. Specific procedures for metals and alloys are given in various handbooks and other references.

Depending on the degree of cold work, annealing temperatures may be higher than the recrystallization temperature. For instance, the recrystallization temperature for copper ranges between 400 and 600°F (200 and 300°C), whereas the annealing temperature ranges from 500 to 1200°F (260 to 650°C) in order to recover the original properties.

To reduce or eliminate residual stresses, the workpiece can be subjected to *stress-relief annealing* or *stress relieving*. The temperature and time required depend on the level of stresses and the material. Cooling rates are generally slow.

If steels are hardened by heat treatment, a process called *tempering* (*drawing*) is used in order to reduce brittleness and increase toughness. Again, the steel is heated to a specific temperature and then cooled at a prescribed rate.

3.6 COLD, WARM, AND HOT WORKING

3.6.1 COLD WORKING

When plastic deformation is carried out at room temperature, it is called *cold working*. With a strain-hardening material, this deformation results in higher strength and hardness, better surface finish, and better dimensional control than working at elevated temperatures. The disadvantages of cold working are that the processes performed require higher forces and energy, and the workpiece material has lower ductility (Fig. 3.24) and may fracture if deformed excessively. Of course, it is possible to anneal the workpiece at an intermediate stage in its deformation history in order to maintain its ductility (*intermediate annealing*).

There are many different methods that can be used to determine whether or not a piece of metal has been cold worked:

a. Observe grain deformation by polishing and etching.

b. Check for anisotropy by taking specimens from different directions and testing them.

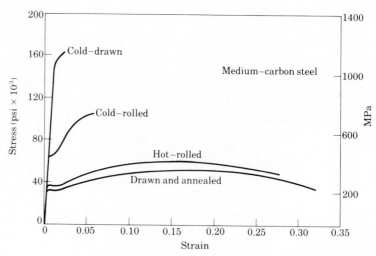

FIGURE 3.24 Effects of cold and hot working on the engineering stress-strain curve for medium-carbon steel. With increasing cold work, the strength increases, and ductility and toughness decrease. Note that the modulus of elasticity remains constant. Gage length = 8 in. (203 mm).

   c. Perform a hardness test and observe indentation profile (Fig. 2.33).

   d. Perform a hardness test and compare with published data on that material. Also, observe any hardness variation at different locations in the part.

   e. Obtain specimens and do tensile tests (or other tests) to determine stress–strain curves. Observe features of these curves, such as the ratio of UTS to yield stress, ductility, etc., and compare with published data.

   f. Anneal some of the specimens and compare test results with those in (e).

   g. Check for residual stresses by sectioning and observing dimensional changes, warping, etc. (Fig. 2.42).

   h. Observe surface appearance, color, oxidation, scale, etc.

   i. Observe forces and energy required in deforming the material, such as by bending or compression.

3.6.2 WARM WORKING

Warm working is carried out at a temperature higher than room temperature but lower than hot working (see below) temperature. This is a compromise between cold and hot working. Warm working has the advantages of reducing the energy required for working the metal and greater precision of the part.

3.6.3 HOT WORKING

If plastic deformation is carried out *above* the recrystallization temperature of the metal it is called *hot working*. This process has the advantages of requiring lower forces because of lower strength and utilizing the greater ductility of the material. Furthermore, coarse grains generally are broken down into smaller grains during hot working. If the original workpiece has any porosity, such as that due to poor casting (see Section 5.11), hot working may eliminate this porosity as a result of deformation and closing of the pores.

    On the other hand, hot working has the disadvantages of poor dimensional control (due to thermal cycling) and the presence of oxide layers on the workpiece due to heating in air (unless carried out in an inert environment). The thickness of this oxide layer can be greatly reduced by proper control of the atmosphere in the furnace and during the transfer of the heated workpiece to the metalworking equipment. The thickness and type of oxide depend on the material, temperature, and time (see also Section 4.2). During plastic deformation, the oxide layers may be disturbed and worked into the material, resulting in poor surface finish. Also, because of temperature and strain gradients, grain size may be nonuniform throughout the part.

    It should also be remembered that, as noted in Section 2.2.7, the strain-rate sensitivity of metals increases with increasing temperature. Thus the benefit obtained by raising the temperature may be counterbalanced by the increased strain-rate sensitivity of the material and hence increased forces (see Fig. 2.17). Note also that,

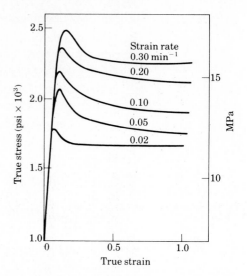

FIGURE 3.25 True stress-true strain curves in compression for pure lead at various strain rates. After an initial hump in the curves, lead behaves much like a perfectly plastic material. This is one reason why lead is used in experiments to verify theoretical studies on stresses in plastic working of metals. *Source:* After S. Kobayashi, *et al., J. Eng. Ind.,* vol. 81, 1959, p. 232.

for metals whose recrystallization temperature is at room temperature, the strain rate effect is already present at room temperature. An example is given for lead in Fig. 3.25.

Another consideration in hot working is the fact that the temperature range for hot working is also a function of deformation rate. The higher the rate of deformation, the higher the temperature necessary for recrystallization to occur. This is due to the fact that at high rates the deformation is completed in a shorter period of time, implying that the temperature required for recrystallization is higher (see Fig. 3.21). This phenomenon is known as *dynamic recrystallization.*

In addition to the factors described above, plastic deformation of steels can be carried out within certain temperature ranges and time limits in order to improve strength. This is known as *thermomechanical treatment* and utilizes time–temperature–transformation diagrams (described in materials science texts). One example is *ausforming* of steels (plastic deformation in the austenite region) where yield strengths of 400,000 psi (2700 MPa) can be obtained.

3.7 FAILURE OF MATERIALS IN DEFORMATION

Failure is one of the most important aspects of material behavior; it directly influences the choice of materials and production methods in manufacturing. Because of the many variables involved, this is a complex area of study. In this section the aspects that are of particular significance to material selection and processing will be described.

There are two types of failure that are of interest in processing materials: *fracture* and separation of the material, either through external or internal cracking, and

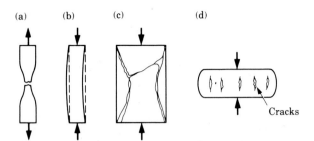

FIGURE 3.26 Types of failure in testing engineering materials. (a) Necking and fracture in tension testing of ductile materials. (b) Buckling of ductile materials under compressive stress. (c) Fracture of brittle materials in compression. (d) Cracking on the barreled surfaces of ductile materials in compression. See also Figs. 6.30 and 6.31. The stresses that cause these cracks are known as secondary tensile stresses.

buckling. These are shown in Fig. 3.26. Fracture is also classified into two general categories: *brittle* and *ductile.*

Whereas failure of materials is generally regarded as undesirable, it should be pointed out that certain products are designed in such a way that failure is essential for their function. Typical examples are food and beverage containers with tops that are removed by tearing the sheet metal along a prescribed path (Fig. 3.27); screw caps for bottles (Fig. 3.28) and caps for plastic milk containers; breakaway lamp posts and

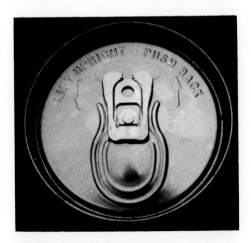

FIGURE 3.27 Scored path on a flip-top aluminum beverage can. This is an example of controlled fracture that takes place by tearing along the scored path.

FIGURE 3.28 A screw cap from a beverage bottle showing fracture along the eight scored sections on the periphery of the cap. During unscrewing, the hoop stresses cause fracture of these sections, allowing the removal of the cap.

runway lights at airports, with reduced sections for easy breakage to minimize damage to vehicles and aircraft; shear pins for machinery; and many machining and material removal processes as detailed in Chapters 8 and 9.

3.7.1 BRITTLE FRACTURE

Brittle fracture occurs with little or no gross plastic deformation preceding the separation of the material into two or more pieces. In tension, this separation usually occurs along *cleavage* planes. Splitting takes place along a crystallographic plane on which the normal tensile stress is maximum (Fig. 3.29). In a polycrystalline metal under tension, the fracture surface (which is perpendicular to the tensile stress, Fig. 3.30a) has a bright granular appearance because of the changes in the direction of the cleavage planes as the crack propagates from grain to grain. An example of the surface of brittle fracture is shown in Fig. 3.31.

   In compression, brittle fracture is more complex and may follow a path that theoretically is at a 45-degree angle to the direction of the applied force. This suggests a maximum shear stress path. An example is shown in Fig. 3.32.

   An example of fracture along a cleavage plane is the cleaving of rock salt or peeling of layers of mica. Tensile stresses normal to the cleavage plane (due to pulling) initiate and control the propagation of fracture. Another example is the behavior of brittle materials such as chalk, gray cast iron, and concrete. In tension, they fail in the manner shown in Fig. 3.30(a). In torsion, they fail along a plane at a 45-degree angle to the axis of twist; that is, along a plane on which the normal tensile stress is maximum (see Fig. 2.50).

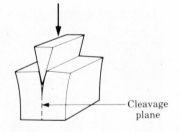

FIGURE 3.29   Fracture by cleavage. This type of fracture is similar to splitting wood with an ax. Note that the crack tip is some distance away from the tip of the wedge. Tensile stresses at the tip of the crack are very high, and cause rapid propagation of cracks in brittle materials. Note, for example, how easily and rapidly a crack in a piece of glass propagates.

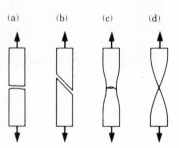

FIGURE 3.30   Types of fracture in uniaxial tension. (a) Brittle fracture in typical engineering (polycrystalline) materials. (b) Shear fracture in ductile single crystals. Compare this figure with Fig. 3.3(a). (c) Ductile cup-and-cone fracture in polycrystalline metals. See also Fig. 3.41. (d) Complete ductile fracture in polycrystalline metals. Note that the reduction of area is 100%.

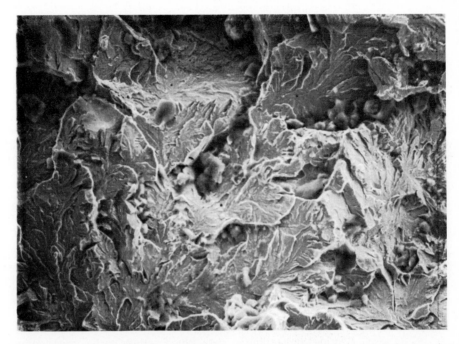

FIGURE 3.31 Typical fracture surface of a metal that has failed by brittle fracture by cleavage. Fracture is transgranular. Magnification: 200×. *Source:* Courtesy of S. L. Meiley, Packer Engineering Associates Inc.

Although there are other factors involved, face-centered cubic metals generally do not fail by brittle fracture, whereas body-centered cubic and some hexagonal close-packed metals fail by cleavage. In general, low temperature and high strain rates promote brittle fracture, especially in certain hcp and most bcc metals. The state of stress can also have a significant role (e.g., hydrostatic pressure, with its beneficial effect on enhancing ductility [Fig. 2.22]).

FIGURE 3.32 Fracture of a solid aluminum-alloy cylindrical specimen in compression between flat dies. Note that the fracture plane is about 45 degrees to the direction of compression. Diameter = 1 in. (25 mm). See also Figs. 6.30 and 6.31.

Defects

Another important factor in fracture is the presence of defects such as external or internal cracks or other stress raisers. When the part is under tension, the tip of the crack is subjected to very high tensile stresses. Under these stresses a crack propagates rapidly because a brittle material has little capacity to dissipate energy, i.e., the area under its stress–strain curve is very small. (The presence of cracks is an important factor in weldments, as described in Section 12.8.)

It has been shown that the tensile strength of a specimen, with a crack perpendicular to the direction of pulling, is related to the length of the crack by the following:

$$\sigma \propto \frac{1}{\sqrt{\text{crack length}}}. \tag{3.5}$$

This observation is essential in explaining the large difference between the compressive and tensile strengths of brittle materials. The difference is on the order of 10 for rocks and similar materials, about 5 for glass, and about 3 for gray cast iron. Under tensile stresses cracks propagate rapidly causing what is known as *catastrophic failure*. However, under compressive stresses cracks cannot form or propagate easily, thus brittle materials exhibit greater strength in compression. In this case, cracks may follow a maximum shear stress path (see Figs. 3.32 and 3.26c), which may then lead to other crack directions.

In polycrystalline metals, the fracture paths most commonly observed are *transgranular* (*transcrystalline* or *intragranular*), where the crack propagates through the grain. *Intergranular* fracture, where the crack propagates along the grain boundaries, is generally observed when the grain boundaries are soft, contain a brittle phase, or have been weakened by liquid or solid metal embrittlement (Fig. 3.33).

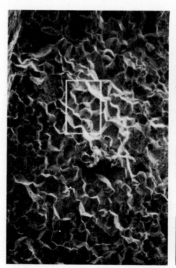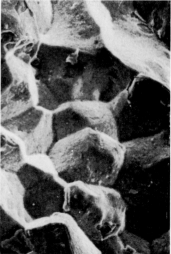

FIGURE 3.33 An example of intergranular fracture, at two different magnifications. Grains and grain boundaries are clearly visible in this micrograph. The fracture path is along grain boundaries (intergranular). Magnification: left, 100×; right, 500×. *Source:* S. L. Meiley.

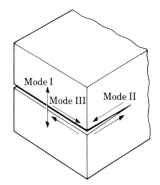

FIGURE 3.34 Three modes of fracture. Mode I has been studied extensively because it is the most commonly observed situation in engineering structures and components. Mode II is rare. Mode III is the tearing process, similar to opening a flip-top can, tearing a piece of paper, or cutting materials with a pair of scissors.

It has been shown that the maximum crack velocity in a brittle material is about 38% of the elastic wave-propagation (or acoustic) velocity of the material. (This velocity is given by the formula $\sqrt{E/\rho}$, where E is the elastic modulus and ρ is the mass density.) Thus, for steel, the maximum crack velocity is 6600 ft/s (2000 m/s).

As can be seen in Fig. 3.34, cracks may be subjected to stresses in different directions. Mode I is tensile stress applied perpendicular to the crack. Modes II and III are shear stresses applied in two different directions. The tearing of paper, cutting of sheet metal with shears, or opening flip-top cans are examples of Mode III fracture.

Stress Corrosion Cracking

An otherwise ductile metal can fail in a brittle manner by *stress—corrosion cracking* (*stress cracking*, or *season cracking*). Parts that are free of any defects after forming may develop cracks, either over a period of time or suddenly (Fig. 3.35). Although the mechanism of stress–corrosion cracking is not fully understood, tensile residual stress and the environment are responsible for this phenomenon (Section 2.7). Cracking may be intergranular or transgranular.

(a)

(b)

FIGURE 3.35 (a) Stress-corrosion cracking of an austenitic stainless steel heat exchanger tube. Length = 12 in. (300 mm). (b) Cross-section of tube. Although not fully understood, important factors in stress corrosion cracking are the degree of inhomogeneity of plastic deformation during manufacturing of the component, the presence of tensile residual stresses, and the environment. *Source:* S. L. Meiley.

The susceptibility of metals to stress–corrosion cracking depends on the type of material, the degree of inhomogeneity of deformation, the magnitude of tensile residual stresses, and the environment. The environment could be corrosive media such as saltwater or other chemicals, or it could be a liquid metal. Brass and austenitic stainless steels are metals commonly susceptible to stress cracking.

The usual procedure to avoid this phenomenon is to *stress-relieve* the part just after it is formed. Full annealing may also be done, but this reduces the strength of the part.

Hydrogen Embrittlement

The presence of hydrogen, as little as 6 ppm (parts per million) by weight, can cause severe embrittlement in many metals, including steels. This phenomenon is known as *hydrogen embrittlement*. Hydrogen may have been introduced into the metal during processing. Minute amounts of oxygen also affect the strength and ductility of some metals.

Fatigue

Another basically brittle type of fracture is *fatigue failure*. Minute external or internal cracks develop at pre-existing flaws in the material. These cracks then propagate, under *cyclic stresses*, and eventually lead to total failure of the part. The fracture surface in fatigue failures is generally characterized by *striations* (also called *beach marks* because of their appearance), as shown in Fig. 3.36. Although this type of failure generally occurs in machine elements and structures, it can also occur in tools and dies in metalworking. The cyclic stresses are due either to the nature of loading or to *thermal stresses*, such as a cool die coming into repeated contact with hot workpieces.

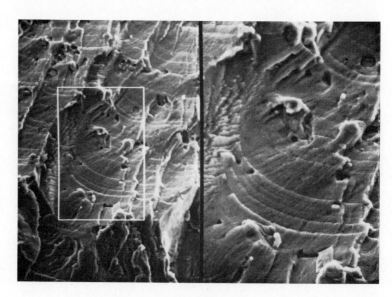

FIGURE 3.36   An example of fatigue fracture. Note the striations (beach marks) on the fracture surface. Such marks on fracture surfaces, such as broken crankshafts, indicate fatigue failure. Most components in machines and equipment fail by fatigue, and not by excessive static loading. Magnification: left, 500×; right, 1000×. *Source:* S. L. Meiley.

● **Illustrative Problem 3.3**

Two pieces of glass rod, one tubular and the other solid and round, are subjected to tension. The rods are of equal length and net cross-sectional area. Which rod is likely to carry the higher tensile load before fracture occurs?

SOLUTION. Glass, like other brittle materials, is very notch sensitive and its strength depends on surface flaws, cracks, and scratches. Hence, the tubular specimen, with its larger surface area, has a greater probability of flaws and a greater number of them. Therefore its strength is likely to be lower than that of the solid, round rod. As discussed in Section 11.8, residual stresses on the surface of glass (due to the nature of manufacturing or preparation of the glass specimens) also play a significant role in the strength of glass. ●

3.7.2 TESTING OF BRITTLE MATERIALS

Many materials, either as workpieces or as tools and dies, are rather brittle. Therefore it is important to determine their properties. Unlike ductile materials, however, it is difficult to prepare specimens from brittle materials. Furthermore, because of their sensitivity to surface defects and notches, it is difficult to clamp the specimens for testing. Alignment of the test specimen can also present some difficulties, resulting in nonuniform stress distribution along its cross-section.

A commonly used method for testing brittle materials is the *bend test*, involving a specimen with a rectangular cross-section supported at its ends. The load is applied at either one or two points; hence, these tests are referred to as 3-*point* or 4-*point* bending, respectively, as shown in Fig. 3.37.

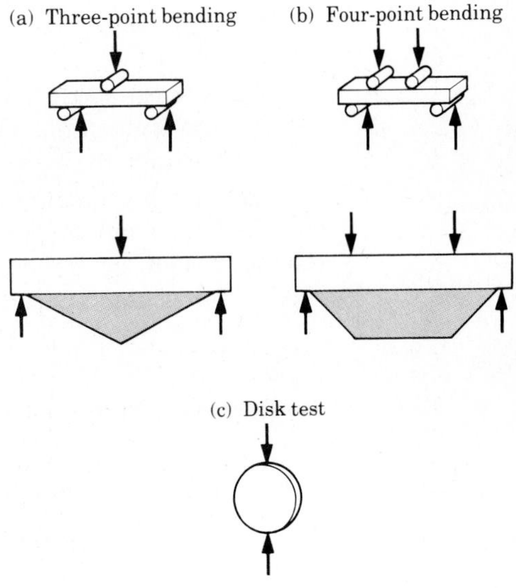

(a) Three-point bending   (b) Four-point bending

(c) Disk test

FIGURE 3.37 Test methods for brittle materials. (a) 3-point bending. (b) 4-point bending. (c) Disk test with diametral compression. Fracture of the disk takes place along a vertical (diametral) line between the arrows. This is a convenient test for materials such as grinding wheels where the round samples are removed with a special rotary cutter.

The tensile stress in these tests is maximal at the bottom surface of the specimen. The stress at fracture is known as the *modulus of rupture* and is obtained from the formula

$$\sigma = \frac{Mc}{I},$$ (3.6)

where M is the bending moment, c is one half of the specimen depth, and I is the moment of inertia of the cross-section.

Note that there is a basic difference between the two loading conditions in Fig. 3.37. In 3-point bending, the maximum stress is at the center of the beam, whereas in the 4-point test the maximum stress is constant between the two loading points. The stress is the same in both cases when all other parameters are maintained. However, there is a greater probability for defects and imperfections to be present in the volume of material between the loading points in the 4-point test than in the much smaller volume under the single load in the 3-point test. This means that the 4-point test is likely to result in a lower modulus of rupture than the 3-point test. This has been verified by experiments. Furthermore, the results of the 4-point test show less scatter than the 3-point test. This is due to the fact that the strength of a small volume of brittle material (such as under the single load in the 3-point test) is subject to greater variations than is the strength of the weakest point in a larger volume (such as in the 4-point test).

Another test for brittle materials (concrete, rocks, carbides) is the *disk test* (Fig. 3.37c). When loaded as shown, fracture initiates internally, causing the disk to split vertically due to a tensile stress. The tensile stress is uniform across the diameter and can be calculated from the formula

$$\sigma = \frac{2P}{\pi dt}$$ (3.7)

where P is the applied load, d is the diameter, and t is the thickness of the disk.

Because of the probabilistic nature of imperfections; their size, distribution, and orientation; and the presence of surface defects, such as scratches and notches, brittle materials generally exhibit a wide scatter in their strength. This necessitates the use of a large number of specimens in order to obtain a statistically significant and reliable strength level for use in design. Generally, the data for each test are plotted on a special type of graph paper.

● **Illustrative Problem 3.4**

A 5-lb block is dropped from a height of 12 in. on a 1-in.-cube-shaped specimen. The specimen is made of a perfectly brittle material with a surface energy of 0.01 in. · lb/in². Estimate the number of pieces into which the specimen will break.

SOLUTION. Assume that each broken piece is a cube of dimension x on each side and that all broken pieces are of equal size. Thus, let

x = lateral dimension of each broken piece

n = number of broken pieces.

Then,

surface area of each broken piece = $6x^2$,

volume of each broken piece = x^3, and

total volume of broken pieces = 1 in^3.

Hence,

$$n = \frac{1}{x^3}.$$

The potential energy of the falling block is dissipated as surface energy. Hence,

$$(5)(12) = (n)(6x^2)(0.01) = \frac{0.06}{x}.$$

Thus,

$x = 0.001$ in.

and

$n = 10^9$ pieces.

Note that, because it represents a very small percentage of the energy involved, the original surface of the specimen (6 in^2) has been ignored in these calculations. Also, although n appears to be a large number (because of the data in this problem) it should be remembered that, as an example, a piece of chalk indeed breaks into a large number of small pieces when stepped upon. ●

3.7.3 DUCTILE FRACTURE

Ductile fracture is associated with plastic deformation prior to failure. In a tension test, highly ductile materials such as gold and lead may neck down to a point and then fail (see Fig. 3.30d). Most metals and alloys neck down to a finite area before failure (Fig. 2.2b).

Ductile fracture generally takes place along planes where the shear stress is maximal. In torsion, for instance, a ductile metal fractures along a plane perpendicular to the axis of twist, that is, the plane on which the shear stress is a maximum. The fracture in shear is a result of extensive slip along slip planes of the grains.

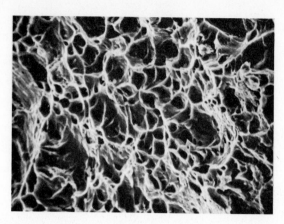

FIGURE 3.38 Surface of ductile fracture in low-carbon steel showing dimples. Fracture is initiated generally at impurities, inclusions, or preexisting voids in the material. *Source:* Courtesy of K.-H. Habig and D. Klaffke, Bundesanstalt für Materialprüfung, Berlin.

Upon close examination of the surface of a ductile fracture (Fig. 3.38) one can observe a fibrous appearance with dimples, as if a number of very small tension tests have been carried out. Failure is initiated with the formation of *microvoids* (usually around small inclusions or pre-existing voids), which then grow and *coalesce* and generate cracks. This process is represented by Fig. 3.39.

Observation of the ductile fracture of a tension-test specimen indicates that fracture starts at the center of the specimen (where the state of stress is hydrostatic [see Illustrative Problem 2.8]) through the growth and coalescence of cavities or voids (Fig. 3.40). The central region then develops a crack, as shown in the midsection of Fig. 3.41. Near the periphery of the necked region, the fracture continues along a surface, at about a 45-degree angle to the axis of the specimen, in a shear-type failure. Because of its appearance, the fracture surface of a tension-test specimen is called *cup-and-cone* fracture.

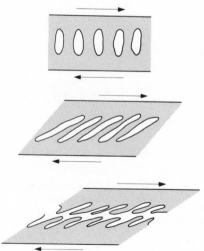

FIGURE 3.39 A schematic illustration of the mechanism of ductile fracture. Note the formation of elongated dimples on the fracture surface. Compare with Fig. 3.38. *Source:* After H. C. Rogers.

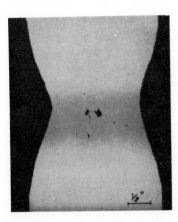

FIGURE 3.40 Void formation at the center of a neck in a tensile test specimen due to the state of hydrostatic tensile stress component (See Fig. P2.3). As the specimen is stretched, the voids coalesce; the net cross-sectional area becomes too small to support the load and the specimen fractures. *Source:* After H. C. Rogers.

FIGURE 3.41 Typical cup-and-cone fracture in a tension test specimen. Not all specimens break symmetrically as in this figure. If broken parts, such as wire, rod, or cable, indicate this type of fracture, the cause is generally overloading. *Source:* Courtesy of R. E. Reed-Hill.

Inclusions are very important in ductility and ductile fracture because of their role in void formation and, hence, in the formability of metals and alloys. Inclusions can be various impurities or second-phase particles such as oxides, carbides, and sulfides. The extent of their influence depends on factors such as their shape and distribution, hardness, and volume fraction.

As can be seen in Figs. 3.42(a) and (b), the greater the volume of inclusions, the lower the ductility. Voids and porosity resulting from processing (e.g., casting) also reduce ductility, Fig. 3.43.

There are two factors that have an effect on void formation. One is the strength of the interface between an inclusion and the matrix. If the bond is strong, there is less tendency for void formation during plastic deformation. The second factor is the hardness of the inclusion. If it is soft (such as manganese sulfide in steels) then it will comply with the overall shape change of the specimen or part during deformation. If it is hard (such as carbides and oxides) then it could lead to void formation, as shown in Fig. 3.44. Hard inclusions may also break up into smaller particles during deformation due to their brittle nature. Inclusions also have significant effects in machining. These are covered in Section 8.11.

(a)

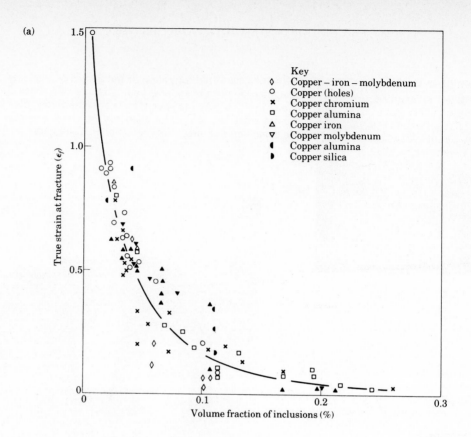

(b)

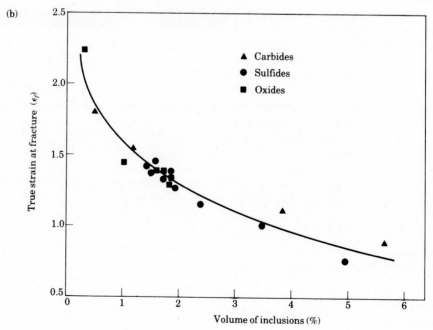

130

FIGURE 3.42 (a) The effect of volume fraction of various second-phase particles on the tensile reduction of area for copper. (b) The effect of volume fraction of natural inclusions on the tensile reduction of area for steel. *Source:* (a) After B. I. Edelson and W. M. Baldwin, *Trans. Q. ASM*, vol. 55, 1962, pp. 230–250. (b) After F. B. Pickering, *Physical Metallurgy and Design of Steels*, App. Sci. Pub., 1978, p. 51.

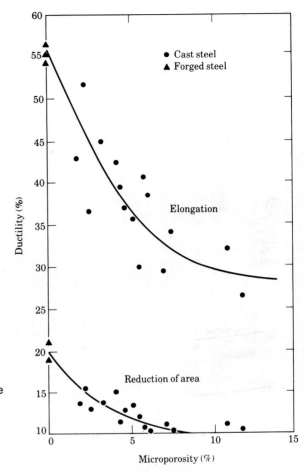

FIGURE 3.43 The effect of microporosity on the ductility of quenched and tempered 1% Cr, 0.25% Mo cast steel. *Source:* After T. Grousatier, *et al.*

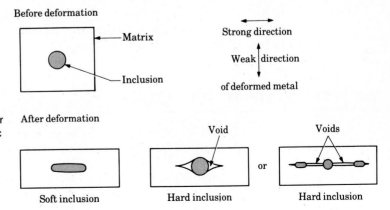

FIGURE 3.44 Schematic illustration of the deformation of soft and hard inclusions, and their effect on void formation in plastic deformation of metals. Note that hard inclusions, because they do not comply with the overall deformation of the ductile matrix, can cause voids.

FIGURE 3.45 Manganese sulfide inclusions (thin dark elongated regions) in AISI 1215 steel. These inclusions are important in machinability of steels, described in Section 8.11.1. Magnification: 100×. *Source:* S. L. Meiley.

The alignment of inclusions during plastic deformation leads to *mechanical fibering*, as shown in Fig. 3.45. For maximum ductility, subsequent processing of such material must involve consideration of the proper direction of working.

The beneficial effect of hydrostatic compression should be reiterated here. It has been shown that, under otherwise identical conditions, the application of hydrostatic pressure increases the ductility (see Fig. 2.22) of metals and nonmetallic materials. Various techniques have been developed for performing deformation processing of materials under a hydrostatic compressive stress state. The most common example is hydrostatic extrusion. In contrast, hydrostatic tension must be avoided because of its adverse effect on formability. Hydrostatic tension can lead to defects in products made by bulk deformation processes (Chapter 6).

3.8 EFFECTS OF TEMPERATURE

The overall effects of temperature on the properties of metals and alloys were described in Section 2.2.6. Ductility generally increases with increasing temperature; however, an important consideration is the presence of any liquid phase in the grain boundaries. Wetting of the grain boundaries weakens them. This wetting can occur in the presence of elements or compounds that melt at a temperature lower than the alloy. This is called *hot shortness*.

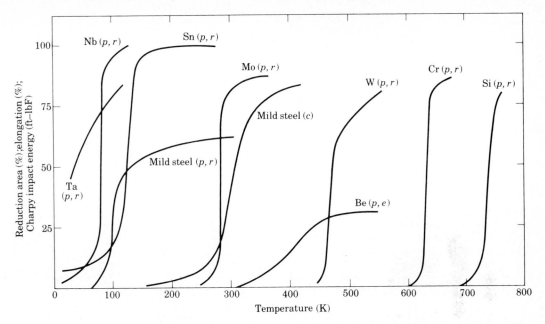

FIGURE 3.46 Transition temperature for various metals. Legend: (*p*, *r*) reduction of area in uniaxial tension; (*p*, *e*) elongation in uniaxial tension; (*c*) Charpy impact test. Note the narrow temperature range across which the behavior of the material undergoes a major transition. *Source:* After A. H. Cottrell, *The Mechanical Properties of Matter*, New York: Wiley, 1964, p. 358.

Transition Temperature

Another important consideration is the *transition temperature*. Body-centered cubic and some hexagonal close-packed metals undergo a sharp change in ductility and toughness across a very narrow temperature range (Fig. 3.46). This rarely occurs in face-centered cubic metals.

The temperature from ductile to brittle behavior is known as the lower transition temperature, and from brittle to ductile behavior as the upper transition temperature. If unspecified, the transition temperature is understood to be the middle of this range.

The transition temperature also depends on such factors as composition, microstructure, grain size, surface finish and geometry of the specimen, and strain rate. High strain rates, abrupt changes in shape, and surface notches raise the transition temperature.

Strain Aging

In strain aging, carbon atoms segregate to dislocations, thereby pinning them and thus increasing the resistance to dislocation movement. The result is increased strength and reduced ductility.

Fig. 3.47 shows the effects of strain aging. The stress–strain curve for low-carbon steel is shown by the curve *abc*, with its upper and lower yield points. If the tension test is stopped at point *e*, the specimen unloaded and then tested again, curve *dec* is

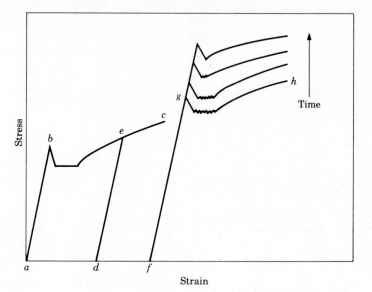

FIGURE 3.47 Effect of strain aging on the stress-strain curve for plain-carbon steel. Strain aging is due to the segregation of carbon atoms to dislocations (Fig. 3.8), pinning them, and increasing the resistance to dislocation movement. The process of migration takes time, hence the increase in strength by time.

obtained. However, if several days pass before the second test is made, curve *fgh* is obtained. Note that the yield stress is higher than before and the upper and lower yield points have reappeared. Instead of taking place over several days at room temperature, this same phenomenon can occur in a few hours at a higher temperature; thus, strain aging can be accelerated.

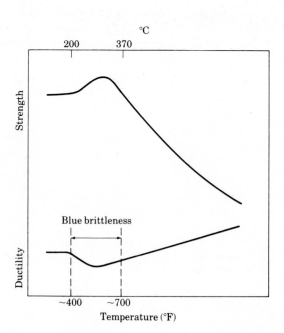

FIGURE 3.48 Blue brittleness in plain-carbon steel. This is an example of accelerated strain aging, i.e., aging above room temperature.

The presence of the upper and lower yield points and the yield-point elongation cause undesirable marks on sheet metal, called *stretcher marks* or *Lueder's lines*. This topic and the general behavior of sheet metal specimens (in contrast to the more common, round tensile-test specimen) are treated in greater detail in Section 7.3.

Blue Brittleness

Blue brittleness in low-carbon steels is another phenomenon where there is a lowering of ductility (and an increase in strength) within a temperature range of about 400 to 700°F (200 to 370°C) (Fig. 3.48). It has been shown that this temperature range increases with increasing strain rate. Blue brittleness is an example of *accelerated strain aging*.

3.9 SIZE EFFECT

The dependence of the properties of a material on its size is known as *size effect*. Note from the foregoing discussions and from Eq. (3.4) that as the size decreases, defects, cracks, imperfections, etc., are less likely to occur. Hence, the strength and ductility of a specimen increase with decreasing size. Figs. 3.49 and 3.50 show this effect for metals and glass fibers.

Whereas these figures relate strength to diameter, the length of the specimen is also important. The greater the length, the greater the probability of defects (Fig. 3.51). As an analogy, a long chain is likely to be weaker than a shorter one since the probability of one of its links being weak increases with the number of links (length of chain).

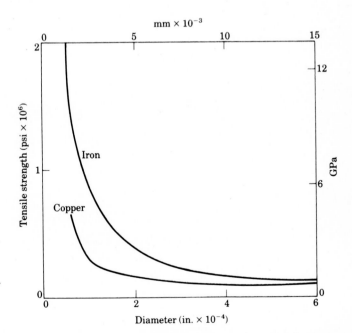

FIGURE 3.49 Tensile strength of iron and copper whiskers as a function of their diameter. This phenomenon is known as size effect; the smaller the size, the greater the strength.

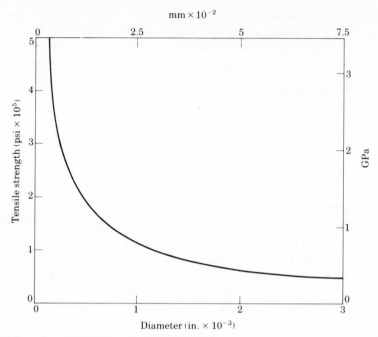

FIGURE 3.50 Tensile strength of glass filaments as a function of their diameter. Because of their high strength, glass fibers are (as are carbon and boron fibers) used extensively in reinforced plastics (See Figs. 10.25 and 10.29.)

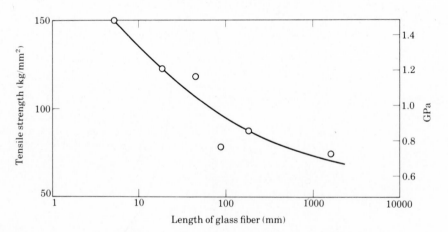

FIGURE 3.51 Tensile strength of glass fibers as a function of fiber length. Diameter = 13 mm. Note the decrease in strength with length, because of the higher probability of defects present in the material with increasing length. This is similar to the observation that a long piece of chain is likely to be less strong than a shorter chain. *Source:* After F. O. Anderegg.

Size effect is also demonstrated by whiskers (see Section 3.2.2), which, because of their small size, are either free of any imperfections or do not contain the type of imperfections that affect their strength. The actual strength in such cases approaches the theoretical strength.

For brittle materials, such as glass, surface conditions are very important, particularly surface flaws, such as cracks and scratches. The mechanical properties of glass are strongly influenced by the method of producing the glass fibers or rods, the thermal history, and the presence of adsorbed layers of water vapor.

3.10 BUCKLING

Buckling, folding, or wrinkling of workpieces during deformation processing is an important problem. In its simplest configuration, Fig. 3.26(b), buckling is a case of elastic instability. The buckling load P is given by *Euler's* formula,

$$P = \frac{n\pi^2 EI}{L^2},\tag{3.8}$$

where E is the modulus of elasticity, I is the least moment of inertia, and L is the effective length of the column. The value of n depends on end conditions. If one end is fixed and the other end is pin-jointed and free to move sideways, then $n = \frac{1}{4}$. If both ends are pin-jointed, $n = 1$, and if both ends are fixed, $n = 4$.

Three examples of buckling in metalworking are:

a. Upsetting the ends of rods to make rivet and bolt heads (Fig. 6.16),

b. Wrinkling of the flange in deep drawing of sheet metal to make cups and beverage cans, (Fig. 7.42) and

c. Folding of thin-walled tubes in reducing the diameter of tubes (Fig. 6.88).

These problems arise from the presence of compressive stresses in the plane of the sheet metal. They can be avoided or minimized by proper selection of the die geometry and other process parameters. The complexity of the geometries involved present considerable difficulties for quantitative studies in buckling in metalworking processes.

3.11 PHYSICAL PROPERTIES

In addition to the strength and ductility characteristics of materials, physical properties also play a significant role, both in processing these materials and during

TABLE 3.3
MECHANICAL AND PHYSICAL PROPERTIES OF WROUGHT METALS AND ALLOYS AT ROOM TEMPERATURE

METAL	DENSITY		MELTING POINT		SPECIFIC HEAT	
	lb/in^3	g/cm^3	°F	°C	BTU/lb · °F	J/kg · °C
Aluminum	0.0975	2.699	1220	660	0.215	900
Aluminum alloys	0.095–0.102	2.63–2.82	890–1210	476–654	0.21–0.22	880–920
Beryllium	0.067	1.854	2332	1278	0.45	1884
Chromium	0.260	7.197	3407	1875	0.11	460
Columbium (niobium)	0.310	8.580	4475	2468	0.065	272
Copper	0.324	8.968	1980	1082	0.092	385
Copper alloys	0.27–0.323	7.47–8.94	1625–2300	885–1260	0.09–0.104	377–435
Iron	0.284	7.86	2798	1537	0.11	460
Steels	0.25–0.33	6.92–9.13	2500–2790	1371–1532	0.107–0.12	448–502
Lead	0.41	11.35	621	327	0.031	130
Lead alloys	0.32–0.41	8.85–11.35	360–620	182–326	0.030–0.045	126–188
Magnesium	0.063	1.744	1202	650	0.245	1025
Magnesium alloys	0.064–0.065	1.77–1.78	1130–1150	610–621	0.25	1046
Molybdenum alloys	0.369	10.21	4730	2610	0.066	276
Nickel	0.322	8.91	2647	1453	0.105	440
Nickel alloys	0.28–0.32	7.75–8.85	2030–2650	1110–1454	0.091–0.13	381–544
Tantalum alloys	0.6	16.6	5425	2996	0.034	142
Titanium	0.163	4.51	3035	1668	0.124	519
Titanium alloys	0.16–0.17	4.43–4.70	2820–3000	1549–1649	0.12–0.13	502–544
Tungsten	0.697	19.29	6170	3410	0.033	138
Zinc	0.258	7.14	786	419	0.092	385
Zinc alloys	0.24–0.26	6.64–7.20	728–977	386–525	0.096	402
Zirconium	0.23	6.37	3366	1852	0.067	280
Zirconium alloys	—	—	—	—	—	—

the useful life of the products. Various physical properties are described below as they relate to manufacturing processes (see Table 3.3).

3.11.1 SPECIFIC HEAT

As noted in Eq. (2.59), the temperature rise in a workpiece, as a result of plastic deformation, is a function of the energy input and the specific heat of the material. The lower the specific heat, the higher the temperature rise.

THERMAL CONDUCTIVITY		YIELD STRESS		ULTIMATE TENSILE STRENGTH		ELONGATION	YIELD STRESS/DENSITY (max)
						in 2 in.	
BTU/hr · ft°F	W/m°C	psi × 10³	MPa	psi × 10³	MPa	(50 mm) %	in. (×10³)
128	222	5	35	13	90	45	50
70–138	121–239	5–80	35–550	13–88	90–600	45–4	800
84	146	27–38	185–260	33–51	230–350	3.5–1	570
39	67	—	—	13	90	0	—
30	52	30	205	40	275	30	95
227	393	10	70	32	220	45	30
17–135	29–234	11–160	76–1100	20–190	140–1310	65–3	500
43	74	6–29	40–200	27–41	185–285	60–3	100
9–30	15–52	30–250	205–1725	60–255	415–1750	65–2	750
20	35	1–2	7–14	2.5	17	50	5
14–27	24–46	2	14	3–8	20–55	50–9	5
89	154	13–15	90–105	23–28	160–195	15–3	240
43–80	75–138	19–44	130–305	35–55	240–380	21–5	675
82	142	115–300	80–2070	135–340	90–2340	40–30	215
53	92	8.5	58	46	320	30	25
7–36	12–63	15–175	105–1200	50–210	345–1450	60–5	550
31	54	70–225	480–1550	80–225	550–1550	40–20	375
10	17	20–80	140–550	40–100	275–690	30–17	500
5–7	8–12	50–200	344–1380	60–210	415–1450	25–7	1250
96	166	80–100	550–690	90–110	620–760	—	145
65	113	—	—	20	140	60	—
61–65	105–113	—	—	20–48	140–330	65–2	—
12	21	10–45	70–310	30–70	205–485	40–25	200
—	—	45–110	310–760	65–120	450–830	27–6	—

3.11.2 THERMAL CONDUCTIVITY

This is an important property with regard to thermal gradients within a work-piece. When heat is generated due to plastic work of deformation or friction, it is generally desirable to conduct this heat away at a high rate; otherwise, a severe rise in temperature may take place. For instance, among all engineering metals and alloys, titanium has the lowest thermal conductivity. Some of the difficulties experienced in machining titanium have been attributed to this. The heat generated is concentrated at the tool tip, where temperature rise is very high, leading to excessive tool wear. Low thermal conductivity can also lead to highly inhomogeneous deformation during plastic working of metals. The influence of thermal conductivity in manufacturing processes is treated in Chapters 5, 6, 8, 9, 10, and 12.

3.11.3 THERMAL EXPANSION

This property, in conjunction with thermal conductivity, plays the most significant role in causing thermal stresses. This is particularly important when, for example, a relatively cool die is brought in contact with a hot workpiece. To reduce thermal stresses, it is desirable to have a combination of high thermal conductivity and low thermal expansion.

Thermal stresses can lead to cracks in many situations, such as in ceramics and in tools and dies made of relatively brittle materials. *Thermal fatigue* is a result of thermal cycling and causes a number of cracks on a surface. *Thermal shock* is the term generally used if cracks develop after a single thermal cycle.

Thermal stresses are caused not only by temperature gradients, but may also be caused by anisotropy of thermal expansion, generally observed in hexagonal close-packed metals and in ceramics. (See Table 11.12.)

The influence of temperature on dimensions and on elastic moduli can be a significant problem in the use of precision instruments and equipment. For instance, a spring will have a different stiffness as its temperature changes. Likewise, a tuning fork or a pendulum will have different frequencies at different temperatures (see Illustrative Problem 3.5).

To alleviate these problems, a family of iron–nickel alloys (known as *Invar* and Invar-type alloys) has been developed with nickel contents ranging from 36 to 49%. These alloys have the interesting properties of very low thermal expansion and a constant modulus of elasticity over a moderate but significant temperature range. This phenomenon is known as the *Invar effect*. These alloys are available in various forms, such as wire, bar, and strip. Because they are *ferromagnetic*, a magnetic field can cause a change in the dimensions and elastic modulus of a component. Hence the environment in which these materials are to be used should also be taken into account.

3.11.4 ELECTRICAL AND MAGNETIC PROPERTIES

Electrical conductivity is of primary importance in certain processes, such as magnetic pulse forming (Section 7.12.3), electrodischarge machining, and electrochemical grinding (Section 9.8.3) where the workpiece should have good electrical conductivity.

Changes in the properties of materials when subjected to electrical or magnetic fields are also important. The *piezoelectric effect* (*piezo*, from the Greek, meaning "to press") is exhibited by some materials, such as certain ceramics and quartz crystals, in which there is a reversible interaction between an elastic strain and an electric field. This property is utilized in making pressure transducers.

Some materials, such as pure nickel and some iron–nickel alloys, expand or contract when subjected to a magnetic field. This phenomenon is called *magnetostriction* and is the principle behind most ultrasonic machining equipment (Section 9.7).

3.11.5 DENSITY

The most significant role that density plays is in the *strength-to-weight ratio* of materials and structures. This is particularly important for weight saving, not only for aircraft and aerospace structures, but for automotive bodies and components and other applications where energy consumption and power limitations are a major concern.

The strength-to-weight ratios of a number of materials are given in Fig. 3.52 and Table 3.3. Note the wide range of these ratios and that the applicable range of temperatures is also quite wide. The temperature range within which a structure is designed to function is, of course, an essential parameter in material selection. Note, for

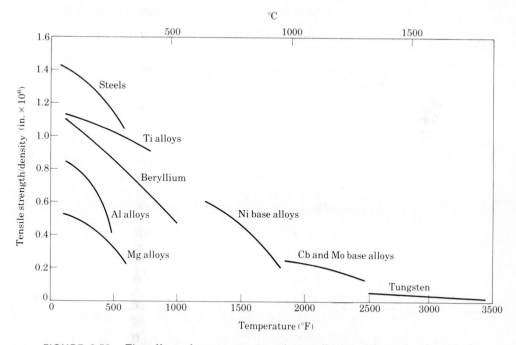

FIGURE 3.52 The effect of temperature on the tensile strength-to-density ratio for various metals and alloys. Note the useful temperature range for these alloys. High ratios are desirable for aircraft and aerospace structures. See also Table 3.3.

instance, that reinforced plastics (Section 10.11) have the narrowest temperature range, whereas graphite and refractory metal alloys have the highest. Although these alloys have a rather low strength-to-weight ratio, they are the only choice when high temperatures are involved. However, the processing of these materials into useful shapes with appropriate mechanical properties presents considerable difficulty, as will become evident in subsequent chapters.

● **Illustrative Problem 3.5**

The natural frequency f of a beam, one end clamped and the other end free, is given by

$$f = 0.56 \sqrt{\frac{EIg}{wL^4}},$$

where E = modulus of elasticity,
$\quad\quad I$ = moment of inertia,
$\quad\quad w$ = weight per unit length,
$\quad\quad L$ = length of beam,
$\quad\quad g$ = gravitational constant.

How does the natural frequency change as the temperature of the beam is increased?

SOLUTION. Assume that the cross-section of the beam is square with dimension h on each side. Then,

$$I = \tfrac{1}{12}h^4 \quad\quad (I \text{ increases with temperature})$$

and

$$w = \frac{\text{weight}}{L} \quad\quad (w \text{ decreases with temperature}).$$

Also, L increases with temperature and E decreases with temperature (Fig. 2.11).

Assuming that thermal expansion is isotropic (which is not the case with some metals and ceramics) and letting the subscript o denote room temperature, the expression for natural frequency can be written as

$$\frac{f}{f_o} = \sqrt{\frac{\dfrac{E}{E_o} \cdot \left(\dfrac{h}{h_o}\right)^4}{\dfrac{w}{w_o} \cdot \left(\dfrac{L}{L_o}\right)^4}}.$$

Since $wL = w_o L_o$ and the ratio of length dimensions is simply given by $(1 + \alpha\Delta T)$, where α is the coefficient of thermal expansion, then

$$\frac{f}{f_o} = \sqrt{\frac{E}{E_o}} (1 + \alpha\Delta T).$$

Whether or not the frequency increases or decreases depends on the rate at which the modulus of elasticity decreases with temperature. As an example, let

$$E_o = 200 \text{ GPa}$$
$$E = 175 \text{ GPa}$$
$$\alpha = 10^{-5}/°C$$
$$\Delta T = 200°C$$

Then,

$$\frac{f}{f_o} = 0.94.$$

Hence the natural frequency will decrease as the beam is heated. ●

3.12 GENERAL PROPERTIES OF FERROUS ALLOYS

By virtue of their relatively low cost and extremely wide range of physical, mechanical, and chemical properties ferrous alloys have extensive applications. Typical examples are springs, sheet metal for appliances and car bodies, cutting tools, I-beams, and fasteners.

Ferrous alloys are available in a wide selection of shapes and sizes, ranging from very thin wires to thick plates for ships and nuclear power plants. As with all materials, the choice of a ferrous alloy for a particular application requires consideration of strength, hardness, impact toughness, fatigue strength, corrosion resistance, machinability, formability, and weldability.

3.12.1 STEELS

Steels (first made in China and Japan around 600–800 AD) are generally divided into the categories of *carbon steels* and *alloy steels* (including *tool steels*). Various alloying elements are added to iron in order to impart certain properties. These are summarized below. The major detrimental effects are stated in parentheses.

Carbon: hardenability, strength, hardness, and wear resistance.

Nickel: strength and toughness; minor effect on hardenability.

Chromium: strength, toughness, hardness, and wear resistance; increases depth of hardness penetration in heat treatment.

Molybdenum: hardenability, wear resistance, toughness; strength, creep resistance, and hardness at elevated temperatures.

Vanadium: strength, abrasion resistance, hardness at elevated temperatures; inhibits grain growth during heat treatment.

Copper: resistance to atmospheric corrosion, improve strength with little loss in ductility. (Can adversely affect surface quality and hot-working characteristics.)

Manganese: hardenability and ductility.

Lead: machinability. (Causes liquid-metal embrittlement.)

Sulfur: machinability. (Lowers impact strength and transverse ductility; impairs surface quality and weldability.)

Silicon: strength, high electrical conductivity; decreases magnetic hysteresis loss.

Phosphorus: strength, hardenability, corrosion resistance, machinability. (Decreases ductility and toughness.)

Boron: hardenability.

Tungsten and Cobalt: strength and hardness at elevated temperatures.

Columbium (Niobium): fine grain size, strength, lowers transition temperature.

Tellurium: machinability of leaded steels.

Zirconium and Cerium: control shape of inclusions (sulfides) and improve toughness in high-strength, low-alloy steels.

Aluminum, Silicon and Calcium: added to steels during solidification to remove oxygen and nitrogen.

Residual elements during the processing of steels can have various effects as outlined below:

Nitrogen: strength, hardness, machinability. (Decreases ductility and toughness.)

Oxygen: slight increase in strength of rimmed steels. (Gross reduction in toughness.)

Hydrogen: (Severe embrittlement of steels.)

Designations

Carbon Steels.   The AISI–SAE designations for carbon and alloy steels have been standardized and are based on four digits, as shown in Table 3.4. The first two digits indicate the alloying elements and their percentage, and the last two digits indicate the carbon content. Thus, for instance, 1112 steel is a resulfurized steel with 0.12% carbon. Typical applications for these steels are given in Tables 3.5 and 3.6.

TABLE 3.4
AISI–SAE DESIGNATIONS FOR STEELS AND THEIR MAJOR ALLOYING ELEMENTS

10xx	Plain carbon steel	48xx	Ni–Mo
11xx	Resulfurized carbon steel	50xx	Cr
12xx	Resulfurized and rephosphorized carbon steel	51xx	Cr
13xx	Mn	52xx	Cr
31xx	Ni–Cr	61xx	Cr–V
33xx	Ni–Cr	81xx	Ni–Cr–Mo
40xx	Mo	86xx	Ni–Cr–Mo
41xx	Cr–Mo	87xx	Ni–Cr–Mo
43xx	Ni–Cr–Mo	88xx	Ni–Cr–Mo
44xx	Mo	92xx	Mn–Si–Cr
45xx	Mo	93xx	Ni–Cr–Mo
46xx	Ni–Mo	94xx	Mn–Ni–Cr–Mo
47xx	Ni–Cr–Mo	98xx	Ni–Cr–Mo

Note: The last two digits indicate the carbon content (e.g., 1020 steel has 0.20% carbon). Leaded steels are identified by the letter "L" inserted between the second and third numerals (e.g., 10L45).

TABLE 3.5
TYPICAL SELECTION OF STEELS FOR VARIOUS APPLICATIONS

PRODUCT	STEEL
Aircraft forgings, tubing, fittings	4140, 8740
Axles	1040, 4140
Ball bearings and races	52100
Bolts	1035, 4042, 4815
Camshafts	1020, 1040
Chains (transmission)	3135, 3140
Coil springs	4063
Connecting rods	1040, 3141, 4340
Crankshafts (forged)	1045, 1145, 3135, 3140
Differential gears	4023
Gears (car and truck)	4027, 4032
Landing gear	4140, 4340, 8740
Lock washers	1060
Nuts	3130
Springs (coil)	1095, 4063, 6150
Springs (leaf)	1085, 4063, 9260, 6150
Tubing	1040
Wire	1045, 1055
Wire (music)	1085

TABLE 3.6

CHARACTERISTICS AND TYPICAL APPLICATIONS OF VARIOUS CARBON AND ALLOY STEELS.
(After K. G. Budinski, in *Wear Control Handbook*, Am. Soc. Mech. Eng., NY, 1980, 946.)

TYPE	AISI GRADES	CONDITION	TYPICAL APPLICATIONS
Low-carbon (carburizing grades)	1117, 1020 1030, 4320 4615, 5015 8620, 9310	Soft	Poor abrasion resistance and metal-to-metal wear resistance, do not hold a cutting edge or sustain high loads. They provide satisfactory service life as pins, guides, shaft, jigs, etc. at very low loads and infrequent use.
		Carburized to >600 HB	Excellent metal-to-metal wear resistance self mated or mated with a tool steel (hardened). Suitable for cams, shafts, gibs, gears, and other sliding devices. Not for cutting devices involving loads that could yield metal under the case. Use lubricated. These steels have fair low stress abrasion resistance but since they do not contain alloy carbides they do not have the abrasion resistance of tool steels.
Medium-carbon (direct hardening)	1040, 1050 1060, 4140 4340, 5060 5150, 6150 8640, 9260 1137, 41L40	Soft	Same as low carbon.
		Hardened and tempered to ~300 HB	Widely used for parts requiring good strength and toughness: shafts, fixtures, forming tools, hand tools, sheared in dies, hammers, etc. Abrasion resistance and metal-to-metal resistance is only fair. Use only at this hardness when toughness and fabricability is more important than wear resistance. 4340 at 350 HB is used for forging dies.
		Hardened and tempered to ~500 HB	Best suited for structural components on tools, levers, piston rods, stops, gears, springs, etc. They do not have the hardenability and alloy carbides of tool steels and maximum hardness is usually in the range of 550–600 HB. Moderate abrasion resistance and good metal-to-metal wear resistance. Wear resistance is insufficient for use in most cutting devices, punches and dies. When nitrided they have much improved metal-to-metal wear resistance.
High-carbon (direct hardening)	52100 1080 1095	Soft	The low stress abrasion and metal-to-metal wear resistance is better than on soft low- and medium-carbon steels. 1080 strip at 450 HB can be used for rule dies.
		Hardened and tempered to >500 HB	52100 at 650–680 HB is a standard steel for rolling element bearings. It also has suitable wear properties for short run dies, embossing rolls, cam followers, gages, etc. 1080 and 1095 steels in strip are widely used for flat springs; cold drawn they are used for many types of wire springs.

A letter prefix may also be added to these digits to indicate the process used in producing the steel. The absence of a prefix indicates open-hearth, B is Bessemer, C is open-hearth carbon steel, and E is electric furnace.

A new numbering system has been developed (for all metals) as a result of a co-operative study by ASTM and SAE, called the Unified Numbering System (UNS). It consists of a letter followed by five digits (Table 3.7).

Carbon steels are generally classified into three groups based on their carbon content: *low-carbon* steels (generally less than 0.30% carbon), *medium-carbon* steels (0.3 to 0.6%) and *high-carbon* steels (more than 0.6% carbon). The general mechanical properties of carbon and alloy steels are given in Tables 3.8 and 3.9.

TABLE 3.7
UNIFIED NUMBERING SYSTEM (UNS) FOR METALS AND ALLOYS

NONFERROUS	UNS SERIES	FERROUS	UNS SERIES
Aluminum	A00001–99999	Specified mechanical property steels	D00001–99999
Copper	C00001–99999	Cast irons	F00001–99999
Rare earth	E00001–99999	AISI-SAE carbon and alloy steels	G00001–99999
Low-melting	L00001–99999	AISI H-steels	H00001–99999
Miscellaneous nonferrous	M00001–99999	Cast steels	J00001–99999
Nickel-base superalloys	N00001–99999	Miscellaneous steels	K00001–99999
Precious metals	P00001–99999	Heat and corrosion resistant	S00001–99999
Refractory	R00001–99999	Tool steels	T00001–99999

TABLE 3.8
TYPICAL MECHANICAL PROPERTIES OF SELECTED CARBON AND ALLOY STEELS IN THE HOT ROLLED, NORMALIZED AND ANNEALED CONDITION

AISI No.*	CONDITION	ULTIMATE TENSILE STRENGTH		YIELD STRENGTH		ELONGA-TION, %	REDUCTION IN AREA, %	HARD-NESS, HB	IZOD IMPACT STRENGTH	
		MPa	psi × 10³	MPa	psi × 10³				J	ft · lb
1020	As-rolled	448	65	330	48	36	59	143	86	64
	Normalized	441	64	346	50	35	67	131	117	86
	Annealed	393	57	294	42	36	66	111	123	91
1040	As-rolled	620	90	413	60	25	50	201	48	36
	Normalized	589	85	374	54	28	54	170	65	48
	Annealed	518	75	353	51	30	57	149	44	32
1060	As-rolled	813	118	482	70	17	34	241	17	13
	Normalized	775	112	420	61	18	37	229	13	9
	Annealed	625	90	372	54	22	38	179	11	8
1080	As-rolled	965	140	586	85	12	17	293	6	5
	Normalized	1010	146	524	76	11	20	293	6	5
	Annealed	615	89	375	54	24	45	174	6	4
1118	As-rolled	521	75	316	45	32	70	149	108	80
	Normalized	477	69	319	46	33	65	143	103	76
	Annealed	450	65	284	41	34	66	131	106	78
3140	Normalized	891	129	599	87	19	57	262	53	39
	Annealed	689	100	422	61	24	50	197	46	34
4340	Normalized	1279	185	861	125	12	36	363	15	11
	Annealed	744	108	472	68	22	49	217	51	37
4820	Normalized	750	109	484	70	24	59	229	109	81
	Annealed	681	98	464	67	22	58	197	92	68
5140	Normalized	792	115	472	68	22	59	229	38	28
	Annealed	572	83	293	42	28	57	167	40	30
6150	Normalized	939	136	615	89	21	61	269	35	26
	Annealed	667	96	412	59	23	48	197	27	20
8620	Normalized	632	91	357	51	26	59	183	99	73
	Annealed	536	77	385	55	31	62	149	112	82
9255	Normalized	932	135	579	84	19	43	269	13	10
	Annealed	774	112	486	70	21	41	229	8	6

* All grades are fine-grained except for those in the 1100 series that are coarse-grained. Heat treated specimens were oil quenched unless otherwise indicated.   Adapted from *ASM Metals Reference Book,* 2 ed., 1983.

TABLE 3.9
TYPICAL MECHANICAL PROPERTIES OF SELECTED CARBON AND ALLOY STEELS IN THE QUENCHED AND TEMPERED CONDITION

AISI NO.*	TEMPERING TEMPERATURE		ULTIMATE TENSILE STRENGTH		YIELD STRENGTH,		ELONGA-TION, %	REDUCTION IN AREA, %	HARDNESS, HB
	°C	°F	MPa	psi × 10³	MPa	psi × 10³			
1040†	205	400	896	130	662	96	16	45	514
	425	800	841	122	634	92	21	57	352
	650	1200	669	97	496	72	28	68	201
1040	205	400	779	113	593	86	19	48	262
	425	800	758	110	552	80	21	54	241
	650	1200	634	92	434	63	29	65	192
1060	205	400	1103	160	779	113	13	40	321
	425	800	1076	156	765	111	14	41	311
	650	1200	800	116	524	76	23	54	229
1080	205	400	1310	190	979	142	12	35	388
	425	800	1289	187	951	138	13	36	375
	650	1200	889	129	600	87	21	50	255
1141	205	400	1634	237	1213	176	6	17	461
	425	800	1165	169	1034	150	12	47	331
	650	1200	710	103	593	86	23	62	217
4340	205	400	1875	272	1675	243	10	38	520
	425	800	1469	213	1365	198	10	44	430
	650	1200	965	140	855	124	19	60	280
5140	205	400	1793	260	1641	238	9	38	490
	425	800	1310	190	1172	170	13	50	365
	650	1200	758	110	662	96	25	66	235
6150	205	400	1931	280	1689	245	8	38	538
	425	800	1434	208	1331	193	10	43	420
	650	1200	945	137	841	122	17	58	282
81B45	205	400	2034	295	1724	250	10	33	550
	425	800	1407	204	1310	190	11	48	405
	650	1200	896	130	793	115	20	55	280
9255	205	400	2103	305	2048	297	1	3	601
	425	800	1606	233	1489	216	8	22	477
	650	1200	993	144	814	118	20	42	285

*All grades are fine-grained except for those in the 1100 series that are coarse-grained. Heat treated specimens were oil quenched unless otherwise indicated.

† Water quenched.

Adapted from *ASM Metals Reference Book*, 2 ed., 1983.

HSLA Steels. In order to improve the strength-to-weight ratio of steels, a number of *high-strength, low-alloy* (HSLA) steels have been developed, particularly for reducing weight in automotive bodies. These steels have a low carbon content and are characterized by a microstructure consisting of fine-grain ferrite and a hard second phase (martensite/austenite). They were developed in the 1930s.

HSLA steels are produced in sheet form by microalloying and controlled rolling. However, their ductility, formability, and weldability is inferior to conventional low-alloy steels. To improve these properties, *dual-phase steels* have been developed (see below) with high tensile strengths and high strain-hardening exponents (n).

A new system of AISI designations for high-strength sheet steel, consisting of three categories, is shown in Table 3.10. The structural quality (S) includes elements of C, Mn, P, and N. Low alloys (X) contain Cb, Cr, Cu, Mo, Ni, Si, Ti, V, and Zr, either singly or in combination. Weathering steels (W) have atmospheric corrosion resistance (approximately four times greater than that of conventional low-carbon steels) and contain Si, P, Cu, Ni, and Cr in various combinations.

Dual-Phase Steels.   Dual-phase steels (D) are specially processed and have high strain-hardening characteristics, which improve their formability. The deoxidation practice for these steels (developed in the late 1960s) is described in detail in Section 5.5. Generally, the formability of sheet steels is identified as follows: F = excellent, K = good, O = fair.

The SAE designations for these steels are similar to those in Table 3.10 with the exception that another letter is added to indicate the carbon content. Thus, O5OXF becomes O5OXLF where L shows the carbon level, in this case L meaning low carbon.

TABLE 3.10
AISI DESIGNATION FOR HIGH-STRENGTH SHEET STEEL

YIELD STRENGTH, psi × 10^3	CHEMICAL COMPOSITION	DEOXIDATION PRACTICE
35	S = structural	F = killed plus
40	quality	sulfide
45		inclusion
50	X = low alloy	control
60		K = killed
70	W = weathering	
80		O = non-killed
100	D = dual phase	
120		
140		

EXAMPLE

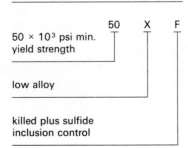

50 × 10^3 psi min. yield strength

low alloy

killed plus sulfide inclusion control

3.12.2 STAINLESS STEELS

Stainless steels, developed in the early 1900s, are generally divided into four groups:

Austenitic (200 and 300 series)

Ferritic (400 series)

Martensitic (400 and 500 series)

Precipitation-hardening stainless steels (P-H)

Stainless steels are characterized by their high chromium content and have a wide range of properties (Table 3.11). (The 300 series also contains high percentages of nickel.) Austenitic stainless steels are nonmagnetic, have excellent corrosion resistance, and harden by cold working. However, with cold work their formability is reduced.

TABLE 3.11
ROOM-TEMPERATURE MECHANICAL PROPERTIES AND TYPICAL APPLICATIONS OF ANNEALED STAINLESS STEELS

AISI TYPE (UNS)	FORM	ULTIMATE TENSILE STRENGTH, psi × 10^3	YIELD STRENGTH, psi × 10^3	ELONGA- TION, %	CHARACTERISTICS AND TYPICAL APPLICATIONS
201 (S20100)	Sheets	95	45	40	High work-hardening rate; low-nickel equivalent of type 301. Flatware; automobile wheel covers, trim.
	Strips	95	45	40	
	Tubing	95	45	40	
202 (S20200)	Sheets	90	45	40	General-purpose low-nickel equivalent of type 302. Kitchen equipment; hub caps; milk handling.
	Strips	90	45	40	
	Tubing	90	45	40	
301 (S30100)	Plates	105	40	55	High work-hardening rate; used for structural applications where high strength plus high ductility is required. Railroad cars; trailer bodies, aircraft structurals; fasteners; automobile wheel covers, trim; pole line hardware.
	Sheets	110	40	60	
	Strips	110	40	60	
	Tubing	105	40	50	
303 (S30300)					Free-machining modification of type 302, for heavier cuts. Screw machine products; shafts; valves; bolts; bushings; nuts.
	Bars	90	35	50	
	Tubing	80	38	53	
303Se (S30323)	Wire	90	35	50	Free-machining modification of type 302, for lighter cuts; used where hot working or cold heading may be involved. Aircraft fittings; bolts; nuts; rivets; screws; studs.
304 (S30400)	Bars	85	35	60	Low-carbon modification of type 302 for restriction of carbide precipitation during welding. Chemical and food processing equipment; brewing equipment; cryogenic vessels; gutters; downspouts; flashings.
	Plates	82	35	60	
	Sheets	84	42	55	
304H (S30409)	Strips	84	42	55	
	Tubing	85	35	50	
	Wire	90	35	60	

continued

TABLE 3.11 (*Continued*)

AISI TYPE (UNS)	FORM	ULTIMATE TENSILE STRENGTH, psi × 10³	YIELD STRENGTH, psi × 10³	ELONGA-TION, %	CHARACTERISTICS AND TYPICAL APPLICATIONS
309 (S30900)	Bars	95	40	45	High temperature strength and scale resistance. Aircraft heaters; heat treating equipment; annealing covers; furnace parts; heat exchangers; heat treating trays; oven linings; pump parts.
	Plates	95	40	45	
	Sheets	90	45	45	
	Strips	90	45	45	
	Tubing	90	45	45	
	Wire	105	70	35	
309S (S30908)					Low-carbon modification of type 309. Welded constructions; assemblies subject to moist corrosion conditions.
310 (S31000)	Bars	95	45	50	Higher elevated temperature strength and scale resistance than type 309. Heat exchangers; furnace parts; combustion chambers; welding filler metals; gas turbine parts; incinerators; recuperators; rolls for roller hearth furnaces.
	Plates	95	45	50	
	Sheets	95	45	45	
	Strips	95	45	45	
	Tubing	95	45	45	
	Wire	105	75	30	
310S (S31008)					Low-carbon modification of type 310. Welded constructions; jet engine rings.
316 (S31600)	Bars	80	30	60	Higher corrosion resistance than types 302 and 304; high creep strength. Chemical and pulp handling equipment; photographic equipment; brandy vats; fertilizer parts; ketchup cooking kettles; yeast tubs.
	Plates	82	36	55	
	Sheets	84	42	50	
	Strips	84	42	50	
	Tubing	85	35	50	
	Wire	80	30	60	
321 (S32100)	Bars	85	35	55	Stabilized for weldments subject to severe corrosive conditions, and for service from 800 to 1600°F (870°C). Aircraft exhaust manifolds; boiler shells; process equipment; expansion joints; cabin heaters; fire walls; flexible couplings; pressure vessels.
	Plates	85	30	55	
	Sheets	90	35	45	
	Strips	90	35	45	
321H (S32109)	Tubing	85	35	50	
	Wire	95	65	40	
410 (S41000)	Bars	75	40	35	General-purpose heat treatable type. Machine parts; pump shafts; bolts; bushings; coal chutes; cutlery; finishing; tackle; hardware; jet engine parts; mining machinery; rifle barrels; screws; valves.
	Plates	70	35	30	
	Sheets	70	45	25	
	Strips	70	45	25	
410S (S41008)	Tubing	75	40	30	
	Wire	75	40	30	
416 (S41600)	Bars	75	40	30	Free-machining modification of type 410; for heavier cuts. Aircraft fittings; bolts; nuts; fire extinguisher inserts; rivets; screws.
	Tubing	75	40	30	
	Wire	75	40	20	
420 (S42000)	Bars	95	50	25	Higher carbon modification of type 410. Cutlery; surgical instruments; valves; wear-resisting parts; glass molds; hand tools; vegetable choppers.
	Wire	95	50	20	
422 (S42200)	Bars	145	125	18	High strength and toughness at service temperatures up to 1200°F (650°C). Steam turbine blades; fasteners.

continued

TABLE 3.11   (*Continued*)

AISI TYPE (UNS)	FORM	ULTIMATE TENSILE STRENGTH, psi × 10³	YIELD STRENGTH, psi × 10³	ELONGA- TION, %	CHARACTERISTICS AND TYPICAL APPLICATIONS
430 (S43000)	Bars	75	45	30	General-purpose nonhardenable chromium type. Decorative trim; nitric acid tanks; annealing baskets; combustion chambers; dishwashers; heaters; mufflers; range hoods; recuperators; restaurant equipment.
	Plates	75	40	30	
	Sheets	75	50	25	
	Strips	75	50	25	
	Tubing	75	40	25	
	Wire	70	40	35	

Adapted from *Metal Progress, 1983 Materials and Processing Databook.*

Most martensitic stainless steels do not contain nickel and harden by heat treatment. These steels have high strength and good ductility. Ferritic stainless steels are not heat treatable.

Stainless steels are most important in applications where corrosion resistance is required. Examples are cutlery, kitchen equipment, surgical instruments, automotive trim, and equipment for the chemical and petroleum industries. Depending on the corrosive media, other materials such as aluminum, nickel, copper, and plastics may also be suitable. Factors such as desirable mechanical and physical properties, temperature range, and methods of fabrication should also be considered in the final selection.

3.12.3 TOOL STEELS

These are specially alloyed steels designed for high strength, impact toughness, and wear resistance at room and elevated temperatures. Table 3.12 shows the general classification of tool steels commonly used in cutting and forming metals. Here, tool means both a cutting tool and a die. These steels are discussed further in Chapters 6, 7, and 8.

3.12.4 CAST IRONS

Four basic types of cast iron are available: white, gray, nodular (ductile), and malleable cast iron. These are described in Section 5.12.

3.13 NONFERROUS METALS AND ALLOYS

Although generally more expensive than ferrous metals, nonferrous metals and their alloys have many important applications because of the wide range of properties they possess (see Table 3.3). These are treated individually in the sections below.

TABLE 3.12
BASIC TYPES OF TOOL STEELS

TYPE	AISI	(UNS)
High-speed	M (molybdenum-base)	T11000
	T (tungsten-base)	T12000
Hot-work	H1 to H19 (chromium-base)	T20810
	H20 to H39 (tungsten-base)	T20820
		T20830
	H40 to H59 (molybdenum-base)	T20840
Cold-work	D (high-carbon, high-chromium)	T30400
	A (medium-alloy air-hardening)	T30100
	O (oil-hardening)	T31500
Shock-resisting	S	T41900
Mold steels	P	T51600
Special purpose	L (low-alloy)	T61200
	F (carbon–tungsten)	—
Water-hardening	W	T72300

3.13.1 ALUMINUM AND ALUMINUM ALLOYS

The important factors in selecting aluminum are its high strength-to-weight ratio, corrosion resistance, high thermal and electrical conductivity, appearance, and good fabrication characteristics. Principal uses of aluminum alloys are in construction, aircraft and other transportation equipment, packaging, and electrical machinery.

A variety of aluminum alloys is available, either as castings or wrought products in many shapes. Wrought alloys of aluminum are of two types: alloys that can be hardened by cold working, and those that can be hardened by heat treatment. Although they vary widely in their properties, techniques have been developed whereby these alloys are machined, formed, and joined with relative ease. Aluminum was first produced in 1825.

Designation
Aluminum alloys are identified by four digits and a temper designation. The major alloying element is identified by the first digit, as follows:

1   commercially pure aluminum, $>99\%$ Al

2   copper

3   manganese

4   silicon

5   magnesium

6   magnesium and silicon

7   zinc

8   other elements

The second digit indicates modifications of the aluminum alloy, and the last two digits indicate the level of purity. Thus, 1090 aluminum is 99.90% pure aluminum.

The temper designations are as follows:

-F   as fabricated

-O   annealed

-H   strain-hardened by cold working

-T   heat-treated

The properties and uses of some of the common aluminum alloys are given in Tables 3.13 and 3.14.

TABLE 3.13
PROPERTIES OF SELECTED ALUMINUM ALLOYS

ALLOY (UNS)	CONDITION	ULTIMATE TENSILE STRENGTH, psi × 10³	YIELD STRENGTH, psi × 10³	ELONGATION IN 2 in.*, %
1100 (A91100)	O	13	5	35–45
1100	H14	18	17	9–20
2011 (A92011)	T3	55	43	15
2024 (A92024)	O	27	11	20–22
2024	T4	68	47	20–19
3003 (A93003)	O	16	6	30–40
3003	H14	22	21	8–16
5052 (A95052)	O	28	13	25–30
5052	H34	38	31	10–14
6061 (A96061)	O	18	8	25–30
6061	T6	45	40	12–17
7075 (A97075)	O	33	15	17–16
7075	T6	83	73	11

* $\frac{1}{16}$ in. thick — $\frac{1}{2}$ in. thick specimen.

TABLE 3.14
MANUFACTURING PROPERTIES AND TYPICAL APPLICATIONS OF WROUGHT ALUMINUM ALLOYS

ALLOY DESIG-NATION	COMMERCIAL FORMS*	CHARACTERISTICS			APPLICATIONS IN PRODUCT DESIGN
		CORROSION RESISTANCE†	*MACHINABILITY†*	*WELDABILITY‡*	
1050	T	A-A	D-C	A-A	Coiled tubing, extruded
1060	ST	A-A	D-C	A-A	Chemical equipment, railroad tank cars
1100	STEBWFO	A-A	D-C	A-A	Sheet metal work, spun hollow ware, tin stock

continued

TABLE 3.14 (*Continued*)

ALLOY DESIG-NATION	COMMERCIAL FORMS*	CHARACTERISTICS			APPLICATIONS IN PRODUCT DESIGN
		CORROSION RESISTANCE†	*MACHINABILITY*†	*WELDABILITY*‡	
1145	O	A-A	—	—	Foil for capacitors, tin stock
1175	C	A-A	—	A-A	Reflector sheet
1235	OT	A-A	—	A-A	Foil for capacitors, tubing
1345	BW	A-A	D-C	A-A	—
1350	STEPBW	A-A	D-C	A-A	Electrical conductors
2011	BW	C-C	A-A	D-D	Screw machine products
2014	STEBF	C-C	C-B	C-B	Truck frames, aircraft structures
2017	BW	C	B	B-C	Screw machine products, fittings
2018	F	C	B	B-C	Aircraft engine cylinder heads and pistons
2024	STEBW	C-C	C-B	C-B	Truck wheels, screw machine products, aircraft structures
2025	F	D	B	B	Forgings, aircraft propellers
2036	S	C	C	B	Auto body sheet
2117	BW	C	C	B	Rivets, redraw rods
2124	S	C-C	C-B	C-B	Aircraft structures (superior fracture toughness)
2218	F	D	B	C	Jet engine impellers and compressor rings, aircraft engine cylinder heads and pistons
2219	STEBF	D	B	A	Structural use at high temperatures (to 600°F, 315°C), high strength weldments for cryogenic and aircraft parts
2419	SEF	C	B	A	Same as 2219 plus high fracture toughness
2618	F	C	B	B-C	Aircraft engines (temperature to 450°F, 230°C)
3003	All forms	A-A	D-C	A-A	Cooking utensils, chemical equipment, pressure vessels, sheet metal work, builders, hardware, storage tanks
3004	S	A-A	D-C	A-A	Sheet metal work, storage tanks, pressure vessels, containers
3005	S	—	—	—	Building products—siding gutters, etc.
3105	S	A-A	D	A-A	Building products—siding gutters, etc.
4032	F	C	B	B	Pistons (low thermal expansion), forgings
4043	W	—	—	—	Welding wire, brazing filler metal, spray gun wire

continued

TABLE 3.14 (*Continued*)

| ALLOY DESIG- NATION | COMMERCIAL FORMS* | CHARACTERISTICS | | | APPLICATIONS IN PRODUCT DESIGN |
		CORROSION RESISTANCE†	*MACHINABILITY*†	*WELDABILITY*‡	
4045	C	—	—	—	Brazing sheet, wire
4047	SW	—	—	—	Brazing sheet, wire
4343	C	—	—	—	Brazing sheet
5005	SBW	A-A	D-C	A-A	Appliances, utensils, architectural trim, electrical conductors
5042	S	—	—	—	Rigid container stock
5050	STBW	A-A	D-C	A-A	Builders hardware, refrigerator trim, coiled tubes
5052	STBWO	A-A	D-C	A-A	Sheet metal work, hydraulic tubes, appliances, bus, truck and marine uses
5056	BWO	A-B	D-C	A-A	Cable sheathing, rivets for magnesium, screen wire, zippers
5083	STEBF	A-B	D-C	A-B	Unfired welded pressure vessels,
5086	STEB	A-B	D-C	A-B	marine, auto, and aircraft parts, cryogenics, TV towers, drilling rigs, transportation equipment, missile components, armor plate
5154	STEBW	A-A	D-C	A-A	Welded structures, storage tanks, pressure vessels, salt water service
5182	S	—	—	—	Rigid container stock
5252	S	A-A	D-C	A-A	Auto and appliance trim
5254	S	A-A	D-C	A-A	Hydrogen peroxide and chemical storage vessels
5356	W	—	—	—	Welding rod, wire, and electrodes
5454	STEB	A-A	D-C	A-A	Welded structures, pressure vessels, marine service tubing
5456	STEBF	A-B	D-C	A-A	High strength welded structures, storage tanks, pressure vessels, marine service
5457	S	A-A	D-C	A-A	Anodized auto and appliance trim (good formability in annealed temper)
5657	S	A-A	D-C	A-A	Anodized auto and appliance trim (good brightness)
6003	C	—	—	—	Cladding for sheets and plates
6005	ET	A	C	A	Trucks and marine structures, railroad cars, furniture
6009	S	B	C	A	Auto body sheet
6010	S	B	C	A	Auto body sheet
6053	BWF	A-A	D-C	A-A	Wire and rods for rivets

continued

TABLE 3.14 (*Continued*)

ALLOY DESIG-NATION	COMMERCIAL FORMS*	CHARACTERISTICS			APPLICATIONS IN PRODUCT DESIGN
		CORROSION RESISTANCE†	*MACHINABILITY*†	*WELDABILITY*‡	
6061	All forms	B-B	D-C	A-A	Heavy-duty structures where corrosion resistance is needed, truck and marine structures, railroad cars, furniture, pipelines, bridge railings, hydraulic tubing
6063	TPE	A-A	D-C	A-A	Pipe, railings, furniture, architectural extrusions, truck and trailer flooring
6066	TEBF	C	B	B	Forgings, extrusions for welded structures
6070	TPEB	B	C	A	Heavy duty welded structures, pipelines
6101	TPEB	A	C-D	A	High strength bus conductors
6151	F	A-B	C	A-B	Moderate strength intricate forgings for machine and auto parts
6201	W	A-B	B-C	A-A	Electrical conductor wire (high strength)
6253	C	—	—	—	Component of clad rod and wire
6262	TEBW	B-B	A-A	B-B	Screw machine products (better corrosion resistance than 2011 or 2017)
6463	E	A-A	D-C	A-A	Architectural and trim extrusions
7001	EB	C	B	D	High strength structures
7005	E	B	—	—	Truck bodies and trailer parts, portable bridges, railroad cars
7049	FES	C	B	D	Structural parts for aircraft
7050	SEF	C	B	C	Aircraft and other structures (resistance to exfoliation, stress-corrosion cracking), good fracture toughness
7072	CS	—	—	—	Fin stock, cladding for sheets, plates, and tubing
7075	STEBWF	C	B-D	D	Aircraft and other structures, keys, hydraulic fittings
7178	SEBW	C	B	D	Aircraft and other structures
7475	S	C	B	D	Aircraft and other structures (good fracture toughness)

* B, bars or rods; C, cladding; E, extrusions; F, forgings or forging stock; O, foil; P, pipe; S, sheets or plates; T, tubing; W, wire.

† Relative ratings in decreasing order of merit—A, B, C, D. Where applicable, ratings for both annealed and hardest tempers are given (for example, A-C).

‡ Weldability: A, generally weldable; B, weldable with special techniques for specific applications; C, limited weldability; D, not weldable. Ratings are given for arc welding. Gas welding and brazeability ratings are the same or differ, up or down, by one rating. Exceptions: most of the 2000 and 7000 alloys.

Data supplied by Alcan Aluminum Corp., Aluminum Assn. Inc., and Aluminum Co. of America.

3.13.2 MAGNESIUM AND MAGNESIUM ALLOYS

Magnesium is the lightest engineering metal available and, hence, its alloys are used where weight is of primary importance. Typical uses are for aircraft and missile components, either in cast or wrought form. Because it is not sufficiently strong in its pure form, magnesium is alloyed with various elements to impart certain properties. Magnesium was first produced in 1808.

Designation

Magnesium alloys are designated as follows in terms of the alloying elements:

A	aluminum	P	lead
E	rare earth	Q	silver
H	thorium	S	silicon
K	zirconium	T	tin
M	manganese	Z	zinc

The general designation is as follows: one or two prefix letters (indicating the principal alloying elements from the list above), two numerals (indicating the percentage of the alloying elements), and a suffix letter indicating minor variations in composition (e.g., AZ31B which has 3% Al and 1% Zn). The condition (temper) of the material is indicated by the same symbols used for aluminum alloys.

The overall mechanical properties of magnesium alloys are generally lower than those for aluminum alloys; they are also more expensive than aluminum alloys (Table 3.15). A variety of these alloys is available with good casting, forming, and machining characteristics. Because they oxidize rapidly, a fire hazard exists when machining or grinding magnesium alloys; precautions must be taken accordingly.

TABLE 3.15
PROPERTIES AND TYPICAL FORMS OF WROUGHT MAGNESIUM ALLOYS

ALLOY	COMPOSITION, %				CONDITION	ULTIMATE TENSILE STRENGTH, MPa	YIELD STRENGTH, MPa	ELONGATION IN 50 mm, %	TYPICAL FORMS
	Al	Zn	Mn	Zr					
AZ31B	3.0	1.0	0.2		F	260	200	15	Extrusions
					H24	290	220	15	Sheet and plates
AZ80A	8.5	0.5	0.2		T5	380	275	7	Extrusions and forgings
HK31A			3Th	0.7	H24	255	200	8	Sheet and plates
ZK60A		5.7		0.55	T5	365	300	11	Extrusions and forgings

3.13.3 COPPER AND COPPER ALLOYS

The general properties of copper and copper alloys are somewhat similar to those of aluminum alloys; in addition, they are among the best conductors of heat and electricity (Table 3.16). Copper is used primarily for electrical, construction, machinery, and transportation components. Copper alloys are available as cast or wrought products. Copper was first produced around 4000 BC.

TABLE 3.16
PROPERTIES AND TYPICAL APPLICATIONS OF WROUGHT COPPER AND BRASSES

TYPE AND UNS NUMBER	NOMINAL COMPOSITION, %	COMMERCIAL FORMS*	ULTIMATE TENSILE STRENGTH, psi × 10³	YIELD STRENGTH, psi × 10³	ELONGATION IN 2 in., %	FABRICATING CHARACTERISTICS AND TYPICAL APPLICATIONS
Oxygen-free electronic (C10100)	99.99 Cu	FRWTPS	32-66	10-53	55-4	Excellent hot and cold workability; good forgeability. Fabricated by coining, coppersmithing, drawing and upsetting, hot forging and pressing, spinning, swaging, stamping. Uses: busbars, bus conductors, waveguides, hollow conductors, lead-in wires and anodes for vacuum tubes, vacuum seals, transistor components, glass to metal seals, coaxial cables and tubes, microwave tubes, rectifiers.
Electrolytic tough pitch copper (C11000)	99.90 Cu. 0.04 O	FRWTPS	32-66	10-53	55-4	Fabricating characteristics same as C10100. Uses: downspouts, gutters, roofing, gaskets, auto radiators, busbars, nails, printing rolls, rivets, radio parts.
Gilding, 95% (C21000)	95.0 Cu, 5.0 Zn	FW	34-64	10-58	45-4	Excellent cold workability, good hot workability for blanking, coining, drawing, piercing and punching, shearing, spinning, squeezing and swaging, stamping. Uses: coins, medals, bullet jackets, fuse caps, primers, plaques, jewelry base for gold plate.

continued

TABLE 3.16 (*Continued*)

TYPE AND UNS NUMBER	NOMINAL COMPOSITION, %	COMMERCIAL FORMS*	ULTIMATE TENSILE STRENGTH, psi × 10³	YIELD STRENGTH, psi × 10³	ELONGATION IN 2 in., %	FABRICATING CHARACTERISTICS AND TYPICAL APPLICATIONS
Commercial bronze, 90% (C22000)	90.0 Cu, 10.0 Zn	FRWT	37-72	10-62	50-3	Fabricating characteristics same as C21000, plus heading and upsetting, roll threading and knurling, hot forging and pressing. Uses: etching bronze, grillwork, screen cloth, weatherstripping, lipstick cases, compacts, marine hardware, screws, rivets.
Jewelry bronze, 87.5% (C22600)	87.5 Cu, 12.5 Zn	FW	39-97	11-62	46-3	Fabricating characteristics same as C21000, plus heading and upsetting, roll threading and knurling. Uses: angles, channels, chain, fasteners, costume jewelry, lipstick cases, compacts, base for gold plate.
Red brass, 85% (C23000)	85.0 Cu, 15.0 Zn	FWTP	39-105	10-63	55-3	Excellent cold workability; good hot formability. Uses: weather-stripping, conduit, sockets, fasteners, fire extinguishers, condenser and heat exchanger tubing, plumbing pipe, radiator cores.
Low brass, 80% (C24000)	80.0 Cu, 20.0 Zn	FW	42-125	12-65	55-3	Excellent cold workability. Fabricating characteristics same as C23000. Uses: battery caps, bellows, musical instruments, clock dials, pump lines, flexible hose.
Cartridge brass, 70% (C26000)	70.0 Cu, 30.0 Zn	FRWT	44-130	11-65	66-3	Excellent cold workability. Fabricating characteristics same as C23000, except for coining, roll threading, and knurling. Uses: radiator cores and tanks, flashlight shells, lamp fixtures, fasteners, locks, hinges, ammunition components, plumbing accessories, pins, rivets.
Yellow brass (C26800, C27000)	65.0 Cu, 35.0 Zn	FRW	46-128	14-62	65-3	Excellent cold workability. Fabricating characteristics same as C23000. Uses: Same as C26000 except not used for ammunition.

continued

TABLE 3.16 (*Continued*)

TYPE AND UNS NUMBER	NOMINAL COMPOSITION, %	COMMERCIAL FORMS*	ULTIMATE TENSILE STRENGTH, psi × 10³	YIELD STRENGTH, psi × 10³	ELONGATION IN 2 in., %	FABRICATING CHARACTERISTICS AND TYPICAL APPLICATIONS
Muntz metal (C28000)	60.0 Cu, 40.0 Zn	FRT	54-74	21-55	52-10	Excellent hot formability and forgeability for blanking, forming and bending, hot forging and pressing, hot heading and upsetting, shearing. Uses: architectural, large nuts and bolts, brazing rod, condenser plates, heat exchanger and condenser tubing, hot forgings.
Free-cutting brass (C36000)	61.5 Cu, 3.0 Pb, 35.5 Zn	FRS	49-68	18-45	53-18	Excellent machinability. Fabricated by machining, roll threading and knurling. Uses: gears, pinions, automatic high speed screw machine parts.
Forging brass (C37700)	59.0 Cu, 2.0 Pb, 39.0 Zn	RS	52 (As extruded)	20	45	Excellent hot workability. Fabricated by heading and upsetting, hot forging and pressing, hot heading and upsetting, machining. Uses: forgings and pressings of all kinds.
Inhibited admiralty (C44300, C44400, C44500)	71.0 Cu, 28.0 Zn, 1.0 Sn	FWT	48-55	18-22	65-60	Excellent cold workability for forming and bending. Uses: condenser, evaporator and heat exchanger tubing, condenser tubing plates, distiller tubing, ferrules.
Naval brass (C46400 to (C46700)	60.0 Cu, 39.25 Zn, 0.75 Sn	FRTS	55-88	25-66	50-17	Excellent hot workability and hot forgeability. Fabricated by blanking, drawing, bending, heading and upsetting, hot forging, pressing. Uses: aircraft turnbuckle barrels, balls, bolts, marine hardware, nuts, propeller shafts, rivets, valve stems, condenser plates, welding rod.

* F, flat products; R, rod; W, wire; T, tube; P, pipe; S, shapes.
Data adapted from Copper Development Association, New York.

The temper designations for these alloys are based on percentage reduction by cold working, such as drawing or rolling (see Tables 6.11 and 6.13).

The most common copper alloys are brasses and bronzes. *Brass* (an alloy of copper and zinc) is one of the earliest alloys developed and has numerous applications, including decoration. *Bronze* is an alloy of copper and tin (Table 3.17).

Copper alloys can acquire many different properties by adding other alloying elements to improve their formability and machinability. Beryllium copper (beryllium bronze) and phosphor bronze have good strength and hardness for applications such as springs and bearings. The color descriptions of wrought copper alloys are given in Table 3.18 to serve as a guide in identifying various copper alloys.

TABLE 3.17
PROPERTIES AND TYPICAL APPLICATIONS OF WROUGHT BRONZES

TYPE AND UNS NUMBER	NOMINAL COMPOSITION, %	COMMERCIAL FORMS*	ULTIMATE TENSILE STRENGTH, psi × 10³	YIELD STRENGTH, psi × 10³	ELONGATION IN 2 in., %	FABRICATING CHARACTERISTICS AND TYPICAL APPLICATIONS
Leaded commercial bronze, nickel-bearing (C31600)	89.0 Cu, 1.9 Pb, 1.0 Ni, 8.1 Zn	FR	37-67	12-59	45-12	Good cold workability; poor hot formability. Uses: electrical connectors, fasteners, hardware, nuts, screws, screw machine parts.
Architectural bronze (C38500)	57.0 Cu, 3.0 Pb, 40.0 Zn	RS	60 (As extruded)	20	30	Excellent machinability and hot workability. Fabricated by hot forging and pressing, forming, bending and machining. Uses: architectural extrusions, store fronts, thresholds, trim, butts, hinges, lock bodies and forgings.
Phosphor bronze, 5% A (C51000)	95.0 Cu, 5.0 Sn, trace P	FRWT	47-140	19-80	64-2	Excellent cold workability. Fabricated by blanking, drawing, bending, heading and upsetting, roll threading and knurling, shearing, stamping. Uses: bellows, bourbon tubing, clutch disks, cotter pins, diaphragms, fasteners, lock washers, wire brushes, chemical hardware, textile machinery, welding rod.

continued

TABLE 3.17 *(Continued)*

TYPE AND UNS NUMBER	NOMINAL COMPOSITION, %	COMMERCIAL FORMS*	ULTIMATE TENSILE STRENGTH, psi × 10³	YIELD STRENGTH, psi × 10³	ELONGATION IN 2 in., %	FABRICATING CHARACTERISTICS AND TYPICAL APPLICATIONS
Phosphor bronze, 10% D (C52400)	90.0 Cu, 10.0 Sn, trace P	FRW	66-147	28 (Annealed)	70-3	Good cold workability for blanking, forming and bending, shearing. Uses: heavy bars and plates for severe compression, bridge and expansion plates and fittings, articles requiring good spring qualities, resiliency, fatigue resistance, good wear and corrosion resistance.
Free-cutting phosphor bronze (C54400)	88.0 Cu, 4.0 Pb, 4.0 Zn, 4.0 Sn	FR	44-75	19-63	50-15	Excellent machinability; good cold workability. Fabricated by blanking, drawing, bending, machining, shearing, stamping. Uses: bearings, bushings, gears, pinions, shafts, thrust washers, valve parts.
Aluminum bronze, 5% (C60800)	95.0 Cu, 5.0 Al	T	60	27	55	Good cold workability; fair hot formability. Uses: condenser, evaporator and heat exchanger tubes, distiller tubes, ferrules.
Low silicon bronze, B (C65100)	98.5 Cu, 1.5 Si	RWT	40-95	15-69	55-11	Excellent hot and cold workability. Fabricated by forming and bending, heading and upsetting, hot forging and pressing, roll threading and knurling, squeezing and swaging. Uses: hydraulic pressure lines, anchor screws, bolts, cable clamps, cap screws, machine screws, marine hardware, nuts, pole-line hardware, rivets, U-bolts, electrical conduits, heat exchanger tubing, welding rod.
Copper nickel, 10% (C70600)	88.7 Cu, 1.3 Fe, 10.0 Ni	FT	44-60	16-57	42-10	Good hot and cold workability. Fabricated by forming and bending, welding. Uses: condensers, condenser plates, distiller tubing, evaporator and heat exchanger tubing, ferrules, salt water piping.

continued

TABLE 3.17 (Continued)

TYPE AND UNS NUMBER	NOMINAL COMPOSITION, %	COMMERCIAL FORMS*	ULTIMATE TENSILE STRENGTH, psi × 10³	YIELD STRENGTH, psi × 10³	ELONGATION IN 2 in., %	FABRICATING CHARACTERISTICS AND TYPICAL APPLICATIONS
Copper nickel, 20% (C71000)	79.0 Cu, 21.0 Ni	FWT	49-95	13-85	40-3	Good hot and cold formability. Fabricated by blanking, forming and bending, welding. Uses: communication relays, condensers, condenser plates, electrical springs, evaporator and heat exchanger tubes, ferrules, resistors.
Nickel silver, 65-10 (C74500)	65.0 Cu, 25.0 Zn, 10.0 Ni	FW	49-130	18-76	50-1	Excellent cold workability. Fabricated by blanking, drawing, etching, forming and bending, heading and upsetting, roll threading and knurling, shearing, spinning, squeezing and swaging. Uses: rivets, screws, slide fasteners, optical parts, etching stock, hollow ware, nameplates, platers' bars.
Nickel silver, 65-18 (C75200)	65.0 Cu, 17.0 Zn, 18.0 Ni	FRW	56-103	25-90	45-3	Fabricating characteristics similar to C74500. Uses: rivets, screws, table flatware, truss wire, zippers, bows, camera parts, core bars, temples, base for silver plate, costume jewelry, etching stock, hollow ware, nameplates, radio dials.

* F, flat products; R, rod; W, wire; T, tube; P, pipe; S, shapes.
Data adapted from Copper Development Association, New York.

3.13.4 NICKEL AND NICKEL ALLOYS

Nickel (discovered in 1751) is a major alloying element and is used extensively in stainless steels and high-temperature alloys (also called *superalloys*, Section 3.15) for such applications as jet engine components, rockets, and nuclear power plants. Nickel alloys have high strength and corrosion resistance at elevated temperatures. Other alloying elements are chromium, cobalt (also called cobalt-base alloys), and molybdenum. The behavior of nickel alloys in machining, forming, casting, and welding can be modified by various alloying elements.

Various nickel alloys have been developed with a range of strength at different temperatures (Table 3.19). *Monel* is a nickel–copper alloy and *Inconel* a nickel–

TABLE 3.18
COLOR DESCRIPTION OF WROUGHT COPPER ALLOYS

UNS NUMBER	ALLOY	COLOR DESCRIPTION
C11000	Electrolytic tough pitch copper	Soft pink
C21000	Gilding, 95%	Red brown
C22000	Commercial bronze, 90%	Bronze gold
C23000	Red brass, 85%	Tan gold
C26000	Cartridge brass, 70%	Green gold
C28000	Muntz metal, 60%	Light brown gold
C61200	Aluminum bronze	Brown gold
C65500	High-silicone bronze, A	Lavender-brown
C70600	Copper-nickel, 10%	Soft lavender
C74500	Nickel silver, 65-10	Gray white
C75200	Nickel silver, 65-18	Silver

TABLE 3.19
PROPERTIES AND TYPICAL APPLICATIONS OF NICKEL ALLOYS (ALL ARE TRADE NAMES)

ALLOY [CONDITION]	PRINCIPAL ALLOYING ELEMENTS, %	ULTIMATE TENSILE STRENGTH, MPa	YIELD STRENGTH, MPa	ELONGA-TION IN 50 mm, %	TYPICAL APPLICATIONS
Nickel 200 [annealed]	None	380–550	100–275	60–40	Chemical and food processing industry, aerospace equipment, electronic parts
Duranickel 301 [age hardened]	4.4 Al 0.6 Ti	1300	900	28	Springs, plastics extrusion equipment, molds for glass, diaphragms
Monel R-405 [hot-rolled]	30 Cu	525	230	35	Screw-machine products, water meter parts
Monel K-500 [age hardened]	29 Cu 3 Al	1050	750	30	Pump shaft, valve stems, springs
Inconel 600 [annealed]	15 Cr 8 Fe	640	210	48	Gas turbine parts, heat-treating equipment, electronic parts, nuclear reactors
Incoloy 800 [annealed]	21 Cr 46 Fe	520–750	205–415	60–30	Heat treating, food processing
Hastelloy C-4 [solution-treated and quenched]	16 Cr 15 Mo	785	400	54	High temperature stability, resistance to stress-corrosion cracking
Hastelloy G [solution-treated]	22 Cr 20 Fe	700	310	62	Chemical applications.
Hastelloy X [solution-treated]	22 Cr 18 Fe	785	360	43	Jet engine parts, industrial furnaces

chrome alloy with a tensile strength of up to 200,000 psi (1400 MPa). A nickel–molybdenum–chromium alloy (*Hastelloy*) has good corrosion resistance and high strength at elevated temperatures. *Nichrome* is an alloy of nickel, chromium, and iron with high oxidation and electrical resistance, and is used for electrical heating elements.

3.13.5   TITANIUM AND TITANIUM ALLOYS

The high strength-to-weight ratio of titanium and its corrosion resistance at room and elevated temperatures make it suitable for aircraft components, naval vessels, and the chemical industry (Table 3.20). Titanium alloys have been developed for service at 1000°F (540°C) for long periods of time, and up to 1400°F (760°C) for short periods. Titanium was discovered in 1791.

Aluminum, vanadium, molybdenum, manganese, and other alloying elements are added to impart certain properties to titanium alloys, such as to improve their workability, strength, and hardenability. The bcc structure of titanium (β-titanium, above 1600°F, 870°C) is quite ductile, whereas its hcp structure is somewhat brittle and is very sensitive to stress corrosion.

3.13.6   LOW-MELTING ALLOYS

The major metals in this category are lead, zinc, and tin.

Lead is selected for its properties of high density, resistance to corrosion, softness, low strength, and good workability. Alloying with various elements gives lead special properties suitable for applications such as piping, collapsible tubing, bearing alloys, cable sheathing, and batteries. Lead was used extensively by the ancient Romans. The word *plumber* comes from *plumbium*, the Latin for lead.

Zinc has two major uses, one for *galvanizing* iron and steel (4.13.1) and the other as an alloy for casting. As a coating, zinc serves as the anode and protects the steel (cathode) from corrosive attack when the coating is scratched or punctured.

Zinc-base alloys are used extensively in casting, especially die casting (see Section 5.7.2), for products such as automotive and appliance components and numerous machine parts. Major alloying elements in zinc are aluminum, copper, and magnesium for strength and dimensional control during casting.

Another use for zinc is in superplastic alloys. Very fine-grained zinc–aluminum is a common example. (See Sections 2.2.7 and 7.13.)

Tin-base alloys generally contain copper, antimony, and lead. These elements control the hardness, strength, corrosion resistance, and frictional properties of these alloys. Because of the low friction coefficient (due to their low shear strength and low adhesion), a typical application of these alloys is for bearing materials (*Babbitt*, developed in the 1830s).

Tin is also used as a protective coating on steel sheet, known as *tin plate*, used for containers for food products. However, unlike galvanized steels, if this coating is

TABLE 3.20
PROPERTIES AND TYPICAL APPLICATIONS OF WROUGHT TITANIUM ALLOYS

NOMINAL COMPOSITION,* %	UNS	CONDITION	ROOM TEMPERATURE				VARIOUS TEMPERATURES					TYPICAL APPLICATIONS
			ULTIMATE TENSILE STRENGTH, psi × 10³	YIELD STRENGTH, psi × 10³	ELON- GATION, %	REDUCTION IN AREA, %	TEMP., F	ULTIMATE TENSILE STRENGTH, psi × 10³	YIELD STRENGTH, psi × 10³	ELON- GATION, %	REDUCTION IN AREA, %	
COMMERCIALLY PURE 99.5 Ti	R50250	Annealed	48	35	30	55	600	22	14	32	80	Airframes; chemical, desalination, and marine parts; plate type heat exchangers; cold spun or pressed parts; platinized anodes; high formability. Also available with low iron content for added corrosion resistance.
LOW-ALLOYED 99.5 Ti, 0.15 Pd	R52250	Annealed	48	35	30	55	600	22	14	32	80	Good corrosion resistance for chemical industry applications where medium is mildly reducing or varies between oxidizing and reducing; high formability.
ALPHA ALLOYS 5 Al, 2.5 Sn	R54520	Annealed	125	117	16	40	600	82	65	18	45	Weldable alloy for forgings and sheet-metal parts such as aircraft engine compressor blades and ducting; steam turbine blades, good oxidation resistance and strength at 600 to 1000°F (315 to 540°C), good stability at elevated temperatures.
5 Al, 2.5 Sn (low O₂)	R54521	Annealed	117	108	16	—	−320 −423	180 229	168 206	16 15	— —	Special grade for high pressure cryogenic vessels operating down to −423°F (−250°C).
NEAR ALPHA 8 Al, 1 Mo, 1 V	R54810	Duplex annealed	145	138	15	28	600 800 1000	115 107 90	90 82 75	20 20 25	38 44 55	Airframe and jet engine parts requiring high strength to 850°F (450°C); good creep and toughness properties; good weldability.
ALPHA-BETA ALLOYS 3 Al, 2.5 V	R56320	Annealed	100	85	20	—	600	70	50	25	—	Aircraft hydraulic tubing, foil; combines strength, weldability, and formability.

continued

TABLE 3.20 (Continued)

NOMINAL COMPOSITION.*%	UNS	CONDITION	ROOM TEMPERATURE — ULTIMATE TENSILE STRENGTH, psi × 10³	YIELD STRENGTH, psi × 10³	ELONGATION, %	REDUCTION IN AREA, %	VARIOUS TEMPERATURES — TEMP. °F	ULTIMATE TENSILE STRENGTH, psi × 10³	YIELD STRENGTH, psi × 10³	ELONGATION, %	REDUCTION IN AREA, %	TYPICAL APPLICATIONS
6 Al, 4 V	R56400	Annealed	144	134	14	30	600, 800, 1000	105, 97, 77	95, 83, 62	14, 18, 35	35, 40, 50	Rocket motor cases; blades and disks for aircraft turbines and compressors; structural forgings and fasteners; pressure vessels; gas and chemical pumps; cryogenic parts; ordnance equipment; marine components; steam turbine blades.
		Solution + age	170	160	10	20	600	142	130	10, 12, 22	28, 35, 45	
6 Al, 4 V (low O₂)	R56401	Annealed	130	120	15	35	−320	220	205	14	—	High pressure cryogenic vessels operating down to −320°F (−195°C).
6 Al, 6 V, 2 Sn	R56620	Annealed	155	145	14	30	600	135	117	18	42	Rocket motor cases, ordnance components; structural aircraft parts and landing gears; responds well to heat treatments; good hardenability.
		Solution + age	185	170	10	20	600	142	130	12	28	
6 Al, 2 Sn, 4 Zr, 6 Mo	R56260	Solution + age	184	170	10	23	600, 800, 1000	148, 138, 123	122, 110, 95	18, 19, 19	55, 67, 70	Components for advanced jet engines.
BETA ALLOYS												
13 V, 11 Cr, 3 Al	R58010	Solution + age	177	170	8	—	600	128	115	19	—	High strength fasteners.
		Solution + age	185	175	8	—	800	160	120	12	—	High strength fasteners, aerospace components, honeycomb panels; good formability, heat treatable.
3 Al, 8 V, 6 Cr, 4 Mo, 4 Zr	R58640	Solution + age	210	200	7	—	400, 600	150, 136	130, 110	20, 17	—	High strength fasteners, torsion bars, aerospace components.
		Annealed	128	121	15	—	600	105	95	22	—	Parts requiring formability and corrosion resistance.
11.5 Mo, 6 Zr, 4.5 Sn	R58030	Solution + age	201	191	11	—	600	131	123	16	—	High strength fasteners, high strength aircraft sheet parts.

* Numbers indicate percentage of alloying elements.

Source: Titanium Div., Martin Marietta Aluminum; Oregon Metallurgical Corp.; RMI Co.; Teledyne Allvac; and Timet.

punctured the steel corrodes because it is anodic to tin (cathode). The low shear strength of the tin coating improves the performance in deep drawing of tin plate.

Another tin alloy is *pewter*, which contains copper and antimony, and is used for tableware and decorative objects. It was developed in the 15th century.

Tin–lead alloys are the common materials for soldering, having a wide range of compositions and different melting points (Section 12.6.2).

3.14 REFRACTORY METALS AND ALLOYS

Refractory metals are molybdenum, columbium (niobium), tungsten, and tantalum (see Tables 3.3 and 3.21). These metals maintain their strength at elevated temperatures more than do most other metals and alloys; hence they are of great interest and use in rocket engines, gas turbines, and various aerospace applications. The temperature range for such service is on the order of 2000 to 4000°F (1100 to 2200°C), where strength and oxidation are of major concern. Although refractory metal elements were discovered about 200 years ago and were used as alloying elements, their use as engineering metals and alloys did not begin until about the 1940s.

3.14.1 MOLYBDENUM

Molybdenum has the general properties of high melting point and modulus of elasticity, good thermal conductivity and resistance to thermal shock. Although it has a very low susceptibility to hydrogen, major disadvantage of molybdenum alloys is their low resistance to oxidation at high temperatures, thus necessitating protective coatings. Principal alloying elements for molybdenum are titanium and zirconium.

Molybdenum is also an important alloying element in casting and wrought alloys (Section 3.12.1).

3.14.2 COLUMBIUM (NIOBIUM)

Columbium possesses good ductility and formability, and has greater oxidation resistance than other refractory metals. Through various alloying elements, columbium alloys can be made to have moderate strength and good fabrication characteristics. These alloys are used in nuclear and superconductor applications.

Columbium is an alloying element in various alloys and superalloys. An example is its use in high-strength, low-alloy (HSLA) steels, which are particularly suited for automotive and similar applications for weight reduction.

3.14.3 TUNGSTEN

Tungsten has the highest melting point of any metal (6170°F, 3410°C). It is characterized by high strength at elevated temperatures, low coefficient of thermal expansion and good thermal conductivity. On the other hand, it has high density, brittleness at low temperatures, and poor oxidation resistance.

Tungsten and its alloys are considered for service at temperatures above 3000°F (1650°C), such as nozzle throat liners in missiles and in the hottest parts of jet and rocket engines.

The filament wire in light bulbs is made of tungsten. It is also used as an electrode

TABLE 3.21
COMPOSITIONS AND PROPERTIES OF REFRACTORY METAL ALLOYS

COMMON DESIGNATION	NOMINAL ALLOY ADDITIONS, %	MOST COMMON PRODUCT FORM	CONDITION*	DUCTILITY CLASS†	TEMPERA-TURE, °F	ULTIMATE TENSILE STRENGTH, psi × 10³
COLUMBIUM ALLOYS						
Unalloyed Cb	None	All	Rx	A	2000	10
Cb-1Zr	1 Zr	All	Rx	A	2000	23
C103 (KBI-3)	10 Hf, 1 Ti, 0.7 Zr	All	Rx	A	2000	27
SCb291	10 Ta, 10 W	Bar, sheet	Rx	A	2000	32
C129	10 W, 10 Hf, 0.1 Y	Sheet	Rx	A	2400	26
FS85	28 Ta, 11 W, 0.8 Zr	Sheet	Rx	A	2400	23
SU31	17 W, 3.5 Hf, 0.12 C, 0.03 Si	Bar, sheet	SP	C	2400	40
MOLYBDENUM ALLOYS						
Unalloyed Mo	None	All	SRA	B-C	1800	52
Doped Mo	K, Si; ppm levels	Wire, sheet	CW	B	3000	30
Low C Mo	None	All	SRA	B	1800	50
TZM	0.5 Ti, 0.08 Zr, 0.015 C	All	SRA	B-C	2400	45
TZC	1.0 Ti, 0.14 Zr, 0.02–0.08 C	All	SRA	B-C	2400	55
Mo-5Re	5 Re	All	SRA	B	3000	2
Mo-30W	30 W	All	SRA	B-C	2000	50
TANTALUM ALLOYS						
Unalloyed Ta	None	All	Rx	A	2400	8
FS61	7.5 W (P/M)	Wire, strip	CW	A	75	165
FS63	2.5 W, 0.15 Cb	All	Rx	A	200	46
Ta-10W	10 W	All	Rx	A	2400	50
KBI-40	40 Cb	All	Rx	A	500	42
TUNGSTEN ALLOYS						
Unalloyed W	None	Bar, sheet, wire	SRA	D	3000	25
Doped W	K, Si, Al; ppm levels	Wire	CW	C	3000	94
W-1 ThO$_2$	1 ThO$_2$	Bar, sheet, wire	SRA	D	3000	37
W-2 ThO$_2$	2 ThO$_2$	Bar, sheet, wire	SRA	D	3000	30
W-3ThO$_2$	3 ThO$_2$	Bar, wire	SRA	D	3000	30
W-4ThO$_2$	4 ThO$_2$	Bar	SRA	D	3000	30
W-15 Mo	15 Mo	Bar, sheet	SRA	D	3000	36
W-50 Mo	50 Mo	Bar, sheet	SRA	D	3000	20
W-3 Re	3 Re	Wire	CW	C	—	—
W-25 Re	25 Re	Bar, sheet, wire	SRA	B	3000	33

*CW, cold worked; Rx, recrystallized; SRA, stress-relief annealed; SP, special thermal processing.

† A, excellent cryogenic ductility; B, excellent room-temperature ductility; C, may have marginal ductility at room temperature; D, normally brittle at room tempera ture.

Source: AMAX Specialty Metals Corp., Edwin S. Bartlett (Battelle Columbus Laboratories), Fansteel Inc., GTE Products Corp., KBI Div. of Cabot Corp., and Schwarkopf Development Corp. (Metallwerk Plansee).

in welding (Chapter 12). Tungsten is an important alloying element in tool and die steels and is used as tungsten carbide, with cobalt as a binder (Section 8.10).

3.14.4 TANTALUM

Tantalum is characterized by a high melting point (5425°F, 3000°C), good ductility, and corrosion resistance. However, it has high density and poor resistance to chemicals except below 300°F (150°C). Tantalum has extensive uses in capacitors and in various electrical and electronic components. Tantalum-base alloys are available in many forms for use in the construction of missiles and aircraft.

3.15 SUPERALLOYS

Superalloys are a series of alloys for high-temperature applications. They are referred to as iron-, cobalt-, or nickel-base superalloys and contain nickel, chrome, cobalt, and molybdenum as major alloying elements. Other elements are aluminum, tungsten, and titanium. These alloys are generally identified by trade names or special numbering systems. Superalloys were developed in the 1940s.

Iron-base superalloys generally contain 32 to 67% iron, 15 to 22% chrome, and 9 to 38% nickel.

Cobalt-base superalloys generally contain 35 to 65% cobalt, 19 to 30% chrome, and up to 35% nickel (Table 3.22).

TABLE 3.22
PROPERTIES AND TYPICAL APPLICATIONS OF SELECTED COBALT-BASE SUPERALLOYS AT 1600°F (870°C) (ALL ARE TRADE NAMES)

ALLOY	CONDITION	ULTIMATE TENSILE STRENGTH, MPa	YIELD STRENGTH, MPa	ELONGATION IN 50 mm, %	TYPICAL APPLICATIONS
AR-13	Cast	290	275	21	High-temperature parts
AR-215	Cast	275	215	59	Nozzle vanes, resistant to hot corrosion
FSX-414	Cast	310	165	23	Gas turbine vanes
Haynes 25 (L-605)	Wrought	325	240	30	Jet engine parts, sheets
Haynes 188	Wrought	420	260	73	Aircraft, chemical, nuclear
Haynes 556	Wrought	300	195	45	Structural, oxidation and creep resistance
MAR-M 302	Cast	275	215	15	Jet engine blades, vanes
MAR-M 322	Cast	550	345	12	Jet engine blades, vanes
S-816	Wrought	360	240	16	Gas turbine blades, bolts, springs

Nickel-base superalloys are the most common and are available in a wide variety of compositions. The range of nickel is from 38 to 76%, up to 27% chrome, and up to 29% cobalt (Table 3.23).

Conventional superalloys have a maximum service temperature of about 1750°F (950°C) for structural applications. The temperatures can be as high as 2200°F (1200°C) for non–load-bearing components. Depending on their composition, super-

TABLE 3.23
PROPERTIES AND TYPICAL APPLICATIONS OF SELECTED NICKEL-BASE SUPERALLOYS AT 1600°F (870°C) (ALL ARE TRADE NAMES)

ALLOY	CONDITION	ULTIMATE TENSILE STRENGTH, MPa	YIELD STRENGTH, MPa	ELONGATION IN 50 mm, %	TYPICAL APPLICATIONS
Astroloy	Wrought	770	690	25	Forgings for high temperature
Hastelloy X	Wrought	255	180	50	Jet engine sheet parts
IN-100	Cast	885	695	6	Jet engine blades and wheels
IN-102	Wrought	215	200	110	Superheater and jet engine parts
IN-792	Cast	840	660	8	Hot corrosion resistance, blades
Inconel 600	Wrought	105	62	80	Heat treating and chemical processing equipment
Inconel 601	Wrought	160	140	92	Heat treating equipment
Inconel 625	Wrought	285	275	125	Aircraft engines and structures, chemical processing equipment
Inconel 718	Wrought	340	330	88	Jet engine and rocket parts
Inconel X-750	Wrought	235	165	47	Gas turbine parts, bolts
MAR-M 200	Cast	840	760	4	Jet engine blades
MAR-M 421	Cast	750	650	6	Hot corrosion resistance
MAR-M 432	Cast	730	605	8	Integrally cast turbine wheels
Rene 41	Wrought	620	550	19	Jet engine parts
Rene 80	Wrought	705	530	12	Turbine blades
Udimet 500	Wrought	660	600	9	Gas turbine parts, sheet, bolt
Udimet 700	Wrought	690	635	27	Jet engine parts
Waspaloy	Wrought	525	515	35	Jet engine parts

alloys have generally good resistance to corrosion, mechanical and thermal fatigue, mechanical and thermal shock, creep, and erosion at elevated temperatures.

Typical applications are for components in aircraft gas turbines; reciprocating engines; rocket engines; tools and dies for hot working of metals; heat-treating equipment; and nuclear, chemical, and petrochemical systems.

SUMMARY

1. The manufacturing properties of metals and alloys depend largely on their mechanical and physical properties. These, in turn, are governed by the crystal structure of metals, imperfections, grain boundaries, grain size, and texture.

2. Metals and alloys can be worked at room, warm, or high temperatures. Their overall behavior, force requirements, and workability depend largely on whether or not the working temperature is below or above the recrystallization temperature.

3. An important factor is fracture of the material. Basic types of fracture are brittle and ductile, intergranular and transgranular. Temperature, strain rate, surface conditions, and state of stress are significant factors in fracture.

4. A variety of metals and alloys is available with a wide range of properties, such as strength, toughness, hardness, ductility, and resistance to oxidation. They can be generally classified as ferrous alloys, nonferrous metals and alloys, refractory metals and their alloys, and superalloys.

5. The selection of a material for a particular application requires careful consideration of many factors. Among these are design considerations, service requirements, long-term effects, compatibility with other materials, environmental attack, and economic factors. (See also Chapter 1.)

BIBLIOGRAPHY

Ashby MF, Jones DRH. *Engineering Materials: An Introduction to their Properties and Applications.* New York: Pergamon, 1980.

Barrett CR, Nix WD, Tetelman AS. *The Principles of Engineering Materials.* Englewood Cliffs, N.J.: Prentice-Hall, 1973.

Byrne JG. *Recovery, Recrystallization and Grain Growth.* New York: Macmillan, 1965.

Caddell RM. *Deformation and Fracture of Solids.* Englewood Cliffs, N.J.: Prentice-Hall, 1980.

Clauser HR. *Industrial and Engineering Materials.* New York: McGraw-Hill, 1975.

Cottrell AH. *The Mechanical Properties of Matter.* New York: Wiley, 1964.

DiBenedetto AT. *The Structure and Properties of Materials.* New York: McGraw-Hill, 1967.

Dieter GE. *Mechanical Metallurgy*, 2d ed. New York: McGraw-Hill, 1976.

Ductility. Metals Park, Ohio: American Society for Metals, 1967.

Flinn RA, Trojan PK. *Engineering Materials and their Applications*, 2d ed. Boston: Houghton Mifflin, 1981.

Gibbons RC, ed. *Woldman's Engineering Alloys*, 6th ed. Metals Park, Ohio: American Society for Metals, 1975.

Gordon JE. *The New Science of Strong Materials or Why You Don't Fall Through the Floor.* Harmondsworth, England: Penguin Books, 1968.

Guy AG. *Essentials of Materials Science.* New York: McGraw-Hill, 1976.

Hertzberg RW. *Deformation and Fracture Mechanics of Engineering Materials*, 2d ed. New York: Wiley, 1983.

Jastrzebski ZD. *The Nature and Properties of Engineering Materials*, 2d ed. New York: Wiley, 1977.

Keyser CA. *Materials Science in Engineering*, 3d ed. Columbus, Ohio: Merrill, 1980.

LeMay I. *Principles of Mechanical Metallurgy.* New York: Elsevier, 1981.

Lynch CT, ed. *Handbook of Materials Science* (3 vols.). Cleveland, Ohio: CRC Press, 1975.

Material Selector. (Annual publication of *Materials Engineering Magazine.* Cleveland, Ohio: Penton/IPC.)

McClintock FA, Argon AS, eds. *Mechanical Behavior of Materials*, Reading, Mass.: Addison–Wesley, 1966.

McLean D. *Mechanical Properties of Metals.* New York: Wiley, 1962.

Nutt MC. *Metallurgy and Plastics for Engineers.* New York: Pergamon, 1977.

Nutt MC. *Principles of Modern Metallurgy*, Columbus, Ohio: Merrill, 1968.

Pascoe KJ. *An Introduction to the Properties of Engineering Materials*, 3d ed. New York: Van Nostrand Reinhold, 1978.

Polakowski NH, Ripling EJ. *Strength and Structure of Engineering Materials.* Englewood Cliffs, N.J.: Prentice-Hall, 1966.

Ralls KM, Courtney TH, Wulff J. *An Introduction to Materials Science and Engineering*, New York: Wiley, 1976.

Reed-Hill RE. *Physical Metallurgy Principles*, 2d ed. New York: Van Nostrand Reinhold, 1973.

Richman NH. *Science of Metals.* Waltham, Mass.: Blaisdell, 1967.

Rosenthal D, Asimow RM. *Introduction to Properties of Materials*, 2d ed. New York: Van Nostrand Reinhold, 1971.

Ruoff AL. *Introduction to Materials Science.* Englewood Cliffs, N.J.: Prentice-Hall, 1972.

Samans CH. *Metallic Materials in Engineering.* New York: Macmillan, 1963.

Sims CT, Hagel WC. *The Superalloys.* New York: Wiley, 1972.

Tegart WJMcG. *Elements of Mechanical Metallurgy.* New York: Macmillan, 1966.

Tetelman AS, McEvily AJ. *Fracture of Structural Materials.* New York: Wiley, 1967.

Tweeddale JG. *The Mechanical Properties of Metals.* New York: Elsevier, 1964.

Van Vlack LH. *Materials for Engineering.* Reading, Mass.: Addison–Wesley, 1982.

Van Vlack LH. *Elements of Materials Science and Engineering*, 4th ed. Reading, Mass.: Addison–Wesley, 1980.

Wulff J, et al. *The Structure and Properties of Materials* (4 vols). New York: Wiley, 1964.

Zackay VF. *High Strength Materials*, New York: Wiley, 1965.

REFERENCES

Aerospace Structural Metals Handbook (5 vols). Columbus, Ohio: Mechanical Properties Data Center, Battelle, 1980.

ASM Metals Reference Book. Metals Park, Ohio: American Society for Metals, 1981.

Brady GS, Clauser HR. *Materials Handbook*, 11th ed. New York: McGraw-Hill, 1977.

Gibbons RC. *Woldman's Engineering Alloys*, 6th ed. Metals Park, Ohio: American Society for Metals, 1979.

Metals Handbook, 8th ed. Vol. 10, *Failure Analysis and Prevention*, Metals Park, Ohio: American Society for Metals, 1975.

Metals Handbook, 9th ed. Vol. 1: *Properties and Selection: Irons and Steels*, 1978; Vol. 2: *Properties and Selection: Nonferrous Alloys and Pure Metals*, 1979, Vol. 3: *Properties and Selection: Stainless Steels, Tool Materials and Special Purpose Metals*, 1980, Metals Park, Ohio: American Society for Metals.

Smithells *Metals Reference Book*, 6th ed. London: Butterworths, 1983.

Periodicals
Materials Engineering

Materials Science and Engineering

Metal Progress

PROBLEMS

3.1. List and explain briefly all the conditions that induce brittle fracture in an otherwise ductile metal.

3.2. A cold-worked piece of metal has been fully annealed. When tested, it is found that it is anisotropic. Explain the probable reason.

3.3. A strip of fully annealed metal is reduced in thickness by cold working from 25 mm to 15 mm. A similar fully annealed strip is reduced from 25 mm to 10 mm. Which one of these cold-worked strips will recrystallize at a lower temperature?

3.4. Explain why the heat content of a cold-worked piece of metal is not equal to the amount of work done in plastic deformation.

3.5. A part fractures while being subjected to plastic deformation at an elevated temperature. List the factors that could have played a role in fracture.

3.6. Estimate the ratio of theoretical shear strength/theoretical tensile strength for the materials listed in Table 2.1. (Ignore those for which the Poisson's ratio is not available.)

3.7. Explain the general process of refining the grain size of a metal by plastic deformation.

3.8. Make a list of materials that would be suitable for: a) a paper clip, b) a buckle for a safety belt, c) door hinges, d) nails, e) a marine propeller, f) a coat bracket, g) a battery cable, h) a gas turbine blade.

4 Surfaces: Measurement, Properties, and Treatment

4.1 INTRODUCTION

In the preceding chapters, materials were treated in terms of their bulk properties, and some descriptions of surfaces as they affect these properties were included. In this chapter surfaces will be treated as distinct entities. Their measurement, properties, and thermal and mechanical treatment will be described. These are important not only in manufacturing processes, but also because they affect the use of products in service.

A surface is the thin layer of material on the boundary between the bulk metal and the environment. Surfaces have unique properties and characteristics that can have significant influence on performance. Physical, chemical, metallurgical, and mechanical properties of surfaces depend not only on the type of bulk metal, but also the environment and processing history of the material or component. The terms *surface integrity* and *surface texture* are used to describe not only surface finish but also other physical and mechanical characteristics and defects that may be present.

In addition to surface properties and their measurement, the subjects of friction, wear, and lubrication (now known as *tribology*) are also treated in this chapter because they are all basically surface phenomena.

4.2 SURFACE STRUCTURE

The surface of a piece of metal is generally composed of several layers, as shown in Fig. 4.1. The bulk metal, or metal substrate, has a structure that depends on its composition and processing history, as described in Chapter 3.

Above this bulk metal is a layer that generally has been plastically deformed and work hardened. The depth and properties of this layer (also known as the *Beilby layer*) depend on a number of factors. The major factors are the magnitude of friction and the type of processing, or operation, to which the surface has been subjected. If, for example, the surface is produced by machining with a dull tool or ground with a dull grinding wheel, this layer will be relatively thick. These effects also depend on the process parameters during machining (Chapter 8). If machined by electrical, chemical, or other techniques (Chapter 9), this layer may undergo phase transformations, or it may undergo melting and resolidification due to high temperatures. Such

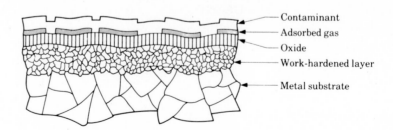

FIGURE 4.1 Schematic illustration of typical surface structure of metals. Surfaces generally have properties that are different from the bulk material. *Source:* After E. Rabinowicz.

a deformed surface also can be produced by high friction and at high speed of sliding, which can melt, burn, or burnish the surface and cause thermal cracks.

Depending on its past history and surface strains produced, this layer may also affect the concentration of defects on the atomic scale, such as dislocations, vacancies, and voids. These defects influence not only the mechanical properties of the surface layers but also their wear resistance.

The work-hardened layer usually contains residual stresses. As described in Section 2.7, residual stresses may be due to nonuniform plastic deformation or to temperature gradients and thermal cycling, or phase transformations. These conditions arise during metalworking operations. Residual stresses in surfaces can have a significant effect on the reactivity of surfaces with chemicals, oxidation, and lubricant behavior. For example, surfaces with tensile residual stresses are more reactive than those with compressive stresses. In addition, residual stresses have a significant effect on the fatigue life and stress–corrosion cracking of components (Section 3.7.1).

Above the work-hardened layer there is, invariably, an oxide layer, unless the metal is processed and kept in an inert (oxygen-free) environment or it is a noble metal. The chemistry and thickness of this layer depend primarily on the material, the environment, and temperature. It also depends on the structure of the work-hardened surface layer.

Most metals are covered with a mixture of oxides and hydroxides because of reactions with air and moisture. For instance, iron has a complex oxide structure with FeO adjacent to the bulk iron, followed by a layer of Fe_3O_4 and then a layer of Fe_2O_3, which is exposed to air. Furthermore, in steels the carbon is preferentially oxidized, leaving a carbon-depleted layer on the surface of the bulk steel right below the FeO layer. For steels, the thickness of the oxide layer (scale) decreases with increasing carbon content, for the same temperature and time.

Aluminum in air has a dense amorphous layer of Al_2O_3 (10 to 20 Å in thickness) with a thick, porous, hydrated layer over it. For instance, a newly scratched or machined piece of aluminum has a bright, shiny appearance, whereas aluminum usually has a dull gray appearance due to the hydroxide layer.

Copper, which has a bright, shiny surface when newly scratched or machined, has a Cu_2O layer next to its bulk. This layer is then covered with a layer of CuO giving it a somewhat dull color. Other metals form various oxides as a result of their reaction with oxygen and water vapor.

The oxide of a metal is generally harder than the base metal, as shown in Table 4.1. Oxides also tend to be brittle and abrasive.

TABLE 4.1
HARDNESS RATIO OF OXIDE/BASE METAL OF CERTAIN ELEMENTS

Molybdenum	0.3	Nickel	2.0
Tantalum	0.6	Lead	20.0
Copper	1.6	Aluminum	70.0
Columbium	1.8	Tin	90.0

Under normal environmental conditions, the oxide layer on metals is generally covered with adsorbed layers of gas and moisture. Finally, the outermost surface of the metal may be covered with contaminants of many kinds unless precautions are taken against them. Typical contaminants are dirt, dust, grease, lubricant residues, cleaning compound residues, and other pollutants from the environment.

It is apparent from the foregoing observations that surfaces, in general, can indeed have considerably different properties and behavior from the bulk metals. This observation is important since the bulk metal generally determines the overall strength and ductility of a component to be formed or used in service, but it is the surface of the component that has direct influence on the following properties:

a. Friction and wear properties in service or in subsequent processing, such as in direct contact with tools and dies,

b. Friction and wear properties at interfaces, and the effectiveness of a lubricant,

c. The appearance of the part and its role in subsequent operations such as painting,

d. Welding, brazing, and soldering characteristics,

e. The initiation of cracks due to surface defects, which could lead to rapid failure of a component, particularly in notch-sensitive materials,

f. Thermal and electrical conductivity at the interface with another body.

Many of the factors described above for metals are also relevant to nonmetallic materials. As will be seen in the rest of this chapter, these factors are of particular significance in manufacturing processes.

4.3 SURFACE FINISH AND ITS MEASUREMENT

All surfaces, regardless of their method of manufacturing and preparation, have unique characteristics referred to as *surface finish* or *texture*. The description of surface texture as a geometric property is quite involved. Certain guidelines have been established for identifying surface texture in terms of generally well-defined and measurable quantities. A typical surface is shown in Fig. 4.2 with reference to which key terms are defined below:

Waviness: the recurrent deviation from a flat surface. It may be caused by deflections of the tools, dies, and the workpiece during processing; warping; nonuniformity in lubrication; vibration; or any mechanical or thermal instability in the system. Waviness is measured in terms of the width and height of the wave.

Roughness: closely spaced, irregular deviations on a scale less than that of waviness. It may be superimposed on waviness. Roughness is expressed in terms of its height, its width, and the distance (length) on the surface along which it is measured.

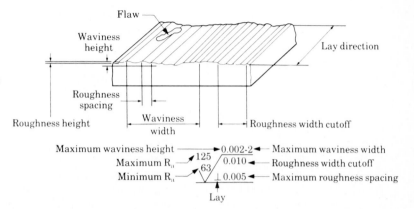

FIGURE 4.2 Standard surface finish terminology and symbols. In spite of these descriptions, most surfaces are difficult to characterize. See, for example, Figs. 4.20, 4.29, 4.30, 8.47, 8.50, 8.61, and 9.21.

Lay: direction of the predominant surface pattern, which is usually discernible by the eye.

Flaws: random irregularities, such as scratches, cracks, holes, depressions, laps, seams, tears, and surface inclusions.

4.3.1 ROUGHNESS MEASUREMENT

Generally, two measures are employed for expressing the surface roughness: the arithmetical average, R_a (also identified as AA or CLA for center-line average), or the root-mean-square average, R_q (RMS).

Based on the schematic illustration shown in Fig. 4.3, the *arithmetical average roughness* is defined as

$$R_a = \frac{a + b + c + d + \cdots}{n} \tag{4.1}$$

and the *root-mean-square average* as

$$R_q = \sqrt{\frac{a^2 + b^2 + c^2 + d^2 + \cdots}{n}}. \tag{4.2}$$

The line AB in Fig. 4.3 is located such that the sum of the areas above the line is equal to the sum of the areas below the line.

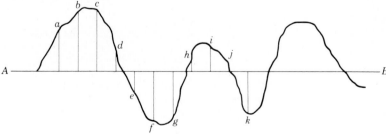

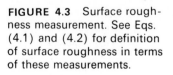

FIGURE 4.3 Surface roughness measurement. See Eqs. (4.1) and (4.2) for definition of surface roughness in terms of these measurements.

Because of its simplicity, the arithmetical average R_a has been adopted internationally and is widely used. However, it can be seen from Eqs. (4.1) and (4.2) that there is a relation between R_q and R_a. For example, it can be shown that for a surface roughness in the shape of a sine wave, R_q is proportional to R_a, i.e., $R_q = (1.11)(R_a)$.

It has been observed experimentally that for most machining processes $R_q = (1.25)(R_a)$ and, approximately, $R_q = (1.45)(R_a)$ for honing. The units that are generally used for surface roughness are microinches (μin.) or micrometers (μm), where 1 μin. $= 0.025$ μm and 1 μm $= 40$ μin.

Surface roughness is usually measured and recorded with a variety of commercially available instruments, called surface *profilometers*. The most common ones employ the travel of a diamond stylus along a straight line over the surface (Fig. 4.4). To highlight the roughness, the profilometer tracings are recorded using a vertical scale (roughness height) that is a few orders of magnitude greater than the horizontal scale. (See Fig. 4.9.)

Because of the finite radius of the tip of the stylus, its path is smoother than the actual surface roughness. The smaller the tip radius and the lower the surface roughness, the closer the path of the stylus to the actual surface profile. The most commonly used stylus diameter is 0.0004 in. (10 μm). A permanent record of the surface profile is made by mechanical and electronic means. The surface roughness is specified on a technical drawing by the symbols shown in Fig. 4.2 and their value is placed to the left of the check mark.

Surface roughness can also be observed directly through an optical or scanning electron microscope. Stereoscopic photographs are particularly useful for a three-dimensional observation of surfaces. Measurement of the surface roughness is also possible by these means.

The measurement and interpretation of surface roughness can be quite complex and even controversial. It is generally not sufficient to describe a surface by its R_a or R_q value alone. Two surfaces may have the same roughness value, as measured by a

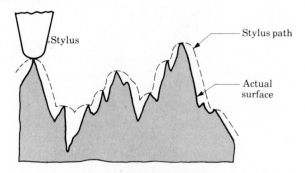

FIGURE 4.4   Path of stylus in surface roughness measurements (broken line), as compared to actual roughness profile. Note that the profile of the stylus path is smoother than the actual profile. *Source*: After D. H. Buckley.

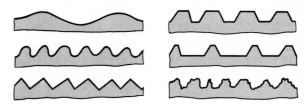

FIGURE 4.5 Various surface profiles that have the same R_a and R_q values. This is an example of the difficulty in describing surfaces by surface roughness values only.

profilometer, but their topography may be quite different, as can be seen in Fig. 4.5. These differences in the actual surface profile can be significant, particularly when evaluating friction, wear, and lubrication mechanisms.

4.3.2 LAY

The symbols denoting the lay of a surface and their meanings are given in Fig. 4.6. The symbol for lay is placed at the lower right of the check mark.

Lay symbol	Interpretation	Examples
—	Lay parallel to the line representing the surface to which the symbol is applied	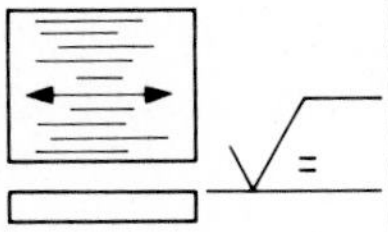
⊥	Lay perpendicular to the line representing the surface to which the symbol is applied	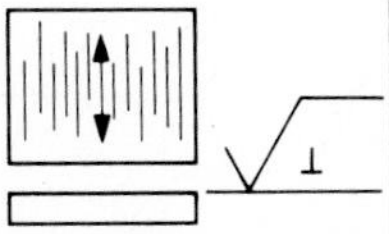
X	Lay angular in both directions to line representing the surface to which symbol is applied	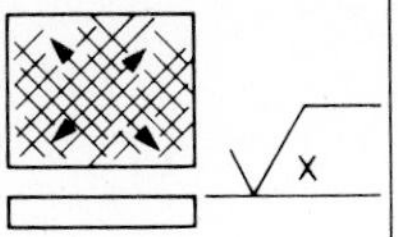
M	Lay multidirectional	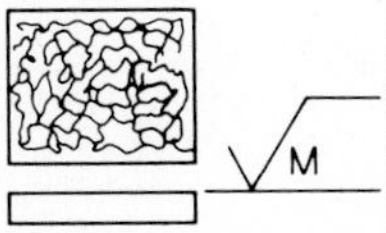

Lay symbol	Interpretation	Examples
C	Lay approximately circular relative to the center of the surface to which the symbol is applied	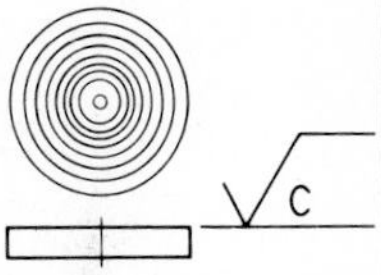
R	Lay approximately radial relative to the center of the surface to which the symbol is applied	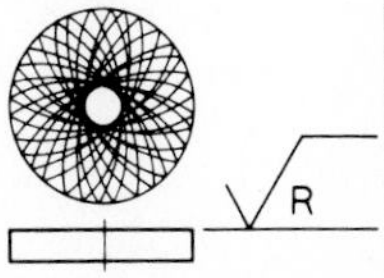
P	Pitted, protuberant, porous, or particulate nondirectional lay	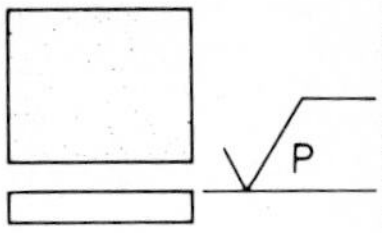

FIGURE 4.6 Lay symbols for engineering surfaces. Lay is defined as directionality in the surface finish of a manufactured part. See, for example, Fig. 8.50.

The symbols used to describe a surface (see Fig. 4.2) specify only the roughness, waviness, and lay. Thus, *flaws* are not included. Whenever important, a special note is included in technical drawings to describe the method of inspection for any surface flaws.

4.4 SURFACE ROUGHNESS IN PRACTICE

The surface finishes required for a number of engineering applications are given in Fig. 4.7. Note that roughness requirements can vary by as much as two orders of magnitude.

Fig. 4.8 shows the range of roughness produced by a number of machining, forming, and casting processes. Note that the range in this chart is as much as three orders of magnitude. Even within a particular process the range of roughnesses

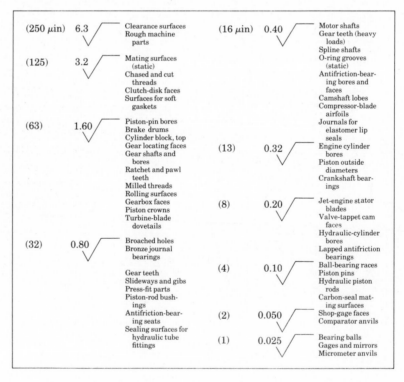

FIGURE 4.7 Typical surface roughness requirements in engineering components. Numbers above the symbols are in micrometers. Note that the range of roughness is two orders of magnitude. See also Figs. 4.8 and 9.12.

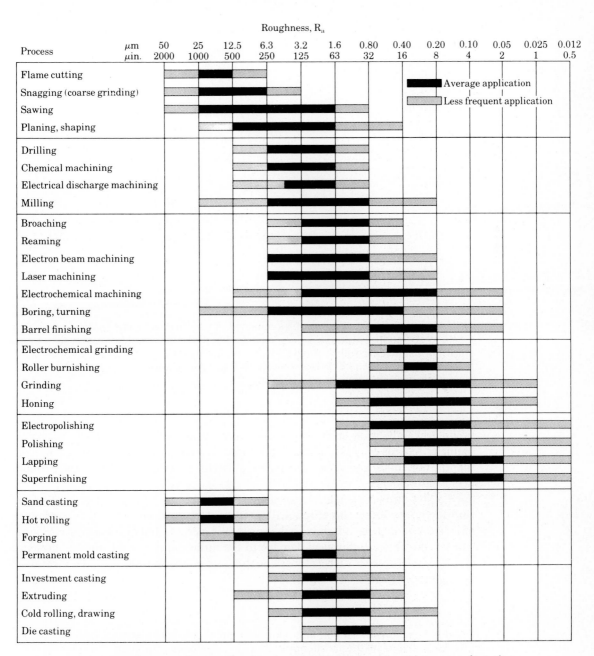

FIGURE 4.8 Range of surface roughness obtained in various manufacturing processes. Note the wide range of roughness within each process. See also Fig. 9.12. *Source:* Courtesy of General Motors.

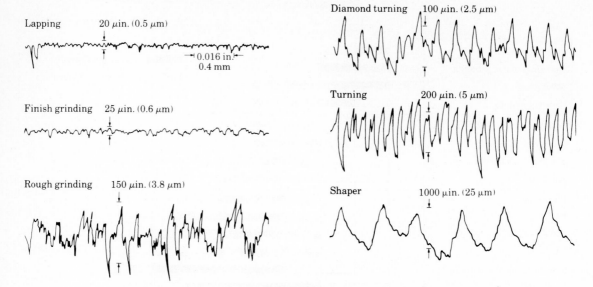

FIGURE 4.9 Typical surface profiles obtained in various machining and surface finishing operations. Note the difference between the vertical and horizontal scales, shown in the first example. Actual surfaces are not nearly as rough as they appear in these illustrations. *Source:* Adapted from *Tool and Manufacturing Engineers Handbook.* New York: McGraw-Hill, 3d ed., 1976, pp. 35–37.

produced can be wide, such as in boring, turning, and grinding. Fig. 4.9 shows profilometer records of typical surfaces produced by various machining processes. Note that the horizontal and vertical scales are different, as is the case in all surface roughness measurements.

Fatigue strength is affected most by the type of process used in generating the surface and its roughness (Figs. 4.10a and b). However, note that the decrease in fatigue strength is due not only to the roughness increase, but also to surface residual stresses, possible metallurgical damage to the surface of the metal, and the presence of surface microcracks such as those due to severe gradients of strain and temperature during processing.

FIGURE 4.10 (a) Reduction in fatigue strength of cast steels with various surface ▶ finishes and ultimate tensile strengths. The effect on fatigue strength is greater for steels with higher strengths. (b) Effect of processing on the endurance limit of steel. See also Fig. 9.29. *Source:* (a) M. R. Mitchell, *J. Eng. Mat. Tech.,* vol. 99, 1977, pp. 329–343. (b) G. C. Noll and C. Lipson, *Proc. Soc. Exp. Stress Anal.,* vol. 3, 1946, pp. 89–101.

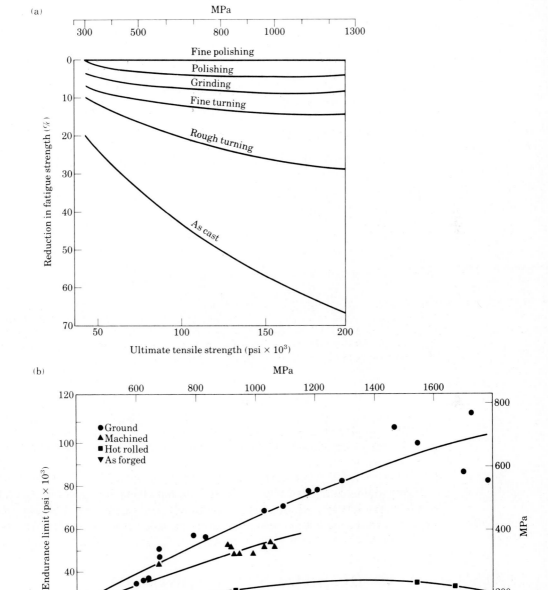

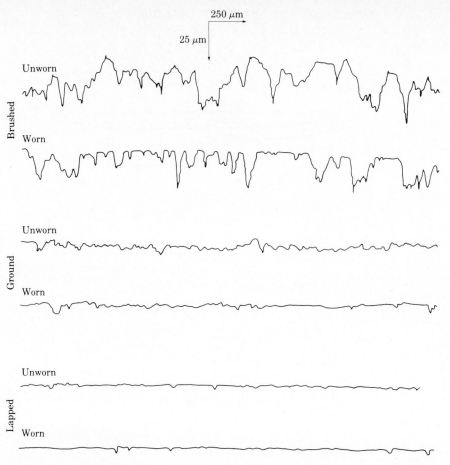

FIGURE 4.11 Changes in the original surface profiles due to wear, as produced by three different processes. Note the flattening of the tops of the peaks as a result of wear. *Source:* E. Wild and K. J. Mack, *Tribology*, vol. 11, 1978, pp. 321–324.

The original surface profile and its texture may be altered as a result of further processing or its use in service. One such example is shown in Fig. 4.11, where it can be seen that the tops of the peaks have been flattened as a result of wear or plastic deformation.

4.5 TESTING TECHNIQUES FOR SURFACE INTEGRITY

A variety of nondestructive and destructive techniques are available for observing and testing surfaces for various defects and flaws. Table 4.2 gives a general listing of the nondestructive techniques available; some of these are widely employed while others are still experimental.

TABLE 4.2
NONDESTRUCTIVE METHODS FOR EVALUATING MATERIALS. (After D. J. Hagemaier,
McDonnell Douglas Corp.)

METHOD	MEASURES OR DETECTS	APPLICATIONS	ADVANTAGES	LIMITATIONS
Acoustic emission	Crack initiation and growth rate Internal cracking in welds during cooling Boiling or cavitation Friction or wear Plastic deformation Phase transformations	Pressure vessels Stressed structures Turbine or gear boxes Fracture mechanics research Weldments Sonic signature analysis	Remote and continuous surveillance Permanent record Dynamic (rather than static) detection of cracks Portable Triangulation techniques to locate flaws	Transducers must be placed on part surface Highly ductile materials yield low amplitude emissions Part must be stressed or operating Test system noise needs to be filtered out
Acoustic-impact (tapping)	Debonded areas or delaminations in metal or nonmetal composites or laminates Cracks in turbine wheels or turbine blades Loose rivets or fasteners	Brazed or adhesive-bonded structures Bolted or riveted assemblies Turbine blades Turbine wheels Composite structures Honeycomb assemblies	Portable Easy to operate May be automated Permanent record or positive meter readout No couplant required	Part geometry and mass influences test results Impactor and probe must be repositioned to fit geometry of part Reference standards required Pulser impact rate is critical for repeatability
Eddy current (100 Hz–10 kHz)	Subsurface cracks around fastener holes in aircraft structure	Aluminum and titanium structures	Detect subsurface cracks not detectable by radiography	Part geometry Will not detect short cracks
Eddy current (10 kHz–6 MHz)	Surface and subsurface cracks and seams Alloy content Heat treatment variations Wall thickness, coating thickness Crack depth Conductivity Permeability	Tubing Wire Ball bearings "Spot checks" on all types of surfaces Proximity gage Metal detector Metal sorting Measure conductivity in % IACS	No special operator skills required High speed, low cost Automation possible for symmetrical parts Permanent record capability for symmetrical parts No couplant or probe contact required	Conductive materials Shallow depth of penetration (thin walls only) Masked or false indications caused by sensitivity to variations, such as part geometry, lift-off Reference standards required Permeability variations

(*continued*)

TABLE 4.2   (*Continued*)

METHOD	MEASURES OR DETECTS	APPLICATIONS	ADVANTAGES	LIMITATIONS
Eddy-sonic	Debonded areas in metal-core or metal-faced honeycomb structures Delaminations in metal laminates or composites	Metal-core honeycomb Metal-faced honeycomb Conductive laminates such as boron or graphite fiber composites Bonded metal panels	Portable Simple to operate No couplant required May be automated	Specimen or part must contain conductive materials to establish eddy-current field Reference standards required Part geometry
Electric current (direct current conduction method)	Cracks Crack depth Resistivity Wall thickness Corrosion-induced wall-thinning	Metallic materials Electrically conductive materials Train rails Nuclear fuel elements Bars, plates, other shapes	Access to only one surface required Battery or dc source Portable	Edge effect Surface contamination Good surface contact required Difficult to automate Electrode spacing Reference standards required
Electrified particle	Surface defects in nonconducting material Through-to-metal pinholes on metal-backed material Tension, compression, cyclic cracks Brittle-coating stress cracks	Glass Porcelain enamel Nonhomogeneous materials such as plastic or asphalt coatings Glass-to-metal seals	Portable Useful on materials not practical for penetrant inspection	Poor resolution on thin coatings False indications form moisture streaks or lint Atmospheric conditions High voltage discharge
Exo-electron emission	Fatigue in metals	Metals	Access to only one surface required Permanent record Quantitative	No surface films or contamination Geometry limitations Skilled technician required
Filtered particle	Cracks Porosity Differential absorption	Porous materials such as clay, carbon, powdered metals, concrete Grinding wheels High-tension insulators Sanitary ware	Colored or fluorescent particles Leaves no residue after baking part over 205°C (400°F) Quickly and easily applied Portable	Size and shape of particles must be selected before use Penetrating power of suspension medium is critical Particle concentration must be controlled Skin irritation

(*continued*)

TABLE 4.2 (*Continued*)

METHOD	MEASURES OR DETECTS	APPLICATIONS	ADVANTAGES	LIMITATIONS
Fluoroscopy (cine-fluorography, kine-fluorography)	Level of fill in containers Foreign objects Internal components Density variations Voids, thickness Spacing or position	Particles in liquid flow Presence of cavitation Operation of valves and switches Burning in small solid-propellant rocket motors	High-brightness images Real-time viewing Image magnification Permanent record Moving subject can be observed	Costly equipment Geometric unsharpness Thick specimens Speed of event to be studied Viewing area
Holography (acoustical-liquid surface levitation)	Lack of bond Delaminations Voids Porosity Resin-rich or resin-starved areas Inclusions Density variations	Metals Plastics Composites Laminates Honeycomb structures Ceramics Biological specimens	No hologram film development required Real-time imaging provided Liquid-surface responds rapidly to ultrasonic energy	Through-transmission techniques only Object and reference beams must superimpose on special liquid surface Immersion test only Laser required
Holography (interferometry)	Strain Plastic deformation Cracks Debonded areas Voids and inclusions Vibration	Bonded and composite structures Automotive or aircraft tires Three-dimensional imaging	Surface of test object can be uneven No special surface preparations or coatings required No physical contact with test specimen	Vibrationfree environment is required Heavy base to dampen vibrations Difficult to identify type of flaw detected
Infrared (radiometers)	Lack of bond Hot spots Heat transfer Isotherms Temperature ranges	Brazed joints Adhesive-bonded joints Metallic platings or coatings; debonded areas or thickness Electrical assemblies Temperature monitoring	Sensitive to 0.85°C (1.5°F) temperature variation Permanent record or thermal picture Quantitative Remote sensing; need not contact part Portable	Emissivity Liquid-nitrogen-cooled detector Critical time-temperature relationship Poor resolution for thick specimens Reference standards required
Leak testing	Leaks: helium, ammonia, smoke, water, air bubbles, radioactive gas, and halogens	Joints: welded, brazed, and adhesive-bonded Sealed assemblies Pressure or vacuum chambers Fuel or gas tanks	High sensitivity to extremely small, tight separations not detectable by other NDT methods Sensitivity related to method selected	Accessibility to both surfaces of part required Smeared metal or contaminants may prevent detection Cost related to sensitivity

(*continued*)

TABLE 4.2   (*Continued*)

METHOD	MEASURES OR DETECTS	APPLICATIONS	ADVANTAGES	LIMITATIONS
Magnetic field	Cracks Wall thickness Hardness Coercive force Magnetic anisotropy Magnetic field Nonmagnetic coating thickness on steel	Ferromagnetic materials Ship degaussing Liquid level control Treasure hunting Wall thickness of nonmetallic materials Material sorting	Measurement of magnetic material properties May be automated Easily detects magnetic objects in nonmagnetic material Portable	Permeability Reference standards required Edge-effect Probe lift-off
Magnetic particle	Surface and slightly subsurface defects; cracks, seams, porosity, inclusions Permeability variations Extremely sensitive for locating small tight cracks	Ferromagnetic materials; bar, forgings, weldments, extrusions, etc.	Advantage over penetrant in that it indicates subsurface defects, particularly inclusions Relatively fast and low cost May be portable	Alignment of magnetic field is critical Demagnetization of parts required after tests Parts must be cleaned before and after inspection Masking by surface coatings
Magnetic perturbation	Cracks Crack depth Broken strands in steel cables Permeability effects Nonmetallic inclusions Grinding burns and cracks under chromium plating	Ferromagnetic metals Broken steel cables in reinforced concrete	May be automated Easily detects magnetic objects in nonmagnetic materials Detects subsurface defects	Requires reference standard Need trained operator Part geometry Expensive equipment
Microwave (300 MHz–300 GHz)	Cracks, holes, debonded areas, etc. in nonmetallic parts Changes in composition, degree of cure, moisture content Thickness measurement Dielectric constant	Reinforced plastics Chemical products Ceramics Resins Rubber Liquids Polyurethane foam Radomes	Between radio waves and infrared in the electromagnetic spectrum Portable Contact with part surface not normally required Can be automated	Will not penetrate metals Reference standards required Horn to part spacing critical Part geometry Wave interference Vibration
Penetrants (dye or fluorescent)	Defects open to surface of parts; cracks, porosity, seams, laps, etc. Through-wall leaks	All parts with non-absorbing surfaces: forgings, weldments, castings, etc. (Note: bleed-out from porous surfaces can mask indications of defects)	Low cost Portable Indications may be further examined visually Results easily interpreted	Surface films, such as coatings, scale, and smeared metal may prevent detection of defects Parts must be cleaned before and after inspection Defect must be open to surface

(*continued*)

TABLE 4.2 (*Continued*)

METHOD	MEASURES OR DETECTS	APPLICATIONS	ADVANTAGES	LIMITATIONS
Radiography (X-rays, film)	Internal defects and variations; porosity; inclusions; cracks; lack of fusion; geometry variations; corrosion thinning Density variations Thickness, gap and position Misassembly Misalignment	Castings Electrical assemblies Weldments Small, thin, complex wrought products Nonmetallics Solid propellant rocket motors Composites	Permanent records; film Adjustable energy levels (5 kv-25 meV) High sensitivity to density changes No couplant required Geometry variations do not affect direction of X-ray beam	High initial costs Orientation of linear defects in part may not be favorable Radiation hazard Depth of defect not indicated Sensitivity decreases with increase in scattered radiation
Radiometry (X-, gamma-, or beta-ray; transmission or backscatter)	Wall thickness Plating thickness Variations in density or composition Fill level in cans or containers Inclusions or voids	Sheet, plate, foil, strip, tubing Nuclear reactor fuel rods Cans or containers Plated parts Composites	Fully automatic Fast Extremely accurate In-line process control Portable	Radiation hazard Beta-ray useful for ultrathin coatings only Source decay Reference standards required
Sonic (less than 0.1 MHz)	Debonded areas or delaminations in metal or nonmetal composites or laminates Cohesive bond strength under controlled conditions Crushed or fractured core Bond integrity of metal insert fasteners	Metal or nonmetal composite or laminates brazed or adhesive-bonded Plywood Rocket motor nozzles Honeycomb	Portable Easy to operate Locates far-side debonded areas May be automated Access to only one surface required	Surface geometry influences test results Reference standards required Adhesive or core thickness variations influence results
Ultrasonic (0.1-25 MHz)	Internal defects and variations; cracks, lack of fusion, porosity, inclusions, delaminations, lack of bond, texturing Thickness or velocity Poisson's ratio, elastic modulus	Wrought metals Welds Brazed joints Adhesive-bonded joints Nonmetallics In-service parts	Most sensitive to cracks Test results known immediately Automating and permanent record capability Portable High penetration capability	Couplant required Small, thin, complex parts may be difficult to check Reference standards required Trained operators for manual inspection Special probes

In the most commonly used destructive technique, *metallography*, samples are cut, polished, etched, and observed under an optical microscope or electron microscope. The samples required are usually much smaller than the part or component being analyzed. The advantages and limitations of nondestructive and destructive techniques are given in Table 4.3.

TABLE 4.3

ADVANTAGES AND LIMITATIONS OF NONDESTRUCTIVE AND DESTRUCTIVE TESTING

NONDESTRUCTIVE TESTING	DESTRUCTIVE TESTING
ADVANTAGES	*ADVANTAGES*
1 Can be done directly on production items without regard to part cost or quantity available, and no scrap losses are incurred except for bad parts	1 Can often directly and reliably measure response to service conditions
2 Can be done on 100% of production or on representative samples	2 Measurements are quantitative, and usually valuable for design or standardization
3 Can be used when variability is wide and unpredictable	3 Interpretation of results by a skilled technician usually not required
4 Different tests can be applied to the same item simultaneously or sequentially	4 Correlation between tests and service usually direct, leaving little margin for disagreement among observers as to meaning and significance of test results
5 The same test can be repeated on the same item	
6 May be performed on parts in service	*LIMITATIONS*
7 Cumulative effect of service usage can be measured directly	1 Can be applied only to a sample, and separate proof that the sample represents the population is required
8 May reveal failure mechanism	2 Tested parts cannot be placed in service
9 Little or no specimen preparation is required	3 Repeated tests of same item are often impossible, and different types of tests may require different samples
10 Equipment is often portable for use in field	
11 Labor costs are usually low, especially for repetitive testing of similar parts	4 Extensive testing usually cannot be justified, because of large scrap losses
	5 May be prohibited on parts with high material or fabrication costs, or on parts of limited availability
LIMITATIONS	6 Cumulative effect of service usage cannot be measured directly, but only inferred from tests on parts used for different lengths of time
1 Results often must be interpreted by a skilled, experienced technician	
2 In absence of proven correlation, different observers may disagree on meaning and significance of test results	7 Difficult to apply to parts in service, and usually terminates their useful life
3 Properties are measured indirectly, and often only qualitative or comparative measurements can be made	8 Extensive machining or other preparation of test specimens is often required
4 Some nondestructive tests require large capital investments	9 Capital investment and manpower costs are often high

Source: ASM Metals Reference Book, 2d ed., 1983.

4.6 FRICTION

Friction is defined as the resistance to relative sliding between two bodies in contact under a normal load. It plays an important role in all metalworking and manufacturing processes because of the relative motion and forces always present between tools, dies, and workpieces.

Friction is an energy-dissipating process that results in generation of heat. The subsequent rise in temperature can have a major effect on the overall operation. Furthermore, because it impedes free movement at interfaces, friction can significantly affect the flow and deformation of the material in many metalworking processes.

However, friction should not always be regarded as undesirable. The presence of friction is necessary for the success or optimization of many operations, the best example of which is the rolling of metals to make sheet and plate. Without friction it would be impossible to roll metals, just as it would be impossible to drive a car on the road.

Various theories have been proposed to explain the phenomenon of friction. An early *Coulomb* model (after C. A. de Coulomb, 1785) states that friction is due to the mechanical interlocking of rough surfaces; this interlocking will require some force to slide the two bodies against each other.

For any model of friction to be valid, it must explain the frictional behavior of two bodies under different loads, speed of relative sliding, temperature, surface conditions, environment, etc., as observed in practice. Consequently, many models have been proposed with varying degrees of success. The most commonly used theory of friction is based on adhesion, because it agrees reasonably well with experimental observations.

4.6.1 ADHESION THEORY OF FRICTION

The adhesion theory of friction (developed by F. P. Bowden and D. Tabor in England) is based on the observation that two clean dry (unlubricated) surfaces, regardless of how smooth they are, contact each other at only a fraction of their apparent area of contact (Fig. 4.12). (This can be verified by different techniques, such as by measuring the electrical resistance between the two contacting bodies.) The static load at the interface is thus supported by the minute asperities in contact with each other. The sum total of the contacting areas is known as the *real* area of contact, A_r. For most engineering surfaces, the average angle that the asperity hill makes with a horizontal line is found to be between 5 and 15 degrees. (Surface roughness illustrations in this chapter are highly exaggerated because of the scales involved.)

Under light loads and with a large real area of contact, the normal stress at the asperity contact is *elastic*. As the load increases, the stresses increase and eventually *plastic* deformation takes place at the *junctions*. Furthermore, with increasing load, the asperities in contact increase in area; also, new junctions are formed with other asperities on the two surfaces in contact with each other. Because of the random height of the asperities, some junctions are in elastic and others in plastic contact.

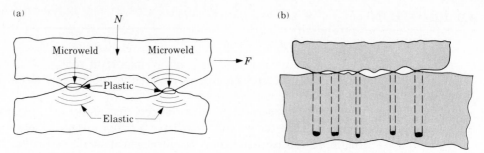

FIGURE 4.12 (a) Schematic illustration of the interface of two contacting surfaces, showing the real areas of contact. (b) Sketch illustrating the proportion of the apparent area to real area of contact. The ratio of the areas can be as high as four to five orders of magnitude.

Adhesion

The intimate contact of asperities creates an *adhesive bond*. The nature of this bond is complex, involving interactions on an atomic scale, and may include mutual solubility and diffusion. The strength of the bond depends on the physical and mechanical properties of the materials in contact, temperature, and the nature and thickness of any oxide film or other contaminants present on the surfaces.

In metalworking processes, the load at the interface is generally high so that the normal stress on the asperity reaches the yield stress. Plastic deformation of the asperity and adhesion then take place at the junctions. In other words, the asperities form *microwelds*. The cleaner the interface, the stronger are the adhesive bonds.

To pull the two surfaces apart requires a certain force, much like pulling apart a welded joint. The coefficient of adhesion is defined as

$$\text{Coefficient of adhesion} = \frac{\text{Tensile force}}{\text{Compressive force applied}}. \tag{4.3}$$

From Fig. 4.13, obtained from an experiment described in the figure caption, note that this coefficient can be quite high with softer metals and that, with few exceptions, it requires a certain finite force to pull the surfaces apart. The more ductile the materials, the greater is the plastic deformation and the stronger is the adhesion of the junctions.

Sliding between the two bodies with such an interface under a normal load N is possible only by the application of a tangential force F. According to the adhesion theory, this is the force required to *shear* the junctions (friction force). The *coefficient of friction* μ at the interface is defined as

$$\mu = \frac{F}{N} = \frac{\tau A_r}{\sigma A_r} = \frac{\tau}{\sigma}, \tag{4.4}$$

where τ is the shear strength of the junction and σ is the normal stress, which, for a plastically deformed asperity, is the yield stress of the asperity. However, since an asperity is surrounded by a large mass of material, the normal stress on an asperity is equivalent to the hardness of the material (see Section 2.5.7). Thus, the coefficient of friction μ can now be defined as

$$\mu = \frac{\tau}{\text{hardness}}. \tag{4.5}$$

Two additional phenomena take place in asperity interactions under plastic stresses: For a strain-hardening material, the peak of the asperity is stronger than the bulk material because of plastic deformation at the junction. Thus, under ideal conditions with clean surfaces, the breaking of the bonds under a tensile load will likely follow a path below or above the geometric interface.

Secondly, a tangential motion at the junction under load will cause *junction growth* (i.e., the contact area increases). This is due to the fact that, according to the yield criteria described in Section 2.8, the effective yield stress of the material de-

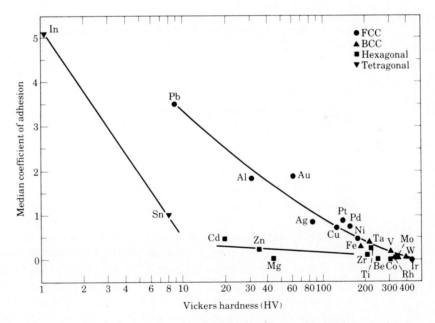

FIGURE 4.13 Influence of hardness and lattice structure of metals on the median coefficient of adhesion. Flat ends of two round bars, one hollowed out, were pressed together and then twisted. Adhesion coefficient is the ratio of the force of pulling the two surfaces apart to the pressing force. Low adhesion coefficients (hence low friction) are generally associated with materials with high elastic modulus, hardness (hence strength), high recrystallization temperature, and high surface energy. Note that hcp metals have low adhesion. Good bearing materials generally have noncubic structures. *Source:* M. E. Sikorski, *J. Basic Eng.*, vol. 85, 1963, pp. 279–285.

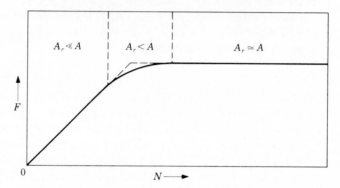

FIGURE 4.14　Schematic illustration of the relation between friction force and normal force. Note that as the real area of contact approaches the apparent area, the friction force reaches a maximum and stabilizes. Most machine components operate in the first region. The second and third regions are encountered in metalworking operations because of the high contact pressures involved between sliding surfaces (i.e., die and workpiece). Relative sliding at the interfaces is an important factor in friction. The tangential motion strips off protective surface barriers, such as oxide layers, adsorbed contaminants and gases, roughens the surfaces, and thus increases adhesion (hence increases friction.)

creases when subjected to a shear stress. Hence, the junction has to grow in area to support the normal load. In other words, the two surfaces tend to approach each other under a tangential force.

If the normal load is increased further, the real area of contact, A_r, in this model ideally continues to increase, especially with soft and ductile metals. In the absence of contaminants or fluids trapped at the interface, the real area of contact eventually reaches the apparent area of contact. This is the maximum contact area that can be obtained. Since the shear strength at the interface is constant, the friction force (shearing force) reaches a maximum and stabilizes (Fig. 4.14).

This condition, known as *sticking* (see also Section. 6.3.1), creates difficulty in defining friction. Sticking in a sliding interface does not necessarily mean complete adhesion at the interface, as in welding; it means that the frictional stress at the surface has reached the shear yield stress k of the material. However, when two clean surfaces are pressed together with sufficiently high forces, cold pressure welding can indeed take place, as described in Section 12.4, and observed in Fig. 4.13.

As the normal load N is increased further, the friction force F remains constant and hence, by definition, the coefficient of friction decreases. This situation, which is an anomaly, indicates that there might be a different and more realistic means of expressing the frictional condition at an interface.

A more recent trend is to define a *friction factor* or *shear factor m* as follows:

$$m = \frac{\tau_i}{k},\tag{4.6}$$

where τ_i is the shear strength of the interface and k is the shear yield stress of the softer material in a sliding pair. (The quantity k is equal to $Y/2$ according to Eq. (2.33) and to $Y/\sqrt{3}$ according to the distortion-energy criterion, where Y is the uniaxial yield stress of the material.)

The value of τ_i is difficult to measure since one has to know the real area of contact, and average values will not be meaningful. However, note that in this definition when $m = 0$, there is no friction and when $m = 1$ complete sticking takes place at the interface. The magnitude of m is independent of the normal force or stress. This is due to the fact that the shear yield stress of a thin layer of material has been shown to be unaffected by the magnitude of the normal stress (see also Fig. 8.14).

Ploughing

If the upper body in the model shown in Fig. 4.12 is harder than the lower one, or if its surface contains protruding hard particles, then as it slides over the softer body it can scratch and produce grooves on the lower surface. This is known as *ploughing* (or plowing) and is an important aspect in frictional behavior; in fact, it can be a dominant mechanism for situations where adhesion is not strong. Such grooves or scratches can readily be seen on metal surfaces or on various machine elements subjected to sliding.

Ploughing may involve two different mechanisms. One is the generation of a groove, whereby the surface is deformed plastically (Fig. 4.15a). The other is the formation of a groove because the cutting action of the upper body generates a chip, as shown in Fig. 4.15(b). Either of these processes involves work supplied by a force that

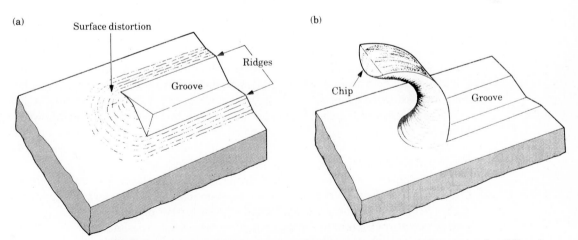

(a)

(b)

FIGURE 4.15 Schematic illustration of (a) ploughing, and (b) cutting in sliding. In ploughing, the material is displaced sideways. In cutting, material is removed from the surface in the form of a chip or sliver, although some ploughing may also be involved. This is an example of abrasive or erosive wear. *Source:* After L. E. Samuels.

manifests itself as friction force. The ploughing force can contribute significantly to friction and to the measured coefficient of friction at the interface. (See also Section 4.8.2.)

4.6.2 COEFFICIENT OF FRICTION

The nature and strength of the interface is the most significant factor in friction. A strong interface requires a high friction force for relative sliding and one with little strength requires a low friction force. Eq. (4.5) indicates that the coefficient of friction can be reduced not only by decreasing the numerator but also by increasing the denominator.

This observation suggests that thin films of low shear strength over a substrate with high hardness is the ideal method for reducing friction. In fact, this is exactly what is achieved with a surface lubricated either with a fluid or solid lubricant.

With no low-shear-strength layer or contaminant present, and under ideal clean conditions with a rigid–perfectly plastic material pair, the coefficient of friction at an asperity can be calculated theoretically from Eq. (4.5) as

$$\mu = \frac{\tau}{\text{hardness}} = \frac{Y/\sqrt{3}}{3Y} \simeq 0.2. \qquad (4.7)$$

If the number of asperities increases, such as with increasing load, the combined shear strength of the junctions may exceed the bulk shear strength of the softer metal. In this case, plastic flow takes place in a sublayer (sticking). The value of the normal stress in Eq. (4.4) then becomes that of the uniaxial compressive yield stress of the softer metal. The coefficient of friction is then

$$\mu = \frac{\tau}{\sigma} = \frac{Y/\sqrt{3}}{Y} = 0.577. \qquad (4.8)$$

Eq. (4.7) can be modified to include a term for ploughing. This term can be estimated roughly to be on the order of 0.1, which should be added algebraically to the value obtained from Eq. (4.7).

The sensitivity of friction to different factors can be illustrated by the following examples.

The coefficient of friction of steel sliding on lead is 1.0, steel on copper is 0.9. However, for steel sliding on copper coated with a thin layer of lead, it is 0.2.

The coefficient of friction of pure nickel sliding on nickel in a hydrogen or nitrogen atmosphere is 5, in air or oxygen it is 3, and in the presence of water vapor it is 1.6. (See Section 4.9 on lubricating mechanisms for further details.)

Coefficients of friction in sliding contact, as measured experimentally, vary from as low as 0.02 to 100, or even higher. This is not surprising in view of the many variables involved in the friction process. In metalworking operations using various lubricants, the range for μ is narrower, as seen in Table 4.4. In cold working of metals, μ ranges

TABLE 4.4
**RANGE OF COEFFICIENTS OF FRICTION IN
METALWORKING PROCESSES AT ROOM
TEMPERATURE USING VARIOUS LUBRICANTS**

PROCESS	WORKPIECE MATERIAL	μ
Deep drawing	Cu, brass, bronze, Al	0.08–0.15
Press forging	Al, Pb	0.2
Rod drawing	Al, steel	0.02–0.20
Sheet drawing	303 stainless steel	0.05–0.15
Strip drawing	Steel, Cu	0.03–0.15
Strip drawing	Ferrous, nonferrous	0.03–0.30
Strip rolling	Al, steel	0.02–0.30
Strip rolling	Al, Cu, brass, steel	0.04–0.10
Tube drawing	Brass, steel	0.03–0.06
Various	Steel	0.05–0.25
Wire drawing	Al, Zn	0.02–0.20
Wire drawing	Brass, steel	0.03–0.10
Wire drawing	Cu, stainless steel	0.07–0.30

roughly from 0.03 to 0.3 (with 0.15 as a good average), whereas in hot working it is generally higher, unless lubricated effectively. In metal cutting operations, μ can be as much as 2, or even higher.

The effects of load, temperature, speed, and environment on the coefficient of friction are difficult to generalize, as each situation has to be studied individually. Relevant details and data are included Chapters 6, 7, 8, 9, 10, and 11.

4.6.3 TEMPERATURE RISE DUE TO FRICTION

Almost all the energy dissipated in overcoming friction is converted into heat. A small portion remains in the bodies as stored energy and some of it goes into surface energy during generation of new surfaces and wear particles.

Frictional heat raises the interface temperature. The magnitude of this temperature rise and its distribution depends not only on the friction force but also on speed, surface roughness, and the physical properties of the materials, such as thermal conductivity and specific heat. The interface temperature increases with increasing friction, speed, and with low thermal conductivity and specific heat of the materials.

One example of the effect of speed on temperature is shown in Fig. 4.16. Note that it is possible for the interface temperature to be high enough to soften and melt the surface. (The temperature cannot, of course, exceed the melting point of the material.) A number of analytical expressions have been obtained to calculate the temperature rise in sliding. However, these are rather involved and are beyond the scope of this text.

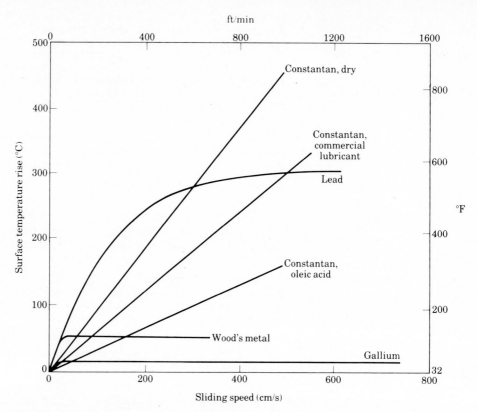

FIGURE 4.16 Temperature rise at the points of contact between various materials and steel. Note that the temperature cannot exceed the melting point of the metal. Wood's metal is an alloy of Bi-Pb-Sn-Cd. Constantan is a Cu-Ni alloy and is used for thermocouples. *Source:* F. P. Bowden and D. Tabor, *The Friction and Lubrication of Solids*, vol. 1, London: Oxford, 1950, pp. 38, 40.

4.7 TECHNIQUES FOR FRICTION MEASUREMENT

The friction coefficients obtained from engineering design reference sources are generally not relevant to metalworking processes. This is because, unlike that in machine elements, the die–workpiece interface in metalworking is subjected to stresses in the plastic range. Furthermore, the magnitude and distribution of these stresses, as well as the temperature, surface conditions, the metalworking fluid, etc., are usually quite different from those in general engineering use and service.

Consequently, the coefficient of friction μ or the friction factor m are determined experimentally either during actual metalworking processes or in simulated tests using small-scale specimens of various shapes. Numerous techniques and instrumentation have been developed for these measurements and the data obtained are extensive.

These techniques generally involve measurement of either forces or dimensional changes of the specimen during plastic deformation.

Because of the difficulties involved in full-scale experimentation on production equipment, simulated tests have been developed and studied extensively. Unfortunately, however, many of these tests do not duplicate the exact conditions of the actual metalworking process, particularly in regard to speeds and part dimensions.

Ring Compression Test

One test for friction measurement that has gained wide acceptance, particularly for the bulk deformation processes treated in Chapter 6, is the *ring compression* test. In this test, a flat ring is compressed plastically between two flat platens (Fig. 4.17a). As its height is reduced, the ring expands outward radially. If friction at the interfaces is zero, both the inner and outer diameters expand as if the ring were a solid disk. With increasing friction, the rate of increase of the inner diameter decreases. For any particular reduction in height, there is a critical value of the coefficient of friction whereby the internal diameter increases (from the original) if μ is low, and decreases if μ is high (Fig. 4.17b).

Several theoretical analyses of this test method have resulted in the development of curves from which the coefficient of friction or the friction factor can be determined by knowing the change in the internal diameter of the specimen (Fig. 4.18).

There is a specific set of curves for each ring geometry. The most common geometry has the ratio 6:3:2 for outer diameter:inner diameter:height of the specimen. Generally the actual size of the specimen is not relevant in these tests. Thus, once

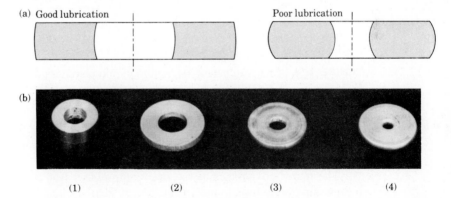

FIGURE 4.17 (a) Effect of lubrication on barreling in ring compression test. With good lubrication, both the inner and outer diameters increase as the specimen is compressed. With poor or no lubrication, friction is high and the inner diameter decreases. The direction of barreling depends on the relative motion of the cylindrical surfaces with respect to the flat dies. (b) Ring compression test results with 50% reduction in height: 1. Original specimen. 2. Low friction. 3. Medium friction. 4. High friction. The change in the diameter of the hole is a measure of friction, as shown in Fig. 4.18. *Source:* A. T. Male and M. G. Cockcroft. *J. Inst. Metals*, vol. 93, 1964–65, p. 39.

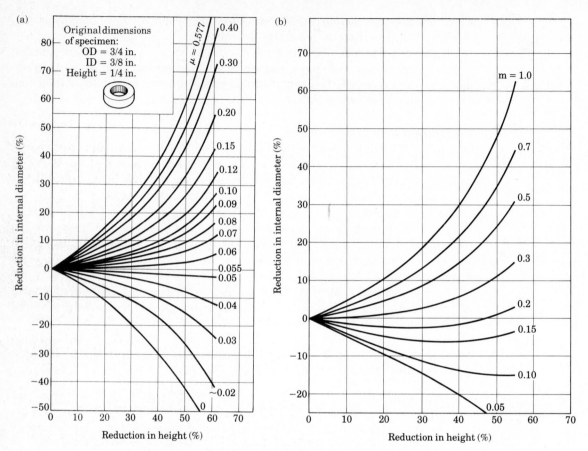

FIGURE 4.18 Charts to determine friction in ring compression tests. (a) Coefficient of friction μ. (b) Friction factor m. Friction is determined from these charts, from the percent reduction in height and by measuring the percent change in the internal diameter of the specimen after compression.

the percent reductions in internal diameter and height are known, μ or m can be determined from these charts.

The major advantages of the ring compression test are that the test does not require any force measurements and that it involves large-scale deformation of the workpiece material, thus simulating actual metalworking operations. Furthermore, this test is simple and can be carried out at different strain rates and temperatures using various lubricants. For testing at elevated temperatures, provisions must be made so that the temperature of the specimen is maintained during the test; otherwise, the ring will chill rapidly when brought in contact with cool platens.

Because of friction, the cylindrical surfaces of the specimen tend to *barrel* (see Fig. 4.17). Although, with effective lubricants, this barreling is minimal, an average diameter can be measured when barreling is present. Also, because of possible anisotropy of the specimen material or other causes, such as lay (see Fig. 4.6) of the surface finish of the dies, the ring may acquire an oval shape after deformation instead of remaining circular. In this case, an average diameter is usually measured.

In addition to the use of this test method for determining the friction for a specific situation of die and workpiece materials, temperature, etc., the test can also be used for rating different metalworking lubricants.

● **Illustrative Problem 4.1**

In a ring compression test, a specimen 10 mm in height with outside diameter (O.D.) = 30 mm and inside diameter (I.D.) = 15 mm is reduced in thickness by 50%. Determine the coefficient of friction μ and the friction factor m if the O.D. after deformation is 38 mm.

SOLUTION. It is first necessary to determine the new I.D. This is obtained from volume constancy as follows:

$$\text{Volume} = \frac{\pi}{4}(30^2 - 15^2)10 = \frac{\pi}{4}(38^2 - \text{I.D.}^2)5.$$

From this, the new I.D. = 9.6 mm.

Thus, the change in internal diameter is

$$\text{Change in I.D.} = \frac{15 - 9.6}{15} \times 100 = 36\% \text{ (decrease).}$$

For a 50% reduction in height and a reduction in internal diameter of 36%, the following values are obtained from Fig. 4.18:

$\mu = 0.21$

$m = 0.72.$ ●

4.8 WEAR

Wear is defined as the loss or removal of material from a surface. Wear has important technologic and economic significance because it changes the geometry of interfaces and may interfere with the overall operation of machinery and equipment. In order to appreciate the importance of wear, note the innumerable parts and components replaced or repaired because they are worn. These may range from regrinding a dull drill to the inspection of a jet engine for worn bearings, seals, and other parts.

Wear and its control are among the most complex subjects to study and to understand, as evidenced by the extensive technical literature on the subject. There are many forms of wear involving several different mechanisms. Even the definition of the types of wear tends to be controversial and not always universally understood or agreed on.

One common classification of wear is by its four basic types: *adhesive*, *abrasive*, *corrosive*, and *fatigue* wear. Other types of wear are *erosion*, *fretting*, *impact*, and *cavitation*. Wear may also be classified as dry or lubricated wear, sliding or rolling contact wear, and wear by plastic deformation or by fracture.

4.8.1 ADHESIVE WEAR

When a tangential force F is applied to the model shown in Fig. 4.12(a) with adhesive bonding (microwelds) of the asperities, shearing can take place either at the original interface or along a path below or above it. The fracture path depends on whether or not the strength of the adhesive bond of the asperities is greater than the cohesive strength (Section 3.2.2) of either of the two bodies.

Because of such factors as strain hardening, diffusion, and mutual solubility, the adhesive bonds are often stronger than the base metals. Thus, during the application of a tangential force, fracture usually occurs in the weaker or softer body, as shown in Fig. 4.19. A wear fragment is thus generated. Although this fragment is attached to the harder (upper) body, it eventually becomes detached during further rubbing at the interface and develops into a loose wear particle.

A typical surface undergoing adhesive wear is shown in Fig. 4.20. The adhesive wear particles in this figure are irregular and have a smeared appearance. However,

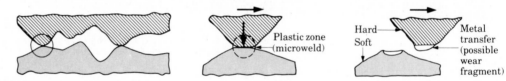

FIGURE 4.19   Schematic illustration of adhesion of material, eventually leading to a wear particle when the metal transferred to the upper body works itself loose under repeated sliding.

FIGURE 4.20   Adhesive metal transfer to a surface, showing wear particles deposited on the surface. The mating body has thus undergone wear by loss of material. Material: 52100 steel, lubricated. *Source:* Courtesy of K.-H. Habig and D. Klaffke, Bundesanstalt für Materialprüfung, Berlin.

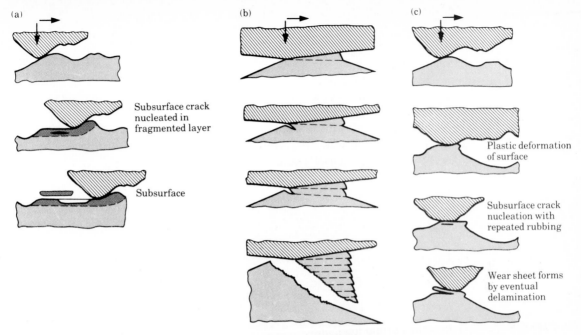

(a)

Subsurface crack
nucleated in
fragmented layer

Subsurface

(b)

(c)

Plastic deformation
of surface

Subsurface crack
nucleation with
repeated rubbing

Wear sheet forms
by eventual
delamination

FIGURE 4.21 Schematic illustration of possible wear mechanisms in sliding contact. (a) Delamination. Note the formation of a subsurface crack that eventually leads to the formation of a wear particle. (b) Delamination. (c) Irregular wear particle formation. *Source:* (a) After N. P. Suh. (b) After E. D. Doyle and D. M. Turley. (c) After T. Kayaba and K. Kato.

these particles can have a variety of shapes depending on the hardness and surface topography of the two bodies. Some of the models proposed for wear particle formation are given in Fig. 4.21. An example of wear particles is shown in Fig. 4.22. Even spherical wear particles can be found. Their formation is generally attributed to the rounding of an original wear particle by rolling between the two surfaces, or to

FIGURE 4.22 Typical wear particles of steel. Note the random size distribution and the large surface-to-thickness ratio of the particles. Wear particles may be of various other shapes, depending on the mechanisms by which they are generated. *Source:* R. Krishna, S. Kalpakjian, and S. Kumar, in *Advanced Techniques in Track/Train Dynamics and Design,* New York: Pergamon, 1978.

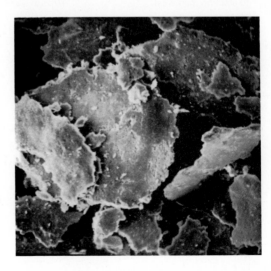

heating of the wear particle, as a result of which it softens and develops a spherical shape due to surface tension.

In more severe cases adhesive wear is also described as *scuffing, smearing, galling,* and *seizure.* The problem of quantifying these surfaces and the wear particles produced is still not resolved. However, based on the probability that a junction between two sliding surfaces would lead to the formation of a wear particle, an expression for adhesive wear has been developed as

$$V = k\frac{LW}{3p},\tag{4.9}$$

where

V is the volume of material removed by wear from the surface,

k is the wear coefficient (dimensionless),

L is the length of travel,

W is the normal load, and

p is the indentation hardness of the softer body.

TABLE 4.5
APPROXIMATE ORDER OF MAGNITUDE FOR WEAR COEFFICIENT k IN AIR

UNLUBRICATED	k	LUBRICATED	k
Zinc on zinc	10^{-1}	52100 steel on 52100 steel	10^{-7} to 10^{-10}
Mild steel on mild steel	10^{-2} to 10^{-3}	Aluminum bronze on hardened steel	10^{-8}
Copper on copper	10^{-2}	Hardened steel on hardened steel	10^{-9}
Stainless steel on stainless steel	10^{-2}		
Copper on mild steel	10^{-3}		
52100 steel on 52100 steel	10^{-3}		
60-40 brass on hardened tool steel	10^{-3}		
Hardened tool steel on hardened tool steel	10^{-4}		
Stellite on tool steel	10^{-5}		
Beryllium copper on tool steel	10^{-5}		
Mild steel on hard steel	10^{-5}		
Stainless steel on tool steel	10^{-5}		
Polytetrafluoroethylene on tool steel	10^{-5}		
Tungsten carbide on mild steel	10^{-6}		
Tungsten carbide on tungsten carbide	10^{-6}		
Polyethylene on tool steel	10^{-7} to 10^{-8}		

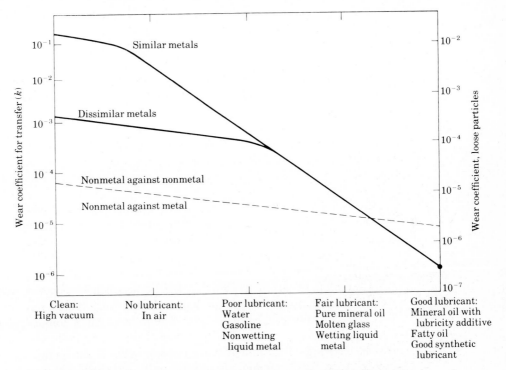

FIGURE 4.23 Typical values of wear coefficient *k* for various material pairs and lubricating conditions. Note the very significant effect of lubrication on wear. *Source:* E. Rabinowicz, *Mat. Sci. Eng.*, vol. 25, 1976, pp. 23–28.

Typical values of *k* for a combination of materials sliding in air are given in Table 4.5. The wear coefficient for the same pair of materials can vary by a factor of 3 depending on whether wear is measured as a loose particle or as a transferred particle (to the other body). Loose particles have a lower *k* value (Fig. 4.23). Likewise, when mutual solubility is a significant parameter in adhesion, similar metal pairs have higher *k* values than dissimilar pairs.

Effect of Surface Films

Thus far the treatment of adhesive wear has been based on the assumption that the surface layers of the two contacting bodies are clean and free of contaminants. Under these conditions, the adhesive wear rate can be very high; this is known as *severe wear*. As described in Section 4.2, however, metal surfaces are almost always covered with oxide layers, with thicknesses generally ranging between 20 and 100 Å. Although this thickness may at first be regarded as insignificant, the oxide layer has a profound effect on wear behavior.

Oxide layers are usually hard and brittle. When such a surface is subjected to rubbing, this layer can have the following effects.

If the load is light and the oxide layer is strongly adhering to the bulk metal, the strength of the junctions between the asperities will be weak and wear will be low. In this situation, the oxide layer acts as a protective film and the type of wear is known as *mild wear*.

The oxide layer can be broken up under high normal loads if the oxide layer is brittle and is not strongly adhering to the bulk metal, or if the asperities are rubbing against each other repeatedly, so that the oxide layer breaks up by fatigue. However, if the surfaces are smooth, the oxide layers are more difficult to break up. When this layer is eventually broken up, a wear particle is formed (Fig. 4.24). In addition, an upper asperity can then form a strong junction with the lower asperity, which is now unprotected by the oxide layer. The wear rate will now be higher until a fresh oxide layer is formed.

The frictional and wear behavior of such an interface thus depend on the rates of oxide layer destruction and oxide layer formation. Although oxide films on metals are formed extremely rapidly in a normal environment, their rate of destruction may also be rapid, or the time available may not be long enough for a sufficiently thick new oxide layer to form.

In addition, a surface exposed to the environment is covered with adsorbed layers of gas and usually with other contaminants. Such films, even if they are very thin (5 Å), generally weaken the interfacial bond strength of contacting asperities by separating the metal surfaces. The net effect of these films is to provide a protective layer by reducing adhesion and thus reducing wear. The magnitude of the effect of adsorbed gases and contaminants depends on many factors. For instance, it has been shown repeatedly that even small differences in humidity can have a profound influence on the wear rate.

● **Illustrative Problem 4.2**

The end of a rod made of 60-40 brass is sliding over the unlubricated surface of hardened tool steel with a load of 200 lb. The hardness of brass is 120 HB. What is the distance traveled to produce a wear volume of 0.001 in^3 by adhesive wear of the brass rod?

SOLUTION. The parameters in Eq. (4.9) for adhesive wear are as follows:

$V = 0.001$ in^3

$k = 10^{-3}$ (from Table 4.5)

$W = 200$ lb

$p = 120$ kg/mm$^2 = 170,700$ lb/in^2.

Thus, the distance traveled is

$$L = \frac{3Vp}{kW} = \frac{(3)(0.001)(170,700)}{(10^{-3})(200)} = 2560 \text{ in.} = 213 \text{ ft.} \quad ●$$

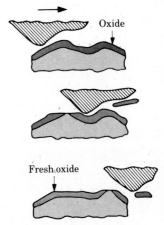

FIGURE 4.24 Schematic illustration of mild wear due to repeated formation and breakdown of surface oxide films. If the rate of oxide film breakdown is higher than the rate of formation, wear increases, and seizure takes place. This leads to what is known as severe wear, with a high *k* value. *Source:* After T. F. J. Quinn.

4.8.2 ABRASIVE WEAR

Abrasive wear is the result of a hard, rough surface, or a surface containing hard, protruding particles, rubbing over a softer, more ductile surface. The mechanism of wear particle removal is basically one of formation of microchips or slivers, thereby producing grooves or scratches on the softer surface (Fig. 4.25). Common examples of abrasive wear are material removal by abrasive papers (emery paper or sandpaper), coarse polishing, and filing.

The shape of the abrasive particles or the surface roughness characteristics of the harder body have a significant effect on abrasive wear. If the particles are dull, a groove or scratch on the surface is produced by a *ploughing* process without necessarily removing material (see Fig. 4.15a). If they are sharp, wear particles are generated by a *cutting* process (see Figs. 4.15b and 9.6).

The *abrasive wear resistance* of metals has been found to be directly proportional to the hardness of materials and inversely proportional to the load. This observation is to be expected because the harder the metal, the less the penetration of the hard particle on a surface under the same load and, hence, the smaller the depth of the

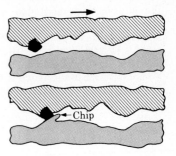

FIGURE 4.25 Schematic illustration of abrasive wear in sliding. This mechanism is similar to processes such as grinding, filing, or sanding. Note the removal of material in the form of a chip or sliver. Longitudinal scratches on a surface are generally indicative of abrasive wear. See also Fig. 4.15.

groove; and the higher the load, the greater the depth of the groove, hence the higher the volume of wear.

Figure 4.26 shows the abrasive wear resistance of a variety of pure metals and steels. The wear resistance in this figure is in relation to a tin–lead alloy. Thus, for example, under otherwise identical conditions, pure tungsten undergoes about 1/60 the volume of wear by abrasion as the tin–lead alloy and pure aluminum about 1/6.

There are two basic types of abrasive wear: 2-body and 3-body wear, as shown in Fig. 4.27. The first type is the basis of *erosive* wear. The abrasive particles (such as sand) are usually carried in a jet of fluid or air, and they remove material from the surface by erosion. The 3-body wear is important in metalworking processes because a lubricant between the die and workpiece may carry with it wear particles (which are generated over a period of time) and cause abrasive wear. Thus, the necessity for proper inspection and filtering of a metalworking fluid is apparent. This type of wear is also significant in the maintenance of various machinery and equipment. There is a possibility that particles (from machining, grinding, the environment, etc.) may contaminate the lubricating system of the machine components and thus scratch and damage the surfaces.

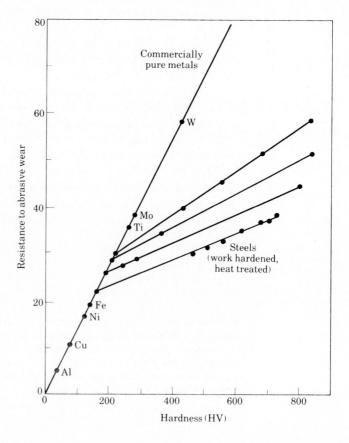

FIGURE 4.26 The effect of hardness on the abrasive wear resistance for commercially pure annealed metals and steels. Wear resistance in this figure is relative to a tin-lead alloy. Thus, under identical conditions, tungsten (W) has an abrasive wear resistance that is about 60 times that of the tin-lead alloy. See also Table 10.2. *Source:* M. M. Kruschov, *Proc. Conf. Lubrication and Wear*, Inst. Mech. Eng., 1957, pp. 655–659.

Two-body abrasion   Three-body abrasion

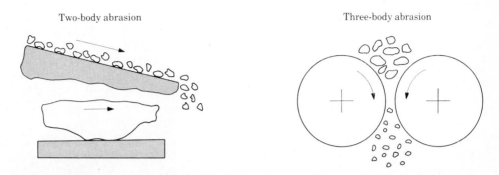

FIGURE 4.27 Schematic illustrations of two-body and three-body abrasive wear. An example of two-body wear is erosion of a surface by hard particles, such as in earth moving or agricultural machinery. An example of three-body wear is crushing stones between two rolls, or wear of dies due to the presence of hard wear particles in a lubricant between the die and the workpiece. *Source:* After H. Czichos.

● **Illustrative Problem 4.3**

A block of metal with a rectangular cross-section is placed on an abrasive belt to reduce its thickness, Fig. P4.1. If the block is supported as shown in the figure, will the removal of material be uniform along the length of the block?

SOLUTION. In order to keep the block in place, the friction force of the abrasive belt has to be balanced by the force applied by the support. The free-body diagram shown in the figure indicates that the reaction force exerted by the table is to the left of the center-line of the block. Thus, the pressure distribution along the bottom of the block (from which material is being removed by abrasion) will be as shown in the figure. Since abrasive wear is proportional to load, this means that wear rate will be increasing from the right to the left of the block. Consequently, the shape of the block will be as shown in the sketch in Fig. P4.1.

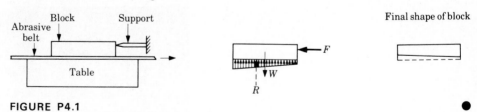

FIGURE P4.1   ●

4.8.3 CORROSIVE WEAR

In this type of wear, also known as *oxidation* wear, the corrosive products (Fig. 4.28) on a surface (due to oxidation) are removed during sliding at the interface, such as by abrasive action. The corrosive wear rate is difficult to quantify because it

FIGURE 4.28   Corrosive wear particles (copper oxide) on copper pin surface, sliding on iron disk. Unlubricated. *Source:* Courtesy of K.-H. Habig and D. Klaffke, Bundesanstalt für Materialprüfung, Berlin.

depends on a number of factors, such as the thickness and strength of the corrosion products, the rate of corrosion, and the mechanism of removal. When the corrosive layer is removed, another layer begins to form and the process of removal and forming is repeated as it is in mild wear.

4.8.4   FATIGUE WEAR

Fatigue wear results when the surface of a material is subjected to cyclic loading, such as in rolling contact. This type of wear is also called *surface fatigue* or *surface fracture wear*. The mechanisms in fatigue wear are complex and involve the initiation of cracks on or below the surface and their propagation (Fig. 4.29). The wear particles are usually formed by *spalling* or *pitting*; that is, two or more cracks meet each other and a small piece of material from the surface breaks off. Depending on the conditions at the surface and the presence of defects, voids, inclusions, and impurities, small- or large-scale surface deformations may precede fatigue wear.

FIGURE 4.29   Surface fatigue of high-speed-steel pin sliding on carbon-steel disk in air. Note the surface cracks perpendicular to the horizontal direction of sliding. Fatigue wear is particularly important in ball or roller bearings. *Source:* Courtesy of Habig and Klaffke, ibid.

In addition to mechanical fatigue due to cyclic loading, another type of wear is by *thermal fatigue*. Here, cracks on the surface are generated by thermal stresses due to thermal cycling, such as a cool die repeatedly contacting hot workpieces. These cracks then join and the surface begins to spall.

4.8.5 GENERAL OBSERVATIONS

It is apparent from the foregoing that wear is a complex phenomenon and that each type of wear is influenced by several factors. Numerous theories have been advanced to explain and interpret the experimental data and observations made regarding wear. Because many factors are involved, it is difficult to quantify wear data. In fact, it is well known that wear data are difficult to reproduce.

The most significant aspects of wear are the nature and geometry of the contacting asperities. The strength of the bond at the asperity junction relative to the cohesive strength of the weaker body is an important factor. It is interesting to note that when a polymer such as PTFE, i.e., *Teflon*, is rubbed against steel, the surface of the polymer (which is much weaker than the steel) is found to be embedded with steel wear particles, Figs. 4.30(a) and (b). Also, when you rub gold with your fingers, wear particles of gold are found on the fingers. The mechanisms for this are not clearly established.

The question also arises as to whether or not there is any direct relation between friction and wear for a pair of metals. Although at first one would expect there to be such a relation, experimental evidence for unlubricated metal pairs indicates that this

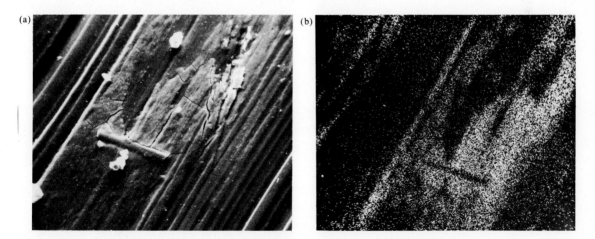

(a)   (b)

FIGURE 4.30 (a) Scanning electron micrograph of PTFE (Teflon) pin surface sliding on steel disk, showing transfer of iron particles to PTFE. (b) Microprobe of PTFE pin surface showing transfer of iron particles (white dots). Experiments have shown that wear particles from a hard surface can be transmitted to a softer surface. *Source:* Courtesy of Habig and Klaffke, ibid.

TABLE 4.6
COEFFICIENT OF FRICTION AND WEAR RATE FOR SOME METAL PAIRS IN PIN-ON-RING TESTS*[*]

| MATERIALS | | COEFFICIENT OF FRICTION, μ | WEAR RATE, cm^3/cm, $\times 10^{-12}$ |
PIN	*RING*		
Mild steel	Mild steel	0.62	157,000
60-40 leaded brass	Tool steel	0.24	24,000
Polytetrafluoroethylene	Tool steel	0.18	2,000
Stellite	Tool steel	0.60	320
Stainless steel, ferritic	Tool steel	0.53	270
Polyethylene	Tool steel	0.65	30
Tungsten carbide	Tungsten carbide	0.35	2

*Halling, J., *Principles of Tribology*. New York: Macmillan, 1975:9.

is not always the case (Table 4.6). These results are not surprising in view of the many factors involved in friction and wear. For instance, ball bearings and races have very little friction yet they can undergo wear after a large number of cycles.

Although wear in general alters the surface topography and may lead to severe surface damage, it can also reduce the surface roughness by removing the peaks of asperities. Thus, under controlled conditions, wear may be regarded as a type of smoothing or polishing process (see Fig. 4.11). The wear rate is very low and the wear particles generated do not interfere significantly with the motion at the interfaces. The *running-in* period of various machines and engines are of this type.

4.8.6 WEAR MEASUREMENT TECHNIQUES

There are various methods of observing and measuring wear:

Visual and tactile inspection,

Measuring dimensional changes and gaging,

Weighing,

Profilometry,

Radiography (wear particles transferred to a surface from an irradiated mating surface),

Analysis of the lubricant for wear particles (spectroscopy),

Monitoring performance and noise level.

The choice of a particular method depends on the accuracy desired and the physical constraints of the system, such as size and difficulty of disassembly (see also Section 8.7.4).

4.9 LUBRICATING MECHANISMS

In manufacturing processes the interface between tools, dies, and workpieces is usually subjected to a wide range of variables. Among the major ones are:

a. Contact pressures, ranging from low values of elastic stresses to multiples of the yield stress of the workpiece material,

b. Relative speeds, ranging from very low (such as in some superplastic metal forming operations [Section 7.13]) to very high speeds (such as in explosive forming, thin wire drawing, grinding, and high-speed metal cutting), and

c. Temperatures, ranging from ambient to almost melting, such as in hot extrusion and squeeze casting.

If two surfaces slide against each other under these conditions, with no protective layers at the interface, friction and wear would be high. To reduce friction and wear, surfaces should be held as far apart as possible. This generally is done by metalworking lubricants, which may be solid, semisolid, or liquid in nature.

Metalworking lubricants are not only fluids or solids with certain desirable physical properties, such as viscosity, but also are chemicals that can react with the surfaces of the tools, dies, and workpieces and alter their physical properties. These are described below.

4.9.1 LUBRICATION REGIMES

There are basically four regimes of lubrication that are relevant to metalworking processes.

a. Thick Film: the surfaces are completely separated by a fluid film whose thickness is about one order of magnitude greater than the surface roughness; thus there is no metal-to-metal contact. The normal load is light and is supported by the hydrodynamic fluid film by the wedge effect due to the relative velocity of the two bodies and the viscosity of the fluid (Fig. 4.31). The coefficient of friction is very low, usually ranging between 0.001 and 0.02.

There is essentially no wear and hence the surfaces retain their original characteristics. Any slight wear is generally due to erosive wear from hard particles or contaminants present in the fluid film. Although hydrodynamic (thick) films are rather rare in metalworking, they can occur in high-speed operations, particularly with high-viscosity lubricants.

b. Thin Film: as the normal load increases, or as the speed and viscosity of the fluid decrease (such as due to a rise in temperature), the film thickness is reduced to about three to five times the surface roughness. There may be some metal-to-metal contact at the higher asperities; this increases friction and leads to slightly higher wear.

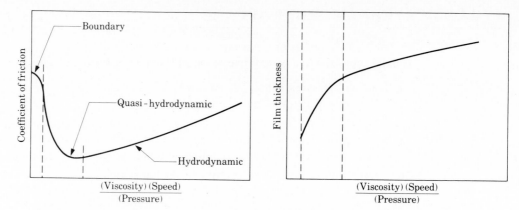

FIGURE 4.31 Stribeck curve describing various lubrication regimes. The three variables in the abscissa are a useful means of describing the conditions at the die-workpiece interface in metalworking processes.

c. Mixed Lubrication: in this situation, a significant portion of the load is carried by the metal-to-metal contact of the asperities and the rest of the load is carried by the pressurized fluid film present in hydrodynamic pockets, such as in the valleys of asperities. The film thickness is less than three times the surface roughness. With proper selection of lubricants, a strongly adhering *boundary* film of a few molecules in thickness can be formed on the surfaces. This film prevents direct metal-to-metal contact and thus reduces wear. Depending on the strength of the boundary film and other parameters, outlined in Sections 4.6 and 4.8, the friction coefficient in mixed lubrication may range up to about 0.4.

d. Boundary Lubrication: the load here is supported by the contacting surfaces, which are covered with a *boundary layer*. Depending on the boundary film thickness and its strength, the friction coefficient ranges from about 0.1 to 0.4. Wear can be relatively high if the boundary layer is destroyed.

   Typical *boundary lubricants* are natural oils, fats, fatty acids, and soaps. Boundary films form rapidly on metal surfaces. As the film thickness decreases and metal-to-metal contact takes place, the chemical aspects of surfaces and their roughness becomes significant. (In thick-film lubrication the viscosity of the lubricant is the important parameter in controlling friction and wear; chemical aspects are not particularly significant, except as they affect corrosion and staining of metal surfaces.)

   A boundary film can *break down* or it can be removed by being disturbed or rubbed off during sliding, or because of desorption due to high temperatures at the interface. The metal surfaces may thus be deprived of this protective layer. The clean metal surfaces then contact each other and, as a consequence, severe wear and scoring can occur. Thus, the adherence of boundary films is an important aspect in lubrication.

Figure 4.31 highlights the importance of viscosity, speed, and pressure on film thickness. Viscosity depends on temperature and pressure and is an important aspect of lubrication.

The role of surface roughness, particularly in mixed lubrication, should be recognized. Roughness can serve to create local reservoirs or pockets for lubricants. The lubricant can be trapped in the valleys of the surface and, because these fluids are incompressible, they can support a substantial portion of the normal load. These pockets also supply lubricant to regions where the boundary layer has been destroyed. There is an optimal roughness for lubricant entrapment.

In metalworking operations, it is generally desirable for the workpiece, not the die, to have the rougher surface, otherwise the workpiece surface will be damaged by the rougher and harder die surface. The recommended surface roughness on most dies is about 15 μin. (0.40 μm).

4.9.2 GEOMETRIC EFFECTS

In addition to surface roughness, the overall geometry of the interacting bodies is an important consideration in lubrication. Figure 4.32 shows the typical metalworking processes of drawing, extrusion, and rolling. The movement of the workpiece into the deformation zone must allow a supply of the lubricant to be carried into the die-workpiece interfaces.

From the analysis of hydrodynamic lubrication, it can be shown that the inlet angle is an important parameter. As this angle decreases more lubricant is entrained; thus, lubrication is improved and friction is reduced (Fig. 4.33). With proper selection of parameters, a relatively thick film of lubricant can be maintained at the die-workpiece interface. A similar situation exists in the compression of a straight solid cylinder between flat platens (Fig. 4.34a). As the approach speed of the platen

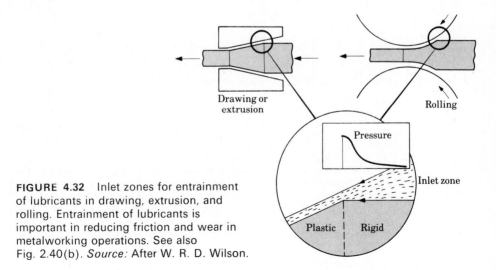

FIGURE 4.32 Inlet zones for entrainment of lubricants in drawing, extrusion, and rolling. Entrainment of lubricants is important in reducing friction and wear in metalworking operations. See also Fig. 2.40(b). *Source:* After W. R. D. Wilson.

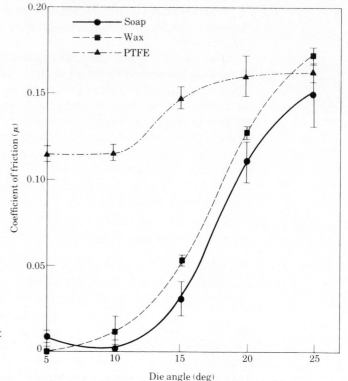

FIGURE 4.33 Effect of die angle and lubrication on friction in strip drawing through a wedge-shaped die. As the die angle decreases, more lubricant is entrained and friction is reduced. *Source*: After W. R. D. Wilson.

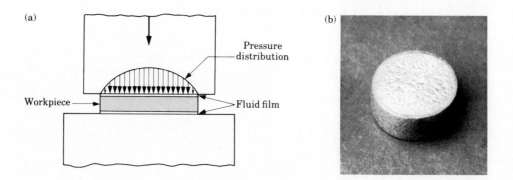

FIGURE 4.34 (a) Pressure distribution at the lubricated interface between a flat die (platen) and a solid cylindrical workpiece. The pressure increases with increasing speed, viscosity, and workpiece diameter, and decreasing fluid film thickness. (b) Rough surface produced in compressing an aluminum specimen, due to the high-viscosity of the lubricant and high speed of compression. The lubricant is trapped at the die-specimen interface. The roughness produced depends on the grain size, similar to those shown in Figs. 3.19 and 3.20.

increases, the lubricant is trapped more and more due to its viscosity; a thick film is thus developed.

However, it should be pointed out that the reduction of friction and wear by entrainment or entrapment of lubricants may not always be desirable. This is because with a thick fluid film the metal surface cannot come in full contact with the die surfaces and, hence, does not acquire the shiny appearance that may be required on the product. (Note, for instance, the shiny appearance of a piece of copper wire.) A thick film of lubricant generates a grainy dull surface on the workpiece; the degree of roughness depends on the grain size of the material (Fig. 4.34b).

In operations such as coining and forging (Chapter 6), trapped lubricants are undesirable because they interfere with the forming process and prevent precise shape generation.

4.10 METALWORKING LUBRICANTS

Metalworking lubricants may be classified into various categories. The most common lubricants are oils, emulsions, synthetic solutions, greases, soaps and waxes, solid lubricants, and glasses. (See also Section 8.12.) There are several variations within each category. Lubricants may also be classified as (and have the characteristics of) liquids, semisolids, and solids.

Because of their diverse chemistry, properties, and characteristics, the study of lubricants is complex. In this section only the general properties of commonly used lubricants will be described briefly. The practice of lubrication and selection of lubricants for individual metalworking processes are treated separately in Chapters 6, 7, 8, and 9. (See also Table 6.7.)

4.10.1 OILS

Oils have high film strength on the surface of a metal, as evidenced by the difficulty of cleaning an oily surface. Although they are effective in reducing friction and wear, they have low thermal conductivity and specific heat (Table 4.7). Thus they are not effective in conducting away the heat generated by friction and plastic deforma-

TABLE 4.7
THERMAL PROPERTIES OF SOME FLUIDS AT ROOM TEMPERATURE

FLUID	SPECIFIC HEAT		THERMAL CONDUCTIVITY	
	$BTU/lb \cdot °F$	$kJ/kg \cdot °C$	$BTU/hr \cdot ft \cdot °F$	$W/m \cdot °C$
Water	1.0	4.19	0.35	0.60
Ethyl alcohol	0.58	2.43	0.105	0.18
Castor oil	0.43	1.80	0.104	0.18
Carbon tetrachloride	0.20	0.84	0.107	0.19

tion in metalworking operations. Furthermore, oils are difficult to dispose of after use. They are also difficult to remove from the surface of a component that may subsequently be painted or welded. (See Table 4.8.)

The sources for oils are mineral (petroleum), animal, vegetable, and fish. Oils may be *compounded* with a variety of additives or with other oils to impart special properties, such as their viscosity–temperature behavior, surface tension, heat resistance, and boundary-layer characteristics. Mineral oils, with or without fillers, used undiluted are known as *neat oils*.

An important lubricant is mineral oil with sulfur or chlorine additives. These are known as *extreme-pressure* (EP) *additives* and may be used either singly or in combination. These oils are also known as *sulfurized* or *chlorinated oils*. They react chemically with metal surfaces and form adherent surface films of metallic sulfides and chlorides. These films have low shear strength and good anti-weld properties. However, they break down at certain temperatures generated at the interface. Phosphorus is another additive in this category, although it is not as effective as sulfur or chlorine and the film breaks down at a lower temperature.

Other additives in mineral oils are animal or vegetable oils and fats. These enhance lubrication by forming metallic soaps on the surface, again reducing friction.

New surfaces are continuously generated in metalworking operations due to plastic deformation. These new surfaces are highly susceptible to chemical reactions, particularly at elevated temperatures. Thus, protective films of sulfides, chlorides, and metallic soaps can be formed on surfaces with relative ease. The formation of these films also depends on the composition of the tool, die, and workpiece materials.

In selecting an oil as a lubricant, the importance of its viscosity–temperature–pressure relationship must be recognized. Temperatures and pressures in metalworking, as stated in Section 4.9, can vary widely not only from process to process and from material to material, but also at different locations of the workpiece during its processing. As noted in Fig. 4.31, lowering the viscosity of a lubricant can have a significant detrimental effect on friction and wear. This is particularly important in view of the fact that in most metalworking processes, unless pressures are low or speeds are high, hydrodynamic or thick-film lubrication is rarely encountered.

4.10.2 EMULSIONS

An emulsion is a mixture of two immiscible liquids. In metalworking lubrication, emulsions are usually a mixture of oil and water. Emulsions (also known as *water-soluble* fluids) are of two types: direct and indirect. In a *direct emulsion*, mineral oil is dispersed in water as very small droplets. In an *indirect emulsion*, water droplets are dispersed in the oil.

Direct emulsions are an important class of metalworking fluids, particularly in high-speed machining of metals, because of their high cooling capacity (due to the presence of water). They are mixed in various proportions of oil and water (1:10, 1:40, etc.). As lubricants, emulsions are not as effective as oils (Table 4.8).

TABLE 4.8
GENERAL CHARACTERISTICS OF METALWORKING FLUIDS. (After E. S. Nachtman in *Metalworking Lubrication*, New York, Am. Soc. of Mech. Eng., 1980, p. 53)

FUNCTION	COMPOUNDED OIL	OIL-IN-WATER EMULSION	SEMISYNTHETIC SOLUTION	SYNTHETIC SOLUTION
1 Provide lubrication at high pressure (boundary lubrication)	1	2	3	4
2 Reduce heat from plastic deformation (heat transfer)	5	2	2	1
3 Provide cushion between workpiece and die to reduce adhesion and pick-up (film thickness)	1	2	3	4
4 Reduce friction between die and workpiece and, thus, heat	1	2	2	1
5 Reduce wear and galling between tool and workpiece (chemical surface activity)	4	3	2	1
6 Flushing action to prevent buildup of scale, dirt (fluid flow)	5	4	3	2
7 Protect surface characteristics; nonstaining	5	2	2	1
8 Minimize processing costs, welding, cleaning, painting	5	2	2	1
9 Minimize environmental effect, air contamination, water recovery	5	2	2	1

Note: 1 = most effective, 5 = least effective.

4.10.3 SYNTHETIC SOLUTIONS

These are chemical fluids containing various inorganic chemicals dissolved in water. A variety of chemical agents are also added to impart different properties. Semi-synthetic solutions are basically synthetic solutions with small amounts of emulsi-fiable oils added. Their lubricating properties are thus better than synthetic fluids.

4.10.4 GREASES, SOAPS, AND WAXES

Greases are solid or semisolid lubricants and are generally composed of soaps, mineral oil, and various fillers and additives. They are adherent to metal surfaces and are highly viscous. Their viscosity depends on the rate of shearing. Although used extensively in lubricating machine elements, greases have somewhat limited use in metalworking.

Soaps are usually reaction products of sodium or potassium salts with fatty acids. Alkali soaps are soluble in water, but other metal soaps are generally insoluble. Metal soaps are commonly used in various metalworking processes. They are effective boundary lubricants and can also form thick-film layers at die–workpiece interfaces, particularly when applied over conversion coatings (Section 4.10.7).

Waxes may be of plant (paraffin) or animal origin. They have complex structures. Compared to greases, waxes are less "greasy" and are more brittle. They have limited use in metalworking operations, except for cold working of copper and (as chlorinated paraffin) for stainless steels and high-temperature alloys.

4.10.5 SOLID LUBRICANTS

The most common solid lubricants are graphite, molybdenum disulfide, glasses, and polymeric films.

Graphite

Graphite is a crystalline form of carbon and has a layered structure, with a c/a ratio of 2.5 (see Fig. 3.1). It is known as a lamellar solid and has highly anisotropic physical and mechanical properties. Graphite is weak in shear along its layers and has a low coefficient of friction in that direction. Thus it can be a good solid lubricant. However, the friction of graphite is low only in the presence of oxygen or water vapor. In a vacuum or an inert-gas atmosphere, the friction of graphite is very high; in fact, it can be quite abrasive. These effects are more pronounced at room temperature than at elevated temperatures.

Graphite may be applied either by rubbing on surfaces or as a colloidal suspension in liquid carriers such as water, oil, or alcohols. It is used widely in a variety of metal-working operations, particularly at elevated temperatures.

Molybdenum disulfide (MoS_2)

This is another widely used lamellar solid lubricant. It has a somewhat similar physical appearance to graphite. Its main difference from graphite is that it decomposes in the presence of oxygen and water vapor. It has a high friction coefficient in ambient environment. Oils are commonly used as carriers for molybdenum disulfide and are used as a lubricant in metalworking operations at room temperature. MoS_2 can also be rubbed on the surfaces of a workpiece.

Metallic and Polymeric Films

Because of their low strength, thin layers of soft metal and polymer coatings are also used as solid lubricants. Typical examples are metals such as lead, indium, cadmium, tin, and silver, and polymers such as PTFE (*Teflon*), polyethylene, and methacrylates. These coatings have limited applications in metalworking operations because they lack strength under high stresses and at elevated temperatures.

The oxides with which almost all metals are covered should be mentioned in this category of solid lubricants. If the oxide of a particular metal has low friction properties and is sufficiently thin, these layers can serve as solid lubricants, particularly at elevated temperatures. Powdered oxides of various metals may also be applied to the surfaces of the workpiece.

4.10.6 GLASSES

Although at room temperature glasses are hard and brittle, they become viscous at elevated temperatures and hence can serve as a liquid lubricant. Their viscosity is a function of temperature (but not of pressure) and depends on the type of glass. Their poor thermal conductivity also makes glasses attractive since they act as a thermal barrier between hot workpieces and relatively cool dies. Typical applications for glass are in hot extrusion and forging.

4.10.7 CONVERSION COATINGS

Lubricants may not always adhere properly to the surfaces of the workpiece to be processed, particularly under the high normal and shear stresses encountered in some operations. This is especially important in bulk deformation processing (forging, extrusion, wire drawing) of steels, stainless steels, and high-temperature alloys.

For these applications, the workpiece surface is treated by a chemical reaction, leaving a somewhat rough and spongy surface that acts as a carrier for the lubricant. *Zinc phosphate* coatings are widely used on carbon and low-alloy steels. Zinc phosphate crystals become plastic under pressure (at the die–workpiece interfaces) and thus follow major surface extensions of the workpiece. For stainless steels and high-temperature alloys, *oxalate* coatings are used. Liquid lubricants or a soap are then applied to the coated surface. The lubricant film adheres to the surface and is not easily scraped off.

4.10.8 ULTRASONIC VIBRATIONS

Significant reductions in friction have been observed in subjecting the die–workpiece interface to ultrasonic vibrations, generally at 20 kHz. It appears that in order to reduce friction with vibrations it is necessary that:

a. The die and the workpiece have a strong tendency for adhesion,

b. The amplitude of vibration be sufficiently high to separate the microwelds at the asperities and expose the surfaces to the lubricant, and

c. The lubricant contain a reactive element to provide the necessary protective layer to the freshly disturbed surface.

This method of reducing friction requires special equipment. Also, considerable experience is needed to properly apply ultrasonic vibrations at appropriate locations in the metalworking system.

4.11 LUBRICANT SELECTION

In addition to the foregoing considerations, selection of a lubricant involves several other factors. Among these are the compatibility of the lubricant with the workpiece and tool and die materials, surface preparation, method of lubricant application, removal of lubricant after use, contamination of the lubricant by other lubricants (such as those used to lubricate the machines), treatment of waste lubricant, storage and maintenance of lubricants, and biologic and ecologic considerations. Brief examples of these factors are given below:

a. While extreme-pressure additives are very important in boundary lubrication, it has been observed, for example, that these lubricants could preferentially attack the cobalt binder in tungsten carbide tools and dies (Section 8.10) and cause changes in the surface roughness and the integrity of die surfaces.

b. The effect of normal stresses on the shear strength (hence frictional resistance) of liquid and solid lubricants must be considered. As Fig. 4.35 indicates, whereas the shear strength (τ_i) of metallic films is not affected by pressure (p), oils and all other materials have a pressure dependency.

c. The chemical reactivity of a lubricant depends on the nature of residual stresses on a surface. Surfaces with tensile residual stresses are more reactive than those with compressive stresses. Also, freshly generated surfaces, such as those by machining and grinding, are more reactive than others.

d. The different functions of a metalworking fluid, whether primarily a lubricant or a coolant, must also be taken into account. Water and water-base lubricants are effective coolants, but not effective lubricants in comparison to oils.

e. In complex metalworking operations, it may be necessary to lubricate certain surfaces of a workpiece, and not others, to induce favorable flow of the material into the die cavities.

f. There are situations where the presence of a lubricant is harmful. If a surface crack is generated (e.g., due to mechanical or thermal fatigue) the fluid is drawn into the crack opening by surface tension. Upon subsequent contact of the bodies, high hydrostatic pressure can develop in this crack (due to the incompressibility of the fluid). This leads to rapid propagation of the crack, resulting in premature surface failure (Section 4.8.4).

g. Metalworking fluids should not leave any harmful residues that may interfere with the operation of the equipment. They should not stain or corrode the workpiece or the equipment. The fluids should be checked periodically for deterioration due to bacterial growth, accumulation of oxides, chips and wear debris, and for general degradation due to temperature and time.

h. Metal surfaces are usually covered with lubricant residues after metalworking operations. These residues should be removed before further processing, such as welding or painting. A variety of cleaning solutions and techniques can be used for this purpose. Among these are alkaline and acid cleaners, petroleum solvents, vapor degreasing, and ultrasonic cleaning.

i. Biologic and ecologic considerations, with their accompanying legal aspects, have acquired great significance. There may be potential health hazards in contact with or inhalation of some metalworking fluids. Recycling of waste fluids and their disposal are other important problems.

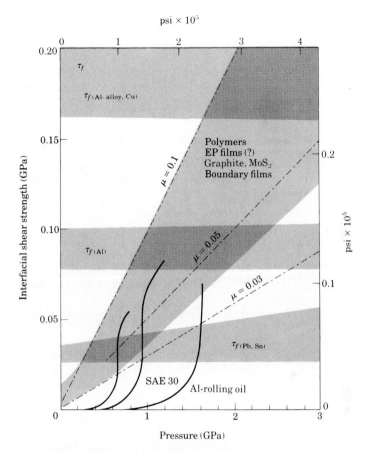

FIGURE 4.35 Interface shear strength of various solid and liquid lubricants as a function of normal pressure. Note that, unlike in lubricants, the shear strength of metallic films does not change with pressure. See also Fig. 8.14. *Source:* After J. A. Schey, in *Metal Forming Plasticity*, Berlin: Springer, 1979, pp. 336–348.

4.12 SURFACE TENSION

An important aspect of lubrication with fluids is their wetting characteristics— how well the fluid spreads itself over the surface of the workpiece in a continuous film, instead of forming droplets. *Wetting* is a phenomenon related to surface tension, which is a manifestation of surface energy.

The molecules in a liquid or solid mass are subjected to equal forces of attraction in all directions. The molecules on the surface of the mass, on the other hand, are subjected to attractive forces from one side only. In order to balance the forces, a net force must act on the surface molecules. This force is called *surface tension*.

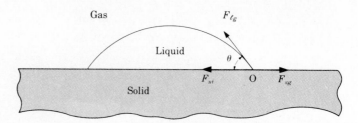

FIGURE 4.36   Surface tension between a liquid, gas, and solid. Wetting of a surface means that the contact angle θ is small. A drop of oil on the surface of water has a small θ, whereas a drop of water on an oily surface has a high θ. Wetting is a characteristic of lubricants that is important in effective lubrication in metalworking processes.

In the example of a drop of liquid on a solid surface (Fig. 4.36), interfacial tensions exist between the surface molecules of solid–liquid (F_{sl}), liquid–gas (F_{lg}), and solid–gas (F_{sg}). The angle θ is called the *contact angle* and is a measure of how well the liquid wets the surface.

The equilibrium of these interfacial tensions in the horizontal direction at point 0 gives

$$F_{sl} + F_{lg} \cos \theta - F_{sg} = 0,$$

or

$$\cos \theta = \frac{F_{sg} - F_{sl}}{F_{lg}}. \tag{4.10}$$

If $F_{sg} > F_{sl}$, the value of $\cos \theta$ is positive and the contact angle is equal to or less than 90 degrees. If $F_{sg} < F_{sl}$ then the contact angle is greater than 90 degrees. An angle of $\theta = 0$ degrees indicates complete wetting and, hence, spreading of the liquid over the surface of the solid.

Note from Eq. (4.10) that spreading or wetting is improved by decreasing F_{lg} or F_{sl}, or both, and by increasing F_{sg}. This can be accomplished either by the addition of *wetting agents* (such as alcohols and glycols) in lubricants or by increasing the temperature. Also, wetting is improved by increasing the roughness of the solid surface.

Although good wetting characteristics of metalworking lubricants is generally an important prerequisite, there may well be situations where preferential wetting may be desirable. For instance, timepiece manufacturers have long recognized the need for nonmigrating (nonwetting) lubricants for pivot points in watches.

Surface tension and wetting are important not only in liquids but also have an important role in grain boundary formation during solidification of polycrystalline metals. (See Sections 3.3 and 5.2.)

● **Illustrative Problem 4.4**

Determine the contact angles for a drop of oil on water and a drop of water on oil.

SOLUTION.   Interfacial tensions between water, oil, and air at room temperature are as follows:

Water–oil 5.3×10^{-2} N/m

Water–air 7.3×10^{-2}

Oil–air 3.1×10^{-2}

Thus, from Eq. (4.10), for a drop of oil on water

$$\cos \theta = \frac{73 - 53}{31} = 0.645$$

or

$$\theta = 50 \text{ degrees}$$

and for a drop of water on oil

$$\cos \theta = \frac{31 - 53}{73} = -0.301$$

or

$$\theta = 108 \text{ degrees.}$$

In this problem it is assumed that the surfaces on which the drops are placed is flat, such as a solid surface covered with a thin layer of water or oil. This problem illustrates why oil spreads over water much more readily than water does on oil. ●

4.13 SURFACE TREATMENTS

Adequate wear resistance of surfaces can be provided either by selecting a material that has the required properties for the particular wear mechanism, or by surface treatment and coatings. Various techniques are available to modify the surface structure and its properties in order to improve wear resistance (Table 4.9). These are described below.

4.13.1 TECHNIQUES FOR SURFACE TREATMENT

Surface Hardening

Traditional surface heat treatment processes, such as *carburizing, carbonitriding, cyaniding, nitriding, flame hardening,* and *induction hardening* are outlined in Table 4.10. Laser beams are also used as a heat source in surface hardening. The hardnesses

TABLE 4.9
SUGGESTED SURFACE TREATMENTS FOR VARIOUS METALS. (After M. K. Gabel and D. M. Donovan in *Wear Control Handbook*, New York, Am. Soc. of Mech. Eng., 1980, p. 349).

METAL	TREATMENT
Aluminum	Chrome plate; anodic coating, phosphate; chromate conversion coating
Beryllium	Anodic coating; chromate conversion coating
Cadmium	Phosphate; chromate conversion coating
Die steels	Boronizing, ion nitriding; liquid nitriding
High-temperature steels	Diffusion
Magnesium	Anodic coating; chromate conversion coating
Mild steel	Boronizing; phosphate; carburizing; liquid nitriding; carbo-nitriding, cyaniding
Molybdenum	Chrome plate
Nickel- and cobalt-base alloys	Boronizing, diffusion
Refractory metals	Boronizing
Stainless steel	Vapor deposition, ion nitriding, diffusion, liquid nitriding, nitriding
Steel	Vapor deposition, chrome plate, phosphate, ion nitriding, induction hardening, flame hardening, liquid nitriding
Titanium	Chrome plate, anodic coating, ion nitriding
Tool steel	Boronizing, ion nitriding, diffusion, nitriding, liquid nitriding
Zinc	Vapor deposition, anodic coating, phosphate, chromate chemical conversion coating

obtained by various processes are shown in Fig. 4.37 for a variety of steels. (See also Table 3.6.)

Surface hardening, as well as some of the other surface-treatment processes described below, induce residual stresses on surfaces. The formation of *martensite* in surface heat-treating processes causes compressive residual stresses. Such stresses are desirable because they improve the fatigue life of components by delaying the initiation of fatigue cracks.

Hard Facing

In these processes, a relatively thick layer of hard metal is deposited on the surface. Typical applications are valve seats, oil-well drilling tools, and metalworking dies. Common techniques are *flame spraying*, *plasma spraying*, and *weld deposition* of hard alloys (Chapter 12). These processes involve diffusion at the interface of the coating and the base metal. Both surface hardening and hard facing generally involve thick layers, as much as 0.060 in. (1.5 mm). This category also includes *electroplating* techniques, such as hard chrome plating for increased wear resistance.

TABLE 4.10
OUTLINE OF HEAT TREATMENT PROCESSES FOR SURFACE HARDENING

PROCESS	METALS HARDENED	ELEMENT ADDED TO SURFACE	PROCEDURE	GENERAL CHARACTERISTICS	TYPICAL APPLICATIONS
Carburizing	Low-carbon steel (0.2% C), alloy steels (0.08–0.2% C)	C	Heat steel at 1600–1750°F (870–950°C) in an atmosphere of carbonaceous gases (gas carburizing) or carbon-containing solids (pack carburizing). Then quench.	A hard, high-carbon surface is produced. Hardness 55 to 65 HRC. Case depth <0.020 to 0.060 in. (<0.5 to 1.5 mm). Some distortion of part during heat treatment.	Gears, cams, shafts, bearings, piston pins, sprockets, clutch plates
Carboni-triding	Low-carbon steel	C and N	Heat steel at 1300–1600°F (700–800°C) in an atmosphere of carbonaceous gas and ammonia. Then quench in oil.	Surface hardness 55 to 62 HRC. Case depth 0.003 to 0.020 in. (0.07 to 0.5 mm). Less distortion than in carburizing.	Bolts, nuts, gears
Cyaniding	Low-carbon steel (0.2% C), alloy steels (0.08 to 0.2% C)	C and N	Heat steel at 1400–1550°F (760–845°C) in a molten bath of solutions of cyanide (e.g., 30% sodium cyanide) and other salts.	Surface hardness up to 65 HRC. Case depth 0.001 to 0.010 in. (0.025 to 0.25 mm). Some distortion.	Bolts, nuts, screws, small gears
Nitriding	Steels (1% Al, 1.5% Cr, 0.3% Mo), alloy steels (Cr, Mo), stainless steels, high-speed tool steels	N	Heat steel at 925–1100°F (500–600°C) in an atmosphere of ammonia gas or mixtures of molten cyanide salts. No further treatment.	Surface hardness up to 1100 HV. Case depth 0.005 to 0.030 in. (0.1 to 0.6 mm) and 0.001 to 0.003 in. (0.02 to 0.07 mm) for high-speed steel.	Gears, shafts, sprockets, valves, cutters, boring bars, fuel-injection pump parts
Boronizing	Steels	B	Part is heated using boron-containing gas or solid in contact with part.	Extremely hard and wear resistant surface. Case depth 0.001–0.003 in. (0.025–0.075 mm).	Tool and die steels
Flame hardening	Medium-carbon steels, cast irons	None	Surface is heated with an oxyacetylene torch, then quenched with water spray or other quenching methods.	Surface hardness 50 to 60 HRC. Case depth 0.030 to 0.25 in. (0.7 to 6 mm). Little distortion.	Gear and sprocket teeth, axles, crankshaft, piston rod, lathe beds and centers
Induction hardening	Same as above	None	Metal part is placed in copper induction coils and is heated by high-frequency current, then quenched.	Same as above.	Same as above.

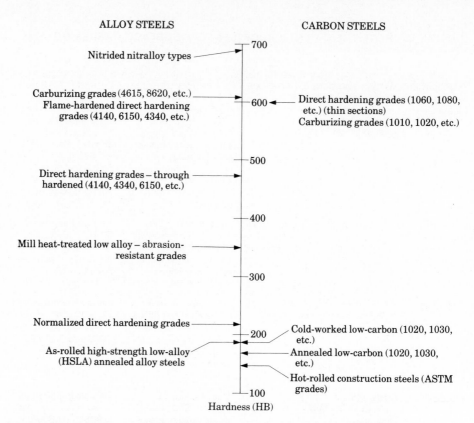

FIGURE 4.37 Hardness ranges obtainable by various thermal and mechanical treatment processes for carbon and alloy steels. Hardness affects not only the strength of metals, but also their frictional and wear characteristics. *Source:* After K. G. Budinski.

Coatings

New developments in coating technology allow the close control of the thickness and the composition of coatings of soft or hard compounds and metals on surfaces (Table 4.11). Typical processes are *chemical vapor deposition* (CVD), *physical vapor deposition* (PVD), *sputtering, pyrolytic decomposition*, and *reactive evaporation*. The coating thickness is much less than in the more traditional techniques and is usually a few micrometers. One of the most useful coatings for metalworking tools is titanium nitride (Section 8.10). The details of these techniques are rather involved and are beyond the scope of this book.

In addition to wear resistance, coatings are also employed to impart certain properties to metal surfaces, such as corrosion and oxidation resistance, appearance, color, and surface preparation for painting.

Metal coatings can be applied by various processes, described below.

a. Hot Dipping: for example, sheet steel is dipped into a bath of molten zinc (*galvanized* steel), tin (*tin cans* for food containers), or aluminum (*aluminized*, for oil-refinery equipment, and various parts for protection at elevated temperatures).

b. Spraying: molten metal is sprayed on a surface with a spray gun (oxyacetylene flame, electric-arc or plasma-arc, at temperatures on the order of 20,000°F [12,000°C], Chapter 12). The surfaces must be clean and rough for improved bond strength. Typical examples are steel structures, storage tanks, and tank cars sprayed with zinc or aluminum, up to 0.010 in. (0.25 mm) thick. The process can also be used for rebuilding worn or undersized parts, such as bearings and shafts. Techniques have also been developed to spray refractory coatings for high-temperature and electrical resistance.

c. Diffusion: (Section 12.4.5), also called cementation.

d. Cladding: metals are covered with a thin layer of corrosion-resistant metal by heating and rolling (Section 12.4.5). Typical examples are steels clad with stainless steel or nickel alloys, or steel wire clad with copper.

e. Electroplating: a metal workpiece (cathode) is plated with another metal (anode) in an *electrolyte* solution.

f. Vapor Deposition: metal vapor is deposited on metals, plastics, glass, and paper by condensation of metal vapor from molten metal.

TABLE 4.11
COATINGS USED FOR VARIOUS FUNCTIONS.
(After S. Ramalingam).

FUNCTION	COATINGS
Reduce wear	Titanium carbide, nitride
Reduce friction	PTFE, molybdenum disulfide
Increase friction	Titanium, bonded abrasives
Improve lubrication	Copper, lead
Increase temperature or load capacity	Electroless nickel
Prevent adhesion	Silver/gold plate
Imbed particles	Indium, lead
Reduce corrosive wear	Chromium plate or diffusion
Retain fluid lubricants	Phosphating, nylon
Rebuild surface	Steel hard surfacing
Reduce surface roughness	Silver plate
Prevent drop erosion	Polyurethane, neoprene
Prevent particle erosion	Cobalt alloy, molybdenum

g. Anodizing: the workpiece is the anode in an electrolytic cell in an acid bath. This is an oxidation process and the part is coated with a hard, porous oxide layer. Organic coatings or dyes of various colors can then be applied to the surfaces with good bond strength. Typical applications are aluminum furniture and utensils, architectural shapes, and automobile trim.

h. Enamels or Porcelains: vitreous (glassy) coatings on metals for resistance to chemicals and service at elevated temperatures.

i. Ceramic Coatings: metal surfaces are sprayed with aluminum oxide or zirconium oxide.

j. Conversion Coatings: see Section 4.10.7.

Another process of coating (actually a metal forming process) is *electroforming*, which is a variation of electroplating. Metal is deposited on a *mandrel* (also called a mold or matrix), which is then removed. Thus, the coating becomes the product. Mandrels can be made of a variety of metallic and nonmetallic materials (which can be made electrically conductive using conductive coatings), and should be physically removable. The electroforming process is suitable for low production runs of intricate parts with expensive metals.

4.13.2 MECHANICAL HARDENING METHODS

To improve the surface properties of finished components, such as shafts with fillets, several techniques are available wherein the surface layer of the metal is work-hardened to a small depth.

a. Shot Peening: the surface is hit repeatedly with steel or cast iron shot (small balls) usually ranging in diameter from 0.005 to 3/16 in. (0.125 to 4.8 mm). Depending on process parameters, the depth of surface deformation can range up to 0.05 in. (1.25 mm). Because it plastically deforms the surface layers, shot peening imparts compressive residual stresses on the surface. Some typical residual stress distributions are shown in Fig. 4.38.

b. Surface Rolling: (Fig. 4.39) the surface of a component is cold worked by a roller instead of shot. The results obtained are similar to shot peening.

c. Explosive Hardening: surfaces are subjected to very high transient pressures. In this technique, the method consists of placing a layer of explosive sheet directly on the metal surface and detonating it. The maximum pressures can be as high as 5×10^6 psi (35,500 MPa) lasting about 2 to 3 μs. It has been observed that large increases in surface hardness can be obtained by this method with very little change (less than 5%) in the shape of the component.

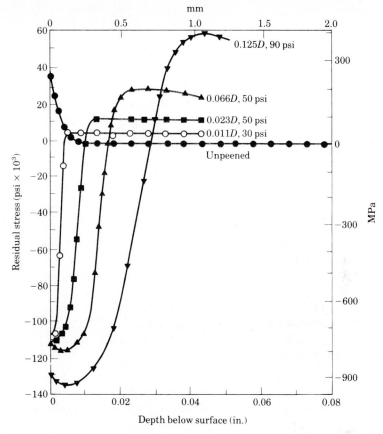

FIGURE 4.38 Effect of shot diameter and air pressure on residual stresses in shot peening of steel, 42 HRC. Note that the residual stresses reverse signs at a rather small depth. See also Figs. 6.93 and 9.10. *D* is shot diameter in inches. *Source:* J. M. Lessels and R. F. Brodrick, Int. Conf. Fatigue in Metals, Inst. Mech. Eng., 1956, p. 621.

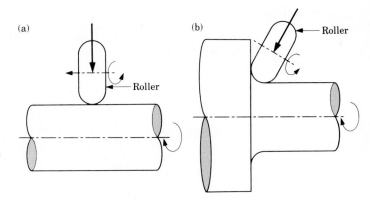

FIGURE 4.39 Surface rolling methods to induce compressive surface residual stresses to improve the fatigue life of a (a) shaft and (b) fillet on a stepped shaft.

SUMMARY

1. In manufacturing processes, surfaces and their properties are as important as the bulk properties of the materials. Surfaces involve not only a particular geometry and appearance, but also are composed of a layer with properties that are generally different from those of the bulk material.

2. Surfaces are exposed to the environment and thus are subject to environmental attack; or they may be in intimate contact with tools and dies (during processing) or other components (during their service life). Consequently, their geometric and material properties can have significant effects, such as in regard to friction and wear.

3. Measurements and descriptions of surfaces and their roughness are a complex problem. Certain standards have been developed and are specified along with other design requirements in manufacturing.

4. Friction and wear are among the most significant factors in processing of materials. In spite of the difficulties involved, considerable progress has been made in understanding these phenomena and in identifying the factors that govern them. Among these are the affinity and reactivity of the two materials in contact, the nature of surface films, the presence of contaminants, and process parameters such as load, speed, and temperature.

5. A wide variety of metalworking fluids is available for specific applications. Their selection and use require careful consideration of many factors in regard to workpiece and die materials and the particular manufacturing process.

6. Surface treatments are used to impart certain physical and mechanical properties, such as resistance to environmental attack and wear and fatigue resistance. Various techniques employed are heat treatment, coatings, surface treatments, and mechanical working of surfaces.

BIBLIOGRAPHY

Friction and Wear

Bowden FP, Tabor D. *The Friction and Lubrication of Solids.* New York: Oxford, vol. I, 1950, vol. II, 1964.

Czichos H. *Tribology—A Systems Approach to the Science and Technology of Friction, Lubrication and Wear.* New York: Elsevier, 1978.

Dowson D. *The History of Tribology.* New York: Longman, 1979.

Engel PA. *Impact Wear of Materials.* New York: Elsevier, 1976.

Friction and Wear Devices, 2d ed. Park Ridge, Ill.: American Society of Lubrication Engineers, 1976.

Glaeser WA, et al., eds. *Wear of Materials*. New York: American Society of Mechanical Engineers, biannual, since 1977.

Halling J. *Principles of Tribology*. New York: Macmillan, 1975.

Kragelskii IV. *Friction and Wear*. Washington, DC: Butterworth, 1965.

Lipson C. *Wear Considerations in Design*. Englewood Cliffs, N.J.: Prentice-Hall, 1967.

Moore DF. *Principles and Applications of Tribology*. New York: Pergamon, 1975.

Moore DF. *The Friction and Lubrication of Elastomers*. New York: Pergamon, 1972.

Neale MJ, ed. *Tribology Handbook*. New York: Butterworth, 1973.

Peterson MB, Winer WO, eds. *Wear Control Handbook*. New York: American Society of Mechanical Engineers, 1980.

Rabinowicz E. *Friction and Wear of Materials*. New York: Wiley, 1965.

Rigney DV, ed. *Fundamentals of Friction and Wear of Materials*. Metals Park, Ohio: American Society for Metals, 1981.

Sarkar AD. *Wear of Metals*. New York: Pergamon, 1976.

Scott D, ed. *Wear*. Volume 13 in *Treatise on Materials Science and Technology*. New York: Academic Press, 1979.

Simons EN. *Metal Wear: A Brief Outline*. London: Frederick Muller Ltd., 1972.

Source Book on Wear Control Technology. Metals Park, Ohio: American Society for Metals, 1978.

Thomas JR, ed. *Rough Surfaces*. New York: Longman Group Ltd., 1982.

Tribology in Metalworking

Bastian ELH. *Metalworking Lubricants*. New York: McGraw-Hill, 1951.

Billett M. *Industrial Lubrication*. New York: Pergamon, 1979.

Braithwaite ER, ed. *Lubrication and Lubricants*. New York: Elsevier, 1967.

Braithwaite ER. *Solid Lubricants and Surfaces*. New York: Macmillan, 1964.

Clauss FJ. *Solid Lubricants and Lubricating Solids*. New York: Academic Press, 1971.

Cutting and Grinding Fluids. Dearborn, Mich.: Society of Manufacturing Engineers, 1967.

Friction and Lubrication in Metal Processing. New York: American Society of Mechanical Engineers, 1966.

Kalpakjian S, Jain S, eds. *Metalworking Lubrication*. New York: American Society of Mechanical Engineers, 1980.

Olds NJ. *Lubricants, Cutting Fluids and Coolants*. Boston: Cahners, 1973.

Schey JA, ed. *Metal Deformation Processes: Friction and Lubrication*. New York: Marcel Dekker, 1970.

Schey JA. *Tribology in Metalworking—Friction, Lubricating and Wear*. Metals Park, Ohio: American Society for Metals, 1983.

Periodicals

American Society of Lubrication Engineers Transactions

Industrial Tribology

Journal of Lubrication Technology (now called *Journal of Tribology*)

Lubrication Engineering

Tribology International

Wear

PROBLEMS

4.1. Give examples from your personal experience or observations in this chapter to confirm the validity of Eq. (4.5).

4.2. It has been observed that, both for lubricated and unlubricated sliding contacts, friction generally decreases with increasing sliding speed. Can you explain this behavior?

4.3. Explain the effect of the included angle of the stylus (conicity) on the shape of its path in Fig. 4.4 (broken lines).

4.4. A 1-in.-high flat ring with an outside diameter of 2 in. and inside diameter of 1 in. is compressed by 50%. It is found that the new outside diameter is 2.45 in. What is your opinion about the magnitude of the friction coefficient?

4.5. Estimate the answer to Illustrative Problem 4.2 if the materials are lubricated.

4.6. What procedure would you suggest if, in a ring compression test, the specimen acquires an oval-shaped inner hole?

4.7. Give the possible reasons why the inner hole of a ring compression specimen may become oval after upsetting.

4.8. Explain the trends that you see in surface finish requirements in Fig. 4.7.

5 Casting Processes

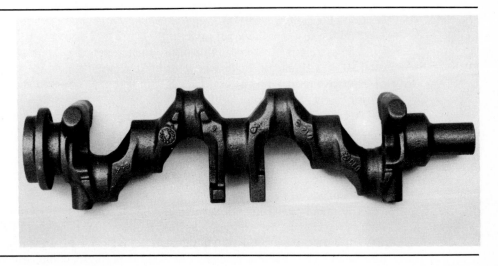

5.1 INTRODUCTION

Casting processes are among the oldest methods of manufacturing various ornamental objects and functional parts. Casting dates back to around 4000 B.C. when copper arrowheads were made. Casting processes basically involve the introduction of a molten metal into a mold cavity, where, upon solidification, the metal takes on the shape of the mold cavity.

The casting process is capable of producing intricate shapes, both internal and external, in a single piece. Very large or hollow parts, which would be very difficult or expensive to make by other methods, can be mass produced economically, using proper casting techniques.

Almost all materials can be cast in, or nearly in, the final shape and dimensions desired so that additional machining, and thus scrap, is reduced. With appropriate control of the parameters involved, parts can be cast with uniform properties throughout. In addition, materials such as gray cast irons have internal damping characteristics that are important in reducing vibrations in machine tools and various other equipment.

The major limitations of casting in comparison with other shaping processes, such as forging, are lower toughness and ductility, and porosity in the cast product.

The variety of parts that can be made by casting is extensive. Typical examples are: frames and housings for machines, structural parts, machine components, engine blocks, crankshafts, pistons and other automotive components, pipe and valves, railroad equipment, statues, and ornamental artifacts. Casting sizes may range from a few millimeters, with weights on the order of a few grams (such as the individual teeth of a zipper), to very large pieces weighing as much as 300 tons.

Regardless of the method of casting, there are certain fundamental aspects that must be well understood and controlled in order to produce parts successfully. The most important of these is the mechanism of solidification of metals. Other important aspects are the preparation of molds, cores, melting and pouring of metals, and cleaning and inspection.

5.2 SOLIDIFICATION OF METALS

When liquid metal is poured into a mold and allowed to cool, a series of complex events takes place. The significant factors are the type of metal (whether it is a pure metal or an alloy), thermal properties (thermal conductivity and specific heat), the geometric relation between volume and surface area of the liquid metal, and the shape of the mold, and the mold material.

5.2.1 PURE METALS

Pure metals have a clearly defined melting (or freezing) point, so solidification takes place at a constant temperature (Fig. 5.1). When the temperature of the liquid metal is reduced to the freezing point, it remains constant while the latent heat of

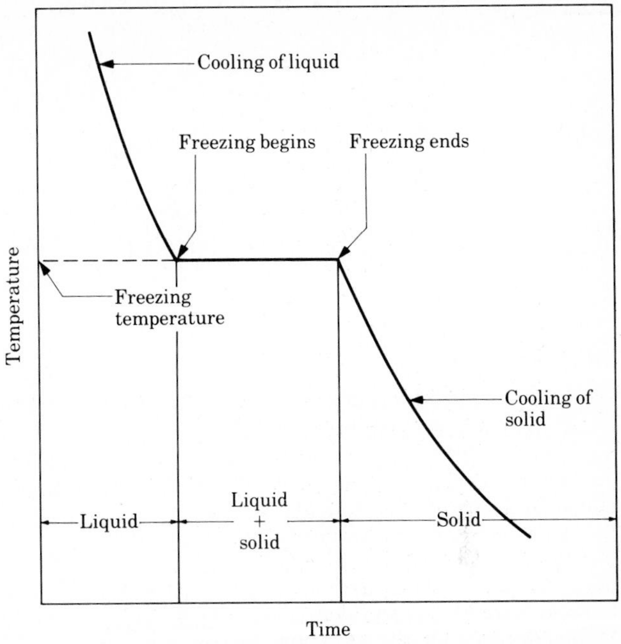

FIGURE 5.1 Cooling curve for the solidification of pure metals. Note that freezing takes place at a constant temperature. This temperature ranges from 621°F (327°C) for lead, 2798°F (1537°C) for iron, and to 6170°F (3410°C) for tungsten. The heat given off during freezing (heat of solidification) is the same as the latent heat of fusion during melting, and the process is reversible.

solidification is given off. At the end of this cycle, complete solidification takes place and the solid metal begins to cool off.

A typical temperature distribution in the mold–liquid metal interface is shown in Fig. 5.2. Heat from the liquid metal is given off through the mold wall and the air. The temperature drop at the air–mold and mold–metal interfaces is due to the

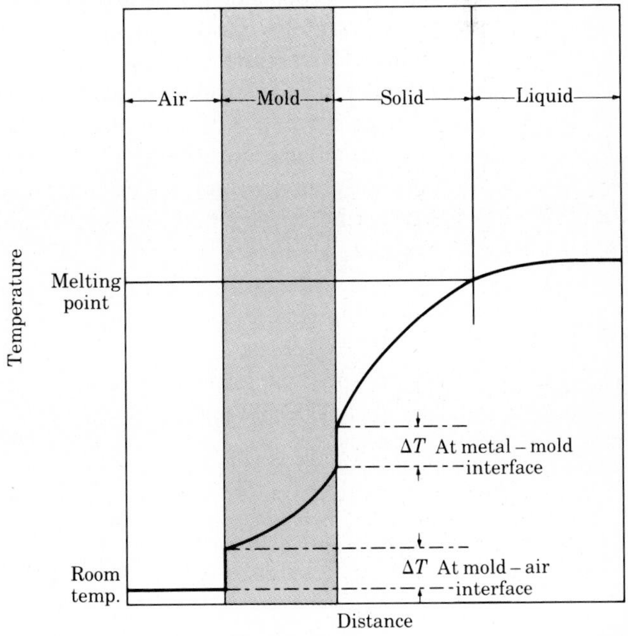

FIGURE 5.2 Temperature distribution at the mold wall and liquid–metal interface during solidification of metals in casting. The sharp temperature drop at the mold–solid metal interface is due to imperfect contact, and at the air–mold interface due to the presence of boundary layers. The rate at which the temperature drops within other regions depends on the thermal conductivity of the materials involved.

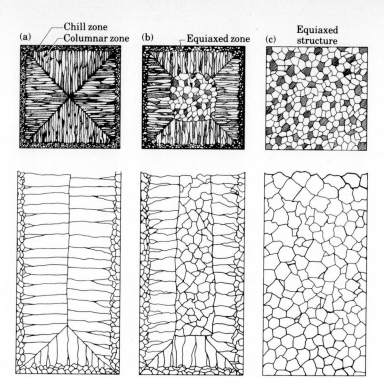

FIGURE 5.3 Schematic illustration of various structures for metals cast in a mold. (a) Pure metals. (b) Solid-solution alloys. A solid solution is a single homogeneous crystalline phase containing two or more chemical elements. (c) Structure obtained in the absence of thermal gradients within the solidifying mass, or using a catalyst to induce heterogeneous nucleation of grains. *Source:* After G. W. Form, J. F. Wallace, J. L. Walker, and A. Cibula.

presence of boundary layers and imperfect contact at these interfaces. The shape of the curves in this figure depends on the thermal properties of the mold material.

The grain structure of a pure metal casting is shown in Fig. 5.3(a). At the mold wall, chilling takes place, resulting in a skin of fine *equiaxed* grains. As solidification continues, *columnar* grains are formed (*dendrites*). They are columnar because only those grains favorably oriented will grow away from the mold wall and toward the center of the solidifying mass. Thus, a preferred orientation develops, and grains that have substantially different orientations are blocked from further growth (Fig. 5.4).

The size and distribution of the overall grain structure depends on the direction and rate of heat flow. Experimental evidence indicates that convection has a strong effect on grain size and that no outer chill zone is observed when convection is absent. Furthermore, the grain size of the solidified metal is larger when convection is reduced.

FIGURE 5.4 Development of preferred texture at a chill face, such as a cool mold. Note that only those grains that are favorably oriented grow away from the cool surface. Grains that have substantially different orientations (e.g., smaller grains in the figure) are blocked off by the larger grains from further growth. (See also Fig. 12.33.)

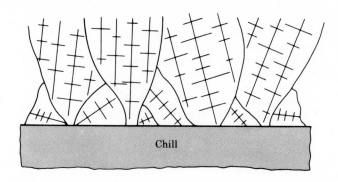

5.2.2 ALLOYS

Unlike pure metals, alloys solidify over a range of temperatures (Fig. 5.5). Solidification begins when the temperature drops below the *liquidus*; it is completed when it reaches the *solidus*. Within this temperature range the alloy is in a mushy or pasty state. Its composition and state are given by the phase diagram of the particular alloy. One example is shown in Fig. 5.6 for a copper–nickel alloy. Here, the solidified alloy is known as a *solid solution* because the alloying element (Cu) is completely dissolved in the base metal (Ni) and each grain has the same composition.

Figure 5.3(b) shows a typical cast structure of a solid solution alloy with an inner equiaxed granular zone. This zone can be extended throughout the casting (Fig. 5.3c) by the addition of a catalyst. This induces heterogeneous nucleation of grains, instead of the usual grain formation from the mold walls and progression toward the center. Another method to obtain hetereogeneous nucleation is to reduce or eliminate thermal gradients within the mold.

As seen in Fig. 5.7(a), dendrites form with three-dimensional arms and branches that eventually interlock with each other. When the alloy is cooled very slowly, each dendrite develops a uniform composition. However, with rapid cooling cored dendrites are formed with surface compositions that are different from that at their centers. The variety of dendritic structures that can be obtained is shown in Fig. 5.7(b), where the effects of time, cooling rate, and alloy type can be observed.

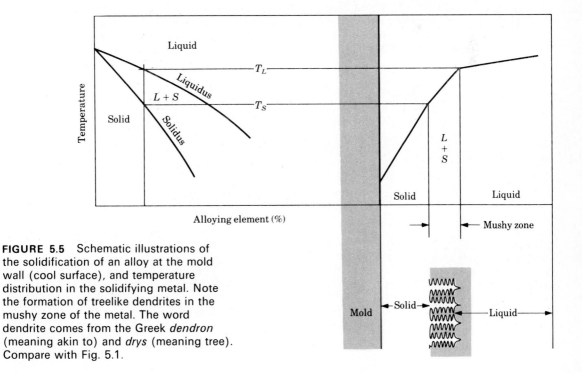

FIGURE 5.5 Schematic illustrations of the solidification of an alloy at the mold wall (cool surface), and temperature distribution in the solidifying metal. Note the formation of treelike dendrites in the mushy zone of the metal. The word dendrite comes from the Greek *dendron* (meaning akin to) and *drys* (meaning tree). Compare with Fig. 5.1.

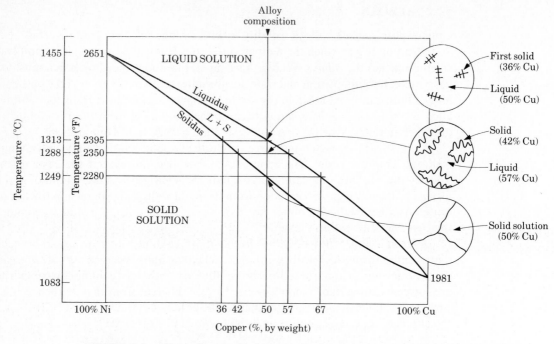

FIGURE 5.6 Phase diagram for nickel-copper alloy system obtained by very slow solidification. The solid solution is a single homogeneous crystalline phase containing copper and nickel. Note that pure nickel and pure copper each have one freezing temperature, as was shown schematically in Fig. 5.1.

As can be seen with sand molds (Fig. 5.7b), the steel with the lowest carbon solidifies with a marked *skin formation* (*short* freezing range). Steels with a higher carbon content show extensive mushy zones during solidification (*long* freezing range). Table 5.1 gives a list of various metals and their freezing ranges.

TABLE 5.1
SOLIDIFICATION CHARACTERISTICS OF METALS AND ALLOYS IN CASTING

SHORT FREEZING RANGE (FREEZING WITH MARKED SKIN FORMATION)	LONG FREEZING RANGE (FREEZING WITH EXTENSIVE PASTY OR MUSHY ZONE)
Aluminum	Aluminum alloys
Aluminum bronzes	Bronzes
Brasses	Magnesium alloys
Copper	Nickel-base alloys
Low-carbon steel	Steels (medium- and high-carbon)

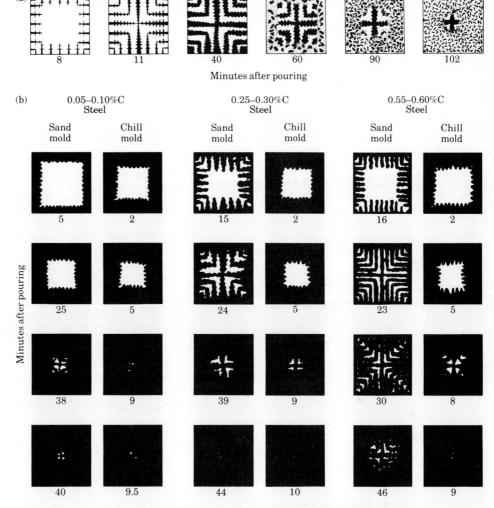

FIGURE 5.7 (a) Solidification mode for gray cast iron. The castings are 7 in. (178 mm) square. Note that after 11 minutes of cooling, dendrites reach each other. However, the casting is still mushy throughout the body (see Fig. 5.5). It takes about two hours for this casting to solidify completely. (b) Solidification of carbon steels in sand and chill (metal) molds. Note the change in the solidification pattern as the carbon content of the steel increases. See also Table 5.1. *Source:* H. F. Bishop and W. S. Pellini, *Foundry*, vol. 80, February, 1952.

5.2.3 SOLIDIFICATION TIME

At the early stage of solidification, a thin skin or shell is formed at the mold wall. The thickness of this solidified skin increases with time toward the center of the liquid metal. With flat mold walls, the thickness of the solid skin has been shown to be proportional to the square root of time. The larger the volume, the lower the rate of

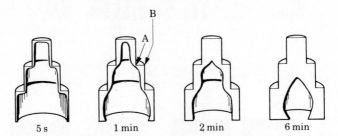

FIGURE 5.8 Solidified skin on a steel casting. The remaining molten metal is poured out at indicated times. Note that at internal corner *A* the casting is thinner than at external corner *B*, which cools faster than corner *A*. Hollow ornamental and decorative objects are made by a similar process (slush casting) that has the freezing pattern shown in this figure. *Source:* After H. F. Taylor, J. Wulff, and M. C. Flemings, *Foundry Engineering*, New York: Wiley, 1959.

cooling. The general formula for the complete solidification time is:

$$\text{Solidification time} = C \text{ (volume/surface area)}^2, \tag{5.1}$$

where *C* is a constant that depends on mold material and temperature. This is known as *Chvorinov's rule* (after N. Chvorinov, 1938) and has been shown to apply well to various geometries. With a more complex development of this basic equation, it can be shown that, for the same volume, a cube freezes faster than a sphere. (See Illustrative Problem 5.1.)

The effect of mold geometry and time can be seen in Fig. 5.8, where the still liquid metal has been removed from the solidified regions. Note that the solidified skin is thinner at internal angles (location *A* in the figure) than at external angles (location *B*). This is due to the fact that cooling is less rapid at internal angles than at external angles.

● **Illustrative Problem 5.1**

Three pieces of castings have the same volume but different shapes. One is a sphere, one a cube, and the other a cylinder with height that is equal to the diameter. Which piece will solidify the fastest and which one the slowest?

SOLUTION.   According to Eq. (5.1) the solidification time will be

$$\text{Solidification time} \propto 1/\text{(surface area)}^2$$

since the volumes are equal. Assume that the volume is unity; the respective surface areas are then determined as follows:

Sphere: $V = \frac{4}{3}\pi r^3$,   hence $r = \left(\dfrac{3}{4\pi}\right)^{1/3}$

$$A = 4\pi r^2 = 4\pi\left(\frac{3}{4\pi}\right)^{2/3} = 4.84.$$

Cube: $V = a^3$, hence $a = 1$

$A = 6a^2 = 6$.

Cylinder: $V = \pi r^2 h = 2\pi r^3$, hence $r = \left(\dfrac{1}{2\pi}\right)^{1/3}$

$A = 2\pi r^2 + 2\pi rh = 6\pi r^2 = 5.54$.

Thus, the solidification times t will be:

$t_{\text{sphere}} = 0.043C$

$t_{\text{cube}} = 0.028C$

$t_{\text{cylinder}} = 0.033C$

Hence, the cube-shaped casting will solidify the fastest and the spherical-shaped casting will solidify the slowest. ●

5.2.4 SHRINKAGE AND POROSITY

Shrinkage or contraction in a casting is due to three factors:

a. Contraction of the liquid metal as it cools,

b. Contraction due to latent heat of fusion (solidification),

c. Contraction of the solidified metal as its temperature drops to room temperature.

For plain-carbon steel, these contractions are typically 1.6 per 100°F (56°C), 3.0, and 7.2% by volume, respectively. Thus, the greatest shrinkage is during cooling of the solidified metal.

The solidification contraction for various cast metals is given in Table 5.2. Note that gray cast iron undergoes expansion during solidification. This is because graphite

TABLE 5.2
SOLIDIFICATION CONTRACTION FOR VARIOUS CAST METALS. (After R. A. Flinn.)

METAL OR ALLOY	VOLUMETRIC SOLIDIFICATION CONTRACTION, %	METAL OR ALLOY	VOLUMETRIC SOLIDIFICATION CONTRACTION, %
Aluminum	6.6	70% Cu–30% Zn	4.5
Al–4.5% Cu	6.3	90% Cu–10% Al	4
Al–12% Si	3.8	Gray iron	Expansion to 2.5
Carbon steel	2.5 to 3	Magnesium	4.2
1% Carbon steel	4	White iron	4 to 5.5
Copper	4.9	Zinc	6.5

has a relatively high specific volume and, when the carbon in the liquid solution precipitates as graphite flakes, it results in a net expansion.

5.2.5 POROSITY

The most undesirable effect of shrinkage is porosity. When the liquid metal is unable to reach the regions where solidification is taking place, a porous region can develop due to the shrinkage of the solidified metal (Fig. 5.9). Micropores can also develop when the liquid metal solidifies and shrinks between dendrites and between dendrite branches.

Porosity due to shrinkage can be reduced or even eliminated by various means. Casting processes must provide for an adequate supply of liquid metal so that cavities

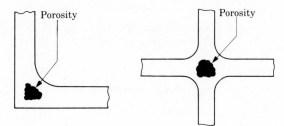

FIGURE 5.9 Typical examples of porosity in castings due to shrinkage of the metal during cooling and solidification. The walls of the pores are rough and angular. Porosity may also be caused by trapped gases in the solidified metal, in which case the walls of the pores (or micropores) are smooth. See also Fig. 5.12(b).

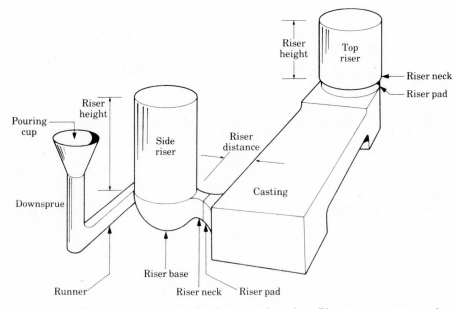

FIGURE 5.10 Schematic illustration of a riser-gated casting. Risers serve as reservoirs of molten metal, supplying additional material to the casting that is undergoing shrinkage during solidification (see Table 5.2); otherwise the casting will be incomplete. *Source:* American Foundrymen's Society.

due to shrinkage during cooling can be avoided. This is best done by *risers* or feeders, which are appendages to the casting. They serve as reservoirs, supplying liquid metal to prevent the formation of pores in the casting (Fig. 5.10).

In order to reduce *piping* in the riser, the surface of the metal may be covered with an *insulating compound* or with an *exothermic mixture* (Fig. 5.11) known as *hot tops*. The latter supplies heat to the riser and prevents its rapid cooling. In addition to risers, other methods for eliminating porosity are by external or internal *chills* (Fig. 5.12) to solidify the metal rapidly, and by changes in design to increase the rate of cooling. The latter method is a means of controlling thermal gradients and obtaining directional solidification.

With alloys, porosity can also be reduced or eliminated by methods that make the mushy region in the casting (i.e., liquid + solid zone $[L + S]$ in Fig. 5.5) as narrow as

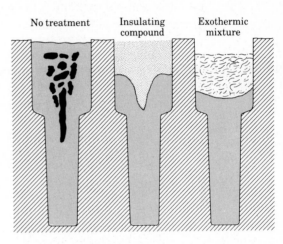

No treatment   Insulating compound   Exothermic mixture

FIGURE 5.11 The use of an insulating compound, or an exothermic (heat producing) mixture, to reduce piping in the riser due to shrinkage. Insulating materials may be refractories or ceramics (see Fig. 11.15). Exothermic materials are typically mixtures of iron oxide and powdered aluminum, such as those used as a source of heat in thermit welding, described in Section 12.5.1. An electric arc may also be used to provide heat to the top.

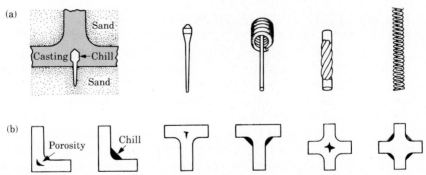

(a) Sand / Casting / Chill / Sand

(b) Porosity / Chill

FIGURE 5.12 Various types of (a) internal and (b) external chills used in casting to eliminate porosity due to shrinkage. Chills increase the cooling rate in the vicinity that they are placed, and thus avoid formation of porosity. They are generally placed in regions, such as intersections as shown, where there is a larger volume of material to cool relative to the volume of material in adjacent regions.

possible. This can be done by making the temperature gradient steep, by using mold materials with high thermal conductivity and by selecting casting alloys with low thermal conductivity and a short solidification-temperature range.

5.2.6 GASES AND POROSITY

Liquid metals have much greater solubility for gases than solids because solubility undergoes an abrupt reduction when solidification occurs. Consequently, when a metal begins to solidify the dissolved gases are expelled from the solution and accumulate in regions of existing porosity, such as in interdendritic areas, or cause microporosity.

Whether microporosity is a result of shrinkage or is due to gases is not easily determined. Generally, if the porosity is spherical and has smooth walls, it is due to gases. If the walls are rough and angular, it is likely to be due to shrinkage between dendrites. Gross porosity, such as those shown in Fig. 5.9, are due to shrinkage.

Porosity is detrimental to the mechanical properties and surface finish of castings and may make them permeable (thus affecting pressure tightness of a cast pressure vessel). It is thus necessary to control or eliminate gases. Hydrogen and nitrogen are commonly found in castings, the former being soluble, to various degrees, in all metals. Dissolved gases may be removed from the molten metal by such methods as flushing or purging with an inert gas, or by melting and pouring the metal in a vacuum. The effect of porosity on ductility is shown in Fig. 3.43.

A vacuum is useful because the amount of dissolved gas is proportional to the square root of the partial pressure of the gas above the melt. If the gas is oxygen, another technique is to introduce elements into the melt to deoxidize the metal. Thus, steel is deoxidized with aluminum or silicon, and copper-base alloys with phosphorus copper.

5.3 FLUID FLOW AND DESIGN CONSIDERATIONS

The introduction of molten metal into a mold and its solidification are basically fluid-flow problems, involving heat transfer, behavior of the liquid metal in the passages of the mold, and the role of thermal gradients on the type of flow. Even before it reaches the mold cavity, the molten metal must be handled carefully to avoid picking up any gas, and to avoid the formation of oxides on surfaces exposed to the environment, or to avoid the introduction of any impurities into the molten metal.

In order to appreciate the importance of these effects, let us first review a basic casting system, as shown in Fig. 5.13. The molten metal is poured through a *pouring basin* or *cup*. It then flows through the *sprue* and *runners* into the mold cavity. As stated earlier, risers serve as reservoirs of molten metal to supply the metal necessary to prevent shrinkage. Although such a gating system appears to be relatively simple, successful casting requires careful design and control of the solidification process to ensure adequate fluid flow in the system.

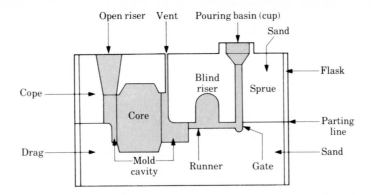

FIGURE 5.13 Schematic illustration of a sand mold showing the gating system, risers, and the core. Cores, which are made of aggregates such as compacted sand, are used to produce internal cavities and recesses in castings, thus reducing subsequent machining operations. Cores are destroyed and removed after the casting has solidified.

Fluid Flow

The molten metal must be introduced to the gating system in such a way that turbulence in the fluid flow is minimized or eliminated, otherwise, mold erosion and gas pickup may occur. *Aspiration* is another problem; it involves air entrapment due to turbulent flow. Aspiration occurs particularly at the sprue and can be reduced substantially by tapering the sprue. Thus, the cross-section of the sprue is made smaller toward the bottom to compensate for the increase in velocity due to gravity of the molten metal. Likewise, the runners, which are channels to the mold cavity, should be streamlined and should be free of sharp corners or sharp changes in cross-section in order to avoid further turbulence.

In ordinary gating systems, *Reynolds* numbers have been found to range between 2000 and 20,000. (In fluid flow, the Reynolds number is the ratio of the inertia to the viscous force. The higher this number the greater the tendency for turbulent flow, and the lower the number the greater the tendency for laminar flow.) Since a Reynolds number of 2000 generally indicates the transition from laminar to turbulent flow, it is difficult to eliminate turbulent flow in casting. However, with proper design of the gating system this turbulence can be minimized.

Another consideration in fluid flow is the elimination of *slag* or *dross*. These are formed as a result of oxidation and by impurities rising to the surface of the molten metal entering the mold. There are various ways to clean the molten metal, such as the use of filters and screens in the pouring of nonferrous alloys. These are generally made of sheet metal, mica, or fiberglass. The proper location and placement of these screens is important for successful cleaning. The slag or dross can also be skimmed (*skim bobs*, Fig. 5.14) with properly designed gating or pouring basins.

Heat Flow

In addition to the basic fluid-flow problems, another important consideration is the heat flow during the complete cycle from pouring to solidification of the casting, and

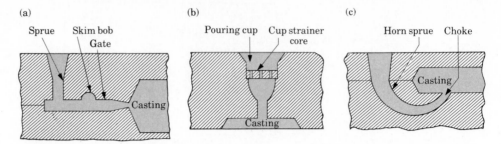

FIGURE 5.14   Types of gates and their features in cleaning the molten metal entering the mold. (a) Parting gate with skim bob to trap slag. (b) Top gate with strainer. (c) Bottom gate with horn sprue. Gating design is important in controlling the rate and turbulence in the molten metal being poured, and the temperature gradient within the casting. *Source:* After L. E. Doyle.

its cooling to room temperature. Heat flow at different locations in the system is a complex phenomenon; it depends on many factors relating to the materials of the casting and the mold and process parameters.

For instance, in the casting of thin sections the metal flow rates must be sufficiently high to avoid premature chilling and solidification. On the other hand, the flow rate must not be so high as to cause excessive turbulence with its detrimental effects on the casting process. The effects of some of these variables are described in Section 5.2 as they relate to the quality of castings.

Fluidity

A commonly used term that combines fluid flow and heat flow characteristics is *fluidity*. This term indicates the capability of the molten metal to flow into the cavities of the system before freezing. Fluidity is related to *viscosity*. However, because comparisons with other metals and alloys is highly desirable, several tests have been developed to quantify fluidity. One example of such tests is shown in Fig. 5.15, where the molten metal is made to flow along a channel at room temperature. The distance of flow of the metal after it stops due to solidification is a measure of fluidity. Obviously this length is a function of the thermal properties of the metal and also of the mold. Furthermore, the length is a function of the nature of the interface between the metal and mold walls. In spite of these considerations, such tests are useful and simulate actual casting situations to a reasonable degree.

From the foregoing, it is apparent that successful casting design (particularly those aspects related to gating and risering) requires careful consideration of many factors, and that fluid flow and heat flow are intimately interrelated. Although many of the present casting practices are based on experience, a great number of analytical and experimental studies have been conducted. These studies are particularly useful in predicting properties and behavior of castings and casting defects. Some details are discussed in Section 5.11, with specific examples.

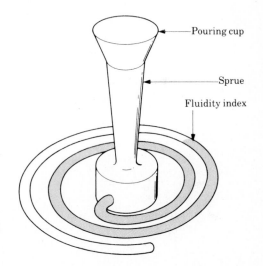

Pouring cup

Sprue

Fluidity index

FIGURE 5.15 A test method for fluidity using a spiral mold. The fluidity index is the length of solidified metal in the spiral. The greater the length, the greater the fluidity of the molten metal. Thus fluidity is akin to viscosity.

5.4 FURNACES AND MELTING PRACTICE

A great variety of furnaces is available in foundries to meet different requirements. The proper choice of a furnace depends on a number of factors. These include:

a. The composition and melting point of the alloy to be melted,

b. Control of the atmosphere to avoid contamination of the metal,

c. Capacity and the rate of melting and delivery,

d. Environmental considerations, such as air pollution and noise,

e. Power supply and availability and cost of fuels, and

f. Economic considerations, such as initial cost, and maintenance and operational costs.

The type of furnaces available for melting metals are shown in Fig. 5.16. The furnaces most commonly used in foundries today are: *cupolas, electric-arc furnaces,* and *induction furnaces.* Furnaces are made with a variety of features to meet specific requirements; choice of a furnace is governed by the factors listed above. Thus, cast iron is melted mostly in cupolas, although electric-arc furnaces are also used. Steel is generally melted in electric-arc furnaces. Other metals and alloys can be melted in a variety of furnaces.

The selection of a furnace depends also on the type of operation, i.e., batch versus continuous melting. In a batch-melting operation, a crucible or hearth of certain capacity is used. In continuous operations, melting and tapping can be done continuously in a cupola.

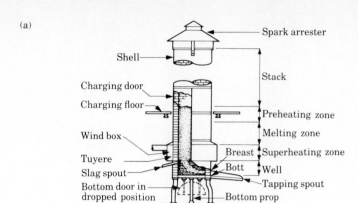

(a)

Spark arrester

Shell

Charging door

Charging floor

Stack

Preheating zone

Melting zone

Wind box

Superheating zone

Tuyere

Breast

Slag spout

Bott

Well

Bottom door in dropped position

Tapping spout

Bottom prop

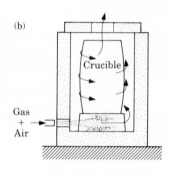

(b)

Crucible

Gas + Air

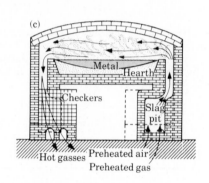

(c)

Metal

Hearth

Checkers

Slag pit

Hot gasses

Preheated air

Preheated gas

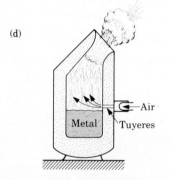

(d)

Air

Metal

Tuyeres

(e)

Refractory cement

Copper induction coils

Molten metal

Crucible

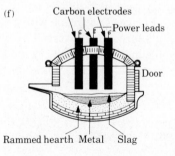

(f)

Carbon electrodes

Power leads

Door

Rammed hearth   Metal   Slag

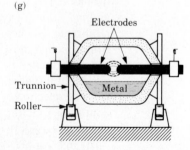

(g)

Electrodes

Trunnion

Metal

Roller

FIGURE 5.16 Types of melting furnaces. (a) Cupola. (b) Crucible. (c) Open-hearth. (d) Converter. (e) Induction. (f) Direct-arc. (g) Indirect-arc. The selection of a furnace for a particular application depends on many technical and economic factors. *Source:* R. A. Flinn, *Fundamentals of Casting*, Reading, Mass.: Addison-Wesley, 1963.

Melting

Melting practice is an important aspect of casting processes as it has a direct bearing on the quality of the castings. Furnaces are charged with *melting stock* composed of metal, alloying elements, and such elements as flux. The metal charge may be composed of commercially pure primary metals and secondary metals that are re-melts of scrap and clean scrapped castings; gates and risers may also be included in the charge.

If the melting points of the alloying elements are low, pure *alloying elements* are added to obtain the desired composition in the melt. If the melting point is high, the alloying element does not mix readily with the low-melting point metals. In this case, lower-melting-point *master alloys* or *hardeners* are used. These generally consist of higher-concentration alloys of one or two of the alloying elements needed. Master alloys should be selected so that they do not have much tendency for gravity segregation. In alloying during melting, the lower-melting-point alloy is generally melted first, then the high-melting-point alloy.

In order to protect the surface of the molten metal against atmospheric reaction and contamination, and also to refine the melt, it must be insulated against heat loss. This is generally done by covering the surface, or mixing the melt, with compounds that form a slag. For cast-steel melts, the composition of slag includes CaO, SiO_2, MnO, and FeO. The slag can be made acid (high SiO_2 content) or basic reducing (high CaO content).

Fluxes are inorganic compounds, such as limestone and dolomite, with other secondary fluxes such as sodium carbonate and calcium fluoride (for cast iron), and borax–silica mixtures (for copper alloys.) These compounds refine the melt by removing dissolved gases and various impurities.

Tapping and pouring the molten metal require careful handling in order to reduce contamination by gases and nonmetallic impurities. The tools, such as ladles, must be clean and free of moisture. Pouring must be done smoothly, as described in Section 5.3 on fluid flow.

5.5 CASTING OF INGOTS

An *ingot* is a simple casting of a specific length and cross-section, and is generally designed for subsequent processing. The ingot is made in a mold that generally has a taper in order to facilitate the removal of the solidified metal. The bottom of the molds may be open or closed, and the taper may be such that the big end is down or up (Fig. 5.17).

Molds are usually made of cupola iron or blast-furnace iron (with 3.5 % carbon) or graphite. Massive iron molds can be made to absorb the heat from the casting. Graphite has a thermal conductivity three times that of steel and is much lighter, but is quite brittle. Thus it is more suitable for smaller ingots. Although iron molds are not cooled, because of their size, some molds, such as those used for casting copper-base alloys, are water cooled and have thin walls.

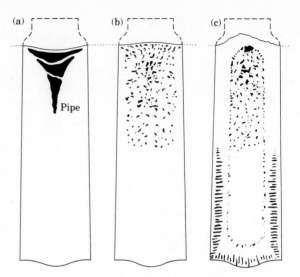

FIGURE 5.17 Typical solidification structures of steel ingots. The dark spots are oxide inclusions, or porosity (blowholes) due to evolved gases during solidification. (a) Killed steel. This is clean steel, with a sound structure, and is used for high-carbon grades of steel. There are no blowholes; however, pipes form on the top (Fig. 5.11), resulting in large cropping losses. (b) Semikilled steel, used for medium-carbon grade steel, as plates and structural members. (c) Rimmed steel, used for low-carbon grades. The surface of this ingot is composed of almost pure iron.

Although the simple shape of an ingot would suggest that casting it is a simple process, this is not the case, as can be appreciated from the discussion in Section 5.2. Unless controlled properly, an ingot can have many imperfections and also porosity; these could play a significant role in further deformation and processing of the ingot. The quality of ingots, especially steel ingots, has been studied extensively and will now be described briefly, particularly as it affects further processing of the ingot.

Killed Steel

Depending on the amount of gas that evolves during solidification, three types of steel ingots can be made: killed, semikilled, and rimmed. *Killed steel* is a fully *deoxidized* steel in which the dissolved oxygen in the liquid metal in the ladle is made to react by the addition into the melt of elements such as aluminum, silicon, ferrosilicon, ferromanganese, and other deoxidizers. These elements have an affinity to oxygen and form metallic oxides; the deoxidation product with aluminum-killed steels is Al_2O_3 and with silicon it is SiO_2.

These oxide inclusions, if sufficiently large, float out of the molten bath and adhere to or are dissolved in the slag. Thus a fully killed steel is free of porosity due to gases. It is also free of *blowholes*, these being large spherical holes near the surface of the ingot. However, an ingot of this type develops a *pipe* (shrinkage cavity) as seen in Fig. 5.17(a). As shown in Fig. 5.11, this defect can be reduced by placing an exothermic compound on the top of the ingot (hot top).

Semikilled Steel

Semikilled steels are *partially deoxidized*. They contain some porosity (generally in the upper central section of the ingot because of gravity) but little or no pipe, thus reducing the scrap (Fig. 5.17b). *Piping* is reduced because it is compensated for by the presence of porosity in that region.

Rimmed Steel

In *rimmed steels*, which generally have a low carbon content, the evolved gases are only partially controlled by the addition of elements such as aluminum. The gas (carbon monoxide) forms blowholes in the mold along the outer rim of the ingot, hence the term *rimmed* (Fig. 5.17c). Blowholes near the surface are called *primary blowholes*; those farther away from the surface are called *secondary blowholes*.

Blowholes are generally not objectionable unless they break through the outer skin of the ingot. Because they are not exposed to oxidation by the atmosphere, these blowholes can later be closed and welded during plastic deformation at elevated temperatures, such as by rolling and forging. Thus, a continuous and homogeneous mass of metal can be obtained. However, if the casting is used without the benefit of plastic deformation, then these porosities can be detrimental to the quality of the casting. (See Fig. 3.43.)

The rimming action in ingots can be controlled by *capping* the ingot. This consists of sealing the top of the mold with a cap to increase the pressure and inhibit the formation of blowholes.

Characteristics

The advantages and limitations of the three types of steels described above can be summarized as follows.

Killed steels have a homogeneous structure with good, uniform properties. They require special techniques and have extensive pipes, and result in relatively poor surface finish after processing.

Semikilled steels have less pipes and residual deoxidizers, and are economical to produce. However, they require plastic working to reduce or eliminate blowholes and their surface finish is poor. Alloy choice is limited and requires close control of deoxidizers.

Rimmed steels have little or no piping and have a ductile skin with good surface finish. However, primary blowholes may break through the skin if they are not controlled properly. These steels also have limitations as to composition of alloys, and segregation of impurities and inclusions toward the center of the ingot.

The casting of ingots of nonferrous metals and alloys is somewhat less complex. The relevant details are given in Section 5.3 on fluid flow.

5.5.1 CONTINUOUS CASTING

The casting of ingots is essentially a *batch* process and the ingot has to be stripped from the mold after solidification. Furthermore, there are problems such as piping, structural variation, and accumulation of slag. These problems can be alleviated and productivity increased by *continuous casting* processes. These processes were first developed for casting of nonferrous metal strips. They are now also used for steel production.

Three techniques for continuous casting are shown in Fig. 5.18. Basically, the molten metal travels through water-cooled molds (cooling chambers) and begins to

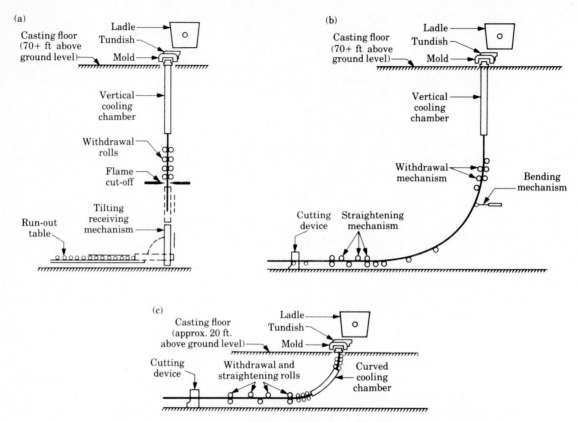

FIGURE 5.18   Three methods of continuous casting of steel. Developed in the early 1950s, continuous casting is an economical method for producing semifinished products such as billets, blooms, ingots, slabs, and tubes of various cross-sections. Unlike other casting processes, the length of the product is not related to the dimensions of the mold. *Source:* Courtesy of U.S. Steel Corporation.

solidify as it travels along a path supported by rollers. The cooling rate is such that the casting develops a solidified skin to support itself. Additional cooling may be provided by water sprays. The molds are generally made of copper with graphite-type or similar lubricants to reduce friction and adhesion at the mold–metal interface. The molds may also be vibrated for this purpose.

As seen in Fig. 5.18, the casting may eventually be cut into desired lengths. The continuously cast metal may also be fed directly into a rolling mill (see Section 6.10) for further reduction in thickness and for shape changes. Square, rectangular, and round shapes of aluminum, copper, and various steels can be economically cast by such techniques.

5.6 CASTING OF SHAPES: EXPENDABLE MOLD

Although ingot casting is an important primary process, an important objective in casting is to produce shapes at or near the final shape of the product. As outlined briefly in Section 5.1, a large variety of shapes can be cast from many different metals and alloys.

Casting processes can be classified according to the molding process, the molding material, and the method of feeding the mold. These processes can also be classified as *expendable-mold* casting and *permanent-mold* casting methods. Expendable molds are made of materials such as sand and various aggregates. Once the casting has solidified, the molds are simply broken up and destroyed to remove the casting. Permanent molds are described in Section 5.7.

Many casting methods employ expendable molds. Among these are four that utilize sand: green-sand molding, shell molding, silicate-bonded sand molds hardened with carbon dioxide, and bonded or unbonded sand with expendable polystyrene patterns. Other methods employing different mold and pattern materials are plaster molding, ceramic molding, and investment casting.

5.6.1 SAND CASTING

This is the traditional method of casting metals in sand molds and has been used for centuries. The majority of castings today are still produced by this method, Fig. 5.13.

Sand is inexpensive and is suitable as mold material because of its resistance to high temperatures. It usually contains clay composed of aluminosilicates, but is also mixed with other clays, such as bentonite and fireclay, for enhanced properties. Important factors in the selection of sand molds are: grain size and shape of the sand, moisture content, content of clay and other additives, hardness, strength, permeability, and collapsibility.

Sand and Sand Molds

The relationship of the strength of sand and its permeability to factors such as grain size and shape, and moisture are shown in Fig. 5.19. *Permeability* is the ability of the sand mold to allow the escape of air, gases, and steam during the casting process. *Collapsibility* is the ability of the sand mold to collapse when the metal casting cools and shrinks in the mold. In this way, hot tearing and cracking of the casting is avoided (Section 5.11).

There are three basic types of sand molds: green-sand mold, skin-dried mold, and dry sand mold. The most common mold material is *green molding sand*. (*Green* means that the sand mold is moist during casting.) This is a mixture of sand (SiO_2), clay (aluminosilicate) as a binder, and water. Other materials are added to impart various properties, such as strength at elevated temperatures, improved thermal stability, and improved surface finish of the castings. Green sand molds are the least

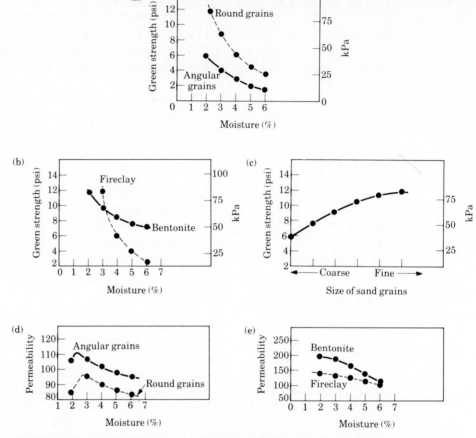

FIGURE 5.19 Effect of moisture, type of clay binder, and shape of sand particles on the strength and permeability of molding sand. Permeability is the ability to allow the escape of gases, air, and steam during solidification; otherwise these elements adversely affect the quality of the castings. *Source:* After L. E. Doyle.

expensive of the sand molds and have the advantages of low distortion; they require less time to reuse, and they have good collapsibility.

Skin-dried molds are usually used for large castings. The surfaces are dried, either in air or with torches, to depths greater than 0.5 in. (12 mm), prior to pouring.

Dry-sand molds are oven dried and are used for medium to large castings. They are stronger than green-sand molds and impart better dimensional accuracy and surface finish. However, the production rate is slower, distortion of the mold is greater, and the castings are more subject to hot tearing due to the lower collapsibility of the mold.

The sand mixture is compacted around the pattern by machines that compress this mixture to between 20 and 80% of its loose bulk density. These molding machines use various motions such as squeezing, jolting, slinging, and blowing.

Patterns

A typical mold for sand casting and the procedures involved are shown in Fig. 5.20; the functions of the various components have already been indicated in Section 5.3. The two halves of the mold, called the *cope* and the *drag*, are made using patterns to obtain the desired shapes. Patterns may be made of wood, plaster, plastic, or metal, depending on the size, shape, dimensional accuracy, and the quantity of castings required. The greater the number of castings desired, the stronger and more durable the pattern material should be. Patterns may also be made of a combination of materials to reduce wear in critical sections of the pattern. Patterns are coated with a parting agent to facilitate their removal after the molds are made.

Pattern design is an important part of the total casting process as it must make provisions for metal shrinkage, ease of removal of the pattern from the sand, and allow favorable fluid flow in the mold cavity. For castings with internal cavities, such as a valve body, *cores* (Figs. 5.13 and 5.20) are employed; these are made of aggregates and are removed from the castings during shakeout and subsequent processing. Core sand should have basically the same properties as molding sand. They are strengthened with bonding agents such as oils and corn flour.

After the two halves of the mold have been shaped and the core has been placed in position (and supported with *chaplets*), the two halves are closed and clamped and weighted down to prevent separation of the mold sections when the metal is poured into the mold cavity.

The design of the *gating system* (pouring basin, sprue, and runners) is important for the proper delivery of the molten metal into the mold cavity (Section 5.3). Turbulence must be minimized, air and gases must be allowed to escape (by means such as vents), and proper temperature gradients must be established to minimize shrinkage and porosity. The design of risers is also important in supplying the necessary liquid metal during the solidification of the casting. After the casting has solidified the mold is broken up and the casting is removed.

The pouring basin may also serve as a riser, as shown in Fig. 5.21(a) for a railroad wheel casting. The figure also illustrates the use of chills.

Almost all metals can be sand cast. As expected, surface finish will reflect the materials used for the mold, and tolerances are not as good as with other casting processes. However, intricate shapes can be cast by this process. Sand casting can be economical for relatively small production runs, and equipment costs are generally low (Table 5.3).

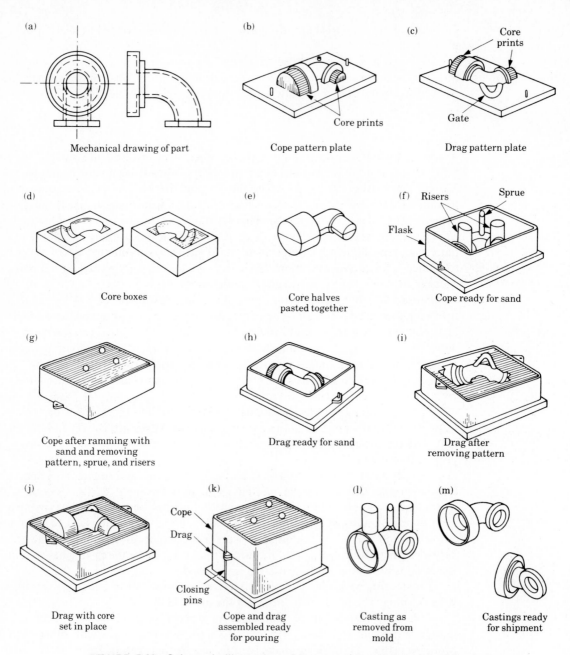

(a) Mechanical drawing of part

(b) Core prints — Cope pattern plate

(c) Core prints — Gate — Drag pattern plate

(d) Core boxes

(e) Core halves pasted together

(f) Risers — Sprue — Flask — Cope ready for sand

(g) Cope after ramming with sand and removing pattern, sprue, and risers

(h) Drag ready for sand

(i) Drag after removing pattern

(j) Drag with core set in place

(k) Cope — Drag — Closing pins — Cope and drag assembled ready for pouring

(l) Casting as removed from mold

(m) Castings ready for shipment

FIGURE 5.20 Schematic illustration of the sequence of operations for a typical part made by sand casting. Sand casting is one of the oldest manufacturing processes. Unlike other casting processes, there is virtually no limit to the size of a sand cast part. *Source:* Courtesy of Steel Founders' Society of America.

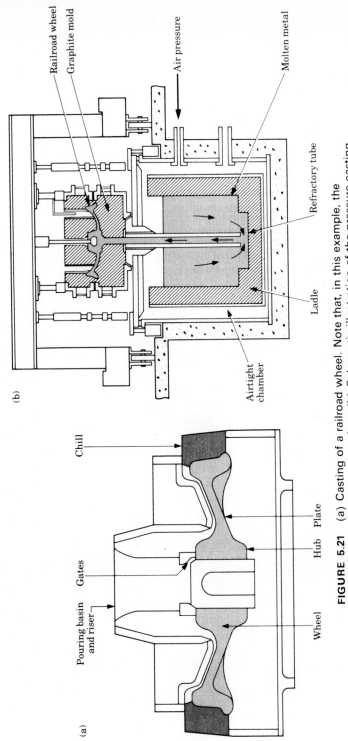

FIGURE 5.21 (a) Casting of a railroad wheel. Note that, in this example, the pouring basin also serves as a riser. (b) Schematic illustration of the pressure casting (pressure pouring, or low-pressure casting) process for manufacturing steel railroad wheels using graphite molds. (See also Fig. 6.27 for the forging methods of manufacturing railroad wheels.) Aluminum engine blocks have also been cast by this process. *Source*: Courtesy of Griffin Wheel Division of Amsted Industries, Inc.

TABLE 5.3
GENERAL CHARACTERISTICS OF CASTING PROCESSES

PROCESS	MATERIALS CAST	SIZE, kg		SURFACE FINISH μin, RMS	POROSITY*	SHAPE COMPLEXITY*
		MINIMUM	MAXIMUM			
Sand	All	0.05	No limit	100–700	5	1–2
Shell	All	0.05	100+	50–150	5	2–3
Plaster	Nonferrous (Al, Mg, Zn, Cu)	0.05	50+	30–70	3	1–2
Investment	All (High-melting pt.)	0.1	100+	50–125	5	1
Permanent mold	All	0.5	300	100–250	2–3	3–4
Die	Nonferrous (Al, Mg, Zn, Cu)	<0.05	50	40–100	1–2	3–4
Centrifugal	All	—	5000+	100–500	1–2	3–4

* Relative rating: 1 best, 5 worst. Cost: L, low; M, medium; H, high.

5.6.2 SHELL MOLDING

In this process, the molding material consists of sand and 2.5 to 4.0% *thermosetting resin* (Section 10.3). The pattern is made of metal heated to between 350 and 700°F (175 and 370°C), coated with a parting agent and then coated with the sand mixture, such as by placing it on a dump box. To complete the curing process, the assembly is then placed in an oven for a short time. The heated metal pattern effects the cure on the pattern-face side, while the curing of the back side of the shell is carried out in an oven for a short time. The shell is then removed from the pattern. Two half shells are made in this way and are then clamped or glued together in preparation for pouring.

Shell molding produces castings with better surface finish and tolerances than sand casting. Complex shapes can be produced with less labor and, although it is a more expensive process, it can be economical because less machining is required and better tolerances are obtained.

5.6.3 CARBON DIOXIDE PROCESS

The mold material in this process is a mixture of sand and 1.5 to 6% *sodium silicate* as the *binder* for the sand. This mixture is then packed around the pattern and is hardened by blowing CO_2 gas through the mixture.

5.6.4 FULL-MOLD PROCESS

When only a few castings are required the expense of making a pattern can be an important factor. In the *expended polystyrene process* (also known as *evaporative casting*), the pattern is made of polystyrene, around which the sand mold is formed.

SECTION THICKNESS, mm		COSTS			PRODUCTION RATE, Pc/hr	SCRAP LOSS*	DIMENSIONAL ACCURACY*
MINIMUM	MAXIMUM	DIE	EQUIPMENT	LABOR			
3	No limit	L	L	L–M	<20	5	5
2	—	L–M	M–H	L–M	<50	2	2
1	—	L–M	M	M–H	<10	2	2
1	75	M–H	L–M	H	<1000	1	1
2	50	M	M	L–M	<60	2	2
0.5	12	H	H	L–M.	<200	2	1
2	100	M	H	L–M	<50	1	3

The sand may be bonded or unbonded. Then, without removing the pattern, the molten metal is poured into the mold; this immediately vaporizes the polystyrene, thus forming the mold cavity. Complex castings of various sizes can be made economically by this process.

5.6.5 PLASTER MOLDING

In this process, the mold is made of *plaster of Paris* (gypsum), with the addition of materials such as asbestos fiber, talc, and silica flour to improve strength and control the time required for setting of the plaster. These components are mixed with water and the resulting *slurry* is poured over the patterns. After setting, the pattern is removed and the mold is dried at around 400°F (200°C). The mold halves are then assembled and the molten metal is poured into the mold to make the casting.

Since there is a limit to the maximum temperature that the plaster mold can withstand (generally about 2200°F [1200°C]) this process is used only for casting of aluminum, magnesium, zinc, and some copper-base alloys. The plaster molding process produces fine details with good surface finish. Because of the good dimensional accuracy obtained (due to the lower shrinkage of the mold), this process and the ceramic-mold and investment casting processes (described below) are known as *precision casting*.

5.6.6 CERAMIC-MOLD CASTING

This is a process similar to plaster molding, except that it uses materials suitable for high-temperature applications. The slurry is a mixture of fine-grained zircon

($ZrSiO_4 $), aluminum oxide ($Al_2O_3 $), and fused silica ($SiO_2 $). These refractory materials are mixed with *bonding agents* and applied to the pattern. After setting, the ceramic shells are removed, backed by fireclay to give strength, and assembled and used as molds.

The resistance to high temperatures of the molding materials allow these molds to be used in casting ferrous and other high-melting-point alloys with good precision and surface finish, and wide range of sizes. However, the process is somewhat expensive.

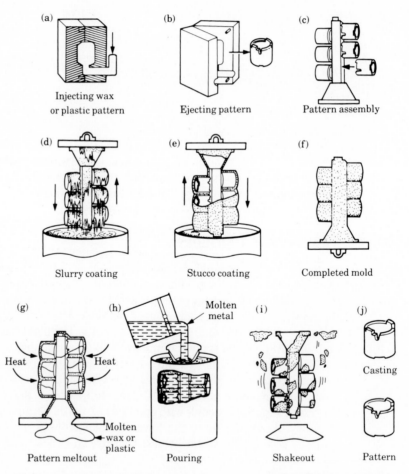

FIGURE 5.22 Schematic illustration of the investment casting (lost wax) process. This process, which dates back to 4000–3000 B.C., produces castings from a variety of materials and with very fine detail. *Source:* Courtesy of Steel Founders' Society of America.

5.6.7 INVESTMENT CASTING

This is an old and commonly used casting process (also known as the *lost-wax process*) that uses a pattern made of *wax*, although *plastics* can also be used. After the pattern is made (Fig. 5.22) it is dipped into a slurry of refractory material, such as very fine silica and liquids including water, ethyl silicate, and acids. Once this initial coating is dried, it is recoated repeatedly to increase thickness. The mold sections are then dried in air and heated to a temperature of 200 to 300°F (90 to 150°C) in an inverted position to melt out the wax.

The mold is then fired to 1200 to 1900°F (650 to 1050°C), depending on the metal to be cast, to drive off the water of crystallization (chemically combined water). After the metal is poured and the casting has solidified, the mold is broken up and the casting is removed. A number of patterns can be joined to make one mold, called a *tree*, thus increasing production rate.

Although this process is costly, it is suitable for casting high-melting-point alloys with good surface finish and close tolerances. It is capable of producing intricate shapes.

A variation of this process is *ceramic-shell casting*, which uses the same type of pattern dipped in ethyl silicate gel and then into a fluidized bed of fine-grained fused silica. It is then dipped into coarser grains for a build-up of additional coatings and thickness to withstand the thermal shock of pouring. The rest of the procedure is similar to investment casting. This process is economical and is used extensively for precision casting.

5.7 CASTING OF SHAPES: PERMANENT MOLD

In all the processes described in the preceding sections, the mold materials were capable of withstanding the temperatures of molten metal, but the mold had to be broken up in order to remove the casting. It is, of course, possible to make molds that do not have to be destroyed after each casting but still can sustain the temperatures of the molten metal. With these molds, high-quantity production can be obtained.

Permanent molds are made by properly selecting mold materials that have sufficient strength and hardness at high temperatures, are resistant to erosion due to fluid metal flow, and are resistant to thermal cycling (thermal fatigue due to continuous heating and cooling each time a casting is poured, solidified, and removed from the mold).

Because metal molds are better heat conductors than expendable molds, the solidifying metal is subjected to a higher rate of cooling, which in turn affects the grain structure and distribution within the casting. This point must be considered when planning castings by permanent-mold techniques. Furthermore, unlike sand and other aggregate mold materials, metal molds do not allow the escape of gases, hence proper venting is essential to avoid gas porosity.

Several casting processes using permanent molds are described below.

5.7.1 PERMANENT-MOLD CASTING

In this process, two halves of a mold are made from such materials as cast iron, steel, bronze, graphite, or refractory metal alloys; the choice depends on the metal to be cast. The mold cavity and gating system are machined. For parts with internal cavities, cores made of metal or an aggregate (which can be broken up) are placed in the mold. In order to increase mold life, the surfaces of the mold cavity are usually coated with refractory slurry or graphite. These coatings also serve as parting compounds and make the removal of the solidified casting easier. Mechanical ejectors may be needed for complex castings. The parting compounds also serve as thermal barriers and control the rate of cooling of the casting.

The molds are clamped together by mechanical means and the molten metal is poured through the gating system. After solidification, the molds are opened and the casting is removed. The process is generally used for nonferrous alloys and cast iron. Steel can be cast in graphite molds as well as metal molds. This process has advantages such a good surface finish, close tolerances, and high production rates. Although equipment cost can be somewhat high, labor costs are relatively low because the process can be mechanized.

Permanent-mold casting can also produce hollow castings. This can be appreciated by reviewing Fig. 5.8, where we note that a solidified skin first develops in a casting and that this skin becomes thicker with time. Thus for ornamental and decorative artifacts, the partly solidified casting can be removed from the molds, inverted, and the remaining molten metal poured out. In this way, only the necessary external skin is preserved. This process is known as *slush-casting*.

In the two processes described above the molten metal flows into the mold cavity by gravity. A more advanced method is the *pressure casting* (pressure *pouring*) process (Fig. 5.21b). The molten metal is forced upward into the mold cavity by air pressure, which is maintained until the casting has solidified. An expendable plunger is also used to seal the hole through which metal is being fed. This process is generally used for high-quality casting of steel (railroad wheels and ingots are typical products). The molds are made of graphite.

5.7.2 DIE CASTING

This process is a further development of permanent-mold casting. The molten metal is forced into the die cavity at pressures ranging between 100 and 100,000 psi (0.7 to 700 MPa). This also means that the dies will tend to part unless clamped together. Thus, die-casting machines are rated according to this clamping force. There are two basic types of die-casting machines: hot-chamber and cold-chamber.

The *hot-chamber* process (Fig. 5.23) involves a piston, which traps a certain volume of molten metal and forces it into the die cavity at pressures of about 2000 psi (15 MPa). Low-melting-point alloys such as lead, tin and zinc are cast by this process. With higher-melting-point alloys there would be problems with the life of various components of the machine.

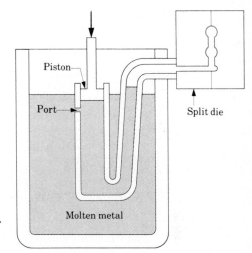

FIGURE 5.23 Hot-chamber die-casting process, generally used with low-melting-point alloys such as lead, tin, and zinc. The pressure of the liquid metal in the die may be on the order of 2000 psi (15 MPa).

In the *cold-chamber* process (Fig. 5.24), the injection cylinder (shot chamber) is filled by ladle with molten metal and the metal is forced into the die cavity at pressures as much as 10 times that in the hot-chamber process. High-melting-point alloys of aluminum, magnesium, and copper are cast by this process.

Die casting has the advantages of high production rates with good strength, high quality of parts, and good surface details and dimensional accuracy. Equipment costs are somewhat high (Fig. 5.25) but labor costs are generally low. One part made by die casting is shown in Fig. 5.26. Properties and typical applications of die casting alloys are given in Table 5.4.

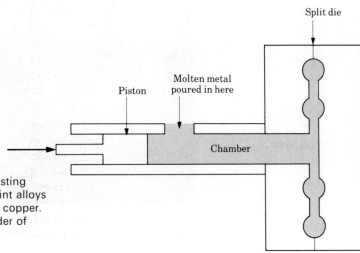

FIGURE 5.24 Cold-chamber die-casting process, used with high-melting-point alloys such as aluminum, magnesium, and copper. The pressures are as much as an order of magnitude higher than those in the hot-chamber process.

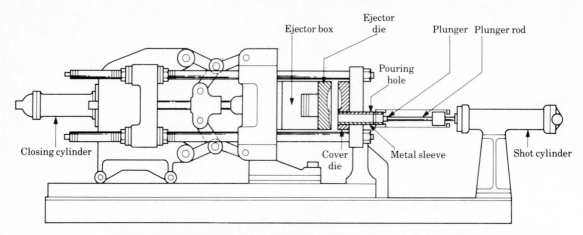

FIGURE 5.25 Schematic illustration of a die-casting machine. These machines are large compared to the size of the casting, because of the high forces required to keep the two halves of the die closed under the pressure of the molten metal in the die cavities.

FIGURE 5.26 An example of a housing made by die casting. Note the fine detail and complex shape of the part that has been cast in one piece. *Source:* Courtesy of J. J. Kolar, American Die Casting Institute.

TABLE 5.4
PROPERTIES AND TYPICAL APPLICATIONS OF COMMON DIE CASTING ALLOYS
(Data from American Die Casting Institute).

ALLOY	ULTIMATE TENSILE STRENGTH, psi $\times 10^3$	YIELD STRENGTH, psi $\times 10^3$	ELONGA- TION IN 2 in., %	APPLICATIONS
Aluminum 380 (3.5Cu-8.5Si)	46	23	2.5	Appliances, automotive
A380 (3.5Cu-8.5Si)	47	23	3.5	components, electrical motor frames and housings, functional and decorative parts
13 (12Si)	43	21	2.5	Complex shapes with
A13 (12Si)	42	19	3.5	thin walls, parts requiring strength at elevated temperatures
Brass 858 (60Cu)	55	30	15	Plumbing fixtures, lock hardware, bushings, ornamental castings
Magnesium AZ91B (9Al-0.7Zn)	34	23	3	Power tools, automotive parts, sporting goods
Zinc No. 3 (4Al)	41	—	10	Automotive parts, office equipment, household utensils, building hardware, toys
5 (4Al-1Cu)	47	—	7	Appliances, automotive parts, building hardware, business equipment

5.8 CENTRIFUGAL CASTING

As the name implies, in this process inertial forces due to rotation are utilized to distribute the molten metal to the mold cavities. There are three types of centrifugal casting: true centrifugal casting, semicentrifugal casting, and centrifuging.

In *true centrifugal* casting, hollow cylindrical shapes, such as pipes, are produced by the technique shown in Fig. 5.27, giving products with uniform quality. The refractory lining can be shaped so that pipes with various outer geometries can be cast, although the inner surface remains cylindrical due to the uniform distribution of the molten metal by centrifugal forces.

An example of *semicentrifugal* casting is shown in Fig. 5.28, where parts with rotational symmetry are cast, such as a wheel with spokes.

In *centrifuging*, mold cavities (of any shape) are placed at a certain radius from the axis of rotation and the molten metal, poured from the center, is forced into the mold by centrifugal forces (Fig. 5.29).

Typical parts made by centrifugal casting are pipes, engine cylinders, large gun barrels, brake drums, and cylinder liners.

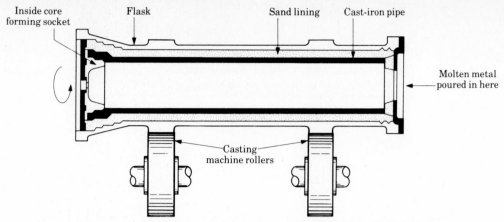

FIGURE 5.27 Schematic illustration of true centrifugal casting. Cast iron pipes of varying cross-section and lamp posts can be made by this process. Outside diameters may be on the order of a few feet. Due to centrifugal forces, the outer skin of the cast parts has higher purity than the inner regions, where the lower-density impurities and nonmetallic inclusions tend to segregate.

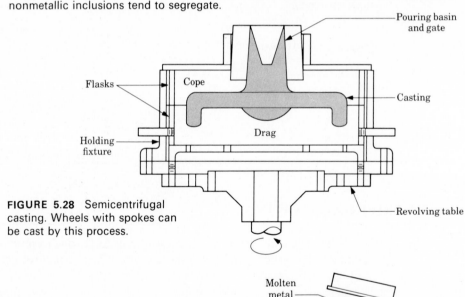

FIGURE 5.28 Semicentrifugal casting. Wheels with spokes can be cast by this process.

FIGURE 5.29 Schematic illustration of centrifuging. The molds are placed at the periphery of the machine and are filled with molten metal by the action of centrifugal forces.

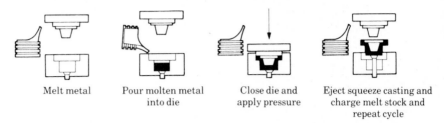

Melt metal | Pour molten metal into die | Close die and apply pressure | Eject squeeze casting and charge melt stock and repeat cycle

FIGURE 5.30 Production sequence for squeeze casting. The metal is melted by induction heating (schematically illustrated in the figure by the coils). This process combines the advantages of casting and forging. It was developed in the 1960s.

5.9 SQUEEZE CASTING

This is a relatively new process and involves solidification under high pressure (Fig. 5.30). The equipment includes a die, a punch, and an ejector pin. The pressure applied by the punch keeps the entrapped gases in solution and the intimate contact at the die–metal interface promotes rapid heat transfer, resulting in a fine microstructure of the casting with good mechanical properties. Parts can be made nearly to net (final) shape, with fine detail on complex geometries, from nonferrous and ferrous wrought metals or casting alloys. Examples are automotive wheels and mortar bodies. The pressures required are less than those for hot or cold forging (Chapter 6).

5.10 CRYSTAL GROWING: DIRECTIONAL SOLIDIFICATION AND MONOCRYSTALS

With the advent of the semiconductor industry, single crystal growing has become a major element in the manufacture of devices such as semiconductors. There are basically two methods of crystal growing. (See also Section 3.2.)

In the *crystal-pulling* method (Fig. 5.31a), the semiconducting material is melted and its surface is brought in contact with a small single-crystal seed. This is then pulled away slowly while being rotated. The liquid metal begins to solidify on the single-crystal seed and continues the crystal structure of the seed. *Dopants* (alloying elements) may be added to the liquid metal to give special properties. Single crystals of germanium and other materials with melting points below 1830°F (1000°C) are made by this process.

In materials with higher melting points, such as silicon, contamination and doping may become a problem. For these materials, single crystals can be grown using the *floating-zone* procedure (Fig. 5.31b). Starting with a rod of polycrystalline silicon resting on a single crystal, a radiofrequency heat source heats these two pieces while moving slowly upward. The single crystal grows upward, maintaining its orientation.

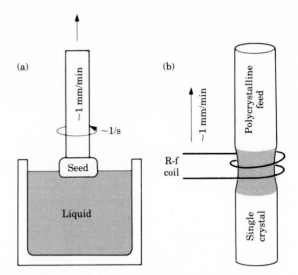

FIGURE 5.31 Two methods of crystal growing. (a) Crystal pulling. (b) Floating zone method. Crystals are especially important in the semiconductor industry. *Source:* L. H. Van Vlack, *Materials for Engineering*, Reading, Mass.: Addison-Wesley, 1982, p. 513.

FIGURE 5.32 Turbine blades manufactured by three different methods. (a) Conventionally cast. (b) Directionally solidified, with columnar grains. (c) Single crystal. The absence of grain boundaries improves the high-temperature strength, ductility, and resistance to thermal shock of single-crystal (monocrystal) blades. *Source:* Courtesy of Pratt and Whitney Commercial Engineering.

Thin wafers are then cut, cleaned, and polished for use in devices such as transistors.

Techniques have been developed to manufacture parts for special applications, the best example being turbine blades and vanes (Fig. 5.32). These blades, because of directional properties (*directional solidification*) or lack of grain boundaries (single crystals or monocrystals), have improved high-temperature strength and creep properties, ductility, and resistance to thermal shock.

5.11 CASTING DEFECTS AND DESIGN CONSIDERATIONS

From the discussion thus far, it is evident that successful casting requires careful balance and control of a large number of factors. The problems involved with gases, fluid flow, and proper gating systems are outlined in Sections 5.2 and 5.3. A major defect in castings is the contraction that follows solidification. If the casting is restrained from shrinking freely, cracking, known as *hot tearing*, occurs (Fig. 5.33). Although many factors are involved, coarse grain size and low-melting-point segregates increase the tendency for hot tearing. The design of the casting and the molds and dies must therefore make provisions for contraction. This is done by proper selection of casting shapes and by preparation of dies that allow for the contraction of the solidifying metal.

The major casting defects (Fig. 5.34) are:

a. *Blows*, *scars* and *blisters*—cavities on the surface of the casting,

b. *Misruns* and *scabs*—incomplete castings,

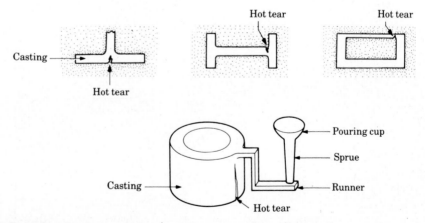

FIGURE 5.33 Examples of hot-tear defects in castings. These defects are due to the inability of the casting to shrink freely during cooling, because of the constraint of the rigid mold. Similar defects are also produced in welding, as shown in Fig. 12.37(b). Exothermic compounds may be used to control cooling at critical sections in a casting to avoid hot tearing (exothermic padding).

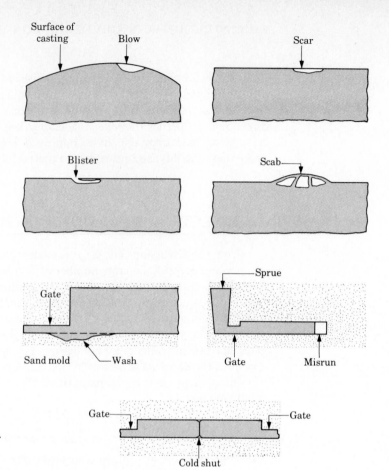

FIGURE 5.34 Some examples of common casting defects. These defects can be minimized or eliminated by proper design of castings, preparation of molds, and control of melting and pouring procedures. *Source:* After J. Datsko.

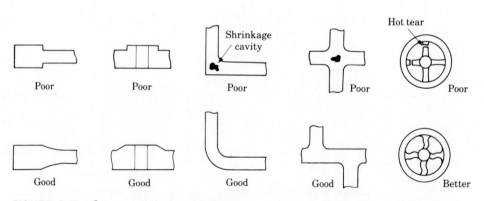

FIGURE 5.35 Suggested design modifications to avoid common defects in castings. Note that good design requires that sharp corners be avoided, section uniformity be maintained, and allowance be made for shrinkage, as in the spokes of the wheel.

   c. *Wash* or *cut*—a projection on the lower surface of the casting due to erosion of the surface of the drag because of fluid flow, and

   d. *Cold shut*—an interface lacking complete fusion due to two streams of liquid metal meeting each other; it is particularly harmful if the interface is covered with an oxide layer.

    A comparison of good and poor casting designs is shown in Fig. 5.35, summarizing some of the problems described in this chapter. In addition, residual stresses in castings (see also Section 2.7) can lead to distortions during subsequent processing, such as by machining portions of a casting for assembly with other components. Thus, stress relieving is necessary for critical applications. (See Section 3.5.)

    The choice of a casting process for a particular part depends on many technical and economic factors, as outlined in Table 5.3. Economic considerations also dictate the selection of appropriate wall thicknesses commensurate with the strength of the cast metal. Fillets and radii should be selected to reduce stress concentrations and to ensure proper fluid flow during the casting process.

5.12 CASTING ALLOYS

    The selection of alloys for casting involves considerations such as mechanical and physical properties, cost, appearance, and special requirements. Casting alloys are classified as ferrous or nonferrous. The aluminum- and magnesium-base alloys are known as light-metal castings. The primary casting properties of various alloys are described below. The mechanical properties of cast metals are given in Fig. 5.36. (See Tables 5.5 and 5.6 and also Tables 3.22 and 3.23.)

TABLE 5.5
TYPICAL APPLICATIONS FOR CASTINGS AND THEIR CHARACTERISTICS

TYPE OF ALLOY	APPLICATION	CASTABILITY	WELDABILITY	MACHINABILITY
Aluminum	Pistons, clutch housings, intake manifolds	E	F	G-E
Copper	Pumps, valves, gear blanks, marine propellers	F-G	F	F-G
Gray iron	Engine blocks, gears, brake disks and drums, machine bases	E	D	G
Magnesium	Crankcase, transmission housings	G-E	G	E
Malleable iron	Farm and construction machinery, heavy-duty bearings, railroad rolling stock	G	D	G
Nickel	Gas turbine blades, pump and valve components for chemical plants	F	F	F
Nodular iron	Crankshafts, heavy-duty gears	G	D	G

E, excellent; G, good; F, fair; VP, very poor; D, difficult.

(*continued*)

TABLE 5.5 (*Continued*)

TYPE OF ALLOY	APPLICATION	CASTABILITY	WELDABILITY	MACHINABILITY
Steel (carbon and low-alloy)	Die blocks, heavy-duty gear blanks, aircraft undercarriage members, railroad wheels	F	E	F
Steel (high-alloy)	Gas turbine housings, pump and valve components, rock crusher jaws	F	E	F
White iron	Mill liners, shot blasting nozzles, railroad brake shoes, crushers and pulverizers	G	VP	VP
Zinc	Door handles, radiator grills, carburetor bodies	E	D	E

E, excellent; G, good; F, fair; VP, very poor; D, difficult.

TABLE 5.6
PROPERTIES AND TYPICAL APPLICATIONS OF CAST NONFERROUS ALLOYS

ALLOYS (UNS)	CONDITION	ULTIMATE TENSILE STRENGTH, MPa	YIELD STRENGTH, MPa	ELONGA-TION IN 50 mm, %	APPLICATIONS
ALUMINUM ALLOYS					
195 (AO1950)	Heat treated	220–280	110–220	8.5–2	Sand castings
319 (AO3190)	Heat treated	185–250	125–180	2–1.5	Sand castings
356 (AO3560)	Heat treated	260	185	5	Permanent mold castings
COPPER ALLOYS					
Red brass (C83600)	Annealed	235	115	25	Pipe fittings, gears
Yellow brass (C86400)	Annealed	275	95	25	Hardware, ornamental
Manganese bronze (C86100)	Annealed	480	195	30	Propeller hubs, blades
Leaded tin bronze (C92500)	Annealed	260	105	35	Gears, bearings, valves
Gun metal (C90500)	Annealed	275	105	30	Pump parts, fittings
Nickel silver (C97600)	Annealed	275	175	15	Marine parts, valves
MAGNESIUM ALLOYS					
AZ91A	F	230	150	3	Die castings
AZ63A	T4	275	95	12	Sand and permanent mold castings
AZ91C	T6	275	130	5	High strength
EZ33A	T5	160	110	3	Elevated temperature
HK31A	T6	210	105	8	Elevated temperature
QE22A	T6	275	205	4	Highest strength

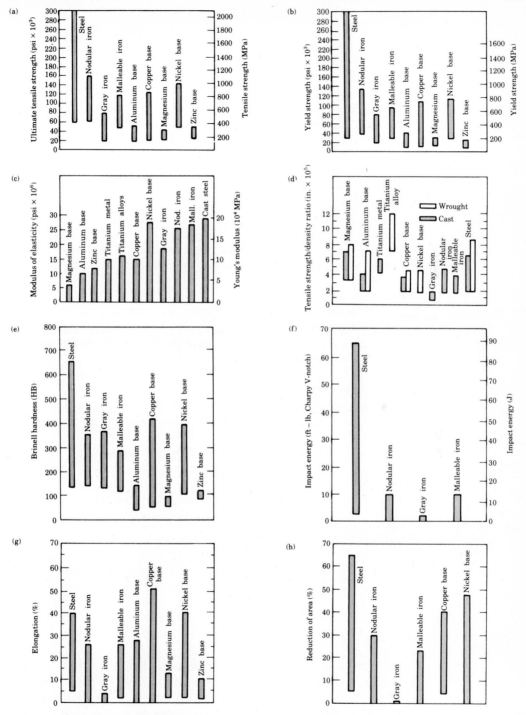

FIGURE 5.36 Mechanical properties for various groups of cast alloys. Compare with various tables of properties in Chapter 3. *Source:* Courtesy of Steel Founders' Society of America.

5.12.1 NONFERROUS CASTING ALLOYS

Aluminum-Base Alloys
This group has a wide range of mechanical properties because of various hardening mechanisms and heat treatment. They have high electrical conductivity, good corrosion resistance to most elements, are nontoxic and lightweight, and have good machinability. These alloys have wide applications, including architectural and decorative use. However, they generally have low resistance to wear and abrasion (except alloys with silicon), and are likely to be attacked by alkali. Their fluidity during casting depends on the presence of oxide and alloying elements.

Magnesium-Base Alloys
This group has the lowest density of all commercial casting alloys. They have good corrosion resistance and moderate strength, depending on heat treatment.

Copper-Base Alloys
Although somewhat expensive, these alloys have the advantages of good electrical and thermal conductivity, corrosion resistance, nontoxicity, and wear properties suitable for bearing materials. Mechanical properties and fluidity are governed by various alloying elements.

Zinc-Base Alloys
This is a low-melting-point alloy group, with good fluidity, and sufficient strength for various structural applications. These alloys are commonly used in die casting.

Nickel- and Cobalt-Base and Other High-Temperature and Refractory Alloys
These alloys have a wide range of properties and require high temperatures for casting. They can, with special techniques, be cast into parts for critical applications, such as jet and rocket engine components. Some of these metals are more suitable for casting than shaping by other manufacturing methods.

5.12.2 STEELS

Because of the high temperatures involved, casting of steels requires considerable knowledge and experience. Steels undergo much shrinkage during solidification, and hence require extensive risering and proper mold design to allow for shrinkage. The high temperatures required also present problems in melting and pouring and the selection of mold materials, particularly in view of the high reactivity of steels with oxygen.

On the other hand, steel castings have more uniform (isotropic) properties than those made by mechanical working. Also, they can be welded without the loss of properties due to the heat generated during welding. Cast weldments have gained great prominence where complex configurations, or the size of the casting, may be factors preventing casting in one piece.

5.12.3 CAST IRONS

Cast irons constitute the largest quantity of all metals cast. They generally possess desirable properties such as wear resistance, hardness, corrosion resistance, and good machinability. These alloys can be cast easily into intricate shapes. Cast irons are alloys of iron, carbon (up to 4%), and silicon (up to 3.5%) (Fig. 5.37).

The term *cast iron* refers to a family of alloys; more specifically, they fall into the following major categories: gray cast iron (or gray iron), white cast iron, malleable iron, and nodular cast iron (or ductile cast iron or spheroidal graphite cast iron). The major characteristics of each of these cast irons are given below. (See also Table 5.7.)

Gray Cast Iron

In gray cast iron the graphite exists in flake form (Fig. 5.38a). It is called *gray* iron because, when broken, the fracture surface has a gray, sooty appearance. The fracture

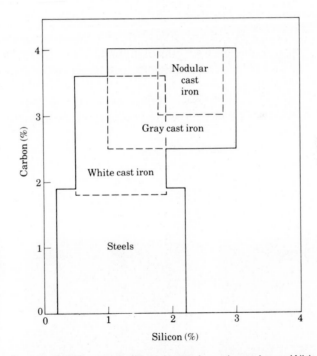

FIGURE 5.37 Range of carbon and silicon in steels and cast irons. White iron can be annealed to produce malleable iron, whereby iron carbide is decomposed into iron and graphite. Silicon is an important element in cast irons and promotes the decomposition of cementite (iron carbide) into ferrite and carbon; otherwise, the presence of large amounts of cementite makes the cast iron extremely hard and brittle, as in white cast iron. Other alloying elements in cast iron and cooling rates have various effects on the microstructure and properties of cast irons.

TABLE 5.7
PROPERTIES AND TYPICAL APPLICATIONS OF CAST IRONS

CAST IRON	TYPE	ULTIMATE STRENGTH, MPa	YIELD STRENGTH, MPa	ELONGATION IN 50 mm, %	TYPICAL APPLICATIONS
Gray	Ferritic	170	140	0.4	Pipe, sanitary ware
	Pearlitic	275	240	0.4	Engine blocks, machine tools
	Martensitic	550	550	0	Wearing surfaces
Nodular (Ductile)	Ferritic	415	275	18	Pipe, general service
	Pearlitic	550	380	6	Crankshafts, highly stressed parts
	Tempered martensite	825	620	2	High-strength machine parts, wear resistance
Malleable	Ferritic	365	240	18	Hardware, pipe fittings, general engineering service
	Pearlitic	450	310	10	Railroad equipment, couplings
	Tempered martensite	700	550	2	Railroad equipment, gears, connecting rods
White	Pearlitic	275	275	0	Wear-resistance, mill rolls

path is along the lengths of graphite flakes. These flakes act as stress raisers; hence, gray cast iron has negligible ductility. However, the presence of the graphite flakes gives this material high damping capacity (to vibrations) due to the internal friction caused by these flakes. Gray cast iron is therefore suitable for machine tools (see Section 8.13 and Fig. 8.62). These castings have relatively few porosities and shrinkage cavities.

Other forms of gray cast iron are called ferritic, pearlitic, and martensitic, and because of the different structures, each has somewhat different properties. Typical uses of gray cast iron are engine blocks, machine bases, bathtubs, electric motor housings, pipes, and wearing surfaces.

(a)   (b)   (c)

FIGURE 5.38   Microstructure of cast irons. (a) Gray cast iron. (b) Malleable iron. (c) Nodular cast iron. Microstructure has a major influence on the strength, toughness, and ductility of cast metals. *Source:* Courtesy of Central Foundry Division, General Motors Corporation.

Gray cast irons are specified by a two-digit designation according to ASTM standards. Thus, Class 20, for example, specifies that the material must have a minimum tensile strength of 20,000 psi (140 MPa).

White Cast Iron

This structure is hard and brittle due to the presence of large amounts of iron carbide (no graphite). It is called white iron because of the white crystalline appearance of the fracture surface. Because of its extreme wear resistance, it is used mainly for such applications as liners for processing abrasive materials, rolls for rolling mills, and railroad brake shoes.

Malleable Iron

In this structure the carbon is present as graphite clusters and not as flakes (Fig. 5.38b). This promotes ductility and shock resistance, hence the term malleable. Malleable iron is obtained by annealing white iron, whereby iron carbide is decomposed into iron and graphite. The main use of malleable iron is for railroad equipment and various hardware.

Malleable irons are specified by a five-digit designation. Thus, for example, 35018 indicates that the yield strength of the material is 35,000 psi (240 MPa) and its elongation is 18% in 2 in.

Nodular (Ductile) Iron

In this structure the graphite is in nodular (spheroid) form, (Fig. 5.38c). This shape permits the material to be somewhat ductile and shock resistant. Nodular cast iron is obtained by the addition of small amounts of magnesium or cerium to the molten metal prior to pouring. Nodular cast iron is used for making machine parts, pipes, and crankshafts (Fig. 5.39).

FIGURE 5.39 Nodular cast iron crankshafts. The casting processes used are shell molding and the full-mold process, although these crankshafts can also be made by sand casting. First put in service in the 1950s, most cars and trucks are now equipped with cast crankshafts. Diesel engine trucks are generally equipped with forged crankshafts (see Fig. 6.26b).
Source: Photograph courtesy of Central Foundry Division, General Motors Corporation.

Nodular irons are specified by a set of two-digit numbers. Thus, for example, 80-55-06 means that the material has a minimum tensile strength of 80,000 psi (550 MPa), a minimum yield strength of 55,000 psi (380 MPa), and 6% elongation in 2 in.

Compacted Graphite Iron

The graphite in this structure is in the form of short, thick flakes with rounded extremities and undulating surfaces. The mechanical and physical properties of this cast iron are intermediate between those of flake graphite and nodular graphite cast irons. Typical applications are disk-brake rotors and diesel-engine heads.

SUMMARY

1. Casting is a fundamental process in that, first, it provides the basic material in the shape of an ingot, which is then mechanically worked into a wrought product, such as plates, and secondly, it is a process capable of producing intricate shapes of widely varying sizes.

2. Casting is a solidification process in which molten metal is poured into a mold and is allowed to cool. Consequently, fluid flow and heat transfer are important parameters, along with the design of the mold to ensure proper flow of the molten metal into the cavities of the mold.

3. Since all metals contract during solidification (except gray cast iron, which expands), cavities can form in the casting due to the fact that metal solidifies first at the mold walls. Cavities can be avoided by using external or internal chills and with sufficient liquid metal supply. Unless controlled, porosity can also result from gases evolving during solidification.

4. Cast structures can be controlled by various means and elements to obtain desired mechanical properties.

5. A wide variety of casting processes have been developed for many applications. For example, turbine blades are now cast either conventionally, by directional solidification, or as single crystals with no grain boundaries, thus improving the high-temperature resistance of the blades.

6. Ferrous and nonferrous alloys are used for casting a wide variety of products, such as engine blocks, machine bases, railroad rolling stock, crankshafts, pump and valve components, transmission housings, and die blocks.

BIBLIOGRAPHY

Angus HT. *Cast Iron: Physical and Engineering Properties*. New York: Butterworths, 1976.

Atlas of Defects in Castings, 2d ed. London: Institute of British Foundrymen, 1961.

Beeley PR. *Foundry Technology*. London: Butterworths, 1972.

Caine JB. *Design of Ferrous Castings*. Des Plaines, Ill.: American Foundrymen's Society, 1963.

Casting Defects Handbook. Des Plaines, Ill.: American Foundrymen's Society, 1972.

Casting Design Handbook. Metals Park, Ohio: American Society for Metals, 1962.

Chalmers B. *Principles of Solidification*. New York: Wiley, 1964.

Cook GJ. *Engineering Castings*. New York: McGraw-Hill, 1961.

Flemings MC. *Solidification Processing*. New York: McGraw-Hill, 1974.

Flinn RA. *Fundamentals of Metal Casting*. Reading, Mass.: Addison–Wesley, 1963.

Heine RW, Loper CR Jr, Rosenthal C. *Principles of Metal Casting*, 2d ed. New York: McGraw-Hill, 1967.

Kay A, Street A. *Die Casting Metallurgy*. Woburn, Mass.: Butterworths, 1982.

Kondic V. *Metallurgical Principles of Founding*. London: Edward Arnold, 1968.

Metals Handbook, 8th ed. Vol. 5, *Forging and Casting*. Metals Park, Ohio: American Society for Metals, 1970.

Mikelonis PJ, ed. *Foundry Technology, Source Book*. Des Plaines, Ill.: American Foundrymen's Society, 1982.

Steel Castings Handbook, 5th ed. Des Plaines, Ill.: Steel Founders' Society of America, 1980.

Walton CF, Opar TJ, eds. *Iron Castings Handbook*. Des Plaines, Ill.: Iron Castings Society, Inc., 1981.

Wieser PF, ed. *Steel Castings Handbook*, 5th ed. Des Plaines, Ill.: Steel Founders' Society of America, 1980.

Winegard WC. *An Introduction to the Solidification of Metals*. London: Institute of Metals, 1964.

Periodicals

British Foundryman

Casting Engineering and Foundry World

Foundry Management and Technology

Foundry Trade Journal

Modern Casting

Transactions of the American Foundrymen's Society

PROBLEMS

5.1. Inspect Table 5.3 and explain qualitatively the reasoning for the various ratings and other data concerning casting processes.

5.2. Describe the important properties of sand in sand casting operations. Explain the effect of various additives in the sand.

5.3. Describe the procedure you would follow to determine whether a defect in a casting is a cavity due to shrinkage or porosity due to gases.

5.4. From your own observations of various products, add other applications for the cast materials listed in Table 5.4.

5.5. A casting may have a different shape from the pattern used to make the mold. Explain the reasons for these differences.

5.6. Explain how you would go about avoiding hot tears in castings.

5.7. It is stated in the caption for Fig. 5.39 that diesel trucks use forged crankshafts. Why could one not use cast crankshafts for this application?

5.8. Using Eq. (5.1) plot the solidification time with respect to the length/diameter ratio of a solidifying mass of constant volume.

5.9. Explain the advantages of pressure casting over other methods.

5.10. Explain why die casting is restricted to nonferrous metals and alloys.

5.11. If you need only five of a certain casting, which process would you use and why?

5.12. Describe your observations of the events depicted in Fig. 5.7.

5.13. Do you think the quality in centrifugal casting will be different from that obtained by other gravity-type casting processes?

5.14. Are there any die-cast parts on your car?

5.15. Explain the advantages of single-crystal turbine blades.

6 Bulk Deformation Processes

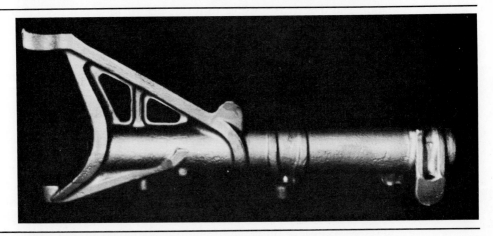

6.1 INTRODUCTION

Deformation processes in manufacturing are those operations that induce shape changes on the workpiece by plastic deformation under forces applied by various tools and dies. It is possible to divide deformation processes into various groups depending on the temperature, size, and shape of the workpiece, and the type of operation. For example, using temperature as a criterion, deformation processes may be divided into three basic categories of cold (room temperature), warm, and hot working. Also, by noting the type of operation, deformation processes may be classified as primary working or secondary working.

Primary working operations are those that take a solid piece of metal (generally from a cast state, such as an ingot) and break it down successively into shapes such as slabs, plates, and billets (see Fig. 1.8). Traditionally, primary working processes are forging, rolling, and extrusion.

Secondary working involves further processing of the products from primary working into final or semifinal products such as bolts, sheet metal parts, and wire.

Such classifications should not be rigid because there are operations belonging to both primary and secondary working. For instance, some forgings result in a product as final as those that result from secondary working. Likewise, in extrusion some products are ready for use without any further processing.

The recent trend has been to classify deformation processes according to the size and shape of the workpiece. Thus, all deformation processes are classified as either *bulk deformation* or *sheet forming*.

Bulk deformation is the processing of workpieces whose surface area-to-volume (or surface area-to-thickness) ratio is relatively small, hence the term *bulk*. In all bulk deformation processing, there is a change in the thickness or cross-section of the workpiece.

In sheet-forming operations the surface-to-thickness ratio is relatively high. In general, the material is subjected to shape changes by various dies. Thickness changes are usually undesirable and, in fact, can lead to failure.

In this chapter four basic bulk deformation processes for metals are described: forging, rolling, extrusion, and drawing of rod and wire. Bulk deformation processing of plastics and nonmetallic materials is covered in Chapters 10 and 11.

6.2 FORGING

Forging denotes a family of processes where plastic deformation of the workpiece is carried out by *compressive forces*. Forging is one of the oldest metalworking operations, dating back to 5000 B.C., and is used in making parts of widely varying sizes and shapes from a variety of metals. Typical parts made by forging today are crankshafts and connecting rods for engines, turbine disks, gears, wheels, bolt heads, hand tools, and a great variety of structural components for machinery and transportation equipment.

TABLE 6.1
HOMOLOGOUS TEMPERATURE RANGES FOR VARIOUS PROCESSES

PROCESS	T/T_m
Cold working	<0.3
Warm working	0.3 to 0.5
Hot working	>0.6

The forging operation can be carried out at room temperature (cold working), or at elevated temperatures, known as warm and hot forging, depending on temperature. The temperature range for these categories is given in Table 6.1 in terms of the homologous temperature, T/T_m, where T_m is the melting point of the workpiece material on the absolute scale. Note that the homologous recrystallization temperature for metals is about 0.5. (See also Fig. 3.17.)

Very simple forgings can be made with a heavy hammer and an anvil by techniques used by blacksmiths for centuries. Usually, though, a set of dies and a press are required.

The three basic categories of forging are: open-die, impression-die, and closed-die forging. These processes are described below.

6.3 OPEN-DIE FORGING

In its simplest form, the *open-die forging* process generally involves placing a solid cylindrical workpiece between two flat dies (platens) and reducing its height by compressing it (Fig. 6.1a). This operation is also known as *upsetting*. The die surfaces may be shaped, such as a conical or curved cavity, thereby forming the ends of the cylindrical workpiece during upsetting.

Under ideal conditions, a solid cylinder deforms as shown in Fig. 6.1(a). This is known as *homogeneous deformation*. Because volume is constant, any reduction in height increases the diameter of the cylinder. The following terms may now be defined for a specimen that has been reduced in height from h_0 to h_1:

$$\text{Reduction in height} = \frac{h_0 - h_1}{h_0} \times 100\%, \tag{6.1}$$

and from Eqs. (2.1) and (2.9), and using absolute values (as is generally the case with bulk deformation processes) of

$$e_1 = \frac{h_0 - h_1}{h_0} \tag{6.2}$$

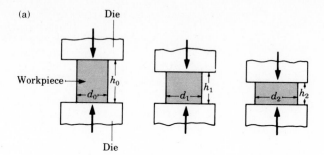

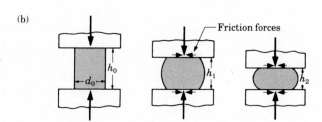

FIGURE 6.1 (a) Ideal deformation of a solid cylindrical specimen compressed between flat frictionless dies. This process is known as upsetting. (b) Deformation in upsetting with friction at the die–workpiece interfaces. The curvature developed on the cylindrical surfaces of the specimen is called barreling. See also Fig. 6.30.

and

$$\epsilon_1 = \ln\left(\frac{h_0}{h_1}\right). \tag{6.3}$$

With a relative velocity v between the platens, the specimen is subjected to a strain rate, according to Eqs. (2.17) and (2.18),

$$\dot{e}_1 = -\frac{v}{h_0} \tag{6.4}$$

and,

$$\dot{\epsilon}_1 = -\frac{v}{h_1}. \tag{6.5}$$

If the specimen is reduced in height from h_0 to h_2, the subscript 1 in the foregoing equations is replaced by subscript 2. Note that the true strain rate $\dot{\epsilon}$ increases rapidly as the height of the specimen approaches zero. Positions 1 and 2 in Fig. 6.1(a) may be regarded as instantaneous positions during a continuous upsetting operation.

Barreling

In actual operations, the specimen develops a *barrel* shape as shown in Fig. 6.1(b). Barreling is primarily due to *frictional* forces at the die–workpiece interfaces that oppose the outward flow of the material at these interfaces.

Barreling also occurs in upsetting *hot* workpieces between *cool* dies. The material at and near the interfaces cools rapidly, while the rest of the specimen is relatively hot.

Because the strength of the material decreases with temperature, the ends of the specimen show a greater resistance to deformation than does the center. Thus, the central portion of the cylinder deforms to a greater extent than do its ends.

In barreling, the material flow within the specimen becomes *nonuniform*, or *inhomogeneous*, as can be seen from the polished and etched section of a barreled cylindrical specimen in Fig. 6.2. Note the stagnant (*dead*) zones at the top and bottom surfaces, which have a wedge effect on the rest of the material, thus causing barreling. This inhomogeneous deformation can also be observed in grid patterns, as shown in Fig. 6.3, which are obtained experimentally using various techniques (Section 6.15.1).

FIGURE 6.2 Grain flow lines in upsetting a solid steel cylinder at elevated temperatures. Note the highly inhomogeneous deformation and barreling. The different shape of the bottom section of the specimen (as compared to the top) is due to the hot specimen resting on the lower cool die before deformation proceeded. The bottom surface was chilled, thus exhibiting greater strength and hence deforming to a lesser extent than the top surface. *Source:* J. A. Schey, *et al.*, IIT Research Institute.

FIGURE 6.3 Schematic illustration of grid deformation in upsetting. (a) Original grid pattern. (b) After deformation, without friction. (c) After deformation, with friction. Such deformation patterns can be used to calculate the strains within a deforming body.

In addition to the single barreling shown in Fig. 6.2, *double barreling* can also be observed. This occurs when friction at the die–workpiece interfaces is high and at large ratios of height-to-cross-sectional area. This phenomenon has been attributed to the fact that, under these conditions, the stagnant zone under the platen is sufficiently remote from the midsection of the workpiece. Thus, the midsection deforms uniformly while the top and bottom portions barrel out, hence the term "double barreling."

Barreling caused by friction can be minimized by using an effective lubricant or ultrasonic vibration of the platens (Section 4.10.8). Also, the use of heated platens or a thermal barrier at interfaces will reduce barreling in hot working.

6.3.1 FORCES AND WORK OF DEFORMATION

In this section expressions for the force required in upsetting will be derived. If friction at the interfaces is zero and the material is perfectly plastic with a yield stress of Y, then the normal compressive stress on the specimen is uniform at a level Y. The force at any height h_1 is then,

$$F = YA_1, \tag{6.6}$$

where A_1 is the cross-sectional area and is obtained from volume constancy. Thus,

$$A_1 = \frac{A_0 h_0}{h_1}.$$

The work of deformation is the product of the volume of the specimen and the specific energy u (Eq. 2.53):

$$\text{Work} = \text{Volume} \int_0^{\epsilon_1} \sigma \, d\epsilon, \tag{6.7}$$

where ϵ_1 is obtained from Eq. (6.3).

If the material is strain hardening, with a true stress–true strain curve given by

$$\sigma = K\epsilon^n, \tag{6.8}$$

then the expression for force at any stage during deformation becomes,

$$F = Y_f A_1, \tag{6.9}$$

where Y_f is the *flow stress* of the material, corresponding to the true strain given by Eq. (6.3) and Fig. 2.51.

The expression for the work done is,

$$\text{Work} = (\text{Volume})(\overline{Y})(\epsilon_1), \tag{6.10}$$

where $\overline{Y}$ is the *average flow stress* and is given by

$$\overline{Y} = \frac{K \int_0^{\epsilon_1} \epsilon^n \, d\epsilon}{\epsilon_1} = \frac{K\epsilon_1^n}{n+1}. \tag{6.11}$$

The calculation of forces with friction at the interfaces requires analytical approaches, as outlined in Section 2.10. A simple method of analysis is the slab method (Section 2.10.1), which, using the distortion-energy criterion, gives the following results for upsetting of rectangular and cylindrical specimens.

Upsetting of Rectangular Workpieces in Plane Strain

For a rectangular workpiece of length $2a$ and height h (with the width remaining constant, hence *plane strain*) (Fig. 6.4), the normal stress p at any location x is given by

$$p = Y'e^{2\mu(a-x)/h} \tag{6.12}$$

(see Section 2.10.1). Thus, the pressure is maximum at the center and diminishes to Y' at the edges. Y' is the stress in plane strain and is equal to $(2/\sqrt{3})Y$ as shown in Section 2.8.3.

For a strain-hardening material, Y' in Eq. (6.12) is replaced by Y'_f. Because of its shape, the pressure-distribution curve in Fig. 6.4 is referred to as the *friction hill*. It can be seen from this expression that the pressure with friction is higher than that with no friction. This is to be expected because the work required to overcome friction must be supplied by the upsetting force.

It can be seen from Eq. (6.12) that the normal stress p increases with increasing friction and a/h ratio. Note that, as expected, this expression reduces to Y' when the coefficient of friction is zero.

The area under the pressure curve in Fig. 6.4 is the upsetting force per unit width of the specimen. This area can be obtained by integration. However, an approximate expression for the average pressure p_{av} can be obtained:

$$p_{av} \simeq Y'\left(1 + \frac{\mu a}{h}\right), \tag{6.13}$$

Again note the significant roles of a/h and friction on the pressure required, especially at high a/h ratios.

The upsetting force F is the product of the average pressure and the contact area and is given by,

$$F = (p_{av})(2a)(\text{width}). \tag{6.14}$$

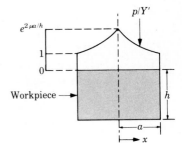

FIGURE 6.4 Normal stress (pressure) distribution in the compression (forging) of a rectangular specimen with sliding friction under conditions of plane strain. The pressure at the left and right boundaries is equal to the uniaxial yield stress of the material in plane strain, Y'.

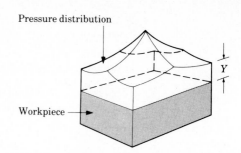

FIGURE 6.5 Normal stress (pressure) distribution in the compression of a rectangular workpiece with sliding friction under conditions of plane stress. Note that the stress at the corners is equal to the uniaxial yield stress of the material, Y.

Upsetting of Rectangular Workpieces in Plane Stress

A rectangular specimen can be upset without being constrained on its sides. It can be shown that, according to the distortion-energy criterion, the normal stress distribution can be given qualitatively by Fig. 6.5. The pressure is Y at the corners because the elements at the corners are in uniaxial compression. There is still a friction hill along the edges of the specimen because of friction along the edges.

The lateral expansion (top view) of the edges of a rectangular specimen in actual plane stress upsetting is shown in Fig. 6.6. Whereas the increase in the length of the specimen is 40%, the increase in width is 230%. This is because, as expected, *the material flows in the direction of least resistance*. The width direction has less frictional resistance than the length. Likewise, a specimen in the shape of an equilateral cube will, after upsetting, acquire a pancake shape. The diagonal direction will expand at a slower rate than the other directions.

Upsetting of Solid Cylinders

For a solid cylindrical specimen of radius r, the normal stress distribution is similar to that shown in Figure 6.4 where the expression for the normal stress p at any radius x is given by

$$p = Ye^{2\mu(r-x)/h}. \tag{6.15}$$

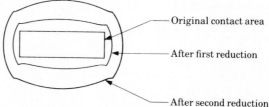

Original contact area

After first reduction

After second reduction

FIGURE 6.6   Increase in contact area of a rectangular specimen (top view) compressed between flat dies with friction. Note that the length of the specimen has increased proportionately to a smaller extent than its width. This is because the material tends to move further in the direction of least frictional resistance. Likewise, a specimen in the shape of a cube acquires the shape of a pancake after deformation with friction.

The average pressure p_{av} can be given approximately as,

$$p_{av} \simeq Y\left(1 + \frac{2\mu r}{3h}\right). \tag{6.16}$$

The upsetting force is thus,

$$F = (p_{av})(\pi r^2). \tag{6.17}$$

For strain-hardening materials, Y, in these expressions, is replaced by the flow stress Y_f. Note that all the expressions are in terms of an instantaneous height h. Thus the force at any h during a continuous upsetting operation must be calculated individually.

In actual forging operations, the value of the coefficient of friction μ in the foregoing equations can be estimated to be 0.05 to 0.1 for cold forging, and 0.1 to 0.2 for hot forging. The exact value depends on the effectiveness of the lubricant. These values can be higher than stated, especially in regions where the workpiece surface is devoid of lubricant. (See also Table 4.4.)

Sticking

It can be noted that the product of μ and p is the frictional stress (surface shear stress) at the interface at any location x from the center of the specimen. As p increases toward the center, μp also increases. However, the value of μp cannot be greater than the shear yield stress k of the material. When $\mu p = k$, *sticking* takes place. (In plane strain the value of k is $Y'/2$.) Sticking does not necessarily mean adhesion at the interface; it is used to describe the fact that, relative to the platen surfaces, the material does not move. (See also Section 8.5.2 and Fig. 8.15.)

For the sticking condition, the normal stress distribution in plane strain can be shown to be

$$p = Y'\left(1 + \frac{a - x}{h}\right). \tag{6.18}$$

It can be seen that the pressure varies linearly with x, as shown in Fig. 6.7.

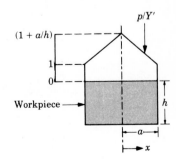

FIGURE 6.7 Normal stress (pressure) distribution in the compression of a rectangular specimen in plane strain and under sticking condition. The pressure at the edges is the uniaxial yield stress of the material in plane strain, Y'. Sticking means that the frictional shear stress on the surface of the specimen is equal to its shear yield stress.

(a)
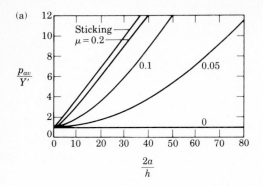

(b)
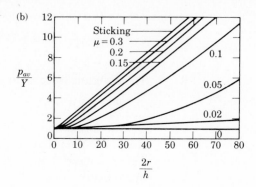

FIGURE 6.8 Ratio of average die pressure to yield stress as a function of friction and aspect ratio of the specimen. (a) Plane-strain compression. (b) Compression of a solid cylindrical specimen. Note that the yield stress for this case is Y, and not Y' as in plane-strain compression in (a). *Source:* (a) After J. F. W. Bishop, *Jr. Mech. Phys. Solids*, vol. 6, 1958, pp. 132–144. (b) Adapted from W. Schroeder and D. A. Webster, *Trans. ASME*, vol. 71, 1949, pp. 289–294.

The effects of friction and the aspect ratio of the specimen (a/h or r/h) on the average pressure p_{av} in upsetting are shown in Fig. 6.8. The pressure is given in a dimensionless form and can be regarded as a pressure-multiplying factor. These curves are convenient to use and again show the significance of friction and the aspect ratio of the specimen on upsetting pressure.

● **Illustrative Problem 6.1**

A cylindrical specimen made of annealed 4135 steel is 6 in. in diameter and 4 in. high. It is upset by open-die forging with flat dies to a height of 2 in. at room temperature. Assuming that the coefficient of friction is 0.2, calculate the force required at the end of the stroke. Use the average-pressure formula.

SOLUTION. The average-pressure formula is, from Eq. (6.16),

$$p_{av} \simeq Y\left(1 + \frac{2\mu r}{3h}\right),$$

where Y is replaced by Y_f because the workpiece material is strain hardening. From Table 2.4, $K = 147,000$ psi and $n = 0.17$.

The absolute value of the true strain is

$$\epsilon_1 = \ln\left(\frac{4}{2}\right) = 0.693.$$

Thus,

$$Y_f = K\epsilon_1^n = (147{,}000)(0.693)^{0.17} = 138{,}000 \text{ psi.}$$

The final height h_1 is 2 in. The radius r at the end of the stroke is found from volume constancy. Thus,

$$\frac{\pi 6^2}{4} \cdot 4 = \pi r_1^2 \cdot 2$$

and

$$r_1 = 4.24 \text{ in.}$$

Therefore,

$$p_{av} \simeq 138{,}000\left(1 + \frac{(2)(0.2)(4.24)}{(3)(2)}\right) = 177{,}000 \text{ psi.}$$

Thus the upsetting force is,

$$F = (177{,}000)\pi \cdot (4.24)^2 = 10^7 \text{ lb} = 5000 \text{ tons (short).}$$

(Note that Fig. 6.8(b) can also be used to solve this problem.) ●

● **Illustrative Problem 6.2**

A block made of a perfectly plastic material with a yield stress Y of 150 MPa has the dimensions of $2a = 0.2$ m, $h = 0.1$ m, and a width of 0.15 m at one instant during plane-strain forging. Assuming there is sticking friction, calculate the peak pressure.

SOLUTION. For this case the applicable formula is Eq. (6.18),

$$p = Y'\left(1 + \frac{a - x}{h}\right),$$

where

$$Y' = 1.15Y = 172.5 \text{ MPa}$$

$$a = 0.1 \text{ m,}$$

$$h = 0.1 \text{ m, and}$$

$$x = 0 \text{ (since the peak pressure is at } x = 0).$$

Thus,

$$p = 172.5\left(1 + \frac{0.1}{0.1}\right) = 345 \text{ MPa.}$$

(Note that a knowledge of the width of the block is not necessary for this problem.)
●

6.4 IMPRESSION-DIE FORGING AND CLOSED-DIE FORGING

In impression-die forging, the workpiece acquires the shape of the die cavities (impressions) while it is being upset between the closing dies. A typical example is shown in Fig. 6.9. Some of the material flows radially outward and forms a *flash*. Because of its high length-to-thickness ratio (equivalent to a high *a/h* ratio), the flash is subjected to high pressures. This in turn means high frictional resistance to material flow in the radial direction in the flash gap. Because high friction encourages the filling of the die cavities, the flash has a significant role in the flow of material in impression-die forging.

Furthermore, if the operation is carried out at elevated temperatures, the flash, because of its high surface-to-thickness ratio, cools faster than the bulk of the workpiece. Thus, the flash exhibits a greater resistance to deformation than the bulk and aids in the filling of the die cavities.

6.4.1 FORCES

Because of the more complex shapes involved, the accurate calculation of forces in impression-die forging is difficult. Depending on its position, each element within the workpiece is generally subjected to different strains and strain rates. Note, for instance, the wide range of strain rates in the specimen shown in Fig. 6.10. The thinner the section, the higher the true strain rate, as observed from Eq. (6.5).

Consequently, the level of strength that the material exhibits at each location will depend not only on the strain and strain rate, but also on the exponents n and m of the workpiece material in Eqs. (2.11) and (2.16).

Because of the many difficulties involved in calculating forces in impression-die forging, certain *pressure-multiplying factors* K_p have been recommended as shown in Table 6.2. These factors are to be used with the expression,

$$F = (K_p)(Y_f)(A), \tag{6.19a}$$

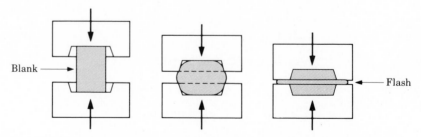

FIGURE 6.9 Schematic illustration of stages in impression-die forging. Note the formation of a flash, which is excess material that is subsequently trimmed off.

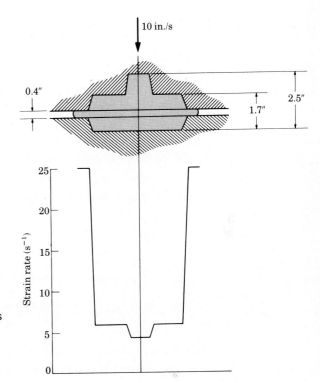

FIGURE 6.10 Strain rate distribution in impression-die forging of an axisymmetric part. Note that, with the same die speed, thinner sections undergo higher strain rates, as indicated by Eqs. (6.4) and (6.5). For strain-rate sensitive materials, these regions will require higher stress levels to continue deformation than the bulk of the forging.

where F is the forging load, A is the projected area of the forging (including the flash), and Y_f is the flow stress of the material (from its true stress–true strain curve) at the strain and strain rate to which the material is subjected.

A typical impression-die–forging load as a function of the stroke of the die is shown in Fig. 6.11. For this axisymmetric workpiece, the force increases gradually as the cavity is filled, Fig. 6.11(b). The force then increases rapidly due to flash formation. In order to obtain the final dimensions and details on the forged part, the dies must close further, with an even steeper rise in the forging load.

TABLE 6.2
RANGE OF K_p VALUES FOR EQ. (6.19a) IN IMPRESSION-DIE FORGING

Simple shapes, without flash	3–5
Simple shapes, with flash	5–8
Complex shapes, with flash	8–12

Note that the flash has a finite contact length with the die called *land*; see Fig. 6.28. This is to ensure that the flash generates sufficient resistance to the outward flow of the material (to aid in die filling) without contributing excessively to the forging load.

- **Illustrative Problem 6.3**

Assume that a forging with a very complex shape has a projected cross-sectional area of 30 in². If the metric tonnage of the forging press is 10,000 what is the maximum flow stress that the material can have? What type of material could this be?

SOLUTION. Since this is a complex forging, the value for K_p in Eq. (6.19) is taken as 12 (from Table 6.2). Thus,

$$Y_f = \frac{F}{K_p A}$$

$$= \frac{(10,000)(2,000)}{(12)(30)} = 55,500 \text{ psi.}$$

By referring to Fig. 2.7, we find that a number of materials could exhibit this level of flow stress. Drawing a horizontal line at a true stress of 55,500 psi, we note that the material could be annealed copper at a true strain of about 0.5, or it could be 1020 steel at a true strain of a little over 0.1. Thus, copper will have to be compressed more than steel to reach the same flow stress level. ●

6.4.2 CLOSED-DIE FORGING

Although not quite correct, the example shown in Fig. 6.11 is also referred to as *closed-die forging*. True closed-die forging examples are shown in Fig. 6.12. No flash is formed and the workpiece is completely surrounded by the dies. Whereas in impression-die forging any excess metal in the die cavity is formed into a flash, this is not the case in closed-die forging. Thus, proper control of the volume of material is essential to obtain a forging of desired dimensions. Undersized blanks in closed-die forging prevent the complete filling of the die; oversized blanks may cause premature die failure or jamming of the dies.

Precision Forging
Precision forging, or *flashless forging*, and similar operations where the part formed is close to the final dimensions of the desired component, are also known as *near–net-shape* production (Fig. 6.13). Any excess material is subsequently removed by various machining processes described in Chapters 8 and 9.

In precision forging, special dies are made and machined to greater accuracy than in ordinary impression-die forging. Precision forging requires a higher capacity forging equipment than other forging processes. Aluminum and magnesium alloys

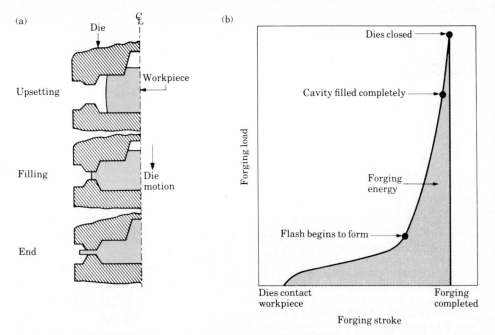

(a)

Die

$\overset{C}{L}$

Workpiece

Upsetting

Die
motion

Filling

End

(b)

Dies closed

Cavity filled completely

Forging load

Forging
energy

Flash begins to form

Dies contact
workpiece

Forging stroke

Forging
completed

FIGURE 6.11 Typical load-stroke curve for closed-die forging. Note the sharp increase in load after the flash begins to form. In hot forging operations, the flash requires high levels of stress because it is thin (i.e., small h) and cooler than the bulk of the forging. *Source:* After T. Altan.

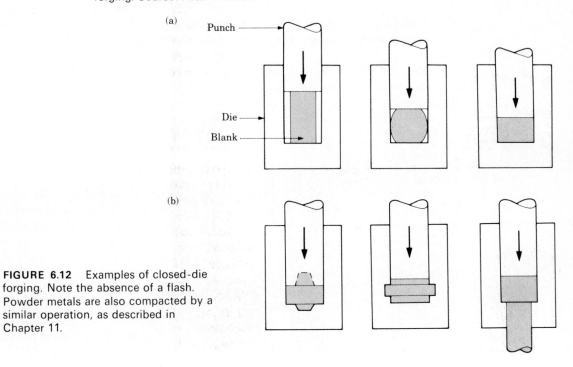

(a)

Punch

Die

Blank

(b)

FIGURE 6.12 Examples of closed-die forging. Note the absence of a flash. Powder metals are also compacted by a similar operation, as described in Chapter 11.

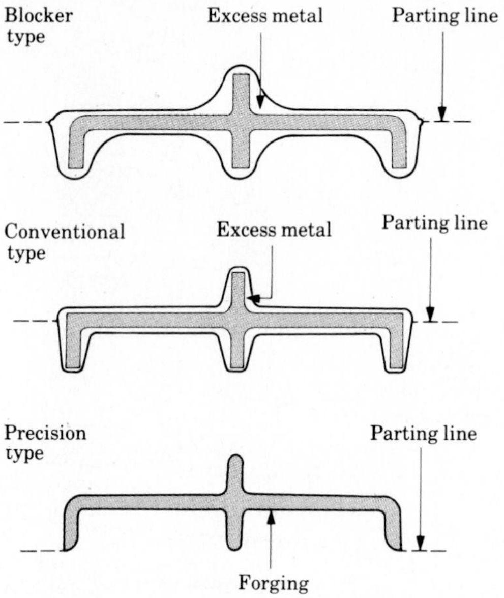

FIGURE 6.13   Three basic types of forging operations. Note that precision forging requires little or no machining to produce the final product. Blocker types require the removal of the largest amount of material. Production of parts that are at or near the final desired dimensions is known as near-net shape production. Forging, squeeze casting, and powder metallurgy techniques are the most commonly used processes in near-net shape forming. Parting line is the location where the two dies meet; see also Fig. 6.28.

are particularly suitable for precision forging because of the low forging loads and temperatures required. Also, they result in little die wear and produce a good surface finish.

Steels and other alloys are more difficult to precision forge. The choice between conventional forging and precision forging requires an economic analysis. Precision forging requires special dies. On the other hand, there is much less machining involved since the part is closer to the desired final shape.

Coining

Another example of closed-die forging is the minting of coins (Fig. 6.14), where the slug is shaped in a completely closed cavity. The pressures required can be as high as five to six times the flow stress of the material in order to produce the fine details of a coin or a medallion. Several coining operations may be necessary in order to obtain full details on some parts. The first coins appeared around 700 B.C. and were further developed by the Romans.

The coining process is also used with forgings and other products to improve surface finish and impart the desired dimensional accuracy (*sizing*) of the products. The pressures are high and usually very little change in shape takes place. Lubricants cannot be tolerated in coining since they can be trapped in the die cavities and prevent the reproduction of fine die surface details.

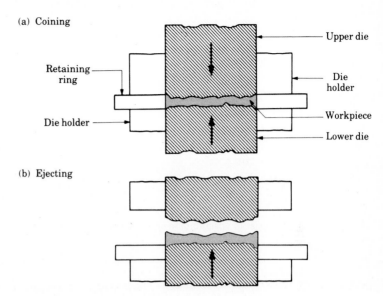

(a) Coining

Retaining ring

Die holder

Upper die

Die holder

Workpiece

Lower die

(b) Ejecting

FIGURE 6.14 Schematic illustration of coining. The earliest coins were made by the Romans using simple tools. This process is also used for decorative items, medallions, patterned tableware, and in sizing parts.

6.5 MISCELLANEOUS FORGING OPERATIONS

The basic forging processes have been described above with some typical examples. (See also Table 6.3.) A number of other forging operations are also carried out in order to impart the desired shape to the products. These are briefly described below.

TABLE 6.3
ADVANTAGES AND LIMITATIONS OF COMMON FORGING PROCESSES*

PROCESS	DESCRIPTION	ADVANTAGES	LIMITATIONS
Open-die forging	Compressive forces (produced by hand tools or mechanical hammers) are applied locally to heated metal stock; little or no lateral confinement is involved. Desired shape is achieved by turning and manipulating workpiece between blows	Simple, inexpensive tools; useful for small quantities; wide range of sizes available; good strength characteristics	Limited to simple shapes; difficult to hold close tolerances; machining to final shape necessary; slow production rate; relatively poor utilization of material; high degree of skill required
Closed-die forging	Compressive forces (produced by a mechanical hammer in a mechanical or hydraulic press) are applied over the entire surface of heated metal stock, forcing metal into a die cavity of desired shape. There are several types of closed-die forgings	Relatively good utilization of material; generally better properties than open-die forgings; good dimensional accuracy; rapid production rate; good reproducibility	High tool cost for small quantities; machining often necessary

(continued)

*Materials Selector. *Materials Engineering Magazine.* Cleveland: Penton/IPC.

TABLE 6.3 (*continued*)

PROCESS	DESCRIPTION	ADVANTAGES	LIMITATIONS
Blocker type	Uses single-impression dies and produces parts with somewhat generalized contours	Low tool costs; high production rates	Machining to final shape necessary; thick webs and large fillets necessary
Conventional type	Uses preblocked workpiece and multiple-impression dies	Requires much less machining than blocker type; rapid production rates; good utilization of material	Somewhat higher tool cost than blocker type
Precision type	Uses minimum draft (often 0°)	Close tolerances; machining often unnecessary; very good material utilization; very thin webs and flanges possible	Requires intricate tooling and elaborate provision for removing forging from tools
Cold heading	Similar to upset forging except metal is cold. Wire up to about 1 in. in diameter is fed to die in punch press and positioned with one end protruding; this end mushrooms out under force of punch and is formed between die and punch face	Good surface strength; alloys used are generally tough, ductile, and crack resistant; excellent surface finish; no scrap loss; rapid production rate	Head volume and shape limited; internal stresses may be left at critical points; size of part limited

Heading. This is basically an upsetting operation performed at the end of a rod (usually round) in order to form a shape with a larger cross-section. Typical examples are the heads of bolts, screws, and nails (Fig. 6.15). An important aspect in the upsetting process during heading is the tendency for buckling if the length-to-diameter ratio is too high. Figure 6.16 gives some general guidelines to avoid buckling in upsetting the ends of rods.

Heading is done on machines called *headers*, which are usually highly automated horizontal machines with high production rates. The operation can be carried out cold, warm, or hot.

Piercing. This is an operation involving indenting the surface of a workpiece with a punch to produce a cavity or an impression (Fig. 6.17). The workpiece may be confined in a die cavity or it may be unconstrained.

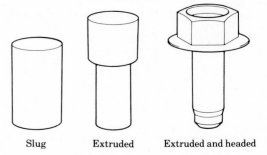

FIGURE 6.15 Stages in heading operation for a bolt blank. Note that the head undergoes basically an upsetting operation. The threads on the bolt are produced either by cutting or forming operations, as shown in Figs. 6.59 and 6.60.

Slug   Extruded   Extruded and headed

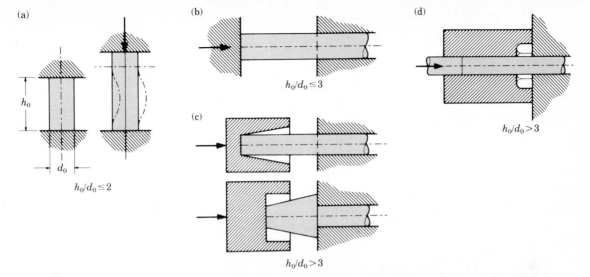

(a)

h_0

d_0

$h_0/d_0 \leq 2$

(b)

$h_0/d_0 \leq 3$

(c)

$h_0/d_0 > 3$

(d)

$h_0/d_0 > 3$

FIGURE 6.16 Limits of buckling in upsetting and heading operations. (a) Upsetting with flat dies. (b) Heading with a flat die. (c) Heading with a conical preformed head. (d) Heading of a bolt in a cold header, a horizontal machine that forms the heads of bolts at very high rates of production. *Source:* Reproduced with permission from J. A. Schey, *Introduction to Manufacturing Processes*, New York, McGraw-Hill, © 1977, p. 115.

The piercing force depends on the punch cross-sectional area and its tip geometry, the flow stress of the material, and the friction at the interfaces. Punch pressures may range from three to five times the flow stress of the material.

The term piercing is also used for the process of cutting of holes with a punch and die (Fig. 7.6).

Hubbing. Hubbing consists of pressing a hardened punch with a particular tip geometry into the surface of a block. This process is used to produce a die cavity, which is then used for subsequent forming operations. The cavity is usually shallow, but for deeper cavities some material may be removed from the surface by machining prior to hubbing.

It has been shown that the pressure required to generate a cavity by hubbing is approximately three times the ultimate tensile strength (UTS) of the material of the

Punch

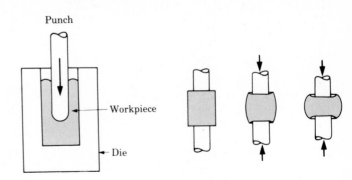

Workpiece

Die

FIGURE 6.17 Examples of piercing operations. The term piercing is also used in punching holes in sheet metal. See also Figs. 6.58 and 7.6.

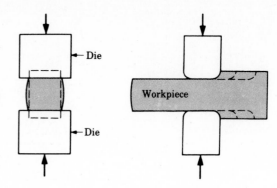

FIGURE 6.18 Schematic illustration of a cogging operation on a rectangular bar. Note that, with simple tools, the thickness and cross-section of a bar can be reduced by multiple cogging operations. Note also the barreling after cogging. Blacksmiths use a similar procedure to reduce the thickness of parts by heating the workpiece and hammering it a number of times at small increments.

block. Thus, the force required is

$$\text{Hubbing force} = 3(\text{UTS})(A), \tag{6.19b}$$

where A is the projected area of the impression. Note that the factor 3 is in agreement with the observations made in regard to hardness of materials, discussed in Section 2.5.7.

Cogging.   In this operation, also called *drawing out*, the thickness of a bar is reduced by successive steps at certain intervals (Fig. 6.18). By this method, a long section of a bar can be reduced in thickness without high forces, since the contact area is small.

Fullering and Edging.   These operations are performed, usually on bar stock, in order to distribute the material in certain regions of a forging (Fig. 6.19). The parts are then formed into final shapes by other forging processes.

Roll Forging.   In this process, the cross-sectional area of a bar is reduced and altered in shape by passing it through a pair of rolls with grooves of various shapes (Fig. 6.20). This operation may also be used to produce a part that is basically the final product, such as tapered shafts, tapered leaf springs, table knives, and numerous tools. Roll forging is also used as a preliminary forming operation, followed by other forging processes. Typical examples are crankshafts and other automotive components.

FIGURE 6.19   (a) Fullering. (b) Edging. In order to distribute the material properly in the die cavities, these operations are often necessary prior to forging complex shapes. See also Figs. 6.25 and 6.26(b).

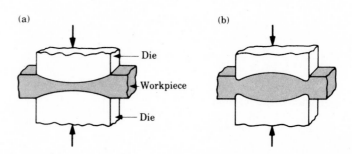

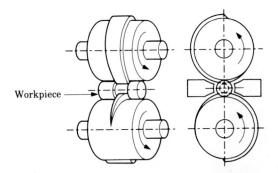

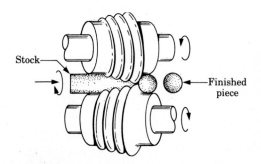

FIGURE 6.20 Schematic illustration of a roll forging (cross rolling) operation. Tapered leaf springs and knives can also be made by this process with specially designed rolls. *Source:* After J. Holub, *Machinery*, vol. 102, Jan. 16, 1963.

FIGURE 6.21 Production of steel balls for bearings by the skew rolling process. Balls for bearings can also be made by the forging process shown in Fig. 6.22.

Skew Rolling. A process similar to roll forging is skew rolling (Fig. 6.21), which is used for making ball bearings. Round wire or rod stock is fed into the roll gap and spherical blanks are formed continuously by the rotating rolls. (Another method of forming blanks for ball bearings is by cutting pieces from a round bar and upsetting them, as shown in Fig. 6.22. The balls are then ground and polished in special machinery.)

Other operations, such as swaging, ring rolling, and extrusion are also considered a part of the overall forging operations. These are covered in other sections in this chapter.

Isothermal Forging

In this process, also known as *hot-die forging*, the dies are heated to the same temperature as that of the hot blank. In this way, cooling of the workpiece is eliminated, the low flow stress of the material is maintained, and material flow within the die cavities is improved. The dies are generally made of nickel alloys (e.g., IN-100) and

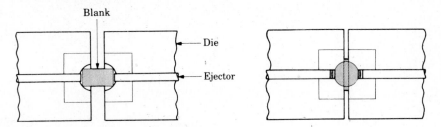

FIGURE 6.22 Production of steel balls by upsetting of a cylindrical blank. Note the formation of flash. The balls are subsequently ground and polished for use in ball bearings and other mechanical components.

FIGURE 6.23 Isothermal forging of a nose wheel for an aircraft. The Ti-6Al-6V-2Sn titanium alloy blank (left) was formed into a wheel in a single stroke of a hydraulic press with a force of 900 tons. Forging temperature was 1650°F (900°C). *Source:* IIT Research Institute.

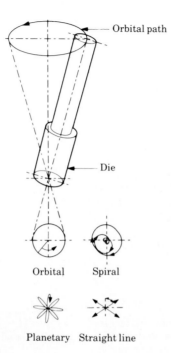

Orbital path

Die

Orbital   Spiral

Planetary   Straight line

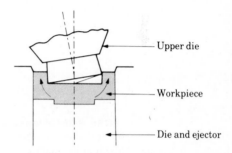

Upper die

Workpiece

Die and ejector

FIGURE 6.24 Schematic illustration of the orbital forging process. Note that the die is in contact only with a portion of the workpiece surface. This process is also called rotary forging, swing forging, or rocking die forging, and can be used for forming bevel gears, wheels, and bearing rings.

complex parts with good dimensional accuracy can be forged in one stroke in hydraulic presses (Fig. 6.23). Isothermal forging is expensive; however, it can be economical for intricate forgings of expensive materials, provided that the quantity required is sufficiently high to justify die costs.

Orbital Forging
This is a relatively new process in which the die moves along an orbital path (Fig. 6.24) and forges or forms the part incrementally.

6.6 FORGING DIES

The proper design of forging dies and the selection of die materials require considerable experience. Important factors concerning forging dies are described below.

6.6.1 DIE DESIGN

The design of forging dies requires a knowledge of the strength and ductility of the workpiece material, its sensitivity to strain rate and temperature, and its frictional characteristics. Die distortion under high forging loads can also be an important consideration in die design, particularly if close tolerances are required.

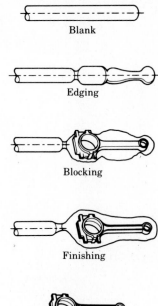

FIGURE 6.25 Stages in forging a connecting rod for an internal combustion engine. Note the amount of flash that is necessary to fill the die cavities properly.

The most important rule in die design is that the workpiece material flows in the direction of least resistance. Thus, it is common practice to distribute the material in such a way as to enable the proper filling of die cavities.

An example of the forging of a connecting rod is shown in Fig. 6.25. Starting with a round bar stock, the bar is first preformed (*intermediate shape*) by some of the techniques described above, and is then formed into the final shape in two additional forging operations, followed by trimming. It is apparent from this illustration that proper distribution of the material is important in order to be able to fill the die cavities and forge the part successfully.

Other examples of intermediate shapes are shown in Fig. 6.26. The selection of these shapes requires experience and involves calculating cross-sectional areas at each location in the forging. The dies used for intermediate forging operations are called *blocker* dies. The care necessary in calculating the volume of the material required to properly fill the die cavities is obvious (see Fig. 6.13). Computer techniques are now being developed to expedite these calculations. The sequence of operations involved in making a railroad wheel is shown in Fig. 6.27.

The terminology used in die design is shown in Fig. 6.28 and the significance of various parameters are described briefly below.

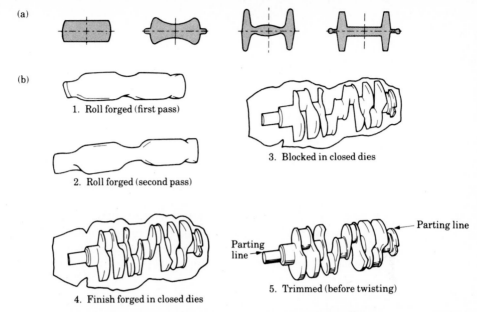

(a)

(b)

1. Roll forged (first pass)

2. Roll forged (second pass)

3. Blocked in closed dies

4. Finish forged in closed dies

5. Trimmed (before twisting)

Parting line

Parting line

FIGURE 6.26 Intermediate stages in forging two different parts. (a) H-section. (b) Crankshaft. These intermediate stages are important for distributing the material and filling the die cavities properly. Crankshafts are also made by casting processes, as shown in Fig. 5.39.

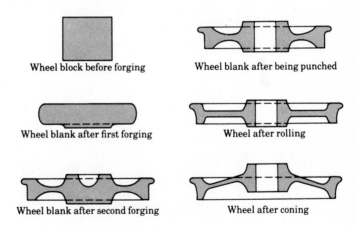

Wheel block before forging

Wheel blank after being punched

Wheel blank after first forging

Wheel after rolling

Wheel blank after second forging

Wheel after coning

FIGURE 6.27 Stages in the manufacture of a railroad wheel. Wheels are also made by casting, as shown in Figs. 5.21(a) and (b). *Source:* Copyright 1964 by United States Steel Corporation.

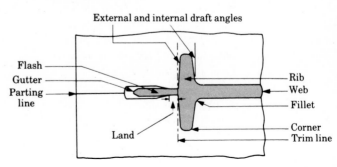

External and internal draft angles

Flash

Gutter

Parting line

Land

Rib

Web

Fillet

Corner

Trim line

FIGURE 6.28 Standard terminology for various features of a typical forging die.

311

Parting Line

This is the line where the two dies meet. For simple symmetrical shapes the parting line is a straight line at the center of the forging; for more complex shapes the line may be offset and may not be in a single plane. The selection of the proper location for the parting line is based on the shape of the part, metal flow, balance of forces, and the flash.

Flash and Gutter

The significance of the flash is described in Section 6.4. After sufficiently constraining lateral flow (due to the length of the land), the flash is allowed to flow into a *gutter*. Thus the extra flash does not increase the forging load unnecessarily. A general guideline for the flash clearance (between the dies) is 3% of the maximum thickness of the forging. The length of the land is usually five times the flash clearance.

Draft Angle

Draft angles are necessary in almost all forgings in order to facilitate the removal of the part from the die. Draft angles usually range between 3 and 10 degrees. Because the forging, upon cooling, shrinks in its radial direction (as well as in other directions), internal draft angles are made larger than external ones. Thus, internal angles are about 7 to 10 degrees and external angles about 3 to 5 degrees.

Radii

The selection of the proper radii for corners and fillets is important to ensure smooth flow of the metal in the die cavity and to improve die life. Small radii are generally not desirable because of their adverse effect on metal flow and their tendency to wear rapidly from stress concentration and thermal cycling. Small radii in fillets can cause fatigue cracking of the dies.

6.6.2 DIE MATERIALS

Because most forgings, particularly large-size ones, are performed at elevated temperatures, general requirements for die materials are as follows: strength and toughness at elevated temperatures, hardenability, mechanical and thermal shock resistance, and wear resistance (particularly abrasive wear resistance, because of the presence of scale on heated forgings).

Selection of die materials depends on factors such as die size, composition and properties of the workpiece, complexity of its shape, forging temperature, type of operation, cost of die material, and the number of forgings required. Heat transfer and distortion of the die are also important factors.

Dies can fail because of factors such as improper heat treatment, improper design, overheating and heat checking (thermal cracks due to temperature cycling), excessive wear, and overloading (see Section 6.25). To reduce breakage and heat checking, dies should be preheated to temperatures of about 300 to 500°F (150 to 260°C).

Common die materials are tool and die steels containing chromium, nickel, molybdenum, and vanadium. (See Tables 6.4, 6.5, and 6.6.)

TABLE 6.4
TYPICAL TOOL AND DIE MATERIALS FOR METALWORKING

PROCESS	MATERIAL
Die casting	H13, P20
Powder metallurgy	
Punches	A2, S7, D2, D3, M2
Dies	WC, D2, M2
Molds for plastics and rubber	S1, O1, A2, D2, 6F5, 6F6, P6, P20, P21, H13
Hot forging	6F2, 6G, H11, H12
Hot extrusion	H11, H12, H13
Cold heading	W1, W2, M1, M2, D2, WC
Cold extrusion	
Punches	A2, D2, M2, M4
Dies	O1, W1, A2, D2
Coining	52100, W1, O1, A2, D2, D3, D4, H11, H12, H13
Drawing	
Wire	WC, diamond
Shapes	WC, D2, M2
Bar and tubing	WC, W1, D2
Rolls	
Rolling	Cast iron, cast steel, forged steel, WC
Thread rolling	A2, D2, M2
Shear spinning	A2, D2, D3
Sheet metals	
Shearing	
Cold	D2, A2, A9, S2, S5, S7
Hot	H11, H12, H13
Pressworking	Zinc alloys, 4140 steel, cast iron, epoxy composites, A2, D2, O1
Deep drawing	W1, O1, cast iron, A2, D2

Notes: Tool and die materials are usually hardened to 55 to 65 HRC for cold working, and 30 to 55 for hot working.

Tool and die steels contain one or more of the following major alloying elements: chromium, molybdenum, tungsten, and vanadium. For further details see references at end of chapter.

For composition and properties of tool and die materials see Tables 6.5, 6.6, 8.7 and 8.8.

TABLE 6.5
COMPOSITION AND TYPICAL APPLICATIONS OF COMMON TOOL AND DIE STEELS

AISI TYPE	COMPOSITION, %						TYPICAL APPLICATIONS
	C	W	Mo	Cr	V	OTHER	
MOLYBDENUM HIGH-SPEED STEELS							
M1	0.85*	1.50	8.50	4.00	1.00	—	Drills, taps, end mills, reamers, milling cutters, hobs, punches, lathe and planar tools, form tools, saws, chasers, broaches, routers, woodworking tools.
M2	0.85; 1.00*	6.00	5.00	4.00	2.00	—	Drills, taps, end mills, reamers, milling cutters, hobs, form tools, saws, lathe and planar tools, chasers, broaches and boring tools.
M3.1	1.05	6.00	5.00	4.00	2.40	—	Drills, taps, end mills, reamers and counterbores, broaches, hobs, form tools, lathe and planar tools, cheeking tools, milling cutters, slitting saws, punches, drawing dies, routers, woodworking tools.
M3.2	1.20	6.00	5.00	4.00	3.00	—	Drills, taps, end mills, reamers and counterbores, broaches, hobs, form tools, lathe and planar tools, cheeking tools, slitting saws, punches, drawing dies, woodworking tools.
M4	1.30	5.50	4.50	4.00	4.00	—	Broaches, reamers, milling cutters, chasers, form tools, lathe and planar tools, cheeking tools, blanking dies and punches for abrasive materials, swaging dies.
M7	1.00	1.75	8.75	4.00	2.00	—	Drills, taps, end mills, reamers, routers, saws, milling cutters, lathe and planar tools, chasers, borers, woodworking tools, hobs, form tools, punches.
M10	0.85; 1.00*	—	8.00	4.00	2.00	—	Drills, taps, reamers, chasers, end mills, lathe and planar tools, woodworking tools, routers, saws, milling cutters, hobs, form tools, punches, broaches.
M33	0.90	1.50	9.50	4.00	1.15	8.00 Co	Drills, taps, end mills, lathe tools, milling cutters, form tools, chasers.
M42	1.10	1.50	9.50	3.75	1.15	8.00 Co	Drills, end mills, reamers, form cutters, lathe tools, hobs, broaches, milling cutters, twist drills, end mills. Hardenable to 67 to 70 HRC.
TUNGSTEN HIGH-SPEED STEELS							
T1	0.75*	18.00	—	4.00	1.00	—	Drills, taps, reamers, hobs, lathe and planer tools, broaches, crowners, burnishing dies, cold extrusion dies, cold heading die inserts, lamination dies, chasers, cutters, taps, end mills, milling cutters.
T5	0.80	18.00	—	4.00	2.00	8.00 Co	Lathe and planar tools, form tools, cutoff tools, heavy duty tools requiring high hot hardness.
T15	1.50	12.00	—	4.00	5.00	5.00 Co	Form tools, lathe and planar tools, broaches, milling cutters, blanking dies, punches, heavy duty tools requiring wear resistance.

*Other carbon contents may be available. Some of these grades can be made with added sulfur to improve machinability. See Section 8.11.

Adapted from AISI Tool Steel Products Manual and producers of tool steels.

TABLE 6.5 (*continued*)

AISI TYPE	COMPOSITION, %						TYPICAL APPLICATIONS
	C	W	Mo	Cr	V	*OTHER*	
CHROMIUM HOT WORK STEELS							
H11	0.35	—	1.50	5.00	0.40	—	Die casting dies, punches, piercing tools, mandrels, extrusion tooling, forging dies, high strength structural components.
H12	0.35	1.50	1.50	5.00	0.40	—	Extrusion dies, dummy blocks, holders, gripper and header dies, forging die inserts, punches, mandrels, sleeves for cold heading dies.
H13	0.35	—	1.50	5.00	1.00	—	Die casting dies and inserts, dummy blocks, cores, ejector pins, plungers, sleeves, slides, extrusion dies, forging dies and inserts.
TUNGSTEN HOT WORK STEELS							
H21	0.35	9.00	—	3.50	—	—	Mandrels, hot blanking dies, hot punches, blades for flying shear, hot trimming dies, extrusion and die casting dies, extrusion and die casting dies for brass, dummy blocks, piercer points, gripper dies, hot nut tools (crowners, cutoffs, side dies, piercers), hot headers.
H24	0.45	15.00	—	3.00	—	—	Punches and shear blades for brass, hot blanking and drawing dies, trimming dies, dummy blocks, hot press dies, hot punches, gripper dies, hot forming rolls, hot shear blades, swaging dies, hot heading dies, extrusion dies.
MOLYBDENUM HOT WORK STEELS							
H42	0.60	6.00	5.00	4.00	2.00	—	Cold trimming dies, hot upsetting dies, dummy blocks, header dies, hot extrusion dies, cold header and extrusion dies and die inserts, hot forming and swaging dies, nut piercers, hot punches, mandrels, chipping chisels.
AIR-HARDENING MEDIUM-ALLOY COLD WORK STEELS							
A2	1.00	—	1.00	5.00	—	—	Thread rolling dies, extrusion dies, trimming dies, blanking dies, coining dies, mandrels, shear blades, slitters, spinning rolls, forming rolls, gages, beading dies, burnishing tools, ceramic tools, embossing dies, plastic molds, stamping dies, bushings, punches, liners for brick molds.

(*continued*)

TABLE 6.5   (*continued*)

AISI TYPE	COMPOSITION, %						TYPICAL APPLICATIONS
	C	W	Mo	Cr	V	*OTHER*	
A9	0.50	—	1.40	5.00	1.00	1.50 Ni	Solid cold heading dies, die inserts, heading hammers, coining dies, forming dies and rolls, die casings, gripper dies. Hot work applications: punches, piercing tools, mandrels, extrusion tooling, forging dies, gripper dies, die casings, heading dies, hammers, coining and forming dies.

HIGH-CARBON, HIGH-CHROMIUM COLD WORK STEELS

AISI TYPE	C	W	Mo	Cr	V	OTHER	TYPICAL APPLICATIONS
D2	1.50	—	1.00	12.00	1.00	—	Blanking dies, cold forming dies, drawing dies, lamination dies, thread rolling dies, shear blades, slitter knives, forming rolls, burnishing tools, punches, gages, knurling tools, lathe centers, broaches, cold extrusion dies, mandrels, swaging dies, cutlery.
D3	2.25	—	—	12.00	—	—	Blanking dies, cold forming dies, drawing dies, lamination dies, thread rolling dies, shear blades, slitter knives, forming rolls, seaming rolls, burnishing tools, punches, gages, crimping dies, swaging dies.
D4	2.25	—	1.00	12.00	—	—	Blanking dies, brick molds, burnishing tools, thread rolling dies, hot swaging dies, wiredrawing dies, forming tools and rolls, gages, punches, trimmer dies, dies for deep drawing.

OIL-HARDENING COLD WORK STEELS

AISI TYPE	C	W	Mo	Cr	V	OTHER	TYPICAL APPLICATIONS
O1	0.90	0.50	—	0.50	—	1.00 Mn	Blanking dies, plastic mold dies, drawing dies, trim dies, paper knives, shear blades, taps, reamers, tools, gages, bending and forming dies, bushings, punches.
O7	1.20	1.75	—	0.75	—	—	Mandrels, slitters, skiving knives, taps, reamers, drills, blanking and forming dies, gages, chasers, brass finishing tools, dental burrs, paper knives, roll turning tools, burnishing dies, pipe threading dies, rubber cutting knives, woodworking tools, hand reamers, scrapers, spinning tools, broaches, blanking and cold forming punches.

TABLE 6.5 (*continued*)

AISI TYPE	COMPOSITION, %						TYPICAL APPLICATIONS
	C	W	Mo	Cr	V	*OTHER*	
SHOCK-RESISTING STEELS							
S1	0.50	2.50	—	1.50	—	—	Bolt header dies, chipping and caulking chisels, pipe cutters, concrete drills, expander rolls, forging dies, forming dies, grippers, mandrels, punches, pneumatic tools, scarfing tools, swaging dies, shear blades, track tools, master hobs.
S2	0.50	—	0.50	—	—	1.00 Si	
S5	0.55	—	0.40	—	—	0.80 Mn, 2.00 Si	Hand and pneumatic chisels, drift pins, forming tools, knock-out pins, mandrels, nail sets, pipe cutters, rivet sets and busters, screw driver bits, shear blades, spindles, stamps, tool shanks, track tools, lathe and screw machine collets, bending dies, punches, rotary shears.
S7	0.50	—	1.40	3.25	—	—	Shear blades, punches, slitters, chisels, forming dies, hot header dies, blanking dies, rivet sets, gripper dies, engraving dies, plastic molds, die casting dies, master hobs, beading tools, caulking tools, chuck jaws, clutches, pipe cutters, swaging dies.
LOW-CARBON MOLD STEELS							
P6	0.10	—	—	1.50	—	3.50 Ni	Heavy duty gears, shafts, bearings, plastic molding dies (hobbed and carburized).
P20	0.35	—	0.40	1.70	—	—	Molds for zinc and plastic articles, holding blocks for die casting dies.
P21	0.20	—	—	—	—	4.00 Ni, 1.20 N	Thermoplastic injection molds, zinc die casting dies, holding blocks for plastic and die casting dies.
WATER-HARDENING STEELS							
W1	0.60; 1.40*	—	—	—	—	—	*Low-carbon:* blacksmith tools, blanking tools, caulking tools, cold chisels, forging dies, rammers, rivet sets, shear blades, punches, sledges. *Medium-carbon:* arbors, beading tools, blanking dies, reamers, bushings, cold-heading dies, chisels, coining dies, countersinks, drills, forming dies, jeweler dies, mandrels, punches, shear blades, woodworking tools. *High-carbon:* glass cutters, jeweler dies, lathe tools, reamers, taps and dies, twist drills, woodworking tools. Vanadium content of W2 imparts finer grain, greater toughness and shallow hardenability.
W2	0.60; 1.40*	—	—	—	0.25	—	

TABLE 6.6
PROCESSING AND SERVICE CHARACTERISTICS OF COMMON TOOL AND DIE STEELS

AISI DESIGNATION	RESISTANCE TO DECARBURIZATION	HARDENING AND TEMPERING				MACHINABILITY	FABRICATION AND SERVICE		
		HARDENING RESPONSE	AMOUNT OF DISTORTION(a)	RESISTANCE TO CRACKING	APPROXIMATE HARDNESS(b), HRC		TOUGHNESS	RESISTANCE TO SOFTENING	RESISTANCE TO WEAR
MOLYBDENUM HIGH-SPEED STEELS									
M1	Low	Deep	A or S, low; O, medium	Medium	60–65	Medium	Low	Very high	Very high
M2	Medium	Deep	A or S, low; O, medium	Medium	60–65	Medium	Low	Very high	Very high
M3 (class 1 and class 2)	Medium	Deep	A or S, low; O, medium	Medium	61–66	Medium	Low	Very high	Very high
M4	Medium	Deep	A or S, low; O, medium	Medium	61–66	Low to medium	Low	Very high	Highest
M7	Low	Deep	A or S, low; O, medium	Medium	61–66	Medium	Low	Very high	Very high
M10	Low	Deep	A or S, low; O, medium	Medium	60–65	Medium	Low	Very high	Very high
M33	Low	Deep	A or S, low; O, medium	Medium	60–65	Medium	Low	Highest	Very high
M42	Low	Deep	A or S, low; O, medium	Medium	65–70	Medium	Low	Highest	Very high
TUNGSTEN HIGH-SPEED STEELS									
T1	High	Deep	A or S, low; O, medium	High	60–65	Medium	Low	Very high	Very high
T5	Low	Deep	A or S, low; O, medium	Medium	60–65	Medium	Low	Highest	Very high
T15	Medium	Deep	A or S, low; O, medium	Medium	63–68	Low to medium	Low	Highest	Highest

TABLE 6.6 *(continued)*

CHROMIUM HOT WORK STEELS

H11	Medium	Deep	Very low	Highest	38–54	Medium to high	Very high	High	Medium
H12	Medium	Deep	Very low	Highest	38–55	Medium to high	Very high	High	Medium
H13	Medium	Deep	Very low	Highest	38–53	Medium to high	Very high	High	Medium

TUNGSTEN HOT WORK STEELS

H21	Medium	Deep	A, low; O, medium	High	36–54	Medium	High	High	Medium to high
H26	Medium	Deep	A or S, low; O, medium	High	43–58	Medium	Medium	Very high	High

MOLYBDENUM HOT WORK STEELS

H42	Medium	Deep	A or S, low; O, medium	Medium	50–60	Medium	Medium	Very high	High

AIR-HARDENING MEDIUM-ALLOY COLD WORK STEELS

A2	Medium	Deep	Lowest	Highest	57–62	Medium	Medium	High	High
A9	Medium	Deep	Lowest	Highest	35–56	Medium	High	High	Medium to high

HIGH-CARBON, HIGH-CHROMIUM COLD WORK STEELS

D2	Medium	Deep	Lowest	Highest	54–61	Low	Low	High	High to very high
D3	Medium	Deep	Very low	High	54–61	Low	Low	High	Very high
D4	Medium	Deep	Lowest	Highest	54–61	Low	Low	High	Very high

OIL-HARDENING COLD WORK STEELS

O1	High	Medium	Very low	Very high	57–62	High	Medium	Low	Medium
O7	High	Medium	W, high; O, very low	W, low; O, very high	58–64	High	Medium	Low	Medium

(continued)

TABLE 6.6 (continued)

AISI DESIGNATION	RESISTANCE TO DECARBURIZATION	HARDENING AND TEMPERING					FABRICATION AND SERVICE		
		HARDENING RESPONSE	AMOUNT OF DISTORTION(a)	RESISTANCE TO CRACKING	APPROXIMATE HARDNESS(b), HRC	MACHINABILITY	TOUGHNESS	RESISTANCE TO SOFTENING	RESISTANCE TO WEAR
SHOCK-RESISTING STEELS									
S1	Medium	Medium	Medium	High	40–58	Medium	Very high	Medium	Low to medium
S2	Low	Medium	High	Low	50–60	Medium to high	Highest	Low	Low to medium
S5	Low	Medium	Medium	High	50–60	Medium to high	Highest	Low	Low to medium
S7	Medium	Deep	A, lowest; O, low	A, highest; O, high	45–57	Medium	Very high	High	Low to medium
LOW-CARBON MOLD STEELS									
P6	High	—	A, very low; O, low	High	58–61(c)	Medium	High	Low	Medium
P20	High	Medium	Low	High	28–37	Medium to high	High	Low	Low to medium
P21	High	Deep	Lowest	Highest	30–40(d)	Medium	Medium	Medium	Medium
WATER-HARDENING STEELS									
W1	Highest	Shallow	High	Medium	50–64	Highest	High(e)	Low	Low to medium
W2	Highest	Shallow	High	Medium	50–64	Highest	High(e)	Low	Low to medium

Adapted from *Tool Steels*, American Iron and Steel Institute, 1978. (a) A, air cool; O, oil quench; S, salt bath quench; W, water quench. (b) After tempering in temperature range normally recommended for this steel. (c) Carburized case hardness. (d) After aging at 510 to 550°C (950 to 1025°F). (e) Toughness decreases with increasing carbon content and depth of hardening.

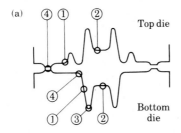

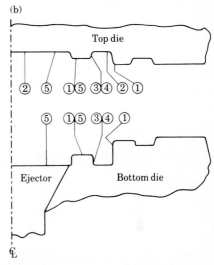

1. Erosion
2. Thermal fatigue
3. Mechanical fatigue
4. Plastic deformation

1. Erosion
2. Pitting (lubricated dies only)
3. Thermal fatigue
4. Mechanical fatigue
5. Plastic deformation

FIGURE 6.29   Types of wear observed in dies in hot forging. Note that a single pair of dies may undergo wear by five different mechanisms, depending on the location on the die surfaces. See Section 4.8 for a description of wear mechanisms. *Source:* (a) After A. Knnappan, *Metal Forming*, vol. 37, No. 1, Dec. 1969–Jan. 1970. (b) After T. A. Dean, in *Metalworking Lubrication*, S. Kalpakjian and S. Jain, eds., ASME, 1980.

Die wear in hot forging may involve a variety of mechanisms, as can be seen from Fig. 6.29. Die wear depends on such factors as the level of stress, temperature, relative motion, and contact time.

Lubrication

Lubrication in forging plays an important role. It affects friction and wear and, consequently, the flow of metal into the die cavities. Lubricants can also serve as a thermal barrier between the hot forging and the relatively cool dies, thus slowing the rate of cooling of the workpiece. Another important role of a lubricant is to serve as a parting agent, that is, to prevent the forging from sticking to the dies. Lubricants can have a significant effect on the wear pattern in forging dies. A wide variety of lubricants can be used in forging. For hot forging, graphite, molybdenum disulfide, and sometimes glass, are commonly used as lubricants. For cold forging, mineral oils and soaps are common lubricants (Table 6.7).

TABLE 6.7
LUBRICANTS COMMONLY USED IN METALWORKING OPERATIONS (From various sources)

MATERIAL AND ALLOY	TEMPERATURE	FORGING	ROLLING	EXTRUSION
Aluminum	Cold	FA + MO, S	FA + MO, MO	D, G, MS, L, S
	Hot	MO + G, MS	FA + CO, FA + E	D, G, PI
Beryllium	Hot	MO + G, J	G	MS, G, J
Copper	Cold	S, E, T	E, MO	S, T, L, Wa, G, MS
	Hot	G	D, E	D, MO + G
Lead	Cold	FO + MO	FA + MO, EM	D, S, T, E, FO
	Hot	—	—	D
Magnesium	Cold	FA + MO, CC + S	D, FA + MO,	D, T
	Hot	W + G, MO + G	MO + FA + EM, D, G	D, PI
Nickel	Cold	—	MO + CL	CC + S
	Hot	MO + G, W + G, GI	W, E	GI, J
Refractory	Hot	GI, G, MS	G + MS	J + GI
Steels	Cold	EP + MO, CC + S	E, CO, MO	CC + S, T, W + MS
(carbon and low alloy)	Hot	MO + G, Sa, GI	W, MO, CO, G + E	GI, G
Steels (stainless)	Cold	CL + MO, CC + S	CL + EM, CL + MO	CC + S, CP, MO
	Hot	MO + G, GI	D, W, E	GI
Titanium	Cold	S, MO	MO, CC + FO, G	CC + G, CC + S
	Hot	W + G, GI, MS	PI, CC, G, MS	J + GI, GI
Zinc	Cold	—	—	D, T
	Hot	—	—	D, T
Zirconium	Cold	—	—	CC + G, CC + MS
	Hot	—	—	J + GI + G

6.7 FORGEABILITY

The forgeability of a metal can be defined as its capability to undergo deformation by forging without cracking. This definition can be expanded to include the flow strength of the metal. Thus, a material with good forgeability is one that can be shaped with low forces without cracking. A number of tests have been developed to measure forgeability, although none is universally accepted. The more common tests are described below.

Upsetting Test
A standard test is to upset a solid cylindrical specimen and observe any cracking on the barreled surfaces (Fig. 6.30). The greater the reduction in height prior to cracking, the greater the forgeability of the metal. These cracks are due to *secondary* tensile stresses on the barreled surfaces. (They are called secondary because no external tensile stress is applied to the material and the stresses develop because of shape changes.)

TABLE 6.7   (*continued*)

ROD AND WIRE DRAWING	SHEET METALWORKING	MACHINING	GRINDING
MO, E, Wa, FA + MO	FO + MO, E, L, Wa, S	D, MO, E, MO + FO, CSN	E, EP
—	G		
G	—	MO, E, CSN	D, E, CSN
FO + E + S, MO, Wa	FO + E, FO + MO, S + Wa	D, E, CSN, MO + FO	CSN, E, MO + FO
—	G	—	—
FO	FO + MO		
—	—		
—	FO + MO	D, MO, MO + FO	D, MO
—	G + MO, S + Wa, T + G,	—	—
CC + S, MO + CL	E, MO + EP, CL, CC + S	MO, E, CSN	CSN, EP
—	G, MS	—	—
G + MS, PI	MS, G	MO, E, EP	EP
S, CC + S	E, S, Wa, FO + MO, PI, CSN	D, MO, E, CSN, EP	CSN, E
—	G	—	—
CC + S, CL + MO	FO + MO, Wa, PI, EP + MO, S	D, MO, E, CSN	CSN, E
—	G	—	—
CC + PI	CL, S, PI, Wa	CSN, EP, MO	CSN, E
—	G, MS	—	—
—	—	D, MO, E, CSN	—
—	—	—	—
—	—	D, E, CSN	—

CC	Conversion coatings	CSN	Chemicals and synthetics	EP	Extreme pressure		
CL	Chlorinated paraffin	D	Dry	FA	Fatty acid		
CO	Compounded oil	E	Emulsion	FO	Fatty oil		
CP	Copper plate						
G	Graphite	L	Lanolin	PI	Polymer	T	Tallow
Gl	Glass	MO	Mineral oil	S	Soap	W	Water
J	Canned or jacketed	MS	Molybdenum disulfide	Sa	Salts	Wa	Wax

(a)

(b)

FIGURE 6.30 Cracks on the barreled surface in upsetting of solid steel cylinders. Diameter is about 2 in. (50 mm) (a) Shear crack. Aspect ratio = 1.5, 60% reduction in height in a hydraulic press at 60 in./min (0.025 m/s). (b) Normal crack. Aspect ratio = 1.0, 60% reduction in a drop hammer. *Source:* After K. M. Kulkarni and S. Kalpakjian.

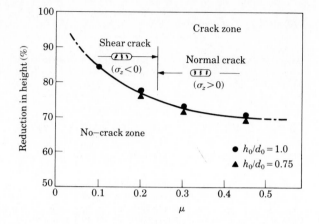

FIGURE 6.31 Cracking in upsetting solid cylindrical AISI 1045 steel specimens, as a function of reduction in height and friction. No cracks are observed below the curve. *Source:* After S. Kobayashi.

Because barreling is a result of friction at the die–workpiece interfaces, friction has a marked effect on cracking in upsetting. As shown in Fig. 6.31, as friction increases the specimen cracks at a lower reduction in height. It is also interesting to note that, whether or not the crack will be at 45 degrees or vertical (Fig. 6.30), depends on the sign of the axial stress σ_z on the barreled surface. If this stress is negative (compressive) the crack is at 45 degrees, and if positive (tensile) it is longitudinal. Depending on the notch sensitivity of the material, any surface defects will also affect the results by causing premature cracking. A typical surface defect is a *seam*: a longitudinal scratch, a string of inclusions, or a fold due to prior working of the material.

Upsetting tests can be performed at various temperatures and strain rates. An optimal range for these parameters can then be specified for forging a particular material. Such a test can serve only as a guideline, since in actual forging the metal is subjected to a different state of stress as compared with a simple upsetting operation.

Hot-Twist Test

In this test a round specimen is twisted continuously until it fails. The test is performed at various temperatures and the number of turns that each specimen undergoes before failure is observed. The optimal forging temperature is then determined from a plot such as that shown in Fig. 6.32. The hot-twist test has been found to be particularly useful in determining the forgeability of steels, although upsetting tests can also be used for this purpose.

It has been observed that small changes in the composition of or impurities in the metal can have a significant effect on forgeability. Note in Fig. 6.32(b) that a slight amount of sulfur in low-carbon steels drastically reduces the number of twists before failure. This is due to hot shortness (Section 3.3.1) and can be improved by increasing the manganese content of the steel.

Forgeability of Various Metals

Based on the results of various tests and observations in practice, the forgeability of several metals and alloys has been determined, as listed below. These are based on

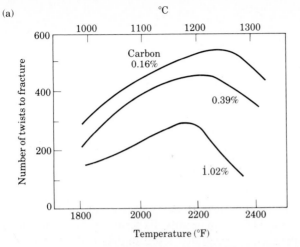

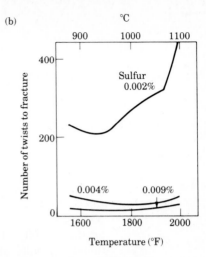

FIGURE 6.32 (a) Effect of carbon content on hot-twist tests for carbon steel. (b) Effect of sulfur on hot-twist tests for low-carbon steel. Note that doubling the amount of sulfur (although still very small) severely reduces the number of twists to fracture, hence its forgeability. *Source:* (a) After C. L. Clark and J. Russ, *Trans. AIMME*, vol. 167, 1946, p. 736. (b) After C. T. Anderson, *et al.*, *Jr. Metals*, vol. 5, 1953, pp. 525–529.

such considerations as the ductility and strength of the metals, forging temperature required, frictional behavior, and the quality of the forgings obtained. The list is in decreasing order of forgeability.

Aluminum alloys
Magnesium alloys
Copper alloys
Carbon and low-alloy steels
Stainless steels
Titanium alloys
Iron-base superalloys
Cobalt-base superalloys
Columbium alloys
Tantalum alloys
Molybdenum alloys
Nickel-base superalloys
Tungsten alloys
Beryllium

Forging Temperature

Typical forging temperature ranges for a variety of metals are given in Table 6.8.

TABLE 6.8
FORGING TEMPERATURE RANGES FOR VARIOUS METALS

	°F	°C		°F	°C
Aluminum alloys	750–850	400–450	Titanium alloys	1400–1800	750–975
Copper alloys	1150–1750	625–950	Refractory alloys	1800–3000	975–1650
Alloy steels	1700–2300	925–1250			

Effect of Hydrostatic Pressure on Forgeability

As described in Section 2.2.8, hydrostatic pressure has a significant beneficial effect on the ductility of metals and nonmetallic materials. Experiments have indicated that the results of room-temperature forgeability tests, outlined above, are improved (that is, cracking takes place at higher strain levels) if the tests are conducted in an environment of high hydrostatic pressure. Although such a test would be difficult to perform at elevated temperatures, techniques have been developed to forge metals in a high compressive environment in order to take advantage of this phenomenon. The pressure-transmitting medium is usually a low-strength ductile metal.

6.8 FORGING DEFECTS

In addition to surface cracking during forging, other defects can occur in forgings due to the material flow patterns in the die cavity. In one example (Fig. 6.33) the flow of the material into the die generates a cavity, known as an *extrusion defect* (see Section 6.17).

In another example (Fig. 6.34) excess material in the web in a forging buckles during forging and develops *laps*. If the web is thick, the excess material flows past the already forged portions of the forging and develops internal cracks (Fig. 6.35). These examples indicate the importance of the proper distribution of the material and the control of flow in the die cavity.

The various radii in the die cavity can have a significant effect on the formation of defects. In the example shown in Fig. 6.36, the material follows a large corner radius

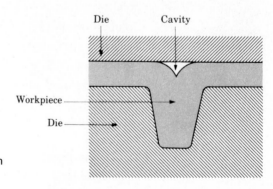

FIGURE 6.33   Defect in forging due to material flow pattern into the die cavity. This defect is known as extrusion defect or piping. The term piping is also used for a similar defect observed in casting ingots, as shown in Fig. 5.11.

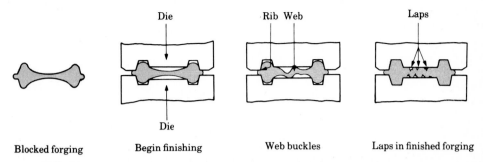

Die · Rib Web · Laps

Die

Blocked forging · **Begin finishing** · **Web buckles** · **Laps in finished forging**

FIGURE 6.34 Laps formed due to buckling of the web during forging. The solution to this problem is to increase the initial web thickness to avoid buckling.

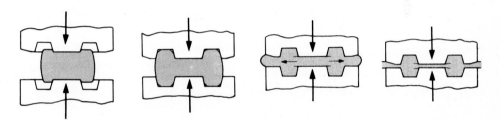

FIGURE 6.35 Internal defects produced in a forging because of an oversized billet. The die cavities are filled prematurely and the material at the center of the part flows past the filled regions as deformation continues.

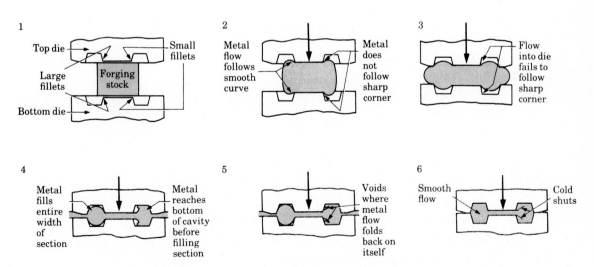

1
Top die · Small fillets
Large fillets · Forging stock
Bottom die

2
Metal flow follows smooth curve · Metal does not follow sharp corner

3
Flow into die fails to follow sharp corner

4
Metal fills entire width of section · Metal reaches bottom of cavity before filling section

5
Voids where metal flow folds back on itself

6
Smooth flow · Cold shuts

FIGURE 6.36 Effect of fillet radius on defect formation in forging. Small fillets (right side of drawings) cause the defects. *Source:* ALCOA.

FIGURE 6.37 Section of a nickel alloy forging showing grain flow pattern. Note the highly nonuniform flow of the material. *Source:* Wyman-Gordon Co.

better than a small radius. With small radii, the material can fold over itself and produce a lap, called *cold shut.*

During the service life of the forged component, such defects can cause fatigue failures and lead to other problems such as corrosion and wear. The importance of inspecting forgings before they are put into service, particularly for critical applications, is thus apparent. (For inspection techniques, see Table 4.2.)

Although it may not be considered a flaw, another important aspect in the quality of a forging is the *grain flow pattern* (Fig. 6.37). There can be situations where the grain flow lines reach a surface perpendicularly, thus exposing the grain boundaries directly to the environment. These are known as *end grains.* They can, in service, be preferentially attacked by the environment and develop a rough surface and act as stress raisers. For critical components, end grains in forgings should be avoided by proper

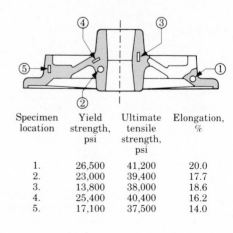

FIGURE 6.38 Mechanical properties of five tensile test specimens taken at various locations and directions in a AZ61 magnesium alloy forging. Note the anisotropy of properties due to inhomogeneous deformation during forging. *Source:* After S. M. Jablonski, *Modern Metals*, vol. 16, 1963, pp. 62–70.

Specimen location	Yield strength, psi	Ultimate tensile strength, psi	Elongation, %
1.	26,500	41,200	20.0
2.	23,000	39,400	17.7
3.	13,800	38,000	18.6
4.	25,400	40,400	16.2
5.	17,100	37,500	14.0

selection of the original workpiece orientation in the die cavity and by control of material flow.

Because the metal flows in various directions in a forging, and also because of temperature variations, the properties of a forging are generally anisotropic. As can be seen in Fig. 6.38, significant variations of strength and ductility are observed in the test specimens taken from different locations and orientations in the forged part.

6.9 FORGING EQUIPMENT

A variety of forging equipment is available with various designs, capacities, speeds, and speed–stroke characteristics, as shown in Table 6.9.

Hydraulic Presses
Hydraulic presses have a constant low speed and are *load-limited*. Large amounts of energy can be transmitted to the workpiece with a constant load available throughout the stroke. Hydraulic presses are thus ideally suited for extrusion-type forging operations. The ram speed can be varied during the stroke. These presses are used for both open-die and closed-die forging operations. The largest hydraulic press in existence has a capacity of 75,000 metric tons (670 MN).

TABLE 6.9
CHARACTERISTICS OF VARIOUS FORGING EQUIPMENT (after T. Altan).

EQUIPMENT	SPEED RANGE		SPEED–STROKE BEHAVIOR
	ft/s	m/s	
Hydraulic press	0.2–1.0	0.06–0.30	
Mechanical press	0.2–5	0.06–1.5	
Screw press	2–4	0.6–1.2	
Gravity drop hammer	12–16	3.6–4.8	
Power drop hammer	10–30	3.0–9.0	
Counterblow hammer	15–30	4.5–9.0	

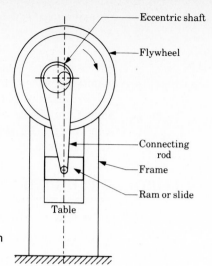

FIGURE 6.39 Schematic illustration of a mechanical press with an eccentric drive. Rotary motion is translated into reciprocating motion. The clutch and brake for this press are located on the eccentric shaft.

Mechanical Presses

Mechanical presses are *stroke limited*. They are basically crank or eccentric types (Fig. 6.39), with speeds varying from a maximum at the center of the stroke to zero at the bottom. The force available depends on the stroke position and becomes extremely high at the bottom dead center. Thus, proper setup is essential to avoid breaking the dies or the equipment. The largest mechanical press has a capacity of 12,000 tons (120 MN).

Screw Presses

Screw presses (Fig. 6.40) derive their energy from a flywheel. The forging load is transmitted through a vertical screw. These presses are *energy limited* and can be used for many forging operations. They are particularly suitable for producing small quantities, for parts requiring precision (such as turbine blades), and for control of ram speed. The largest screw press has a capacity of 16,000 tons (160 MN).

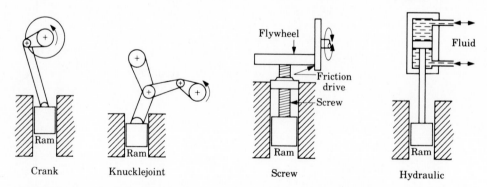

FIGURE 6.40 Schematic illustration of various types of presses used in metalworking. The choice of the press is an important factor in the overall operation.

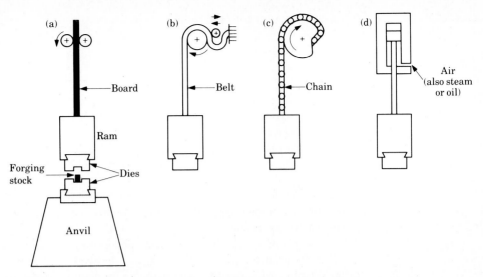

FIGURE 6.41 Schematic illustration of various types of gravity-drop hammers. (a) Board drop. (b) Belt drop. (c) Chain drop. (d) Air drop. *Source:* After T. Altan.

Hammers

Hammers (Fig. 6.41) derive their energy from the potential energy of the ram, which is then converted to kinetic energy; thus they are *energy limited*. The speeds are high; therefore the low forming times minimize the cooling of the hot forging, thus allowing the forging of complex shapes, particularly with thin and deep recesses. To complete the forging, several blows may have to be made on the part. In power hammers (developed in the 1830s) the ram is accelerated in the downstroke by steam or air, in addition to gravity. The highest energy available in hammers is 850,000 ft · lb (1.15 MN · m).

Counterblow Hammers

Counterblow hammers have two rams that simultaneously approach each other to forge the part. They are generally of the mechanical–pneumatic or mechanical–hydraulic type. These machines transmit less vibration to the foundation. The largest counterblow hammer has a capacity of 900,000 ft · lb (1.2 MN · m).

Equipment Selection

The selection of forging equipment depends on such factors as the size and complexity of the forging, the strength of the material and its sensitivity to strain rate, and the degree of deformation required. As a guideline, presses are generally preferred for aluminum, magnesium, beryllium, bronze, and brass. Hammers are preferred for copper, steels, titanium, and refractory alloys. Production rate is also a consideration in equipment selection. The number of strokes per minute ranges from a few for hydraulic presses to as much as 300 for power hammers. (See also Table 6.9.)

6.10 ROLLING

Rolling, first developed in the late 1500s, is the process of reducing the thickness or changing the cross-section of a workpiece by compressive forces exerted by a pair of *rotating rolls*. The basic operation in this category is flat rolling (or simply rolling), where the rolled products are flat plates and sheets. Plates are used for structural applications such as bridges, ships, boilers, and nuclear vessels. Sheets (generally 1/4 in. [6 mm] or less in thickness) are used for a great variety of applications as described in Chapter 7. Typical examples are sheet metal for automotive and other transportation equipment, appliances, beverage cans, containers, welded tubing, and office and kitchen equipment.

The basic rolling process has been developed into various other operations. These are shape rolling (structural shapes such as I-beams), ring rolling (bearing races, ring-gear blanks, ring-shaped structural components), thread rolling (threaded fasteners such as screws and bolts), and rotary tube piercing (seamless tubing).

Because of the diversity of the products obtained, particularly in making flat plates, sheet, and foil, rolling is an important process and has been studied extensively. The original material to be rolled is usually an *ingot* from a foundry (see Section 5.5 and Fig. 1.8). This ingot is first rolled hot, and the coarse-grained, brittle, and porous structure of the cast ingot is broken down into a *wrought structure* with greater ductility and finer grain size.

The product from the first rolling operation is a *bloom* (which usually has a square cross-section at least 6 in. [150 mm] on the side) or a *slab* (rectangular). Blooms are processed further by shape rolling into structural shapes. Slabs are rolled into *plates* and *sheets*. *Billets* are usually square with smaller cross-sectional area than blooms. They are then rolled into shapes such as round *rods* and *bars* of various cross-sections.

6.11 FLAT ROLLING

A schematic illustration of the flat rolling process is shown in Fig. 6.42. A strip of thickness h_0 enters the roll gap and is reduced to h_f by the rotating (powered) rolls. The surface speed of the roll is V_r. In order to keep the volume rate of metal flow constant, the velocity of the strip must increase as it moves through the roll gap (similar to fluid flow through a converging channel). At the exit of the roll gap, the velocity of the strip is V_f (Fig. 6.43). Since V_r is constant along the roll gap, there is sliding between the roll and the strip.

There is one point along the arc of contact where the two velocities are the same. This is known as the *neutral point* or *no-slip point*. To the left of this point, the roll moves faster than the workpiece, and to the right the workpiece moves faster than the roll. Because of friction at the interfaces, the frictional forces, which oppose motion, act on the strip surfaces as shown in Fig. 6.43.

(Top roll removed)

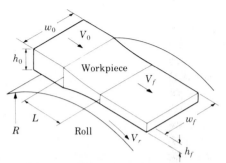

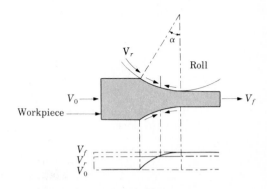

FIGURE 6.42 Schematic illustration of the flat rolling process. A greater volume of metal is formed by rolling than by any other metalworking process.

FIGURE 6.43 Relative velocity distribution between roll and strip surfaces. Note the difference in the direction of frictional forces.

Note the similarity between Figs. 6.43 and 6.1(b). The friction forces oppose each other at the neutral point. In upsetting, these forces are equal to each other because of the symmetry of the operation. In rolling, the frictional force on the left of the neutral point must be higher than the force on the right. In this way, there is a net frictional force to the right, which then enables the strip to be pulled into the roll gap and make the rolling operation possible. Furthermore, the net frictional force and the surface velocity of the roll must be in the same direction in order to supply work to the system. This indicates that the neutral point should be toward the exit in order to satisfy these requirements.

Forward slip in rolling is defined in terms of the exit velocity of the strip V_f and the surface speed of the roll V_r as

$$\text{Forward slip} = \frac{V_f - V_r}{V_r} \tag{6.20}$$

and is a measure of the relative velocities involved.

6.11.1 FORCES AND STRESSES

Although the deformation zone in the roll gap is subjected to a state of stress similar to that in upsetting, the calculation of forces and stress distribution in flat rolling is more involved because of the curved surface of contact. In addition, the material at the exit is strain hardened, so the flow stress at the exit is higher than that at the entry.

The stresses on an element in the entry and exit zones, respectively, are shown in Fig. 6.44. Note that the only difference between the two elements is the direction of the friction force. Using the *slab method* of analysis for plane strain, described in

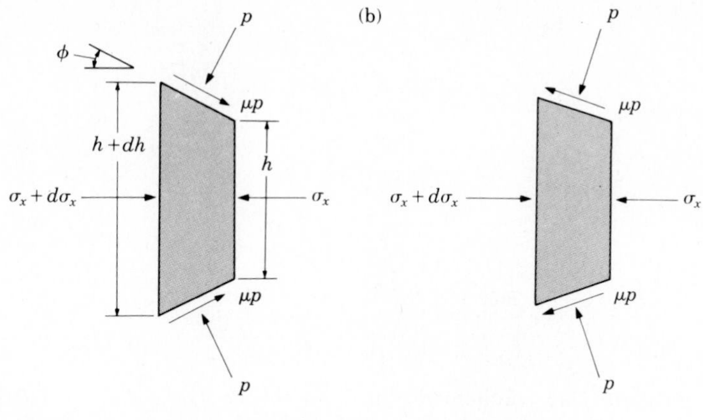

FIGURE 6.44 Stresses on an element in flat rolling (see also Fig. 2.53). (a) Entry zone. (b) Exit zone.

Section 2.10.1, it can be shown that the normal stress (pressure) distribution along the arc of contact is as follows:

In the entry zone,

$$p = Y' \frac{h}{h_0} e^{\mu(H_0 - H)} \qquad (6.21)$$

and in the exit zone,

$$p = Y' \frac{h}{h_f} e^{\mu H}, \qquad (6.22)$$

where H is a convenient parameter and is defined as,

$$H = 2\sqrt{\frac{R}{h_f}} \tan^{-1}\left(\sqrt{\frac{R}{h_f}} \cdot \phi\right). \qquad (6.23)$$

and H_0 is the value of H at $\phi = 0$.

Note from these expressions that the pressure p is a function of h and the angular position ϕ along the arc of contact. At entry, $\phi = \alpha$ and at exit, $\phi = 0$. For a strain-hardening material, the flow stress Y_f in these expressions corresponds to the strain that the material has undergone at that particular location in the roll gap. (Although the stresses in rolling are compression and, hence, negative signs should be used according to Section 2.8, it is the practice in rolling and in other bulk deformation analyses to ignore the negative signs.)

These expressions also indicate that the pressure increases with increasing strength of the material, increasing coefficient of friction, and increasing R/h_f ratio. Note that the R/h_f ratio in rolling is equivalent to the a/h ratio in upsetting.

The theoretical pressure distribution in the roll gap, for a non–strain-hardening material, is shown in Fig. 6.45. Note the similarity of this curve to Fig. 6.4 (it is a *friction hill*). Also note that the neutral point shifts toward the exit as friction decreases. This is to be expected because, when friction approaches zero, the rolls begin to slip instead of pulling the strip in. This means that the neutral point must approach

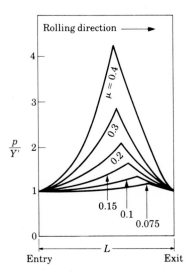

FIGURE 6.45 Pressure distribution in the roll gap as a function of the coefficient of friction (see also Fig. 2.54). The area under the curve is the roll separating force per unit width of the strip. Note that, as friction increases, the neutral point shifts toward the entry. Without friction, the rolls slip and the neutral point shifts completely toward the exit.

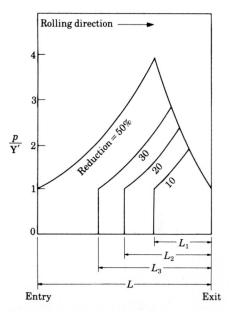

FIGURE 6.46 Pressure distribution in the roll gap as a function of reduction in thickness. Note the increase in the area under the curves with increasing reduction in thickness, thus increasing the roll-separating force.

the exit. (This situation is similar to the spinning of automobile wheels when accelerating on wet pavement.)

The effect of reduction in thickness of the strip on the pressure distribution is shown in Fig. 6.46. As reduction increases, the length of contact in the roll gap increases. This in turn increases the peak pressure. The curves shown are theoretical; actual pressure distributions, as determined experimentally, have smoother curves with their peaks rounded off.

The area under these pressure–contact length curves is the *roll force per unit width* of the strip (i.e., *specific roll force, F*). This area can be calculated graphically once the points are plotted. This force can also be calculated from the expression,

$$F = \int_0^{\phi_n} pR \, d\phi + \int_{\phi_n}^{\alpha} pR \, d\phi. \tag{6.24}$$

A simpler method of calculating roll force is to multiply the contact area with an average contact stress. Thus,

$$F = Lwp_{av}, \tag{6.25}$$

where L is the arc of contact and w is the width of the strip. The dimension L can be approximated as the projected area and is given by the expression,

$$L = \sqrt{R\Delta h},$$ (6.26)

where R is the roll radius and Δh is the difference between the original and final thicknesses of the strip (*draft*).

The estimation of p_{av} depends on the h/L ratio, where h is now the average thickness of the strip in the roll gap. For large h/L ratios (small reductions and/or small roll diameters), the rolls act as indenters in a hardness test. Friction is not significant and p_{av} is obtained from Fig. 2.39. Small h/L ratios (large reductions and/or large roll diameters) are equivalent to high a/h ratios, as in Fig. 6.4. Thus, friction is predominant and p_{av} is obtained from Fig. 6.8(a). For strain-hardening materials the appropriate flow stresses must be calculated.

As a rough approximation and for low frictional conditions, Eq. (6.25) can be simplified as,

$$F = Lw\overline{Y},$$ (6.27)

where $\overline{Y}$ is the average flow stress (see Fig. 2.51) of the material in the roll gap.

6.11.2 TORQUE AND POWER REQUIREMENTS

The roll torque T for each roll can be calculated analytically from the expression,

$$T = \int_{\phi_n}^{\alpha} \mu p R^2 \, d\phi - \int_0^{\phi_n} \mu p R^2 \, d\phi.$$ (6.28)
$$\quad\;\;_{\text{(entry)}} \qquad\qquad _{\text{(exit)}}$$

Note that the negative sign indicates the change in direction of the friction force at the neutral point. Thus, if the frictional forces are equal to each other, the torque is zero.

The torque in rolling can also be estimated by assuming that the roll-separating force F acts in the middle of the arc of contact, i.e., a moment arm of $0.5L$, and that F is perpendicular to the plane of the strip. (Whereas $0.5L$ is a good estimate for hot rolling, a value of $0.4L$ has been found to be a better estimate for cold rolling.)

The torque per roll T is then given by,

$$T = \frac{FL}{2}.$$ (6.29)

The power required per roll is,

$$\text{Power} = T\omega$$ (6.30)

where $\omega = 2\pi N$ and N is the revolutions per minute of the roll. Thus, for a set of two powered rolls,

$$\text{hp} = \frac{2\pi FLN}{33,000},$$ (6.31)

where F is in lb and L in ft. Also,

$$kW = \frac{2\pi FLN}{60,000},$$ (6.32)

where F is in newtons and L in meters.

● **Illustrative Problem 6.4**

A 9 in. wide 6061-O aluminum strip is rolled from a thickness of 1.00 in. to 0.80 in. If the roll radius is 12 in. and the roll rpm is 100, calculate the horsepower required for this operation.

SOLUTION.   The power needed for a set of two rolls is given by Eq. (6.31),

$$hp = \frac{2\pi FLN}{33,000},$$

where F is given by Eq. (6.27) and L by Eq. (6.26). Thus,

$$F = Lw\bar{Y}$$

$$L = \sqrt{R\Delta h}.$$

We now have

$$L = \sqrt{(12)(1.0 - 0.8)} = 1.55 \text{ in.}$$

$$w = 9 \text{ in.}$$

For 6061-O aluminum, $K = 30,000$ psi and $n = 0.2$ as obtained from Table 2.4. The true strain in this operation is

$$\epsilon_1 = \ln\left(\frac{1.0}{0.8}\right) = 0.223.$$

Thus, from Eq. (6.11)

$$\bar{Y} = \frac{(30,000)(0.223)^{0.2}}{1.2} = 18,500 \text{ psi},$$

and

$$F = (1.55)(9)(18,500) = 258,000 \text{ lb.}$$

Therefore,

$$hp = \frac{(2\pi)(258,000)(0.13)(100)}{33,000} = 639 \text{ hp}$$

$$= 477 \text{ kW.} \quad ●$$

6.11.3 FORCES IN HOT ROLLING

Because ingots and slabs are usually hot rolled, the calculation of forces and torque in hot rolling is important. However, there are two major difficulties. One is the proper estimation of the coefficient of friction μ at elevated temperatures (see below). The other is the strain-rate sensitivity of metals at high temperatures (Section 2.2.7).

The average strain rate in flat rolling can be obtained by dividing the strain by the time that it takes for an element to undergo this strain in the roll gap. The time can be approximated as L/V_r. Thus,

$$\dot{\bar{\epsilon}} = \frac{V_r}{L} \ln\left(\frac{h_0}{h_f}\right). \tag{6.33}$$

The flow stress Y_f of the material corresponding to this strain rate must first be obtained and then substituted in the proper equations. These calculations are approximate. Variations in μ and temperature within the strip in hot rolling further contribute to the difficulties in calculating forces accurately.

6.11.4 FRICTION

It is necessary to have some friction in rolling because the rolls cannot pull the strip into the roll gap without it. On the other hand, as we have seen, forces and power requirements increase with increasing friction.

In cold rolling, the coefficient of friction μ usually ranges between 0.02 and 0.3, depending on the materials and lubricants used. The low ranges for the coefficient of friction are obtained with effective lubricants and regimes approaching hydrodynamic lubrication, such as in cold rolling of aluminum at high speeds.

In hot rolling, the coefficient of friction may range from about 0.2, with effective lubrication, to as high as 0.7, indicating sticking, which usually occurs with steels, stainless steels, and high-temperature alloys.

The maximum possible *draft* (i.e., $h_0 - h_f$) in flat rolling can be shown to be a function of friction and roll radius as follows:

$$\Delta h_{max} = \mu^2 R. \tag{6.34}$$

Hence, the higher the friction and the larger the roll radius, the greater the maximum draft. As expected, the draft is zero when there is no friction. The maximum value of the angle α (*angle of acceptance*) in Fig. 6.43 is geometrically related to Eq. (6.34). From the simple model of a block sliding down an inclined plane, it can be seen that,

$$\alpha_{max} = \tan^{-1}\mu. \tag{6.35}$$

If α_{max} is larger than this value, the rolls begin to slip because the friction is not high enough to pull the material through the roll gap.

6.11.5 FRONT AND BACK TENSION

Although the roll force F can be reduced by various means, such as lower friction, smaller roll radii, smaller reductions, and higher workpiece temperatures, an effective method is to reduce the apparent compressive yield stress of the material by applying longitudinal tension. Recall from yield criteria (Section 2.8) that if tension is applied to a strip (Fig. 6.47), the yield stress normal to the strip surface will be reduced and hence the roll pressure will decrease.

Tensions in rolling can be applied either at the entry (*back tension* σ_b) or at the exit of the strip (*front tension* σ_f) or both. Eqs. (6.21) and (6.22) are then modified as follows to include the effect of tension:

For the entry zone:

$$p = (Y' - \sigma_b) \frac{h}{h_0} e^{\mu(H_0 - H)} \tag{6.36}$$

and for the exit zone:

$$p = (Y' - \sigma_f) \frac{h}{h_f} e^{\mu H}. \tag{6.37}$$

Depending on the relative magnitudes of the tensions applied, the neutral point may shift, as can be seen in Fig. 6.47. This shift will affect the pressure distribution, torque, and power requirements in rolling. Front tension is controlled by the torque on the coiler (*delivery* reel) around which the rolled sheet is coiled. The back tension is controlled by a braking system in the uncoiler (*payoff* reel). Special instrumentation is available for these controls. Tensions are particularly important in rolling thin, high-strength materials because they require high roll forces.

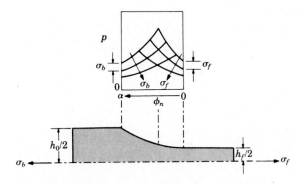

FIGURE 6.47 Pressure distribution as a function of front and back tension. Note the shifting of the neutral point, and the reduction in the area under the curves with increasing tension.

6.11.6 ROLL DEFLECTIONS

Roll forces tend to bend the rolls, as shown in Fig. 6.48(a), with the result that the strip is thicker at its center than at its edges (*crown*). The usual method of avoiding this problem is to grind the rolls so that their diameter at the center is slightly larger than at the edges. This is known as *camber*. In rolling sheet metals the camber is generally less than 0.01 in. (0.25 mm) on the radius.

When properly designed, such rolls produce flat strips as shown in Fig. 6.48(b). However, a particular camber is correct only for a certain load and width of strip. In hot rolling, uneven temperature distribution in the roll can also cause variation in its diameter along the length of the roll. In fact, camber can be controlled by varying the location of the coolant (lubricant) on the rolls in hot rolling.

Forces also tend to flatten the rolls elastically, much like the flattening of tires on automobiles. In rolling, flattening of the rolls means that the roll radius is larger and hence a larger contact area for the same reduction in thickness. Thus, the roll force F increases.

It can be shown that the new (distorted) roll radius R' is given by,

$$R' = R\left(1 + \frac{CF}{h_0 - h_f}\right). \tag{6.38}$$

In this formula C is 1.6×10^{-4} in.^{2}/klb (2.3×10^{-2} mm^2/kN) for steel rolls and 3.15×10^{-4} (4.57×10^{-2}) for cast iron rolls. F is the roll force per unit width of strip, klb/in. (kN/mm). The higher the elastic modulus of the roll material the less the roll will distort. It can also be seen from Eq. (6.38) that reducing the roll force, such as by using an effective lubricant or taking smaller reductions, also reduces roll flattening.

It can be seen from Eq. (6.38) that R' cannot be determined directly since it is a function of the roll force F, which itself is a function of the roll radius. The solution is obtained by trial and error. (Note that R in Eqs. 6.23, 6.24, and 6.28 should be replaced by R' when roll flattening occurs to a significant degree.)

(a)

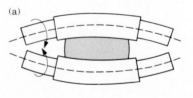

(b)

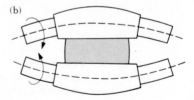

FIGURE 6.48 (a) Bending of straight cylindrical rolls because of the roll-separating force. (b) Bending of rolls, ground with camber, that produce a sheet with uniform thickness during rolling.

6.12 FLAT ROLLING PRACTICE

The initial breakdown of an ingot by hot rolling converts the coarse-grained, brittle, and porous cast structure to a *wrought* structure with finer grains and greater ductility. The reduction in thickness per pass in practice depends on the type of workpiece material and its ductility. For high-strength alloys reductions are usually small, whereas for soft and nonferrous metals they are high.

Temperatures in hot rolling are similar to those in hot forging (Section 6.7). The advantages of hot rolling are the lower forces needed and the higher ductility of the metal. Cold rolling, on the other hand, requires higher forces but it produces a much better surface finish, allows better control of tolerances, and produces plates and sheet with higher strength. Cold rolling produces anisotropic sheets, as shown in Fig. 3.23, and as further described in Section 7.9.

Although rolling plates and sheets with high width-to-thickness ratios is essentially a process of plane strain, with smaller ratios, such as a square cross-section, the width increases considerably during rolling. This increase in width is known as *spreading* (Fig. 6.49). Spreading decreases with increasing width-to-thickness ratios of the entering material, increasing friction, and increasing ratios of roll radius-to-strip thickness.

The thickness of sheet is identified by a *gage number* (Table 6.10); the smaller the number the thicker the sheet. There are different numbering systems depending on the sheet metal. Rolled sheets of copper and brass are also identified by thickness changes by rolling, as shown in Table 6.11. Nonferrous metals such as aluminum foil used in wrapping candy and cigarettes, can be rolled to a thickness on the order of 0.0008 in. (0.02 mm).

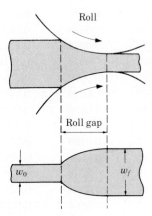

FIGURE 6.49 Spread (increase in width) in flat rolling. A similar observation can be made when dough is rolled with a rolling pin.

TABLE 6.10
STANDARD SHEET AND WIRE GAGES FOR VARIOUS MATERIALS
(Dimensions in inches)

GAGE NO.	ALUMINUM AND BRASS — BROWN & SHARP GA.	STEEL SHEETS — MFRS. STD. GA.	GALVANIZED SHEET	STAINLESS STEEL	COPPER SHEET — STUBBS GA.	STEEL WIRE — STEEL WIRE GA.	MUSIC WIRE
6-0's	0.5800	—	—	—	—	0.4615	0.004
5-0's	0.5165	—	—	—	—	0.4305	0.005
4-0's	0.4600	—	—	—	—	0.3938	0.006
3-0's	0.4096	—	—	—	—	0.3625	0.007
2-0's	0.3648	—	—	—	—	0.3310	0.008
0	0.3249	—	—	—	—	0.3065	0.009
1	0.2893	—	—	—	0.300	0.2830	0.010
2	0.2576	—	—	—	0.284	0.2625	0.011
3	0.2294	0.2391	—	—	0.259	0.2437	0.012
4	0.2043	0.2242	—	—	0.238	0.2253	0.013
5	0.1819	0.2092	—	—	0.220	0.2070	0.014
6	0.1620	0.1943	—	—	0.203	0.1920	0.016
7	0.1443	0.1793	—	—	0.180	0.1770	0.018
8	0.1285	0.1644	0.1681	0.1719	0.165	0.1620	0.020
9	0.1144	0.1495	0.1532	0.1562	0.148	0.1483	0.022
10	0.1019	0.1345	0.1382	0.1406	0.134	0.1350	0.024
11	0.0907	0.1196	0.1233	0.1250	0.120	0.1205	0.026
12	0.0808	0.1046	0.1084	0.1094	0.109	0.1055	0.029
13	0.0720	0.0897	0.0934	0.0938	0.095	0.0915	0.031
14	0.0641	0.0747	0.0785	0.0781	0.083	0.0800	0.033
15	0.0571	0.0673	0.0710	0.0703	0.072	0.0720	0.035
16	0.0508	0.0598	0.0635	0.0625	0.065	0.0625	0.037
17	0.0453	0.0538	0.0575	0.0563	0.058	0.0540	0.039
18	0.0403	0.0478	0.0516	0.0500	0.049	0.0475	0.041
19	0.0359	0.0418	0.0456	0.0438	0.042	0.0410	0.043
20	0.0320	0.0359	0.0396	0.0375	0.035	0.0348	0.045
21	0.0285	0.0329	0.0366	0.0344	0.032	0.0317	0.047
22	0.0253	0.0299	0.0336	0.0313	0.028	0.0286	0.049
23	0.0226	0.0269	0.0306	0.0281	0.025	0.0258	0.051
24	0.0201	0.0239	0.0276	0.0250	0.022	0.0230	0.055
25	0.0179	0.0209	0.0247	0.0219	0.020	0.0204	0.059
26	0.0159	0.0179	0.0217	0.0188	—	0.0181	0.063
27	0.0142	0.0164	0.0202	0.0172	—	0.0173	0.067
28	0.0126	0.0149	0.0187	0.0156	—	0.0162	0.071
29	0.0113	0.0135	0.0172	0.0141	—	0.0150	0.075
30	0.0100	0.0120	0.0157	0.0125	—	0.0140	0.080

TABLE 6.10 (*continued*)

GAGE NO.	ALUMINUM AND BRASS	STEEL SHEETS	GALVANIZED SHEET	STAINLESS STEEL	COPPER SHEET	STEEL WIRE	
	BROWN & SHARP GA.	*MFRS. STD. GA.*			*STUBBS GA.*	*STEEL WIRE GA.*	*MUSIC WIRE*
31	0.0089	0.0105	0.0142	—	—	0.0132	0.085
32	0.0080	0.0097	0.0143	—	—	0.0128	0.090
33	0.0071	0.0090	—	—	—	0.0118	0.095
34	0.0063	0.0082	—	—	—	0.0104	0.100
35	0.0056	0.0075	—	—	—	0.0095	0.106
36	0.0050	0.0067	—	—	—	0.0090	0.112
37	0.0045	0.0064	—	—	—	0.0085	0.118
38	0.0040	0.0060	—	—	—	0.0080	0.124

TABLE 6.11
HARDNESS TERMINOLOGY FOR COLD-ROLLED COPPER AND BRASS SHEETS

TERMINOLOGY	DESIGNATION	INCREASE IN GAGE NUMBER	REDUCTION IN THICKNESS, %
Annealed (dead soft)		0	0
$\frac{1}{4}$ hard	H01	1	11
$\frac{1}{2}$ hard	H02	2	21
$\frac{3}{4}$ hard	H03	3	29
Hard	H04	4	37
Extra hard	H06	6	50
Spring	H08	8	60
Extra spring	H10	10	69
Special spring	H12	12	75
Super spring	H14	14	80

6.12.1 DEFECTS IN ROLLED PRODUCTS

Successful rolling practice requires a balance of many factors including material properties, process variables, and lubrication. Defects may be on the surfaces of the rolled plates and sheets, or they may be structural defects within the material.

Surface defects may be due to inclusions and impurities in the material, scale, rust, dirt, roll marks, and other causes related to the prior treatment and working of the material. In hot rolling blooms, billets, and slabs, the surface is usually conditioned prior to rolling by various means, such as by torch (*scarfing*), to remove scale.

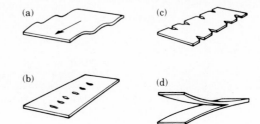

FIGURE 6.50 Schematic illustration of typical defects in flat rolling. (a) Wavy edges. (b) Zipper cracks in center of strip. (c) Edge cracks. (d) Alligatoring.

Structural defects are those that distort or affect the integrity of the rolled product. Some typical defects are shown in Fig. 6.50. Wavy edges are due to the bending of the rolls; the edges of the strip are thinner than the center. Since the edges elongate more than the center, they buckle because they are restrained from expanding freely.

The cracks shown in Fig. 6.50(b) and (c) are usually due to low ductility and to barreling. *Alligatoring* is a complex phenomenon due to inhomogeneous deformation of the material during rolling or to defects in the original cast ingot such as piping. (See Fig. 5.17.)

6.12.2 RESIDUAL STRESSES

Residual stresses can be generated in rolled sheets and plates due to inhomogeneous plastic deformation in the roll gap (Fig. 6.51). Small-diameter rolls or small reductions tend to plastically work the metal at its surfaces (similarly to shot peening or roller burnishing). This generates compressive residual stresses on the surfaces and tensile stresses in the bulk. (See also Figs. 4.38 and 6.93b.)

On the other hand, large-diameter rolls and high reductions tend to deform the bulk to a greater extent than the surfaces. This is due to the frictional constraint at the surfaces along the arc of contact. This situation generates residual stresses that are opposite to the previous case, as shown in Fig. 6.51.

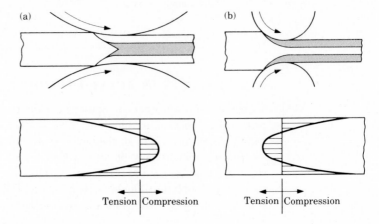

FIGURE 6.51 The effect of roll radius on the type of residual stresses developed in flat rolling. (a) Large rolls, or large reduction in thickness. (b) Small rolls, or small reduction.

(a)

(b)

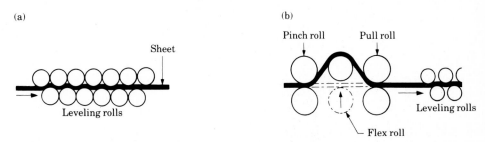

FIGURE 6.52 Schematic illustration of methods of roller leveling. These processes are used to flatten rolled sheets.

6.12.3 ROLLER LEVELING

A rolled sheet may not be sufficiently flat for subsequent use or processing. To improve *flatness*, the strip is passed through a series of leveling rolls (which are usually driven individually) where it is flexed in opposite directions. Figure 6.52 shows two different roll arrangements for roller leveling; various other roller arrangements are also available. The minimum sheet thickness that can be flattened by this method is about 0.020 in. (0.5 mm).

6.13 ROLLING EQUIPMENT AND PRACTICE

A wide variety of rolling equipment is available with a number of roll arrangements. The major types are shown in Fig. 6.53. *Two-high* or *three-high* (developed in the mid 1800s) rolling mills are used for initial breakdown passes on cast ingots with roll diameters ranging from 24 to 55 in. (600 to 1400 mm).

It is desirable to use small-diameter rolls because the smaller the roll radius the lower the roll force. However, small rolls deflect under the roll forces and have to be supported by other rolls (Fig. 6.53c and d).

The range of the dimensions in rolling is extremely wide. The rolled product can be as wide as 200 in. (5000 mm) and as thin as 0.0001 in. (0.0025 mm) obtained in a cluster mill.

The *cluster mill* (*Sendzimir*) is particularly suitable for cold rolling thin strip of high-strength metals. One arrangement for the rolls is shown in Fig. 6.54. The *work roll* (smallest roll) can be as small as 1/4 in. (6 mm) in diameter and is usually made of tungsten carbide for rigidity, strength, and wear resistance.

Stiffness in rolling mills is important for controlling dimensions. Mills can be highly automated with rolling speeds as high as 5000 ft/min (25 m/s). Flat rolling can also be carried out with front tension only, using idling rolls (*Steckel rolling*). For this case, the torque T on the roll (Eq. 6.28) is zero, assuming frictionless bearings.

Requirements for roll materials are mainly strength and resistance to wear. Three common roll materials are cast iron, cast steel, and forged steel (Table 6.4). For hot

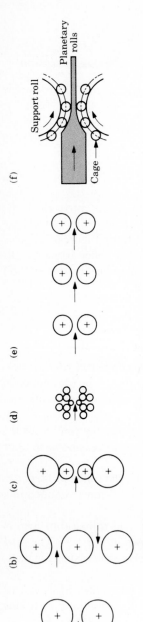

FIGURE 6.53 Schematic illustration of various roll arrangements. (a) 2-high. (b) 3-high. (c) 4-high. (d) Cluster. (e) Tandem rolling with three stands. (f) Planetary.

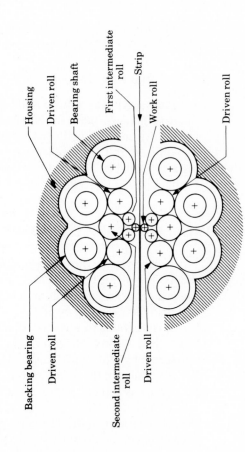

FIGURE 6.54 Schematic illustration of a cluster (Sendzimir) mill. These mills are very rigid and are used in rolling thin sheets of high-strength materials, with good control of dimensions.

rolling, roll surfaces are generally rough and may even have notches or grooves in order to pull the metal through the roll gap at high reductions. Rolls for cold rolling are ground to a fine finish and, for special applications, are also polished.

Hot rolling of ferrous alloys is usually carried out without a lubricant, although graphite may be used. Aqueous solutions are used to cool the rolls and to break up the scale on the workpiece. Nonferrous alloys are hot rolled with a variety of compounded oils, emulsions, and fatty acids. Cold rolling is carried out with low-viscosity lubricants, including mineral oils, emulsions, paraffin, and fatty oils. (See Table 6.7.)

6.14 MISCELLANEOUS ROLLING PROCESSES

In this section non–flat-rolling operations are described.

6.14.1 SHAPE ROLLING

Straight structural shapes such as bars of various cross-sections, channel sections, I-beams, and railroad tracks are rolled by passing the stock through a number of pairs of specially designed rollers (Fig. 6.55). These processes were first developed in the late 1700s. The original material is usually a *bloom* (Fig. 1.8). Designing a series of rolls (*roll-pass design*) requires experience in order to avoid defects and hold tolerances, although some of these defects may also be due to the material rolled (Fig. 6.56). The material elongates as it is reduced in cross-section. However, since for a shape such as a channel the reduction is different in different locations within the section, elongation is not uniform. This can cause warping or cracking of the product. Airfoil shapes can also be produced by shape-rolling techniques.

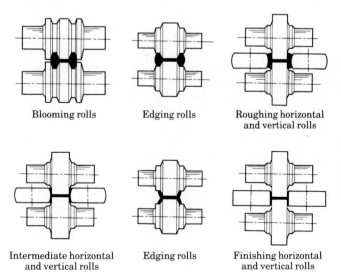

Blooming rolls   Edging rolls   Roughing horizontal
and vertical rolls

Intermediate horizontal   Edging rolls   Finishing horizontal
and vertical rolls   and vertical rolls

FIGURE 6.55 Stages in shape rolling of an H-section. Various large structural sections, such as I-beams, are made by this process.

FIGURE 6.56 Internal defect developed during shape rolling of a steel rail. These defects are generally attributed to the presence of impurities and inclusions in the material which, under adverse states of stress and flow pattern during deformation, lead to fracture.

6.14.2 RING ROLLING

In this process, a small-diameter thick ring is expanded into a larger-diameter ring with a smaller cross-section. This is done by placing the ring between two rolls, one of which is driven (Fig. 6.57). The thickness is reduced by bringing the rolls closer as they rotate. The reduction in thickness is compensated for by an increase in the diameter of the ring.

A typical procedure for producing a *seamless ring* is shown in Fig. 6.58. A great variety of cross-sections can be ring rolled with shaped rolls. This process can be carried out at room or at elevated temperatures, depending on the size and strength of the product.

The advantages of ring rolling, compared with other processes for making the same part, are: short production runs, material savings, close tolerances, and favorable grain flow direction. Typical applications of ring rolling are large rings for rockets and turbines, gearwheel rims, ball and roller bearing races, flanges and reinforcing rings for pipes, and pressure vessels.

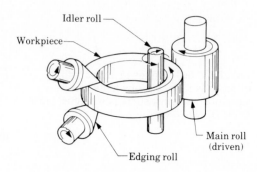

FIGURE 6.57 A horizontal ring rolling operation. The reduction in the thickness of the ring is compensated for by the increase in ring diameter. Large bearing races and turbine components are made by this process.

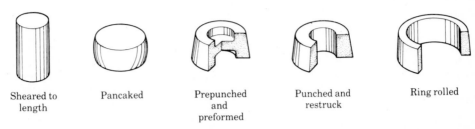

Sheared to length | Pancaked | Prepunched and preformed | Punched and restruck | Ring rolled

FIGURE 6.58   Stages in the manufacturing of a ring with varying wall thickness using a combination of processes. The race for a large tapered bearing can be made by these processes.

6.14.3 THREAD AND GEAR ROLLING

This is a cold-forming process by which threads are formed on round rods or work-pieces by passing them between reciprocating or rotating dies (Fig. 6.59). Typical products are screws, bolts, and similar threaded parts.

With flat dies the threads are rolled on the rod or wire with each stroke of the reciprocating die. Production rates are very high, depending on the diameter of the product. With small diameters the rates can be as high as eight per second and with larger diameters (as much as 1 in. [25 mm]) about one per second.

Two- or three-roller thread-rolling machines are also available. In another design (Fig. 6.59b), threads are formed with a rotary die with production rates as high as 80 per second.

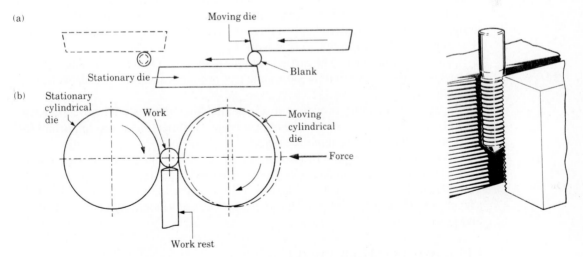

FIGURE 6.59   Thread rolling processes. (a) Flat dies. (b) Two-roller dies. These processes are used extensively in making threaded fasteners at high rates of production.

(a)

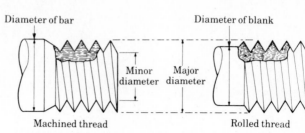

Machined thread   Rolled thread

FIGURE 6.60 (a) Schematic illustration of machined and rolled threads. Note the increase in the blank diameter for the rolled thread. (b) Grain flow lines in machined and rolled threads. Unlike machining, which cuts through the grains of the metal, rolled threads have improved strength due to cold working.

(b)

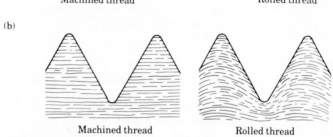

Machined thread   Rolled thread

The thread-rolling process has the advantages of generating threads without any metal loss and with improved strength due to cold working. The surface finish is very smooth and the process induces compressive residual stresses on the surfaces, thus improving fatigue life. The product is superior to that made by thread cutting and is used in the production of almost all externally threaded fasteners. (See also Section 12.2.)

Because of volume constancy in plastic deformation, a rolled thread requires a smaller diameter round stock to produce the same major diameter as a machined thread (Fig. 6.60). Also, whereas machining removes material by cutting through the grain flow lines of the material, rolled threads have a grain flow pattern that improves the strength of the thread due to cold working.

Thread rolling can also be carried out internally with a fluteless forming tap. The process is similar to external thread rolling and produces accurate threads with good strength.

In all thread-rolling processes it is essential that the material have sufficient ductility and that the rod or wire be of proper size. Lubrication is also important for good surface finish and to minimize defects.

Spur and helical gears are also produced by cold-rolling processes similar to thread rolling. The process may be carried out on solid cylindrical blanks or on pre-cut gears. Helical gears can also be made by a direct extrusion process, using specially shaped dies. Cold rolling of gears has many applications in automatic transmissions and power tools.

6.14.4 ROTARY TUBE PIERCING

This is a hot-working process for making long, thick-walled seamless tubing by the technique shown in Fig. 6.61. The principle of the process is based on the observa-

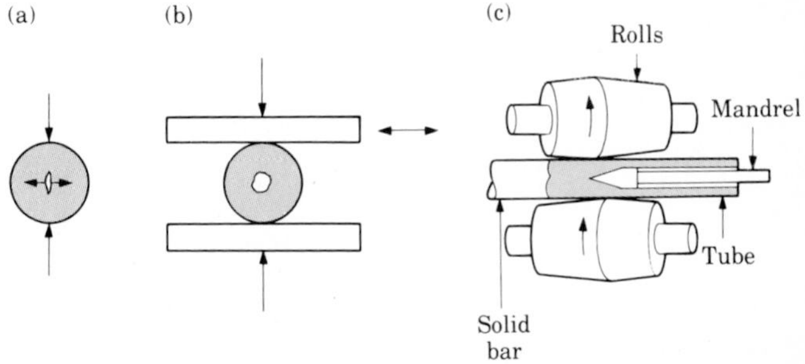

FIGURE 6.61 Cavity formation by secondary tensile stresses in a solid round bar and its utilization in the rotary tube piercing process. This is the principle of the Mannesmann mill for seamless tubemaking.

tion that when a round bar is subjected to radial compression in the manner shown in Fig. 6.61(a), tensile stresses develop at the center of the rod. When subjected to cycling compressive stresses as shown in Fig. 6.61(b), a cavity begins to form at the center of the rod.

In *rotary tube piercing* (*Mannesmann* process, developed in the 1880s) this process is carried out by an arrangement of rotating rolls, as shown in Fig. 6.61(c). The axes of the rolls are skewed in order to pull the round bar through the rolls by their rotary action. A mandrel assists the operation by expanding the hole and sizing the inside diameter of the tube. Because of the severe deformation that the metal undergoes in this process, it is necessary to use high-quality bars free of defects.

6.15 EXTRUSION

In the basic extrusion process, developed in the late 1700s for lead pipe, a round billet is placed into a chamber and is forced through a die opening by a ram. The die may be round or of various other shapes. There are four basic types of extrusion: direct, indirect, hydrostatic, and impact (Fig. 6.62).

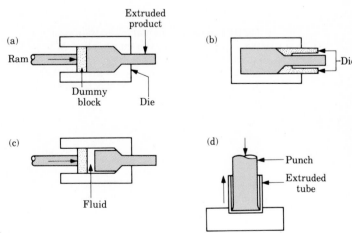

FIGURE 6.62 Types of extrusion.
(a) Direct. (b) Indirect.
(c) Hydrostatic. (d) Impact extrusion.

Direct extrusion (*forward* extrusion) is similar to forcing the paste through the opening of a toothpaste tube. The billet slides relative to the container wall; the wall friction increases the ram force considerably.

In *indirect extrusion* (*reverse*, *inverted*, or *backward* extrusion), the die moves toward the billet; thus, except at the die, there is no relative motion at the billet–container interface.

In *hydrostatic* extrusion, the chamber is filled with a fluid that transmits the pressure to the billet, which is then extruded through the die. There is no friction along the container walls.

Impact extrusion is a form of indirect extrusion and is particularly suitable for hollow shapes.

Extrusion processes can be carried out hot or cold. Because a chamber is involved, each billet is extruded individually and hence extrusion is basically a *batch* process. *Cladding* by extrusion can also be carried out with coaxial billets (such as copper clad with silver), provided that the flow stresses of the two metals are similar.

6.15.1 METAL FLOW IN EXTRUSION

Because the billet is forced through a die, with a substantial reduction in its cross-section, the metal flow pattern in extrusion is an important factor in the overall process. A common technique for investigating the flow pattern is to half the round billet lengthwise and mark one face with a grid pattern (Fig. 6.63). The two halves are then placed together in the container (they may also be fastened together or *brazed* to keep the two halves intact) and extruded. They are then taken apart and inspected.

(a) Solid billet

(b) Billet sectioned longitudinally and one cut face marked with a grid

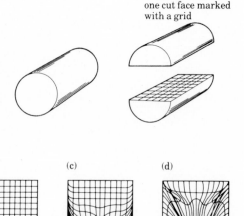

FIGURE 6.63 Schematic illustration of placing grid lines in the mid-section of a round billet to observe metal flow pattern in extrusion. The two halves are then placed in the extrusion chamber and extruded. See Fig. 6.64.

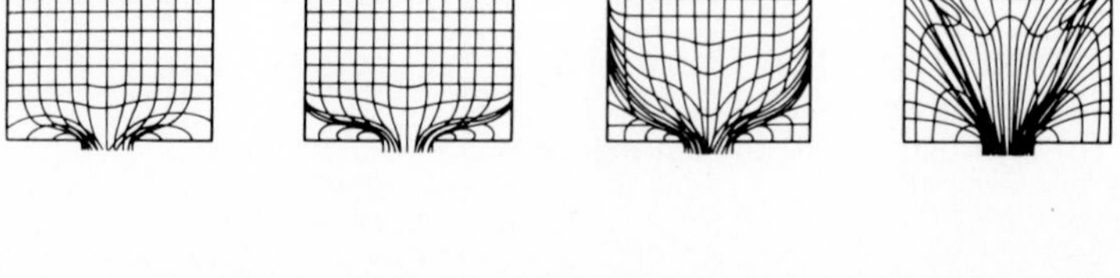

(a)   (b)   (c)   (d)

FIGURE 6.64 Schematic illustration of four different types of metal flow in direct extrusion. (See Section 6.15.1.)

Figure 6.64 shows some typical results in direct extrusion with square dies. The conditions under which these different flow patterns are obtained are outlined below.

a. The most homogeneous flow pattern is obtained when there is no friction at the billet–container–die interfaces (Fig. 6.64a). This type of flow occurs when the lubricant is very effective, or with indirect extrusion.

b. A second pattern is obtained when there is little or no friction at the container wall, but some friction at the die surfaces. A small volume of the metal remains stationary (Fig. 6.64b) (*dead-metal zone*) at the corners of the container during extrusion.

c. When friction along all interfaces is high, the dead-metal zone increases, resulting in a third type of flow pattern. Note that there is a high-shear area as the material flows into the die exit, somewhat like a funnel (Fig. 6.64c). This could indicate that the billet surfaces (with their oxide layer and lubricant) could enter this high-shear zone and be extruded, causing defects in the extruded product.

d. In the fourth type of flow the high-shear zone extends further back (Fig. 6.64d). This can be due to two factors: high container-wall friction, which retards the flow of the billet, and materials whose flow stress drops rapidly with increasing temperature. In hot working, the material near the container walls cools rapidly and hence increases in strength. Thus the material in the central regions flows toward the die more easily than that at the outer regions. As a result, a large dead-metal zone forms and the flow is inhomogeneous. This flow pattern leads to a defect known as a *pipe* or *extrusion defect* (Section 6.17.2).

It is apparent from these typical examples that two factors that greatly influence metal flow in extrusion are the frictional conditions at billet–container–die interfaces and thermal gradients in the billet.

6.15.2 FORCES

The ram force in direct extrusion can be calculated as described below for different situations.

Ideal Deformation
The *extrusion ratio* R is defined as,

$$R = \frac{A_0}{A_f},$$ (6.39)

where A_0 is the billet cross-sectional area and A_f is the area of the extruded product (Fig. 6.65). The absolute value of the true strain is then,

$$\epsilon_1 = \ln\left(\frac{A_0}{A_f}\right) = \ln\left(\frac{L_f}{L_0}\right) = \ln R,$$ (6.40)

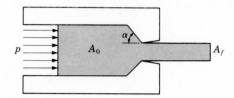

FIGURE 6.65   Variables in the direct extrusion process.

where L_0 and L_f are the lengths of the billet and the extruded product, respectively.

For a perfectly plastic material with a yield stress Y, the energy dissipated in plastic deformation per unit volume, u, is,

$$u = Y\epsilon_1. \tag{6.41}$$

Hence the work done on the billet is,

$$\text{Work} = (A_0)(L_0)(u). \tag{6.42}$$

This work is supplied by the ram force F, which travels a distance L_0. Thus,

$$\text{Work} = FL_0 = pA_0L_0, \tag{6.43}$$

where p is the *extrusion pressure* at the ram. Equating the work of plastic deformation to the external work done, we find

$$p = u = Y \ln\left(\frac{A_0}{A_f}\right) = (Y)(\ln R). \tag{6.44}$$

Note that, for strain-hardening materials, Y should be replaced by the average flow stress $\overline{Y}$ (Fig. 2.51).

Ideal Deformation and Friction

Equation (6.44) pertains to ideal deformation without any friction. Based on the slab method of analysis, it can be shown that when friction at the die–billet interface is included (but not the container-wall friction) the pressure p is given by,

$$p = Y\left(1 + \frac{\tan \alpha}{\mu}\right)[(R)^{\mu \cot \alpha} - 1]. \tag{6.45}$$

An estimate of p can also be obtained by assuming that, because of the dead zone, material flow in the container takes place along a 45-degree angle. If the frictional stress along this 45-degree "die" is assumed to be the yield stress in shear k of the material (where $k = Y/2$), it can be shown that,

$$p = 1.7Y \ln R. \tag{6.46}$$

In this analysis, the force required to overcome friction at the billet–container interface has been neglected. Assuming that the frictional stress is equal to the shear yield stress of the material k, the additional ram pressure required due to friction, p_f,

FIGURE 6.66 Schematic illustration of typical extrusion pressure as a function of ram travel. (a) Direct extrusion. (b) Indirect extrusion. The pressure in direct extrusion is higher, because of frictional resistance in the chamber as the billet moves toward the die.

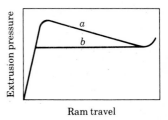

can be obtained as,

$$(p_f)\left(\frac{\pi D_0^2}{4}\right) = \pi D_0\, kL$$

or,

$$p_f = k\frac{4L}{D_0} = Y\frac{2L}{D_0}, \tag{6.47}$$

where L is the length of the billet remaining in the container. Thus Eq. (6.46) becomes,

$$p = Y\left(1.7 \ln R + \frac{2L}{D_0}\right). \tag{6.48}$$

For strain-hardening materials, Y in these expressions should be replaced by $\overline{Y}$.

Note that as the billet is extruded further, L decreases and thus the ram force decreases (Fig. 6.66), whereas in indirect extrusion the ram force is not a function of billet length.

Actual Forces
The derivation of analytical expressions, including friction, die angle, and redundant work due to inhomogeneous deformation of the material (Section 2.9), can be difficult. Furthermore, there are difficulties in estimating the coefficient of friction, the flow stress of the material, and the redundant work in a particular operation. Consequently, a convenient *empirical* formula has been developed:

$$p = Y(a + b \ln R), \tag{6.49}$$

where a and b are constants determined experimentally. Approximate values for a and b are 0.8 and 1.2 to 1.5, respectively.

Again, note that for strain-hardening materials Y is replaced by $\overline{Y}$, which is the average flow stress of the material when undergoing the strain given by Eq. (6.40).

Optimum Die Angle
The die angle has an important effect on forces in extrusion. This can be understood from the following observations:

a. The *ideal* work of deformation is independent of the die angle (Fig. 6.67) since it is only a function of the extrusion ratio.

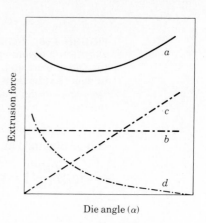

FIGURE 6.67 Schematic illustration of extrusion force as a function of die angle. (a) Total force. (b) Ideal force. (c) Force required for redundant deformation. (d) Force required to overcome friction. Note that there is an optimum die angle where the total extrusion force is a minimum.

b. The *frictional* work increases with decreasing die angle because the length of contact at the billet–die increases, thus requiring more work.

c. The *redundant* work due to inhomogeneous deformation increases with the die angle.

Since the total ram force is the sum of these three components, there is an angle where this force is a *minimum* (Fig. 6.67). Unless the behavior of each component as a function of the die angle is known, the determination of this *optimum angle* is difficult. Furthermore, the optimum die angle also depends on other factors, as described in Section 6.17.

Forces in Hot Extrusion

Because of the strain-rate sensitivity of metals at elevated temperatures, forces in hot extrusion are difficult to calculate.

It can be shown that the average true strain rate $\dot{\bar{\epsilon}}$ is given by,

$$\dot{\bar{\epsilon}} = \frac{6V_0 D_0^2 \tan \alpha}{D_0^3 - D_f^3} \ln R, \tag{6.50}$$

where V_0 is the ram velocity. Note from this equation that for high extrusion ratios $(D_0 >> D_f)$ and for $\alpha = 45$ degrees, as may be the case with a square die (thus developing a dead zone) and poor lubrication, the strain rate reduces to

$$\dot{\bar{\epsilon}} = \frac{6V_0}{D_0} \ln R. \tag{6.51}$$

The effect of ram speed and temperature on extrusion pressure is shown in Fig. 6.68. As expected, pressure increases rapidly with ram speed, especially at elevated temperatures.

As extrusion speed increases, the rate of work done per unit time also increases. Because work is converted into heat, the heat generated at high speeds may not be dissipated at a sufficiently high rate. The subsequent rise in temperature can cause

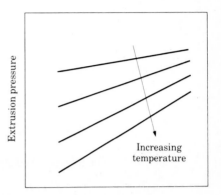

FIGURE 6.68 Schematic illustration of the effect of temperature and ram speed on extrusion pressure. Compare with Fig. 2.13.

incipient melting of the workpiece material and cause defects. Circumferential surface cracks, due to *hot shortness*, may also develop; in extrusion this is known as *speed cracking*. These problems can be eliminated by reducing the extrusion speed.

A convenient parameter that is used to estimate forces in extrusion is an experimentally determined *extrusion constant K_e*, which includes various factors and is given by,

$$p = K_e \ln R. \tag{6.52}$$

Figure 6.69 gives some typical values of K_e for a variety of materials.

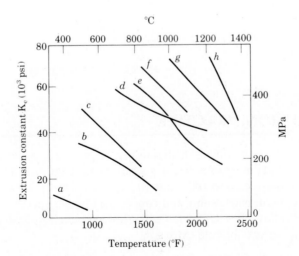

FIGURE 6.69 Extrusion constant K_e for various materials as a function of temperature. Note the range of temperature for various materials. a. 1100 aluminum, b. Copper, c. 70–30 brass, d. Beryllium, e. Cold-rolled steel, f. Stainless steel, g. Molybdenum, h. Chromium. *Source:* After P. Loewenstein, ASTME Paper SP63-89.

● **Illustrative Problem 6.5**

A copper billet 5 in. in diameter and 10 in. long is extruded at 1500°F at a speed of 10 in./s. Using square dies and with poor lubrication, estimate the force required in this operation if the final diameter is 2 in.

SOLUTION. The extrusion ratio is

$$R = \frac{5^2}{2^2} = 6.25.$$

The average true strain rate is, from Eq. (6.51),

$$\dot{\bar{\epsilon}} = \frac{(6)(10)}{(5)} \ln 6.25 = 22/s$$

The flow stress of copper at hot working temperatures is estimated to be (Table 2.6)

$$\sigma = C\dot{\epsilon}^m = (19,000)(22)^{0.06} = 22,870 \text{ psi.}$$

Assuming that $\bar{Y} = \sigma$, then from Eq. (6.48),

$$p = \bar{Y}\left(1.7 \ln R + \frac{2L}{D_0}\right)$$
$$= (22,870)\left[(1.7)(1.83) + \frac{(2)(10)}{(5)}\right]$$
$$= 162,630 \text{ psi.}$$

Hence

$$F = (p)(A_0) = (162,630)\frac{(\pi)(5)^2}{4}$$
$$= 3.2 \times 10^6 \text{ lb} = 1600 \text{ tons (short).} ●$$

6.16 EXTRUSION PRACTICE

A number of materials can be extruded to a wide range of cross-sectional shapes and dimensions. Extrusion ratios can range from about 10 to 100 or over. Ram speeds may be up to 100 ft/min (0.5 m/s). Generally, slower speeds are preferable for aluminum, magnesium, and copper, and higher speeds for steels, titanium, and refractory alloys. Presses for hot extrusion are generally hydraulic and horizontal, and usually vertical for cold extrusion.

6.16.1 HOT EXTRUSION

In addition to the strain-rate sensitivity of the material at elevated temperatures, hot extrusion requires other special considerations. Cooling of the billet in the container can result in highly inhomogeneous deformation. Furthermore, because the billet is heated prior to extrusion, it is covered with an oxide layer (unless heated in an inert atmosphere). Because of different frictional properties this can affect the flow of the material and may produce an extrusion covered with an oxide layer. In order to avoid this problem, the diameter of the dummy block ahead of the ram (Fig. 6.62a) is made a little smaller than that of the container. A thin cylindrical shell (*skull*), composed mainly of the oxidized layer, is thus left in the container and the extruded product is free of oxides.

Lubrication

Lubrication is important in hot extrusion. For steels, stainless steels, and high-temperature materials glass is an excellent lubricant, first developed in France in the early 1940s. It maintains its viscosity at elevated temperatures, has good wetting characteristics, and it acts as a thermal barrier between the billet and the container and the die, thus minimizing cooling. A circular glass pad is usually placed at the die entrance. This pad softens and melts away slowly as extrusion progresses and forms an optimal die geometry. The viscosity–temperature index of the glass is an important factor in this application.

Solid lubricants such as graphite and molybdenum disulfide are also used in hot extrusion. Nonferrous metals are usually extruded without a lubricant, although graphite may be used. (See Table 6.7.)

For materials that have a tendency to stick to the container and the die, the billet can be enclosed in a *jacket* of a softer metal such as copper or mild steel (*canning*). In addition to acting as a low-friction interface, canning prevents the contamination of the billet by the environment, or the billet material from contaminating the environment if the material is toxic or radioactive. The canning technique is also used for processing metal powders. (See Section 11.2.3.)

Dies

Die materials for hot extrusion are usually hot-work die steels (Table 6.4). To extend die life, coatings may be applied to the dies. Designing dies requires considerable experience because of the wide variety of products extruded. Typical die designs are shown in Fig. 6.70. Dies with angles of 90 degrees (square dies or *shear dies*) can also be used in extrusion of nonferrous metals, especially aluminum.

Tubing is also extruded with a ram fitted with a mandrel (Fig. 6.71). For billets with a pierced hole, the mandrel may be attached to the ram. If the billet is solid, then it must first be pierced in the container by the mandrel.

The complexity of an extrusion is a function of the ratio of the perimeter to the cross-sectional area of the part, known as the *shape factor*. Thus, a solid, round extrusion is the simplest shape. Hollow shapes (Fig. 6.72) can also be extruded by *welding-chamber* methods using various special dies known as *spider*, *porthole*, and *bridge* dies. The metal flows around the arms of the die into strands, which are then rewelded under the high die pressures at the exit. This process is suitable only for aluminum

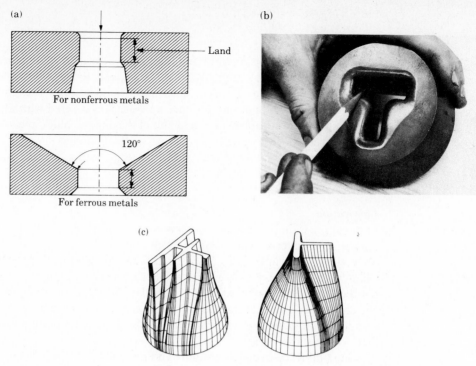

(a)

Land

For nonferrous metals

120°

For ferrous metals

(b)

(c)

FIGURE 6.70 Typical extrusion die configurations. (a) For nonferrous and ferrous materials. (b) T-shaped extrusion die made of W-Cr hot-work die steel and used with glass lubricant. Although die design is usually based on experience, analytical techniques are being developed to design dies to streamline the metal flow pattern, as shown in (c), and produce defect-free extrusions, as well as minimize the energy of deformation. *Source:* (b) Jones & Laughlin Steel Corp. (c) After H. L. Gegel, J. S. Gunasekera, and S. M. Doraivelu.

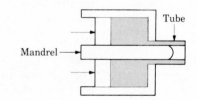

Tube

Mandrel

FIGURE 6.71 Extrusion of a seamless tube. The hole in the billet may be prepunched or pierced, or it may be generated during extrusion.

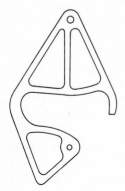

FIGURE 6.72 An example of extrusion that requires special dies to produce the five hollow sections in the part. This component, made of 6063-T6 aluminum, is a ladder lock for aluminum extension ladders. It is 5/16 in. (8 mm) thick, 9 in. (230 mm) long, and weighs 9.5 lb/ft (14 kg/m). Other than sawing the individual pieces from a long extrusion, no additional machining is required. Extrusion has been shown to be the most economical method for making this part.

TABLE 6.12
EXTRUSION TEMPERATURE RANGES FOR VARIOUS METALS

	°F	°C
Lead	400–500	200–250
Aluminum and its alloys	700–900	375–475
Copper and its alloys	1200–1800	650–975
Steels	1600–2400	875–1300
Refractory alloys	1800–4000	975–2200

and some of its alloys because of their capacity for pressure welding. Lubricants cannot be used as they prevent rewelding during extrusion.

Typical temperature ranges for hot extrusion are given in Table 6.12.

6.16.2 COLD EXTRUSION

Cold extrusion is a general term often denoting a combination of processes, such as direct and indirect extrusion and forging (Fig. 6.73). Many materials can be extruded into various configurations, with the billet either at room temperature or at a temperature of a few hundred degrees.

Cold extrusion has gained wide acceptance in industry because of the following advantages over hot extrusion:

a. Improved mechanical properties due to strain hardening, provided that the heat generated by plastic deformation and friction does not recrystallize the extruded metal.

b. Good control of tolerances, thus requiring a minimum of machining operations,

c. Improved surface finish, provided that lubrication is effective,

d. Lack of oxide layers,

e. High production rates and economics.

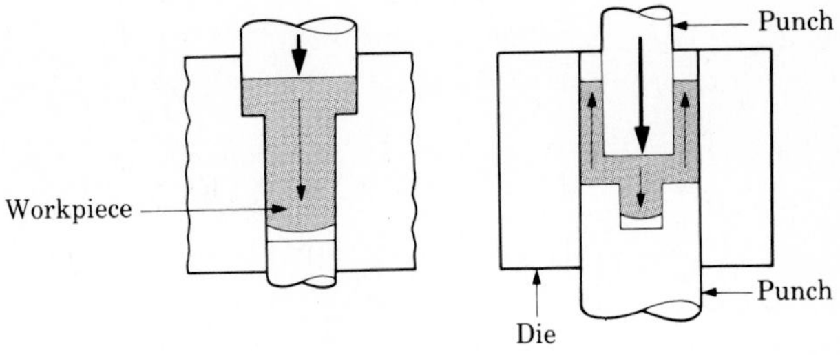

FIGURE 6.73 Examples of cold extrusion. Arrows indicate the direction of material flow. These parts may also be considered as forgings, as shown in Fig. 6.12.

(a)   (b)

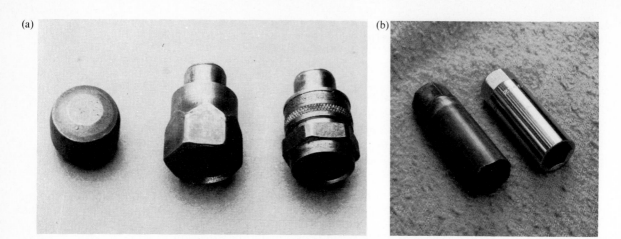

FIGURE 6.74   (a) Production of the metal part of an automotive spark plug by cold extrusion and machining. (b) Automotive socket produced by cold extrusion. Left: as extruded, right: finished product.

On the other hand, the stresses on the tooling in cold extrusion are very high, especially with steel workpieces. The stress levels on the tooling are on the order of the hardness of the material, that is, at least three times its flow stress. The design of tooling and selection of appropriate tool materials are therefore critical to success in cold extrusion. Examples of cold extrusion are shown in Fig. 6.74.

The hardness of tooling usually ranges between 60 and 65 HRC for the punch and 58 to 62 HRC for the die. Punches are a critical component as they must have not only sufficient strength, but also toughness, wear, and fatigue resistance. (See Table 6.4.)

Lubrication is also critical, especially with steels, because of the generation of new surfaces and the possibility of seizure between the metal and the tooling due to the breakdown of lubrication. The most effective lubrication is phosphate conversion coatings on the workpiece and soap (or wax in some cases) as the lubricant. (See Table 6.7 and Section 4.10.7.)

Temperature rise in cold extrusion is an important factor, especially at high extrusion ratios. The temperature may be sufficiently high to initiate and complete the recrystallization process of the cold-worked metal, thus reducing the advantages of cold working (Section 3.5).

6.16.3 IMPACT EXTRUSION

This is a process that is often included in the category of cold extrusion and is similar to indirect extrusion (Fig. 6.62b). The punch descends at a high speed on the *blank*

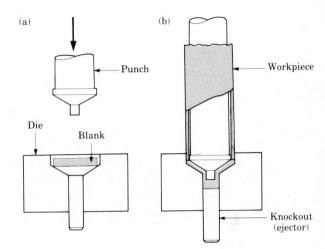

FIGURE 6.75 Impact extrusion of a collapsible tube, such as those used for tooth paste, made of low-melting-point alloys. This operation is also known as the Hooker process.

(*slug*), which is then extruded upward. The thickness of the extruded tubular section is a function of the clearance between the punch and the die cavity.

The impact-extrusion process usually produces tubular sections whose wall thicknesses are small in relation to their diameter. This ratio can be as small as 0.005. The concentricity between the punch and the blank is important for uniform wall thickness.

A typical example of impact extrusion is the production of collapsible tubes, such as for toothpaste (Fig. 6.75). The punch travel is determined by the setting of the press. Other shapes produced by impact extrusion are shown in Fig. 6.76. A variety of nonferrous metals are impact extruded in this manner, using vertical presses at production rates as high as two parts per second.

FIGURE 6.76 Examples of impact extruded products made of nonferrous metals. Some of these parts may also be made by other processes, such as casting and machining. The choice of the process depends on a number of technical and economic considerations, as described in Chapters 1 and 13.

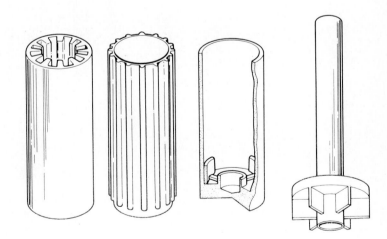

6.16.4 HYDROSTATIC EXTRUSION

This is a process, first proposed in 1893, whereby the extrusion pressure is supplied by the fluid surrounding the billet (Fig. 6.62c). Hydrostatic extrusion is similar to direct extrusion, with the exception that there is no container-wall friction and the pressure in the chamber supplies some fluid to the die surfaces, thus reducing friction significantly. The process can be carried out at room temperature using oils (such as castor oil) or at elevated temperatures using glass, heat-resisting grease, metal oxide or salts as the pressure-transmitting medium for temperatures above 900°F (500°C).

Brittle materials can be extruded successfully by hydrostatic extrusion because of increased ductility due to the hydrostatic state of stress (Section 2.2.8). Ductility is improved even further by extruding the part into a second pressurized chamber that has a lower pressure (*fluid-to-fluid extrusion*). Various other techniques have been developed for hydrostatic extrusion, including continuous extrusion.

Ductile materials have been extruded successfully into a variety of shapes, including wire extrusion, as shown in Fig. 6.77, and copper pipes. Extrusion ratios are up to about 200 for industrial applications, although much higher ratios have been obtained experimentally with ductile nonferrous metals. In spite of the success obtained in experimental studies, hydrostatic extrusion has had limited industrial applications, largely because of the complex nature of tooling and because of long cycle times.

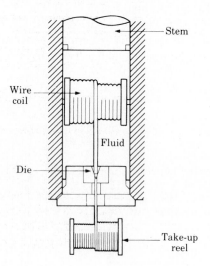

FIGURE 6.77 Reduction of the cross-sectional area of wire by hydrostatic extrusion. Note that this is a continuous operation. Because of practical difficulties, hydrostatic extrusion has not become a significant manufacturing process.

6.17 DEFECTS IN EXTRUSION

There are three principal defects in extrusion, as described below.

6.17.1 SURFACE CRACKING

If the extrusion temperature, friction, or extrusion speed are too high, surface temperatures rise significantly and can lead to surface cracking and tearing (*fir-tree cracking* or *speed cracking*). These cracks are intergranular, are usually due to *hot shortness* (Section 3.3.1) and occur especially with aluminum, magnesium, and zinc alloys. This defect is also observed with other metals, such as the molybdenum alloy extrusion shown in Fig. 6.78. This situation can be avoided by using lower temperatures and speeds.

Surface cracking may also occur at low temperatures and has been attributed to periodic sticking of the extruded product along the die land (see Fig. 6.70a). When the product being extruded sticks to the die land, the extrusion pressure increases rapidly. Shortly thereafter, the product moves forward again and the pressure is released. The cycle is then repeated.

6.17.2 EXTRUSION DEFECT

As described in Section 6.15.1, the type of metal flow observed in Fig. 6.64(d) tends to draw surface oxides and impurities toward the center of the billet, much like a funnel (Fig. 6.79). This defect is known as *extrusion defect, pipe, tailpipe,* or *fish-tailing.* A considerable portion of the material can be rendered useless as an extruded product because of it, by as much as one third of the length of the extrusion.

FIGURE 6.78 Cracking of molybdenum alloy, extruded at 4000°F (2200°C) with an extrusion ratio of 8. This defect has been attributed to hot shortness. *Source:* After V. DePierre.

FIGURE 6.79 Tailpipe (extrusion defect) in an extruded refractory metal alloy. *Source:* After V. DePierre.

This defect can be reduced by modifying the flow pattern to a less inhomogeneous one, such as by controlling friction and minimizing temperature gradients. Another method is to machine the surface of the billet prior to extrusion to eliminate scale and impurities. The extrusion defect can also be avoided by using a dummy block (Fig. 6.62a) that is smaller in diameter than the container, thus leaving a thin shell along the container wall as extrusion progresses.

6.17.3 INTERNAL CRACKING

The center of an extruded product can develop cracks (variously known as *center-burst, center cracking, arrowhead fracture,* or *chevron cracking*), as shown in Fig. 6.80. These cracks are attributed to a state of hydrostatic tensile stress (also called *secondary tensile stresses*) at the centerline of the deformation zone in the die. This situation is similar to the necked region in a uniaxial tensile-test specimen (Section 2.2).

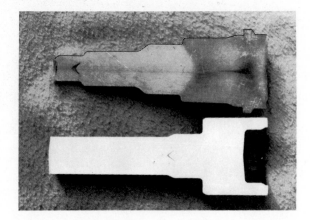

FIGURE 6.80 Chevron cracking in round steel bars during extrusion. Unless inspected by various techniques (see Table 4.2), such internal defects may remain undetected and possibly cause failure of the part in service.

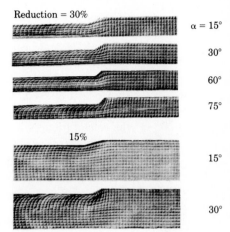

FIGURE 6.81 Grid deformation in indirect plane-strain extrusion of clay at various die angles and thickness reductions. Note that with increasing die angle, the upper surface layers of the extruded part are deformed to a greater extent than the bulk.

Studies have indicated that the major variables affecting hydrostatic tension are the die angle, extrusion ratio (reduction in cross-sectional area), and friction. The situation can be best understood by observing the extent of inhomogeneous deformation in extrusion. The experimental results shown in Fig. 6.81 indicate that, for the same reduction, the larger the die angle the more inhomogeneous the deformation across the part. Note how severely the surface layers are deformed at large die angles, which may include a dead zone.

In addition to the die angle, another factor in internal cracking is the die contact length. The smaller the die angle the longer the contact length. This situation is similar to a hardness test with a flat indenter. The size and depth of the deformation zone (see Fig. 2.37) increases with increasing contact length. This is illustrated in Fig. 6.82.

As shown in Fig. 2.40, an important parameter is the h/L ratio. It can be seen that the higher this ratio, the more inhomogeneous the deformation. High ratios mean small reductions and large die angles. Inhomogeneous deformation indicates that the center of the billet is not in a fully plastic state (it is rigid). This is because the plastic

FIGURE 6.82 Deformation zone in extrusion showing rigid and plastic zones. Note that the plastic zones do not meet, thus leading to chevron cracking. The same observations are also made in drawing round bars through conical dies, and drawing flat sheet plate through wedge-shaped dies. *Source:* After B. Avitzur.

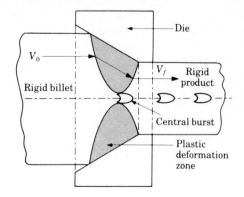

deformation zones under the die contact lengths do not reach each other (Fig. 6.82). Likewise, small reductions and high die angles retard the flow of the material at the surfaces, while the central portions are more free to move through the die.

The situations described above (high h/L ratios) have been shown to generate hydrostatic tensile stresses in the center of the billet, causing the type of defects shown in Fig. 6.80. These defects form more readily in materials with impurities, inclusions, and voids since they serve as nucleation sites for defect formation. As for the role of friction, it appears that high friction in extrusion delays the formation of these cracks.

Such cracks have also been observed in tube extrusion (Fig. 6.71) and in spinning of tubes (Fig. 7.76). The cracks appear on the inside surfaces for the reasons described above.

In summary, the tendency for center cracking increases with increasing die angles and levels of impurities, and decreases with increasing extrusion ratio. These observations are also valid for drawing of rod and wire, described in Sections 6.19 and 6.23.

6.18 EXTRUSION EQUIPMENT

The basic equipment for extrusion is a hydraulic press. These presses are mostly horizontal. They are designed for a variety of extrusion operations. Crank-type mechanical presses are also used for cold extrusion and piercing, and for mass production of steel tubing. (See also Fig. 6.40.) The largest hydraulic press for extrusion has a capacity of 16,000 tons (160 MN).

6.19 ROD AND WIRE DRAWING

This is an operation where the cross-sectional area of a bar is reduced by pulling it through a converging die (Fig. 6.83). The die opening may be any shape. Wire drawing involves smaller-diameter materials than rod drawing, with sizes as small as 0.001 in. (0.025 mm). The drawing process, which was an established art by the 11th century, is somewhat similar to extrusion with the exception that, in drawing, the product is under tension, whereas in extrusion the billet is under compression.

Rod and wire drawing are usually finishing processes. The product is either used

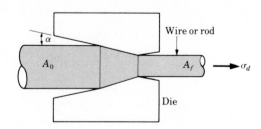

FIGURE 6.83   Variables in drawing round rod or wire.

as produced or is further processed into other shapes, usually by bending or machining. Rods are used for a variety of applications, such as small pistons, tension-carrying structural members, shafts, and spindles, and as the raw material for fasteners such as bolts and screws. Wire and wire products have a wide range of applications, such as electrical and electronic equipment and wiring, cables, springs, musical instruments, fencing, bailing, wire baskets, and shopping carts.

6.20 DRAWING OF ROUND SECTIONS

6.20.1 DRAWING FORCES AND STRESSES

The major variables in the drawing process are the reduction in cross-sectional area, the die angle, and friction (Fig. 6.83).

Ideal Deformation
For a round rod or wire, the drawing stress σ_d for the simplest case of ideal deformation (no friction or redundant work) can be obtained by the same approach as that in extrusion, described in Section 6.15.2. Thus,

$$\sigma_d = Y \ln\left(\frac{A_0}{A_f}\right). \tag{6.53}$$

Note that this expression is the same as Eq. (6.44) and that it also represents the energy per unit volume u.

For strain-hardening materials Y is replaced by an average flow stress $\overline{Y}$ in the deformation zone (Fig. 6.84). Thus, for a material that exhibits the true stress–true strain behavior of

$$\sigma = K\epsilon^n,$$

the average flow stress $\overline{Y}$ is given by,

$$\overline{Y} = \frac{K\epsilon_1^n}{n+1}.$$

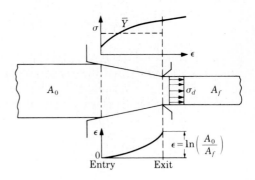

FIGURE 6.84 Variation in strain and flow stress in the deformation zone in drawing. Note that the strain increases rapidly toward the exit. This is because when the exit diameter is zero, the true strain reaches infinity.

The drawing force F is then,

$$F = \overline{Y}A_f \ln\left(\frac{A_0}{A_f}\right). \tag{6.54}$$

It can be seen that the greater the reduction in cross-sectional area and the stronger the material, the higher the drawing force.

Ideal Deformation and Friction
The effect of friction in drawing is to increase the drawing force, since work has to be supplied to overcome friction. Using the slab method of analysis (Section 2.10.1) it can be shown that the drawing stress σ_d is given by the expression,

$$\sigma_d = Y\left(1 + \frac{\tan\alpha}{\mu}\right)\left[1 - \left(\frac{A_f}{A_0}\right)^{\mu\cot\alpha}\right]. \tag{6.55}$$

Investigations have shown that even though this expression does not include the redundant work, it is in good agreement with experimental data for small die angles and for a wide range of reductions.

Redundant Work of Deformation
Depending on the die angle and reduction, the material in drawing undergoes inhomogeneous deformation, much as it does in extrusion. This means that the redundant work of deformation also has to be included in the expression for the drawing stress. One expression is as follows:

$$\sigma_d = Y\left\{\left(1 + \frac{\tan\alpha}{\mu}\right)\left[1 - \left(\frac{A_f}{A_0}\right)^{\mu\cot\alpha}\right] + \frac{2}{3}\alpha^2\left(\frac{1-r}{r}\right)\right\}, \tag{6.56}$$

where r is the fractional reduction of area and α is the die angle in radians.

The first term in Eq. (6.56) represents the ideal and frictional work components, and the second term the redundant work component. Note that this last term is, as expected, a function of the die angle. The higher the angle, the greater the inhomogeneous deformation and, hence, the greater the redundant work.

Another expression for the drawing stress for small die angles and including all three components of work is

$$\sigma_d = Y\left[\left(1 + \frac{\mu}{\alpha}\right)\ln\left(\frac{A_0}{A_f}\right) + \frac{2}{3}\alpha\right]. \tag{6.57}$$

The last term in this expression is the redundant-work component, whereby this work increases linearly with the die angle, as shown in Fig. 6.67.

Because redundant deformation is a function of the h/L ratio (see Fig. 2.40), an inhomogeneity factor Φ can replace the last term in Eq. (6.57). For drawing of round sections this factor is given approximately as,

$$\Phi = 1 + 0.12\left(\frac{h}{L}\right). \tag{6.58}$$

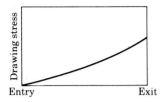

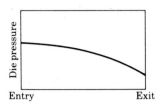

FIGURE 6.85 Variation in the drawing stress and die contact pressure along the deformation zone. Note that as the drawing stress increases, the die pressure decreases. This can be observed from the yield criteria, described in Section 2.8.

A simple expression for the drawing stress is then given by

$$\sigma_d = \Phi Y \left(1 + \frac{\mu}{\alpha}\right) \ln\left(\frac{A_0}{A_f}\right). \tag{6.59}$$

The foregoing expressions for the drawing stress, while not always agreeing with experimental data, give a reasonably good approximation of the stresses required for wire drawing. More importantly, they identify the effects of the various parameters involved. With good lubrication the value of μ in these equations ranges from about 0.03 to 0.1.

6.20.2 DIE PRESSURE

The die pressure p at any diameter along the die contact length can be conveniently obtained from the expression,

$$p = Y_f - \sigma, \tag{6.60}$$

where σ is the tensile stress in the deformation zone at any diameter. (Thus, σ is equal to σ_d at the exit and zero at entry). Y_f is the flow stress of the material at any diameter corresponding to that strain.

This expression is based on yield criteria for an element subjected to the stress system shown in Fig. 2.44(b), where the compressive stresses in the two principal directions are both equal to p. Note that since the tensile stress increases toward the exit, the die pressure drops toward the exit. The tensile stress and die pressure distributions are shown qualitatively in Fig. 6.85.

● **Illustrative Problem 6.6**

A round rod of annealed 302 stainless steel is being drawn from a diameter of 10 mm to one of 8 mm at a speed of 0.5 m/s. Assume that the frictional and redundant work is 40% of the ideal work of deformation. (a) Calculate the power required in this operation. (b) Calculate the die pressure at the die exit.

SOLUTION.   (a) The true strain in this operation is

$$\epsilon_1 = \ln\left(\frac{10^2}{8^2}\right) = 0.446.$$

From Table 2.4, $K = 1300$ MPa and $n = 0.30$. Thus,

$$\bar{Y} = \frac{K\epsilon_1^n}{n+1} = \frac{(1300)(0.446)^{0.30}}{1.30} = 785 \text{ MPa.}$$

From Eq. (6.54)

$$F = \bar{Y}A_f \ln\left(\frac{A_0}{A_f}\right),$$

where,

$$A_f = \frac{(\pi)(0.008)^2}{4} = 5 \times 10^{-5} \text{ m}^2.$$

Hence,

$$F = (785)(5 \times 10^{-5})(0.446) = 0.0175 \text{ MN}$$

and

$$\text{Power} = (F)(V_f) = (0.0175)(0.5) = 0.00875 \text{ MN} \cdot \text{m/s} = 0.00875 \text{ MW}$$
$$= 8.75 \text{ kW}$$

$$\text{Actual power} = (1.4)(8.75) = 12.25 \text{ kW}$$

(b) From Eq. (6.60)

$$p = Y_f - \sigma,$$

where Y_f represents the flow stress of the material at the exit (for this question). Thus

$$Y_f = K\epsilon_1^n = (1300)(0.446)^{0.30} = 1020 \text{ MPa,}$$

and σ in this equation is the drawing stress σ_d. Hence, using the actual force

$$\sigma_d = \frac{F}{A_f} = \frac{(1.4)(0.0175)}{0.00005} = 490 \text{ MPa}$$

Therefore the die pressure at the exit is

$$p = 1020 - 490 = 530 \text{ MPa.} \quad \bullet$$

6.20.3 DRAWING AT ELEVATED TEMPERATURES

At elevated temperatures, the flow stress of metals is a function of the strain rate. It can be shown that the average true strain rate $\dot{\bar{\epsilon}}$ in the deformation zone in drawing can be given by,

$$\dot{\bar{\epsilon}} = \frac{6V_0}{D_0} \ln\left(\frac{A_0}{A_f}\right). \tag{6.61}$$

(Note that this expression is the same as Eq. 6.51.)

In a particular drawing operation at an elevated temperature, the average strain rate is first calculated, from which the flow stress of the material is determined and substituted for Y in the foregoing equations for drawing stress and die pressure.

6.20.4 OPTIMUM DIE ANGLE

It was shown in Section 6.15.2 that, due to the various effects of the die angle on the three work components (ideal, friction, and redundant work), there is an optimum die angle where the extrusion force is a minimum. Because drawing involves a similar type of deformation, there is an optimum die angle in rod and wire drawing also, as shown in Fig. 6.67. Figure 6.86 shows a typical example from an experimental study. The optimum die angle for the minimum force increases with reduction. Note that the optimum angles in drawing are rather small.

6.20.5 MAXIMUM REDUCTION PER PASS

Note from all the expressions that with greater reduction the drawing stress increases. There is obviously a limit to the magnitude of the drawing stress. Thus, if it reaches the yield stress of the material, the product will simply continue to yield further as it leaves the die. This is not acceptable since the product will undergo further deformation. Thus, the maximum possible drawing stress can only be equal to the yield stress of the exiting material.

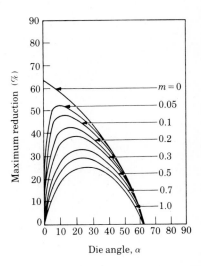

FIGURE 6.86 The effect of reduction in cross-sectional area on the optimum die angle in drawing copper wire. Note that the optimum angle increases with reduction. *Source:* After J. G. Wistreich.

In the ideal case of a perfectly plastic material with a yield stress Y, the limiting condition is,

$$\sigma_d = Y \ln\left(\frac{A_0}{A_f}\right) = Y, \qquad (6.62a)$$

or

$$\ln\left(\frac{A_0}{A_f}\right) = 1.$$

Hence,

$$\frac{A_0}{A_f} = e$$

and therefore,

$$\text{Maximum reduction per pass} = \frac{A_0 - A_f}{A_0} = 1 - \frac{1}{e} = 0.63 = 63\%. \qquad (6.62b)$$

The effects of friction and die angle on maximum reduction per pass are similar to those shown in Fig. 6.67. Because both friction and redundant work increase the drawing stress, the maximum reduction per pass will be lower than the ideal. In other words, the exiting material must have a larger cross-sectional area to sustain the higher drawing forces.

These effects are shown in Fig. 6.87 using the friction factor m, described in Section 4.6.1. As for the effect of strain hardening, it can be seen that with strain hardening, the exiting material is stronger than the rest. Hence, the maximum reduction per pass increases.

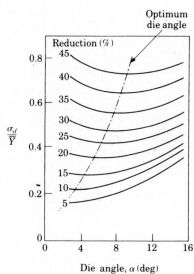

FIGURE 6.87   Effect of friction factor m and die angle on maximum possible reduction in wire drawing. A value of $m = 1$ indicates complete sticking (see Section 4.6.1). Note that the maximum possible reduction is 63%, as derived in Eq. (6.62b). *Source:* After B. Avitzur.

● **Illustrative Problem 6.7**

Obtain an expression for the maximum reduction per pass for a material with a true stress–true strain curve of $\sigma = K\epsilon^n$. Ignore friction and redundant work.

SOLUTION. From Eq. (6.53) for this material

$$\sigma_d = \bar{Y} \ln\left(\frac{A_0}{A_f}\right) = \bar{Y}\epsilon_1,$$

where

$$\bar{Y} = \frac{K\epsilon_1^n}{n+1}$$

and σ_d, for this problem, can have a maximum value equal to the flow stress at ϵ_1

$$\sigma_d = Y_f = K\epsilon_1^n.$$

Hence Eq. (6.62a) can be written as

$$K\epsilon_1^n = \frac{K\epsilon_1^n}{n+1}\,\epsilon_1,$$

or

$$\epsilon_1 = n + 1.$$

Since $\epsilon_1 = \ln(A_0/A_f)$ and maximum reduction $= (A_0 - A_f)/A_0$, these expressions reduce to

Maximum reduction per pass $= 1 - e^{-(n+1)}$. ●

Note that when $n = 0$ (perfectly plastic material) this expression reduces to Eq. (6.62b). As n increases the maximum reduction per pass increases.

6.21 DRAWING OF FLAT STRIP

Whereas drawing of round sections is axisymmetric, flat-strip drawing with high width-to-thickness ratios can be regarded as a *plane strain* problem. The dies are wedge-shaped and thus the process is somewhat similar to rolling of wide strips. There is little or no change in the width of the strip during drawing. This process, although not of any industrial significance, is the fundamental deformation mechanism in ironing, as described in Section 7.9.1.

The treatment of this subject, as far as forces and maximum reductions are concerned, is similar to that for round sections. The drawing stress for the ideal condition is given by (with Y' as the yield stress of the material in plane strain),

$$\sigma_d = Y' \ln\left(\frac{t_0}{t_f}\right), \tag{6.63}$$

where t_0 and t_f are the original and final thicknesses of the strip, respectively. The effects of friction and redundant deformation in strip drawing are similar to those for round sections. For instance, Eq. (6.56) can be used for strip drawing by replacing the cross-sectional areas by the thicknesses.

The maximum reduction per pass is obtained by equating the drawing stress (Eq. 6.63) to the uniaxial yield stress of the material. Thus, it can be shown that

$$\text{Maximum reduction per pass} = 1 - \frac{1}{e^{\sqrt{3}/2}} = 0.58 = 58\%. \tag{6.64}$$

6.22 DRAWING OF TUBES

Tubes produced by extrusion or other processes can be reduced in thickness or diameter (*tube sinking*) by the tube-drawing processes illustrated in Fig. 6.88. A variety of mandrels are used for different purposes. Shape changes can also be imparted by using dies and mandrels with various profiles.

Drawing forces, die pressures, and maximum reduction per pass in tube drawing can be calculated with approaches similar to those described for round rods.

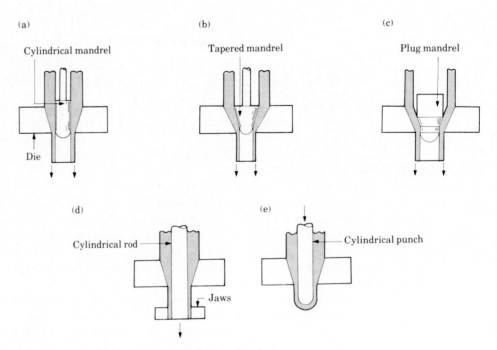

FIGURE 6.88   Various methods of tube drawing.

6.23 DRAWING PRACTICE

Successful drawing operations require careful selection of process parameters and consideration of many factors. These are reviewed in the sections below.

6.23.1 DIE DESIGN, EQUIPMENT, AND MATERIALS

A typical die design for drawing, with its characteristic features, is shown in Fig. 6.89. Die angles usually range from 6 to 15 degrees. The purpose of the land is to size and set the final diameter of the product. Also, when the die is reground after use, the land maintains the exit dimension of the die opening.

A rod or wire is fed into the die by first *pointing* it by swaging (forming the tip of the rod into a conical shape, Section 6.24). After placing it into the die, the tip is clamped into the jaws of the wire-drawing machine and the rod or wire is drawn continuously through the die.

Equipment

There are two types of equipment used in drawing. A *draw bench* is similar to a long horizontal tensile testing machine. It is used for single draws of straight large–cross-section rods and tubes for lengths up to 100 ft (30 m) with a hydraulic or chain-drive mechanism. Rod and wire of smaller cross-sections are drawn by a *bull block* (*capstan*), which is a rotating drum around which the wire is wrapped. The tension in this setup provides the force required in drawing the wire.

Reductions in cross-sectional area per pass range from about 10 to 45%; usually, the smaller the cross-section the smaller the reduction per pass. Reductions per pass higher than 45% may result in breakdown of lubrication and deterioration of surface finish. Light reductions may also be made (*sizing pass*) on rods to improve surface finish and dimensional accuracy.

In most wire drawing operations the wire passes through a series of dies (*tandem drawing*). In order to avoid excessive tension in the exiting wire, it is wound one or two turns around a capstan between each pair of dies. The speed of the capstan is adjusted so that it supplies not only tension, but also a small back tension to the wire entering the next die. Back tension reduces the die pressure and extends die life.

The terminology for drawn copper and brass wire is given in Table 6.13.

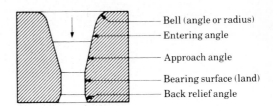

FIGURE 6.89 Terminology for a typical die for drawing round rod or wire.

Bell (angle or radius)
Entering angle
Approach angle
Bearing surface (land)
Back relief angle

TABLE 6.13
HARDNESS TERMINOLOGY FOR COLD-DRAWN COPPER AND BRASS WIRE

TERMINOLOGY	DESIGNATION	REDUCTION IN DIAMETER, %
$\frac{1}{4}$ hard	H01	11
$\frac{1}{2}$ hard	H02	21
$\frac{3}{4}$ hard	H03	29
Hard	H04	37
Extra hard	H06	50
Spring	H08	60
Extra spring	H10	69
Special spring	H12	75
Super spring	H14	80

Drawing Practice

Drawing of large sections can be done at elevated temperatures. In cold drawing, because of strain hardening, intermediate annealing between passes may be necessary to maintain sufficient ductility. Steel wires for springs and musical instruments are made by a heat-treatment process that precedes or follows the drawing operation (*patenting*). These wires have ultimate tensile strengths as high as 700,000 psi (4800 MPa) with tensile reduction of area of about 20%.

Dies

Die materials are usually alloy tool steels, carbides, or diamond (see Table 6.4). A diamond die (used for drawing of fine wires) may be a single crystal or a poly-crystalline diamond in a metal matrix. Carbide and diamond dies are made as inserts or nibs, which are then supported in a steel casing, as shown in Fig. 6.90.

In addition to rigid dies, a set of idling rolls are also used in drawing of rods or bars of various shapes. This arrangement, Fig. 6.91, is known as *Turk's head* and is more

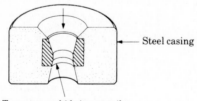

Steel casing

Tungsten carbide insert (nib)

FIGURE 6.90 Tungsten carbide die insert in a steel casing. Carbides have high compressive strength, hardness, and wear resistance, hence are suitable as die materials. See Table 8.7.

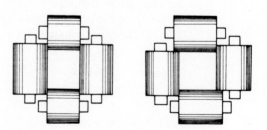

FIGURE 6.91 Arrangement of rolls in a Turk's head. Different cross-sections can be obtained by rearranging the rolls.

FIGURE 6.92 Schematic illustration of typical wear pattern in a wire drawing die.

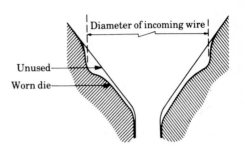

versatile than ordinary dies since the rolls can be adjusted to various positions for different products.

Drawing speeds depend on the material and the cross-sectional area. They may be as low as 30 ft/min (0.15 m/s) for heavy sections, to as high as 10,000 ft/min (50 m/s) for very fine wire. Temperature can rise substantially at high drawing speeds.

A typical wear pattern on a drawing die is shown in Fig. 6.92. Note that die wear is the highest at entry. Since the die pressure is highest in this region, it may be partially responsible for wear. But it has been shown that there are other factors involved. These are: variation in the diameter of the entering wire and vibration (thus subjecting the entry contact zone to fluctuating stresses), and the presence of abrasive scale on the surface of the entering wire.

6.23.2 LUBRICATION

Proper lubrication is essential in rod, tube, and wire drawing. (See Table 6.7.) The two basic types are dry and wet drawing.

In *dry drawing* the surface of the wire is coated with various lubricants, depending on its strength and frictional characteristics. A common lubricant is soap. The rod to be drawn is first surface treated by *pickling*. This removes the surface scale that could lead to surface defects and, being quite abrasive, would considerably reduce die life. The soap is picked up by the wire as it goes through a box filled with soap powder (stuffing box).

With high-strength materials such as steels, stainless steels, and high-temperature alloys, the surface of the rod or wire may be coated either with a softer metal or with a *conversion coating*. Copper or tin can be chemically deposited on the surface of the metal. This thin layer of softer metal acts as a solid lubricant during drawing.

Conversion coatings (Section 4.10.7) may consist of sulfate or oxalate coatings on the rod; these are then typically coated with soap, as a lubricant. Polymers are also used as solid lubricants, such as in drawing of titanium.

In *wet drawing*, the dies and the rod are completely immersed in a lubricant. Typical lubricants are oils and emulsions containing fatty or chlorinated additives, and various chemical compounds.

6.23.3 DEFECTS AND RESIDUAL STRESSES

Defects in drawing are similar to those observed in extrusion, especially *center cracking* (see Section 6.17.3). An additional defect is the formation of *seams*, which are longitudinal scratches or folds in the material. Such defects can open up during subsequent forming operations such as by upsetting, heading, thread rolling, or bending of the rod or wire. A variety of surface defects can also result from improper selection of process parameters and lubrication.

Because of inhomogeneous deformation, a cold-drawn rod, wire, or tube usually contains residual stresses. A typical example is given in Fig. 6.93, showing a wide range of residual stresses within the rod in three principal directions. Note that, for very light reductions, the surface residual stresses are compressive. Light reductions are equivalent to shot peening or surface rolling, thus improving fatigue life. (See also Figs. 4.38 and 4.39.)

In addition to fatigue, residual stresses can also be significant in stress–corrosion cracking over a period of time, and in warping of the component when a layer is subsequently removed, such as by machining.

(a)

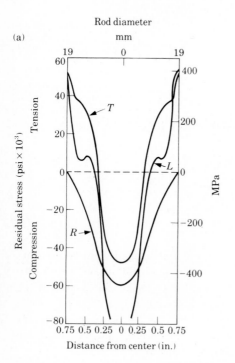

(b)

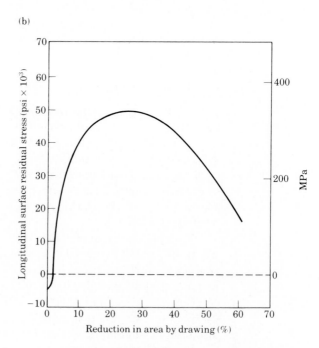

FIGURE 6.93   (a) Residual stresses in cold-drawn AISI 1045 carbon steel round rod. (b) Surface residual stress as a function of reduction by drawing. Note that very light reductions induce compressive surface stresses. See also Fig. 4.38. *Source:* (a) After E. S. Nachtman, ASME Paper 55-SA-49. (b) After W. Linicus and G. Sachs, 1932.

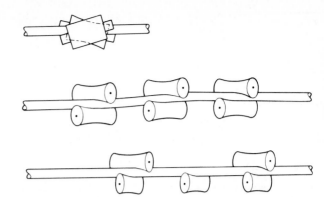

FIGURE 6.94 Schematic illustration of roll straightening of round rods. Different roll arrangements may be used in this technique. See also Fig. 6.52.

Rods and tubes that are not sufficiently straight (or if supplied in a coiled form) can be straightened by passing them through an arrangement of pairs of rolls placed at different axes (Fig. 6.94). The rolls subject the product to a series of bending and unbending operations, similar to the method shown in Fig. 6.52.

6.24 SWAGING

In this process, also known as *rotary swaging* or *radial forging*, a solid rod or a tube is reduced in diameter by the reciprocating radial movement of two or four dies (Fig. 6.95). The die movements are generally obtained through a set of rollers in a

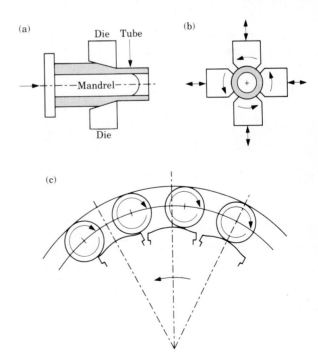

FIGURE 6.95 Schematic illustration of the swaging process. (a) Side view. (b) Front view. (c) Schematic illustration of roller arrangement, curvature on the four radial hammers (that give motion to the dies), and the radial movement of a hammer as it rotates over the rolls.

cage. The internal diameter and the thickness of the tube can be controlled with or without mandrels as shown in Fig. 6.96. Mandrels can also be made with longitudinal grooves (similar in appearance to a splined shaft); thus internally shaped tubes can be swaged (Fig. 6.97). The rifling in gun barrels is made by swaging a tube over a mandrel with spiral grooves. The deformation of a metal tube in swaging it over a hexagonal mandrel is shown in Fig. 6.98.

The swaging process is usually limited to a diameter of about 2 in. (20 mm), although special machinery has been built to swage gun barrels of larger diameter. The length of the product in this process is not limited, except by the length of the mandrel (if needed). Die angles are usually only a few degrees, and may be compound —i.e., the die may have more than one angle, for more favorable material flow during swaging. Lubricants are used for improved surface finish and longer die life. The process is generally carried out at room temperature. Parts produced by swaging have improved mechanical properties and good dimensional accuracy.

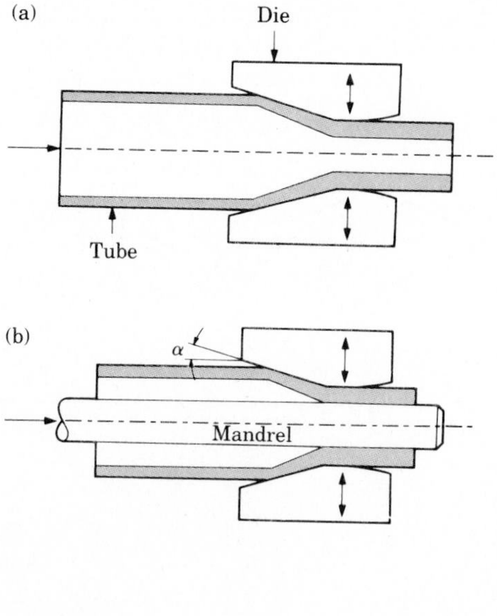

FIGURE 6.96   Reduction of outer and inner diameters of tubes by swaging. (a) Free sinking without a mandrel. The ends of solid bars and wire are tapered (pointing) by this process in order to feed the material into the conical die. (b) Sinking on a mandrel. Coaxial tubes of different materials can also be swaged in one operation.

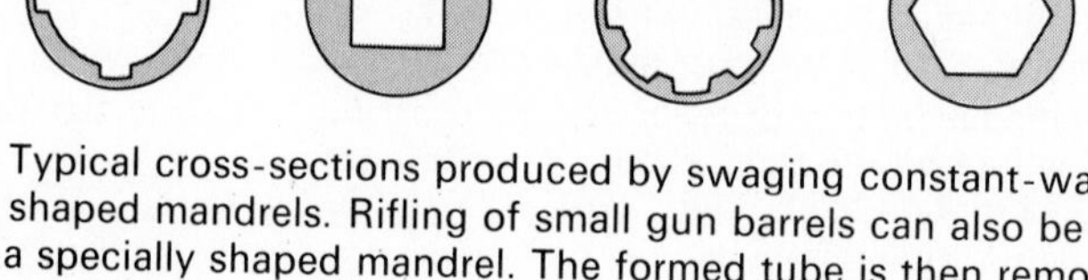

FIGURE 6.97   Typical cross-sections produced by swaging constant-wall-thickness tube blanks on shaped mandrels. Rifling of small gun barrels can also be made by swaging using a specially shaped mandrel. The formed tube is then removed by slipping it off the mandrel.

FIGURE 6.98 Grid flow pattern in a copper tube swaged over a hexagonal mandrel. The curvature developed in the originally straight radial lines is due to the rotation of the dies relative to the workpiece (counterclockwise in this figure, as also shown in Fig. 6.95c).

6.25 TOOL AND DIE FAILURES IN METALWORKING

Failure of tools and dies in metalworking operations are generally due to one or more of the following causes:

a. Design: die design involves not only the proper selection of tool and die materials, but also dimensions. In order to withstand the forces in processing, a die must have proper cross-sectional areas and clearances. Sharp corners, radii and fillets, and sudden changes in cross-section can have deleterious effects on die life.

b. Heat Treatment: to obtain improved hardness, wear resistance, and strength tool and die steels are usually heat treated (Table 4.10). Improper heat treatment is one of the most common causes of failure. Particularly important is the condition and composition of the die surfaces. Thus, the proper selection of temperatures for heat treatment, quenching media, quenching practice, tempering procedures, and handling are important.

c. Finishing Operations: after heat treatment, tools and dies are subjected to grinding or chemical and electrical machining processes (Chapter 9) to obtain the desired surface finish and dimensional accuracy. The grinding process, if not controlled properly, can cause surface damage to the dies due to excessive heat and can also produce tensile residual stresses, thus affecting fatigue life. Scratches on the surface can act as stress raisers. Likewise, a commonly used process such as electrical discharge machining can cause surface damage and cracks, unless process parameters are controlled carefully.

d. Handling and Use: the proper handling, installation, and alignment of tools and dies are also important. Overloading of tools and dies can cause premature failure.

Tool and die materials such as carbides, ceramics, and diamond, in spite of their hardness and resistance to abrasion, are susceptible to cracking and chipping due to impact forces or thermal gradients. (See Table 8.7.) Surface preparation and finish are thus important. Even the metalworking fluid—i.e., lubricant—can have an adverse effect on tool and die materials, such as the tendency for sulfur and chlorine additives in fluids to leach away the cobalt binder in tungsten carbide tools and dies.

Failure of tools and dies is invariably associated with the initiation and propagation of cracks. Consequently, an understanding of the *fracture toughness* of tool and die materials is essential. This subject, which is beyond the scope of this text, deals with the relationship of cracks, and their propagation, to a large number of material and process parameters. Among the major factors involved are the manufacturing and processing history of the tool and die materials, type and orientation of flaws and inclusions, heat treatment, surface finish, type of loading, and the state of stress.

SUMMARY

1. In bulk deformation processes the workpiece using dies and various other tooling is subjected to major changes in its dimensions. The products of bulk deformation processing, such as plates, sheet, forgings, rod, wire, and extruded shapes are then processed by various other methods described in subsequent chapters.

2. The bulk properties, as well as the surface properties, of the workpiece material are important, since the material is in contact with dies under pressures sufficiently high to cause yielding. The strain-hardening capability, as well as its strain-rate sensitivity, of the material is significant, particularly if high temperatures are involved.

3. The basic bulk deformation processes are forging, rolling, extrusion, and rod and wire drawing. In these operations, the important factors are the forces required to carry out the operation, the workability of the material, the final properties of the product, surface characteristics, defects that may be introduced during processing, selection of tool and die materials, and lubrication.

4. As in other metalworking operations, the process parameters that are important in bulk deformation processes are workpiece temperature, deformation rate, reduction per pass, and frictional conditions at the die–workpiece interfaces.

5. It is desirable for the material to be worked in as high a compressive environment as possible (approaching hydrostatic compression) in order to avoid defects and surface cracking. In this regard, deformation geometry is important, as it governs the depth of the plastic zone in the workpiece. This geometry can be controlled by proper die design and by process parameters.

BIBLIOGRAPHY

General Introductory Texts

Alting L. *Manufacturing Engineering Processes.* New York: Marcel Dekker, 1982.

Amstead BH, et al. *Manufacturing Processes,* 7th ed. New York: Wiley, 1977.

DeGarmo EP. *Materials and Processes in Manufacturing,* 5th ed. New York: Macmillan, 1979.

Doyle LE, et al. *Manufacturing Processes and Materials for Engineers*, 2d ed. Englewood Cliffs, N.J.: Prentice-Hall, 1969.

Lindberg RA. *Processes and Materials of Manufacture*, 3d ed. Boston: Allyn and Bacon, 1983.

Niebel BW, Draper AB. *Product Design and Process Engineering*. New York: McGraw-Hill, 1974.

Sachs G. *Fundamentals of the Working of Metals*. New York: Pergamon, 1954.

General Introductory Texts with Some Analytical Treatment

Alexander JM, Brewer RC. *Manufacturing Properties of Materials*. New York: Van Nostrand Reinhold, 1963.

Altan T, Oh S and Gegel H. *Metal Forming—Fundamentals and Applications*. Metals Park, Ohio, American Society for Metals, 1983.

Cook NH. *Manufacturing Analysis*. Reading, Mass.: Addison–Wesley, 1966.

Crane FAA. *Mechanical Working of Metals*. New York: Macmillan, 1964.

Dieter GE. *Mechanical Metallurgy*, 2d ed. New York: McGraw-Hill, 1976.

Harris JN. *Mechanical Working of Metals*. New York: Pergamon, 1983.

Parkins RN. *Mechanical Treatment of Metals*. London: Allen and Unwin, 1968.

Rowe GW. *Elements of Metalworking Theory*. London: Edward Arnold, 1979.

Rowe GW. *Principles of Industrial Metalworking Processes*. London: Edward Arnold, 1977.

Schey JA. *Introduction to Manufacturing Processes*. New York: McGraw-Hill, 1977.

Advanced Texts

Avitzur B. *Handbook of Metal-Forming Processes*. New York: Wiley, 1983.

Avitzur B. *Metal Forming: Processes and Analysis*. New York: McGraw-Hill, 1968.

Backofen WA. *Deformation Processing*. Reading, Mass.: Addison–Wesley, 1972.

Blazynski TZ. *Metal Forming: Tool Profiles and Flow*. New York: Halsted Press, 1976.

Ford H, Alexander JM. *Advanced Mechanics of Materials*, 2d ed. New York: Halsted Press, 1977.

Hoffman O, Sachs G. *Introduction to the Theory of Plasticity for Engineers*. New York: McGraw-Hill, 1953.

Hosford WF, Caddell RM. *Metal Forming, Mechanics and Metallurgy*. Englewood Cliffs, N.J.: Prentice-Hall, 1983.

Johnson W, Mellor PB. *Engineering Plasticity*. New York: Van Nostrand Reinhold, 1973.

Lippmann H, ed. *Engineering Plasticity: Theory of Metal Forming Processes* (2 vols.). New York: Springer, 1977.

Lippmann H, ed. *Metal Forming Plasticity*. New York: Springer, 1979.

Slater RA. *Engineering Plasticity: Theory and its Application to Metal Forming Processes*. New York: Halsted Press, 1974.

Thomsen EG, Yang CT, Kobayashi S. *Mechanics of Plastic Deformation in Metal Processing*. New York: Macmillan, 1964.

Forging

Altan T, et al. *Forging: Equipment, Materials and Practices*. Columbus, Ohio: Battelle Memorial Institute, Metals and Ceramics Information Center, 1973.

Feldman HD. *Cold Forging of Steel*. New York: Chemical Publishing Co., 1962.

Forging Design Handbook. Metals Park, Ohio: American Society for Metals, 1972.

Geleji A. *Forge Equipment, Rolling Mills and Accessories* (in English). Budapest: Academiai Kiado, 1967.

Jenson JE, ed. *Forging Industry Handbook*. Cleveland, Ohio: Forging Industry Association, 1970.

Kyle PE. *The Closed Die Forging Process*. New York: Macmillan, 1954.

Metals Handbook, 8th. ed., Vol. 5, *Forging and Casting*. Metals Park, Ohio: American Society for Metals, 1970.

Open Die Forging Manual, 3d ed. Cleveland, Ohio: Forging Industry Association, 1982.

Sabroff AM, Boulger FW, Henning HJ. *Forging Materials and Practices*. New York: Van Nostrand Reinhold, 1968.

Thomas A. *DFRA Forging Handbook: Die Design*. Sheffield, England: Drop Forging Research Association, 1980.

Watkins MT. *Metal Forming*. I. *Forging and Related Processes*. New York: Oxford, 1975.

Rolling

Elements of Rolling Practice, 2d ed. Sheffield, England: The United Steel Companies, Ltd., 1963.

Larke EC. *The Rolling of Strip, Sheet, and Plate*, 2d ed. London: Chapman and Hall, 1963.

Roberts WL. *Cold Rolling of Steel*. New York: Marcel Dekker, 1978.

Roberts WL. *Hot Rolling of Steel*. New York: Marcel Dekker, 1983.

Roll Pass Design. Sheffield, England: The United Steel Companies, Ltd., 1960.

Starling CW. *The Theory and Practice of Flat Rolling*. London: University of London Press, 1962.

Tselikov AI, Smirnov VV. *Rolling Mills*. New York: Pergamon, 1965.

Underwood LR. *The Rolling of Metals*, Vol. 1. New York: Wiley, 1950.

Wusatowski Z. *Fundamentals of Rolling*. New York: Pergamon, 1969.

Extrusion

Alexander JM, Lengyel B. *Hydrostatic Extrusion*. London: Mills and Boon, 1971.

Everhart JE. *Impact and Cold Extrusion of Metals*. New York: Chemical Publishing Co., 1964.

Laue K, Stenger H. *Extrusion—Processes, Machinery, Tooling*. Metals Park, Ohio: American Society for Metals, 1981.

Pearson CE, Parkins RN. *The Extrusion of Metals*, 2d ed. New York: Wiley, 1961.

Wire Drawing

Bernhoeft CP. *The Fundamentals of Wire Drawing*. London: The Wire Industry Ltd., 1962.

Dove AB, ed. *Steel Wire Handbook*, Vol. 1, 1968, Vol. 2, 1969, Vol. 3, 1972, Vol. 4, 1980. Branford, Conn.: Wire Association, Inc.

Tassi OJ, ed. *Nonferrous Wire Handbook* (2 vols.). Branford, Conn.: Wire Association International, Inc., 1977 and 1981.

Pomp A. *The Manufacture and Properties of Steel Wire*. London: The Wire Industry Ltd., 1954.

Tool and Die Failures

Kalpakjian, S., ed. *Source Book on Tool and Die Failures*. Metals Park, Ohio: American Society for Metals, 1982.

Lubrication

See Bibliography on Tribology in Metalworking in Chapter 4.

Periodicals

American Machinist

International Journal of Production Research

Journal of Applied Metalworking

Journal of Engineering for Industry

Journal of Engineering Materials and Technology

Journal of the Iron and Steel Institution

Journal of Mechanical Working Technology

Manufacturing Engineering

Metal Process

Metallurgia

Production

Production Engineer

Wire

Wire Industry

Wire Journal International

PROBLEMS

6.1. Plot the force vs. reduction-in-height curve in compressing a cylindrical, annealed copper specimen 1 in. high and 1 in. in diameter, up to a reduction of 75%. Ignore friction.

6.2. Plot the same curve in Problem 1 above for a coefficient of friction of 0.2. Ignore barreling.

6.3. A manufacturer is successfully hot forging a certain part using material supplied by Company A. A new supply of material is obtained from Company B, with the same nominal composition of the major alloying elements as the previous material. However, it is found that the new forgings are cracking even though the same procedure is followed as before. Can you offer an explanation?

6.4. A thin circular disk of soft metal is being compressed between two round punches with flat ends. All parts have the same diameter. Assuming there is no friction and that the metal is not strain hardening:
   a. make a rough sketch of the shape of the metal disk after it has been compressed to a fraction of its height,
   b. explain the change, if any, in the magnitude of the punch force as deformation progresses. Ignore any temperature effects.

6.5. Explain why there might be a change in the density of a forged product.

6.6. To determine forgeability, a hot twist test is performed on a round bar 25 mm in diameter and 300 mm long. It is found that it underwent 200 turns before it fractured. Calculate the shear strain at the outer surface of the bar at fracture. (See Eq. 2.21.)

6.7. Describe the role of surface oxide layers on the original workpiece in its behavior in an impression-die–forging operation.

6.8. Derive an expression for the average pressure in plane strain compression with sticking friction, based on Eq. (6.18).

6.9. What is the value of the coefficient of friction when, for plane strain compression, the forging load with sliding friction is equal to the load with sticking friction? Use average pressure formulas.

6.10. In an impression-die forging with flash, identify the regions where lubrication should be applied and those regions where it should be minimal.

6.11. Explain the changes that occur in the roll pressure distribution in rolling if one of the rolls is idling, i.e., power is shut off to that roll.

6.12. Explain how the roll diameter and reduction per pass affect the residual stress magnitudes and distributions in rolling.

6.13. In rolling of a flat strip, explain the conditions under which the neutral point can be shifted toward the entry zone.

6.14. List and explain the methods you would use to reduce the roll-separating force in rolling.

6.15. Explain the advantages and limitations of using small-diameter rolls.

6.16. What would be the answer to Illustrative Problem 6.4 if the material is annealed 304 stainless steel?

6.17. Calculate the force required in direct extrusion of 1100-O aluminum from a diameter of 6 in. to one of 2 in. Assume that the redundant work is 40% of the ideal work of deformation and that the friction work is 25% of the total work.

6.18. Explain the different ways die geometry affects the extrusion process.

6.19. Explain why glass is a good lubricant for hot extrusion. Would it also be suitable for impression-die forging with deep recesses? Explain any limitations.

6.20. It has been stated that the extrusion ratio, die geometry, extrusion speed, and temperature all affect extrusion pressure. Explain why this is so.

6.21. Calculate the power required in Illustrative Problem 6.6 if the material is annealed 70–30 brass.

6.22. Explain the factors that affect centerburst in extrusion and drawing of a rod.

6.23. Explain why hydrostatic extrusion is particularly suitable for brittle materials.

6.24. In wire and rod drawing the maximum die pressure is at the die entry. Explain why.

6.25. Using Eq. (6.55), make a plot similar to Fig. 6.86 for the following conditions: $K = 70$ MPa, $n = 0.25$, $\mu = 0.04$.

7 Sheet-Metal Forming Processes

7.1 INTRODUCTION

In the preceding chapter, it was noted that workpieces could be characterized by the ratio of their surface area to their volume or thickness. Sheet forming, unlike bulk deformation processes, involves workpieces with a high ratio of surface area to thickness.

A sheet thicker than 1/4 in. (6 mm) is generally called a *plate*. Although relatively thick plates, such as those used in boilers, bridges, ships, and nuclear power plants may have smaller ratios of surface area to thickness, sheet forming usually involves relatively thin materials.

The products made by sheet-forming processes include a large variety of shapes and sizes, ranging from simple bends to double curvatures with shallow or deep recesses. Typical examples are metal desks, appliance bodies, hubcaps, aircraft panels, beverage cans, car bodies, and kitchen utensils. Thus sheet forming, which is also called *pressworking*, is among the most important of metalworking processes. It dates back to as early as 5000 B.C. when household utensils, jewelry, and other objects were made by hammering and stamping metals such as gold, silver, and copper.

In this chapter, the characteristic features of sheet metals and their formability will first be reviewed. The major processes of sheet forming, the products, and the equipment involved will then be described. The processing of polymer sheet is covered in Section 10.12.

Sheet metal is produced by a rolling process, as described in Section 6.10. If the sheet is thin, it is generally coiled; if thick, it is available as flat sheets or plates, which may have been decoiled and flattened. Before a sheet-metal part is formed, a blank of suitable dimensions is first removed from a large sheet. This is usually done by a shearing process described below. (See also, Section 12.11.)

7.2 SHEARING

The shearing process involves cutting sheet metal by subjecting it to shear stresses, usually between a *punch* and a *die* much like a paper punch (Fig. 7.1). The punch and die may be any shape; they may be circular or straight blades similar to a pair of

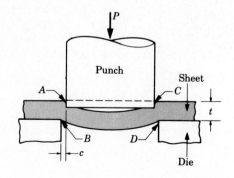

FIGURE 7.1 Schematic illustration of the shearing process with a punch and die. This is a common method of producing holes and various openings in sheet metals.

scissors. The major variables in the shearing process are: the punch force P, the speed of the punch, lubrication, surface condition and materials of the punch and die, their corner radii, and the clearance between the punch and the die.

7.2.1 SHEARING MECHANISM

The overall features of typical sheared edges for the two sheared surfaces (the slug and the sheet) are shown in Fig. 7.2. Note that the edges are neither smooth nor perpendicular to the plane of the sheet. The *clearance c* (Fig. 7.1) is the major factor determining the shape and quality of the sheared edge. As shown in Fig. 7.3, as clearance increases, the edges become rougher and the zone of deformation becomes larger. The material is pulled into the clearance area, and the edges of the sheared zone become more and more rounded. In fact, if the clearance is too large, the sheet metal is bent and subjected to tensile stresses, instead of undergoing a shearing deformation.

Note also the formation of a *burr* in Fig. 7.2. Burr height increases with increasing clearance and increasing ductility of the metal. Tools with dull edges are also a major factor in burr formation. The height, shape, and size of the burr can have a significant

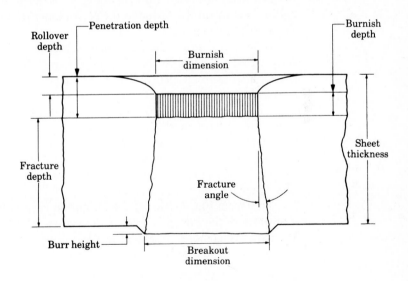

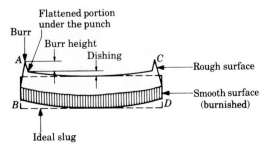

FIGURE 7.2   Characteristic features of a punched hole and the punched slug.

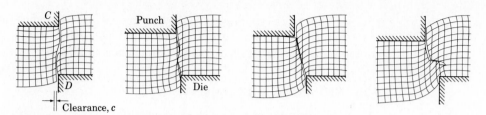

FIGURE 7.3   Effect of clearance c between the punch and die on the deformation zone in shearing. As clearance increases, the material tends to be pulled into the die, rather than being sheared. In practice, clearances usually range between 2 and 10% of the thickness of the sheet.

effect in many subsequent forming operations, such as in flanging, where a burr could lead to cracks.

Observing the shearing mechanism reveals that shearing usually starts with the formation of cracks on both the top and bottom edges of the workpiece (A and B in Fig. 7.1). These cracks eventually meet and complete separation takes place. The rough fracture surface in the slug in Fig. 7.2 is due to these cracks. The smooth, shiny, and burnished surface is due to the contact and rubbing of the sheared edge against the walls of the die.

In the slug shown, the burnished surface is in the lower region since this is the section that rubs against the die wall. On the other hand, inspection of the sheared surface on the sheet itself reveals that the burnished surface is on the upper region in the sheared edge and is due to rubbing against the punch.

It has been observed that the ratio of the burnished to rough areas on the sheared edge increases with increasing ductility of the sheet metal; it decreases with increasing material thickness and clearance. The punch travel required to complete the shearing process depends on the maximum shear strain that the material can undergo before fracture. Thus, a brittle metal or one that is highly cold worked, requires little travel of the punch to complete shearing.

Note from Fig. 7.4 that the deformation zone is subjected to high shear strains. The width of this zone is found to depend on the rate of shearing, i.e., punch speed. With increasing punch speed, the heat generated due to plastic deformation is confined to a smaller zone (approaching a narrow adiabatic zone) and the sheared surface is smoother.

Punch Force

The punch force P is basically the product of the shear strength of the sheet metal and the cross-sectional area being sheared. However, friction between the punch and the workpiece can increase this force substantially. Since the sheared zone is subjected to cracks, plastic deformation, and friction, the punch force–stroke curves can have various shapes. One typical curve for a ductile material is shown in Fig. 7.5. The area under this curve is the *total work* done in shearing.

(a)

(b)

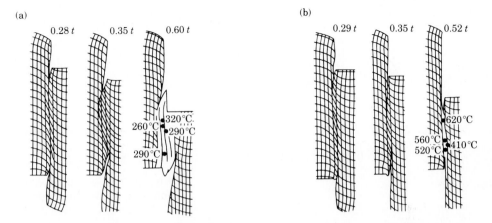

FIGURE 7.4 Deformation and temperature rise in the shearing zone. The temperature was measured by thermocouples. (a) Punching at slow speed. (b) High speed. Note that the deformation is confined to a narrow zone in high-speed shearing, and the temperature is higher than in slow-speed shearing. *Source:* After N. Yanagihara, H. Saito, and T. Nakagawa. Numbers above the figures indicate punch penetration.

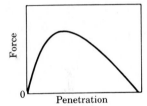

FIGURE 7.5 Typical punch-penetration curve in shearing. The area under the curve is the work done in shearing. The shape of the curve depends on the process parameters and material properties.

An approximate empirical formula for estimating the *maximum punch force P* is given by

$$P = 0.7(\text{UTS})(t)(L), \tag{7.1}$$

where UTS is the *ultimate tensile strength* of the sheet metal, t is its thickness, and L is the total length of the sheared edge. Thus for a round hole of diameter D, we have $L = \pi D$.

7.2.2 SHEARING PRACTICE

Shearing Operations

A variety of operations are based on the shearing process. For example, *punching* (also called *piercing*) is the term used when the slug is discarded (Fig. 7.6), whereas in *blanking* the slug is the part needed. Other shearing operations are called *perforating* (a number of punched holes), *slitting*, *parting*, *notching*, and *lancing* (Fig. 7.7). These processes are also known as *die-cutting*.

Another cutting operation is *nibbling*. This is performed on a nibbler, a machine where a punch (round or triangular in cross-section) moves up and down rapidly

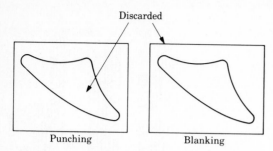

FIGURE 7.6 Punching (piercing) and blanking.

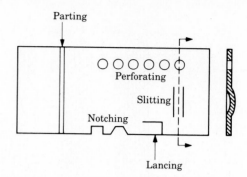

FIGURE 7.7 Examples of perforating, slitting, notching, lancing, and parting of sheet metal.

into the die opening. The sheet metal is fed through the gap and a cut, composed of a number of overlapping holes (similar to making an elongated hole or slot with a round paper punch), is made. Sheets can be cut along any desired path by manual control. This process is economical for small production runs since no special dies are required.

Clearances

Because the formability of the sheared part may be influenced by the quality of its sheared edges, the control of clearance is important (Section 7.5.2). In practice, clearances generally range between 2 and 10% of the thickness t of the sheet; the thicker the sheet, the larger the clearance. Thicker blanks require even larger clearances, such as 10%. However, the smaller the clearance, the better the quality of the sheared edge.

Fine Blanking

Very smooth and square edges can be produced by *fine blanking*. One basic design is shown in Fig. 7.8a. A V-shaped stinger (impingement) locks the sheet metal tightly in place and prevents the type of distortion shown in Fig. 7.3. The fine-blanking process involves very small clearances on the order of 1% and, generally, slow punching speeds. The operation is usually carried out on triple-action hydraulic presses, i.e., the movements of the punches, pressure pad, and the die are controlled individually. Fine blanking usually involves parts with holes punched simultaneously with blanking (Fig. 7.8b).

Shearing Practice

Note in Fig. 7.1 that the surfaces of the punch and die are flat. This means that the punch force builds up rapidly during shearing because the entire cross-sectional area is being sheared at one time. It is possible to control the area being sheared at any moment by making the punch and die surfaces *beveled*, as shown in Fig. 7.9. (This is similar to the punch geometry of a paper punch.) This geometry is particularly suitable

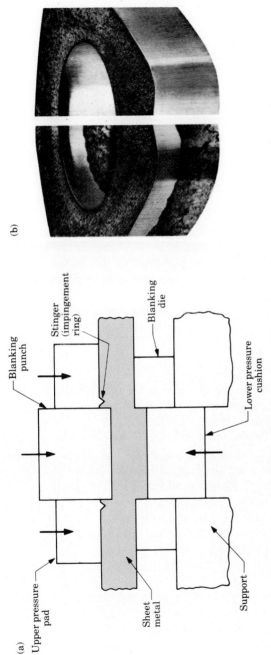

(b)

(a)

FIGURE 7.8 (a) Schematic illustration of set-up for fine blanking. (b) Comparison of sheared surfaces by conventional (left), and fine-blanking (right) techniques. Clearances in fine blanking are on the order of 1% of the thickness of the sheet. *Source:* Courtesy of Feintool U.S. Operations.

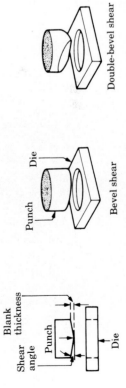

FIGURE 7.9 Examples of the use of shear angles on punch and die in shearing operations. The tip of the punch in an ordinary paper punch has a similar configuration.

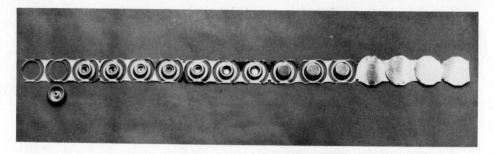

FIGURE 7.10   An example of forming a part with transfer dies. This part is the top piece for an aerosol spray can. Note that a different forming operation is performed at each step, and that the part is attached to the strip until the last operation.

for shearing thick blanks because it reduces the force required at the beginning of the stroke and also reduces the resulting noise level.

If clearances are small, there is significant friction at the interfaces and the parts being sheared may bind to the tools. In fact, the burnished regions indicate such a binding. To remove the parts, a stripper plate is generally used. Various types of stripping devices are available. (See also Fig. 7.40.)

For high production rate of parts requiring multiple operations, punching, blanking, or other operations may be done with *progressive dies*. The sheet metal is fed from a coil strip and a different operation is performed with each stroke of a series of punches (Fig. 7.10).

Several operations on the same piece may also be done in one stroke with *compound dies*. These operations are somewhat slower and the dies are costlier than individual operations. *Transfer dies* may also be used where the sheet metal undergoes different operations in each station of the press.

Shearing operations can also be carried out with a pair of circular blades (*slitting*), similar to those in a can opener, which follow a straight cut or any other desired path. A slit edge normally has a burr which may be removed by rolling the edge over to smooth the burr. (See hemming in Fig. 7.28 and beading in Fig. 7.34.) Slitting equipment may be the driven type (where the blades are powered as in an electric can opener), or the pull-through type (where the strip is pulled through idling blades).

Burrs from shearing operations may be removed by various processes such as filing or vibratory and barrel finishing (Section 9.6).

Tool and Die Materials

Tool and die materials for shearing are generally tool steels; for high production rates, carbide dies are used (Table 6.4). Lubrication is important in reducing tool and die wear, and improving edge quality. The basic machine is known as a *punch press* and is similar to those shown in Figs. 6.39 and 6.40. Rubber pads may also be used in blanking of sheet metal (with a setup similar to that in Fig. 7.57), where the rubber pad serves as the die.

7.2.3 MISCELLANEOUS METHODS OF CUTTING SHEET METAL AND PLATES

In addition to the shearing processes described above, there are other techniques for cutting sheet metals. The sheet or plate may be cut with a saw, such as a band saw. This is a chip removal process, and is described in Chapter 8. In another process, called *shaving*, the extra material from a rough sheared edge is trimmed by cutting (similar to removing a thin layer of wood with a chisel).

Flame cutting is another common method, particularly for thick steel plates, as described in Section 12.11. This process has wide applications in building ships and heavy structural components.

In *friction sawing*, a disk or blade rubs against the sheet or plate at surface speeds up to 25,000 ft/min (130 m/s) for disks, and 15,000 ft/min (80 m/s) for blades. In this process, the frictional energy is converted into heat, which then rapidly softens a narrow zone of the metal. The continuous movement of the disk or blade then pulls the softened metal out of the cutting zone. To help remove the material, some tools have teeth or notches. The friction sawing process is more suitable for ferrous alloys, as nonferrous metals have a tendency to adhere to the disk or blade and interfere with the cutting operation.

● **Illustrative Problem 7.1**

Estimate the force required in punching a 1-in.–diameter hole through a $\frac{1}{4}$-in.–thick annealed titanium alloy Ti-6Al-4V sheet.

SOLUTION. The punch force is given by Eq. (7.1)

$$P = 0.7(\text{UTS})(t)(L),$$

where UTS for this titanium alloy is found to be (from Table 3.20) 144,000 psi. Thus,

$$P = 0.7(144,000)(0.25)(\pi)(1)$$
$$= 79,170 \text{ lb.} \quad ●$$

7.3 SHEET-METAL CHARACTERISTICS

Sheet metals are generally characterized by a high ratio of surface area to thickness. Forming of sheet metals is carried out generally by tensile forces in the plane of the sheet; otherwise the application of compressive forces could lead to buckling, folding, and wrinkling of the sheet.

In bulk deformation processes such as forging, rolling, extrusion, and wire drawing (Chapter 6) there is an intentional change in the thickness or the lateral dimensions of the workpiece. However, in most sheet-forming processes any thickness change is due to the stretching of the sheet under tensile stresses (Poisson's ratio). Thickness decreases should generally be avoided as they could lead to necking and failure.

Since the basic mechanisms of all sheet-forming processes are stretching and bending, there are certain factors that have significant influence on the overall operation. The major ones are: elongation, yield-point elongation, anisotropy, grain size, residual stresses, springback, and wrinkling. These are treated in the sections below.

7.3.1 ELONGATION

Although sheet-forming operations rarely involve simple uniaxial stretching, the observations made in regard to simple tensile testing can be useful in understanding the behavior of sheet metals. Recall from Fig. 2.2 that a specimen subjected to tension first undergoes uniform elongation (which corresponds to the UTS, after which necking begins). This elongation is then followed by additional nonuniform elongation until fracture takes place (postuniform elongation). Since the material is being stretched in sheet forming, high uniform elongation is desirable for good formability (Section 7.4).

It was shown in Section 2.2.3 that, for a material whose true stress–true strain curve can be represented by the equation

$$\sigma = K\epsilon^n, \tag{7.2}$$

the strain at which necking begins (*instability*) is given by

$$\epsilon = n. \tag{7.3}$$

Hence, the true uniform strain in a simple stretching operation (uniaxial tension) is numerically equal to the strain-hardening exponent n. A large n indicates large uniform elongation and, hence, it is desirable for sheet forming.

Necking of a sheet specimen generally takes place at an angle ϕ to the direction of tension, as shown in Fig. 7.11(a) (*localized necking*). For an isotropic sheet specimen in simple tension, the Mohr circle for this situation is constructed as follows (Fig. 7.11b). The strain ϵ_1 is the longitudinal strain, and ϵ_2 and ϵ_3 are the lateral strains. Since the Poisson's ratio in the plastic range is 0.5, the lateral strains have the value $-\epsilon_1/2$. The narrow neck band in Fig. 7.11(a) is in *plane strain* along its length because it is constrained by the material above and below the neck.

The angle ϕ can now be determined from the Mohr circle by a rotation (either clockwise or counterclockwise) of 2ϕ from the ϵ_1 position (Fig. 7.11b). This angle is about 110 degrees; thus the angle ϕ is about 55 degrees. Note that although the length of the neck band remains essentially constant during the test, its thickness decreases (due to volume constancy) and the specimen eventually fractures. The angle ϕ will be different for materials that are anisotropic in the plane of the sheet.

Whether necking is *localized* or *diffuse* (Fig. 7.11c) depends on the strain rate sensitivity m of the material, as given by the equation

$$\sigma = C\dot{\epsilon}^m. \tag{7.4}$$

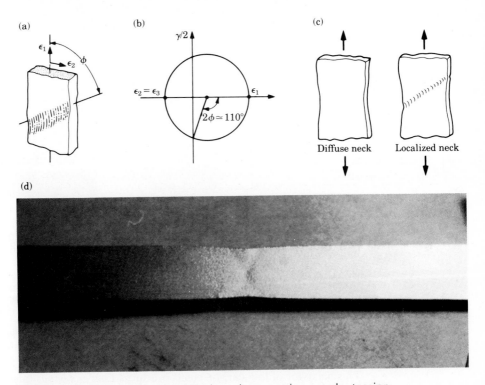

FIGURE 7.11 (a) Localized necking in a sheet specimen under tension. (b) Determination of the angle of neck from Mohr's circle for strain. (c) Schematic illustrations for diffuse and localized necking, respectively. (d) Localized necking in an aluminum strip stretched in tension. Note the double neck.

As described in Section 2.2.7, the higher the value of m, the more diffuse the neck becomes. An example of localized necking on an aluminum strip in tension is shown in Fig. 7.11(d). Note the double localized neck, i.e. ϕ can be in the clockwise and counterclockwise position in Fig. 7.11(a).

In addition to uniform elongation and necking, the *total* elongation (such as 2 in. [50 mm] gage length of a tension-test specimen) is also a significant factor in the formability of sheet metals. Note that total elongation is the sum of uniform elongation and postuniform elongation. Uniform elongation is governed by the strain-hardening exponent n, whereas postuniform elongation is governed by the strain-rate–sensitivity index m. The higher the m value, the more diffuse the neck and hence the greater the postuniform elongation before fracture. Thus, the total elongation of the material increases with increasing values of both n and m.

7.3.2 YIELD-POINT ELONGATION

Low-carbon steels exhibit a behavior called *yield-point elongation*, with upper and lower yield points, as shown in Fig. 7.12a. This behavior indicates that after the

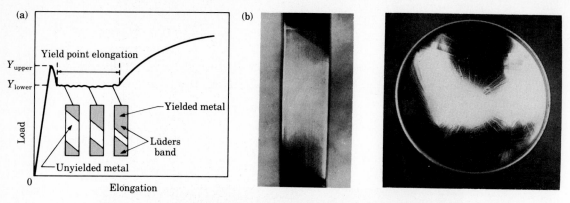

FIGURE 7.12 (a) Yield point elongation and Lueder's bands in tension testing. (b) Lueder's bands in low-carbon steel sheet. Compare with Fig. 7.12a. (c) Stretcher strains at the bottom of a steel can for household products. *Source:* (b) Courtesy of R. B. Liss, Caterpillar Tractor Co.

material yields, it stretches further in certain regions in the specimen with no increase in the lower yield point, while other regions have not yet yielded. When the overall elongation reaches the yield-point elongation, the entire specimen has been deformed uniformly. The magnitude of the yield-point elongation depends on the strain rate; with higher rates, the elongation generally increases.

This behavior of low-carbon steels produces *Lueder's bands* (also called *stretcher strain marks* or *worms*) on the sheet, as shown in Fig. 7.12(b). These are elongated depressions on the surface of the sheet and can be objectionable in the final product because of surface appearance. They can also cause difficulty in subsequent coating and painting operations. Stretcher strain marks on the curved bottom of steel cans for common household products are shown in Fig. 7.12(c); aluminum cans do not exhibit this behavior.

The usual method of avoiding this problem is to eliminate or to reduce yield-point elongation by reducing the thickness of the sheet 0.5 to 1.5% by cold rolling, known as *temper rolling* or *skin rolling*. Because of strain aging (Section 3.8), the yield-point elongation reappears after a few days at room temperature, or after a few hours at higher temperatures. Thus, the sheet metal should be formed within a certain time limit (anywhere from one to three weeks for rimmed steel, Section 5.5) in order to avoid the reappearance of stretcher strains.

7.3.3 ANISOTROPY

Another important factor influencing sheet-metal forming is *anisotropy,* or *directionality,* of the sheet metal. Anisotropy is acquired during the thermomechanical processing history of the sheet. Recall from Section 3.4 that there are two types of anisotropy: *crystallographic anisotropy* (due to preferred orientation of the grains) and *mechanical fibering* (due to the alignment of impurities, inclusions, voids, etc., throughout the thickness of the sheet during processing).

Anisotropy may be present not only in the plane of the sheet, but also in its thickness direction. The former is called *planar* anisotropy and the latter *normal* or *plastic* anisotropy. These are discussed in detail in Section 7.9.

7.3.4 GRAIN SIZE

Grain size of the sheet metal is important for two reasons: first, because of its effect on the mechanical properties of the material, and secondly, because of its effect on the surface appearance of the formed part. The coarser the grain, the rougher the surface appearance (*orange peel*, Fig. 3.19). An ASTM grain size of No. 7 or finer is preferred for general sheet metal forming. (See also Sections 3.3 through 3.5.)

7.3.5 RESIDUAL STRESSES

Residual stresses can be present in sheet metal parts because of the nonuniform deformation of the sheet during forming. When disturbed, such as by removing a portion of it, the part may distort.

Tensile residual stresses on surfaces can also lead to *stress–corrosion cracking* of sheet-metal parts (Fig. 7.13) unless they are properly stress relieved (Section 3.5).

7.3.6 SPRINGBACK

Sheet-metal parts, because they are generally thin and are subjected to relatively small strains, are likely to undergo considerable *springback*. This is particularly

FIGURE 7.13 Stress-corrosion cracking in a deep-drawn brass part for a light fixture. The cracks developed over a period of time. Brass and austenitic (300 series) stainless steels are among metals that are susceptible to stress-corrosion cracking.

significant in bending and other sheet-forming operations where the bend radius-to-thickness ratio is high, such as in automotive panels. This subject is treated in detail in Section 7.5.3.

7.3.7· WRINKLING

Although in sheet forming the metal is generally subjected to tensile stresses, the method of forming may be such that compressive stresses are developed in the plane of the sheet. An example is the wrinkling of the flange in deep drawing due to circumferential *compression* (Section 7.9). Other terms that are used to describe similar phenomena are *buckling*, *folding*, and *collapsing*.

The tendency for wrinkling increases with the unsupported or unconstrained length or surface area of the sheet metal, decreasing thickness, and nonuniformity of the thickness of the sheet. Such defects may also be initiated by lubricants that are trapped or distributed nonuniformly between the surface of the sheet and the dies.

7.4 FORMABILITY OF SHEET METALS

Sheet-metal *formability* is generally defined as the ability of the metal to undergo the desired shape change without failure, such as by necking or tearing. Although not generally practiced, it is also possible to include in this definition the strength of the metal: the stronger it is, the lower is its formability.

The formability of sheet metals has been of great interest because of its technologic significance. This section describes the different methods that are used in predicting formability.

Cupping Tests

Since sheet forming is basically a biaxial stretching process, the earliest tests developed to determine or to predict formability were cupping tests, such as the *Erichsen* and *Olsen* tests (stretching) and the *Swift* and *Fukui* tests (drawing, see Section 7.9).

A typical cupping test is the Erichsen test, shown in Fig. 7.14. The sheet-metal specimen is clamped over a circular flat die with a load of 1000 kg. A 20-mm–diameter

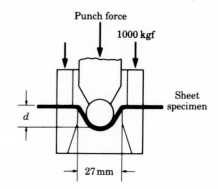

FIGURE 7.14 The Erichsen cupping test to determine formability of sheet metal. The sheet metal under the steel ball is subjected to biaxial stretching. The greater the distance *d* before failure, the greater the formability of the material. Ball diameter = 20 mm.

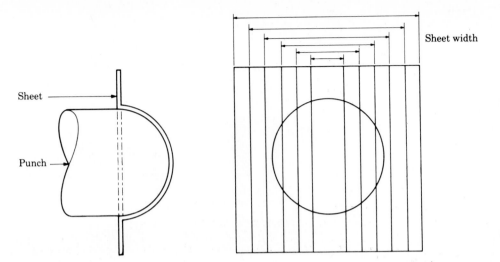

FIGURE 7.15 Schematic illustration of the bulge test on sheet specimens with different widths, clamped at the edges. The narrower the specimen, the more uniaxial is the stretching. A large square specimen stretches biaxially under the hemispherical punch. See also Fig. 7.16.

steel ball is then hydraulically pushed into the sheet metal until a crack appears on the stretched specimen. The distance d, measured in millimeters, is known as the Erichsen number. The greater the value of d, the greater the formability of the sheet.

Cupping tests measure the capability of the material to be stretched before fracturing and are relatively easy to perform. On the other hand, they do not simulate the exact conditions of actual forming operations since the stretching under the ball is axisymmetric.

Equal (balanced) biaxial stretching of sheet metals is also performed in the *bulge test*, which has been used extensively to simulate sheet-forming operations. In this test, a circular blank is clamped at its periphery and is bulged by hydraulic pressure, thus replacing the punch in Fig. 7.14. The process is one of pure stretch forming and there is no friction involved as there is in using a punch.

Forming Limit Diagrams

An important development in testing formability of sheet metals is the construction of *forming limit diagrams* (FLD). In these tests, the sheet blank is marked with a grid pattern of circles or similar patterns, using chemical etching or photoprinting techniques. The blank is then stretched over an unlubricated punch and the deformation of the circles is observed and measured in regions where necking and tearing have occurred. For improved accuracy of measurement, the circles are made as small as practical.

In order to develop unequal biaxial stretching, the specimens are prepared with varying widths (Fig. 7.15). Thus, a square specimen produces equal (balanced) biaxial stretching under the punch, whereas a specimen with a small width appoaches uniaxial stretching. After a series of such tests is performed on a particular sheet metal

403

FIGURE 7.16   Bulge test results on steel sheets of various widths. The first specimen (left) stretched further, before cracking, than the last specimen. From left to right, the state of stress changes from uniaxial to biaxial stretching. *Source:* Courtesy of R. W. Thompson, Inland Steel Research Laboratories.

(Fig. 7.16), the boundaries between failed and safe regions are plotted on the forming limit diagram (Fig. 7.17). Another example of a failed part with round and square grid patterns is shown in Fig. 7.18.

In the forming limit diagram the *major* and *minor strains* (engineering) are obtained as follows: Note in Fig. 7.17 that after stretching the original circle has deformed into an ellipse. The major axis of the ellipse represents the major direction and magnitude of stretching. The major strain plotted in Fig. 7.17 is the engineering strain (%) in this direction. Likewise, the minor axis of the ellipse represents the magnitude of the stretching *or* shrinking in the transverse direction. Thus the minor strain can be negative or positive. (The major strain is always positive since forming of sheet metal takes place by stretching.)

For example, if we draw a circle in the center of a sheet-metal tensile-test specimen and then stretch it, the minor strain will be negative. This is due to the Poisson's ratio (i.e., the specimen becomes narrower as it is stretched). This can be easily observed by experimenting with a wide rubber band. On the other hand, if we place a circle on a spherical rubber balloon and inflate it, the minor strain is positive and equal in magnitude to the major strain. The circle simply becomes a larger circle.

By observing the difference in surface area between the original circle and the ellipse, one can also determine whether or not the thickness of the sheet has changed. If the area of the ellipse is larger than the original circle the sheet has become thinner (because volume remains constant in plastic deformation).

Friction at the punch–metal interface can also be an important factor in the test results. With well-lubricated interfaces, the strains are more uniformly distributed over the punch. Depending on the notch sensitivity of the sheet metal, surface scratches, deep gouges, and blemishes can reduce formability and cause premature

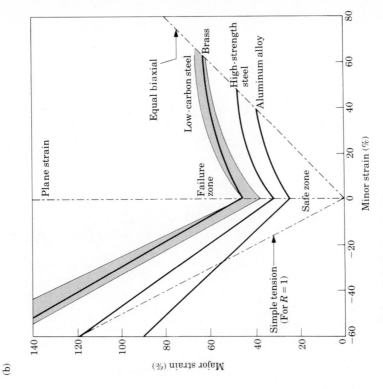

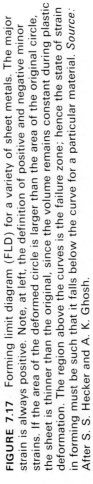

(a)

(b)

FIGURE 7.17 Forming limit diagram (FLD) for a variety of sheet metals. The major strain is always positive. Note, at left, the definition of positive and negative minor strains. If the area of the deformed circle is larger than the area of the original circle, the sheet is thinner than the original, since the volume remains constant during plastic deformation. The region above the curves is the failure zone; hence the state of strain in forming must be such that it falls below the curve for a particular material. *Source:* After S. S. Hecker and A. K. Ghosh.

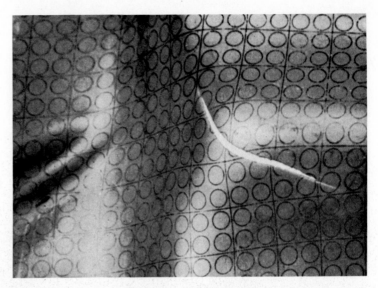

FIGURE 7.18 An example of using grid marks (circular and square) to determine the magnitude and direction of surface strains in sheet metal forming. Note that the crack (tear) is generally perpendicular to the major (positive) strain. *Source:* After S. P. Keeler.

tearing and failure during testing and in actual forming operations. With effective lubrication the coefficient of friction in sheet metal forming generally ranges from about 0.05 to 0.1 for cold forming, and from 0.1 to 0.2 for elevated temperatures.

From Fig. 7.17 we observe, as expected, that different materials have different forming limit diagrams. The higher the curve, the better the formability of the material. Also note that for the same minor strain, say 20%, a compressive minor strain is associated with a higher major strain before failure than a tensile (positive) minor strain. In other words, it is desirable for the minor strain to be negative, i.e., shrinking in the minor direction. As we shall see in Section 7.6.3, special tools have been designed for forming sheet metals that take advantage of the beneficial effect of negative minor strains on extending formability.

The effect of sheet-metal thickness on forming limit diagrams is to raise the curves in Fig. 7.17. Thus, the thicker the sheet, the higher its formability curve, hence the more formable it is. On the other hand, in actual forming operations, a thick blank may not bend as easily around small radii (Section 7.5.2). The possible effect of the rate of deformation on the forming limit diagrams should also be assessed for each material. Extensive studies are being carried out in developing new test methods to predict the behavior of metals in sheet forming operations.

● **Illustrative Problem 7.2**

A thin-walled spherical shell made of the aluminum alloy shown in Fig. 7.17 is being expanded by internal pressure. If the original diameter is 200 mm, what is the maximum diameter to which it can be expanded safely?

SOLUTION. Since the material is being stretched in a state of equal biaxial tension, we find from Fig. 7.17 that the maximum allowable engineering strain is about 40%. Thus,

$$e = \frac{\pi D_f - \pi D_0}{\pi D_0} = 0.40$$

$$= \frac{D_f - 200}{200} = 0.40.$$

Hence,

$$D_f = 280 \text{ mm.} \quad \bullet$$

7.5 BENDING OF SHEET AND PLATE

One of the most common metalworking operations is *bending*. This process is used not only to form parts such as flanges, curls, seams, and corrugations, but also to impart stiffness to the part by increasing its moment of inertia (note, for instance, that a long strip of metal is much less rigid when flat than when it is formed into the shape of a gutter.) In this section, the mechanics of bending operations and the practices that are involved are described.

7.5.1 MECHANICS OF BENDING

The terminology used in bending is shown in Fig. 7.19. In bending, the outer fibers of the material are in tension and the inner fibers are in compression. Theoretically, the strains at the outer and inner fibers are equal in magnitude and are given by the equation

$$e_o = e_i = \frac{1}{(2R/T) + 1}, \tag{7.5}$$

where R is the bend radius and T is the thickness.

FIGURE 7.19 Bending terminology. The bend radius is measured to the inner surface of the bend. Note that the length of the bend is the width of the sheet. Also note that the bend angle and the bend radius (sharpness of the bend) are two different variables.

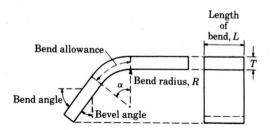

Experimental evidence indicates that while this equation holds reasonably well for the inner fiber strain e_i, actual values of e_o are considerably higher than e_i. This is due to the shifting of the neutral axis toward the inner surface. The length of bend (Fig. 7.19) is smaller in the outer region than in the inner region. (This can easily be observed by bending a rectangular eraser. See also Fig. 7.21c.) The difference between the outer and inner strains increases with decreasing R/T ratio, i.e., sharper bends.

7.5.2 MINIMUM BEND RADIUS

It is apparent from Eq. (7.5) that as the R/T ratio decreases, the tensile strain at the outer fiber increases and that the material may crack after a certain strain is reached. The radius R at which a crack appears on the outer surface of the bend is called the *minimum bend radius*. The minimum radius to which a part can be bent safely is normally expressed in terms of its thickness, such as $2T$, $3T$, $4T$, etc. Thus, a $3T$ bend radius indicates that the smallest radius to which the sheet can be bent, without cracking, is 3 times its thickness. The minimum bend radius for a variety of materials has been determined experimentally and is available in various handbooks. Some typical results are given in Table 7.1.

Studies have also been conducted to establish a relationship between the minimum R/T ratio and a given mechanical property of the material. One such analysis is based on the following assumptions:

a. The true strain at cracking on the outer fiber in bending is equal to the true strain at fracture ϵ_f of that material in a simple tension test.

b. The material is homogeneous and isotropic,

c. The sheet is bent in a state of plane stress, i.e., its L/T ratio (Fig. 7.19) is small.

TABLE 7.1
MINIMUM BEND RADIUS FOR VARIOUS MATERIALS AT ROOM TEMPERATURE

MATERIAL	CONDITION	
	SOFT	HARD
Aluminum alloys	0	6T
Beryllium copper	0	4T
Brass, low-leaded	0	2T
Magnesium	5T	13T
Steels		
austenitic stainless	0.5T	6T
low-carbon, low-alloy and HSLA	0.5T	4T
Titanium	0.7T	3T
Titanium alloys	2.6T	4T

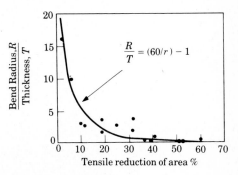

FIGURE 7.20 Relationship between the ratio of bend radius-to-sheet thickness and tensile reduction of area for a variety of materials. Note that sheet metal with a reduction of area of about 50% can be bent and flattened over itself without cracking. See also Fig. 7.28. *Source:* After J. Datsko and C. T. Yang, *Jr. Eng. Ind.,* vol. 82, 1960, pp. 309–314.

The true strain at fracture in tension is given by

$$\epsilon_f = \ln\left(\frac{A_0}{A_f}\right) = \ln\left(\frac{100}{100 - r}\right),$$

where r is the percent reduction of area in a tension test (Eq. 2.7). From Section 2.2.2,

$$\epsilon_0 = \ln(1 + e_o) = \ln\left(1 + \frac{1}{(2R/T) + 1}\right) = \ln\left(\frac{R + T}{R + (T/2)}\right).$$

Equating the two expressions and simplifying, we obtain

$$\text{Minimum } \frac{R}{T} = \frac{50}{r} - 1. \tag{7.6}$$

Some experimental data are shown in Fig. 7.20. The curve that best fits the experimental data is found to be

$$\text{Minimum } \frac{R}{T} = \frac{60}{r} - 1. \tag{7.7}$$

Observe that the R/T ratio approaches zero (complete bendability, i.e., the material can be folded over itself) at a tensile reduction of area of 50%. It is interesting to note that this is the same value obtained in spinnability of metals, described in Section 7.11, where a material with 50% reduction of area is found to be completely spinnable.

To increase the bendability of a metal, one may increase its tensile reduction of area either by *heating* or by the application of *hydrostatic pressure,* as described in Section 2.2.8. Other techniques may also be employed to increase the compressive environment in bending, such as applying compressive forces in the plane of the sheet during bending to minimize tensile stresses in the outer fibers of the bend area.

As the length of the bend (dimension L in Fig. 7.19) increases, the state of stress at the outer fibers changes from uniaxial stress to a biaxial stress. This is due to the fact that L tends to become smaller due to stretching of the outer fibers, but it is constrained by the material around the bend area. Since biaxial stretching tends to reduce ductility (i.e., strain to fracture), it is found that as the length L increases, the minimum

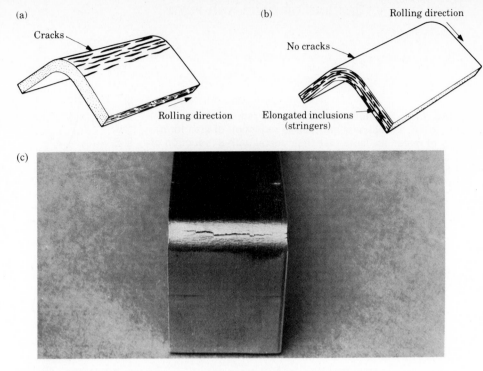

Cracks

Rolling direction

(b)

Rolling direction

No cracks

Elongated inclusions
(stringers)

(c)

FIGURE 7.21 (a) and (b) The effect of elongated inclusions (stringers) on cracking as a function of the direction of bending with respect to the original rolling direction of the sheet. This example shows the importance of the direction of cutting workpieces from large sheets that are subsequently bent to make a product. (c) Cracks on the outer radius of an aluminum strip bent to an angle of 90 degrees.

bend radius increases. However, at a length of about $10T$, there is no further increase in the minimum bend radius and a plane strain condition is fully developed.

It has also been observed that, as the R/T ratio decreases, narrow sheets (smaller length of bend) crack at the edges and wider sheets crack at the center, where the biaxial stress is the highest. Bendability also depends on the edge condition of the sheet being bent. Since rough edges are points of stress concentration, bendability decreases with rougher edges.

Another factor that is significant in edge cracking is the amount and shape of inclusions in the sheet metal and the amount of cold working that the edges undergo during shearing. Inclusions in the form of *stringers* are more detrimental than globular-shaped inclusions. The removal of the cold-worked regions (by machining or heat treating) greatly improves the resistance to edge cracking.

Thus, anisotropy of the sheet is also an important factor in bendability. As depicted in Fig. 7.21, cold rolling of sheets produces anisotropy due to the alignment of impurities, inclusions, and voids (mechanical fibering). Thus, the transverse ductility is reduced, Fig. 7.21(c). In bending such a sheet, caution should be used in cutting the blank in the proper direction from the rolled sheet, although this may not always be possible in practice.

7.5.3 SPRINGBACK

Since all materials have a finite modulus of elasticity, plastic deformation is followed by *elastic recovery* upon removing the load. In bending, this recovery is known as *springback*. As shown in Fig. 7.22, the final bend angle after springback is smaller and the final bend radius is larger than before. (This can be easily observed by bending a piece of wire.) Springback occurs not only in flat sheets or plate, but also in bars, rod, and wire of any cross-section.

A quantity characterizing springback is the *springback factor* K_s, which is determined as follows: Since the bend allowance (Fig. 7.19) is the same before and after bending, the following relationship is obtained for pure bending,

$$\text{Bend allowance} = \left(R_i + \frac{T}{2} \right)\alpha_i = \left(R_f + \frac{T}{2} \right)\alpha_f.$$

From this relationship, K_s is defined as

$$K_s = \frac{\alpha_f}{\alpha_i} = \frac{(2R_i/T) + 1}{(2R_f/T) + 1}, \tag{7.8}$$

where R_i and R_f are the initial and final bend radii, respectively.

We note from this expression that the springback factor K_s depends only on the R/T ratio. A springback factor of $K_s = 1$ indicates no springback and $K_s = 0$ indicates complete elastic recovery (Fig. 7.23), such as in the leaf spring of an automobile.

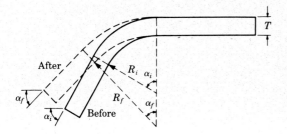

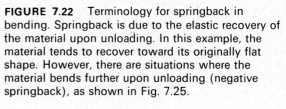

FIGURE 7.22 Terminology for springback in bending. Springback is due to the elastic recovery of the material upon unloading. In this example, the material tends to recover toward its originally flat shape. However, there are situations where the material bends further upon unloading (negative springback), as shown in Fig. 7.25.

FIGURE 7.23 Springback factor K_s for various materials. (a) 2024-0 and 7075-0 aluminum. (b) Austenitic stainless steels. (c) 2024-T aluminum. (d) 1/4 hard austenitic stainless steels. (e) 1/2 hard to full-hard austenitic stainless steels. See Table 6.11 for the definition of hardness in this context. *Source:* After G. Sachs, *Principles and Methods of Sheet-Metal Fabricating.* Reinhold, 1951, p. 100.

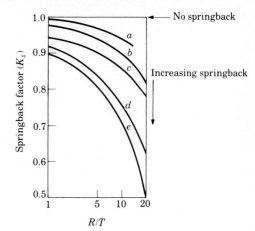

From Fig. 2.8 recall that the amount of elastic recovery depends on the stress level and the modulus of elasticity E of the material. Thus, elastic recovery increases with the stress level and with decreasing elastic modulus. Based on this observation, an approximate formula has been developed to estimate springback,

$$\frac{R_i}{R_f} = 4\left(\frac{R_i Y}{ET}\right)^3 - 3\left(\frac{R_i Y}{ET}\right) + 1, \tag{7.9}$$

where Y is the yield stress of the material at 0.2% offset (Fig. 2.2).

Negative Springback

The springback observed in Fig. 7.22 is called positive springback. Under certain conditions it is also possible to have a negative springback. In other words, the bend angle becomes larger after the bend has been completed and the load is removed. This phenomenon is generally associated with V-die bending (Fig. 7.24).

The development of negative springback can be explained by observing the sequence of deformation in Fig. 7.24. If we remove the piece at stage (b), it will undergo positive springback. At stage (c) the ends of the part are touching the male punch. Note that between stages (c) and (d) the part is actually being bent in the opposite direction to that between (a) and (b). Note also the lack of conformity of the punch radius and the inner radius of the part in (b) and (c). In (d), however, the radii are the same.

Upon unloading, the part in stage (d) will springback inward because it is being *unbent* from stage (c) both at the tip of the punch and in the arms of the part. This inward (negative) springback can be greater than the positive springback that would result upon unloading from stage (c) because of the large strains that the material is undergoing in the bend area in stage (b).

Some experimental results with round wires using three different materials are shown in Fig. 7.25. Note that positive springback increases with increasing R/D ratio, where D is the wire diameter. (This result is similar to the R/T effect in Fig. 7.23.) However, springback becomes negative at low R/D ratios due to the fact that low R/D ratios indicate less conformity at the punch radius–wire interface in Fig. 7.24.

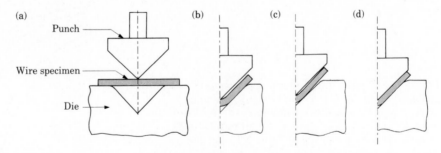

FIGURE 7.24   Schematic illustration of the stages in bending round wire in a V-die. This type of bending can lead to negative springback; this does not occur in air bending, shown in Fig. 7.27. *Source:* After K. S. Turke and S. Kalpakjian, *Proc. NAMRC III,* 1975, pp. 246–262.

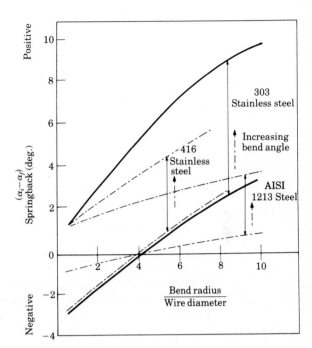

FIGURE 7.25 Range of positive and negative springback for various materials (with the same modulus of elasticity) as a function of the ratio of bend radius-to-wire diameter. *Source:* After K. S. Turke and S. Kalpakjian, *Proc. NAMRC III*, 1975, pp. 246–262.

Compensation for Springback

In practice, compensation for springback is usually accomplished by *overbending* the part. Several trials may be necessary to obtain the desired result. Another method is to *coin* the bend area by subjecting it to high localized stresses between the tip of the punch and the die surface. This is known as *bottoming* the punch. Because springback decreases with decreasing yield stress (all other parameters remaining the same), bending may also be carried out at elevated temperatures to reduce springback. Another method is *stretch bending* where the part is subjected to tension while being bent in a die.

● **Illustrative Problem 7.3**

A 20-gage steel sheet is bent to a radius of 0.5 in. Assuming that its yield stress is 40,000 psi, calculate the radius of the part after it is bent.

SOLUTION. The appropriate formula is Eq. (7.9) where

$R_i = 0.5$ in.

$Y = 40,000$ psi

$E = 29 \times 10^6$ psi

$T = 0.0359$ in. (from Table 6.10)

Thus,

$$\frac{R_i Y}{ET} = \frac{(0.5)(40,000)}{(29 \times 10^6)(0.0359)} = 0.0192$$

and

$$\frac{R_i}{R_f} = 4(0.0192)^3 - 3(0.0192) + 1$$
$$= 0.942.$$

Hence

$$R_f = \frac{0.5}{0.942} = 0.531 \text{ in.} \quad \bullet$$

7.5.4 FORCES

Bending forces can be estimated by assuming that the process is one of simple bending of a rectangular beam. Thus, the bending force is a function of the strength of the material, the length and thickness of the part, and the die opening W, as shown in Fig. 7.26. Excluding friction, the general expression for the *maximum bending force P* can be given by

$$P = k\frac{YLT^2}{W}, \qquad (7.10)$$

where the factor k ranges from about 0.3 for a wiping die, 0.7 for a U-die, and 1.3 for a V-die (Fig. 7.26), respectively.

This expression can be simplified to

$$P = \frac{(UTS)LT^2}{W} \qquad (7.11)$$

for a V-die and can be used as a general guide for other die geometries. Equation (7.11) has been found to apply well to situations where the punch radius and sheet thickness are small as compared to die opening W.

The bending force is also a function of punch travel. It increases from zero to a maximum level and may decrease as the bend is completed. The force then increases sharply as the punch bottoms in the case of die bending. In *air bending*, or *free bending*, (Fig. 7.27), the force does not increase again after it begins to decrease.

7.6 MISCELLANEOUS BENDING OPERATIONS

In this section common industrial bending operations are described.

7.6.1 PRESS-BRAKE FORMING

Small pieces can be bent easily with simple fixtures by using a press. Long (20 ft [7 m] or more) and relatively narrow pieces of sheet metal or plates are usually

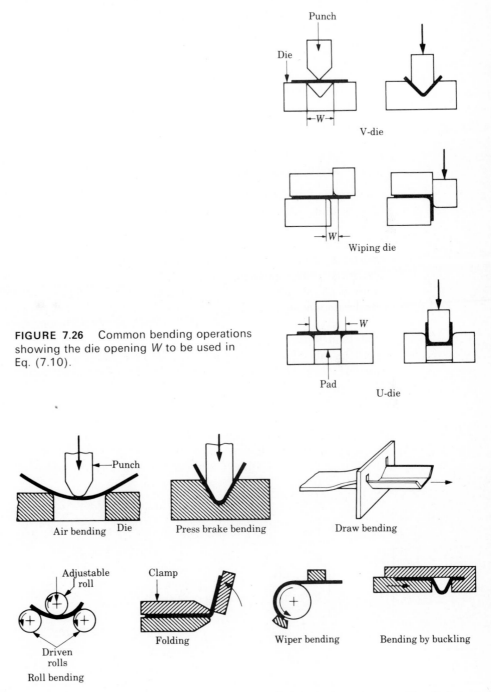

FIGURE 7.26 Common bending operations showing the die opening W to be used in Eq. (7.10).

FIGURE 7.27 Examples of various types of bending operations. *Source:* After K. Lange.

bent in a *press brake*. This machine uses long dies in a mechanical or hydraulic press and is suitable for small-quantity production runs. The tooling is simple and adaptable to a wide variety of shapes, such as those shown in Fig. 7.28.

Many ferrous and nonferrous metals are bent by press-brake forming. For materials with lower ductility or for those requiring high forces, both the dies and the sheet metal may be heated. Die materials may range from hardwood (for low-strength materials and small-quantity runs) to carbides. For most applications, carbon-steel or gray-iron dies are used (Table 6.4).

7.6.2 ROLL FORMING

In press-brake forming, the dies have finite lengths and each piece is bent individually. For bending continuous lengths of sheet metal and for large-quantity production runs, the roll-forming process is used.

In this process, also known as *contour roll forming* or *cold roll forming*, the metal is bent in stages by passing it through a series of rolls (Fig. 7.29). Typical products are channels, gutters, pipes and tubing (with lock seams or welded seams), siding, and various panels and frames (Figs. 7.30 and 7.31). The length of the part is limited only by the amount of material supplied from the coiled stock. The parts are usually sheared and stacked continuously. The sheet thickness generally ranges from about 0.005 to 0.75 in. (0.125 to 20 mm).

The rolls are generally made of carbon steel or gray iron and may be chromium plated for improved wear resistance of the rolls and better surface finish of the product. Lubricants may also be used to improve roll life and surface finish and also to cool the rolls and the workpiece. Forming speeds are generally below 300 ft/min (0.5 m/s), although they can be much higher in certain applications.

Proper design and sequencing of the rolls (which are generally mechanically driven) requires much experience and includes consideration of tolerances, springback, tearing, and buckling. The material is generally formed from the center outward.

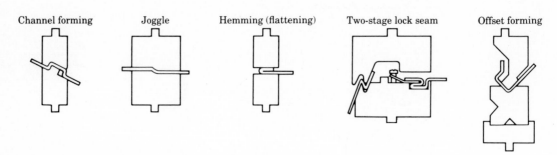

| Channel forming | Joggle | Hemming (flattening) | Two-stage lock seam | Offset forming |

FIGURE 7.28 Schematic illustration of various bending operations.

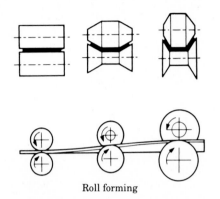

Roll forming

FIGURE 7.29 Schematic illustration of the roll forming process. This is an economic method of forming continuous shapes, such as channels and gutters.

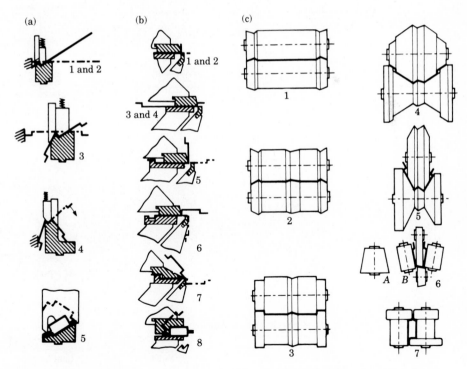

FIGURE 7.30 Schematic illustration of the manufacture of a door frame by three different methods. (a) Die bending. (b) Folding. (c) Roll forming, or contour rolling. *Source:* After K. Lange.

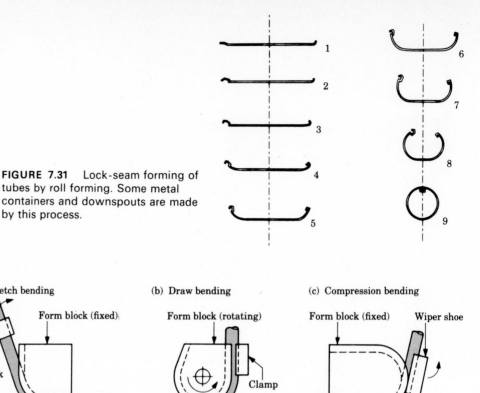

FIGURE 7.31 Lock-seam forming of tubes by roll forming. Some metal containers and downspouts are made by this process.

1
2
3
4
5
6
7
8
9

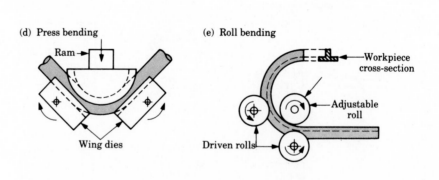

(a) Stretch bending

Form block (fixed)

Chuck

Workpiece   Chuck

(b) Draw bending

Form block (rotating)

Clamp

Pressure bar

(c) Compression bending

Form block (fixed)   Wiper shoe

Clamp

(d) Press bending

Ram

Wing dies

(e) Roll bending

Workpiece cross-section

Adjustable roll

Driven rolls

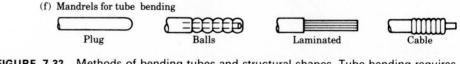

(f) Mandrels for tube bending

Plug   Balls   Laminated   Cable

FIGURE 7.32 Methods of bending tubes and structural shapes. Tube bending requires internal mandrels (or filling with particulate materials, such as sand) to keep the tube from collapsing. A tube tends to collapse during bending due to the longitudinal compressive stresses along the inner radius of the bend. *Source:* After L. E. Doyle.

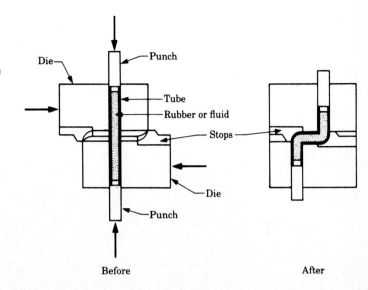

FIGURE 7.33 A method of forming a tube with sharp angles using axial compressive forces. Compressive stresses are beneficial in forming operations because they delay fracture. Note that the tube is supported internally with rubber or fluid to avoid collapsing during forming. *Source:* After J. L. Remmerswaal and A. Verkaik, *Int. Conf. Manufacturing Technology,* ASME, 1967, pp. 1171–1181.

7.6.3 BENDING OF TUBES AND SECTIONS

Since bending involves compressive strains at the inner fibers of a part, bending of tubes and hollow sections requires special tooling to avoid buckling and folding. The simplest and oldest method of bending tubes or pipes is to pack them with loose particles, such as sand, and then bend them. This technique prevents the tube from buckling by pressurizing the inside of the tube and distributing the compressive strains uniformly. After bending, the sand is shaken out of the bent tube. Tubes can also be plugged with various flexible internal mandrels, as shown in Fig. 7.32. This figure also illustrates various bending techniques for tubes and sections.

The beneficial effect of forming metals under a high compressive environment is demonstrated in the bending of a tube with relatively sharp corners (Fig. 7.33). In this operation, the tube is subjected to longitudinal compressive stresses, which reduce the stresses in the outer fibers in the bend area, thus improving the bendability of the material.

7.6.4 MISCELLANEOUS FORMING OPERATIONS

In this section various additional forming operations are described. (See also Section 12.2.)

Beading

Two examples of beading are shown in Figs. 7.34(a) and (b). In the first process, the edge of the sheet metal is bent into the cavity of a die. The bead gives stiffness to the part by increasing the moment of inertia at the edges. The bead also prevents sharp

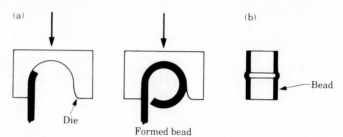

FIGURE 7.34 Examples of beads. (a) At edges, by edge rolling. (b) In tubes, by bulging. Beads act as stiffeners in sheet metal products. Such beads can be found on cans for food and various other products.

edges from coming into contact with hands, and it improves the appearance of the part.

The beading shown in Fig. 7.34(b) can be performed either on flat panels (to improve stiffness) or on cylindrical containers. An example of the latter is food cans, where beading prevents collapsing of the walls due to development of a vacuum while cooling after the can is sealed.

Flanging

This is a process of bending the inner or outer edges of sheet metals. In the *dimpling* operation (Fig. 7.35a), a hole is punched and then expanded into a flange, or a shaped punch pierces the sheet metal and expands the hole (Fig. 7.35b). The stretching of the hole subjects the edges to high tensile strains, which could lead to cracking and tearing of the flange. As the ratio of flange diameter to hole diameter increases, the strains increase proportionately. The larger the hole, the less the strain for the same flange depth.

The ends of tubes are flanged by a similar process, as shown in Fig. 7.35(c). When the angle of bend is less than 90 degrees (conical), such as in fittings, the process is called *flaring*. The condition of the edges is also important in this operation; the rougher the edge, the greater the tendency for cracking.

Flanging is also done on flat sheets, as shown in Fig. 7.35(d). In *shrink flanging*, the flange is subjected to compressive hoop stresses, which, if excessive, cause wrinkling of the edges of the flange. (This can be demonstrated by placing a piece of paper on top of a round cup and wrapping it around the cup. Note how the circumference of the paper develops wrinkles. Also note the wrinkles on the periphery of paper plates.) The tendency for wrinkling increases with decreasing radius of curvature of the flange. In *stretch flanging*, the edges of the flange are subjected to tensile stresses and, if excessive, can lead to cracking.

Hemming

In this process, also called *flattening*, the edge of the sheet metal is folded over itself (Fig. 7.28). This is done to increase the stiffness of the part, to improve its appearance, and to avoid sharp edges.

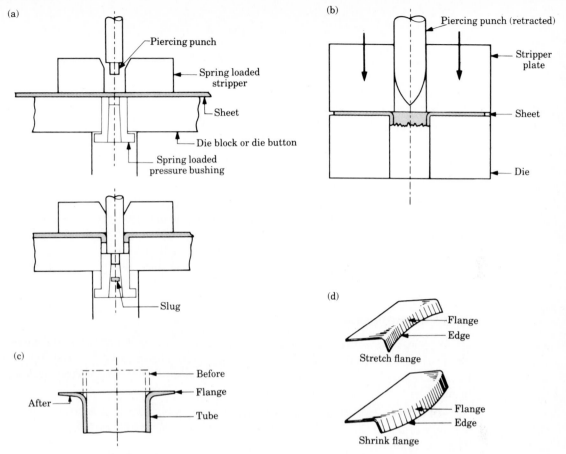

FIGURE 7.35 Various sheet-metal forming operations. (a) Dimpling. (b) Piercing of sheet metal to form a flange. Note that, unlike dimpling, no metal is removed in this piercing operation. (c) Flanging of a tube. Note the thinning of the edges of the flange due to circumferential expansion. (d) Examples of stretch flanging and shrink flanging on sheet metal. Flanges are used not only for assembly with other components at the edges, but they also act as stiffeners.

Seaming

In seaming, two edges of sheet metal are attached by hemming. A single seam is shown in Fig. 12.4(a). Double seams for watertight joints are also made by this process.

7.7 STRETCH FORMING

In this process, the sheet metal is clamped at its edges and stretched over a die or form block, which moves upward (or downward, depending on the arrangement). The process is used primarily for aerospace and automotive applications with a variety of materials. The parts may have single or double curvatures such as in aircraft-wing–skin panels, automobile door panels, and window frames.*

The forming limits in stretch forming can be obtained from the forming limit diagrams, described in Section 7.4, and depend on the biaxial strains to which the material is subjected. In many operations, the blank is a rectangular sheet and is clamped along the pair of narrower edges, thus allowing the material to shrink in width. Careful control of the amount of stretching is an important aspect of this process. (See Illustrative Problem 7.4.)

Stretch forming cannot produce parts with sharp contours or re-entrant corners (depressions on the surface). Although this process is generally a low-volume–production method, it is versatile and economical. Die materials are generally made of wood, zinc alloys, steel, or plastics. Most applications require little or no lubricant. Various accessory equipment can be used in conjunction with this basic process. Among these is forming with both male and female dies while the part is being stretched.

● **Illustrative Problem 7.4**

A 15-in.-long workpiece (Fig. P7.1) is stretched by a force F until $\alpha = 20$ degrees. The original cross-sectional area is 0.5 in^2. The material has a curve $\sigma = 100,000\epsilon^{0.3}$. Find the total work done, ignoring end effects or bending. What is α_{max} before necking begins?

SOLUTION. This situation is equivalent to stretching a workpiece from 15 in. to a length of $a + b$. For $\alpha = 20$ degrees, we find that the final length $L_f = 16.8$ in. The true strain then is

$$\epsilon = \ln\left(\frac{L_f}{L_o}\right) = \ln\left(\frac{16.8}{15}\right) = 0.114.$$

The work done per unit volume is

$$u = \int_0^{0.114} \sigma \cdot d\epsilon = 10^5 \int_0^{0.114} \epsilon^{0.3}\, d\epsilon = 10^5 \left.\frac{\epsilon^{1.3}}{1.3}\right]_0^{0.114}$$

$$= 4570 \text{ in} \cdot \text{lb/in}^3.$$

The volume of the workpiece is

$$V = (15)(0.5) = 7.5 \text{ in}^3.$$

Hence,

$$\text{Work} = (u)(V) = 34,275 \text{ in.} \cdot \text{lb.}$$

---

* Aluminum skins for Boeing 767 and 757 aircraft are made by stretch forming, with a tensile force of 2 million lb (9 MN). The skins are 40 ft by 8.3 ft by 1/4 in. (12 m by 2.5 m by 6.4 mm).

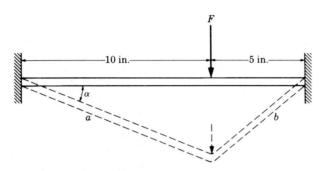

FIGURE P7.1

The necking limit for uniaxial tension is given by Eq. (7.3), the problem being similar to a stretch-forming operation. Thus,

$$L_{max} = L_o e^n$$
$$= 15e^{0.3} = 20.2 \text{ in.}$$

Hence $a + b = 20.2$ in. From similarities in triangles, we obtain

$$a^2 - 10^2 = b^2 - 5^2$$
$$a^2 = b^2 + 75.$$

Also,

$$a + b = 20.2.$$

Hence $a = 12$ in. and $b = 8.2$ in. Therefore,

$$\cos \alpha = \frac{10}{12} = 0.833$$

or

$$\alpha_{max} = 33.6 \text{ degrees.} \quad \bullet$$

7.8 BULGING OF TUBES

In this process, the diameter of the tube is expanded in one or more sections by a flexible *plug* (rubber or polyurethane), hydraulic pressure, or by segmented mechanical dies. Typical products are coffee or water pitchers, barrels, and beads on drums (Fig. 7.34b).

The basic process of bulging involves placing a tubular, conical, or curvilinear part in a split-female die and expanding it, such as with the use of a flexible plug (Fig. 7.36). The punch is then retracted and the part is removed by opening the dies. For

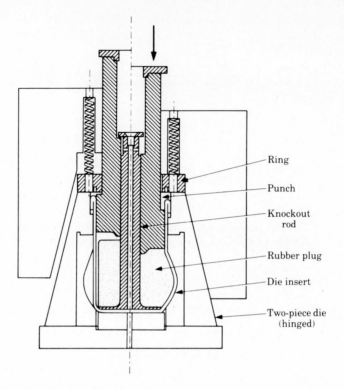

Ring

Punch

Knockout rod

Rubber plug

Die insert

Two-piece die (hinged)

FIGURE 7.36 Bulging of a tubular component with rubber or polyurethane plug. Water pitchers and similarly shaped containers can be made by this method.

parts with complex shapes, the plug may be shaped instead of being cylindrical, in order to apply greater pressure in some locations than in others.

The major advantage of using polyurethane plugs is that they are resistant to abrasion, wear, and lubricants. Furthermore, they do not damage the surface finish of the part. *Hydraulic pressure* can also be used in these operations, although it requires sealing and hydraulic controls. *Segmented dies*, whereby bulging is carried out by expanding the dies mechanically, are relatively inexpensive and can be used for large-quantity production runs.

The formability in bulging processes can be enhanced by the application of longitudinal compressive stresses, as shown in the example in Fig. 7.37 (similar to Fig. 7.33).

An interesting application of bulging is in the manufacture of bellows (Fig. 7.38). After the tube is bulged in several equidistant locations, it is compressed axially to collapse the bulges, thus forming bellows. The tube material must of course be able to undergo the large strains due to bending during the collapsing process.

Tubular blanks may also be expanded by mechanical means using special tooling. Figure 7.39 shows the production of a rear-axle housing from a tubular blank. A slot is first made in the center of the tube. The ends of the tube are then swaged (Section 6.24). The center section is heated and expanded by inserting special tooling through the slot. Various other processes are also used in completing the part, as shown in the figure.

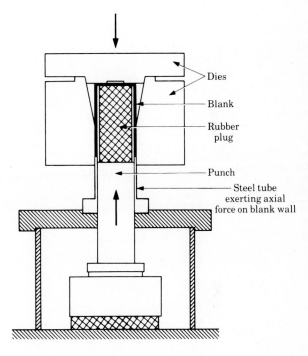

FIGURE 7.37 Bulging of a tubular component using longitudinal compressive stresses to enhance formability. *Source:* After J. L. Remmerswaal and A. Verkaik, *Int. Conf. Manufacturing Technology*, ASME, 1967, pp. 1171–1181.

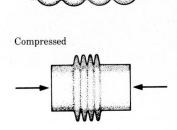

FIGURE 7.38 Sequence of processes for manufacturing metal bellows. A straight tube is first bulged by internal pressure. It is then compressed longitudinally to collapse the bulged sections and thus form bellows. Such products are useful for flexible connectors, such as for gas lines for appliances.

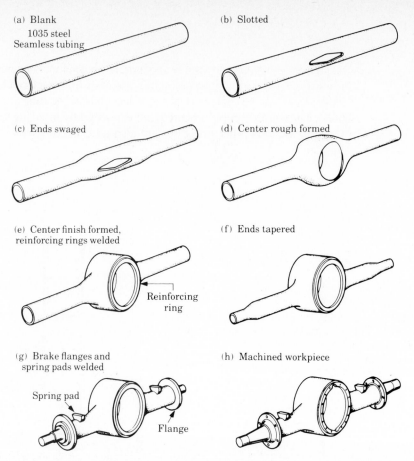

(a) Blank
1035 steel
Seamless tubing

(b) Slotted

(c) Ends swaged

(d) Center rough formed

(e) Center finish formed, reinforcing rings welded

Reinforcing ring

(f) Ends tapered

(g) Brake flanges and spring pads welded

Spring pad

Flange

(h) Machined workpiece

FIGURE 7.39 Manufacturing operations in the production of a rear-axle housing from an AISI 1035 steel tubular blank. *Source:* ASM Committee on Bending of Bars and Tubing, "Bending and Forming of Tubing," *Metals Handbook*, 8th ed., vol. 4, Lyman, Taylor, ed., American Society for Metals, 1969, p. 316.

7.9 DEEP DRAWING

In this process, a flat sheet-metal blank is formed into a cylindrical or box-shaped part by means of a punch that presses the blank into the die cavity (Fig. 7.40). Although the process is generally called deep drawing, meaning deep parts, the basic operation also produces parts that are shallow or have moderate depth.

Deep drawing, first developed in the 1700s, has been studied extensively and has become an important metalworking process. Typical parts produced are beverage cans, pots and pans, containers of all shapes and sizes, sinks, and automobile panels.

7.9.1 MECHANICS OF DEEP DRAWING

The basic parameters in deep drawing a cylindrical cup are shown in Fig. 7.41. A circular sheet blank, with a diameter D_0 and thickness t_0, is placed over a die opening with a corner radius of R_d. The blank is held in place with a blankholder, or hold-down ring, with a certain force. A punch, with a diameter of D_p and a corner radius R_p, moves downward and pushes the blank into the die cavity, thus forming a cup. The significant variables in deep drawing are:

a. Properties of the sheet metal,

b. The ratio of blank diameter to punch diameter,

c. The clearance between the punch and the die,

d. Punch and die corner radii,

e. Blankholder force,

f. Friction and lubrication at the punch, die, and workpiece interfaces, and

g. Speed of the punch.

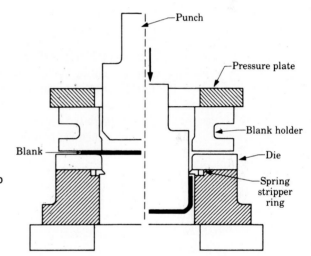

FIGURE 7.40 Schematic illustration of the deep drawing process. This is the first step in the basic process by which aluminum beverage cans are produced today. The stripper ring facilitates the removal of the formed cup from the punch.

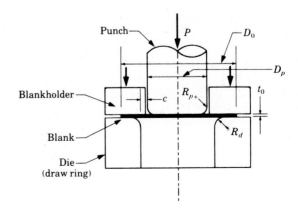

FIGURE 7.41 Variables in deep drawing of a cylindrical cup. Only the punch force in this illustration is a dependent variable; all others are independent variables, including the blankholder force.

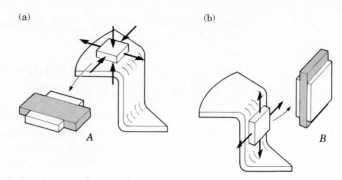

(a)   (b)

A   B

FIGURE 7.42 Deformation of elements in the flange (a) and the cup wall (b) in deep drawing of a cylindrical cup.

Stresses

At an intermediate stage during the deep drawing operation, the workpiece is subjected to the states of stress shown in Fig. 7.42. On element *A* in the blank, the radial tensile stress is due to the blank being pulled into the cavity, and the compressive stress normal to the element is due to the blankholder pressure. With a free-body diagram of the blank along its diameter, it can be shown that the radial tensile stresses lead to compressive hoop stresses on element *A*. (It is these hoop stresses that tend to cause the flange to wrinkle during drawing, thus requiring a blankholder under a certain force.) Under this state of stress, element *A* contracts in the hoop direction and elongates in the radial direction.

The cup wall, which is already formed, is subjected principally to a longitudinal tensile stress, as shown in element *B*. The punch transmits the drawing force *P* (see Fig. 7.41) through the walls of the cup and to the flange that is being drawn into the cavity. The tensile hoop stress on element *B* is due to the fact that the cup may be held tightly on the punch due to its contraction under tensile stresses in the cup wall. (A thin-walled tube, when subjected to tension, becomes smaller in diameter, as can be shown from the generalized Hooke's law equations.) Thus, element *B* tends to elongate in the longitudinal (axial) direction with no change in its width, since it is constrained by the rigid punch.

An important aspect in this operation is the determination of how much stretching and how much pure drawing is taking place (Fig. 7.43). Note that either with a high blankholder force or with the use of *draw beads* (Figs. 7.44 and 7.45), the blank can be prevented from flowing freely into the die cavity. The deformation of the sheet metal takes place mainly under the punch and the sheet begins to stretch, eventually resulting in necking and tearing. Whether necking is localized or diffuse depends on the following:

a. The strain rate sensitivity index *m* of the sheet metal (the higher the *m* value the more diffuse the neck; see Sections 2.27 and 7.3.1),

b. Geometry of the punch, and

c. Lubrication.

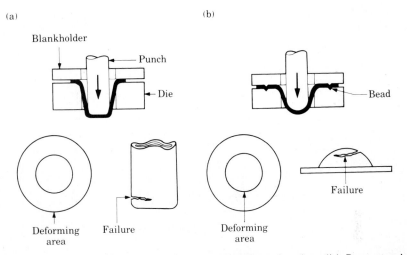

FIGURE 7.43 Examples of drawing operations. (a) Pure drawing. (b) Pure stretching. The bead prevents the sheet metal from flowing freely into the die cavity, as also shown in Figs. 7.44 and 7.45.

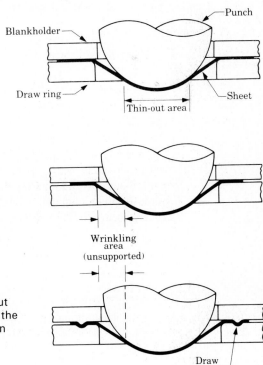

FIGURE 7.44 Drawing with a hemispherical punch with and without beads. Note the unsupported area of the sheet metal. If excessive, this area can develop wrinkles.

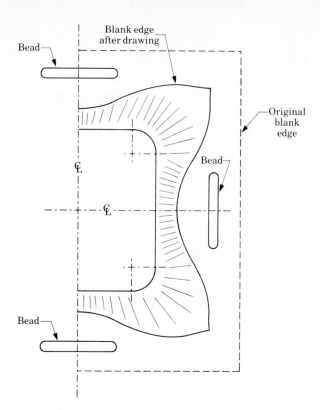

FIGURE 7.45 Metal flow in drawing of a box-shaped part. Beads retard and control the flow of the sheet metal into the die cavity. The design of dies for drawing of irregular-shaped parts requires considerable experience, as analytical studies on forming of such shapes is very limited.

On the other hand, a low blankholder force will allow the blank to flow freely into the die cavity, whereby the blank diameter is reduced as drawing progresses. This is referred to as *pure drawing*; the deformation of the sheet is mainly in the flange, and the cup wall is subjected only to elastic stresses. However, these stresses increase with increasing D_0/D_p ratio and can eventually lead to failure when the cup wall cannot support the load required to draw in the flange (Fig. 7.43a). Also note that, in pure drawing, element A tends to increase in thickness as it moves toward the die cavity because it is being reduced in diameter.

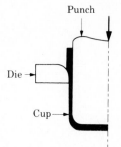

FIGURE 7.46 Schematic illustration of the ironing process. Note that the cup wall is thinner than at its bottom. All beverage cans without seams (known as 2-piece cans) are ironed, generally in three steps, after being drawn into a cup by deep drawing. (Cans with separate tops and bottoms are known as 3-piece cans.)

Ironing

If the thickness of the sheet as it enters the die cavity is greater than the clearance between the punch and the die, then the thickness will be reduced. This is known as *ironing*; it produces a cup with constant wall thickness (Fig. 7.46). Thus, the smaller the clearance the greater the ironing. Obviously, because of volume constancy, an ironed cup will be longer than one produced with a large clearance.

7.9.2 DEEP DRAWABILITY (LIMITING DRAWING RATIO)

The *limiting drawing ratio* (LDR) is defined as the maximum ratio of blank diameter to punch diameter that can be drawn without failure, i.e., D_0/D_p. Many attempts have been made to correlate this ratio with various mechanical properties of the sheet metal.

In an ordinary deep drawing process, failure generally occurs by thinning in the cup wall under high longitudinal tensile stresses. By following the movement of the material into the die cavity (Fig. 7.42), we note that the material should be capable of undergoing a reduction in width (by being reduced in diameter), yet it should resist thinning under the longitudinal tensile stresses in the cup wall.

The ratio of width (w) to thickness (t) strain (Fig. 7.47) is

$$R = \frac{\epsilon_w}{\epsilon_t} = \frac{\ln\left(\dfrac{w_0}{w_f}\right)}{\ln\left(\dfrac{t_0}{t_f}\right)}, \tag{7.12}$$

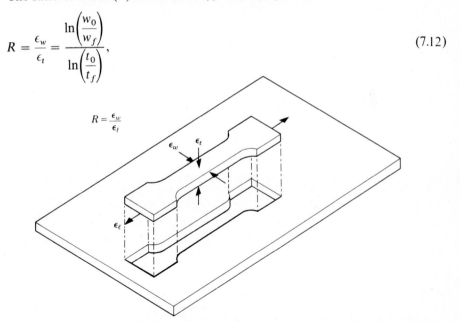

FIGURE 7.47 Definition of normal anisotropy ratio R in terms of width and thickness strains in a tensile-test specimen cut from a rolled sheet. Note that the specimen can be cut in different directions with respect to the length (i.e., rolling direction) of the sheet.

where R is known as the *normal anisotropy* of the sheet metal (also known as *plastic anisotropy* or *strain ratio*). The subscripts o and f refer to the original and final dimensions, respectively. An R value of one indicates that the width and thickness strains are equal to each other, i.e., the material is isotropic.

Since sheet metals are generally thin compared to their surface area, there is a possibility of error in the measurement of small thicknesses. Equation (7.12) can be modified, based on volume constancy, to the following expression:

$$R = \frac{\ln\left(\dfrac{w_o}{w_f}\right)}{\ln\left(\dfrac{w_f \ell_f}{w_o \ell_o}\right)}, \tag{7.13}$$

where ℓ refers to the gage length of the sheet specimen.

To calculate R, the final length and width in a test specimen are generally measured at an elongation of 15 to 20%, or for materials with lower ductility, below the elongation where necking begins.

Rolled sheets generally have planar anisotropy. Thus, the R value of a specimen cut from a rolled sheet (Fig. 7.47) will depend on its orientation with respect to the rolling direction of the sheet. In this case, an average R value, $\bar{R}$, is calculated as follows:

$$\bar{R} = \frac{R_0 + 2R_{45} + R_{90}}{4}, \tag{7.14}$$

where the subscripts 0, 45, and 90 refer to angular orientation (in degrees) of the test specimen with respect to the rolling direction of the sheet. Thus, an isotropic material has an $\bar{R}$ value of one. Some typical $\bar{R}$ values are given in Table 7.2. Although hexagonal close-packed metals usually have high $\bar{R}$ values, the low value for zinc is due to its high c/a ratio in the crystal lattice (see Fig. 3.1 and Table 3.1).

Experiments with sheet steels have shown that $\bar{R}$ is related to the average modulus of elasticity, $\bar{E}$, as shown in Fig. 7.48. Thus, a high value of $\bar{R}$ is associated with a high

TABLE 7.2
TYPICAL RANGE OF AVERAGE NORMAL ANISOTROPY RATIO $\bar{R}$ FOR VARIOUS SHEET METALS

Zinc	0.2
Hot-rolled steel	0.8–1.0
Cold-rolled rimmed steel	1.0–1.35
Cold-rolled aluminum-killed steel	1.35–1.8
Aluminum	0.6–0.8
Copper and brass	0.8–1.0
Titanium	4–6

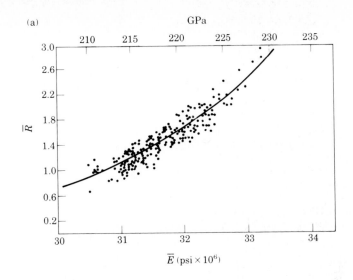

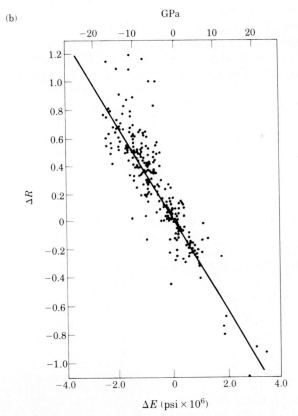

FIGURE 7.48 (a) Relationship between average normal anisotropy $\bar{R}$ and the average modulus of elasticity $\bar{E}$ for steel sheet. (b) Relationship between planar anisotropy ΔR and planar variation in modulus of elasticity ΔE. *Source:* After P. R. Mould and T. R. Johnson, Jr., *Sheet Met. Ind.*, vol. 50, 1973, p. 328.

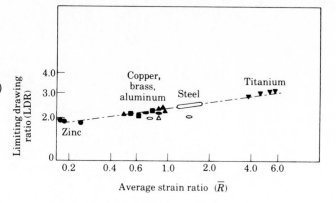

FIGURE 7.49 Effect of average normal anisotropy $\bar{R}$ on limiting drawing ratio (LDR) for a variety of sheet metals. Zinc has a high c/a ratio (see Table 3.1 and Fig. 3.1), whereas titanium has a low ratio. *Source:* After M. Atkinson, *Sheet Metal Industries,* vol. 44, 1967, p. 167.

value of $\bar{E}$. The average modulus, $\bar{E}$, is determined in the same manner as $\bar{R}$ is. Techniques and equipment are now available to measure the modulus of elasticity of sheet specimens by observing their natural frequencies (see also Illustrative Problem 3.5).

The direct relationship between $\bar{R}$ and LDR, as determined experimentally, is shown in Fig. 7.49. It has been shown that, in spite of its scatter, no other mechanical property of sheet metal indicates as consistent a relationship to LDR as $\bar{R}$ does. For an isotropic material and based on ideal deformation, the maximum LDR is equal to $e = 2.718$.

The *planar anisotropy* of a sheet, ΔR, can also be defined in terms of directional R values as follows,

$$\Delta R = \frac{R_0 - 2R_{45} + R_{90}}{2}, \tag{7.15}$$

which is the difference between the average of the R values in the 0- and 90-degree directions to rolling and the R value at 45 degrees. ΔR has been found to be related to ΔE, as shown in Fig. 7.48 where ΔE is determined in the same manner as ΔR is in Eq. (7.15) with R replaced by E.

FIGURE 7.50 Earing in drawn steel cups, due to planar anisotropy of the sheet metal. Earing leads to material waste since the eared portion has to be trimmed off and discarded. *Source:* Courtesy of J. S. Kallend, Illinois Institute of Technology.

Earing

Planar anisotropy is the cause for the *ears* that form in drawn cups (Fig. 7.50), producing a wavy edge. The number of ears produced may be two, four, or six. The height of the ears increases with increasing ΔR. When $\Delta R = 0$, no ears are formed. Ears are objectionable because they have to be trimmed off, resulting in waste of material.

Deep drawability is thus enhanced with a high $\bar{R}$ and a low ΔR value. Generally, however, sheet metals with a high $\bar{R}$ value also have a high ΔR value. Attempts are being made to develop textures in sheet metals to improve drawability. The controlling parameters in processing of the metals have been found to be: alloying elements, additives, processing temperatures, annealing cycles after processing, thickness reduction in rolling, and cross rolling (biaxial) of plates in processing them into sheets.

● **Illustrative Problem 7.5**

Estimate the limiting drawing ratio (LDR) that you would expect from a sheet metal, which, when stretched by 23% in length, decreases in thickness by 10%.

SOLUTION. From volume constancy of the test specimen, we have

$$w_0 t_0 \ell_0 = w_f t_f \ell_f \qquad \text{or} \qquad \frac{w_f t_f \ell_f}{w_0 t_0 \ell_0} = 1.$$

From the information given,

$$\frac{\ell_f - \ell_0}{\ell_0} = 0.23 \qquad \text{or} \qquad \frac{\ell_f}{\ell_0} = 1.23$$

and

$$\frac{t_f - t_0}{t_0} = -0.10 \qquad \text{or} \qquad \frac{t_f}{t_0} = 0.90.$$

Hence

$$\frac{w_f}{w_0} = 0.903.$$

From Eq. (7.13), we obtain

$$R = \frac{\ln\left(\dfrac{w_0}{w_f}\right)}{\ln\left(\dfrac{t_0}{t_f}\right)} = \frac{\ln 1.107}{\ln 1.111} = 0.965.$$

Assuming that the sheet has planar isotropy, then $R = \bar{R}$ and from Fig. 7.49 we obtain

$$\text{LDR} = 2.4 \quad ●$$

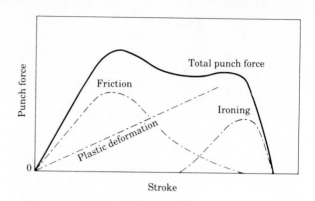

FIGURE 7.51 Schematic illustrations of the variation of punch force with stroke in deep drawing, showing different components of the punch force. Note that ironing does not begin until after the punch has traveled a certain distance and the cup is formed partially.

7.9.3 FORCES

The *punch force P* supplies the work required in deep drawing. This work, as in other deformation processes, is composed of: ideal work of deformation, redundant work, friction work and, when present, the work required for ironing (Fig. 7.51).

Because of the many variables involved in this operation, and due to the fact that deep drawing is not a steady-state process, the punch force is difficult to calculate. Although various expressions have been developed, a simple and very approximate formula for the punch force P is given by

$$P = \pi D_p t_0 (\text{UTS}) \left(\frac{D_0}{D_p} - 0.7 \right). \tag{7.16}$$

Note that Eq. (7.16) does not include friction, the punch and die corner radii, or the blankholder force. However, this empirical equation makes rough provision for these factors.

The punch force is supported basically by the cup wall. If this force is excessive, *tearing* will occur as shown in Fig. 7.43(a). Note that the cup has been drawn to a considerable depth before failure occurs. This can be expected from Fig. 7.51, where we observe that the punch force does not reach a maximum until after the punch has traveled a certain distance.

It has been established that punch corner radius and die radius (if they are greater than 10 times the sheet thickness) do not affect the maximum punch force significantly.

7.9.4 DRAWING PRACTICE

In view of the complex interaction of various factors, certain guidelines have been established for successful deep drawing practice.

The *blankholder pressure* is generally 0.7 to 1.0% of the sum of the yield and ultimate tensile strength of the sheet metal. Too high a blankholder force increases the punch load (because of friction) and leads to tearing of the cup wall. On the other hand, if it is too low wrinkling will occur in the flange.

(a)   (b)

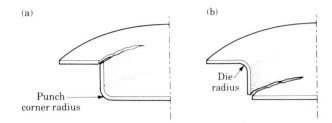

Punch corner radius   Die radius

FIGURE 7.52 Effect of die and punch radii in deep drawing on fracture of a cylindrical cup. (a) Die radius too small. The die radius should generally be 5 to 10 times the sheet thickness. (b) Punch corner radius too small. Because friction between the cup and the punch aids in the drawing operation, excessive lubrication of the punch is detrimental to drawability.

Clearances generally range between 7 and 14% greater than the original sheet thickness. The choice of clearance depends on the thickening of the cup wall (which is a function of the drawing ratio). As the clearance decreases, ironing increases. If the clearance is too small, the blank may simply be pierced and sheared by the punch (Fig. 7.1).

The *corner radius* of the punch and the die are important. If they are too small they can cause fracture at the corners (Fig. 7.52). If they are too large, wrinkling of the unsupported area takes place (see Fig. 7.44). Wrinkling in this region (and from the flange area) causes a defect on the cup wall called *puckering*.

Draw beads (see Figs. 7.44 and 7.45) are useful in controlling the flow of the blank into the die cavity. They are especially necessary in drawing box-shaped and non-symmetric parts (see Fig. 7.45). The proper design and location of draw beads requires considerable experience. Draw beads also help in reducing the blankholder forces required.

Redrawing

Containers or shells that are too difficult to draw in one operation are generally *redrawn*, as shown in Fig. 7.53(a). Another process is *reverse* drawing, shown in Fig. 7.53(b), where the metal is subjected to bending in the opposite direction to its original bending configuration. This reversal in bending results in *strain softening* and is another example of the Bauschinger effect described in Section 2.3.1. This operation requires lower forces and the material behaves in a more ductile manner.

Drawing Without a Blankholder

Deep drawing may be carried out without a blankholder, provided that the sheet metal is relatively thick to prevent wrinkling. The dies are specially contoured for this operation, two examples being shown in Figs. 7.54 and 7.55. An approximate limit for drawing without a blankholder is given by

$$D_0 - D_p < 5t_0. \tag{7.17}$$

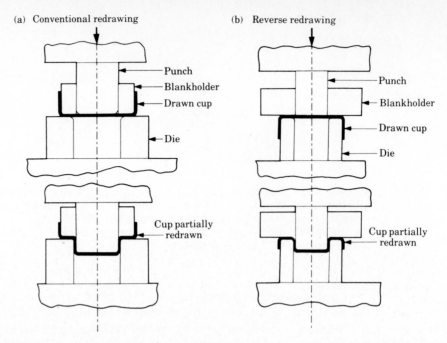

FIGURE 7.53 Reducing the diameter of drawn cups by redrawing operations. (a) Conventional redrawing. (b) Reverse redrawing. Small-diameter deep containers undergo many drawing and redrawing operations.

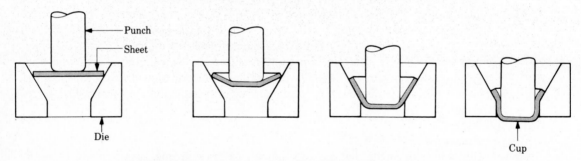

FIGURE 7.54 Deep drawing without a blank holder, using a die with a conical profile. The sheet metal should be relatively thick for this process to be successful; otherwise it will wrinkle at the periphery (similar to the wrinkles on the periphery of paper plates or along the edges of thin aluminum trays for food products).

It is apparent from this expression that the process can be used with thin materials for shallow draws. Although the punch stroke is greater, a major advantage of this process is the reduced cost of tooling and equipment.

Various Operations
A large variety of parts are also made with shallow or moderate depths. Some of these involve drawing or stretching, or a combination of these operations. Parts may

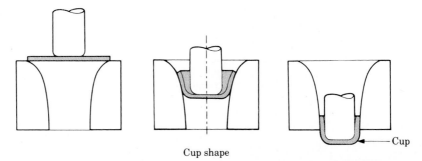

Cup shape

Cup

FIGURE 7.55   Deep drawing without a blank holder using a tractrix die profile. Tractrix is a special curve, the construction for which can be found in texts on analytical geometry or in handbooks.

be embossed with male and female dies or by other means. *Embossing* is an operation consisting of a number of very shallow draws (made with matching dies) on a sheet. The process is used principally for decorative purposes.

Lubrication

Lubrication in deep drawing is important in lowering forces, increasing drawability, reducing wear of the tooling, and reducing defects in parts. In general, lubrication of the punch should be held to a minimum, as friction between the punch and the cup improves drawability. For general applications, commonly used lubricants are mineral oils, soap solutions, and heavy-duty emulsions. For more difficult applications, coatings, wax, and solid lubricants are used (Table 6.7).

Tooling and Equipment

The most common tool materials for deep drawing are tool steels and alloyed cast iron. Other materials, including carbides and plastics, may also be used depending on the particular application (Table 6.4).

A double-action hydraulic press is usually used for deep drawing, although mechanical presses are also used. The double-action press controls the punch and blankholder independently, and forms the part at a constant speed. Punch speeds generally range between 20 and 60 ft/min (0.1 and 0.3 m/s). Speed is generally not important in drawability, although slower speeds are used for high-strength metals.

7.10 RUBBER FORMING

In all the processes described in the preceding sections the dies are made of solid materials. It is possible to replace one of these dies with a flexible material such as a rubber or polyurethane membrane.

The simplest example of rubber forming is shown in Fig. 7.56 for bending operations. Note that the solid female die has been eliminated, thus reducing the cost

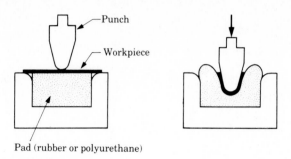

FIGURE 7.56 Bending with a rubber or polyurethane pad. The pad, which replaces the female die, also protects the outer surfaces of the part from damage.

Pad (rubber or polyurethane)

of the operation substantially. Furthermore, the outer surface of the sheet is prevented from being damaged or scratched during bending since it is not in contact with a hard metal (die) surface.

There are four basic types of rubber-forming operations: Guerin, Verson–Wheelon, Marform, and Hydroform processes.

The *Guerin* process (Fig. 7.57) has rubber or a similar flexible material attached to the upper ram that descends and forms the part over a male die or dies. Pressures are usually on the order of 1500 psi (10 MPa). This process is used extensively in the aircraft industry, particularly in making parts with shallow flanges. For more complicated shapes, such as sharp bends and deep indentations, auxiliary devices are used to increase the pressure in specific areas of the part.

The main feature of the *Verson–Wheelon* process (Fig. 7.58) is the use of a soft rubber bag under hydraulic pressure. The forming pressure is four to five times greater than that in the Guerin process; thus it is suitable for more complicated and deeper shapes.

In the *Marform* process (Fig. 7.59) the blankholder pressure is regulated automatically through a pressure-control valve. This process is suitable for deeper draws and causes less wrinkling than in the two preceding operations. The equipment is a package unit that can be installed on a hydraulic press.

In the *Hydroform* process (Fig. 7.60) the pressure over the rubber membrane is controlled throughout the forming cycle, with maximum pressures up to 15,000 psi

FIGURE 7.57 The Guerin process. Used extensively in the aircraft industry, this process is particularly useful in making parts with shallow flanges.

Pad (rubber or polyurethane)

Die

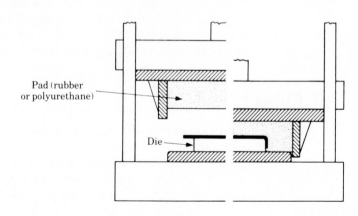

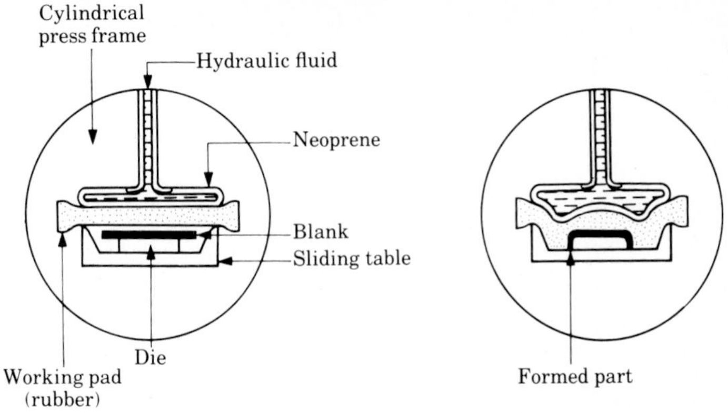

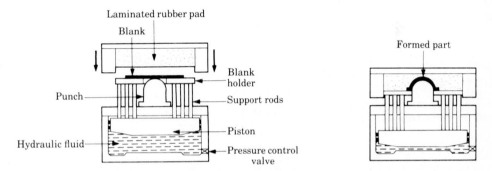

FIGURE 7.58 The Verson–Wheelon process. This process is suitable for making deeper flanges and more complex shapes than the Guerin process.

FIGURE 7.59 The Marform process. The equipment is a package unit and can be installed in a hydraulic press.

(100 MPa). This allows close control of the part during forming. Deeper draws are obtained in the Hydroform process than in conventional deep drawing. This is due to the fact that the pressure around the rubber membrane forces the cup against the punch. The friction at the punch–cup interface reduces the longitudinal tensile stresses in the cup and thus delays fracture. This process is particularly suitable for deep drawing of relatively thick parts.

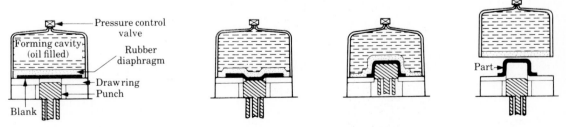

FIGURE 7.60 The Hydroform process. Steel plates 0.5 in. (12.7 mm) thick can be formed by this process. Note that, unlike ordinary deep drawing, the dome pressure forces the cup against the punch.

Advantages

When selected properly, rubber-forming processes have the advantage of low tooling cost, flexibility and ease of operation, low die wear, and no damage to the surface of the sheet metals. It is thus possible to form parts with laminated sheets of various nonmetallic materials or coatings (see Section 7.16). Polyurethanes are used widely in these processes because of their resistance to abrasion, to cutting by sharp edges of the sheet metal, and to fatigue.

7.11 SPINNING

Spinning involves the forming of axisymmetric parts over a rotating mandrel with the use of rigid tools or rollers. There are three basic types of spinning processes: conventional (or manual), shear, and tube spinning. The equipment for all these processes is basically similar to a lathe (Fig. 8.64) with various special features. Spinning processes are described below.

7.11.1 CONVENTIONAL SPINNING

In this process, a circular blank of flat or preformed sheet metal is held against a rotating mandrel while a rigid tool deforms and shapes the material over the mandrel (Fig. 7.61a). The tools may be actuated either manually or by a hydraulic mechanism. The operation involves a sequence of passes (Fig. 7.62) and requires considerable skill. This process has been found to be quite difficult to study analytically.

Typical shapes made by conventional spinning processes are shown in Fig. 7.63. This process is particularly suitable for conical and curvilinear shapes, which would

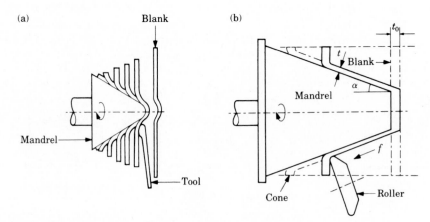

FIGURE 7.61   Schematic illustration of spinning processes. (a) Conventional spinning. (b) Shear spinning. Note that in shear spinning the diameter of the spun part is, unlike conventional spinning, the same as that of the blank.

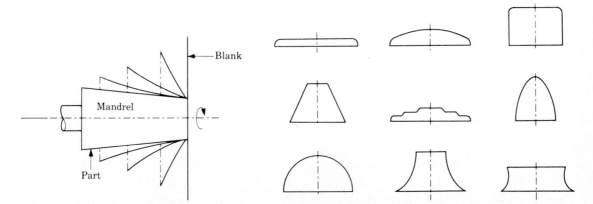

FIGURE 7.62 Stages in conventional spinning of a tubular component from a flat, circular metal disk. This operation requires considerable skill to prevent the part from collapsing or buckling during spinning.

FIGURE 7.63 Typical shapes produced by the conventional spinning process. Circular marks on the external surfaces of components usually indicate that the parts have been made by spinning. Examples are aluminum kitchen utensils and light reflectors.

otherwise be difficult or uneconomical to form by other methods. Part diameters may range up to 20 ft (6 m). Although most spinning is performed at room temperature, thick parts or metals with low ductility or high strength require spinning at elevated temperatures.

Tooling costs in spinning are relatively low. However, because the operation requires multiple passes to form the final part (see Fig. 7.62), it is economical for small production runs only. (See also Fig. 13.12.)

7.11.2 SHEAR SPINNING

In this process (also known as *power spinning*, *flow turning*, *hydrospinning*, or *spin forging*), an axisymmetrical conical or curvilinear shape is generated by keeping the diameter of the part constant (see Fig. 7.61b). Parts typically made by this process are rocket-motor casings and missile nose cones. Although a single roller can be used, two rollers are desirable to balance the radial forces acting on the mandrel.

In shear spinning over a conical mandrel the thickness t of the spun part is simply

$$t = t_0 \sin \alpha. \tag{7.18}$$

Spinnability

An important factor in shear spinning is the spinnability of the metal, i.e., the smallest thickness to which a part can be spun without fracture. To determine spinnability, a simple test method has been developed (Fig. 7.64), where a circular blank is spun over an ellipsoid mandrel, according to Eq. (7.18). Since the thickness is eventually reduced to zero, all materials will fail at some thickness t_f.

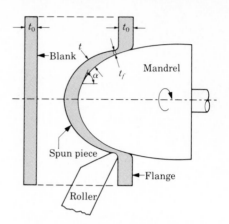

FIGURE 7.64 Schematic illustration of a shear-spinnability test. As the roller advances, the part thickness is reduced. The reduction in thickness at fracture is called the maximum spinning reduction per pass. *Source:* After R. L. Kegg, *Jr. Eng. Ind.,* vol. 83, 1961, pp. 119–124.

Some of the specimens subjected to this test are shown in Fig. 7.65. The data on a variety of materials are given in Fig. 7.66. The maximum spinning reduction in thickness is defined as

$$\text{Maximum reduction} = \frac{t_0 - t_f}{t_0} \times 100\% \qquad (7.19)$$

and is plotted against the tensile reduction of area of the material. Note that if a metal has a tensile reduction of area of about 50% or greater, it can be reduced in thickness by 80% by spinning in one pass. For less ductile materials the spinnability is lower.

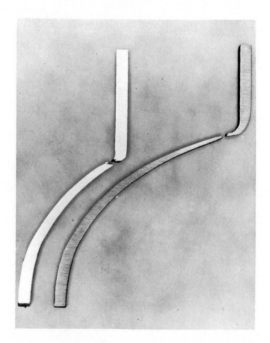

FIGURE 7.65 Specimens from shear-spinnability tests. (a) 2024-T36 aluminum. (b) Copper. *Source:* S. Kalpakjian, *Jr. Eng. Ind.,* vol. 83, 1961, pp. 478–484.

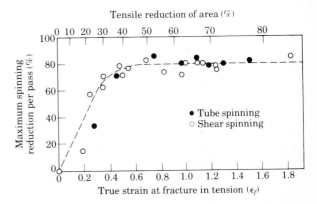

FIGURE 7.66 Experimental data showing the relationship between maximum spinning reduction per pass and the tensile reduction of area of the original material. Note that once a material has about 50% reduction of area in a tension test, further increase in the ductility of the original material does not improve its spinnability. *Source:* S. Kalpakjian, *Jr. Eng. Ind.*, vol. 86, 1964, pp. 49–54.

Process variables such as feed and speed do not appear to have a significant influence on spinnability.

It can be seen from Fig. 7.65 that the first specimen has failed in *shear* under the roller, whereas the second failed in *tension* after the reduction was completed. Once a material is sufficiently ductile, failure takes place by the high tensile stresses in the spun section. This type of failure is similar to the maximum reduction per pass in rod or wire drawing described in Section 6.20.5.

It is noted in Figs. 7.20 and 7.66 that maximum bendability and spinnability both coincide with a tensile reduction of area of about 50%. Any further increase in the ductility of the original material does not improve its formability. For materials with low ductility, spinnability can usually be improved by forming at elevated temperatures.

Forces

It can be shown that in spinning, the force that is primarily responsible for supplying energy is the tangential force, F_t. For an ideal situation in the shear spinning of a cone, this force can be shown to be

$$F_t = ut_0 f \sin \alpha, \tag{7.20}$$

where u is the specific energy of deformation, Eq. (2.54). It is given by the area under the true stress–true strain curve corresponding to a true strain related to the shear strain by

$$\epsilon = \frac{\gamma}{\sqrt{3}} = \frac{\cot \alpha}{\sqrt{3}} \tag{7.21}$$

(see Eq. 2.52). Actual forces can be as much as 50% higher than those given by Eq. (7.20) due to factors such as redundant work and friction.

Spinning Practice

Large parts, up to about 10 ft (3 m), can be spun to close tolerances by shear spinning. There is little waste of material and the operation is completed in a relatively short time. If carried out at room temperature, the part has a higher yield strength than the original material, but lower ductility and toughness. A great variety of shapes can be spun with relatively simple tooling, generally made of tool steel. Because of the large plastic deformation involved, the process generates considerable heat. This is usually carried away by a coolant-type fluid applied during spinning.

7.11.3 TUBE SPINNING

In this process, tubes are reduced in thickness by spinning them on a mandrel using rollers. The operation may be carried out externally or internally (Fig. 7.67). Also, the part may be spun forward or backward, similar to a drawing or a backward extrusion process. In either case, the reduction in wall thickness results in a longer tube.

Typical parts made by this process are shown in Fig. 7.68. Note that various internal and external profiles can be generated by changing the path of the roller during its travel along the mandrel. Tube spinning can be used for making pressure

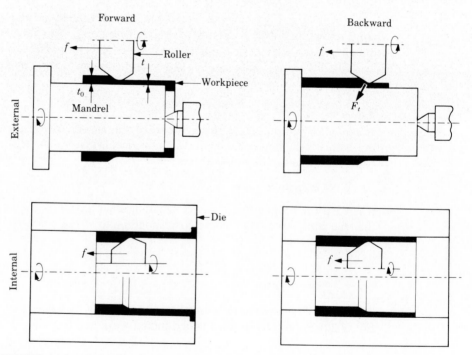

FIGURE 7.67   Examples of external and internal tube spinning and the variables involved.

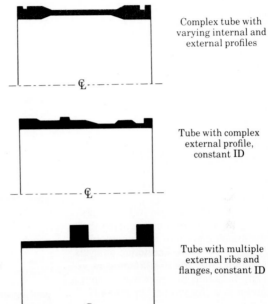

Complex tube with varying internal and external profiles

Tube with complex external profile, constant ID

Tube with multiple external ribs and flanges, constant ID

FIGURE 7.68 Typical shapes produced by tube spinning processes. Such parts can be manufactured by spinning at a relatively high production rate and with little waste of material.

vessels, automotive components, and rocket and missile parts. It can also be combined with shear spinning, such as in the compressor shaft for an aircraft engine shown in Fig. 7.69.

Spinnability

Spinnability in this process is determined by a test method similar to shear spinning (Fig. 7.70). It has been shown that maximum reduction per pass in tube spinning is related to the tensile reduction of area of the material (see Fig. 7.66), with results very similar to shear spinning. Some of the parts spun in this test setup are shown in Fig. 7.71. Note that, as in shear spinning, ductile metals have failed in tension after the reduction in thickness has taken place, whereas the less ductile metal has failed in the deformation zone under the roller.

Forces

The ideal tangential force F_t in forward tube spinning can be expressed as

$$F_t = \overline{Y}\Delta t f, \tag{7.22}$$

where Δt is $(t_0 - t)$ (see Fig. 7.67) and $\overline{Y}$ is an average flow stress of the material. For backward spinning, the ideal tangential force is roughly twice that given by Eq. (7.22). Due to friction and redundant work of deformation, actual forces have been found to be about twice those obtained by this equation.

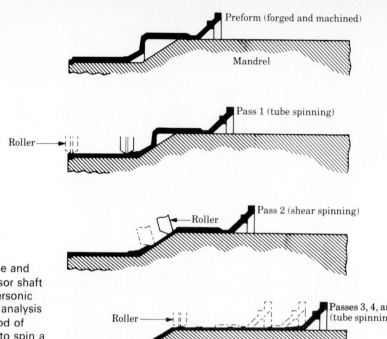

Preform (forged and machined)

Mandrel

Roller → Pass 1 (tube spinning)

Roller — Pass 2 (shear spinning)

Roller → Passes 3, 4, and 5 (tube spinning)

$\mathcal{C_L}$

~40in.(1m)

FIGURE 7.69 Stages in tube and shear spinning of a compressor shaft for the jet engine of the supersonic Concorde aircraft. Economic analysis indicated that the best method of manufacturing this part was to spin a preformed (forged and machined) tubular blank.

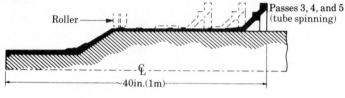
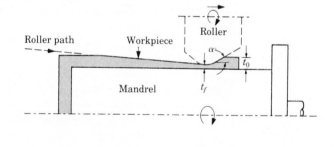

Roller path   Workpiece   Roller

α t_0

Mandrel t_f

FIGURE 7.70 Schematic illustration of a tube-spinnability test. As the roller advances, the wall thickness is reduced and the part eventually fractures, as shown in Fig. 7.71. *Source:* S. Kalpakjian, *Jr. Eng. Ind.*, vol 86, 1964, pp. 49–54.

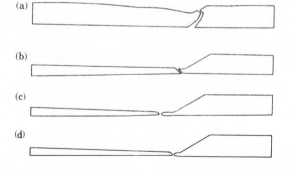

(a)

(b)

(c)

(d)

FIGURE 7.71 Schematic illustration of specimens from tube-spinnability tests. (a) 2024-T4 aluminum. (b) 6061-T6 aluminum. (c) Copper. (d) Mild steel. See also Fig. 7.65. *Source:* S. Kalpakjian, *Jr. Eng. Ind.*, vol. 86, 1964, pp. 49–54.

7.12 HIGH-ENERGY-RATE FORMING

Sheet-metal–forming processes that use chemical, electrical, or magnetic sources of energy are described in this section. These processes are called *high-energy-rate* because the energy is released in a very short time. The basic operations are explosive forming, electrohydraulic forming, and magnetic-pulse forming.

7.12.1 EXPLOSIVE FORMING

The most common explosive forming process (first proposed in the early 1900s) is shown in Fig. 7.72. The workpiece is clamped over a die, the air in the die cavity is evacuated, and then the workpiece lowered into a tank filled with water. An explosive charge is then placed at a certain height and detonated. The rapid conversion of the explosive into gas generates a shock wave. The pressure of this wave is sufficiently high to form the metal.

The peak pressure p, generated in water, is given by the expression

$$p = K\left(\frac{\sqrt[3]{W}}{R}\right)^a, \tag{7.23}$$

where

p = peak pressure in pounds per square inch,

K = a constant that depends on the explosive (e.g., 21,600 for TNT),

W = weight of the explosive in lb,

R = distance of the explosive from the workpiece (standoff) in feet, and

a = a constant, generally taken as 1.15.

An important factor in determining peak pressure is the *compressibility* of the energy-transmitting medium (e.g., water) and its *acoustic impedance*, i.e., the product of mass density and sound velocity in the medium. The lower the compressibility and

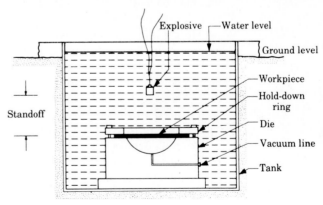

FIGURE 7.72 Schematic illustration of the explosive forming process. Although explosives are generally used for destructive purposes, their energy can be controlled and utilized in forming large parts that would otherwise be difficult or expensive to produce by other methods.

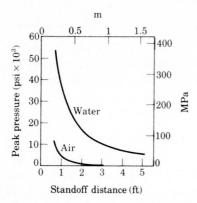

FIGURE 7.73 Influence of the standoff distance and type of energy-transmitting medium on the peak pressure obtained using 4 lb (1.8 kg) of TNT. To be effective, the pressure-transmitting medium should have high density and low compressibility. In practice, water is a commonly used medium.

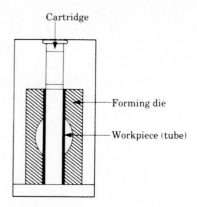

FIGURE 7.74 Schematic illustration of the confined method of explosive bulging of tubes. Thin-walled tubes of nonferrous metals can be formed to close tolerances by this process.

the higher the density of the medium, the higher the peak pressure (Fig. 7.73). The distance between the water level and the explosive should not be too small, otherwise the energy is dissipated rapidly to the environment. Detonation speeds are typically 22,000 ft/s (6700 m/s), while the speed at which the metal is formed is estimated to be on the order of 100 to 600 ft/s (30 to 200 m/s).

Explosive-Forming Practice

A great variety of shapes can be formed explosively, provided that the material is sufficiently ductile at high strain rates. Depending on the number of parts to be formed, dies may be made of aluminum alloys, steel, ductile iron, zinc alloys, reinforced concrete, wood, plastics, and composite materials. The final properties of parts made by this process are basically the same as those made by conventional methods. Safety is an important aspect in these operations.

Since it only needs one die and the process is versatile, with no limit to the size of the workpiece, explosive forming is particularly suitable for small-quantity production runs of large parts. Steel plates 1 in. (25 mm) thick and 12 ft (3.6 m) in diameter have been formed by this method. Tubes of 1 in. (25 mm) wall thickness have also been bulged by explosive-forming techniques.

Another explosive-forming method is shown in Fig. 7.74, which uses a cartridge as the source of energy; no other energy-transmitting medium is used. The process can be used for bulging and expanding of thin-walled tubes. Die wear and die failure can be a significant problem in this operation.

● **Illustrative Problem 7.6**

Calculate the peak pressure in water when using 0.1 lb of TNT at a standoff of 1 ft. Is this pressure sufficiently high for forming sheet metals?

SOLUTION. Using Eq. (7.23) we find

$$p = (21,600)\left(\frac{\sqrt[3]{0.1}}{1}\right)^{1.15}$$

$$\simeq 9000 \text{ psi}$$

This pressure would be sufficiently high to form sheet metals. Note in Illustrative Problem 2.7, for instance, that the pressure required to expand a thin-walled spherical shell of a material similar to soft aluminum alloys is only 400 psi. Also note that a process such as Hydroform (Section 7.10) has a maximum hydraulic pressure in the dome of about 15,000 psi. Other rubber-forming processes use pressures from about 1500 to 7500 psi. Thus, the pressure obtained in this problem is sufficient for most sheet-forming processes. ●

7.12.2 ELECTROHYDRAULIC FORMING

In this process, also called *underwater-spark* or *electric-discharge forming*, the source of energy is a spark from electrodes connected with a thin wire (Fig. 7.75). The energy is stored in a bank of condensers charged with direct current. The rapid discharge of this energy through the electrodes generates a shock wave, which then forms the part.

This process is essentially similar to explosive forming, except that it uses a lower level of energy and is used with smaller workpieces. It is also a safer operation.

7.12.3 MAGNETIC-PULSE FORMING

In this process, the energy stored in a capacitor bank is discharged rapidly through a magnetic coil. In a typical example, a ring-shaped coil is placed over a tubular workpiece to be formed (collapsed) over another solid piece to make an integral part (Fig. 7.76).

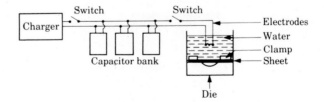

FIGURE 7.75 Schematic illustration of the electrohydraulic forming process.

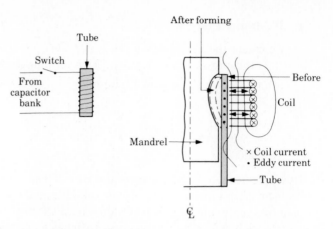

FIGURE 7.76 Schematic illustration of the magnetic pulse forming process. The part is formed without physical contact with any object.

The magnetic field produced by the coil crosses the metal tube, thus generating eddy currents in the tube. This current, in turn, produces its own magnetic field. The forces produced by the two magnetic fields oppose each other; thus there is a repelling force between the coil and the tube. The high forces generated collapse the tube over the inner piece.

The higher the electrical conductivity of the workpiece, the higher is the magnetic force. It is not necessary for the metal to have any special magnetic properties.

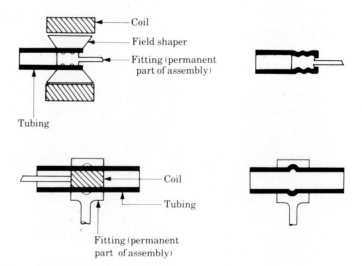

FIGURE 7.77 Typical operations performed by magnetic pulse forming. Flat coils can also be used to make embossed parts, such as metal ash trays.

FIGURE 7.78   Aluminum tube compressed over a hexagonal plug by the magnetic pulse forming process. *Source:* General Dynamics.

Magnetic-pulse forming is used for a variety of operations, such as swaging of thin-walled tubes over rods, cables, plugs, etc., and for bulging and flaring (Figs. 7.77 and 7.78). Flat coils are also made for forming of flat sheet metal, such as embossing and shallow drawing operations.

7.13 SUPERPLASTIC FORMING

The superplastic behavior of certain very-fine-grained alloys (normally less than 10 to 15 μm), where very large elongations (up to 2000 %) are obtained at certain temperatures and low strain rates, was described in Section 2.2.7. These alloys, such as Zn-Al and titanium, can be formed into complex shapes with common metalworking or polymer processing techniques (Section 10.12).

The high ductility and relatively low strength of superplastic alloys present the following advantages:

a. Lower strength of tooling, because of the low strength of the material at forming temperatures, hence, lower tooling costs,

b. Forming of complex shapes in one piece, with fine detail and close tolerances, and elimination of secondary operations,

c. Weight and material savings due to the formability of the materials, and

d. Little or no residual stresses in the formed parts.

The limitations of superplastic forming are:

a. The material must not be superplastic at service temperatures, and

b. Because of the extreme strain-rate sensitivity of the superplastic material, it must be formed at sufficiently low rates (typically at strain rates of 10^{-4} to 10^{-2}/s). Forming times range anywhere from a few seconds to several hours. Thus, cycle times are much longer than conventional forming processes. Superplastic forming is therefore a batch-forming process.

Superplastic alloys (particularly Zn-22Al and Ti-6Al-4V) can be formed by bulk deformation processes, such as compression molding, closed-die forging, coining, hubbing, and extrusion. Sheet forming of these materials can also be done using operations such as thermoforming, vacuum forming, and blow molding. Structural components such as those for aircraft are also made by a combination of superplastic forming and diffusion bonding (Section 12.4.5). Sheet metals suitable for these operations are Ti-6Al-4V, 7475 aluminum (modified 7075), Inconel 100 and Incoloy 718 nickel alloys, and iron-base, high-carbon alloys.

Commonly used die materials in superplastic forming are low-alloy steels, cast tool steels, ceramics, graphite, and plaster of Paris. Selection depends on the forming temperature and strength of the superplastic alloy.

7.14 MISCELLANEOUS SHEET-FORMING PROCESSES

A forming method utilizes *gas mixtures* in a closed container. When ignited, the pressures generated are sufficiently high to form parts. The principle is similar to the generation of pressure in an internal-combustion engine. However, this method is not commonly used in practice.

The expansion of a chemical or liquid medium can also be achieved by using *liquified gases*, such as liquid nitrogen. When allowed to reach room temperature in an enclosed container, the liquid nitrogen expands and develops pressures sufficiently high to form sheet-metal parts.

Peen Forming
This is a process whereby curvatures on thin sheet metals are generated by *shot peening* one surface of the sheet. This is done with cast iron or steel shot, discharged either from a rotating wheel or with air blast from a nozzle.

In this technique, the surface of the sheet is subjected to compressive stresses, which tend to expand the surface layer. Since the material below the peened surface remains rigid, the surface expansion causes the sheet to develop a curvature. The process also induces compressive residual stresses on the final part (just as shot peening does) with improved fatigue strength of the sheet. (See Fig. 4.38.)

The peen-forming technique has been used by the aircraft industry in generating smooth and complex curvatures on aircraft wing skins. Wing panels 80 ft (25 m) long have been formed by this method using steel shot about 0.1 in (2.5 mm) in diameter at speeds of 200 ft/s (60 m/s).

7.15 DENT RESISTANCE OF SHEET-METAL PARTS

In certain applications involving sheet-metal parts, such as automotive body panels, appliances, and office furniture, an important consideration is the dent resistance of the sheet-metal panel.

It is to be expected that the factors significant in dent resistance are the yield stress Y, thickness t, and the shape of the panel. It has been shown that *dent resistance* can be defined as

$$\text{Dent resistance} \propto \frac{Y^2 t^4}{S}, \tag{7.24}$$

where S is the panel stiffness, which is defined as

$$S = (E)(t^a)(\text{shape}) \tag{7.25}$$

where the value of a ranges from 1 to 2 for most panels. As for shape, the smaller the curvature (hence the flatter the panel), the greater the dent resistance because of its flexibility. Thus, dent resistance increases with increasing strength and thickness, and decreases with increasing elastic modulus E and stiffness, and decreasing curvature.

Dents are usually caused by *dynamic* forces, such as those due to falling objects or other objects that hit the sheet-metal panel. In typical automotive panels, for instance, impact velocities range up to 150 ft/s (45 m/s). This suggests that it is the dynamic yield stress (high strain rate), rather than the static yield stress, that is the significant strength parameter. Denting under quasistatic forces could also be important.

For materials whose yield stress increases with strain rate, denting requires higher energy levels than under static conditions. Furthermore, it has been observed that dynamic forces tend to cause more *localized* dents than static forces which tend to spread the dented area. It should also be noted that, because a portion of the energy goes into elastic deformation, the modulus of resilience of the sheet metal (Eq. 2.5) is an additional factor to be considered.

7.16 COATED SHEET METAL

Sheet metals, especially steel, are available precoated with a variety of organic coatings, films, and laminates. Coated sheet metals are used primarily for appearance and eye appeal and also offer corrosion resistance. Coatings are applied to the coil stock on continuous lines. Coating thicknesses generally range from 0.0001 to 0.008 in. (0.0025 to 0.2 mm) on flat surfaces.

Coatings are available with a wide range of properties such as flexibility, durability, hardness, resistance to abrasion and chemicals, color, texture, and gloss. Coated sheet metal is subsequently formed into various products such as TV cabinets, appliance housings, paneling, shelving, residential siding, and metal furniture. Table 7.3 gives the characteristics and typical applications of coated sheet metal. (See also Chapter 10 for the description and properties of the organic materials used for coatings.)

TABLE 7.3
CHARACTERISTICS AND TYPICAL APPLICATIONS OF COATINGS FOR SHEET METAL

COATING	CHARACTERISTICS	APPLICATIONS
Acrylic film laminates	Extremely weatherable, thick, mar-resistant; available in many colors and patterns.	Wide use in building applications.
Amine alkyds	Highly decorative properties; exterior durability; economical; not suitable for severe forming operations. Most widely used tough coating.	Drapery hardware, metal containers, swimming pool sides, indoor and outdoor furniture.
Epoxies	Properties range from hard and brittle to flexible and elastic. Excellent toughness, flexibility, and heat and abrasion resistance.	Sanitary coating, metal container linings, adhesives, closures; excellent primer under exterior topcoats.
Fluorocarbons	Excellent formability, color retention, resistance to solvents, and chalking.	Curtain wall, residential siding, industrial building components.
Fluorocarbon film laminates	Long life and durability.	Exterior siding, exterior building product applications.
Metal-to-metal laminates	Consist normally of one heavier component of structural value and one lighter gage, more precious or functional component on one or both sides of the structural base.	Decorative trim; applications for chemical resistance, noise abatement, acoustical insulation; metal furniture.
Phenolics	Good stain resistance; excellent acid and solvent resistance.	Beer can interior linings, closures, metal container linings.
Plastisols and organosols	Excellent scuff and mar resistance, good formability, good decorative properties, high chemical resistance. May be applied in heavy film thicknesses, permitting embossing and printing.	TV cabinets, appliance housings, interior wall paneling, protective linings for metal containers, metal furniture, commercial and industrial building panels.
Polyesters	Flexibility and durability ranging from excellent to marginal.	Exterior and interior uses.
Silicone modified alkyds, polyesters, and acrylics	Outstanding exterior durability, superior non-chalking property, and gloss retention. Limited in flexibility.	Exterior applications, especially industrial building sheet and residential siding.
Solution vinyls	Outstanding flexibility and exterior durability suitable for severe forming operations.	Exterior siding, metal window frames, tubing, TV cabinets, auto dashboard panels, metal furniture.

TABLE 7.3   (*continued*)

COATING	CHARACTERISTICS	APPLICATIONS
Strippable coatings	Widely used as protective coatings (sometimes temporary) for polished or coated metals during forming, shipping, and installing.	Protection of exterior building panels and siding that is to be cut or formed at the building site; new and replacement automobile trim and other parts, appliance housings.
Thermoset acrylics	Excellent stain, abrasion, mar resistance, and durability. Formability is limited.	Exterior building products, appliance housings, cabinets, lighting fixtures.
Vinyl-alkyds	A compromise between the economy of alkyds and the flexibility of vinyls; used where forming is involved and cost is a factor.	Roof decking and any application where a straight alkyd would be used but more-than-normal fabrication is involved.
Vinyl film laminates	Largest selection of colors and patterns; can be permanently bonded to the adhesive coated metal to withstand all types of metal forming. Abrasion and stain resistance.	TV cabinets, humidifiers, dehumidifiers, store fixtures and shelving, automotive instrument panels, copy machine covers, water coolers.
Vinyl foams	Can be supplied already foamed on coil for flat panel application or coated in plastisol form for foaming after fabrication of article. Can be topped with PVC, acrylic, and other films.	Thermal and acoustical insulation, safety padding and packaging, noise reduction in metal furniture and wall partitions.

Source: National Coil Coaters Association.

SUMMARY

1. Sheet-metal forming processes generally involve workpieces that have a high ratio of surface area to thickness. Furthermore, unlike bulk deformation processes, in sheet forming, the material is generally prevented from being reduced in its cross-sectional area in order to avoid necking and fracture.

2. Sheet forming generally involves the application of tensile stresses in the plane of the sheet by the use of various tools and dies. Because of the large surfaces involved, friction can be a significant factor in the overall operation as far as forces and formability is concerned.

3. The surfaces of the parts formed by these processes are generally not subjected to further processing, except surface coatings, painting, and joining. Consequently, surface finish can be a significant factor. Lubrication is also an important parameter, as it directly affects surface finish. Surface conditions are important since many sheet-metal parts are subsequently joined or welded.

4. The important material parameters are the capacity of the sheet metal to stretch uniformly (hence the desirability for a high strain-hardening exponent n) and

its resistance to thinning (hence high normal anisotropic properties). Various tests have been developed to predict the formability of sheet metals in actual processing.

5. Because of the low thickness-to-length ratios in most sheet-forming operations, buckling and wrinkling are significant problems. These can be avoided by proper design of tools and dies, and minimizing the unsupported length of the material during processing.

6. Since sheet metals are generally thin compared to the shapes to which they are formed, springback is a significant problem, especially in bending.

7. The forces and energy required in sheet forming can be transmitted to the workpiece not only through solid tools and dies, but also by means such as flexible rubber or polyurethane members, and through electrical, chemical, and magnetic means. Unlike in bulk deformation processes, forces in sheet forming are not particularly significant as far as energy requirements are concerned. A wide variety of processes are available for forming sheet metals into complex shapes.

BIBLIOGRAPHY

Benjamin WP. *Plastic Tooling*. New York: McGraw-Hill, 1972.

Blickwede, D. *New Knowledge about Sheet Steel*. Metals Park, Ohio: American Society for Metals, 1970.

Bruno EJ, ed. *High-Velocity Forming of Metals*, rev. ed. Dearborn, Mich.: Society of Manufacturing Engineers, 1968.

Crane EV. *Plastic Working of Metals and Non-metallic Materials in Presses*, 3d ed. New York: Wiley, 1964.

Davies RS, Austin ER. *Developments in High Speed Metal Forming*. New York: Industrial Press, 1970.

Eary DF, Reed EA. *Techniques of Pressworking Sheet Metal*, 2d ed. Englewood Cliffs, N.J.: Prentice-Hall, 1974.

Ezra AA. *Principles and Practice of Explosive Metalworking*. London: Industrial Newspapers Ltd, 1973.

Grainger JA. *Flow Turning of Metals*. Brighton, England: The Machinery Publishing Co, 1969.

Jevons JD. *The Metallurgy of Deep Drawing and Pressing*, 2d ed. London: Chapman and Hall, 1941.

Koistinen DP, Wang N-M, eds. *Mechanics of Sheet Metal Forming*, New York: Plenum, 1978.

Metals Handbook, 8th ed. Vol. 4, *Forming*. Metals Park, Ohio: American Society for Metals, 1969.

Sachs G. *Principles and Methods of Sheet Metal Fabricating*, 2d ed. New York: Van Nostrand Reinhold, 1966.

Source Book on Forming of Steel Sheet. Metals Park, Ohio: American Society for Metals, 1976.

Springborn RK, ed. *Cold Bending and Forming Tube and Other Sections.* Dearborn, Mich: Society of Manufacturing Engineers, 1966.

Watkins MT. *Metal Forming II: Pressing and Related Processes.* New York: Oxford, 1975.

Willis J. *Deep Drawing*, London: Butterworths, 1954.

(See also, General References in Chapter 6.)

Periodicals

American Machinist

Journal of Applied Metalworking

Journal of Engineering for Industry

Journal of Mechanical Working Technology

Manufacturing Engineering

Metal Progress

Sheet Metal Industries

PROBLEMS

7.1. Explain why the springback factor in Fig. 7.23 decreases with increasing R/T ratio.

7.2. Derive Eq. (7.5).

7.3. In deep drawing of a cylindrical cup (Fig. 7.42) is it always necessary that there be tensile circumferential stresses on the element in the cup wall?

7.4. In Illustrative Problem 7.4, calculate the work done by the force–distance method.

7.5. What would be the answer to Illustrative Problem 7.4 if the tip of the force F is fixed to the strip, while maintaining the same lateral position?

7.6. Assume that you are carrying out a sheet-forming operation and you find that the material is not sufficiently ductile. What would you suggest to improve the ductility?

7.7. Explain why cupping tests (such as Erichsen) do not always predict the behavior of the sheet metal in actual forming operations.

7.8. Calculate the press force needed in punching 5052-O aluminum foil, 0.1 mm, in the shape of a square hole, 30 mm on each side.

7.9. Estimate the limiting drawing ratios (LDR) for the materials listed in Table 7.1.

7.10. Describe the role of beads in the pressworking of sheet metal.

7.11. What would be the LDR for a material that is perfectly isotropic?

7.12. Estimate the bending force required for a $\frac{1}{8}$ in. thick and 10 in. wide titanium alloy Ti-5Al-2.5Sn in a V-die with a width of 8 in.

7.13. For the same material and thickness in Problem 7.12 above, calculate the force required in deep drawing, with a blank diameter of 12 in. and punch diameter of 9 in.

7.14. In Illustrative Problem 7.6, calculate the amount of TNT required to develop a pressure of 15,000 psi.

7.15. Describe what you observe in Fig. 7.3.

7.16. Inspect the bottom of cans for household products and beverages. Do you see the type of patterns shown in Fig. 7.12(c) on aluminum can bottoms?

7.17. Explain the advantages of rubber-forming processes for sheet metal.

7.18. We note that the curve in Fig. 7.66 levels off after a certain tensile reduction of area. Can you offer an explanation?

8 Material-Removal Processes: Cutting

8.1 INTRODUCTION

The shapes and components produced by various casting and forming processes presented in the preceding chapters often require further finishing operations before the product is ready for use. These subsequent operations usually consist of material removal by cutting and various other machining processes.

Material-removal processes include a wide variety of operations. Unlike in casting and forming, the tooling and equipment (called *machine tools*) are capable of generating a wide range of shapes with relatively minor changes in tooling and adjustments of the machine tools.

There are a number of reasons for specifying machining processes in a manufacturing operation:

a. Closer dimensional control or tolerances may be required than are available from casting or forming processes.

b. Special surface characteristics or texture may be required on all or part of the surfaces of the product.

c. The part may have external and internal geometric features that cannot be produced by other processes.

d. It may be more economical to machine the part than to obtain the finished product by other manufacturing processes.

e. Some parts have to be heat treated for such purposes as hardness and wear resistance. Heat-treated parts may require additional surface-finishing operations for the desired dimensions and tolerances.

Against these advantages, material-removal processes have certain limitations:

a. The removal process inevitably wastes some material.

b. It generally takes a longer time to remove a volume of material than to shape the part by forming or other processes.

c. Unless carried out properly, material-removal processes can have adverse effects on the surface quality and properties of the product.

d. Removal processes are generally energy, capital, and labor intensive as described in this chapter and in Chapter 13.

In spite of these limitations, machining processes and equipment are indispensable to manufacturing technology. Since the development of early lathes around the 1800s, machining processes have undergone continuous progress leading to new techniques that use electrical, chemical, and optical sources of energy, and modern equipment.

In this chapter all basic processes of material removal by cutting and the equipment involved are described. The basic mechanics of each cutting process will first be analyzed and the significant variables will be identified. Material-removal processes by abrasive, chemical, electrical, and other means are described in Chapter 9. The economics of machining operations are treated in Chapter 13.

8.2 MECHANICS OF CHIP FORMATION

All traditional machining processes by cutting, such as turning on a lathe, drilling, or grinding, remove material from a surface by producing chips. It has been observed experimentally that the basic mechanics are essentially the same for all cutting processes and that they can be idealized by the two-dimensional model shown in Fig. 8.1. This is known as *orthogonal* cutting.

8.2.1 ORTHOGONAL CUTTING MODEL

In this model, a tool with a certain *rake angle* α (positive as shown in Fig. 8.1) and a small *clearance angle* moves along the workpiece at a velocity V and a depth of cut t_o. A chip is generated ahead of the tool by shearing the material continuously along the *shear plane*. This plane makes an angle ϕ with the surface of the workpiece and is called the *shear angle*. Below the shear plane the material is not deformed; it is only subjected to elastic stresses. Above the shear plane we have the chip, already formed, climbing up the face of the tool as cutting continues.

The thickness of the chip, t_c, can be determined by knowing t_o, α, and ϕ. The ratio of t_o to t_c is known as the *cutting ratio*, r, which can be expressed as,

$$r = \frac{t_o}{t_c} = \frac{\sin \phi}{\cos(\phi - \alpha)}, \tag{8.1}$$

or

$$\tan \phi = \frac{r \cos \alpha}{1 - r \sin \alpha}. \tag{8.2}$$

Note that only when $\phi = \alpha$ is the length of the shear plane equal to the chip thickness. It has been observed that the chip thickness is always greater than the depth of cut, hence r is always less than unity.

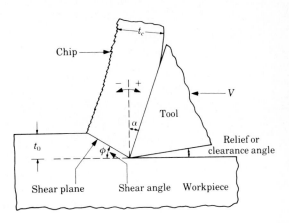

FIGURE 8.1 Schematic illustration of the orthogonal cutting process with a tool. Among various process parameters, the cutting speed V, rake angle α, and the depth of cut t_0 are independent variables, whereas the chip thickness t_c and shear angle ϕ are dependent variables.

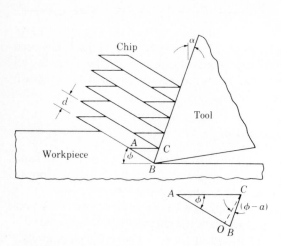

FIGURE 8.2 Schematic illustration of the basic mechanism of chip formation in cutting. This model is similar to a deck of cards sliding along each other. All chip-type machining processes, such as drilling, grinding, and filing, involve this mechanism of material removal.

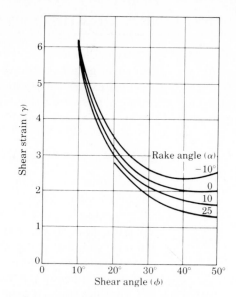

FIGURE 8.3 Shear strain in cutting as a function of shear angle and rake angle. These curves are plots of Eq. (8.3). Note that the shear strains can be on the order of 5 or higher.

The shearing process to which the material is subjected during cutting can be likened to a deck of cards sliding against each other, as shown in Fig. 8.2. Based on the definition of shear strain in Fig. 2.1(c), the shear strain γ in cutting can be expressed as,

$$\gamma = \frac{AB}{OC} = \frac{AO}{OC} + \frac{OB}{OC}$$

or

$$\gamma = \cot \phi + \tan(\phi - \alpha). \tag{8.3}$$

This relationship is shown in Fig. 8.3. It can be seen that large shear strains are associated with low shear angles and low, or negative, rake angles (Table 8.1). Shear strains as high as five or larger have been observed in actual cutting operations.

From the orthogonal model, the velocity of the chip V_c and the velocity V_s at which shearing is taking place along the shear plane can be determined with the use of the diagram shown in Fig. 8.4.

Since mass continuity must be maintained, and assuming that the width of the chip is the same as the workpiece width, then,

$$V t_o = V_c t_c$$

or

$$V_c = Vr. \tag{8.4}$$

TABLE 8.1(a)
ORTHOGONAL METAL CUTTING DATA. (After E. G. Thomsen.)
(Material: 4130 steel, t_o = 0.0025 in., w = 0.475 in., V = 90 ft/min, tool: high-speed steel.)

α	ϕ	γ	μ	β	F_c, lb	F_t, lb	u_t, $\frac{\text{in.-lb}}{\text{in}^3}$	u_s	u_f	$\frac{u_f}{u_t}$, %
25°	20.9°	2.55	1.46	56	380	224	320	209	111	35
35	31.6	1.56	1.53	57	254	102	214	112	102	48
40	35.7	1.32	1.54	57	232	71	195	94	101	52
45	41.9	1.06	1.83	62	232	68	195	75	120	62

TABLE 8.1(b)
ORTHOGONAL METAL CUTTING DATA. (After M. E. Merchant.)
(Material: 9445 steel, t_o = 0.037 in., w = 0.25 in., tool: cemented carbide.)

α	V, ft/min	ϕ	γ	μ	β	F_c	F_t	u_t	u_s	u_f	$\frac{u_f}{u_t}$, %
+10	197	17	3.4	1.05	46	370	273	400	292	108	27
	400	19	3.1	1.11	48	360	283	390	266	124	32
	642	21.5	2.7	0.95	44	329	217	356	249	107	30
	1186	25	2.4	0.81	39	303	168	328	225	103	31
−10	400	16.5	3.9	0.64	33	416	385	450	342	108	24
	637	19	3.5	0.58	30	384	326	415	312	103	25
	1160	22	3.1	0.51	27	356	263	385	289	96	25

Also,

$$V_s = \frac{V \cos \alpha}{\cos(\phi - \alpha)}. \tag{8.5}$$

This relationship can also be obtained by observing, in Fig. 8.4, that,

$$\frac{V}{\cos(\phi - \alpha)} = \frac{V_s}{\cos \alpha} = \frac{V_c}{\sin \phi}$$

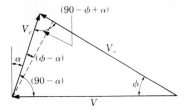

FIGURE 8.4 Velocity relationships in orthogonal cutting. Note that the chip has the smallest velocity.

The *shear strain rate* is the ratio of V_s to the thickness d of the sheared element (shear zone), or,

$$\dot{\gamma} = \frac{V_s}{d}.$$

(8.6)

Experimental evidence indicates that d is on the order of 10^{-3} to 10^{-4} in. (10^{-2} to 10^{-3} mm). This means that, even at low cutting speeds, the shear strain rate is very high, on the order of 10^3 to 10^6/s. A knowledge of the shear strain rate is essential because of its effects on the strength and ductility of the material, and the chip morphology (chip form).

8.2.2 OBSERVATIONS OF ACTUAL CHIPS IN CUTTING

When we observe actual chip formation under different cutting conditions, we find that there are significant deviations from the ideal model described above. Some of the typical chip morphologies observed in practice are shown in Fig. 8.5 and are described below.

a. Total deformation of the material (to form a continuous chip) does not take place instantaneously in a shear plane, but rather in a shear zone. This is known as the *primary shear zone*. This type of chip is closest to the ideal model and can be observed in practice with some metals at high cutting speeds and/or high rake angles.

b. This is a continuous chip, where *a secondary shear zone* develops at the tool–chip interface, with little relative motion of the chip surface with respect to the tool face. The secondary shear zone becomes deeper as friction at the tool–chip interface increases.

c. Shear does not occur in a plane but rather in a zone with curved boundaries. Note that the lower boundary is below the machined surface, which undergoes severe surface strains. Soft metals machined at low speeds produce this type of shear zone.

d. Depending on the sharpness of the tool tip and frictional conditions at the tool–chip interface, a *built-up edge* (BUE) forms at the tip of the tool. This edge is composed of deposited layers of material from the workpiece and, in effect, alters the tool shape (Section 8.3.1).

e. *Inhomogeneous* strain in a continuous chip, showing zones of low and high shear strain. Materials with low thermal conductivity (such as titanium) and flow stress that decreases sharply with temperature exhibit this behavior.

f. This is a *discontinuous* chip. It is composed of segments that may be firmly or loosely attached to each other (Section 8.3.2).

Variations of the basic types outlined above are also observed under various cutting conditions or with different material properties. Some of these chip formations

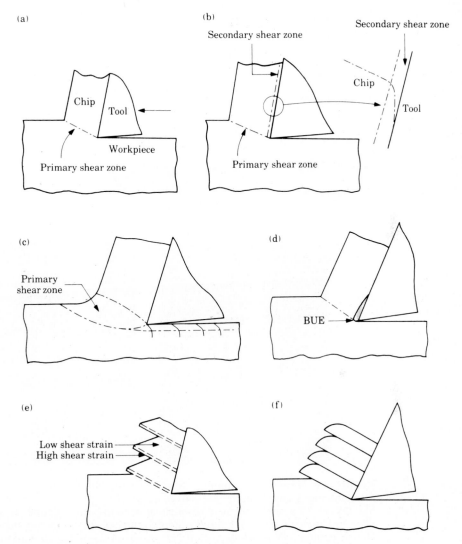

FIGURE 8.5 Basic types of chips observed in metal cutting. See text for details. *Source:* After M. C. Shaw.

can easily be observed by scraping the surface of a stick of butter or wax, or similar materials, with a knife. By changing the temperature of the butter and the angle at which the knife is held, different types of chips can be produced.

Micrographs of four basic types of chips are shown in Fig. 8.6(a) through (e). An additional observation can be made in cutting operations: The chip begins to curl away from the tool face some distance away from the tip of the tool. Thus the tool–chip interface has a finite contact length (Section 8.3.3).

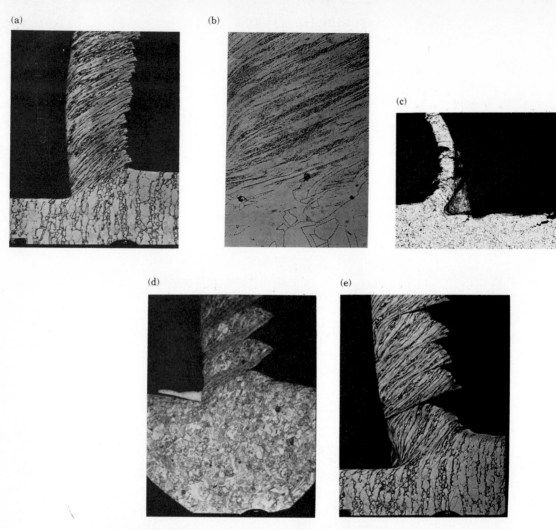

FIGURE 8.6 Photomicrographs of basic types of chips obtained in cutting metals. (a) Continuous chip in cutting 60–40 brass at 670 ft/min (3.4 m/s) at a depth of cut of 0.008 in. (0.2 mm). (b) Secondary shear zone in cutting copper at 330 ft/min (1.7 m/s). (c) Built-up edge in cutting sintered tungsten. (d) Inhomogeneous (serrated) chip in cutting 321 stainless steel at 250 ft/min (1.25 m/s) at a depth of cut of 0.008 in. (0.2 mm). (e) Discontinuous chip in cutting 60–40 brass at 50 ft/min (0.25 m/s). Note the small built-up edge at the root of the chip. *Source:* Parts (a), (b), (d), (e) courtesy of P. K. Wright, Carnegie-Mellon University. Part (c), A. J. Moser and S. Kalpakjian.

8.2.3 VARIABLES IN CUTTING

In the two-dimensional cutting process described above there are a number of independent and dependent variables. The basic independent variables are the following:

a. Tool material and its metallurgical condition;

b. Tool geometry, e.g., rake angle, clearance angle, nose radius, and its surface characteristics;

c. Workpiece material and its condition;

d. Workpiece temperature;

e. Cutting conditions, e.g., speed, feed, and depth of cut;

f. Cutting fluid; and

g. Characteristics of the machine tool, e.g., stiffness, power, and damping.

Dependent variables are those influenced by changes in the independent variables. These are as follows:

a. The shear angle;

b. The chip form or morphology.

c. Forces and energy dissipated in the cutting process;

d. Temperature rise as a result of the cutting operation, since the work done is converted into heat;

e. Wear of the tool; and

f. The surface finish and surface texture obtained after machining.

A complete analysis of such an operation requires the establishment of quantitative relationships between all the independent and dependent variables. In view of the number of factors involved and the variety of materials and conditions available, this is a complex problem.

Consider, for example, the following questions: If the surface finish being produced by cutting is rough and unacceptable, which of the independent variables do we change first? For instance, do we change the rake angle of the tool? If so, do we increase it or decrease it? Do we use a cutting fluid and, if so, what kind? If it is found that the tool wears and becomes dull rapidly, what do we change? Do we cut the material at a slower speed or do we decrease the depth of cut? If the cutting tool begins to vibrate (chatter), what should be done to eliminate the vibration?

Numerous questions of a similar nature can be asked. These are significant and must be resolved in order to carry out a machining process efficiently and economically. In the rest of this chapter, the major cutting processes are analyzed and, to aid us in answering such questions, the necessary relationships are established.

8.2.4 FORCES IN ORTHOGONAL CUTTING

In performing a cutting operation on a piece of metal, it is found that the force on the tool is R (*resultant force*) as shown in Fig. 8.7. This force can be resolved into two components: a *cutting* component F_c, which is in the direction of the cutting velocity, and a *thrust* component F_t, which is in the direction normal to the cutting velocity or to the machined surface.

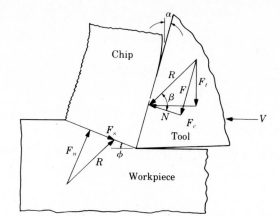

FIGURE 8.7 Force relationships in orthogonal cutting. The resultant force R is resolved into the shear plane, tool face, and in the direction of cutting. The energy required for cutting is supplied by the cutting force F_c.

We note that F_c is in the direction of the cutting velocity V. This is as it should be since, in order to supply the energy required to carry out the cutting process, the cutting force and the tool movement must be in the same direction. The power consumed in cutting is the product of F_c and V.

It can be seen that the thrust force does not do any work on the system. However, the toolholder and the machine tool must be sufficiently stiff to prevent or minimize any deflection due to the thrust force. For instance, if the thrust force is high, or if the machine tool is not sufficiently rigid, the tool will be pushed away from the surface being machined, thus reducing the depth of cut. The direction of F_t can, under certain circumstances, be opposite to that shown in Fig. 8.7 as will be shown in Section 8.5.3.

The resultant force R can also be resolved into two components on the tool face: a *friction* force F at the tool–chip interface, and a *normal* force N. The ratio of F to N is the coefficient of friction μ at the interface, and the angle β is known as the *friction angle*.

It can be shown that μ can be expressed as,

$$\mu = \frac{F_t + F_c \tan \alpha}{F_c - F_t \tan \alpha}.$$ (8.7)

The resultant force R is transmitted to the shear plane and can also be resolved into two components: a *shear* force F_s acting along the shear plane, and a *normal* force F_n acting perpendicular to the shear plane.

8.3 CHIP MORPHOLOGY

A chip, as obtained from an orthogonal cutting process, is generally wider than it is thick. It has two surfaces: one that is in contact with the tool face, and the other from the original surface of the workpiece. Upon inspection of the chip, we find that

TABLE 8.2
STRENGTH AND HARDNESS OF CHIPS IN TURNING MILD STEEL (After K. Nakayama).

	Original material	45°	35°	27°	10°	10°	10°
Rake Angle, α		45°	35°	27°	10°	10°	10°
Feed (f), mm/rev		0.30	0.30	0.20	0.20	0.20	0.20
Cutting speed (V), m/min		50	50	168	168	168	76
Cutting fluid		Soluble oil	None	None	Soluble oil	None	None
Shear strain, γ		1.1	1.7	2.1	2.9	3.1	4.0
Tensile strength (UTS), kg/mm^2	46	75	84	91	92	93	95
Vickers hardness number (HV)	209	272	289	302	320	314	325
HV/UTS	4.5	3.6	3.4	3.3	3.5	3.4	3.4

the tool side of the chip surface is shiny because it has been rubbing along the tool face under high pressure and temperature. This is called a *burnished* surface.

The other surface of the chip has a jagged steplike appearance (see Figs. 8.1 and 8.6a). This is due to the basic mechanism of chip formation shown in Fig. 8.2. This side of the chip is not in contact with any solid body and, hence, it maintains its surface texture as generated. Typical cross-sections of chips are shown in Fig. 8.6.

As a result of strain hardening (due to the shear strains to which it is subjected), the chip usually becomes harder, less ductile, and stronger than the original workpiece material. The increase in hardness and strength of the chip depends on the shear strain. As shown in Table 8.2, as the rake angle decreases, the shear strain increases and the chip is stronger and harder. With increasing strain, the material tends to behave like a rigid, perfectly plastic body.

8.3.1 BUILT-UP EDGE

The *built-up edge* (BUE) is often encountered in metal cutting operations. This edge starts with a layer of workpiece material that adheres to the tip of the tool (see Fig. 8.8). The edge then begins to grow with the additional deposition of layers as cutting progresses. As the BUE becomes larger, it becomes unstable and eventually breaks up. Part of this edge is carried away by the chip and the rest is deposited randomly on the workpiece surface. The process of BUE formation and destruction is then repeated.

BUE formation can be likened to walking over muddy terrain. Mud sticks to the shoes, changing the shape of the bottom of the shoe. It grows in size, and then may fall off and the process is repeated.

The BUE is one of the most important factors affecting surface finish in machining (Figs. 8.6(c) and 8.8). Other factors influencing surface finish are the condition of the cutting edge of the tool, feed marks, and chatter of the tool during cutting, as described in Sections 8.9 and 8.13.

As can be seen in Figs. 8.6(c) and 8.8, a BUE, in effect, changes the tool geometry. Note, for instance, the large tip radius and the roughness of the surface generated.

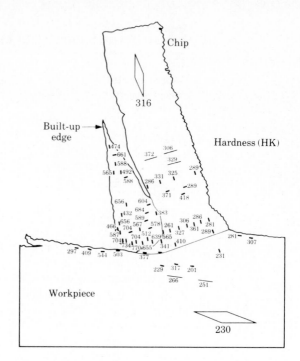

FIGURE 8.8 Hardness distribution in the cutting zone for AISI 3115 steel. Note that some regions in the built-up edge are as much as three times harder than the bulk metal. The built-up edge becomes, in effect, the cutting tool. *Source:* N. Zlatin and M. E. Merchant, Trans ASME, Special Pamphlet, 1947.

Also, due to strain hardening and successive layers of material deposited on the BUE, the hardness of the BUE has increased by threefold from the original hardness of the bulk workpiece material.

A tenacious BUE can significantly change the cutting geometry and be detrimental to the cutting process and to surface finish, Fig. 8.9. On the other hand, a thin, stable BUE is generally regarded as desirable because it protects the surface of the tool.

Mechanism of Built-Up Edge Formation

The exact mechanism of formation of the BUE is not yet clearly understood. However, investigations have identified two distinct mechanisms that contribute to this formation. One is the adhesion of the workpiece material to the rake face of the tool. The strength of this bond depends on the affinity of the workpiece and tool materials. The other mechanism is the growth of the adhered metal layers to form a BUE. (See also Section 4.6.1.)

One of the important factors in forming the BUE is the ability of the workpiece material for strain hardening; the higher the strain-hardening exponent, the greater the tendency for BUE formation. Experimental evidence indicates that as the cutting speed increases the BUE decreases or is eliminated. It is not clear whether the speed, as such, plays a significant role (such as by increasing the strain rate), or whether its effect is felt through the increase in temperature with speed. The latter appears to be in contradiction with the observation that higher temperatures improve adhesion, since as speed increases, temperature increases but BUE decreases. (See Section 8.6.)

In addition to the factors outlined above, it has been shown that the tendency for BUE formation can be reduced by decreasing the depth of cut, increasing the rake angle, using a tool with a small tip radius, and using an effective cutting fluid.

8.3.2 DISCONTINUOUS CHIPS

The mechanism of discontinuous chip formation has proved to be complex. Brittle materials produce such chips, primarily because they do not have the capacity to undergo the shear strains in cutting. Discontinuous chips are generally composed of loosely connected segments or, in some cases, unconnected segments.

Various other factors that have been identified as having an effect on discontinuous chip formation are: depth of cut, cutting speed, and impurities and hard particles in the workpiece material. Discontinuous chip formation may be influenced by these factors as described below.

Impurities and hard particles act as nucleation sites for cracks (Fig. 3.44), thereby producing discontinuous chips. A large depth of cut increases the probability of such defects being present in the cutting zone. Higher cutting speeds mean higher temperatures, hence the material is more ductile and has less tendency to form discontinuous chips.

Another factor in discontinuous-chip formation is the magnitude of the compressive stresses on the shear plane. Recall from Fig. 2.31 that the maximum shear strain

(a) (b)

FIGURE 8.9 Surface finish obtained in cutting with a built-up edge. (a) Surface of AISI 5130 steel bar (hardness, 241 HB) turned on a lathe. Magnification: 15X. (b) Surface of AISI 1018 steel (143 HB) in face milling. 15X. *Source:* Courtesy of Metcut Research Associates Inc.

at fracture increases with increasing compressive stress. If the normal stress is not sufficiently high, the material is unable to undergo the shear strain required to form a continuous chip.

Shear localization (Fig. 8.6d) also produces discontinuous chips (also known as *inhomogeneous* or *serrated* chips). This is generally associated with high localized strains, high strain rate (which is a function of the cutting speed) and low thermal conductivity of the workpiece material.

Cutting fluids (Section 8.12) also have an influence on chip morphology. An effective cutting fluid can reduce friction at the tool–chip interface and thus increase the shear angle. This, in turn, means lower shear strains and less tendency for discontinuous chip formation.

The stiffness of the toolholder and the machine tool can also have an effect on discontinuous chip formation. Low stiffness means that deflections are present due to the cutting forces. This implies that elastic energy is being stored in the system. If the tool moves across weak regions during the cut, this energy may be released suddenly, leading to vibrations and instabilities in the cutting zone. (See also Section 8.13.)

8.3.3 CHIP CURL

Although a common phenomenon observed in all cutting operations with metals and nonmetallic materials, the reasons for *chip curl* are still not clearly understood. Among possible factors contributing to chip curl are: nonuniform stress distribution and inhomogeneity of the material across the shear plane, thermal effects in the cutting area, strain-hardening characteristics, the secondary shear zone, and the segmentation of the chip (nonuniform straining).

In an orthogonal cutting operation, the radius of the chip is found to be small at the beginning of the cut and increases as the cut progresses. In addition to being dependent on the workpiece material, other process variables also have an effect on chip curl. Generally, the radius of curvature has been found to decrease (the chip becomes more curly) with decreasing depth of cut, increasing rake angle, and with low friction at the tool–chip interface.

Chip Breakers

Long continuous chips can be a hazard in machining. This is because such chips tend to become entangled around the cutting area and the tool post, thus interfering with the operation.

The usual procedure to avoid this situation is to decrease the radius of curvature of the chip and break it by bending. This is done by a *chip breaker*, which can be a piece clamped to the rake face of the tool (Fig. 8.10), or it can be an integral part of the tool as manufactured. A large variety of cutting tools with chip-breaking features is available.

In interrupted cutting operations, such as milling, chip breakers are generally not necessary since the chips already have finite lengths due to the intermittent nature of the cutting operation. (See Fig. 8.32e.)

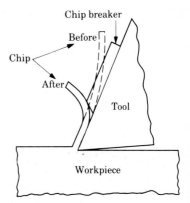

FIGURE 8.10 The action of a chip breaker. The chip breaker reduces the radius of curvature of the chip. Some cutting tools, called inserts, are made with a built-in chip breaker. See also Figs. 8.34 and 8.35(c).

8.4 SHEAR-ANGLE RELATIONSHIPS

Because the shear angle and the shear zone have great significance in the mechanics of cutting, a great deal of effort has been expended to determine the relationship of the shear angle to material properties and process variables.

One of the earliest analyses is based on the assumption that the shear angle adjusts itself so that the cutting force is a minimum, or so that the maximum shear stress occurs in the shear plane. From the force diagram in Fig. 8.7 the following relationships can be obtained:

$$F_c = R \cos(\beta - \alpha) \tag{8.8}$$

and

$$F_s = R \cos(\phi + \beta - \alpha). \tag{8.9}$$

The area of the shear plane is given by,

$$A_s = \frac{wt_0}{\sin \phi}, \tag{8.10}$$

where w is the width of the cut. Thus the shear stress in the shear plane is,

$$\tau = \frac{F_s}{A_s} = \frac{F_c \sec(\beta - \alpha) \cos(\phi + \beta - \alpha) \sin \phi}{wt_0}. \tag{8.11}$$

Assuming that β is independent of ϕ, the shear angle corresponding to the maximum shear stress is then found by differentiating this expression with respect to ϕ and equating it to zero. Hence,

$$\frac{d\tau}{d\phi} = \cos(\phi + \beta - \alpha)\cos \phi - \sin(\phi + \beta - \alpha)\sin \phi = 0.$$

Thus,

$$\tan(\phi + \beta - \alpha) = \cot \phi = \tan(90° - \phi)$$

or

$$\phi = 45° + \frac{\alpha}{2} - \frac{\beta}{2}. \qquad (8.12)$$

Note that this expression indicates that, as the rake angle decreases and/or as the friction at the tool–chip interface increases, the shear angle decreases and the chip is thus thicker. This is to be expected since decreasing α and increasing β tend to lower the shear angle.

A second method of determining ϕ is based on slip-line analysis (Section 2.10.2) and on the following assumptions:

a. The shear plane AB in Fig. 8.11(a) is a plane of maximum shear stress.

b. There are no shear or normal stresses along AC, as it should be, since the chip beyond AC is unconstrained.

c. The material is rigid, perfectly plastic.

The appropriate Mohr's circle construction for this shear zone is shown in Fig. 8.11(b). It can be seen that the angle η is given by,

$$\eta = 45° - \beta.$$

Since, from geometry,

$$\eta = \phi - \alpha,$$

the expression for the shear angle is,

$$\phi = 45° + \alpha - \beta. \qquad (8.13)$$

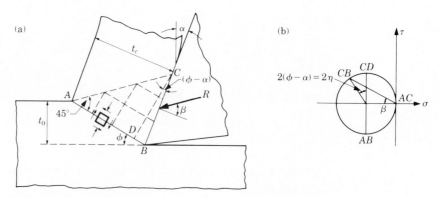

FIGURE 8.11   (a) Slip-line field in orthogonal cutting. Line AB is the shear plane along which the shear stress is assumed to be a maximum. (b) Mohr's circle for stresses in the cutting zone. Point CB on the circle represents the normal and shear stresses at the tool–chip interface, respectively.

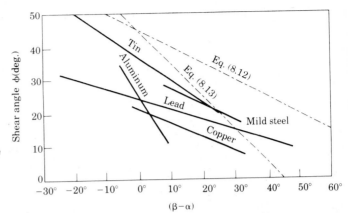

FIGURE 8.12 Comparison of experimental and theoretical shear angle relationships. More recent analytical studies have resulted in better agreement with experimental data.

Note that this expression is similar to Eq. (8.12) and indicates the same trends, although numerically it gives different values.

In another study, the following simple relationships have been proposed, which could serve as a guide for estimating the shear angle for practical purposes:

$$\phi = \alpha \qquad \text{for} \quad \alpha > 15°, \tag{8.14}$$

$$\phi = 15° \qquad \text{for} \quad \alpha < 15°. \tag{8.15}$$

A number of other expressions have been obtained for the shear angle based on various models and different assumptions. However, many of these expressions have generally been found not to be in good agreement with experimental data over a wide range of conditions (Fig. 8.12) largely because shear rarely occurs in a thin plane. There is, on the other hand, qualitative agreement in that the shear angle always decreases with increasing $(\beta - \alpha)$. Recent and more comprehensive studies appear to accurately predict the shear angle analytically, especially for continuous chips. (See also Table 8.1 for additional data.)

8.5 CUTTING ENERGY AND FORCES

In this section, the energy dissipated in a cutting operation is described in terms of the forces and other process parameters involved.

8.5.1 SPECIFIC ENERGIES

Refer to Fig. 8.7 and note that the *total power* input in cutting is,

Power $= F_c V.$

Letting the width of the cut be w, then the *total energy per unit volume* of material removed (*specific energy*), u_t, is

$$u_t = \frac{F_c V}{w t_o V} = \frac{F_c}{w t_o}. \tag{8.16}$$

In other words, u_t is simply the ratio of the cutting force to the projected area of the cut.

From Figs. 8.4 and 8.7 it can also be seen that the power required to overcome friction at the tool–chip interface is the product of F and V_c, or, in terms of frictional specific energy, u_f,

$$u_f = \frac{FV_c}{wt_oV} = \frac{Fr}{wt_o \cdot} = \frac{(F_c \sin \alpha + F_t \cos \alpha)r}{wt_o}. \tag{8.17}$$

Likewise, the power required for shearing along the shear plane is the product of F_s and V_s. Hence the specific energy for shear, u_s, is given by,

$$u_s = \frac{F_s V_s}{wt_o V}. \tag{8.18}$$

The total specific energy, u_t, is the sum of the two,

$$u_t = u_f + u_s. \tag{8.19}$$

It should be pointed out that there are two other sources of energy in cutting. One is the *surface energy* (Section 4.12) due to the formation of new surfaces when a layer of material is removed by cutting. However, this represents a very small amount of energy as compared to the shear and frictional energies involved. The other source is the energy associated with the *momentum change* as the metal crosses the shear plane. (This is similar to the forces involved in a turbine blade due to momentum changes of the fluid or gas.) Although in ordinary metal-cutting operations the momentum energy is negligible, this energy can be significant at very high cutting speeds (above 25,000 ft/min [125 m/s] or so).

Some experimental data on specific energies are given in Table 8.1. Note that, as the rake angle increases, the frictional specific energy remains more or less constant, whereas the shear specific energy is reduced rapidly. Thus, the ratio u_f/u_t increases considerably as α increases.

This trend can also be predicted by obtaining an expression for this ratio as follows:

$$\frac{u_f}{u_t} = \frac{FV_c}{F_c V} = \frac{R \sin \beta}{R \cos(\beta - \alpha)} \cdot \frac{Vr}{V} = \frac{\sin \beta}{\cos(\beta - \alpha)} \cdot \frac{\sin \phi}{\cos(\phi - \alpha)}. \tag{8.20}$$

Since it has been observed experimentally that as α increases, both β and ϕ increase (Table 8.1), then inspection of Eq. (8.20) indicates that the ratio u_f/u_t should also increase with α.

It is apparent that u_f and u_s are related to each other. Although u_f is not required for the cutting action to take place, it affects the magnitude of u_s. This is because as friction increases, the shear angle decreases. A decreasing shear angle, in turn, increases the magnitude of u_s.

The calculation of all the parameters involved in these specific energies in cutting presents considerable difficulties. There are good theoretical computations available, but they are difficult to perform. The reliable prediction of cutting forces and energies is,

TABLE 8.3
**APPROXIMATE POWER
REQUIREMENTS IN VARIOUS
CUTTING OPERATIONS** (at spindle
motor, corrected for 80% efficiency).

MATERIAL	UNIT POWER, hp · min/in.³*
Aluminum alloys	0.16–0.4
Cast irons	0.6–2.0
Copper alloys	0.5–1.2
High-temperature alloys	1.2–3.1
Magnesium alloys	0.16–0.2
Nickel alloys	1.8–2.5
Refractory alloys	1.4–3.5
Stainless steels	1.1–1.9
Steels	1.0–3.4
Titanium alloys	1.1–1.5

* Multiply by 2.73 to obtain $W \cdot s/mm^3$.

therefore, still based largely on experimental data (Table 8.3 and Fig. 8.13). These data should serve as a useful guide. The wide range of values in Table 8.3 can be attributed to differences in strength within each material group, and other variables such as friction and operating conditions.

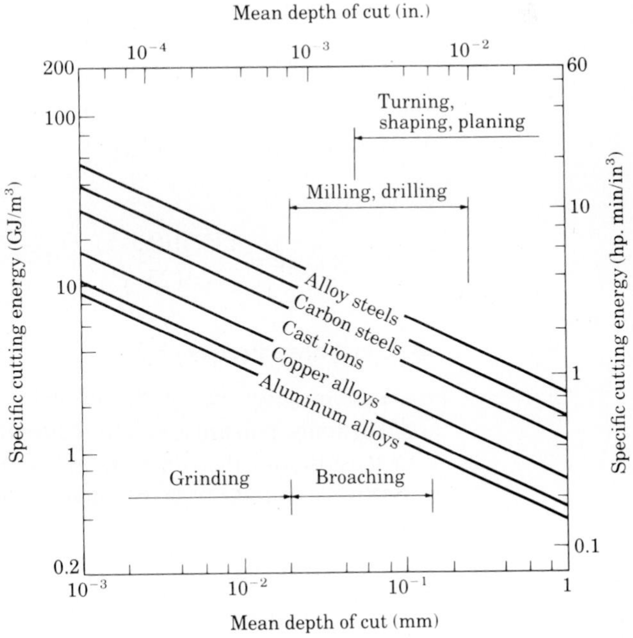

FIGURE 8.13 Approximate values for the specific energy in cutting. Note that the energy required increases with the strength of the material and decreasing depth of cut. See also Table 8.3. *Source:* After G. Boothroyd, *Fundamentals of Metal Machining and Machine Tools*, New York, McGraw-Hill, 1975, p. 57.

● **Illustrative Problem 8.1**

An orthogonal cutting process is being carried out, where $t_o = 0.005$ in., $V = 400$ ft/min, $\alpha = 10$ degrees, and width of cut $= 0.25$ in. It is observed that: $t_c = 0.009$ in., $F_c = 125$ lb, and $F_t = 50$ lb.

Calculate the percentage of the total energy that goes into overcoming friction at the tool–chip interface.

SOLUTION.   The percentage can be expressed as:

$$\frac{\text{Friction energy}}{\text{Total energy}} = \frac{FV_c}{F_c V} = \frac{Fr}{F_c},$$

where

$$r = \frac{t_o}{t_c} = \frac{5}{9} = 0.555,$$

$$F = R \sin \beta,$$

$$F_c = R \cos(\beta - \alpha),$$

and

$$R = \sqrt{F_t^2 + F_c^2} = \sqrt{50^2 + 125^2} = 135 \text{ lb}.$$

Thus

$$125 = 135 \cos(\beta - 10),$$

from which

$$\beta = 32 \text{ degrees}$$

and

$$F = 135 \sin 32° = 71.5 \text{ lb}.$$

Hence

$$\text{Percentage} = \frac{(71.5)(0.555)}{125} = 0.32 = 32\%. \quad ●$$

● **Illustrative Problem 8.2**

You are given two pieces of 304 stainless steel rods, each 0.500 in. in diameter and 6 in. long. You are asked to reduce the diameters to 0.480 in., one piece by pulling it in tension and the other by machining it on a lathe in one pass. Calculate the respective amounts of work involved and explain the difference in the energies dissipated.

SOLUTION.   The work done in pulling the rod is

$$W_{\text{tension}} = (u)(\text{volume}),$$

where

$$u = \int_0^{\epsilon_1} \sigma \, d\epsilon.$$

The strain is found from

$$\epsilon_1 = \ln\left(\frac{0.500}{0.480}\right)^2 = 0.0816.$$

From Table 2.4, the following values for K and n are obtained for this material:

$K = 185{,}000$ psi,

$n = 0.45.$

Thus

$$u = \frac{K\epsilon_1^{n+1}}{n+1} = \frac{(185{,}000)(0.0816)^{1.45}}{1.45}$$

$$= 3370 \text{ in} \cdot \text{lb/in}^3$$

and

$$W_{\text{tension}} = (3370)(\pi)(0.25)^2(6) = 3970 \text{ in.} \cdot \text{lb.}$$

From Table 8.3 an average value for the specific energy in machining stainless steels is obtained as 1.5 hp $\cdot$ min/in.^{3}. The volume of material machined is,

$$\frac{\pi}{4}[(0.5)^2 - (0.480)^2](6) = 0.092 \text{ in.}^3.$$

The specific energy, in appropriate units, is

Specific energy $= (1.5)(33{,}000)(12)$

$$= 594{,}000 \text{ in.} \cdot \text{lb/in.}^3.$$

Thus, the machining work is

$$W_{\text{machining}} = (594{,}000)(0.092) = 54{,}650 \text{ in.} \cdot \text{lb.}$$

Note that the work in machining is about 14 times that for tension. The difference between the energies can be explained by noting that tension requires very little straining and involves no friction. Machining involves friction and the material removed, even though small in volume, has undergone much higher strains than the bulk material in tension. Assuming an average shear strain of 3 (from Tables 8.1 and 8.2), which is equivalent to an effective strain of 1.7 (Eq. 2.52), we note that, in machining, the material is subjected to a strain that is 21 times that in tension. These differences explain why machining consumes much more energy than reducing the diameter of this rod by stretching. It should be noted, however, that as the diameter of the rod decreases, the difference between the two energies becomes smaller, assuming that the same depth of material is to be removed. ●

8.5.2 STRESSES IN THE SHEAR ZONE

The stresses in the shear plane and at the tool–chip interface are analyzed below, by assuming that they are uniformly distributed. The forces in the shear plane can be resolved into shear and normal forces and stresses.

The average shear stress in the shear plane is given by

$$\tau = \frac{F_s}{A_s} \tag{8.21}$$

and the average normal stress by

$$\sigma = \frac{F_n}{A_s}, \tag{8.22}$$

where A_s is the area of the shear plane, namely,

$$A_s = \frac{wt_o}{\sin \phi}.$$

Figure 8.14 gives some data pertaining to these average stresses. In this figure, the rake angle is a parameter and the shear plane area is increased by increasing the depth of cut. We can make the following observations from these curves:

a. The shear stress on the shear plane is independent of the rake angle,

b. The normal stress on the shear plane decreases with increasing rake angle, and consequently,

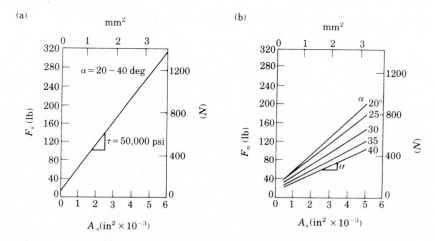

FIGURE 8.14 Shear force and normal force as a function of shear plane area and rake angle for 85–15 brass. Note that the shear stress in the shear plane is constant, regardless of the magnitude of the normal stress. Thus, normal stress has no effect on the shear flow stress of the material. *Source:* After S. Kobayashi and E. G. Thomsen, *Jr. Eng. Ind.,* vol. 81, 1959, pp. 251–262.

FIGURE 8.15 Schematic illustration of the distribution of normal and shear stresses at the tool–chip interface (rake face). Note that, whereas the normal stress increases continuously towards the tip of the tool, the shear stress reaches a maximum and remains at that value (known as sticking). See also Fig. 6.7.

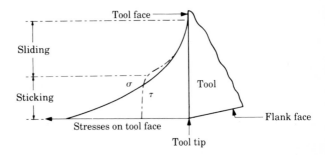

c. It can then be concluded that the normal stress in the shear plane has no effect on the magnitude of the shear stress. This phenomenon has also been verified by other mechanical tests. (However, normal stress has a strong influence on the magnitude of the shear strain in the shear zone. As noted in Fig. 2.31, the maximum shear strain to fracture increases with the normal compressive stress.)

The determination of the stresses on the rake face of the tool presents considerable difficulties. One problem is the accurate determination of the length of contact at the tool–chip interface. This length has been shown to increase with decreasing shear angle, indicating that the contact length is a function of rake angle, cutting speed, and friction at the tool–chip interface. Another problem is that the stresses are not uniformly distributed on the rake face.

Studies with photoelastic techniques have shown that the actual stress distribution is qualitatively as shown in Fig. 8.15. Note that the stress normal to the tool face is maximal at its tip and decreases rapidly toward the end of the contact length.

The shear stress has a similar trend with the exception that it levels off at approximately the center of the contact length. This behavior indicates that sticking is taking place, whereby the shear stress has reached the shear yield strength of the material. Such a sticking region has been observed on some chips. (Sticking regions can be observed in various other metal forming processes also.) See also Section 4.6.1.

8.5.3 OBSERVATIONS ON FORCES IN THE CUTTING ZONE

In addition to being a function of the strength of the workpiece material, forces in cutting are also influenced by other variables. Extensive data are available. In one set of data, Table 8.1, note that the cutting force F_c increases with increasing depth of cut, decreasing rake angle, and decreasing speed.

The effect of cutting speed can, from the analysis of the data in Table 8.1(b), be attributed to the fact that as speed decreases the shear angle decreases and the friction coefficient increases. Both of these effects increase the cutting force.

Another factor that can have a significant influence on the cutting force is the tip radius of the tool. The larger the radius (i.e., the duller the tool) the higher the force. Experimental evidence indicates that, for depths of cut on the order of five times the tip radius or higher, the effect of dullness on the cutting forces is negligible.

Note from Fig. 8.16 that, if the depth of cut is very small compared to the nose radius, the tip radius will in effect determine the rake angle. Even though the tool may have a positive rake angle, the effective angle is actually highly negative. This is why dull tools cannot remove thin layers of material. Such dull tools may actually rub against the surface without removing any material. This causes smearing and burnishing of the workpiece surface and possibly leads to damage. However, the tools could also impart compressive residual stresses on the surface. (See also flank wear, Section 8.7.1.)

Direction of Thrust Force

Referring to Fig. 8.7, note that the cutting force must always be in the direction of cutting in order to supply energy to the system. Although the thrust force does not contribute to the total work done, it is important to know its magnitude. For instance, if the toolholder is not sufficiently stiff, the thrust force will simply deflect the tool and reduce the depth of cut. Thus, tolerances cannot be held.

Also note from Fig. 8.7 that the thrust force is downward. We will now see that it is possible for this force to act upward. First, we express the thrust force as follows:

$$F_t = R \sin(\beta - \alpha) \tag{8.23}$$

or

$$F_t = F_c \tan(\beta - \alpha). \tag{8.24}$$

Because the magnitude of F_c is always positive (say, as shown in the figure), the sign of F_t can be either positive or negative, depending on the relative magnitudes of β and α. It is seen that when $\beta > \alpha$, the sign of F_t is positive (downward) and when $\beta < \alpha$, it is negative (upward). It is thus possible to have an upward thrust force at high rake angles and/or with low friction at the tool–chip interface.

The situation can be visualized by observing in Fig. 8.7 that when $\mu = 0$, $\beta = 0$ and the resultant force R coincides with the force N. In such a case, R will have an upward thrust force component. Also note that for $\alpha = 0$ and $\beta = 0$, the thrust force is zero.

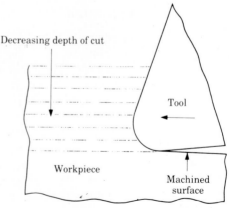

FIGURE 8.16 Schematic illustration of a dull tool (large tool tip radius) in orthogonal cutting, and the dependence of the effective rake angle on depth of cut. Note that at small depths of cut the rake angle is effectively negative. Chips may not be formed, since the tool tends to ride over the surface of the workpiece, producing a burnished surface, causing temperature rise, and possible surface damage.

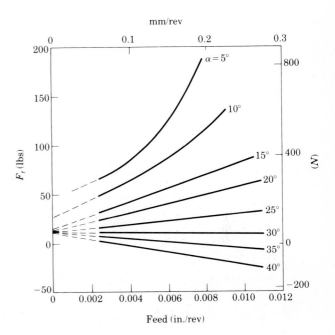

FIGURE 8.17 Thrust force as a function of rake angle and feed in orthogonal cutting of AISI 1112 cold-rolled steel. Note that at high rake angles the thrust force is negative. A negative thrust force has important implications in the design of machine tools, and in controlling the stability of the cutting process. *Source:* S. Kobayashi and E. G. Thomsen, *Jr. Eng. Ind.* vol. 81, 1959, pp. 251–262.

These observations have been verified experimentally, as shown in Fig. 8.17. The influence of the depth of cut is obvious, in that as t_o increases R must also increase so that F_c will also increase. This supplies the additional energy required to remove the extra material produced by the increased depth of cut.

The change in direction and magnitude of the thrust force can play a significant role. Within a certain range of operating conditions, it leads to instability problems in machining, particularly if the machine tool is not stiff enough. However, such high rake angles are rarely used in machining most metals.

8.6 TEMPERATURE

The mechanical energy dissipated in cutting is converted into heat. The main sources of heat are in the shear zone and at the tool–chip interface. Heat is also generated with dull or worn tools when the lower part of the tool tip rubs against the surface of the workpiece.

The interaction of these heat sources, combined with the geometry of the cutting zone, results in a complex temperature distribution, as shown in Fig. 8.18. The distribution depends on many factors, such as specific heat and thermal conductivity, cutting speed, depth of cut, and the type of cutting fluid used, if any.

Various studies have been made to investigate temperatures in cutting, based on heat transfer and dimensional analysis using experimental data. Although it is observed in Fig. 8.18 that there are considerable temperature gradients in the cutting

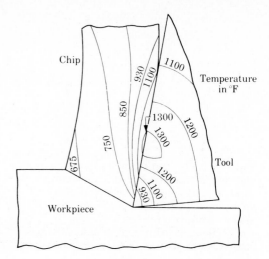

FIGURE 8.18 Typical temperature distribution in the cutting zone. Note that the maximum temperature is about half way up the face of the tool, and that there is a steep temperature gradient across the thickness of the chip. Some chips may become red hot, causing safety hazards to the operator and thus requiring safety guards. *Source:* After G. Vieregge, Werkstatt und Betrieb, No. 11, 1953, p. 696.

zone, a simple but approximate expression for the mean temperature for orthogonal cutting is given by,

$$T = \frac{1.2 Y_f}{\rho c} \sqrt[3]{\frac{V t_o}{K}}, \tag{8.25}$$

where

T = mean temperature of tool–chip interface (°F),

Y_f = flow stress of the material (psi),

V = cutting speed (in./s),

t_o = depth of cut (in.),

ρc = volumetric specific heat of the workpiece (in. $\cdot$ lb/in.^{3} $\cdot$ °F), and

K = thermal diffusivity of the workpiece (in.^{2}/s). Thermal diffusivity is the ratio of thermal conductivity to volumetric specific heat.

Equation (8.25) indicates that temperature increases with the strength of the workpiece material, cutting speed, and depth of cut. Also, the lower the thermal conductivity of the workpiece material, the higher the temperature. It has been shown that the thermal properties of the tool are relatively unimportant compared to those of the workpiece. Because some of the parameters in Eq. (8.25) depend on temperature, it is important that appropriate values be used that are compatible with the predicted temperature range.

Based on Eq. (8.25), an expression for the mean temperature T in turning (on a lathe) is given by,

$$T \propto V^m f^n, \tag{8.26}$$

where m and n are constants, V is the cutting speed, and f is the feed of the tool (see Fig. 8.24).

Approximate values for m and n are as follows:

TOOL	m	n
Carbide	0.2	0.125
High-speed steel	0.5	0.375

It has been observed that the depth of cut has negligible influence on the mean temperature for depths exceeding twice the tip radius of the tool.

Experimental techniques have been developed to measure the local temperatures on the tool surfaces. One example of experimental results (using thermocouples) is given in Fig. 8.19. Note that the maximum temperature is at a point away from the tip of the tool and that it increases with cutting speed.

The temperature generated in the shear plane is a function of the specific energy for shear, u_s, and the specific heat of the material. Hence, temperature rise will be highest in cutting materials with high strength and low specific heat, as seen from Eq. (2.59).

The temperature rise at the tool–chip interface is a function of the coefficient of friction. Flank wear (Section 8.7.1) is also a source of heat because of rubbing of the tool on the machined surface.

Cutting speed has a significant influence on temperature. As the speed increases, there is little time for the heat to be dissipated and hence temperature rises. The

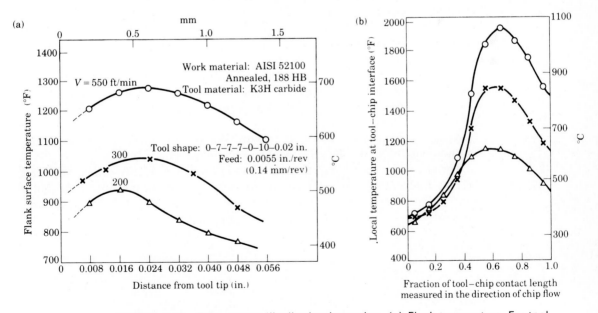

FIGURE 8.19 Temperature distribution in turning. (a) Flank temperature. For tool shape, see Fig. 8.33. (b) Tool-chip interface temperature. Note that the rake face temperature is higher than at the flank surface. *Source:* After B. T. Chao and K. J. Trigger, *Jr. Eng. Ind.*, vol. 83, 1961, pp. 496–504.

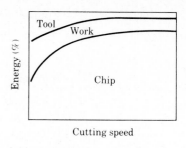

FIGURE 8.20 Typical energy distribution as a function of cutting speed. Note that most of the cutting energy is carried away by the chip (in the form of heat), particularly as speed increases. For dimensional accuracy during cutting, it is important not to allow the workpiece temperature to rise significantly.

chip is a good heat sink, in that it carries away most of the heat generated. (This action of the chip is similar to *ablation*, where layers of metal melt away from a surface, thus carrying away the heat.) As cutting speed increases, a larger proportion of the heat is carried away by the chip, as seen in Fig. 8.20.

8.7 TOOL WEAR

We have seen that tool surfaces are subjected to forces, temperature, and sliding. As described in Section 4.8, these are conditions that induce wear.

Because of its effects on the quality of the machined surface and the economics of machining (Section 13.3.5), the study of tool wear is one of the most important and complex aspects of machining operations. Whereas cutting speed is an independent variable, the forces and temperatures generated are dependent variables and are functions of a large number of parameters.

Similarly, wear depends on tool and workpiece materials (their physical, mechanical, and chemical properties), tool geometry, cutting fluid properties, and various other operating parameters. The types of wear on a tool depend on the relative role of these variables.

Analytical studies on tool wear present considerable difficulties and therefore our knowledge of wear is based largely on experimental data.

The basic wear behavior of a tool is shown in Fig. 8.21 (for a two-dimensional cut) and Figs. 8.22 and 8.23 (for a three-dimensional cut, where the position of the tool is shown in Fig. 8.24). The various regions of wear are identified as follows: *flank wear*, *crater wear*, *nose wear*, and *chipping* of the cutting edge.

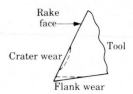

FIGURE 8.21 Schematic illustration of crater and flank wear on a cutting tool. Wear of tools is one of the most important aspects in cutting and has a major influence on the economics of the overall machining operation.

(a)

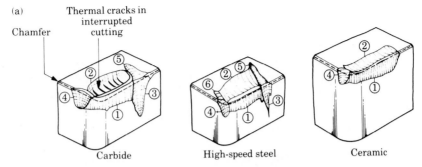

1. Flank wear (wear land)
2. Crater wear
3. Primary groove (outer diameter groove or wear notch)
4. Secondary groove (oxidation wear)
5. Outer metal chip notch
6. Inner chip notch

(b)

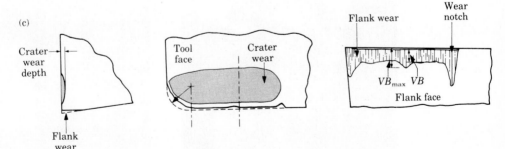

High-speed steel tool, thermal softening and plastic flow

Ceramic tool, chipping and fracture

1. Flank wear
2. Crater wear
3. Failure face
4. Primary groove
5. Outer metal chip notch
6. Plastic flow around failure face

(c)

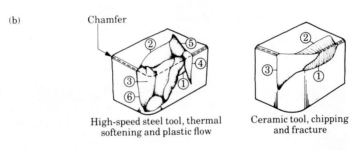

FIGURE 8.22 (a) Types of wear observed in cutting tools. The thermal cracks shown are usually observed in interrupted cutting operations, such as in milling.
(b) Catastrophic failure of tools. (c) Features of tool wear in a turning operation. The symbol VB indicates average flank wear. *Source:* (b) After V. C. Venkatesh. (c) From International Standards Organization, ISO.

The tool profile can be altered by these various wear and fracture processes, and this change is bound to influence the cutting operation. In addition to these wear processes, plastic deformation of the tool can also take place to some extent, such as in softer tools at elevated temperatures. Gross chipping of the tool is called catastrophic failure, whereas wear is generally a gradual process.

Because of the complex interactions between material and process variables, the interpretation of these wear patterns and study of the wear mechanisms must be carried out cautiously.

8.7.1 FLANK WEAR

Flank wear has been studied extensively and is generally attributed to the following:

a. Sliding of the tool along the machined surface causing adhesive and/or abrasive wear, depending on the materials involved; and

b. Temperature, because of its influence on tool material properties.

Following a classic and extensive study by F. W. Taylor, published in 1907, the following relationship was established for cutting various steels:

$$VT^n = C, \qquad\qquad (8.27)$$

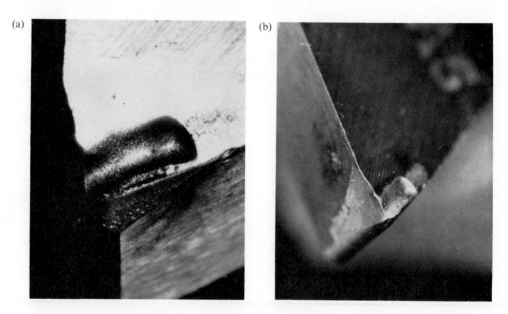

(a)   (b)

FIGURE 8.23   (a) Crater wear on rake face (top surface) and flank wear (side front view) on a carbide insert, used in cutting sintered tungsten on a lathe. (b) Crater wear and chipping (lower left side) on a carbide insert, used in machining tungsten. *Source:* A. J. Moser and S. Kalpakjian.

FIGURE 8.24 Terminology in a turning operation on a lathe, where *f* is the feed (in./revolution or mm/revolution) and *d* is the depth of cut. Note that feed in turning is equivalent to the depth of cut in orthogonal cutting (Fig. 8.1), and depth of cut in turning is equivalent to width of cut in orthogonal cutting. (Compare with Fig. 8.1.)

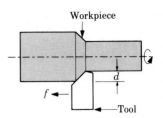

where V is the cutting speed, T is the time that it takes to develop a flank wear land (VB in Fig. 8.22c) of certain dimensions, n is an exponent that depends on cutting conditions, and C is a constant. Equation (8.27) is a simple version of the many relationships that were developed by Taylor among the variables involved.

Each combination of workpiece and tool material, and each cutting condition has its own n value and a different constant C. The range of n values that have been determined experimentally is given in Table 8.4.

Recommended Wear Land

The time T in Eq. (8.27) is called *tool life* and is measured in minutes. The recommended maximum *wear land* (VB) after which the tool has to be changed or resharpened is generally as follows:

a. High-Speed Steel Tools: 0.060 in. (1.5 mm) for turning and face milling, 0.020 in. (0.5 mm) for end milling, 0.015 in. (0.4 mm) for drilling, and 0.006 in. (0.15 mm) for reaming.

b. Carbide Tools, roughing: 0.030 in. (0.8 mm).

c. Carbide Tools, finishing: 0.015 in. (0.4 mm), down to 0.006 in. (0.15 mm) for reaming.

The maximum allowable wear land may be made smaller than those shown above for better accuracy and surface finish.

The recommended cutting speed for a high-speed steel tool is generally the one that gives a tool life of 60 to 120 min, and for carbide tools 30 to 60 min.

Tool Life Curves

Typical *tool life curves* are shown in Fig. 8.25. Note the rapid decrease in tool life with increasing speed, and the strong influence of the condition of the workpiece material.

TABLE 8.4
RANGE OF *n* VALUES FOR EQ. (8.27).

High-speed steels	0.08–0.2
Cast alloys	0.1–0.15
Carbides	0.2–0.5
Ceramics	0.5–0.7

Because of the influence of temperature on the physical and mechanical properties of materials, it is to be expected that wear is strongly influenced by temperature. Experimental investigations have shown that there is indeed a direct relation between flank wear and temperature generated during cutting (Fig. 8.26).

While cutting speed has been found to be the most significant process variable in tool life, depth of cut and feed rate are also important. Thus, Eq. (8.27) can be modified as follows,

$$V T^n d^x f^y = C, \tag{8.28}$$

where d is the depth of cut (Fig. 8.24) and f is the feed rate (inches per revolution or millimeters per revolution).

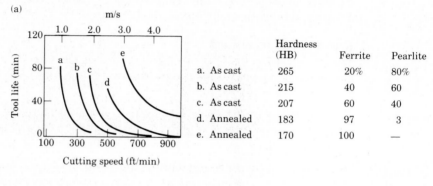

		Hardness (HB)	Ferrite	Pearlite
a.	As cast	265	20%	80%
b.	As cast	215	40	60
c.	As cast	207	60	40
d.	Annealed	183	97	3
e.	Annealed	170	100	—

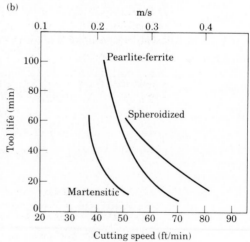

FIGURE 8.25 Effect of workpiece microstructure on tool life in turning. Tool life is given in terms of the time (in minutes) that it takes to reach a flank wear land of specified dimension (Fig. 8.21). (a) Ductile cast iron. (b) Steels, with identical hardness. Note the rapid decrease in tool life as the cutting speed increases.

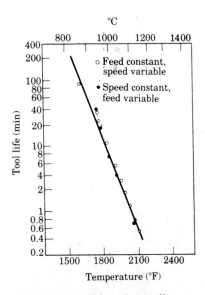

FIGURE 8.26 Relationship between measured temperature during cutting and tool life (flank wear). Note that high cutting temperatures severely reduce tool life. See also Eq. (8.26). *Source:* After H. Takeyama and Y. Murata.

Work material: Heat-resistant alloy
Tool material: Tungsten carbide
Tool life criterion: 0.024 in. (0.6 mm) flank wear

The exponents x and y must be determined experimentally for each cutting condition. Taking, as an example, $n = 0.15$, $x = 0.15$, and $y = 0.6$ as typical values encountered in practice, we can see that, in decreasing order of importance, are cutting speed, feed rate, and depth of cut.

Equation (8.28) can be rewritten as,

$$T = C^{1/n}V^{-1/n}d^{-x/n}f^{-y/n} \tag{8.29}$$

or

$$T \simeq C^7 V^{-7}d^{-1}f^{-4}. \tag{8.30}$$

For a constant tool life, the following observations can be made from Eq. (8.30):

a. If the feed rate or the depth of cut is increased the cutting speed must be decreased, and vice versa, and

b. Depending on the exponents, a reduction in speed can then result in an increase in the volume of the material removed, due to the increased feed rate and/or depth of cut.

The effect of cutting speed on the volume of metal removed, with depth of cut and feed remaining constant, can be seen by analyzing Fig. 8.25(a). First we must remember that the tool has to be changed, or resharpened, after it reaches a particular wear land (i.e., tool life). As an example, cutting at 200 ft/min, for a tool life of 40 min, and looking at material condition (a) in the figure, we note that the tool travels a distance of $40 \times 200 = 8000$ ft. If we now cut at 400 ft/min the tool life is about 6 min, hence the tool travels $6 \times 400 = 2400$ ft.

Because, at a constant feed and depth of cut, the volume of material removed is directly proportional to the distance of tool travel, it is possible that increasing the speed can result in a reduction in the volume of metal removed between tool changes. This observation has important economic significance (Section 13.3.5). For rough cutting operations, the feed should be set at the maximum possible value on the machine tool.

The following observations concerning flank wear may also be made. In obtaining the n value for Eq. (8.27) experimentally, such as those shown in Fig. 8.25, caution should be exercised in using these equations beyond the regions in which they are applicable. On a log–log plot, tool-life curves may be linear over a relatively narrow range of cutting speeds, but are rarely found to be linear over a wide range.

Furthermore, it has been observed experimentally that, at very low cutting speeds, the exponent n can become negative. Thus, when the curves in Fig. 8.25 are extended to the left, they may actually reach a maximum and then curve downward. In other words, it is possible to obtain the same tool life at two different cutting speeds.

The wear *groove* or *notch* on the tools shown in Figs. 8.22(a) and (c) has been attributed to the fact that this region is the boundary where the chip is no longer in contact with the tool. This boundary (also called *depth-of-cut line*) oscillates due to inherent variations in cutting and accelerates the wear process, possibly by fatigue. Furthermore, this grooved region is also in contact with the original surface layer of the workpiece (prior to the cut being taken). Since a machined surface is likely to have a thin oxidized and/or work-hardened layer (Fig. 4.1), this too could contribute to this localized wear.

Scale and oxide layers on a workpiece increase the wear of the tool since these layers are rather abrasive. In such cases, it is desirable to have a depth of cut that is greater than the thickness of the oxide film or the highly cold-worked layer. (See Table 4.1 and Fig. 4.1.)

As can be seen in Fig. 8.25, it is necessary to specify the condition of the material, such as its heat treatment or degree of prior cold work. Note the great difference in tool life for different microstructures. Heat treatment is important; for instance, ferrite has a hardness of about 100 HB, pearlite 200 HB, and martensite 300 to 500 HB. Impurities and hard constituents in the material are also important considerations as they increase tool wear, mostly by abrasion.

● **Illustrative Problem 8.3**

In Table 8.4 we note that the n value ranges from 0.08 to 0.7. To what factors can this difference be attributed?

SOLUTION.   From the Taylor tool-life equation we find that

$$T \propto \frac{1}{V^{1/n}}.$$

Thus n is a measure of the sensitivity of the tool life to the cutting speed. The smaller the value of n, the greater the sensitivity.

Although wear is a complex phenomenon, we know from Chapter 4 that hardness is an important factor in wear. The harder the tool, the lower the adhesive and abrasive-wear rates. We also know that, in wear, the affinity of the mating surfaces to bond and weld is also very important.

Since temperature affects the hardness of the tools (Fig. 8.52), we must also recognize the importance of the temperature rise in metal cutting. In this regard, we should note the values of the coefficient m in Section 8.6 and in Eq. (8.26) for various tool materials. Note also the role of the coefficient of friction at the tool–chip interface. We must also recognize the role of thermal conductivity of various tool materials (see Table 8.7) in dissipating the heat generated.

We may now list and rate three tool materials in the following order: hardness at high temperature/friction and affinity at tool–chip interface/thermal conductivity/m value. In relative terms of low and high, we have:

High-speed steel   low/high/high/high

Tungsten carbide   high/low/high/low

Ceramics   high/low/low/—

Combining these observations, we find that, qualitatively, the order and range for n values should be as shown in Table 8.4.   ●

8.7.2 CRATER WEAR

The most significant factors in *crater* wear are temperature and the degree of chemical affinity between the tool and the workpiece. The factors affecting flank wear also influence crater wear. It can be seen from Figs. 8.15 and 8.18 that the rake face of the tool is subjected to high levels of stress and temperature, in addition to sliding at relatively high speeds. Peak temperatures, for instance, can be on the order of 2000°F (1100°C) (Fig. 8.19b).

It is interesting to note that the location of maximum crater wear generally coincides with the location of the maximum temperature. Experimental evidence indicates that there is a direct relation between crater-wear rate and tool–chip interface temperature (Fig. 8.27). Note the very sharp increase in crater wear after a certain temperature range has been reached.

The cross-section of the tool–chip interface in cutting steel at high speeds is shown in Fig. 8.28. Note the location of the crater-wear pattern and the discoloration of the tool (loss of temper) as a result of high temperatures. Note also how well the discoloration profile agrees with the temperature profile shown in Fig. 8.18.

The effect of temperature on crater wear has been described in terms of a diffusion mechanism (the movement of the atoms across the tool–chip interface). Diffusion depends on the tool–workpiece material combination and on temperature, pressure,

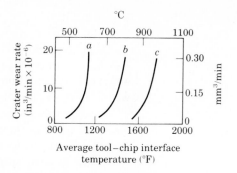

FIGURE 8.27 Relation between crater wear rate and average tool–chip interface temperature in turning. (a) High-speed steel tool. (b) C-1 carbide. (c) C-5 carbide. Note that crater wear increases rapidly within a narrow range of temperature. *Source:* After K. J. Trigger and B. T. Chao.

and time. As these quantities increase, the diffusion rate increases. Unless these factors are favorable, crater wear will not take place by diffusion. Wear will then be due to some other factors, such as those outlined for flank wear.

It should be noted that high temperatures may also cause softening and plastic deformation of the tool due to the decrease in yield strength with temperature. This type of deformation is generally observed in machining high-strength materials.

The various patterns of wear for a carbide tool in cutting austenitic stainless steel are shown in Fig. 8.29. A BUE forms at low speeds. As the cutting speed is increased, crater wear and plastic deformation of the cutting edge of the tool take place due to increased temperature.

Due to the complex interaction of all the parameters involved, the wear behavior of cutting tools cannot be generalized. Each particular set of parameters will result in a specific type of wear on the cutting tool.

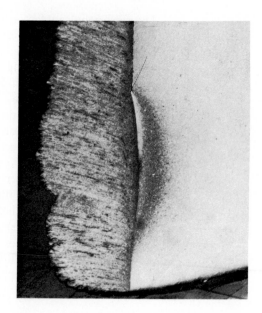

FIGURE 8.28 Interface of chip (left) and rake face of tool (right), and crater wear in cutting AISI 1004 steel at 585 ft/min (3 m/s). The discoloration of the tool is an indication of high temperature (loss of temper). Note how the crater wear pattern coincides with the discoloration pattern. Compare this pattern with the temperature distribution shown in Fig. 8.18. *Source:* Courtesy of P. K. Wright.

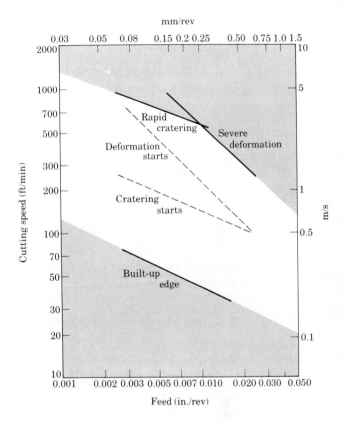

FIGURE 8.29 Plastic deformation and wear of a steel-cutting grade carbide tool as a function of cutting speed and feed. Note that deformation and crater wear take place at higher speeds, indicating higher temperatures. Charts like this are useful in determining the optimum range of feed and speed in cutting operations. *Source:* After P. A. Dearnley and E. M. Trent.

8.7.3 CHIPPING

Chipping is the term used to describe the breaking away of a piece from the cutting edge of the tool. The chipped pieces may be very small (microchipping or macrochipping), or they may involve relatively large fragments (gross chipping or fracture). Unlike wear, which is a more gradual process, chipping is a phenomenon that results in a sudden loss of tool material.

Two main causes of chipping are mechanical shock (impact by interrupted cutting, such as in milling, Fig. 8.37) and thermal fatigue. Chipping may occur in a region in the tool where a small crack or defect already exists. Thermal cracks (Fig. 8.22a) are due to the thermal cycling of the tool in interrupted cutting. Thermal cracks are generally perpendicular to the cutting edge (Fig. 8.22a).

High positive rake angles can contribute to chipping due to the small included angle of the tool tip. It is also possible that crater wear may progress toward the tool tip and weaken it, thus causing chipping. Chipping or fracture can be reduced by selecting tool materials with high impact and thermal-shock resistance. (See Table 8.8.)

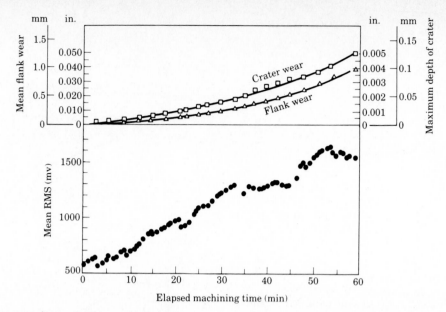

FIGURE 8.30 Relationship between mean flank wear, maximum crater wear, and acoustic emission (noise generated during cutting) as a function of machining time. This technique is being developed as a means for monitoring wear rate in various cutting processes without interrupting the operation. *Source:* After M. S. Lan and D. A. Dornfeld, *Proc. NAMRC-X*, 1982, pp. 305–311.

8.7.4 WEAR-MEASUREMENT TECHNIQUES

Tool-wear measuring and sensing techniques fall into two categories: direct and indirect. The first category involves the direct optical measurement of wear, such as by observing changes in tool profile or workpiece dimensions. Other methods are by observing the rake-face side of the chip for crater-wear particles, and measuring wear by radioactive techniques. (See also Section 4.8.6.)

Indirect methods of wear measurement involve the correlation of wear with process variables such as forces, temperature rise, surface finish and integrity, vibrations, and sound (acoustic) emission.

The most common and reliable technique is the direct observation and optical measurement of wear on the tool. However, this requires that the cutting operation be stopped. Since this involves the interruption of the steady-state nature of the cut, and also influences the economics of the operation, it is desirable to monitor the rate of tool wear on-line. Indirect methods may be used for this purpose, but there are problems involved concerning reliability and calibration. Cutting forces can also be monitored as an indication of tool wear, such as in drilling and tapping (Section 8.8.4).

One recent development is the *acoustic emission* technique, which utilizes a piezoelectric transducer attached to a toolholder. The transducer picks up signals, which are acoustic emissions resulting from the stress waves generated during cutting. Experimental studies have shown that the acoustic emission, as measured by the mean root-mean square (RMS) of the signal, increases with increasing wear (Fig. 8.30). This technique has thus far been used in turning and drilling.

Considerable variability has been observed in wear tests in machining. Reproducibility of wear data is difficult in view of the many factors that affect wear (Section 4.8). Among these factors are metallurgic and other property variations in the materials involved, the machine tool used, application of cutting fluids, environmental conditions, and inaccuracies in controlling process parameters and wear-measurement techniques.

8.8 MECHANICS OF VARIOUS CUTTING OPERATIONS

The majority of cutting operations in practice are three-dimensional. The basic difference between two-dimensional (orthogonal) and three-dimensional (*oblique*) cutting is shown in Fig. 8.31.

In orthogonal cutting, the tool edge is perpendicular to the movement of the tool. An example in practice is the *cutting-off*, or *parting*, process shown in Fig. 8.32(a). This is an operation to separate a piece of metal from a longer piece, or to reduce the diameter of a piece at a certain section along its length.

In oblique cutting, the cutting edge is at an angle i, called the *inclination angle* (Fig. 8.31). The chip flows up the rake face of the tool at an angle α_c (*chip flow angle*), measured in the plane of the tool face. The angle α_n is known as the *normal rake angle* (which is a basic geometric property of the tool) and is the angle between the normal line oz to the workpiece surface and the line oa on the tool face.

The workpiece material approaches the tool at a velocity V, and leaves the surface (as a chip) with a velocity V_c. The *effective rake angle* α_e is calculated in the plane of these two velocities. Assuming that the chip flow angle α_c is equal to the inclination angle i (which is found experimentally to be approximately correct), it can be shown that the effective rake angle α_e is given by,

$$\alpha_e = \sin^{-1}(\sin^2 i + \cos^2 i \sin \alpha_n). \tag{8.31}$$

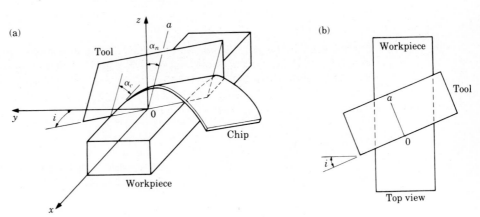

FIGURE 8.31   (a) Schematic illustration of cutting with an oblique tool. (b) Top view showing the inclination angle *i*. Most machining processes involve oblique cutting.

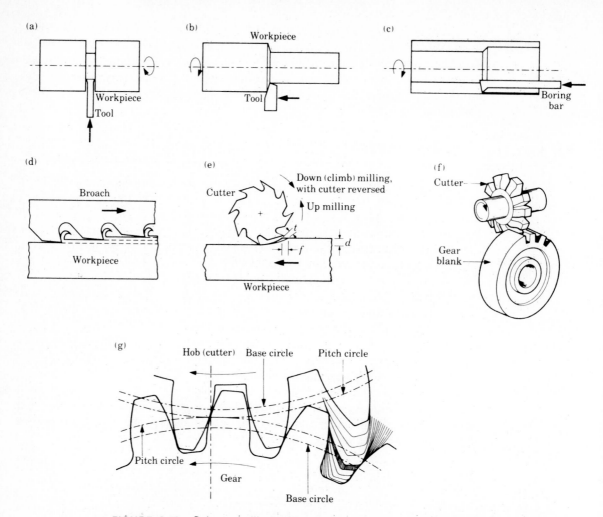

FIGURE 8.32 Schematic illustration of various cutting operations. (a) Cutting off. (b) Turning. (c) Boring. (d) Broaching. (e) Milling. (f) Gear cutting (forming). (g) Gear cutting with a hob (generating).

Since both i and α_n can be measured directly, the effective rake angle can be calculated. As i increases, the effective rake angle increases; hence, the shear strain decreases and the chip becomes thinner and longer.

Although numerous studies have been made regarding the geometry of various oblique cutting operations, many of these analyses are rather lengthy and too complex to be presented here.

Some of the common cutting operations (Fig. 8.32) and the appropriate tool geometries are described below. The machine tools for these processes are described in Section 8.14.

8.8.1 TURNING (CUTTING ON A LATHE)

The most common oblique cutting tool is the *single-point lathe tool* (Fig. 8.33a), commonly made of materials such as tool steels, high-speed steels, and carbides.

Tool Geometry

Note that the tool has two cutting edges (side cutting edge angle, SCEA, and end cutting edge angle, ECEA), and that it has a number of other features equivalent to the rake angle and clearance angle shown in Fig. 8.1. These tools are described by a standardized nomenclature as shown in Fig. 8.33(b).

The effect of tool geometry on chip movement in turning is shown in Fig. 8.34. Note the effect of oblique cutting on chip formation and breakage. By changing the angles on a lathe tool, the chip can be made to flow in various directions.

For each group of workpiece material (ferrous, nonferrous, nonmetallic), there is generally a set of optimal tool angles in turning (Table 8.5) and various other cutting operations. Rake angles are important in controlling the direction of chip flow and in the strength of the tool tip. Whereas positive angles improve the cutting operation, they may, depending on the tool material, cause premature failure due to the small included angle of the tool tip.

Side rake angle is more important than back rake angle, although the latter usually controls the chip-flow direction. Relief angles control the interference and rubbing at the tool–workpiece interface. Other angles in Fig. 8.33 have various effects

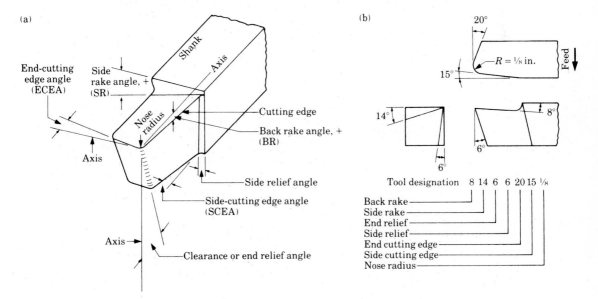

FIGURE 8.33 (a) Terminology for a right-hand cutting tool. Right hand means that the tool travels from right to left, as in normal turning operations on a lathe. (b) Tool designation and symbols.

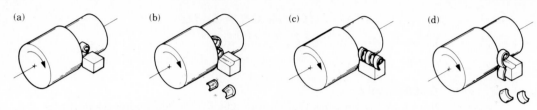

FIGURE 8.34 Types of chips produced in turning with various tool geometries.
(a) Chip is tightly curled against the surface being machined. (b) Chip hits the
workpiece and breaks. (c) Continuous chip moving away from the workpiece. The
direction of movement can be controlled by varying the tool angles. (d) Chip hits the
tool shank and breaks. Long continuous chips are generally not desirable because they
tend to interfere with the cutting operation by getting entangled around the tool post.
Source: After G. Boothroyd.

on chip formation and tool strength. Nose radius affects surface finish and the
strength of the tool tip. However, large nose radii can lead to tool chatter.

Tool Inserts

Cutting tools made of carbides and ceramics are manufactured as *inserts* (Fig. 8.35a).
They are available in a large variety of materials and shapes. These tools are called
indexable inserts because they have multiple cutting edges. (These tools were previously called *throw-away* inserts as they generally are not reground or reused.)
Thus, a square insert has eight and a triangular one has six cutting edges.

TABLE 8.5
GENERAL RECOMMENDATIONS FOR TURNING TOOLS

MATERIAL	HIGH-SPEED STEEL AND CAST ALLOY TOOLS					CARBIDE TOOLS				
	BACK RAKE	SIDE RAKE	END RELIEF	SIDE RELIEF	SIDE AND END CUTTING EDGE	BACK RAKE	SIDE RAKE	END RELIEF	SIDE RELIEF	SIDE AND END CUTTING EDGE
Aluminum and magnesium alloys	20	15	12	10	5	0	5	5	5	15
Copper alloys	5	10	8	8	5	0	5	5	5	15
Steels	10	12	5	5	15	−5	−5	5	5	15
Stainless steels	5	8–10	5	5	15	−5–0	−5–5	5	5	15
High-temperature alloys	0	10	5	5	15	5	0	5	5	45
Refractory alloys	0	20	5	5	5	—	—	5	5	15
Titanium alloys	0	5	5	5	15	−5	−5	5	5	5
Cast irons	5	10	5	5	15	−5	−5	5	5	15
Thermoplastics	0	0	20–30	15–20	10	0	0	20–30	15–20	10
Thermosets	0	0	20–30	15–20	10	0	15	5	5	15

Inserts are also provided with *chip breakers*; they have a standard terminology and standard dimensions for identifying the type of chip-breaker geometry (Fig. 8.35c). Inserts are used for turning, as well as for other cutting operations, as described below.

Boring

Boring operations (Fig. 8.32c) use tools similar to those used in turning. However, because a long boring bar must be used on the inside surfaces and must reach the full length of the workpiece, tool deflections can be a significant problem. The boring bar must be sufficiently stiff (such as by using materials with high elastic modulus, e.g., carbides) to avoid vibrations and chatter, which would lead to poor surface finish and cause premature tool failure. (See Section 8.13.)

Modifications of the basic boring process include *gun drilling* and *trepanning*. In gun drilling, the thrust force of the boring tool tip is balanced by bearing pads sliding

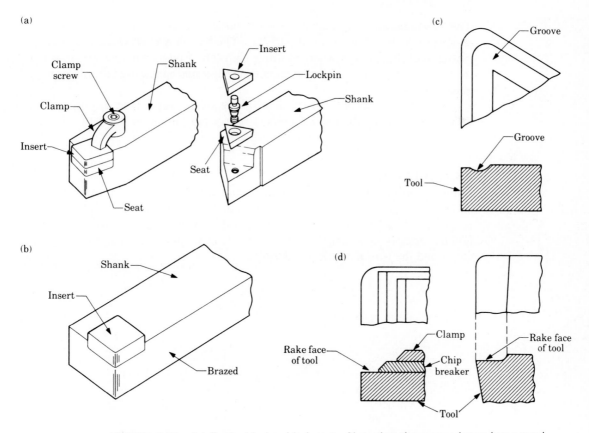

FIGURE 8.35 (a) Typical indexable inserts. Note that the square insert has a total of eight cutting edges, and the triangular insert has six. (b) Insert brazed on a tool shank. (c) Groove-type chip breaker on an insert. (d) Obstruction type chip breaker.

along the inside surface of the hole. In the *trepanning* technique, an annular ring of material is removed leaving the core intact, a process similar to cutting a round hole in wood with a hole saw.

Other Operations

In addition to turning straight, conical, or curved pieces and boring, other operations that can be carried out on a lathe are *drilling* and *reaming* (Section 8.8.4), *parting* or *cutting off* (Fig. 8.32a), *facing* and *knurling*. In facing, a flat surface is produced at the end of the part, such as a bar. Knurling is an operation where a regularly shaped roughness is generated on a cylindrical surface. A typical example is metal knobs. Knurling is done with hardened rollers (with their surface being a replica of the profile to be generated), which are pressed against the rotating workpiece. The surface roughness is produced by a forming, rather than cutting, process.

8.8.2 BROACHING

This is an operation for machining large surfaces (such as cast-iron engine blocks), and grooves on outside or inside surfaces. The *broach* is in effect a linear multitooth cutting tool (Fig. 8.32d), where multiple chips are produced in one stroke. Broaches can also be made round with circular cutting edges and can be used to enlarge holes by machining.

Note that the teeth of a broach are successively deeper (Fig. 8.32d), so that the total depth of material removed in one stroke is the sum of the depth of cut of all the teeth. The terminology for a broach is given in Fig. 8.36.

8.8.3 MILLING

Milling is an important machining operation with a variety of configurations (Fig. 8.37). A *milling cutter* is a multitooth tool and produces a number of chips in one revolution. Unlike other processes described thus far, the thickness of the chip in milling varies along its length because of the relative longitudinal motion between the cutter and the workpiece (Fig. 8.32e).

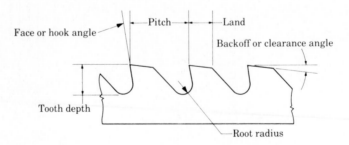

FIGURE 8.36   Terminology for a broach. (See also Fig. 8.32d.)

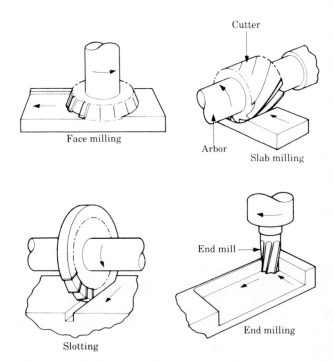

FIGURE 8.37 Various milling operations. Milling is one of the most versatile cutting processes.

Face milling

Slab milling

Cutter

Arbor

Slotting

End mill

End milling

This is described as the *chip depth of cut* or *undeformed chip thickness t* and is given by,

$$t \simeq 2f\sqrt{\frac{d}{D}}, \tag{8.32}$$

where f is the feed per tooth, d is the depth of cut, and D is the cutter diameter.

An important consideration in milling is the direction of cutter rotation. In *up* milling, the maximum thickness of the chip is at the end of the cut, and in *down* (*climb*) milling, it is at the beginning of the cut.

Thus, cutting in climb milling starts with the chip at its thickest location. Because of the resulting high contact forces, it is necessary to have a rigid setup and to eliminate backlash in the system. Climb milling has the advantage of holding the workpiece in place due to the downward force (particularly with thin workpieces), and has a smoother operation with less chatter. However, it is not suitable in machining workpieces with hard scale, such as hot-worked metals and castings, because the abrasive surface scale damages the teeth of the cutter during initial engagement.

In up milling, the workpiece has a tendency to be pulled upward. The advantages are that tooth engagement is not a function of surface variations, and contamination or scale on the surface of the workpiece does not affect tool life.

The spiral teeth on some of the cutters in Fig. 8.37 are for the purpose of reducing the contact load resulting from the sudden engagement of the full length of the cutting

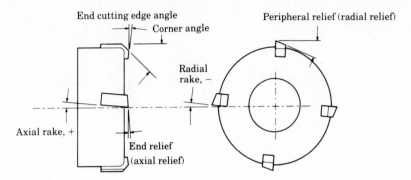

FIGURE 8.38   Terminology for a face milling cutter. A great variety of milling cutters is available with many cutting teeth.

edge. This makes the operation smoother and reduces the contact load. Another arrangement in a face-milling cutter using inserts is shown in Fig. 8.38, which also includes the appropriate terminology for these cutters.

8.8.4   DRILLING, REAMING, AND TAPPING

Hole-making processes are among the most important operations in manufacturing. Two basic processes of hole generating are punching (Section 7.2) and drilling.

An ordinary, standard (*chisel point*) *twist drill*, shown in Fig. 8.39, is characterized by a geometry in which the normal rake angle and the velocity of the cutting edge are a function of their distance from the center of the drill.

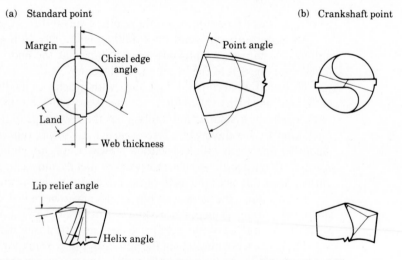

FIGURE 8.39   Terminology for standard point and crankshaft point drills.

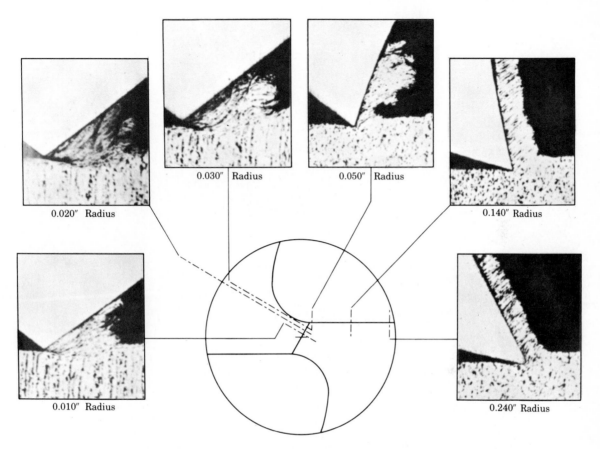

FIGURE 8.40 Cutting conditions at various radial positions in drilling. Note that the drill has an increasingly positive rake angle away from the center. This composite photograph is obtained by a quick-stop mechanism. The drill is removed and the workpiece is sectioned at different locations, polished, and etched. *Source:* Courtesy of A. W. Haggerty, Cincinnati Milacron, Inc.

Figure 8.40 shows the cross-sections of the drill–workpiece interface at different radial distances. Note that near the center, the rake angle is highly negative, but toward the outer radius of the drill, it becomes more and more positive, producing a smooth continuous chip (Fig. 8.41).

The material at the center of the hole is subjected to a deformation by wedge displacement and produces a rather coarse chip. This is a region where the cutting speed is low, the rake angle is highly negative and, hence, the shear angle is very low. These factors contribute to an increase of the thrust force (along the axis of the drill), and may cause the drill to "walk" on the workpiece surface because the drill does not have a centering action.

FIGURE 8.41 Chip formation with a standard chisel-point drill. Compare with Fig. 8.40. *Source:* Courtesy of A. W. Haggerty, Cincinnati Milacron, Inc.

To start a hole properly, the drill can be guided either by fixtures (such as a bushing), or by making a small starting hole with a center drill (Fig. 8.42). Various other drilling operations are also shown in this figure. General recommendations for drill geometry are given in Table 8.6.

Reaming is an operation used to obtain a more dimensionally accurate hole after the hole is first made by a drill or another process. A reamer is similar to a long milling cutter (end mill) (Figs. 8.42 and 8.43).

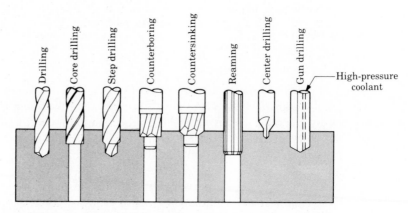

FIGURE 8.42 Various types of drills and drilling operations. Hole-making operations are an important aspect of manufacturing. Holes are made generally for purposes of assembly of various components.

TABLE 8.6
**GENERAL RECOMMENDATIONS FOR DRILL GEOMETRY FOR
HIGH-SPEED STEEL TWIST DRILLS**

MATERIAL	POINT ANGLE	LIP RELIEF ANGLE	CHISEL EDGE ANGLE	HELIX ANGLE	POINT
Aluminum alloys	90–118	12–15	125–135	24–48	Standard
Magnesium alloys	70–118	12–15	120–135	30–45	Standard
Copper alloys	118	12–15	125–135	10–30	Standard
Steels	118	10–15	125–135	24–32	Standard
High-strength steels	118–135	7–10	125–135	24–32	Crankshaft
Stainless steels, low strength	118	10–12	125–135	24–32	Standard
Stainless steels, high strength	118–135	7–10	120–130	24–32	Crankshaft
High-temp. alloys	118–135	9–12	125–135	15–30	Crankshaft
Refractory alloys	118	7–10	125–135	24–32	Standard
Titanium alloys	118–135	7–10	125–135	15–32	Crankshaft
Cast irons	118	8–12	125–135	24–32	Standard
Plastics	60–90	7	120–135	29	Standard

Tapping generates internal threads in a hole with the use of a tap, which is basically a threading tool (Fig. 8.44).

Tool-Life Measurement

Whereas tool life in turning and milling is generally measured by the time it takes to reach a wear land of specified dimensions, the tool life of drills and taps is usually measured by the number of holes drilled or tapped.

The procedure consists of drilling or tapping a number of holes and recording the torque or thrust force. After a number of holes are drilled or tapped, the forces

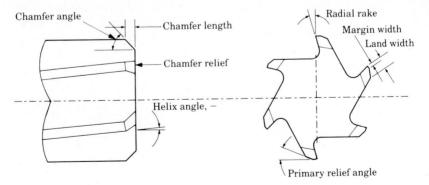

FIGURE 8.43 Terminology for a reamer. Reamers are generally used for enlarging or finishing holes to accurate dimensions.

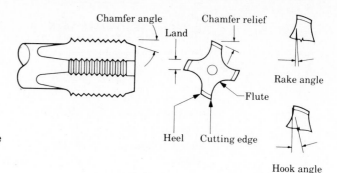

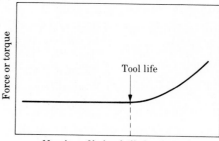

FIGURE 8.44 Terminology for a tap. Taps are used for threading holes. Large holes are threaded on a lathe or similar machine tools.

FIGURE 8.45 Determination of drill (or tap) life by observing the rise in force, or torque, as a function of the number of holes drilled (or tapped). Drill life may also be determined by observing acoustic emissions or vibrations during drilling (see also Fig. 8.30).

begin to increase due to wear (Fig. 8.45). The life of the tool is then specified as the number of holes drilled or tapped when this transition begins to take place. Acoustic-emission techniques and recording vibrations and analyzing the signature are also recent methods used in monitoring wear in drilling. (See also Section 8.7.4.)

8.8.5 FORMING AND GENERATING

A machined shape can be produced either by forming or by generating.

In the *forming* process, the shape of the cutting tool is reproduced on the workpiece. In *generating*, the shape that is produced on the workpiece depends not only on the shape of the cutting tool but also on the relative motion between the tool and the workpiece.

The cutting of gears is a good example of these different methods. In form cutting of gears, the cutter has the shape of the space between the teeth (Fig. 8.32f). Thus, the cutter is basically a milling cutter, such as those shown in Fig. 8.37. The precision of the tooth profile depends on the accuracy of the cutter. Also, form cutting can only be done on gear teeth that have constant width, such as spur or helical gears, but not on bevel gears.

In *gear generating*, the generating tool (cutter) can be considered as one of the gears in a conjugate pair, and the other as the gear blank (Fig. 8.32g). The required relative motion between the shafts of the cutter and the gear blank are obtained by means of a train of gears. The cutter gear reciprocates and is fed radially into the gear blank on a machine that is basically a shaper (Section 8.14.4). Most high-quality gears are made by the generating process.

Shaping of spur and helical gears may be done with a *hob*, which is basically a helical cutter. Both the gear blank and the hob rotate during the cut.

Gears may also be *shaved*, whereby the cutter removes small amounts of metal from the gear teeth. This method produces gear teeth with improved surface finish and profile accuracy. For greater accuracy, the gear teeth may also be ground and honed.

Bevel gears are made on special machines, usually with two half-tooth cutters which cut the inner surfaces of two adjacent teeth. Gears may also be shaped by *rolling*, as described in Section 6.14.3.

8.8.6 SAWING AND FILING

Sawing is a cutting operation where the tool is a series of small teeth on the saw. The width of cut is narrow and the process is used on all materials that are machinable by other cutting processes.

Typical saw teeth and saw-blade designs are shown in Fig. 8.46. In order to prevent the saw from binding and rubbing during cutting, the teeth are offset so that the width of cut (*kerf*, see also Fig. 12.43) is greater than the width of the blade. At least two or three teeth should always be in contact with the workpiece in order to prevent snagging. Thus, the thinner the stock being sawed, the finer the teeth of the saw should be, i.e., the greater the number of teeth per unit length of the saw.

The basic types of saws are *hack* saws (straight blades), *circular* saws (called *cold* saws for cutting metal), and *band* saws (long, flexible blades). Saws are made from a variety of steels. Carbide-tipped steel blades are available, as are diamond-edged blades. Saws are made in a variety of tooth form and spacing, blade thickness, width, and sizes. Sawing can also be done using thin abrasive wheels, or by electrical and chemical removal processes, as described in Chapter 9.

Friction sawing is a process where a disk or blade rubs against the workpiece at speeds up to 25,000 ft/min (125 m/s). The frictional energy is converted into heat, which then rapidly softens a narrow zone on the workpiece. The action of the disk or

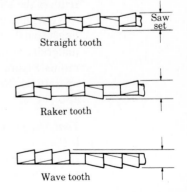

FIGURE 8.46 Various types of saw teeth. The teeth are staggered to provide clearance for the saw blade, otherwise it could bind and prevent free movement of the saw. Blades may be straight or circular. For sawing hard and abrasive materials, blade teeth may be made of carbide inserts brazed on the saw blade. Blades may also be coated with synthetic diamond particles for cutting hard and brittle materials, such as glass and stones.

Straight tooth

Saw set

Raker tooth

Wave tooth

blade (which are sometimes provided with teeth or notches) pulls the softened metal out of the cutting zone. The friction-sawing process is suitable for a variety of ferrous metals, but nonferrous metals tend to adhere to the blade.

Filing is small-scale removal of material from a surface. Files are made of hardened steels and are available in a variety of cross-sections (flat, round, half round, square, triangular, etc.) and with many tooth forms.

Rotary files and *burrs* are also available. These are cutters in the shape of cones, cylinders, spheres, and other shapes, with various tooth profiles. Their action is similar to that of reamers, except that they remove small amounts of material at very high speeds, with small-diameter burrs running at 45,000 rpm. Burrs are also used in cleaning and beveling the edges of sheet metal prior to resistance welding to make containers for food, beverage, and household products (Fig. 12.5c).

Although sawing and filing may be done by hand, various sawing and filing machines are available with automatic features for large-quantity production.

8.9 SURFACE FINISH AND INTEGRITY

Because machining is basically a finishing process with specified dimensions, tolerances, and surface finish, the type of surface that a machining operation generates and its characteristics are of great importance in manufacturing. As described in Chapter 4, a surface is not only a geometric entity but also a layer with its own properties and structure. Various terms and techniques pertaining to surface roughness, waviness, etc., their measurement, and effects on properties are described in Sections 4.3 and 4.4.

The factors that have an influence on surface finish and integrity in cutting processes are described below:

Built-Up Edge (BUE)

The presence of a BUE greatly influences surface finish because it changes the tool profile. The beneficial effect of higher cutting speeds in reducing or eliminating the BUE should be recognized. Other factors concerning material and process variables also have an effect on surface finish. Reducing friction at the tool–chip interface also reduces the tendency for BUE formation.

Ceramic and diamond tools (see Section 8.10) generally produce better surface finish, partly because of their much lower tendency to form BUE than other tool materials. Some of the effects of the BUE are similar to those described below for nose radius. Figure 8.47 shows some examples of various surface finishes obtained in cutting operations. Note the considerable damage that some surfaces have undergone. See also Fig. 8.9.

Tool Sharpness

If the tool is excessively dull compared to the depth of cut (Fig. 8.16), it will cause rubbing over the machined surface and generate heat and residual stresses (Fig. 8.48). These can cause surface damage, such as by tearing and cracking. Surfaces with

(a) 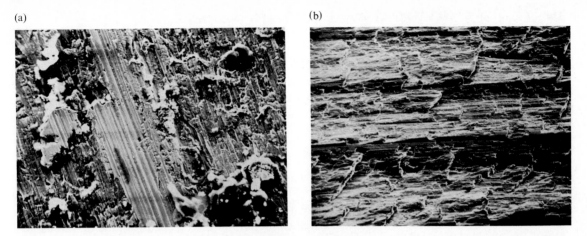 (b)

FIGURE 8.47 Typical surfaces produced by various cutting operations, as observed with a scanning electron microscope. (a) Surface of steel produced by turning. (b) Surface of steel produced by shaping (Fig. 8.69). See also Fig. 8.9. *Source:* Courtesy of J. T. Black and S. Ramalingam.

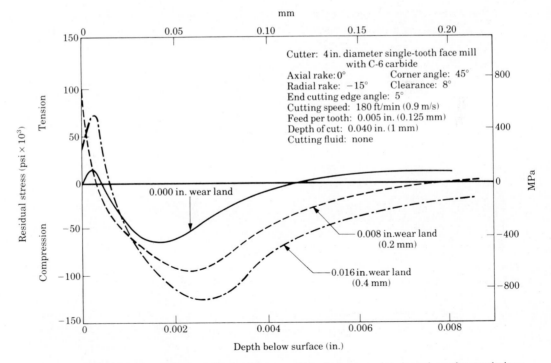

FIGURE 8.48 Effect of flank wear on residual stresses developed on the workpiece surface in milling AISI 4340 steel, 52 HRC. Note that the maximum compressive stresses occur at very small depths below the surface of the workpiece. See also Figs. 4.38, 6.93, and 9.10. *Source:* After W. P. Koster and J. B. Kohls, SME Paper IQ72–207, 1972.

cracks, tears, and tensile residual stresses have lower fatigue strength and corrosion resistance. Depending on the operating conditions, a dull tool (equivalent to a large nose radius in Fig. 8.16) can work harden a layer of the surface considerably. If the temperature is high enough, phase transformations can also take place, thus affecting surface integrity.

Nose Radius and Feed Marks

As the tool moves across the workpiece, it leaves a spiral profile on the machined surface, known as feed marks (Figs. 8.49 and 8.50). The higher the feed rate f, the rougher is the surface. Likewise, the smaller the nose radius, the rougher the surface is. (The nose radius in these figures should not be confused with the tip radius, that

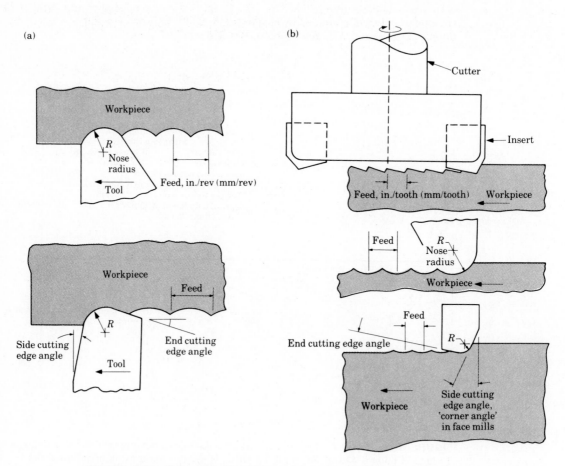

FIGURE 8.49 Schematic illustration of theoretical surfaces produced by (a) turning and (b) face milling tools.

(a)

(b)

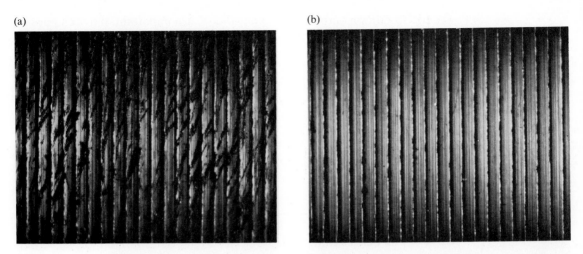

FIGURE 8.50 Feed marks in turning sintered tungsten. (a) Carbide insert. Roughness along the feed marks is due to built-up edge. See also Fig. 8.6(c). (b) Ceramic insert. Note that the surface is smoother than in (a). *Source:* A. J. Moser and S. Kalpakjian.

is dullness, in Fig. 8.16.) A general guideline is to make the nose radius at least three times the feed rate. Note that the depth of cut has no effect on the surface finish. These are purely geometric considerations and do not include other surface-altering effects by BUE or dullness of the tool.

In some operations it is possible to reduce roughness due to feed marks. One such example is given in Fig. 8.51 for face milling. An adjustable wiper blade removes the roughness generated by the cutting tool.

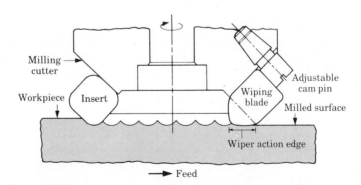

FIGURE 8.51 Wiper blade (micro-finish blade) for smoothing feed marks in face milling. *Source:* Adapted from *Machinability Data Handbook,* 3d ed., Metcut Research Associates Inc.

The edge condition of the cutting of the tool is also important. If it is chipped or uneven (Figs. 8.22 and 8.23b), the surface finish will be influenced adversely.

Vibration and Chatter

Although this subject is treated in some detail in Section 8.13, it should be recognized that, if the tool vibrates or chatters during the cut, it will affect the surface finish. This is due to the fact that a tool, vibrating at a certain amplitude, changes the dimensions of the cut. Excessive chatter can also cause chipping and premature failure of the more brittle cutting tools, such as ceramics and diamond.

8.10 CUTTING-TOOL MATERIALS

We have seen that a cutting tool is subjected to static and variable forces, varying temperatures and speeds, and wear. Accordingly, the major requirements for a cutting tool material are the following:

a. *Hardness*, especially hot hardness, so that the hardness and strength of the tool are maintained at the temperatures encountered in cutting operations;

b. *Toughness*, so that the sudden loading of the tool, as in interrupted cutting operations, does not chip or fracture the tool;

c. *Wear resistance*, so that the tool lasts a reasonably long time before it needs to be resharpened or changed; and

d. *Chemical stability* (or inertness) with respect to the workpiece material.

The selection of a proper tool material depends on a number of conditions: the type of cutting operation, whether it is a *roughing* or *finishing* cut; the workpiece material, its state, shape, and surface condition; and the stiffness of the machine tool and its power (Tables 8.7 through 8.11).

A wide variety of cutting-tool materials is now available. These are divided into the following general categories: carbon tool steels, high-speed steels, cast alloys, cemented carbides, ceramics, cubic boron nitride, and diamond (Table 8.7.)

8.10.1 CARBON TOOL STEELS

This is the oldest type of steel for cutting metals (now rarely used). They are relatively inexpensive and are shock resistant (Table 8.8). These steels can be heat-treated to obtain a wide range of hardness, are easily formed and ground to shape, and they maintain a sharp cutting edge if excessive abrasion and high temperatures are avoided. Alloying elements can be introduced into these steels to improve hardenability with less distortion; manganese is the major alloying element (Section 3.12).

Carbon tool steels were used widely for drills, taps, broaches, and reamers; however, they do not have sufficient hot hardness to be used for high-speed–machining

TABLE 8.7
TYPICAL PROPERTIES OF TOOL MATERIALS

PROPERTY	HIGH-SPEED STEELS	CAST ALLOYS	CEMENTED CARBIDES		CERAMICS	CUBIC BORON NITRIDE	DIAMOND
			WC	TiC			
Hardness	83–86 HRA	82–84 HRA 46–62 HRC	90–95 HRA 1800–2400 HK	91–93 HRA 1800–3200 HK	91–95 HRA 2000–3000 HK	4000–5000 HK	7000–8000 HK
Compressive strength psi × 10³ MPa	600–650 4100–4500	220–335 1500–2300	600–850 4100–5850	450–560 3100–3850	400–650 2750–4500	1000 6900	1000 6900
Transverse rupture strength psi ×10³ MPa	350–700 2400–4800	200–300 1380–2050	150–375 1050–2600	200–275 1380–1900	50–135 345–950	105 700	200 1350
Impact strength in.-lb J	12–70 1.35–8	3–11 0.34–1.25	3–12 0.34–1.35	7–11 0.79–1.24	<1 <0.1	— —	— —
Modulus of elasticity psi ×10⁶ GPa	30 200	— —	75–100 520–690	45–65 310–450	45–60 310–410	125 850	120–150 820–1050
Density g/cm³ lb/in³	8.6 0.31	8–8.7 0.29–0.31	10–15 0.36–0.54	5.5–5.8 0.2–0.22	4–4.5 0.14–0.16	3.48 0.13	3.5 0.13
Volume of hard phase, %	7–15	10–20	70–90	—	100	95	95
Melting or decomposition temperature °F °C	2370 1300	— —	2550 1400	2550 1400	3600 2000	2400 1300	1300 700
Thermal conductivity, W/m.°C	—	—	42–125	17	29	13	—
Coefficient of thermal expansion, ×10⁻⁶/°C	12	—	4–6.5	7.5–9	6–7.5	4.8	—

TABLE 8.8
GENERAL CHARACTERISTICS OF CUTTING-TOOL MATERIALS. THESE TOOL MATERIALS HAVE A WIDE RANGE OF COMPOSITIONS AND PROPERTIES, THUS OVERLAPPING CHARACTERISTICS EXIST IN MANY CATEGORIES OF TOOL MATERIALS. (Komanduri R. *Kirk-Othmer Encyclopedia of Chemical Technology*, 3d ed. New York: Wiley, 1978.)

	CARBON AND LOW/MEDIUM ALLOY STEELS	HIGH SPEED STEELS	CAST COBALT ALLOYS	CEMENTED CARBIDES	COATED CARBIDES	CERAMICS	POLYCRYSTALLINE CUBIC BORON NITRIDE	DIAMOND
Hot hardness			increasing →					
Toughness			increasing →					
Impact strength			increasing →					
Wear resistance			increasing →					
Chipping resistance			increasing →					
Cutting speed			increasing →					
Depth of cut	light to medium	light to heavy	light to heavy	light to heavy	light to heavy	light to heavy	light to heavy	very light for single crystal diamond
Finish obtainable	rough	rough	rough	good	good	very good	very good	excellent
Method of processing	wrought	wrought, cast, HIP* sintering	cast and HIP sintering	cold pressing and sintering	CVD†	cold pressing and sintering or HIP sintering	high pressure—high temperature sintering	high pressure—high temperature sintering
Fabrication	machining and grinding	machining and grinding	grinding	grinding		grinding	grinding and polishing	grinding and polishing
Thermal shock resistance	increasing →							
Tool material cost	increasing →							

* Hot isostatic pressing.

† Chemical vapor deposition.

TABLE 8.9
OPERATING CHARACTERISTICS OF CUTTING TOOL MATERIALS. (Komanduri, R. *Kirk-Othmer Encyclopedia of Chemical Technology.* 3d ed. New York: Wiley, 1978.)

TOOL MATERIALS	MACHINING OPERATION AND (CUTTING SPEED RANGE)	MODES OF TOOL WEAR OR FAILURE	LIMITATIONS
Carbon steels	Tapping, drilling, reaming (low speed)	Buildup, plastic deformation, abrasive wear, microchipping	Low hot hardness, limited hardenability, and limited wear resistance
Low/medium alloy steels	Tapping, drilling, reaming (low speed)	Buildup, plastic deformation, abrasive wear, microchipping	Low hot hardness, limited hardenability, and limited wear resistance
High-speed steels	Turning, drilling milling, broaching (medium speed)	Flank wear, crater wear	Low hot hardness, limited hardenability, and limited wear resistance
Cemented carbides	Turning, drilling, milling, broaching (medium speed)	Flank wear, crater wear	Cannot use at low speed due to cold welding of chips and microchipping
Coated carbides	Turning (medium to high speed)	Flank wear, crater wear	Cannot use at low speed due to cold welding of chips and microchipping
Ceramics	Turning (high speed to very high speed)	Depth-of-cut line notching, microchipping, gross fracture	Low strength, low thermomechanical fatigue strength
Cubic boron nitride	Turning, milling (medium to high speed)	Depth-of-cut line notching, chipping, oxidation, graphitization	Low strength, low chemical stability at higher temperature
Diamond	Turning, milling (high to very high speed)	Chipping, oxidation, graphitization	Low strength, low chemical stability at higher temperature

TABLE 8.10
GENERAL GUIDELINES FOR TOOL MATERIAL SELECTION FOR MACHINING OPERATIONS.
(After V. A. Tipnis, in *Wear Control Handbook, ASME.* 1980: 891.)

MACHINING OPERATION	WORKPIECE	TOOL MATERIAL							CONVENTIONAL CUTTING SPEED RANGES (ft/min)
		HIGH-SPEED STEEL	CARBIDE	CAST ALLOY	COATED CARBIDE	CERAMIC, CERMET	CUBIC BORON NITRIDE	DIAMOND	
Turning	Cast irons, Carbon steels, Alloy steels, alloy cast irons	×	×	×	×	×	×		Low to very high. High-temperature alloys: 10—20 Aluminum and magnesium: up to 2500 Thermoplastics: 250—400 Thermosets: 400—1000
	Aluminum, brass		×	×	×			×	
	Nickel-base, titanium	×	×	×	×	×			
	High-silicon aluminum				×			×	
	Plastics, composites		×	×				×	
Milling	Cast irons, carbon steels	×		×					Low to medium for end milling. Medium to very high for face milling
	Alloy steels, alloy cast irons		×	×					
	Aluminum, brass	×	×	×					
	Nickel-base, titanium		×	×					
	High-silicon aluminum, Plastics, composites		×						

TABLE 8.10 (Continued)

MACHINING OPERATION	WORKPIECE	TOOL MATERIAL							CONVENTIONAL CUTTING SPEED RANGES (ft/min)
		HIGH-SPEED STEEL	CARBIDE	CAST ALLOY	COATED CARBIDE	CERAMIC, CERMET	CUBIC BORON NITRIDE	DIAMOND	
Drilling, Reaming, Tapping	Cast irons, carbon steels	x	x						Low to medium. High-temperature alloys: 15–20 Aluminum and magnesium: 350 Plastics: 150–400
	Alloy steels, alloy cast irons	x	x						
	Aluminum, brass	x	x						
	Nickel-base, titanium	x	x						
	High-silicon aluminum	x	x						
	Plastics, composites	x	x						
Gear Cutting, Hobbing Broaching	Cast irons, carbon steels	x	x	x					Low to medium
	Alloy steels, alloy cast irons	x	x						
	Aluminum, brass	x							
	Nickel-base, titanium	x	x						
	High-silicon aluminum	x							
	Plastics, composites	x							

Note: Low speed: < 100 ft/min (<0.5 m/s)
Medium speed: 100–500 ft/min (0.5–2.5 m/s)
High speed: 500–1000 ft/min (2.5–5 m/s)
Very high speed: >1000 ft/min (>5 m/s)

TABLE 8.11
ISO CLASSIFICATION OF CEMENTED CARBIDE TOOLS ACCORDING TO USE

MAIN GROUPS OF CHIP REMOVAL			GROUPS OF APPLICATION			DIRECTION OF INCREASE IN CHARACTERISTIC	
SYMBOL	CATEGORIES OF MATERIALS TO BE MACHINED	DISTIN-GUISHING COLOR	TOOL DESIGNA-TION	WORKPIECE MATERIAL	CONDITIONS	OF CUT	OF CARBIDE
P	Ferrous metals with long chips	BLUE	P 01	Steel, steel castings	Finish turning and boring, high cutting speeds, small chip section, accuracy of dimensions and fine finish, vibration-free operation	Increasing speed / Increasing feed	Wear resistance / Toughness
			P 10	Steel, steel castings	Turning, copying, threading and milling, high cutting speeds, small or medium chip sections		
			P 20	Steel, steel castings Malleable cast iron with long chips	Turning, copying, milling, medium cutting speeds and chip sections, planing with small chip sections		
			P 30	Steel, steel castings Malleable cast iron with long chips	Turning, milling, planing, medium or low cutting speeds, medium or large chip sections, and machining in unfavorable conditions*		
			P 40	Steel Steel castings with sand inclusion and cavities	Turning, planing, slotting, low cutting speeds, large chip sections, with the possibility of large cutting angles for machining in unfavorable conditions* and work on automatic machines		
			P 50	Steel Steel castings of medium or low tensile strength, with sand inclusion and cavities	For operations demanding very tough carbides, turning, planing, slotting, low cutting speeds, large chip sections, with the possibility of large cutting angles for machining in unfavorable conditions* and work on automatic machines		

M	Ferrous metals with long or short chips and non-ferrous metals	YELLOW		M 10	Steel, steel castings, manganese steel Gray cast iron, alloy cast iron	Turning, medium or high cutting speeds. Small or medium chip sections	↑ Increasing speed — Increasing feed	Wear resistance — Toughness	
				M 20	Steel, steel castings, austenitic or manganese steel, gray cast iron	Turning, milling. Medium cutting speeds and chip sections			
				M 30	Steel, steel castings, austenitic steel, gray cast iron, high temperature resistant alloys	Turning, milling, planing. Medium cutting speeds, medium or large chip sections			
				M 40	Mild free cutting steel, low tensile steel Non-ferrous metals and light alloys	Turning, parting off, particularly on automatic machines	↓	↓	
K	Ferrous metals with short chips, non-ferrous metals and non-metallic materials	RED		K 01	Very hard gray cast iron, chilled castings of over 85 Shore, high silicon aluminum alloys, hardened steel, highly abrasive plastics, hard cardboard, ceramics	Turning, finish turning, boring, milling, scraping	↑ Increasing speed — Increasing feed	Wear resistance — Toughness	
				K 10	Gray cast iron over 220 HB, malleable cast iron with short chips, hardened steel, silicon aluminum alloys, copper alloys, plastics, glass, hard rubber, hard cardboard, porcelain, stone	Turning, milling, drilling, boring, broaching, scraping			
				K 20	Gray cast iron up to 220 HB, non-ferrous metals— copper, brass, aluminum	Turning, milling, planing, boring, broaching, demanding very tough carbide			
				K 30	Low hardness gray cast iron, low tensile steel, com-pressed wood	Turning, milling, planing, slotting, for machining in unfavorable conditions* and with the possibility of large cutting angles			
				K 40	Soft wood or hard wood Non-ferrous metals	Turning, milling, planing, slotting, for machining in unfavorable conditions* and with the possibility of large cutting angles	↓	↓	

* Raw material or components in shapes which are awkward to machine, casting or forging skins, variable hardness etc., variable depth of cut, interrupted cut, work subject to vibrations.

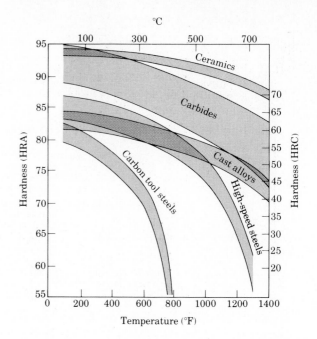

FIGURE 8.52 Hardness of various tool materials as a function of temperature. Hot hardness (hardness at elevated temperatures) is an important property of cutting tools and die materials. The wide range within each curve is due to the variety of compositions available for that group of tool material.

operations, where there is a substantial rise in temperature in the cutting zone. Note, for instance, how rapidly the hardness of these steels decreases with increasing temperature (Fig. 8.52).

8.10.2 HIGH-SPEED STEELS

These are the most highly alloyed tool steels and maintain their hardness, strength, and sharp cutting edge at high operating temperatures (consequently high cutting speeds, therefore the term *high-speed*). They were first developed in the early 1900s. The alloying elements and their role are described in Section 3.12. With suitable procedures, these steels can be fully hardened with little danger of distortion or cracking. (See Tables 6.5 and 6.6.)

There are two basic types of high-speed steels: *tungsten-type* (T series) and *molybdenum-type* (M series). The T series contain 12 to 20% tungsten, with chromium, vanadium, and cobalt as the other major alloying elements. The M series contain 3.5 to 10% molybdenum, with chromium, vanadium, tungsten, and cobalt as the other alloying elements. They generally have higher abrasion resistance than the T series and lesser distortion in heat treatment; also, they are less expensive. The M series constitute about 95% of all high-speed steels produced in the United States.

High-speed steel tools are used in a variety of machining operations requiring complex tool geometries, such as drills, reamers, taps, and gear cutters. Because of their toughness, they are especially suitable for high positive-rake-angle tools and for low-powered machine tools with low stiffness.

High-speed steel tools can be coated with titanium carbide and titanium nitride, by physical vapor deposition techniques, for improved performance (see *coated carbides* below).

8.10.3 CAST ALLOYS

These cutting-tool materials, which were introduced in 1915, have compositions of the following elements with the specified ranges: 38 to 53% cobalt, 30 to 33% chromium and 10 to 20% tungsten. Because of their high hardness (58 to 64 HRC) they have high wear resistance and maintain their hardness at elevated temperatures. (Fig. 8.52).

Cast-alloy tools, which are cast and ground into any desired shape, are not as tough as high-speed steels and are sensitive to shock loading; hence, they are less suitable for interrupted cutting than high-speed steels. They are recommended for deep continuous roughing operations at relatively high feed rates and speeds, as much as twice those possible with high-speed steels. Cast-alloy tool performance is in between that of high-speed steels and carbides. Cutting fluids are not necessary and are usually used only to obtain a special surface finish. These tools are used only in special applications.

8.10.4 CEMENTED CARBIDES

Carbide tool materials, first introduced in the 1930s, are basically composed of carbides with a binder, and are also known as *cemented carbides*. Because of their high hardness over a wide range of temperatures, high modulus of elasticity, high thermal conductivity, and low thermal expansion, they are among the most important tool materials.

There are three general groups of carbides for machining operations:

a. Tungsten Carbide (WC): in its basic composition this is made of tungsten carbide with cobalt as the binder. It is used for cutting nonferrous abrasive materials, cast irons, and nonmetallic materials. Other tungsten carbides also contain titanium carbide, tantalum carbide, and niobium carbide to improve crater-wear resistance and hot hardness. These are used for machining steels.

b. Titanium Carbide (TiC): in this tool material the binder is composed of nickel and molybdenum. It has higher wear resistance but is not as tough as tungsten carbide. TiC is suitable for cutting hard materials (mainly steels and cast iron) and for cutting at high speeds, i.e., high temperature (Figs. 8.53 and 8.54).

c. Coated Carbides: these generally consist of a tungsten carbide base coated (by chemical vapor deposition, CVD, Section 4.13.1) with a very thin layer (2×10^{-4} to 3×10^{-4} in., 5 to 9 μm) of TiC, titanium nitride (TiN), or aluminum oxide (Al_2O_3). These coatings can also be applied to tool and die steels. Multilayer coatings are also becoming available.

Titanium carbide coatings are effective for abrasive wear resistance, and titanium nitride for preventing adhesion at the tool–chip interface, thus reducing pickup or galling.

Titanium nitride coating is a more recent important development and it greatly improves tool life. However, these tools do not perform as well at low cutting speeds since the coating can be removed or chipped off due to adhesion. Proper lubrication is therefore important. High-speed steel tools with various titanium nitride coatings are now available; they have a yellow-gold color. The flank-wear surfaces can be reground after use since regrinding does not remove the coating on the rake face.

Because of its low thermal conductivity, aluminum oxide is a good thermal barrier and is, therefore, also suitable as a coating on tools for high speed and high feed-rate operations. Coated carbides are now being used for machining operations, with increases in tool life by as much as fivefold over uncoated carbides.

The amount of cobalt binder has a significant effect on the properties of tungsten carbide (Fig. 8.55). Note that with increasing cobalt content (decreasing volume of hard phase), the strength, hardness, and wear resistance decrease while the transverse rupture strength (in a bend test, Fig. 3.37) increases. Consequently, the toughness of carbides also increases with increasing cobalt content.

Inserts

Carbide *inserts* are available in a great variety of shapes, such as square, triangular, and round. They are generally clamped or brazed on the tool shank (see Fig. 8.35). Because of the difference in coefficients of thermal expansion, brazing should be done carefully to avoid cracking or warping. The strength of an insert depends on its shape (Fig. 8.56); the smaller the included angle the lower the strength. This is also generally true for all cutting tools.

In order to improve the edge strength and prevent chipping, a small *chamfer* is produced along the edges of the insert, called K-land or T-land in practice. The edge

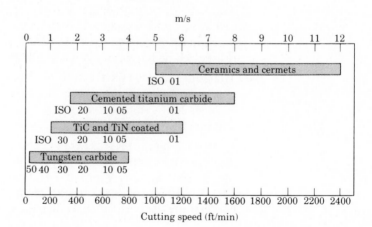

FIGURE 8.53 Approximate cutting speed range for optimum use of various cutting tool materials. ISO designations are shown for various groups of cutting tools. Cermets consist of aluminum oxide and titanium carbide. Coated cutting tools presently account for over 40% of machining applications and are developing rapidly.

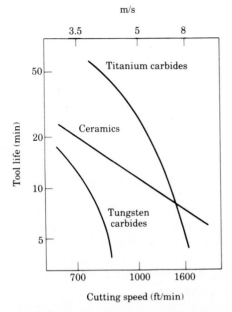

FIGURE 8.54 Tool life curves for various tool materials in medium and light turning operations, as a function of cutting speed. Note how the curve for ceramics crosses over the curve for titanium carbides as speed, hence temperature, increases.

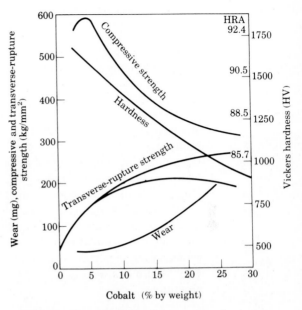

FIGURE 8.55 Effect of cobalt content in tungsten carbide tools on mechanical properties. Note that hardness is directly related to compressive strength and hence, inversely, with wear.

may also be honed to a radius of about 0.001 in. (0.025 mm) to improve its strength and avoid chipping.

Stiffness of the machine tool is of great importance in using carbide tools. Light feeds, low speeds, and chatter are harmful. To take full advantage of carbide tools, stiff, high speed, powerful machine tools have been built for high production rates. Cutting fluids are generally not needed, but if used they should be applied in large quantities and continuously to minimize the heating and quenching of the tool in interrupted cutting operations.

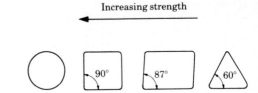

FIGURE 8.56 Relative strength of four different shapes of inserts. The cutting edge is that shown by the angles.

8.10.5 CERAMICS (OXIDES)

These tool materials, first used in the 1940s, consist primarily of fine-grained aluminum oxide (Al_2O_3) particles, which are bonded together; thus they are 100% hard phase. Minor additions of elements, such as titanium oxide and titanium carbide, help obtain optimal properties. Included in this group of materials are *borides* and *cermets*. A typical cermet is 70% Al_2O_3 and 30% TiC.

Ceramic tools have very high abrasion resistance and hot hardness (Fig. 8.57), and have less tendency to adhere to metals during cutting, i.e., less tendency to form a built-up edge. Consequently, good surface finish is obtained. However, they lack toughness and premature tool failure can result by chipping or catastrophic failure. New developments in ceramic tools include inserts consisting of silicon nitride and aluminum oxide, pressed and sintered at 3300°F (1800°C).

Ceramic cutting inserts (available in shapes similar to carbides) are effective in very-high-speed, uninterrupted cutting operations, such as finishing or semifinishing by turning. To reduce thermal shock, cutting fluids should not be used. Tool and setup geometry are important. Negative rake angles are generally used to avoid chipping. Tool failure can also be reduced by increasing the stiffness of machine tools and mountings.

8.10.6 CUBIC BORON NITRIDE (CBN)

Next to diamond this is the hardest material presently available. The cutting tool is made by bonding a layer of polycrystalline cubic boron nitride to a cemented carbide substrate. While the carbide provides the shock resistance, the CBN layer provides very high wear resistance and edge strength. Unlike diamond, CBN is chemically

FIGURE 8.57 Range of properties for different tool materials. New developments include coated carbides, polycrystalline synthetic diamonds, and tools consisting of pressed and sintered silicon nitride and aluminum oxide. See also Tables 8.7 and 8.8.

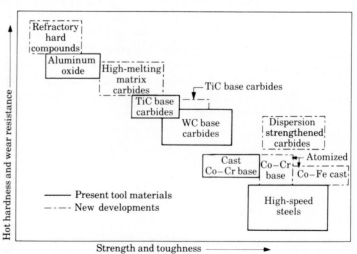

inert to iron and nickel at elevated temperatures. It also has oxidation resistance at high temperature. It is therefore particularly suitable for machining hardened ferrous and high-temperature alloys. CBN is also used as an abrasive (Chapter 9).

8.10.7 DIAMOND

The general properties of diamond are its extreme hardness, high abrasion resistance, very low coefficient of friction, low coefficient of thermal expansion, and high thermal conductivity. Diamond is used when good surface finish and dimensional accuracy are required, particularly with soft nonferrous alloys and abrasive nonmetallic materials.

Because diamond is brittle, tool shape is important. Low rake angles are normally used to provide a stronger cutting edge. Special care should be given to the proper mounting and crystal orientation of the diamond in order to obtain optimal use. The abrasive wear resistance of diamond varies by at least tenfold depending on the crystal orientation. Wear in diamond also takes place by microchipping due to thermal stresses and oxidation, and transformation to carbon due to frictional heating.

Diamond tools can be used satisfactorily at almost any speed, but are suitable only for light, uninterrupted finishing cuts. In order to minimize fracture, the diamond must be resharpened as soon as it becomes dull. Because of the strong chemical affinity, diamond is not used for machining ferrous metals and nickel alloys.

Although single crystal diamonds, such as the diamond on a ring, have been used in the past, a recent development (1970s) is *polycrystalline diamond* tools. These are very small synthetic crystals fused together by a high-pressure, high-temperature process and bonded to a cemented carbide substrate. Thus, the random orientation of the diamond crystals prevents the propagation of cracks through the structure. The carbide substrate provides the necessary toughness for shock loading of the tool. Diamond is also used as an abrasive in grinding and polishing operations (Chapter 9).

8.11 MACHINABILITY

Machinability of a material is generally defined in terms of three factors:

a. Forces and power consumption,

b. Tool wear, and

c. Surface finish and integrity.

Thus, a material with good machinability is one requiring low power consumption, with low tool wear, and producing a good surface finish with no surface damage. An additional parameter sometimes included in machinability is chip curl. Long, thin curled chips, if they cannot be broken up with a chip breaker (see Section 8.3.3), can interfere with the operation by becoming entangled in the cutting area. Thus, a material producing such a chip is less machinable than one with breakable or discontinuous chips.

Because of the complex nature of the relation between these parameters, it is difficult to establish quantitative relationships to define the machinability of a material. In actual production, tool life and surface integrity are generally considered to be the most important considerations in machinability.

The number of variables involved and the variety of materials that are machined preclude quantitative ratings for machinability. It is advisable to refer to machining recommendations (see bibliography for this chapter) that are based on extensive testing, practical experience, and data collected in manufacturing industries.

Machinability can also be viewed in terms of the type of chip formation and the nature of the shear zone. Referring to Fig. 8.5, it is obvious that a large shear zone, high friction at the tool–chip interface, and a built-up edge are detrimental to good machinability. Thus, a very ductile material that has a tendency to adhere to the tool face is likely to produce a poor surface finish. This has been observed to be true with such materials as iron, pure aluminum, copper, and stainless steels, regardless of their strength.

Mechanical and physical properties also play a role in the energy consumption and the temperatures generated during cutting. For instance, titanium is difficult to machine partly because of the high temperatures generated due to its poor thermal conductivity, and partly because of its tendency to adhere to the cutting tool.

The strength of the workpiece material also has an effect on machinability, with the additional consideration of how the strength was obtained, i.e., by cold working or heat treatment. More importantly, it is the composition and the presence of any hard phases in the material that affect tool wear, as described in the section below. Machinability of materials can be significantly improved by altering their composition with various additives as described below.

8.11.1 MACHINABILITY OF STEELS

Because steels are among the most important engineering materials, their machinability has been studied extensively. Machinability of steels has been improved (*free machining*) mainly by adding lead (*leaded steels*) and sulfur (*resulfurized steels*). The manner in which these additives influence the cutting process is not fully understood. One mechanism may be that they produce films of low shear strength, and thus reduce the friction in the secondary shear zone at the tool–chip interface.

Leaded Steels
Lead additives are in the form of a dispersion of fine lead particles. Lead is insoluble in iron, copper, aluminum, and their alloys. During the cutting process these particles are sheared and spread over the tool–chip interface. The low shear strength of lead makes it act as a solid lubricant.

This behavior has been verified by the presence of high concentrations of lead on the tool side (smooth bright side) of the chips in machining leaded steels. In addition to this mechanism in the secondary shear zone, the presence of lead in the primary shear zone could also reduce the strength of the shear plane, thus reducing forces and power consumption. Lead may be used in either nonsulfurized or resulfurized steels.

Resulfurized Steels

The role of sulfur in steels appears to be through formation of manganese sulfide inclusions (second-phase particles), which act as stress raisers in the primary shear zone (Fig. 8.58). The shape, orientation, distribution, and volume fraction of these inclusions have significant influence on machinability. Although still controversial, it has also been suggested that sulfur may impart low-shear-strength characteristics to steel and reduce friction in the secondary shear zone.

Elements such as tellurium and selenium (both chemically similar to sulfur) in resulfurized steels also act as inclusion modifiers, increasing their hardness, and aiding in stress raising.

Calcium Deoxidized Steels

Another important development is *calcium deoxidized steels*, in which oxide flakes of calcium aluminosilicate (CaO, SiO_2, and Al_2O_3) are formed. These flakes, in turn, affect the secondary shear zone through a decrease in strength and friction. Although this mechanism is not yet fully understood, experimental evidence indicates that crater wear is reduced, especially at higher speeds, with appropriate selection of tool materials.

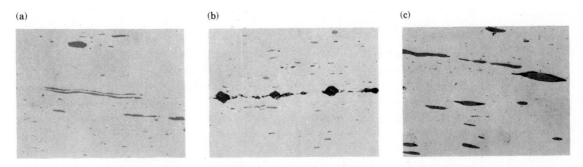

(a) (b) (c)

FIGURE 8.58 Photomicrographs showing various types of inclusions in low-carbon, resulfurized free-machining steels. (a) Manganese sulfide inclusions in AISI 1215 steel. (b) Manganese sulfide inclusions and glassy manganese silicate-type oxide (dark) in AISI 1215 steel. (c) Manganese sulfide with lead particle as tails in AISI 12L14 steel. *Source:* Courtesy of H. Yaguchi, Inland Steel Company.

Other Elements in Steels

The presence of aluminum and silicon is always harmful in steel. These elements combine with oxygen in steel and form aluminum oxides and silicates, which are hard and abrasive, thus increasing tool wear and reducing machinability (Fig. 8.58).

Carbon, manganese, and phosphorus have various effects on steels depending on their composition. Other alloying elements, which improve the properties of steels, such as nickel, chromium, molybdenum, and vanadium, generally reduce machinability. The effect of boron is negligible.

The role of gaseous elements (such as oxygen, hydrogen, and nitrogen) on machinability has not been clearly established. Any effect that they might have would depend on the presence and quantity of other alloying elements in the steels.

In selecting elements to improve machinability, the possibility of detrimental effects of these elements on the properties and strength of the machined product in service must also be considered. At elevated temperatures, for instance, lead causes embrittlement of steels, known as *hot shortness* (see Section 3.3.1). At room temperature lead has no significant effect on the mechanical properties of steels.

As for sulfur, unless there is sufficient manganese, it can cause a severe reduction in the hot workability of steels (see Fig. 6.32b). At room temperature, the mechanical properties of resulfurized steels depend on the orientation of the deformed manganese sulfide inclusions (see Figs. 3.45 and 8.58) with respect to the direction of the tensile stress (mechanical fibering). See also Figs. 3.16(a) and (b) and 3.44.

8.11.2 MACHINABILITY OF VARIOUS MATERIALS

In this section brief guidelines concerning the machinability of various metals and nonmetallic materials are presented. It is beyond the scope of this text to treat this subject in great detail, as numerous reference books are available as listed in the bibliography at the end of this chapter. (See also, Tables 8.5 through 8.11 for recommendations for tool geometry, tool materials, and speeds. Some general guidelines for tool life are given in Fig. 8.59.)

Aluminum and Its Alloys

The softer grades of aluminum alloys have a tendency for BUE formation. High cutting speeds, high rake angles, and high clearance angles are recommended. Wrought alloys with high silicon content and cast aluminum alloys may be abrasive and hence require harder tool materials.

Beryllium

Similar to cast irons (see below) but more abrasive and toxic, hence requiring machining in a controlled environment.

Cast Irons

Gray cast irons are generally machinable; however, they are very abrasive. Free carbides in the casting reduce machinability and cause tool chipping or fracture, thus

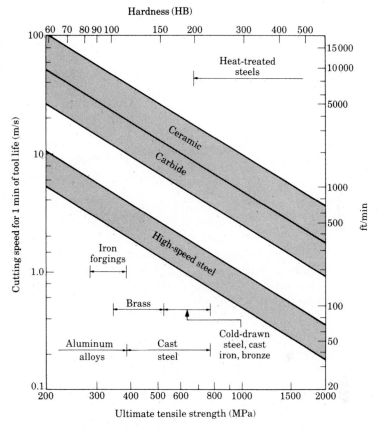

FIGURE 8.59 Approximate cutting speed for one-minute tool life, showing the effects of tool and workpiece materials and cutting speed. Note the rapid decrease in tool life with increasing strength of the workpiece material. *Source:* After G. Boothroyd, *Fundamentals of Metal Machining and Machine Tools*, New York, McGraw-Hill, 1975, p. 147.

requiring tools with higher toughness. Nodular and malleable cast irons are machinable with hard-grade tool materials.

Cobalt-Base Alloys
Very abrasive and highly work hardening. These alloys require sharp and abrasion-resistant tool materials and low speeds and feeds.

Copper
Ductile and work hardening; can be difficult to machine. Cast copper alloys are easy to machine. Brasses (copper–zinc alloys) are easy to machine, especially with the addition of lead. Bronze (copper–tin alloy) is more difficult to machine than brass.

Magnesium
Very easy to machine, with good surface finish and tool life. Care should be exercised during machining because of its high oxidation rate and the danger of fire.

Molybdenum
Ductile and work hardening. Requires sharp tools for good surface finish.

Nickel-Base Alloys
Work hardening, abrasive, and very strong up to high temperatures. Forces and temperatures are high. These alloys are difficult to machine.

Stainless Steels
Austenitic (300 series) steels have machining characteristics similar to nickel-base alloys and are therefore difficult to machine. Chatter could be a problem, so rigid equipment is required. Ferritic stainless steels (also 300 series) have good machinability. Martensitic (400 series) steels are abrasive and tend to form a BUE; they also require tool materials with high hot hardness and crater-wear resistance. Precipitation-hardening stainless steels are strong and abrasive; they require hard, abrasion-resistant tool materials.

Steels
Leaded and resulfurized steels are relatively easy to machine, but as the carbon content increases, machinability decreases. However, plain low-carbon steels (less than 0.15 % C) can produce poor surface finish due to BUE. Cast steels are somewhat more abrasive, but have machinability similar to wrought steels. Tool and die steels are very difficult to machine and usually require annealing prior to machining.

Tantalum
Very work hardening, ductile, and soft. Produces poor surface finish and has high tool wear. Requires an effective cutting fluid.

Titanium
The poor thermal conductivity of titanium and its alloys (lowest of all engineering metals) causes significant temperature rise and BUE. Requires an effective cutting fluid.

Tungsten
Brittle, strong, and very abrasive. Machinability is improved greatly at elevated temperatures.

Zirconium
Good machinability; however, it requires a coolant-type cutting fluid because of the danger of explosion and fire.

The machinability of various nonmetallic materials is outlined below.

Graphite
Very abrasive. It requires hard tools with very sharp edges and resistance to abrasion.

Reinforced Plastics
Very abrasive if reinforced with glass fibers (see Section 10.11); difficult to machine. They require careful removal of machining debris to avoid contact with and inhaling of fibers.

Thermoplastics
Low thermal conductivity, low elastic modulus, and low softening temperature. Consequently, machining of thermoplastics requires tools with positive rake (up to 20 degrees) and high relief (clearance) angles, small depths of cut and feed, relatively high speeds, and proper support of the workpiece for dimensional control. Tools (generally high-speed steels or carbide) should be sharp and abrasion resistant. Cooling of the cutting zone may be necessary. This can usually be done with a jet of air, vapor mist, or water-soluble oils; these fluids also remove chips from the cutting area.

Residual stresses may develop during machining of thermoplastics. To relieve residual stresses, parts are annealed at temperatures ranging from 175 to 315°F (80 to 160°C) for a period of time. Parts should be cooled to room temperature slowly and uniformly.

Thermosetting Plastics
Brittle and sensitive to thermal gradients during cutting. Their machinability is generally similar to thermoplastics.

8.11.3 HOT MACHINING

Metals and alloys that are difficult to machine at room temperature can be machined more easily at elevated temperatures (*hot machining*). Cutting forces are lowered and tool life can be increased. However, there are problems in heating the workpiece and controlling the temperature. Investigations have shown that, except in isolated cases, hot machining offers no significant advantage over machining at room temperature with appropriate tools and cutting fluids. In *laser-assisted machining*, the heat source is a laser focused to an area just ahead of the cutting tool.

8.11.4 AUTOMACHINING

This is a term that describes cutting of metals using single-point tools made of the workpiece material. The cutting tool is not subjected to any additional treatment or coatings. When used on a lathe, it is found that continuous chips are obtained, with good surface finish, on materials such as 2011-T3, 2024-T3, and 301 stainless steel. The surface finish is fair in automachining of mild steel and poor in copper.

The mechanism by which automachining is possible has been determined experimentally to be due to the formation of a BUE. It is the BUE tip that acts as the cutting edge; the chip does not contact the original tool at its rake face during cutting. Since the hardness of the BUE can be considerably higher than that of the workpiece material (see Fig. 8.8), cutting is actually performed by the harder BUE. Although it is not likely to be of industrial significance, automachining is nevertheless an interesting observation in metal cutting.

8.11.5 GENERAL GUIDELINES

It is apparent that many factors should be taken into account in the selection of tools and process parameters in the machining of a particular material. Close observation of the wear rate and chipping of the tool is important. Also, the tools must be changed at the proper time.

If a poor surface finish is produced, the cause must be determined properly, e.g., whether due to BUE, tool wear, chipping, or chatter due to inadequate stiffness of the setup or the machine tool (Section 8.13). The feed rate, depth of cut, and cutting speed will depend on the capacity of the machine tool. The less stiff the machine, the lower these parameters must generally be in order to obtain an acceptable surface finish and good tool life.

8.12 CUTTING FLUIDS

The primary functions of a cutting fluid are to:

a. Reduce friction and wear and thus improve tool life, surface finish, and reduce forces and energy consumption,

b. Cool the cutting zone and thus reduce temperature and distortion,

c. Wash away the chips,

d. Protect the newly machined surfaces from corrosion.

A cutting fluid can have interchangeable roles as a coolant or a lubricant. Its effectiveness depends on a number of factors:

a. Temperatures encountered,

b. Cutting speed,

c. Type of machining operation, and

d. The method of application.

The role of a cutting fluid depends on the range of variables involved in machining. For instance, as the cutting speed increases in an operation such as turning or milling, the temperature increases and thus cooling of the cutting area is of great importance. On the other hand, if the speed is low, such as in broaching and tapping, lubrication is the important factor in reducing the tendency for BUE formation and improving surface finish.

General lubrication mechanisms are treated in Section 4.9. The mechanisms by which cutting fluids influence machining operations are described below.

8.12.1 CUTTING-FLUID ACTION

In view of the high pressures and relative sliding at the tool–chip interface, the question has been raised as to how a cutting fluid penetrates this interface to affect the cutting mechanism. An early model postulates that the fluid is drawn into this interface by the capillary action of the interlocking network of surface asperities (Fig. 8.60).

Because of the small size of this capillary network, it is desirable for the cutting fluid to have a small molecular size and proper wetting (surface tension) characteristics (Section 4.12). More recent studies, using a transparent sapphire cutting tool, have shown that:

a. The cutting fluid appears to gain access to the tool–chip interface by seeping into the diverging gap from the sides of the chip, rather than by movement against the flow of the chip.

b. The lubricating capability of the fluid arises from the formation of films of low shear strength, and the necessary chemical reactions take place due to the high re-activity of the freshly generated surfaces (tool side of the chip). These reactions are obtained with fluids containing additives such as chlorine and sulfur. If the additive is a fatty oil, it forms metallic soaps, again reducing the friction and the tendency for the chip to adhere to the tool face.

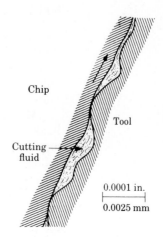

FIGURE 8.60 Schematic illustration of tool–chip interface, showing capillary passages allowing the cutting fluid to enter the interface. Recent experiments have shown that cutting fluids enter the interface through the sides (perpendicular to the page), rather than downward against the movement of the chip. *Source:* After M. E. Merchant.

It is also possible that, with sufficiently high temperatures, cutting fluids are converted to a gaseous state, thus further penetrating the tool–chip interface because of small molecular size. It should be pointed out, however, that these processes require time. Hence, as the cutting speed increases, there is less likelihood that an effective capillary action will take place. With high speeds, the role of a cutting fluid becomes mainly one of cooling the cutting area.

In addition to the cooling and lubricating functions of a cutting fluid, suggestions have been made concerning the possible effect of the fluid in reducing the shear strength of the workpiece material (*Rebinder effect*). Numerous experiments have been conducted to verify the existence of such an effect.

Although it is well known that fluids can have an influence on the mechanical properties of materials, the relevance of these observations to metal cutting is difficult to appraise. This is largely due to the difficulty of isolating the events taking place in the primary and secondary shear zones in the cutting area. When observed, the Rebinder effect has been found to depend on the particular material or alloy being machined.

There are situations where the use of cutting fluids can be detrimental. For example, in interrupted cutting operations, such as milling, the cooling action of the cutting fluid extends the range of temperature (amplitude) to which the cutter teeth are subjected. This can lead to thermal cracks due to thermal cycling.

Another example of the effect of cutting fluids is based on the observation that the cutting fluid can, under certain circumstances, cause the chip to curl into a very small radius. This concentrates the stresses and heat closer to the tip of the tool and reduces tool life.

The properties of various lubricants are discussed in Section 4.10. A general guide to the selection of cutting fluid in cutting operations is given in Table 6.7.

8.12.2 APPLICATION METHODS

The most common method of application is *flood cooling* with rates ranging from 3 gal/min (10 liters/min) for single-point tools to 60 gal/min (225 liters/min) per cutter for multiple-tooth cutters. In some operations, such as gun drilling and end milling, high-pressure systems at 100 to 2000 psi (0.7 to 14 MPa) are used to wash away the chips.

Mist cooling is another method of application, particularly with water-base fluids, at 10 to 80 psi (0.07 to 0.6 MPa) air pressure. Although it requires venting (to prevent inhaling by the machine operator) and has limited cooling capacity, mist cooling supplies the fluid to inaccessible areas and provides better visibility of the workpiece being machined.

The third method of application is manual. The lubricants used are generally solid lubricants, such as graphite, molybdenum disulfide, soaps, and waxes. This method has very limited applications.

8.12.3 EFFECTS OF CUTTING FLUIDS

The selection of a cutting fluid should also include various other considerations as outlined below.

Effects on Workpiece Material

The selection of a fluid should, in certain cases, be dictated by whether or not the machined component will be subjected to possible environmental attack and to high service stresses, thus possibly leading to stress–corrosion cracking (see Section 3.7.1). This is particularly important when using cutting fluids with sulfur and chlorine additives.

Attention must also be paid to how well the machined parts have been cleaned and washed in order to remove any residue of cutting fluids. Additional considerations are staining of the machined surfaces by various cutting fluids, especially on copper and aluminum.

Effects on Machine Tools

Just as a cutting fluid can have an adverse effect on the workpiece material, it can have the same effect on the machine tool and its various components, such as bearings and slideways. The choice of the fluid must therefore include considerations of compatibility with the materials used in the machine tool structure.

Biologic and Environmental Effects

Because the machine-tool operator is always in the general area of a cutting fluid, the effects of contact with the fluid should be of primary concern. In addition to fumes, smoke, and odors, the fluid can also cause severe reactions on the skin and various parts of the operator's body.

Considerable effort and progress have been made in studying these problems and in ensuring safety in the use of cutting fluids. An additional important factor is the effect of the fluid on the environment, particularly in regard to its degradation and ultimate disposal. These practices must comply with federal, state, and local codes and ordinances. (See also Section 4.11.)

● **Illustrative Problem 8.4**

A machining operation is being carried out with an effective cutting fluid. Explain the effects on the mechanics of the cutting operation and the total energy consumption if the fluid is shut off.

SOLUTION.   An effective cutting fluid is a good lubricant. Thus, when the fluid is shut off, the friction at the tool–chip interface will increase. The following chain of events then takes place:

1. Fluid is shut off.

2. Friction at the tool–chip interface increases (β increases).

3. The shear angle decreases.

4. The shear strain increases.

5. The chip is thicker.

6. A built-up edge is likely to form.

As a consequence:

1. The shear energy in the primary zone increases.

2. The friction energy in the secondary zone increases.

3. Hence, the total energy increases.

4. Surface finish is likely to deteriorate.

5. The temperature in the cutting zone increases, hence tool wear increases.

6. Tolerances may be difficult to maintain due to the increased temperature and expansion of the workpiece during machining. ●

8.13 VIBRATION AND CHATTER IN MACHINING

Vibration and chatter of a cutting tool are complex phenomena, which, if uncontrolled, can lead to premature tool failure. This is particularly important with brittle tool materials such as ceramics, some carbides, and diamond. In addition, vibrations affect the machined surface and its integrity. If excessive, vibrations may even damage machine tools. Furthermore, the noise generated may be objectionable, particularly if it is high frequency.

There are two basic types of vibration in machining: *forced* vibration and *self-excited* vibration.

8.13.1 FORCED VIBRATION

This is generally due to some periodic applied force present in the machine, such as from gear drives, imbalance of the machine tool components or misalignments, motors and pumps, or other equipment nearby. In a machining process such as milling, or turning a shaft with a keyway or a splined shaft, forced vibrations are also caused by the periodic entry and exit of the cutting tool, since the process is one of interrupted cutting.

The basic remedy for forced vibration is to isolate or remove the forcing element. If the forcing frequency is at or near the natural frequency of a component of the machine-tool system, then one of the frequencies may be raised or lowered. The amplitude of vibration can be reduced by increasing the stiffness or damping of the system.

Although changing the cutting process parameters generally does not appear to have much influence on forced vibrations, changing the cutting speed can sometimes help. Changing the cutting forces, especially the thrust force, also can help. This can be done by modifying the tool geometry. (See Section 8.5.3.)

FIGURE 8.61 Chatter marks on the surface of a workpiece in turning. Chatter is an important problem in machining because of its detrimental effects on surface finish and tool life. *Source:* Courtesy of Carboloy Systems Department, General Electric Company.

8.13.2 SELF-EXCITED VIBRATION

These vibrations, called *chatter*, are due to the interaction of the dynamics of the chip-removal process and the structural dynamics of the machine tool. The excited vibrations are usually very high in amplitude and may cause damage to the machine tool, as well as lead to premature tool failure. Chatter typically begins with a disturbance in the cutting zone, such as lack of homogeneity in the workpiece material or its surface condition and geometry, or a change in frictional conditions at the tool–chip interface.

The most important type of self-excited vibration is called *regenerative chatter*. This results from the tool cutting a surface that has roughness or disturbances left from the previous cut (Fig. 8.61). Because of the resulting fluctuations in cutting forces, the tool is subjected to vibrations and the process continues repeatedly, hence the term regenerative.

Self-excited vibrations can generally be controlled by increasing the dynamic stiffness of the system and by damping. *Dynamic stiffness* is defined as the ratio of the amplitude of the force to the amplitude of vibration. It should be noted that a machine tool has different stiffnesses at different frequencies. Consequently, changes in operating parameters can also influence chatter.

8.13.3 GENERAL GUIDELINES

Although the solution to some of the chatter problems may require detailed analysis of all the components involved, there are some general guidelines that could help solve many of the problems. These are outlined below.

Turning

Minimize tool overhang, increase overall stiffness (either by proper design or by using materials with higher elastic modulus), support workpiece rigidly, modify tool geometry (such as by reducing the length of the cutting edge in contact with the work-piece and reducing the nose radius), change the stock removal rate, and change operating parameters such as cutting speed, feed rate, and depth of cut.

Boring

Similar to those in turning, with increased emphasis on the stiffness of the boring bar, possibly using carbides with elastic moduli 2.5 to 3 times that of steels.

Milling

Emphasis is on changing tool geometry, reducing the number of teeth or using a cutter with random spacing of teeth, mounting cutters as close to the spindle as possible, and changing operating conditions.

Various studies have indicated that the tendency for a particular workpiece material to chatter is proportional to the cutting forces and the depth and width of cut. Consequently, for a given material, the tendency to chatter generally increases as its hardness increases. Thus, aluminum and magnesium alloys have less tendency to chatter than martensitic and precipitation-hardening stainless steels, nickel alloys, and high-temperature and refractory alloys. However, an important factor is the chip morphology (see Figs. 8.5 and 8.6); continuous chips generally do not cause chatter, whereas discontinuous chips have a tendency to cause chatter.

8.13.4 DAMPING

Damping can be an important factor in machine-tool vibration and chatter. There are three sources of damping: internal, joints in the machine tool, and external.

Internal damping is due to the energy loss in materials during vibration. Thus, for instance, steel has less damping than gray cast iron (Fig. 8.62). (One can observe this difference by tapping these materials and listening to the sound.) The high damping in

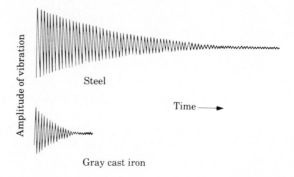

FIGURE 8.62 Relative damping capacity of steel and gray cast iron. These are the two most common materials used for machine tool construction. Damping is an important factor in reducing chatter. Because they dissipate energy, internal damping of materials is also called internal friction.

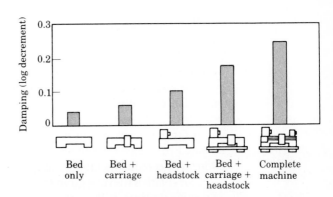

FIGURE 8.63 Damping of vibrations as a function of the number of components on a lathe. Joints between different components dissipate energy because of friction or due to the presence of lubricants. Thus, the greater the number of joints, the greater the damping. The term log decrement is a measure of how rapidly the amplitude of vibration decays with time. *Source:* After J. Peters, CIRP Annals, 1965.

cast iron is due to the energy absorption by the graphite flakes. Hence, with all other parameters constant, vibrations decay faster in a cast iron machine tool structure than in one made of steel. (Internal damping of materials is also called *internal friction.*)

The second source of damping, although less significant, is the bolted joints in a machine tool. Small relative movements dissipate energy by friction in dry joints. In joints where oil is present, the internal friction of the oil layers during relative movements dissipates energy, thus contributing to damping. This type of damping is cumulative because of the number of joints in a machine tool (Fig. 8.63).

The third source of damping is an external one, similar to a shock absorber. Tuned, damped vibration absorbers have been developed and installed on machine tools.

8.14 MACHINE TOOLS

The general characteristics of the machine tools used in cutting operations are reviewed in this section. The equipment can be classified in different categories depending on their function and the accessories involved. The major categories are listed and described below.

8.14.1 LATHES

Lathes are generally considered to be the oldest machine tools and were developed in the 1750s. The basic operations carried out on a lathe are turning, boring, and facing.

The most common lathe (*engine lathe*) is shown in Fig. 8.64. The basic components are the bed, headstock, tailstock, and carriage. One end of the workpiece is clamped in a chuck or a collet, which rotates in the headstock. The other end of the workpiece is supported by the tailstock. A cutting tool, attached to a tool post and to the carriage, removes material by traveling along the bed. The cutting operation is performed at a certain desired peripheral speed of the workpiece, feed rate, and depth of cut (see Fig. 8.24).

With appropriate tools, accessories, and attachments, various other operations can also be carried out on a lathe. These are thread cutting, drilling, knurling, polishing, and generating external or internal, tapered or contoured surfaces on axisymmetric parts.

The size of a lathe is specified by its *swing* (the maximum diameter of the workpiece that can be machined) and by the maximum distance between centers of the headstock and tailstock.

A further development is the *automatic lathe*. The movements and controls are actuated by various mechanical means, such as cams and numerical and computer control, thus reducing labor or requiring less skilled labor. (See Chapter 13.)

Turret lathes (developed in the 1850s) carry out multiple cutting operations, such as turning, boring, drilling, thread cutting, and facing. Various cutting tools, usually up to six (because of the hexagonal turret, Fig. 8.65) are installed on a turret. The

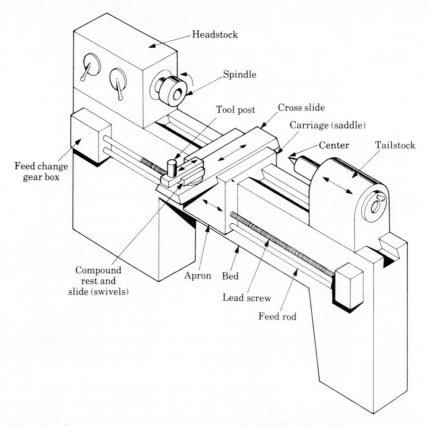

FIGURE 8.64 Schematic illustration of a lathe. Lathes are among the most common machine tools on which a variety of cutting operations can be carried out, such as facing, boring, drilling, and internal and external threading. Capacities of lathes range from fractional horsepower to more than 200 hp (150 kW) for vertical boring mills shown in Fig. 8.66.

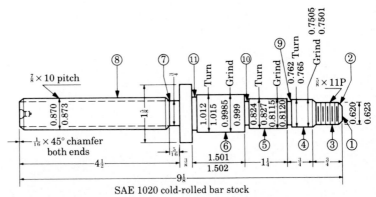

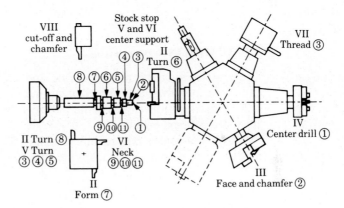

FIGURE 8.65 A typical part made on a turret lathe and the cutting components on the turret. After one operation is completed, the turret rotates and performs the next operation. Note the various operations performed and the tolerances obtained. All dimensions are in inches (1 in. = 25.4 mm). *Source:* Warner and Swasey.

operation is quite versatile. It may be operated by hand or automatically and, once set up properly, it does not require skilled labor. There are a great variety of turret lathes available for different purposes.

Screw machines (the first screw lathe was developed in the 1770s) are designed for high-production-rate machining of screws and similar parts. All the operations are performed automatically with tools attached to a special turret. The bar stock is fed automatically after each screw has been machined to a finished product. These machines may be equipped with single or multiple spindles to reduce the cycle time per piece.

8.14.2 BORING MACHINES

Although boring operations can be carried out on a lathe, boring machines are used for large workpieces. These machines are either horizontal or vertical (Fig. 8.66), and are available with a variety of features for boring and facing operations.

Jig borers are high-precision, vertical boring machines and are used for making jigs and fixtures, and for boring parts held in a jig.

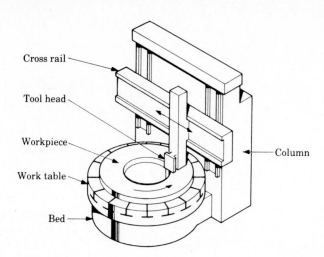

FIGURE 8.66 Schematic illustration of a vertical boring mill. Boring mills are available to handle workpieces over 40 ft (12 m) in diameter. *Source:* After G. Boothroyd.

8.14.3 MILLING MACHINES

These are among the most versatile and useful machine tools because they are capable of performing a variety of cutting operations. There is a large selection of milling machines (developed in the 1860s) available with numerous features. The basic types are outlined below.

Knee-and-column type milling machines (Fig. 8.67) are the most common and are used for general-purpose machining. The spindle, to which the milling cutter is at-

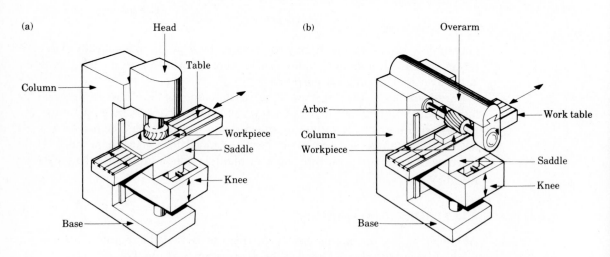

FIGURE 8.67 Schematic illustration of knee-and-column type milling machines. (a) Vertical spindle. (b) Horizontal spindle. Milling machines are among the most versatile machine tools. See also Fig. 8.37. Capacities of milling machines range up to 75 hp (55 kW). *Source:* After G. Boothroyd.

tached, may be horizontal (for slab milling, Fig. 8.37), or vertical (for face milling). In universal milling machines the table on which the workpiece is mounted can be swiveled on a horizontal plane for milling complex shapes such as helices.

Bed-type milling machines have the table mounted on a bed (replacing the knee) and can move only in the longitudinal direction. These machines are simple and rigidly constructed and are used for high-quantity production. The spindles may be horizontal or vertical and can be of duplex or triplex types (that is, with two or three spindles for simultaneous machining of two or three surfaces on a part). Other types of milling machines are also available, such as the planer type for heavy work and machines for special purposes.

Among accessories for milling machines, one of the most widely used is the *universal dividing head*. This is a fixture that rotates the workpiece to specified angles between individual machining steps. Typical uses are in milling parts with polygonal surfaces and in machining gear teeth.

8.14.4 PLANERS AND SHAPERS

These machine tools are generally used for machining large flat surfaces although other shapes, such as grooves, can also be machined. In a *planer*, the workpiece travels under the cross-rail equipped with cutting tools (Fig. 8.68), whereas in a *shaper* the tool travels and the workpiece is stationary (Fig. 8.69).

Because of the size of the workpieces involved, planers are among the largest machine tools. Shapers may be horizontal or vertical and are also used for machining notches and keyways.

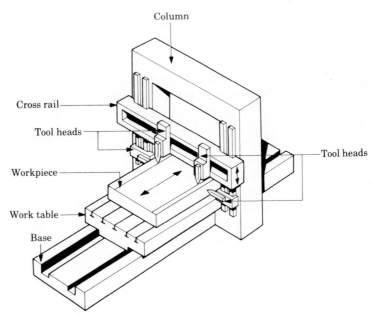

FIGURE 8.68 Schematic illustration of a planer. Very large workpieces can be machined on planers. Carbide tools must be lifted from the workpiece upon the return stroke to avoid chipping of the tool. Capacities range up to 150 hp (110 kW). *Source:* After G. Boothroyd.

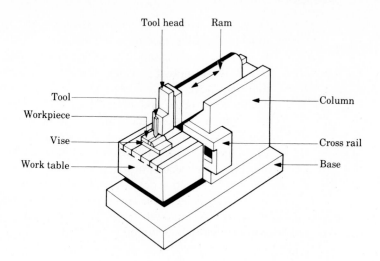

FIGURE 8.69 Schematic illustration of a horizontal shaper. Shapers may also be vertical. Capacities are up to about 20 hp (15 kW).

8.14.5 BROACHING MACHINES

These machines push or pull broaches (Fig. 8.32d), either horizontally or vertically, to generate various external or internal surfaces. They are of relatively simple construction and are actuated hydraulically and have only linear motion of the broach. Because of the great variety of shapes and parts involved, broaching machines are manufactured in a wide range of designs and sizes.

8.14.6 DRILLING MACHINES

These machines are used for drilling holes, tapping, counterboring, reaming, and other general-purpose boring operations. They are usually vertical, the most common one being a drill press (Fig. 8.70). Drilling machines are manufactured in a wide variety of sizes, from simple bench type units to radial drills and large production machines with multiple spindles.

8.14.7 MACHINE TOOL STRUCTURES

The design and construction of machine tools are important aspects of manufacturing engineering. Today there are stringent requirements for very high precision in manufactured products, often with materials that are difficult to machine and with precise specifications concerning surface finish and integrity.

The proper design of machine-tool structures requires a careful analysis of materials available for construction, their forms and properties, the dynamics of the particular machining process, power requirements, and the required range of process variables.

Stiffness and damping are very important considerations. Stiffness involves both the dimensions of the structural components and the elastic moduli of the materials used. Damping involves the nature of the joints in the structure, as well as the type of materials used.

The C-frame structure has been a commonly used design because of easy accessibility to tools and workpieces. However, it is not as stiff as a box-type structure. On the other hand, with the advances made in numerical control, automation and robotics, the importance for accessibility has diminished.

Efficient use of materials and cost considerations require that light-weight or minimum-weight designs be considered. Rather than using traditional designs, with large masses for machine-tool structures (generally made of castings), high dynamic stiffnesses can be achieved with light-weight structures having high static stiffness. This also leads to a high natural frequency of the structure.

Fabrication processes, such as welding or mechanical fastening of individual components (Chapter 12), are the logical choices for light-weight designs. However, this approach to fabrication increases costs for labor as well as for material. Furthermore, since steels are the likely choice for such structures, the benefit of the higher damping capacity of castings is sacrificed.

On the other hand, designs can incorporate frictional joints between various members of the machine tool structure. The orientation of these joints with respect to the sources of vibration should also be included in the overall analysis of the structure.

Regardless of the design, no structure is completely rigid since it is constructed from materials with finite moduli of elasticity. Therefore, in machining critical

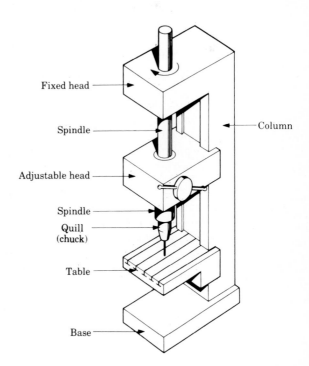

FIGURE 8.70 Schematic illustration of a drill press. Drilled holes, plain or tapped, are used for assembling parts, or for providing passages for air or fluids in a variety of machines and equipment. In radial drilling machines, the head is installed on a radial arm thus permitting drilling of holes at various distances from the column.

components, deflections in the machine-tool structure, either due to forces or thermal gradients in the machine, may be an important problem. Automatic compensation for these deflections requires the use of sophisticated sensing devices.

Deviations of a cutting tool from its prescribed path are due not only to deflections under cutting forces, but also to tool wear, location of the cut on a particular workpiece, or to thermal expansions and contractions in various locations of the total system.

SUMMARY

1. Material-removal processes by cutting involve the use of single-point or multipoint cutters, which remove material by generating chips. Unlike in most deformation processes, described in Chapters 6 and 7, cutting processes are capable of dimensional accuracies, surface finish, and shape complexities that make it an almost necessary operation in the final production of a part.

2. Cutting processes traditionally involve turning, milling, drilling, boring, broaching, reaming, tapping, sawing, and filing. The mechanism of chip formation in all these processes is essentially the same.

3. Chips are generated by a shearing process at the tip of the cutting tool. Chip morphology depends not only on the type of material being cut, but also on processing parameters such as speed, rake angle, depth of cut, feed, and cutting fluids. The type of chip generated is important, particularly as it affects surface finish and integrity. An important factor is the presence of a built-up edge.

4. It is important to study the mechanics of chip formation in order to understand the interrelationship among parameters such as forces, tool wear, surface finish and integrity, vibrations, and various process parameters.

5. Machine tool characteristics can also be significant not only for tolerances and dimensional control of the part, but also concerning vibration and chatter during cutting. Two important parameters are damping and the stiffness of the system.

6. A wide variety of cutting-tool materials have been developed for specific applications. These materials have a wide range of mechanical and physical properties, such as toughness, impact strength, hot hardness, and resistance to wear and chipping.

BIBLIOGRAPHY

General

Armarego EJA, Brown RH. *The Machining of Metals.* Englewood Cliffs, N.J.: Prentice-Hall, 1969.

Boothroyd G. *Fundamentals of Metal Machining and Machine Tools.* New York: McGraw-Hill, 1975.

Brierly RG, Siekmann HJ. *Machining Principles and Cost Control.* New York: McGraw-Hill, 1964.

Gillespie LK. *Deburring, Capabilities and Limitations.* Dearborn, Mich.: Society of Manufacturing Engineers, 1976.

Influence of Metallurgy on Hole Making Operations. Metals Park, Ohio: American Society for Metals, 1978.

Kobayashi A. *Machining of Plastics.* New York: McGraw-Hill, 1967.

Koch P. *Wood Machining Processes.* New York: Ronald, 1964.

Koenigsberger F. *Design Principles of Metal-Cutting Machine Tools.* New York: Pergamon, 1964.

Koenigsberger F, Tlusty J. *Machine Tool Structures.* New York: Pergamon, 1970.

Kronenberg M. *Machining Science and Application.* New York: Pergamon, 1966.

Machining Difficult Alloys. Metals Park, Ohio: American Society for Metals, 1962.

Machining with Carbides and Oxides. Dearborn, Mich.: Society of Manufacturing Engineers, 1962.

Metals Handbook, 8th ed. Vol. 3, *Machining.* Metals Park, Ohio: American Society for Metals, 1967.

Mills B, Redford AH. *Machinability of Engineering Materials*, Applied Science Pub., 1983.

Shaw MC. *Metal Cutting Principles*, 3d ed. Cambridge, Mass.: The Technology Press of The Massachusetts Institute of Technology, 1954.

Swinehart HJ, ed. *Gun Drilling, Trepanning and Deep Hole Machining.* Dearborn, Mich.: Society of Manufacturing Engineers, 1967.

Tobias SA. *Machine Tool Vibration*, Glasgow, Scotland: Blackie and Son, 1965.

Trent EM. *Metal Cutting*, 2d ed. Woburn, Mass.: Butterworth, 1984.

Wilson W, ed. *Machining the Space-Age Metals.* Dearborn, Mich.: Society of Manufacturing Engineers, 1965.

Woldman NE, Gibbons RC. *Machinability and Machining of Metals.* New York: McGraw-Hill, 1951.

Zorev NN. *Metal Cutting Mechanics.* New York: Pergamon, 1966.

(See also General References in Chapter 6.)

Tool Materials

Bhattacharyya A, Ham I. *Design of Cutting Tools.* Dearborn, Mich.: Society of Manufacturing Engineers, 1969.

Burns J, ed. *Diamond Abrasives and Tools.* New York: Macmillan, 1964.

Cutting Tool Material Selection. Dearborn, Mich.: Society of Manufacturing Engineers, 1968.

Cutting Tool Materials. Metals Park, Ohio: American Society for Metals, 1981.

Kalpakjian S, ed. *New Developments in Tool Materials and Applications.* Chicago: Illinois Institute of Technology, 1977.

Kalpakjian S, ed. *Source Book on Tool and Die Failures.* Metals Park, Ohio: American Society for Metals, 1982.

King AG, Wheildon WM. *Ceramics in Machining Processes*. New York: Academic Press, 1966.

Komanduri R, ed. *Advances in Hard Material Tool Technology*. Pittsburgh: Carnegie Press, 1976.

Metals Handbook, 9th ed. Vol. 3, *Properties and Selection: Stainless Steels. Tool Materials and Special Purpose Metals*. Metals Park, Ohio: American Society for Metals, 1980.

Properties and Selection of Tool Materials. Metals Park, Ohio: American Society for Metals, 1975.

Roberts GA, Carry RA. *Tool Steels*, 4th ed. Metals Park, Ohio: American Society for Metals, 1980.

Wilson R. *Metallurgy and Heat Treatment of Tool Steels*. New York: McGraw-Hill, 1975.

Cutting Fluids

Cutting and Grinding Fluids. Dearborn, Mich.: Society of Manufacturing Engineers, 1967.

Olds NJ. *Lubricants, Cutting Fluids and Coolants*. Boston: Cahners, 1973.
(See also Bibliography in Chapter 4.)

Recommendations for Machining

Machinery's Handbook, 21st ed. New York: Industrial Press, 1980.

Machining Data Handbook, 3d ed. (2 vols.). Cincinnati: Machinability Data Center, 1980.

Metals Handbook, 8th ed. Vol. 3, *Machining*. Metals Park, Ohio: American Society for Metals, 1967.

Tool and Manufacturing Engineers Handbook, 4th ed. Vol. 1, *Machining*. Dearborn, Mich.: Society of Manufacturing Engineers, 1983.

Periodicals

American Machinist

Cutting Tool Engineering

International Journal of Machine Tool Design and Research

International Journal of Production Research

Journal of Engineering for Industry

Machine and Tool Blue Book

Machinery

Manufacturing Engineering

Modern Machine Shop

Production

Production Engineer

PROBLEMS

8.1. Are the following statements correct for orthogonal cutting?
   a) For the same shear angle, there are two rake angles that give the same cutting ratio.
   b) For the same depth of cut and rake angle, the type of cutting fluid used has no influence on chip thickness.
   c) If the cutting speed, shear angle, and rake angle are known, the chip velocity can be calculated.
   d) The function of a chip breaker is to increase the curvature of the chip.

8.2. In metal cutting, it has been observed that the shear strains obtained are higher than those calculated from the properties of the material. To what factors would you attribute this difference?

8.3. Describe all the effects of cutting fluids on chip formation.

8.4. Can you offer an explanation as to why the maximum temperature in cutting is located at about the middle of the tool–chip interface?

8.5. Explain why in Illustrative Problem 8.2 it is stated at the end that the difference between the two energies involved becomes smaller as the rod diameter decreases. Note that the depth of cut (0.010 in. on the radius) is kept constant.

8.6. Explain the effect of material properties and operating conditions on inhomogeneous (serrated) chip formation in metal cutting (Fig. 8.6d).

8.7. It has been observed that the cutting force F_c in Fig. 8.7 increases with increasing depth of cut and decreasing rake angle. Explain why.

8.8. Under what conditions would you discourage the use of cutting fluids?

8.9. What are the effects of performing a cutting operation with a dull tool tip?

8.10. It is generally undesirable to allow temperatures to rise too high in cutting operations. Explain why this is so.

8.11. Explain the characteristics of up milling and down (climb) milling.

8.12. Describe and explain the trends that you see in Table 8.2.

8.13. To what factors do you attribute the large difference in the specific energies in Table 8.3?

8.14. Explain why the same tool life may be obtained at two different cutting speeds.

8.15. Inspect Table 8.7 and explain which tool materials would not be particularly suitable for interrupted cutting operations and why.

8.16. Explain the disadvantages in a cutting operation where the type of chip produced is like that shown in Fig. 8.5(c).

8.17. We note in Fig. 8.25 that tool life can be almost infinite at low cutting speeds. Would you recommend that all machining be done at low speeds? Explain the disadvantages.

8.18. How would you explain the effect of cobalt content on the properties of carbides as shown in Fig. 8.55?

9 Material-Removal Processes: Abrasive, Chemical, Electrical, and High-Energy Beams

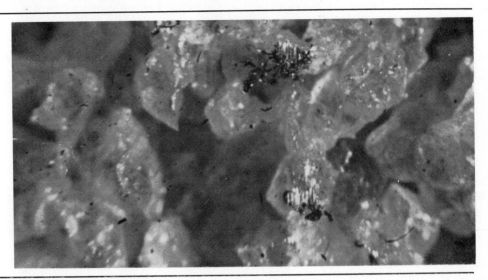

9.1 INTRODUCTION

In all the cutting processes described in Chapter 8, it was noted that the tool is made of a certain material and that it has a clearly defined geometry. Furthermore, the cutting process is carried out by chip removal, the mechanics of which are reasonably well understood.

There are many situations in manufacturing, however, where the workpiece material is either too *hard*, too *brittle*, or its *shape* is difficult to produce with sufficient accuracy by any of the cutting methods described in the preceding chapter. This chapter describes all machining processes where chip removal is performed either on a very small scale (grinding) or where material is removed by mechanisms other than in chips.

The processes described in this chapter use various sources of energy, including chemical, electrical, and optical. These processes are used widely in manufacturing industries, either for their technical capabilities or because of the economics of the overall operation.

9.2 ABRASIVE PROCESSES

In Chapter 8 material-removal processes employed single-point tools (turning) or multipoint tools (milling) of known geometry. Typical tool materials included high-speed steels, carbides, and ceramics.

In abrasive removal processes, the tool is made of abrasive particles of irregular geometry. The most common example is the use of a *grinding wheel* where the particles are held together by a suitable bond. Other examples are belt grinding or sanding with coated abrasives (sandpaper, emery paper), and processes such as honing, lapping, buffing, polishing, shot blasting, and ultrasonic machining.

An abrasive is a hard particle with sharp edges capable of removing material from a softer surface by scratching (or abrasive wear, Section 4.8.2), which is basically a machining process involving chip removal. A basic difference between cutting with a tool and grinding lies in the self-sharpening capability of abrasive grains, which fracture periodically (Section 9.4.4).

The *abrasives* commonly used in grinding and related processes are aluminum oxide (Al_2O_3), silicon carbide (SiC), cubic boron nitride (CBN), and diamond.

Table 9.1 gives the hardness of various materials. It is noted that all four abrasives listed above are considerably harder than hardened steels or cutting-tool materials such as tungsten carbide.

Abrasives, as they are used in manufacturing processes, are very small in size as compared to cutting tools. Size is identified by *grit number* (Fig. 9.1) with 10 being very coarse, 100 fine, and 500 very fine. The sharp edges of the grits, or grains, and their

TABLE 9.1
KNOOP HARDNESS FOR VARIOUS MATERIALS AND ABRASIVES

Common glass	300–500	Zirconium carbide	2100
Flint, quartz	800–1100	Titanium nitride	2000
Zirconium oxide	1000	Titanium carbide	1800–3200
Hardened steels	700–1300	Silicon carbide	2100–3000
Emery, garnet, topaz	1350	Boron carbide	2800
Tungsten carbide	1800–2400	Cubic boron nitride	4000–5000
Aluminum oxide	2000–3000	Diamond	7000–8000

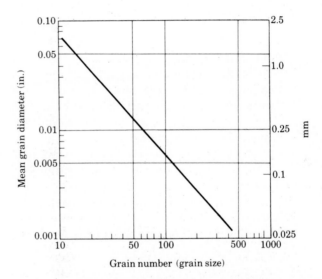

FIGURE 9.1 Relationship between abrasive grain size (number) and its mean diameter. Grain size up to 24 is considered coarse, above 220 very fine. Grain sizes as fine as 1000 may be used for finishing operations such as polishing.

small size allow the removal of material from the workpiece surface in very small quantities.

Abrasives also have high elastic modulus (about 50×10^6 psi [3.5×10^5 MPa]), which means that they do not distort much under the forces encountered in grinding. However, in bonded abrasives described in Section 9.3, the bond has a low modulus and determines the overall stiffness of the product such as a grinding wheel.

It is apparent from the foregoing observations that abrasives have the following special applications:

a. Removal of material from hard or heat-treated surfaces,

b. Very small volume of material removal, and

c. Close control of dimensions and tolerances.

Because of these important characteristics, abrasive methods of material removal are used in such critical applications as grinding ball bearings and roller bearings and their races, gears, cams, crankshafts, pistons, valves and cylinders, rolling-mill rolls, tools, dies and cutters, precision components for instruments, and cutting off of metallic and nonmetallic parts.

9.2.1 ABRASIVES

Abrasives found in nature that are useful in manufacturing are emery, corundum (aluminum oxide), quartz, garnet, and diamond. However, natural abrasives have been found to contain unknown and variable amounts of impurities and to possess nonuniform properties that make their performance somewhat unreliable. Consequently, aluminum oxide and silicon carbide, which are the most common abrasives, are now almost completely manufactured synthetically in electric furnaces.

Aluminum oxide (developed in 1893) is obtained by the fusion of bauxite, iron filings, and coke. *Silicon carbide* (1891) is made with silica sand, petroleum coke, and small amounts of sodium chloride and sawdust. These abrasives are then crushed and graded into various sizes by passing them through standard screens.

Both natural and synthetic diamond (developed in 1955) are also used for abrasive processes. These are known as *industrial* or *synthetic diamonds*. Synthetic diamond is produced by using very high pressures and temperatures. *Cubic boron nitride* (1962), another abrasive, is also produced synthetically by similar techniques.

9.3 BONDED ABRASIVES

Because each abrasive grit (or grain) removes a very small amount of material at a time, reasonable rates of removal can be obtained only by having a large number of grains acting together. This is done by using *bonded abrasives*, typically in the form of a grinding wheel (developed in the mid 1800s with natural abrasives).

The structure of a bonded-abrasive wheel is shown in Fig. 9.2. The grains are spaced at some distance from each other and are held together by bonds, which are like posts between the grains. The most commonly used bonded abrasives are vitrified and resin bonded. Other bonding agents are silicates, shellac, rubber, oxychloride, and metal. Selection of proper bond materials should include factors such as strength, flexibility, and sensitivity to temperature and grinding fluids.

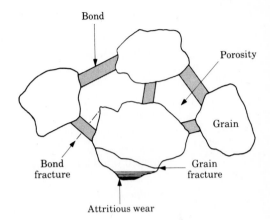

FIGURE 9.2 Schematic illustration of a bonded abrasive (grinding wheel) and its wear and fracture patterns. The type of grain, porosity, and the type, strength, and percentage of the bond can be controlled to produce grinding wheels with different properties and grinding characteristics.

9.3.1 BOND TYPES

Vitrified Bond

This bond is essentially a *glass* and is also called a *ceramic*. The raw materials are feldspar and clays, which are mixed with the abrasives, moistened, and molded into shapes under pressure. These products (known as "green") are then fired slowly up to a temperature of about 2300°F (1260°C) to fuse the glass. They are then cooled slowly (to avoid thermal cracking), finished to size, inspected, and tested. The abrasive grains are thus held together by individual posts of glass.

Vitrified bonds produce wheels that are strong, brittle, stiff, porous, and resistant to oil, acids, and water, and are used for a wide variety of applications. However, because they are brittle, they lack resistance to mechanical and thermal shock. Some vitrified wheels are strengthened by steel backing plates or cups, which are also used to mount the wheels on grinding machines.

Resinoid Bonds

These bonding materials are *thermosetting resins* (Sections 10.3 and 10.9) with a wide range of compositions and properties. Because the bond is an organic compound, wheels with resinoid bonds are also called *organic wheels*. They are manufactured by mixing the abrasive with liquid or powdered phenolic resins and other additives, pressing into shape, and then curing at temperatures of about 350°F (175°C).

Because of the lower elastic modulus of thermosetting resins, resinoid-bonded wheels are more flexible than vitrified wheels. Their strength can be reduced by grinding fluids, thus the compatibility of the wheel with fluids should be considered in the use of these wheels.

Resinoid-bonded wheels can be *reinforced* with one or more layers of fiberglass mats of various mesh sizes. The role of this reinforcement is to prevent the disintegration of the wheel, should it break for some reason, rather than to improve its strength. Large-diameter resinoid wheels can also be strengthened internally with one or more steel rings, such as those used in foundries.

Silicate Bond

The sodium silicate bond is essentially water glass. It is mixed with the abrasive grains, molded, dried, and baked. The strength of silicate bonds is lower than that of vitrified bonds, so its applications are very limited.

Shellac Bond

Flake shellac (the refined form of a resinous substance secreted by insects) is mixed with the abrasive grains, molded, and baked to form a bond. Shellac bonds are suitable for applications where both strength and flexibility are required.

Rubber Bond

This is the most flexible bond used with abrasives. Crude rubber, sulfur, and the abrasive grains are mixed together, rolled into sheets, cut into circles and then heated under pressure to vulcanize the rubber. Very thin wheels can be made with rubber bonds for applications such as cut-off wheels (similar to circular saws and used for cutting bars and structural members).

Oxychloride Bond

This bond is composed of the oxide and chloride of magnesium which are mixed with the abrasive grains and set at room temperature. It has limited use and because grinding fluids attack it, wheels made with oxychloride bond must be used dry.

Metal Bonds

Metal bonds are primarily used with diamond and cubic boron nitride. With powder metallurgy techniques (Chapter 11) the abrasive grains are bonded, usually in a bronze matrix, to the periphery of the wheel (to depths of 0.25 in. [6 mm] or less). The bulk of the wheel may be made of steel, bronze, or plastic.

9.3.2 GRADE

The grade of a bonded abrasive is a measure of the *strength* of the bond; hence grade includes not only the type of bond but also the amount of the bond. Because strength and hardness are related, the grade is also referred to as *hardness* of a bonded abrasive. Thus, a hard wheel has a strong bond and/or has a large amount of bonding material between the grains. A soft wheel has a lower strength bond and/or less bond material.

9.3.3 STRUCTURE

This is a measure of the *porosity* of the bonded abrasive, i.e., spacing between the grains (see Fig. 9.2). It ranges from dense to open. Porosity is needed to provide clearance for the chips. The structure of a wheel also determines the number of cutting points per unit area on the wheel; the greater the porosity, the fewer the number of cutting points.

9.3.4 DESIGNATION

Bonded abrasives are marked with a standardized system of letters and numbers indicating the type of abrasive, grain size, grade, structure, and bond type. Figure 9.3 shows the marking system for aluminum oxide and silicon carbide bonded abrasives, and Fig. 9.4 for diamond and cubic boron nitride.

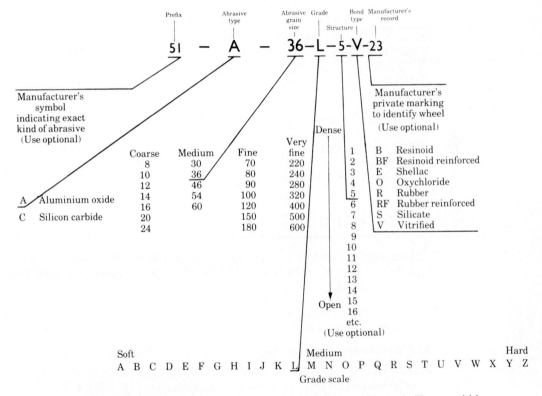

FIGURE 9.3 Standard marking system for aluminum oxide and silicon carbide bonded abrasives. Note that the use of a number identifying the structure may be optional. See also Table 9.4.

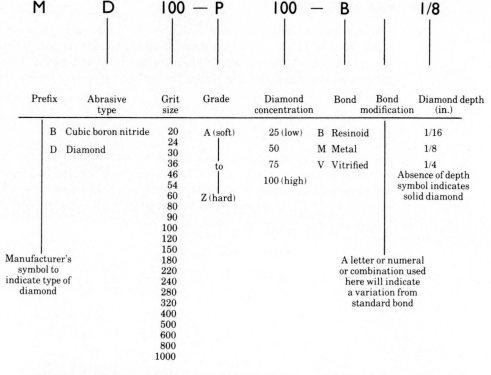

FIGURE 9.4 Standard marking system for cubic boron nitride and diamond bonded abrasives. These two abrasives are the hardest materials available and are manufactured synthetically under high pressure and temperature.

9.4 MECHANICS OF GRINDING

Grinding is basically a chip-removal process where the cutting tool is an individual abrasive grain. The major differences between the action of a single grain and a single-point tool are:

a. The individual grain has an irregular geometry. It is spaced randomly along the periphery of the wheel, Fig. 9.5.

b. There are variations in the radial positions of the grains.

c. The average rake angle of the grains is highly negative, as much as −60 degrees or even lower. Consequently the shear angles are very low (see Fig. 8.1).

d. The cutting speeds are very high, typically 6000 ft/min (30 m/s), Table 9.2.

An example of chip formation by an abrasive grain is shown in Fig. 9.6. Note the negative rake angle, the low shear angle, and the small size of the chip. A variety of

FIGURE 9.5 The grinding surface of an abrasive wheel (A46–J8V) showing grains, porosity, wear flats on grains (see also Fig. 9.9), and metal chips from the workpiece adhering to the grains. Note the random distribution and shape of the abrasive grains. Magnification: 50×.

TABLE 9.2
TYPICAL RANGE OF SPEEDS FOR ABRASIVE PROCESSES

	ft/min	m/s
Grinding		
Workpiece	30–200	0.15–1
Wheel	4500–6500	22.5–32.5
Buffing	6000–12,000	30–60
Polishing	5000–7500	25–37.5

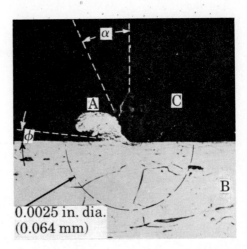

0.0025 in. dia.
(0.064 mm)

FIGURE 9.6 Grinding chip being produced by a single abrasive grain. (A) chip, (B) workpiece, (C) abrasive grain. Note the large negative rake angle of the grain and, consequently, the small shear angle. Compare this figure with those in Fig. 8.6. *Source:* Courtesy of M. E. Merchant.

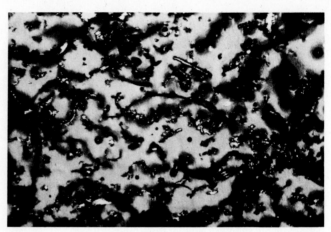

FIGURE 9.7 Typical chips produced in grinding. Chips can be collected easily by holding an adhesive tape against the sparks from a grinding wheel. Most chips are thin and long. However, some chips may become spherical because they melt and resolidify.

metal chips can be observed in grinding, as can be seen in Fig. 9.7. (Chips are easily collected by holding a piece of adhesive tape against the sparks of grinding wheel.)

9.4.1 FORCE AND ENERGY REQUIREMENTS

The mechanics of grinding and the variables involved can best be studied by analyzing the surface grinding operation (Fig. 9.8). A grinding wheel of diameter D is removing a layer of metal at a depth d, known as the *wheel depth of cut*. An individual grain on the periphery of the wheel is moving at a tangential velocity V (*up grinding*, as shown in Fig. 9.8) and the workpiece is moving at a velocity v. The grain is removing a chip whose undeformed thickness (*grain depth of cut*) is t and the undeformed length is ℓ.

It can be shown that, for $v << V$, the *undeformed chip length ℓ* is approximately

$$\ell = \sqrt{Dd}. \tag{9.1}$$

It can also be shown that the *undeformed chip thickness t* is given by

$$t = \sqrt{\frac{4v}{VCr}} \sqrt{\frac{d}{D}}, \tag{9.2}$$

where C is the number of cutting points per unit area on the periphery of the wheel, and r is the ratio of the chip width to average undeformed chip thickness.

Experimental observations indicate the value of C to be roughly on the order of 10^2 to 10^3 per square inch (0.1 to 10 per square mm); the finer the grain size of the wheel, the larger this number. The magnitude of r has been estimated to be between 10 and 20 for most grinding operations. If we substitute typical values for a grinding operation in the foregoing equations, it can be seen that ℓ and t are very small quantities (see Illustrative Problem 9.1).

Relative Grain Force

If it is assumed that the force on the grain (cutting force F_c in Fig. 8.7) is proportional to the cross-sectional area of the undeformed chip, it can be shown that the relative grain force is given by

$$\text{Relative grain force} \propto \frac{v}{VC}\sqrt{\frac{d}{D}}. \tag{9.3}$$

The actual force is the product of the relative grain force and the strength of the metal being ground.

Specific Energy

The specific energy consumed in producing a grinding chip can be considered to consist of the following three components:

$$u = u_{\text{chip}} + u_{\text{ploughing}} + u_{\text{sliding}}, \tag{9.4}$$

where u_{chip} is the specific energy required for chip formation by plastic deformation and $u_{\text{ploughing}}$ is the specific energy required for ploughing, which is plastic deformation without chip removal, similar to Fig. 4.15(a). The last term, u_{sliding}, can best be understood by observing the grain in Fig. 9.9. The grain develops a *wear flat* as a result of the grinding operation (similar to flank wear, Fig. 8.21). The wear flat slides along the surface being ground and, because of friction, requires energy for sliding. The larger the wear flat, the higher the grinding force. (Other effects of wear flats are described below.)

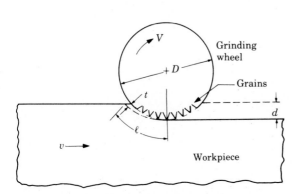

FIGURE 9.8 Variables in the surface grinding operation. In actual grinding, the wheel depth of cut d and contact length l are much smaller than the wheel diameter D. The dimension t is known as the grain depth of cut. In practice d is on the order of 0.0005 to 0.003 in. (0.01 to 0.07 mm).

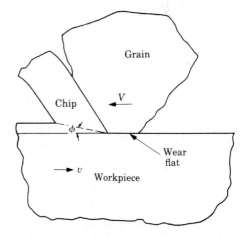

FIGURE 9.9 Schematic illustration of chip formation by an abrasive grain. Compare with Figs. 8.1 and 9.6. Note the negative rake angle, small shear angle, and the wear flat on the grain.

Typical specific energy requirements in grinding are given in Table 9.3. Note that these energy levels are much higher than those in cutting operations with single-point tools (see Table 8.3). This difference has been attributed to the following factors:

a. Size Effect: As stated above, the size of grinding chips is quite small as compared to chips from other cutting operations, by about two orders of magnitude. As described in Section 3.9, the smaller the size of a piece of metal, the greater its strength. Thus, grinding involves higher specific energy than cutting.

b. Wear Flat: Because a wear flat requires frictional energy for sliding, this energy can contribute substantially to the total energy consumed. The size of the wear flat in grinding is much larger than the grinding chip, unlike in metal cutting by a single-point tool where flank wear land (Fig. 8.21) is small compared to the size of the chip.

c. Chip Morphology: Since the average rake angle of a grain is highly negative (Fig. 9.6), the shear strains are very large. This indicates that the energy for plastic deformation to produce a grinding chip is higher than in other cutting processes. Furthermore, ploughing (see Fig. 4.15a) consumes energy without contributing to chip formation.

● **Illustrative Problem 9.1**

Estimate the undeformed chip length and undeformed chip thickness for a typical surface grinding operation.

SOLUTION. The formulas for undeformed length and thickness, respectively, are:

$$\ell = \sqrt{Dd}$$

and

$$t = \sqrt{\frac{4v}{VCr}} \sqrt{\frac{d}{D}}.$$

From Table 9.2 let us select the following values:

$$v = 0.5 \text{ m/s}$$

$$V = 30 \text{ m/s}$$

and also, assume that

$$d = 0.05 \text{ mm}$$

$$D = 200 \text{ mm}$$

TABLE 9.3
**APPROXIMATE UNIT POWER REQUIREMENTS
FOR SURFACE GRINDING**

WORKPIECE MATERIAL	HARDNESS	UNIT POWER,* hp · min/in.³
Aluminum	150 HB	2.5–10
Cast iron (class 40)	215 HB	4.5–22
Low-carbon steel (1020)	110 HB	5 –25
Titanium alloy	300 HB	6 –20
Tool steel (T15)	67 HRC	6.5–30

* Multiply by 2.73 to obtain $W \cdot s/mm^3$.

and let

$C = 2$ per mm^2, and

$r = 15$.

Then,

$$\ell = \sqrt{(200)(0.05)} = 3.2 \text{ mm} = 0.126 \text{ in.}$$

and

$$t = \sqrt{\frac{(4)(0.5)}{(30)(2)(15)}} \sqrt{\frac{0.05}{200}} = 0.006 \text{ mm} = 2.3 \times 10^{-4} \text{ in.}$$

Because of plastic deformation, the actual length of the chip will be shorter, and the thickness greater than these values (see Fig. 9.6). ●

9.4.2 TEMPERATURE

The work expended in grinding is mainly converted into heat. The surface temperature rise ΔT has been found to be a function of the ratio of the total energy input to the surface area ground. Thus, in surface grinding, if w is the width and L is the length of the surface area ground, then,

$$\Delta T \propto \frac{uwLd}{wL} \propto ud. \tag{9.5}$$

If size effect is introduced (see Figs. 3.49 and 3.50) and it is assumed that u varies inversely with the undeformed chip thickness t, then the temperature rise is,

$$\Delta T \propto \frac{d}{t} \propto d^{3/4} \sqrt{\frac{VC}{v}} \sqrt{D}. \tag{9.6}$$

The peak temperatures in chip generation during grinding may be as high as 3000°F (1650°C). However, the time involved in producing a chip is extremely short, on the order of microseconds, hence melting may or may not occur. Because the chips carry away much of the heat generated (as in metal cutting, Fig. 8.20), only a fraction of the heat generated is conducted to the workpiece.

Experiments indicate that as much as one half the chip-formation energy is conducted to the workpiece. (This is higher than in metal cutting.) The heat generated due to sliding and ploughing is conducted mostly into the workpiece.

Sparks

The sparks observed in grinding metals are actually glowing chips. They glow because of the exothermic reaction of the hot chips with oxygen in the atmosphere. Sparks have not been observed with any metal that is ground in an oxygen-free environment. The color, intensity, and shape of the sparks depend on the composition of the metal being ground.

If the heat generated due to exothermic reaction is sufficiently high, it is possible for the chip to melt and, due to surface tension, acquire a round shape and solidify as a shiny spherical particle. (Such chips can be seen in Fig. 9.7.) Observation of these particles under scanning electron microscopy has revealed that they are hollow and have a fine dendritic structure, indicating that they were once molten and resolidified. (It has been suggested that some of the spherical particles may also be produced by plastic deformation and rolling of chips at the grit-workpiece interface.)

9.4.3 EFFECTS OF TEMPERATURE

Temperature rise in grinding can have significant effects on surface properties and residual stresses. Temperature gradients distort the part because of differential thermal expansion and contraction. Furthermore, if the heat generated is allowed to be conducted into the workpiece, it expands the part being ground and hence it is difficult to control tolerances.

Tempering

Excessive temperature rise due to grinding can cause tempering and softening of the surfaces of steel components. Grinding process parameters must therefore be chosen carefully so that excessive temperature rise is avoided. The use of grinding fluids is an effective means of controlling temperatures. (See Section 8.12 and Table 6.7.)

Burning

If the temperature is excessive the surface may burn. Burning produces a bluish color on steels, which is an indication of oxidation at high temperatures. A burn may not be objectionable in itself. However, the surface layers may undergo metallurgical transformations, with martensite formation in high-carbon steels due to rapid cooling. This is known as *metallurgical burn*.

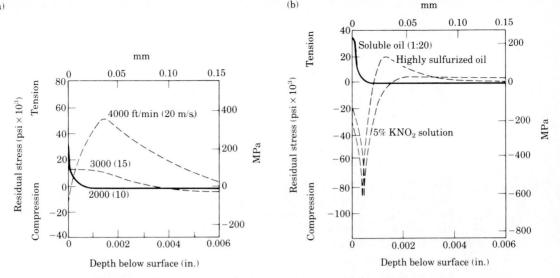

FIGURE 9.10 Residual stresses developed on the workpiece surface in grinding tungsten. (a) Effect of wheel speed. (b) Effect of grinding fluid. Tensile residual stresses on a surface are detrimental to the fatigue life of ground components. The variables in grinding can be controlled to minimize residual stresses. This is known as low-stress grinding. *Source:* After N. Zlatin, *et al.*, 1963.

High temperatures in grinding may also lead to thermal cracking of the surface of the workpiece, known as *heat checking.* Cracks are usually perpendicular to the grinding direction; however, under severe grinding conditions, parallel cracks may also develop.

Residual Stresses

Temperature change and gradients within the workpiece are mainly responsible for residual stresses in grinding. Other contributing factors are the physical interactions of the abrasive grain in chip formation and the sliding of the wear flat along the workpiece surface, causing plastic deformation of the surface. Two examples of residual stresses in grinding are shown in Fig. 9.10, demonstrating the effects of wheel speed and the type of grinding fluid used. The method and direction of the application of grinding fluid can also have a significant effect on residual stresses.

Because of the deleterious effect of tensile residual stresses on fatigue strength (Fig. 4.10), process parameters should be chosen carefully. Residual stresses can usually be lowered (*low stress* or gentle grinding) by using softer grade wheels (free cutting wheels), lower wheel speeds, and higher work speeds.

9.4.4 WEAR OF BONDED ABRASIVES

The overall wear of a bonded abrasive takes place by three distinct mechanisms: attritious wear, fracture of the grain, and fracture of the bond.

Attritious Wear

The sharp cutting edges of an abrasive grain become dull by *attrition*, developing the wear flat shown in Fig. 9.9. This type of wear is due to the interaction of the grain with the workpiece material. Complex physical and chemical reactions take place between the two materials. These include oxidation of the grain surface due to high temperatures, diffusion, chemical degradation or decomposition, fracture at a microscopic scale, melting, and plastic flow.

Attritious wear is low when the two materials are chemically inert with respect to each other, thus lowering the tendency for reaction and adhesion. For example, the rate of attritious wear of aluminum oxide on steel is much lower than that for silicon carbide. This is because aluminum oxide is relatively inert with respect to iron. On the other hand, silicon carbide can be dissolved in molten iron.

Thus, the selection of the type of abrasive for low attritious wear is based on the reactivity of the grain and the workpiece and mechanical properties, such as the relative hardness and toughness. The environment and the type of grinding fluid used also have an effect on grain–workpiece interactions.

Grain Fracture

Abrasive grains are brittle and, hence, their fracture characteristics in grinding are important. If the wear flat from attritious wear is excessive, the grain becomes dull and cutting is inefficient. It is desirable for the grain to fracture or fragment at a moderate rate so that new sharp cutting edges are produced. (This is equivalent to breaking a dull piece of chalk in order to be able to draw fine lines on the board.)

Friability

Friability describes the fracture behavior of abrasives. High friability indicates low strength or low fracture resistance of the grain. Thus, a highly friable grain fragments more easily under the dynamic forces in a grinding operation than one with low friability. The shape and size of the grain also have an effect on friability. Block-shaped particles are less friable than those that are platelike. Also, because of size effect, smaller particles are stronger and hence less friable than larger ones.

Aluminum oxide has lower friability than silicon carbide. This indicates that aluminum oxide has less tendency to fragment (or fail by microchipping) during grinding. Selection of a material for a particular application should also include consideration of the attritious wear rate. A grain–workpiece combination with high attritious wear and low friability indicates a dull grain with a large wear flat; cutting is inefficient and surface damage is likely to occur. The following combinations are generally recommended:

a. Aluminum Oxide: with steels, ferrous alloys, and alloy steels

b. Silicon Carbide: with cast iron, nonferrous metals, and hard and brittle materials such as carbides, ceramics, marble, and glass

c. Diamond: with cemented carbides and some hardened steels

d. Cubic Boron Nitride: with steels and cast irons at 50 HRC or above, and high-temperature superalloys.

Bond Fracture

The strength of the bond (*grade*) is also a significant parameter in grinding. If the bond is too strong, dull grains cannot be dislodged and cutting becomes inefficient. On the other hand, if the bond is too weak the wear rate of the wheel is too high, tolerances may not be held, and the operation becomes uneconomical. In general, softer bonds are recommended for materials that are difficult to grind and for reducing residual stresses and thermal damage. Hard-grade wheels are used for softer metals and to remove large amounts of material at high rates.

Dressing

Dressing is the process of conditioning worn grains on the surface of a bonded abrasive in order to produce sharp new grains. This is done when the wheel becomes dull due to excessive attritious wear (*glazing*, due to the shiny appearance of the wheel surface), or when the wheel becomes *loaded*. Loading is a condition where the porosities on the surface of the wheel become filled or clogged with chips or other materials during grinding. A loaded wheel cuts very inefficiently, generates frictional heat, and causes surface damage.

Dressing is done by various methods. In one method, a specially shaped diamond, or diamond cluster, is moved across the width of the grinding face of a rotating wheel and removes a layer of abrasives from the wheel surface. This can be done either dry or wet, depending on whether the wheel will be used dry or wet, respectively, in practice.

In another very coarse method a set of star-shaped steel disks is pressed against the wheel and removes material from the surface by crushing the grains. Dressing can also be done with abrasive sticks or with other abrasive wheels. Dressing techniques and the rate at which the surface of the wheel is dressed is a significant factor affecting grinding forces and surface finish. A finely dressed wheel produces a fine surface finish.

Dressing is also done to generate a certain shape or form on a grinding wheel with the purpose of grinding profiles on workpieces. One example is thread grinding. Dressing is also a means of *truing* the wheel (making the wheel into a true circle).

9.4.5 GRINDING RATIO

The *grinding ratio G* is defined as,

$$G = \frac{\text{volume of metal removed}}{\text{volume of wheel wear}}. \tag{9.7}$$

In a particular grinding operation this ratio depends on many factors, such as the type of wheel, the workpiece material, the grinding fluid, and process parameters such as depth of cut and speeds. *G* ratios in practice vary over a wide range, from about 2 to 200 and higher.

A particular grinding wheel may act soft or hard, regardless of its grade, just as a pencil acts soft when writing on rough paper (wear rate is high) and acts hard on smooth paper. In grinding, this behavior is due to the force acting on the grain. Thus, the greater the force the greater the tendency for grain fragmentation or bond fracture to occur, hence the softer the wheel acts.

From Eq. (9.3) it is evident that this force increases with work speed v, and wheel depth of cut d, and decreases with increasing wheel speed V, number of cutting points C (more dense structure), and wheel diameter D. Thus, a wheel in surface grinding acts soft when v and d increase, or when V, C, and D decrease. This, in turn, affects the grinding ratio G.

It is not always desirable to try to obtain a high G ratio with a longer wheel life in mind, as this could indicate dulling of the grains and possible surface damage. A lower G value may be quite acceptable when an overall economic analysis justifies the same.

● **Illustrative Problem 9.2**

A surface-grinding operation is being carried out with the wheel running at a constant speed. Will the wheel act soft or hard as the wheel wears down over a period of time?

SOLUTION.   Referring to Eq. (9.3)

$$\text{Relative grain force} \propto \frac{v}{VC}\sqrt{\frac{d}{D}},$$

we note that the only parameter that changes by time in this operation is the wheel diameter D (assuming that d remains constant and the wheel is dressed periodically). Hence, as D becomes smaller by time, the relative grain force increases. Therefore, the wheel acts softer. Some grinding machines are equipped with variable-speed spindle motors to accommodate these changes, and also to make provision for different diameter wheels. ●

9.5 GRINDING PRACTICE

Grinding can be carried out in a variety of wheel–workpiece configurations, as shown in Fig. 9.11. The basic types are: surface, cylindrical, internal, and centerless grinding. The feed may be along the surface of the workpiece (*traverse* grinding, or *cross-feeding*) or it may be into the workpiece (*plunge* grinding). Shaping the surface of the grinding wheel by dressing techniques generates various shapes on the workpiece surface. This process is similar to gear-tooth forming by cutting (Fig. 8.32f). The surface finish and tolerances obtained in various grinding and other removal processes are given in Fig. 9.12.

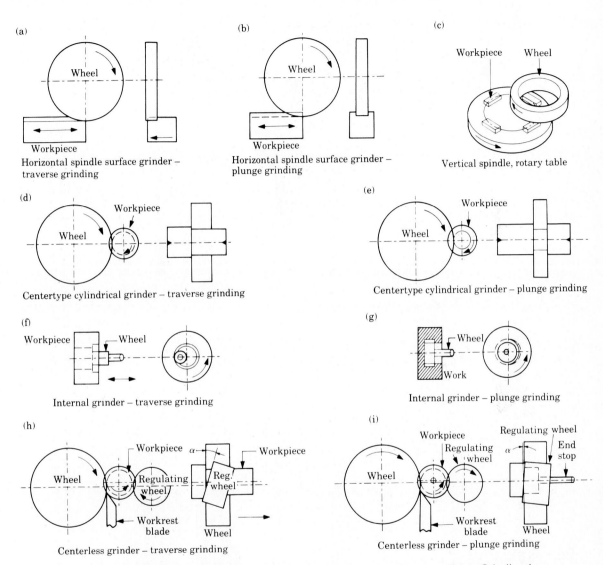

(a) Horizontal spindle surface grinder – traverse grinding

(b) Horizontal spindle surface grinder – plunge grinding

(c) Vertical spindle, rotary table

(d) Centertype cylindrical grinder – traverse grinding

(e) Centertype cylindrical grinder – plunge grinding

(f) Internal grinder – traverse grinding

(g) Internal grinder – plunge grinding

(h) Centerless grinder – traverse grinding

(i) Centerless grinder – plunge grinding

FIGURE 9.11 Schematic illustration of various grinding operations. Grinding is a versatile and important finishing operation, although it can also be used for large-scale removal operations. *Source:* Adapted from *Machinability Data Handbook*, 3d ed., 1980, Metcut Research Associates Inc.

A great variety of grinders is available for different operations. Because of the precision generally required in grinding, special machines have been built with numerous features for automatic loading of the workpieces, clamping, cycling, and wheel dressing.

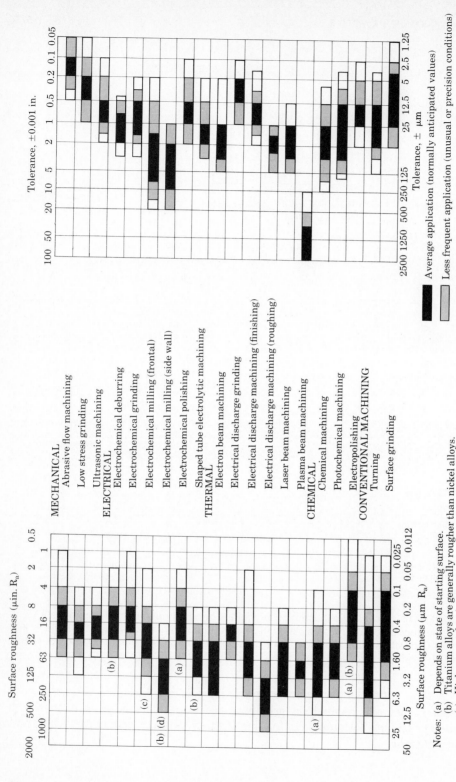

FIGURE 9.12 Surface roughness and tolerances obtained in various material removal processes. Note the wide range of roughness within each machining process. See also Fig. 13.4. *Source:* Compiled from data in *Machinability Data Handbook*, 3d ed., by permission of the Machinability Data Center. © 1980 by Metcut Research Associates Inc.

Tolerance, ±0.001 in.

Tolerance, ± µm

■ Average application (normally anticipated values)

▨ Less frequent application (unusual or precision conditions)

☐ Rare (special operating conditions)

MECHANICAL
  Abrasive flow machining
  Low stress grinding
  Ultrasonic machining
ELECTRICAL
  Electrochemical deburring
  Electrochemical grinding
  Electrochemical milling (frontal)
  Electrochemical milling (side wall)
  Electrochemical polishing
  Shaped tube electrolytic machining
THERMAL
  Electron beam machining
  Electrical discharge grinding
  Electrical discharge machining (finishing)
  Electrical discharge machining (roughing)
  Laser beam machining
  Plasma beam machining
CHEMICAL
  Chemical machining
  Photochemical machining
  Electropolishing
CONVENTIONAL MACHINING
  Turning
  Surface grinding

Surface roughness (µin. R$_a$)

Surface roughness (µm R$_a$)

Notes: (a) Depends on state of starting surface.
     (b) Titanium alloys are generally rougher than nickel alloys.
     (c) High current density areas.
     (d) Low current density areas.

9.5.1 GRINDING OPERATIONS AND EQUIPMENT

Surface Grinding

Surface grinding is performed with a horizontal-spindle or a vertical-spindle wheel where a number of parts can be ground at the same time. It is one of the most common grinding operations (Fig. 9.13).

The workpiece is secured on a magnetic chuck, which is attached to the work table. A straight wheel (Type 1, Fig. 9.14a) is mounted on a horizontal spindle. Grinding is done with the table reciprocating in the longitudinal direction and feeding laterally after each stroke (cross feed).

The size of a surface grinder is identified by the dimensions of the surface that can be ground on that machine. In addition to the design shown in Fig. 9.13, other designs include grinders with vertical spindles (using cup wheels [Fig. 9.14d]), and rotary tables for grinding a number of pieces in one operation (Fig. 9.11).

Cylindrical Grinding

In this operation (see Fig. 9.11), the external cylindrical surface and the shoulders of a workpiece are ground. Examples include crankshafts, axles, spindles, and rolls for rolling mills. Threads and workpieces with two or more diameters (*plunge grinding*) are also ground on these machines. In cylindrical grinding the workpiece reciprocates along its axis, although for large and long workpieces the grinding wheel reciprocates. The latter design is called a *roll grinder*.

Cylindrical grinders are identified by the maximum diameter and length of the workpiece that can be ground. In universal grinders, both the workpiece and the wheel axis can be swiveled around a horizontal plane, thus permitting the grinding of tapered shafts and similar parts.

Internal Grinding

In this operation a small wheel grinds the inside diameter of the part, such as bearing races or bushings. The workpiece is held inside a rotating chuck in the headstock and the wheel rotates at 30,000 rpm or higher. Internal grinders also have features whereby the headstock can be swiveled on a horizontal plane to grind tapered holes.

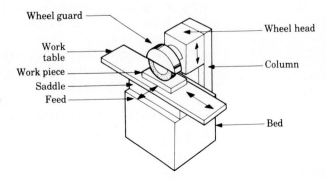

FIGURE 9.13 Schematic illustration of a surface grinder. Workpieces are usually held in place with a magnetic chuck. Note the wheel guard to protect the operator in case of wheel breakage. These machines may also be equipped with fixtures to apply grinding fluids during the operation. The fluid should be turned on after the wheel is turned on, and it should be turned off before stopping the wheel.

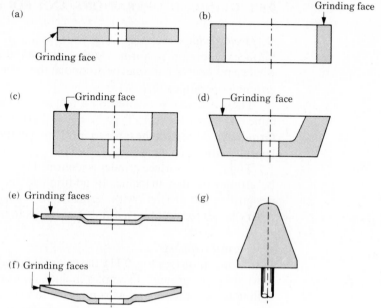

FIGURE 9.14 Examples of some of the common types of grinding wheels. Note that each wheel has a specific grinding surface. Grinding on other faces is improper and unsafe. (a) Type 1, straight. (b) Type 2, cylinder. (c) Type 6, straight cup. (d) Type 11, flaring cup. (e) Type 27, depressed center. (f) Type 28, depressed center. (g) Mounted. Mounted wheels are available in a large variety of shapes and are generally used with small hand-held grinders.

Centerless Grinding

In this process the cylindrical workpiece, which may be a short rod or a long bar, is supported on a workrest blade and is ground between two wheels. Grinding is done by the larger wheel; the smaller wheel (*regulating wheel*) has a rubber bond and regulates the axial movement of the workpiece. This is done by tilting the regulating wheel and running it at speeds of about 1/20 of the larger wheel.

Centerless grinding is a continuous production process requiring little skill. Typical parts ground are roller bearings, piston pins and similar components. Internal centerless grinders, where the workpiece is supported by three rolls, are also available.

Other Grinding Operations

A variety of other special-purpose grinders are available, including tool and cutter grinders, multiple-wheel plunge grinders, cam and drill grinders. For grinding large castings in foundries, swing-frame grinders are available. Rough grinding of castings is called *snagging*. This is usually done on floorstand grinders using wheels as large as 36 in. (915 mm) in diameter.

Portable grinders (either air or electrically driven, or with a flexible shaft connected to a motor) are also available for operations such as grinding off of weld beads, cutting-off operations, and polishing.

9.5.2 ABRASIVE MACHINING

Although grinding has traditionally been associated with small rates of material removal and precision finishing operations, it can also be used for large-scale removal operations such as milling, shaping, and planing. *Abrasive machining* is the term used to describe such processes, where the wheel depth of cut d can be on the order of 1/4 in. (6 mm). Grinders with capacities up to 300 hp (225 kW) are available for this purpose.

Surface finish is of secondary importance. The wheels are mostly resin bonded and are rarely dressed. The overall economics, and its competitive position with other metal-cutting processes, have indicated that abrasive machining can be economical for specific applications.

9.5.3 GRINDING WHEELS

Some of the more commonly used grinding wheels are shown in Fig. 9.14. The surfaces that are to be used for grinding are indicated by an arrow.

Wheel speeds are specified in terms of maximum surface speeds. Recommended speeds usually range from 1500 ft/min (7.5 m/s) for titanium, to 6500 ft/min (32.5 m/s) for steels. However, recently speeds have been increased to as high as 16,000 ft/min (80 m/s) with specially made wheels. Work speeds usually range between 30 and 200 ft/min (0.15 and 1 m/s). (See Table 9.2.)

Typical recommendations for grinding wheels for a variety of operations are given in Table 9.4.

Safety

Because grinding wheels are brittle and are run at high speeds during use, certain procedures must be followed in their handling, storage, and use. Grinding wheels should be properly stored, and protected from extremes of environment. They should be visually inspected for any cracks prior to installing them on spindles. They should be balanced and mounted on spindles with flanges of proper design and dimensions.

TABLE 9.4
TYPICAL RECOMMENDATIONS FOR GRINDING WHEELS

Bolts (centerless grinding)	A80-O6V
Carbide tools (roughing)	C60-18V
	D100-R75B
Crankshafts (diesel)	A54-O5V
Drills (sharpening)	A60-L5V
Glass tubing (cutoff)	C46-J5V
Granite	C16-M4B
Knives (sharpening)	A120-G8V
Nylon	A36-L8V
Pistons, cast iron	A46-18V
Razor blades (final finishing)	A800-K7B
Steel tubing (cutoff)	A80-W10R
Valves, automotive (refacing)	A80-J5V
Welds, stainless steel (portable grinders)	A24-T6R

Grinding wheels should be used according to their specifications and should not be abused. Protection of the operator and wheel guarding are important. (See ANSI Standard B7.1 for details.)

9.5.4 GRINDING FLUIDS

The functions of grinding fluids are similar to those for cutting fluids (Section 8.12). Although grinding can be performed dry, the use of a fluid prevents temperature rise and improves the surface finish and the efficiency of the operation. Grinding fluids are typically water-base emulsions (for general grinding) and oils (for thread grinding). (See Table 6.7.)

Grinding fluids may be applied as a stream (*flood*) or as *mist*, which is a mixture of fluid and air. Because of high surface speeds, it is believed that there is probably an airstream or air blanket around the periphery of the rotating wheel that prevents the fluid from reaching the cutting zone. For these cases, special nozzles have been designed with the fluid applied under high pressure in the proper direction.

9.5.5 GRINDING CHATTER

Chatter is particularly critical in grinding as it adversely affects surface finish and wheel performance. Vibrations in grinding may be attributable to bearings, spindles, unbalanced wheels, and external sources of vibrations. The grinding process can itself cause vibrations due to regenerative chatter (Section 8.13).

The analysis of chatter in grinding requires considerations similar to those described in Section 8.13 for cutting. Thus, the important parameters are stiffness and damping. Additional factors unique to grinding are inhomogeneities in the grinding wheel, dressing techniques, uneven wheel wear, and the gyroscopic action of the wheel during grinding.

These parameters produce characteristic chatter marks on ground surfaces. It is often possible to locate the source of the problem by carefully studying these marks. General guidelines have been established to reduce the tendency for chatter in grinding. Among these are the use of soft-grade wheels, frequent dressing of the wheel, changes in dressing techniques, reduction in material removal rate, and more rigid support of the workpiece.

9.6 MISCELLANEOUS FINISHING OPERATIONS

Abrasive grains are used in a variety of other applications as described below.

Coated Abrasives
These consist of abrasive grains deposited on materials such as paper, cloth, and similar flexible backings (Fig. 9.15). Typical examples are sandpaper, emery cloth, and sanding disks.

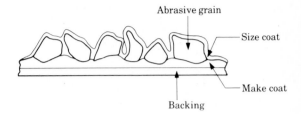

FIGURE 9.15 Schematic illustration of the structure of a coated abrasive. Sandpaper (developed in the 16th century) and emery cloth have this type of structure. Coated abrasives are available in the form of sheets or continuous belts for high-production work.

The grains in coated abrasives are more pointed than those used in grinding wheels. They are deposited electrostatically perpendicular to the plane of the backing. Coated abrasives are available as sheets and belts, and they normally have an open structure. They are used widely in woodworking, finishing flat or curved surfaces on metallic and nonmetallic parts, and polishing of metallographic specimens. The surface finish obtained depends mainly on grain size.

Abrasive Belts

In addition to their traditional use for finishing operations, coated *abrasive belts* are used for high-rate material-removal processes, in some cases replacing conventional grinding operations. Belt speeds range between 2500 and 5500 ft/min (12 and 28 m/s). Specific energy requirements are on the order of 5 to 7 hp $\cdot$ min/in.^{3} (14 to 19 W $\cdot$ s/mm^3) for low-carbon steels. Machines for abrasive-belt operations require proper support of the belt and are rigidly constructed.

Honing

Honing is used primarily for surface finishing of the inside of holes. The honing tool consists of shaped aluminum oxide or silicon carbide sticks mounted on a mandrel, which rotates in the hole with a reciprocating motion. The surface finish obtained is very fine, depending on the type of abrasive, speed, and pressure applied. A fluid is necessary to remove chips and to keep temperatures low.

Superfinishing is similar to honing. The pressure applied is very light. The motion of the hone is short and is controlled in such a manner that the grains do not travel along the same path on the workpiece.

Lapping

Lapping is a finishing operation used on flat or cylindrical surfaces. The lap is usually made of cast iron, copper, leather, or cloth. The abrasive particles are either embedded in the lap, or they may be carried through a fluid. Using fine abrasives, up to size 900, very smooth surface finish and tolerances can be obtained. Depending on the hardness of the workpiece, lapping pressures range from 1 to 20 psi (0.007 to 0.14 MPa).

Polishing

Polishing is a process of producing a smooth, lustrous surface finish. The mechanism by which polishing is accomplished has been debated for some time. It appears that two of the basic mechanisms involved are fine-scale abrasion and smearing, softening, or melting of surface layers.

The first mechanism involves removal of material with fine abrasive particles between the polisher and the workpiece surface. The second mechanism is accomplished by the high temperatures generated due to friction, which softens the surface of the workpiece and produces a smeared surface layer. Polishing is done with cloth disks or belts coated with powders or fine abrasives, such as aluminum oxide or diamond. Mirrorlike finishes can also be obtained on metal surfaces by *electropolishing*. This process is the reverse of electroplating and, because there is no mechanical contact, it is particularly suitable for polishing irregular shapes.

Buffing is similar to polishing, with the exception that very fine abrasives are used on disks made of cloth or hide and running at high speed (Table 9.2).

Vibratory and Barrel Finishing

These processes are used for improving the surface finish and removing burrs from relatively small parts. In this batch-type operation, specially shaped abrasive pellets are placed in a container together with the parts. The container is then either vibrated or tumbled. The impact of individual abrasives and metal particles removes sharp edges and burrs.

Various Finishing Operations

Other finishing or cleaning operations, for less stringent surface-finish requirements, are shot blasting (abrasive-jet machining), abrasive-flow machining, and wire brushing.

In *shot blasting*, the abrasive particles are usually sand propelled with a high-velocity jet of air onto the surface of the material. This process is particularly useful in cutting, stripping, cleaning, and deburring operations on metallic and nonmetallic materials.

In *abrasive-flow machining*, the abrasive grains are mixed in a viscous matrix, which is then forced through the openings and passages in a part. The movement of the matrix under pressure erodes burrs and sharp corners.

Wire brushing involves a rotary wire brush that rotates at high speed and produces a surface composed of fine longitudinal scratches on both external and internal surfaces.

9.7 ULTRASONIC MACHINING

In ultrasonic machining, material is removed from a surface by microchipping or erosion with abrasive particles. The tip of the tool (Fig. 9.16a) vibrates at low amplitude (0.002 to 0.005 in. [0.05 to 0.125 mm]) and at high frequency (20 kHz). This, in turn, transmits a high velocity to fine abrasive grains between the tool and the surface of the workpiece.

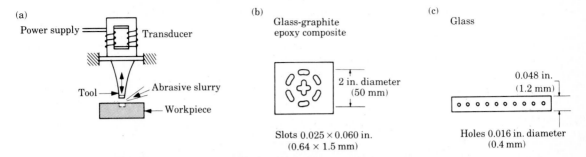

(a) Power supply — Transducer

Tool — Abrasive slurry

Workpiece

(b) Glass-graphite epoxy composite

2 in. diameter (50 mm)

Slots 0.025 × 0.060 in. (0.64 × 1.5 mm)

(c) Glass

0.048 in. (1.2 mm)

Holes 0.016 in. diameter (0.4 mm)

FIGURE 9.16 (a) Schematic illustration of the ultrasonic machining process where material is removed by microchipping and erosion. (b) and (c) Typical examples of holes produced by ultrasonic machining. Note the dimensions of cut and the type of workpiece materials.

The grains are usually boron carbide; aluminum oxide or silicon carbide are also used. Grain size ranges from 100 (for roughing) to 1000 (for finishing). The grains are in a water slurry with concentrations ranging from 20 to 60% by volume. The slurry also carries away the debris from the cutting area.

Mechanics

Microchipping in ultrasonic machining is possible because the stresses produced by particles impacting on a surface are high. This is because the time of contact between the particle and the surface is very short (10 to 100 microseconds) and the area of contact is very small.

It has been shown that the time of contact t_o can be expressed as,

$$t_o \simeq \frac{5r}{c_o}\left(\frac{c_o}{v}\right)^{1/5}, \tag{9.8}$$

where r is the radius of a spherical particle, c_o is the elastic wave velocity in the workpiece ($c_o = \sqrt{E/\rho}$), and v is the velocity with which the particle strikes the surface.

The force F of the particle on the surface is obtained from the rate of change of momentum. Thus,

$$F = \frac{d(mv)}{dt}, \tag{9.9}$$

where m is the mass of the particle.

The average force $\bar{F}$ due to a particle striking the surface and rebounding is given by,

$$\bar{F} = \frac{2mv}{t_o}. \tag{9.10}$$

Substitution of numerical values in Eq. (9.10) indicates that even small particles can exert significant forces and, because of the very small contact area, produce very high stresses. In brittle materials, these stresses are sufficiently high to cause microchipping and erosion of the surface.

Applications

Ultrasonic machining is best suited for hard, brittle materials, such as ceramics, carbides, glass, precious stones, and hardened steels. The tip of the tool is usually made of low-carbon steel and undergoes wear. It is attached to a transducer through the toolholder. With fine abrasives, tolerances of 0.0005 in. (0.0125 mm) or better can be held in this process. Fig. 9.16(b) and (c) show two applications of ultrasonic machining.

● **Illustrative Problem 9.3**

Explain what change, if any, takes place in the magnitude of the impact force of a particle in ultrasonic machining as the temperature of the workpiece is increased.

SOLUTION.   The force of a particle is given by Eq. (9.10). For this problem, m and v are constant. Inspecting the expression for the contact time t_o in Eq. (9.8), we note that it can now be written as,

$$t_o \propto \frac{1}{c_o^{4/5}} \propto \frac{1}{E^{2/5}}.$$

Since, with increasing temperature, the modulus of elasticity E decreases (Fig. 2.11), t_o increases. Therefore, the impact force decreases according to Eq. (9.10).   ●

9.8 CHEMICAL AND ELECTRICAL MACHINING

The machining processes described thus far remove material by chip formation, abrasion, or fracture by microchipping. The mechanical energy required to perform these operations is supplied through either static or dynamic forces using tools of various shapes.

There are situations, however, where these processes are not satisfactory or economical for reasons such as:

a. The hardness and strength of the material are high, i.e., typically above 400 HB.

b. The shape of the part is complex, such as a die cavity in a block of carbide or die steel, or a small-diameter deep hole.

c. The workpiece is too flexible or too delicate to support the cutting forces imposed by conventional machining processes. Examples are thin shells, slender and fragile parts, or components that are difficult to clamp (such as honeycomb structures).

d. Better control of tolerances, part distortion, surface finish, and burr-free performance is required.

e. There are adverse temperature or residual stress effects that may result from conventional machining processes.

In the 1940s these factors led to the development of chemical and electrical means of material removal, as described below. These processes, when selected and applied properly, offer significant economic and technical advantages over traditional methods of machining.

9.8.1 CHEMICAL MACHINING (CM)

In this process material is removed from a surface by dissolution using chemical reagents. There are two types of chemical machining: milling and blanking.

In *chemical milling*, material is removed to produce large shallow cavities on plates, forgings, extrusions, and castings for the purpose of overall weight reduction without sacrificing stiffness. Tapered parts on flat sheet, plate, or disks can also be made by this process.

Selective attack by the chemical reagent on different areas of a workpiece is controlled by removable masking (Fig. 9.17a) or by partial immersion in the reagent. Masking with tapes or paint is a common practice. Elastomers (rubber, neoprene) or plastics (polyvinyl chloride, polyethylene, polystyrene) are also used as masking materials.

An example of tapering is shown in Fig. 9.17(b). A solid circular metal disk (workpiece) of constant thickness is rotated in the reagent while moving its spindle gradually upward. Different curvatures can be obtained by varying the vertical speed with time.

In *chemical blanking* a portion of the material is completely removed from the workpiece. Thus, the process is similar to blanking, with the exception that material is

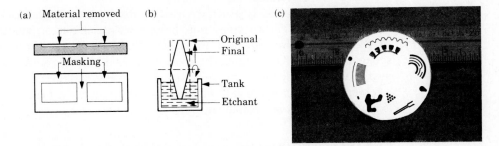

FIGURE 9.17 Examples of chemical machining. (a) Material removed from the surfaces of a plate to reduce its weight while maintaining stiffness. (b) A tapered disk. (c) An example of chemical machining (photochemical) on steel sheet 0.004 in. thick (0.1 mm). Note the fine detail and small size of holes produced. This part is 2 3/8 in. (60 mm) in diameter. *Source:* Newcut, Newark, NY.

removed by chemical dissolution rather than by shearing with a punch and die (Fig. 7.1). Typical applications are *etching* of printed circuit boards, decorative panels, and thin stampings.

Chemicals used for etching in chemical machining depend on the workpiece material. Sodium hydroxide is usually used for aluminum, solutions of hydrochloric and nitric acids for steels, and iron chloride for stainless steels. The workpieces must be cleaned prior to machining. Temperature control and stirring during machining improve the uniformity of material removal. After machining, the parts should be washed thoroughly.

Chemical machining has been applied to a wide variety of metals at depths of cut up to about 0.5 in. (12 mm). Surface damage in chemical milling involves preferential etching and intergranular attack. The typical removal rate is 0.001 in./min (0.025 mm/min). The range of surface finish obtained by this and other processes is given in Fig. 9.12. Tolerances are approximately 10% of the depth removed.

Electropolishing and *photoetching* (Fig. 9.17c) are modifications of chemical machining.

9.8.2 ELECTROCHEMICAL MACHINING (ECM)

In this process, electrolytes dissolve the reaction products formed on the workpiece by electrochemical action; thus, the process is the reverse of electroplating. The electrolyte is pumped at a high rate through the passages in the tool (Fig. 9.18a).

The tool is usually made of brass, copper, or bronze; a variety of other metals can also be used. The tool does not undergo any wear. The electrolyte is usually sodium chloride mixed in water. Sodium nitrate ($NaNO_3$) is also used.

A direct-current power supply in the range of 5 to 25 V maintains current densities, which, for most applications, range between 1000 and 5000 A/in^2 (1.5 and 7.8 A/mm^2) of active cutting surface. For a given material, the penetration rate is proportional to the current density. Surface finish is improved and removal rate is increased with increasing current density.

The ECM process is generally used in machining complex cavities in high-strength materials. The workpiece must be an electrical conductor. Small round or odd-shaped holes are also machined (Fig. 9.18b). ECM is not suited for producing sharp square corners or flat bottom. The process leaves a burr-free surface; in fact, it can also be used as a deburring process, as shown in Fig. 9.18(c). ECM does not cause any thermal damage to the part. Machines with current capacities as high as 40,000 A and as small as 5 A are available.

9.8.3 ELECTROCHEMICAL GRINDING (ECG)

This process is a combination of electrochemical machining and conventional grinding. The equipment is similar to a grinder except that the wheel is a rotating cathode disk with abrasive particles on its periphery, and a flow of electrolyte (usually sodium nitrate) is provided for chemical machining (Fig. 9.19a).

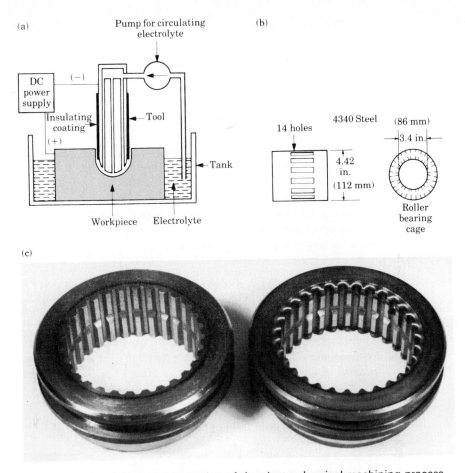

FIGURE 9.18 (a) Schematic illustration of the electrochemical machining process. (b) Roller bearing cage made by ECM. (c) Deburred part for automotive transmissions. Left, as machined; right, as deburred by ECM. *Source:* Anocut.

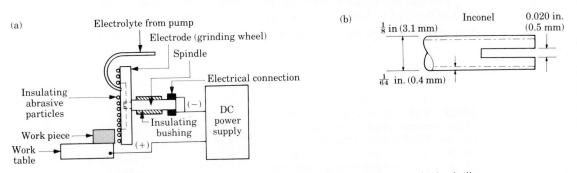

FIGURE 9.19 (a) Schematic illustration of the electrochemical grinding process. Material is removed by a combination of electrochemical and abrasive actions. (b) Thin slot produced in nickel-alloy tubular part by electrochemical grinding.

The wheel is metal-bonded with diamond or aluminum oxide abrasives; it rotates at a surface speed of 4000 to 7000 ft/min (20 to 35 m/s). The abrasives serve as insulators between the wheel and the workpiece and continuously remove electrolytic products from the working area.

The majority of metal removal in ECG is by electrolytic action with currents ranging between 50 and 3000 A. Final cuts are usually done mostly by the grinding action to produce a surface with good finish and dimensional control. Thus wheel wear is very low.

The ECG process is suited for applications similar to milling, grinding, cutting off, sawing, and tool and cutter sharpening (Fig. 9.19b). It is not adapted to cavity-sinking operations such as die making. It has been successfully applied to carbides and a variety of high-strength alloys.

9.8.4 ELECTRICAL-DISCHARGE MACHINING (EDM)

The principle of this process is based on erosion of metal by *spark discharges*. Two charged electrodes are the tool and the workpiece, connected to a direct-current power supply as shown in Fig. 9.20. When the potential difference is sufficiently high, a transient spark discharge takes place through the dielectric fluid, removing a minute amount of metal from the workpiece surface (Fig. 9.21). The discharge is repeated at rates between 50 and 500 kHz. Voltages usually range between 50 and 300 V and currents from 0.1 to 500 A.

The dielectric fluid has the following functions:

a. Acts as an insulator until the potential is sufficiently high,

b. Carries away the debris in the tool-workpiece gap, and

c. Provides a cooling medium.

The most common dielectric fluids are mineral oils. Kerosene, and distilled and deionized water are also used in specialized applications.

The gap between the tool and the workpiece is critical; hence, the downward feed of the tool is done by servocontrol. The rate of metal removal depends on the current density and the physical properties of the workpiece. For a given material,

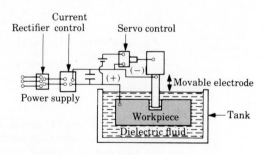

FIGURE 9.20 Schematic illustration of the electrical discharge machining process. EDM is one of the more commonly used modern machining processes. It has numerous applications, such as producing cavities in very large automotive dies.

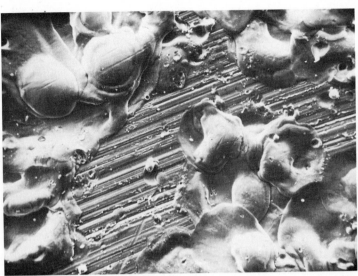

FIGURE 9.21 Craters produced by electrical discharges on a previously ground surface (longitudinal marks). Note the rough surface produced, consisting of molten and resolidified metal. Unless subjected to further finishing operations, such a surface can severely reduce the fatigue strength of parts produced by EDM. See Fig. 9.29. *Source:* Courtesy of J. T. Black and S. Ramalingam.

the frequency of discharges or the energy per discharge is usually varied to control the removal rate.

The removal rate and surface roughness in EDM increase with increasing current and decreasing frequency of sparks. Removal rates generally range from 0.005 to 1.5 in³/h (0.08 to 25 cm³/h). Higher rates are possible but they produce a rough finish with poor surface integrity. Thus, finishing cuts are made at low removal rates.

The melting point and latent heat of melting are important physical properties that determine the volume of metal removed per discharge (Fig. 9.22). Since the process does not involve mechanical energy, hardness or strength of the workpiece does not have any significant effect on removal rate.

Tools

Tools for EDM are usually made of brass, copper, copper-tungsten, or graphite. The tools are shaped by a variety of processes such as forming, casting, powder metallurgy, and machining. Tool wear is an important factor since it affects tolerances and the shape produced in the workpiece. It has been shown that tool wear is related to the melting points of the materials involved (Fig. 9.23). Thus, graphite has the highest wear resistance. Note the very wide range of the wear ratio. Tool wear in electrical-discharge machining of some steels is minimized by reversing the polarity and using copper tools, known in practice as *no-wear* EDM.

Applications

The EDM process is widely used with numerous applications such as producing die cavities for large automotive components, small-diameter (0.002 in. [0.05 mm)] deep holes using tungsten wire as the electrode (Fig. 9.24a and b), narrow slots, and

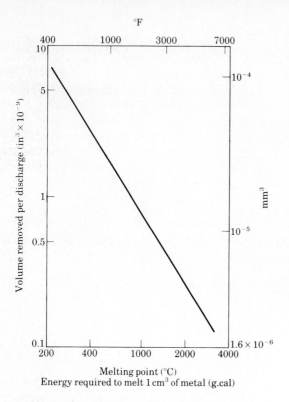

FIGURE 9.22 Effect of melting point of the workpiece material on volume removal rate in EDM. Mechanical properties do not have any significant effect on removal rate in EDM, hence the process can be used on hard and strong materials, provided they are electrically conducting.

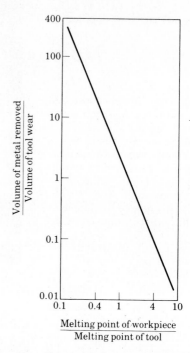

FIGURE 9.23 Relationship between tool wear and melting points of the workpiece and the tool in EDM. Tool wear can be minimized by changing the polarity of the tool, known as no-wear EDM. The ordinate in this figure is similar to the grinding ratio G in Eq. (9.7). Note that this ratio can be less or more than one in EDM.

in the production of shapes on structures such as honeycombs. The process can be used on any material with electrical conductivity.

There are several variations of EDM such as sawing, grinding, and wire cutting. In sawing, a setup similar to a band or circular saw (with no teeth) is used with the same electrical circuit as in EDM. Narrow cuts can be made at high rates of metal removal. Because cutting forces are negligible, the process can be used on thin or delicate components.

In grinding with EDM, the wheel is made of graphite or brass, and has no abrasives. Material is removed by repetitive spark discharges. Figure 9.25 shows an application of electrical-discharge grinding.

Traveling-Wire EDM

A variation of EDM is electrical-discharge wire cutting (Fig. 9.26), often called *traveling-wire* EDM. In this process, a slowly moving wire travels along a prescribed path on the workpiece and cuts the metal, much like a band saw, with sparks acting like cutting teeth.

The wire material is usually brass, copper, or tungsten and is typically about 0.01 in. (0.25 mm) in diameter; thus, narrow cuts are possible. The wire is usually used only once, as it is relatively inexpensive. It travels (at speeds ranging from 0.1 to 6 in./s [2.5 to 150 mm/s]) so that a constant diameter is maintained during the cut. This process is used in cutting plates, as thick as 6 in. (150 mm), and for making punches, dies, and die components from hardened tool materials.

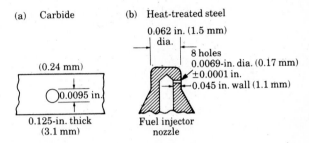

FIGURE 9.24 Typical parts produced by electrical discharge machining. Note the type of workpiece materials used (hard and strong), size of the holes produced, and the tolerances obtained in the fuel injection nozzle in (b). This nozzle has a number of holes on the periphery of the part.

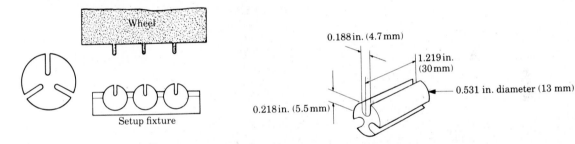

FIGURE 9.25 An example of grooves produced in parts made by the electrical discharge grinding process. The absence of forces in this process allows the machining of closely spaced thin slots. Cast irons are usually not suitable, because the sand inclusions on the surface of the workpiece can damage the wheel.

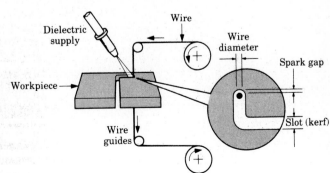

FIGURE 9.26 Schematic illustration of the traveling-wire EDM process (also called electrical-discharge wire cutting). This operation is similar to cutting on a band saw, with spark discharges acting like cutting teeth. As much as 50 hours of accurate cutting can be performed with one reel of wire, which is discarded after use.

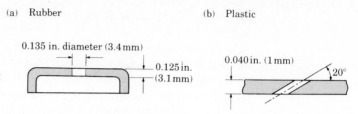

FIGURE 9.27 Holes produced in nonmetallic materials (rubber and plastic) by laser beam machining. Since there is no force involved in this process, the lack of stiffness of the part is not a significant factor.

(a) Rubber

(b) Plastic

0.135 in. diameter (3.4 mm)

0.125 in. (3.1 mm)

0.040 in. (1 mm)   20°

9.9 HIGH-ENERGY–BEAM MACHINING

The high-energy densities of lasers and electron beams are put to use in some machining operations. These beams are also used in welding, hence the descriptions and illustrations in Section 12.5.3 are basically interchangeable with the processes described below.

Laser-Beam Machining (LBM)

In the laser-beam–machining process, the source of energy is a *laser* (an acronym for *l*ight *a*mplification by *s*timulated *e*mission of *r*adiation) that focuses optical energy on the surface of the workpiece. Energy densities suitable for machining range from 10^4 to 10^6 kW/in² (10 to 10^3 kW/mm²). This highly focused, high-density energy beam melts and evaporates the metal. The surface produced by LBM is usually rough and has a heat-affected zone (Section 12.7) that may have to be removed or treated for critical applications.

Important physical parameters in LBM are reflectivity of the workpiece surface, thermal conductivity, specific heat, and latent heat of melting and evaporation. The lower these quantities, the higher the efficiency of laser-beam machining. The process is used on a variety of metallic and nonmetallic materials, and for small-scale cutting operations, such as slitting and drilling holes (Fig. 9.27) as small as 0.0002 in. (0.005 mm) with depth-to-diameter ratios of 50 to 1. (See also Section 12.5.3.)

Laser beams are also used in combination with a gas stream, such as oxygen, nitrogen, or argon. This is known as a *laser beam torch* and is used in cutting thin sheet materials.

In electron-beam machining the source of energy is high-velocity electrons striking a surface (Section 12.5.3). Its applications are similar to laser-beam machining. *Plasma beams* (see Fig. 12.21) are also used in cutting metals at high speeds.

9.10 WATER-JET MACHINING

High-pressure jets of water are also used in various processes such as cutting and deburring, in addition to other operations such as descaling, cleaning, degreasing, and nuclear decontamination. Because of the change in momentum, a water jet can produce forces on solid surfaces that are sufficiently high for these operations. For some of these applications, a pressure of 1000 psi (7 MPa) at the nozzle is sufficient.

An application of water-jet machining in cutting a variety of metallic and nonmetallic materials is shown in Fig. 9.28. The water jet acts like a saw and cuts a very narrow groove (*kerf*, see Figs. 9.26 and 12.43) in the material. Although pressures as high as 200,000 psi (1400 MPa) can be generated, a pressure level on the order of 60,000 psi (400 MPa) is used for efficient operation. A variety of materials (polymers, fabrics, wood products, paper, leather, insulation materials, brick, rubber, etc.) in thicknesses ranging from 0.03 in. (0.8 mm) to 1 in. (25 mm) or more can be cut by this method. The advantages of the process are that cuts can be started at any location without the need for a drilled hole, no heat is involved, wetting of the workpiece material is minimal, no deformation of the rest of the workpiece takes place (hence the process is useful for pliable materials), and the burr produced is minimal. Jet nozzle diameters usually range between 0.004 and 0.012 in. (0.1 to 0.3 mm).

In surface finishing operations abrasives may also be added to the water jet to increase the rate of material removal from the surface. Multiple nozzles may also be used for operations involving large surfaces or a number of parts.

FIGURE 9.28 An example of water jet machining. The letter K was machined from a 6-mm thick plastic sheet, with a water jet at 57,000 psi (400 MPa) with a nozzle diameter of 0.004 in. (0.1 mm). This operation can be performed on ceramics, brick, leather, paper, textiles, printed circuit board, and a variety of other materials.
Source: Sugino USA, Inc.

SUMMARY

1. Machining processes involve not only single- or multi-point tools, but also other methods. Among these are abrasive, chemical, electrical, and high-energy–beam machining. The hardness of the workpiece material is significant to some extent in abrasive processes, but not in the others. This is because these processes rely on mechanisms that do not involve mechanical properties such as strength, hardness, or ductility of the material. Rather, they involve physical, chemical, and electrical properties.

2. Abrasive processes are capable of producing smooth surfaces with very fine accuracy and tolerances. They are basically finishing operations, although abrasives are also used in large-scale material removal operations.

3. Chemical and electrical methods of machining are particularly suitable for difficult-to-work materials. They do not involve forces, significant temperatures, residual stresses, or the mechanical properties of the material. However, their effect on surface integrity must be observed carefully as these processes can produce surfaces that have undergone considerable damage, particularly in electrical-discharge machining, thus reducing fatigue life (Fig. 9.29).

4. The overall choice of machining processes depends not only on technical considerations, but also on economic factors. Depending on the particular process, machining a workpiece can involve considerable time, thus contributing significantly to the overall cost of the product.

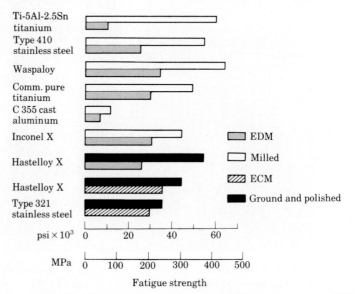

FIGURE 9.29 Comparison of fatigue strength for several alloys machined by various processes. To improve fatigue strength, various finishing operations can be performed, including shot peening and surface rolling. (See Figs, 4.38 and 4.39). *Source:* Pratt and Whitney.

BIBLIOGRAPHY

Abrasive Processes

Farago FT. *Abrasive Methods Engineering*. Metals Park, Ohio: American Society for Metals, 1976.

Grinding Stresses: Cause, Effect and Control. Cleveland: The Grinding Wheel Institute (no date).

Lewis KB. *The Grinding Wheel*, rev. ed. Cleveland: The Grinding Wheel Institute, 1959.

Shaw MC, ed. *New Developments in Grinding*. Pittsburgh: Carnegie Press, 1972.

(See also Bibliography in Chapter 8.)

Chemical and Electrical Machining Processes

DeBarr AE, Oliver DA. *Electrochemical Machining*. New York: American Elsevier, 1968.

Faust CL, ed. *Fundamentals of Electrochemical Machining*. Princeton, N.J.: Electrochemical Society, 1971.

Harris WT. *Chemical Milling: The Technology of Cutting Materials by Etching*. New York: Oxford, 1976.

Livshits AV. *Electro-Erosion Machining of Metals*. London: Butterworths, 1960.

Machining Data Handbook, 3d ed., vol. 2. Cincinnati: Machinability Data Center, 1980.

McGeough JA. *Principles of Electrochemical Machining*. London: Chapman and Hall, 1974.

Source Book on Applications of the Laser in Metalworking. Metals Park, Ohio: American Society for Metals, 1979.

Springborn, RK, ed. *Non-Traditional Machining Processes*. Dearborn, Mich.: Society of Manufacturing Engineers, 1967.

Wilson JF. *Practice and Theory of Electrochemical Machining*. New York: Wiley, 1971.

Cutting Fluids

Cutting and Grinding Fluids. Dearborn, Mich.: Society of Manufacturing Engineers, 1967.

Old, NJ. *Lubricants, Cutting Fluids and Coolants*. Boston: Cahners, 1973.

(See also Bibliography in Chapter 4.)

Recommendations for Machining

Machinery's Handbook, 21st ed. New York: Industrial Press, 1980.

Machining Data Handbook, 3d ed., 2 vols. Cincinnati: Machinability Data Center, 1980.

Metals Handbook, 8th ed., vol. 3. *Machining*, Metals Park, Ohio: American Society for Metals, 1967.

Tool and Manufacturing Engineers Handbook, 4th ed., vol. 1, *Machining*. Dearborn, Mich.: Society of Manufacturing Engineers, 1983.

Periodicals

American Machinist

Cutting Tool Engineering

International Journal of Machine Tool Design and Research

Journal of Engineering for Industry

Machinery

Manufacturing Engineering

Production

Production Engineer

PROBLEMS

9.1. The grinding ratio *G* depends on the following parameters: type of grinding wheel, hardness of workpiece, depth of cut, wheel speed, workpiece speed, and type of grinding fluid used. Explain why.

9.2. Describe the trends that you see in Table 9.4.

9.3. Based on the topics covered in Chapters 8 and 9, suggest several removal processes for concrete. Consider its composition and the properties of the elements in the aggregate.

9.4. Make a list of material-removal processes that may be suitable for the following materials: (a) glass, (b) cast iron, (c) thermoplastics, (d) thermosets, (e) diamond, and (f) annealed copper.

9.5. Give all possible technical and economic reasons why the material-removal processes described in this chapter might be preferred, or even necessary, over those described in Chapter 8.

9.6. Describe the various methods you would use to determine the number of active cutting points per unit surface area on the periphery of a grinding wheel.

9.7. Explain why mechanical properties are not significant in some of the removal processes described in this chapter.

9.8. Explain why ultrasonic machining is not suitable for soft, ductile materials.

9.9. Calculate the average impact force on a steel plate by a spherical aluminum oxide particle of grit size 200 dropped from a height of 1 ft. (See Table 8.7 for density.)

9.10. Explain all the effects that a wear flat has on grinding.

9.11. Why are synthetic abrasives superior to natural abrasives?

9.12. It is generally recommended that a grinding wheel be of a soft grade in grinding hardened steels. Explain why this is so.

9.13. Explain the factors involved in selecting the appropriate type of abrasive for a particular grinding operation.

9.14. What processes would you recommend in die sinking in a die block?

9.15. Explain why the specific energy of material removal in grinding is higher than in other cutting operations.

10 Plastics: Properties and Processing Methods

10.1 INTRODUCTION

Plastics are an important class of materials with an extremely wide range of mechanical, physical, and chemical properties. The first plastics (natural polymers) were produced in the 1860s; however, modern plastics technology (synthetic polymers) began in the 1920s. These materials are called *plastics*, from the Greek *plastikos* meaning they can be molded and shaped.

Plastics are characterized by the following properties: low density, low strength and elastic modulus, low thermal and electrical conductivity, high chemical resistance, and high coefficient of thermal expansion. They can be cast, formed, machined, and joined into different shapes and are available in a wide variety of properties, colors, and opacities. Also included in this group of materials are rubbers and elastomers.

Because of their many unique properties, plastics have increasingly replaced metallic components in numerous applications. This trend is explained by considerations with respect to service requirements, strength-to-weight ratio, design and cost of the material, and ease of manufacturing.

With various additives and reinforcements, plastics are used in a wide range of consumer and industrial products. Some of the major applications are in automotive, electrical and electronic products such as integrated circuits, mechanical equipment, food and beverage containers, packaging, signs, housewares, textiles, safety equipment, toys, appliances, and optical equipment.

In this chapter the structure and characteristics of plastics are described, together with their engineering properties, processing methods, and applications.

10.2 THE STRUCTURE OF PLASTICS

Plastics are composed of *polymer* molecules and various additives. Polymers are long-chain molecules (also called giant molecules or *macromolecules*), which are formed by polymerization; that is, linking and cross-linking of different monomers.

Monomers

A *monomer* is the basic building block of polymers. The word *mer* (from the Greek *meros*, meaning part) indicates the smallest repetitive unit, similar to the term unit cell in crystal structures. Thus, *polymer* means *many mers* or units. Monomers are organic materials. They include carbon atoms joined in covalent bonds (electron sharing) with other atoms such as hydrogen, oxygen, nitrogen, fluorine, chlorine, silicon, and sulfur.

A typical monomer is the ethylene molecule shown in Fig. 10.1(a). These molecules can be made to attach themselves to other ethylene molecules by a chemical reaction involving heat, pressure, and a catalyst. In this reaction, the double bonds between the carbon atoms open and the molecules arrange themselves in a long line (Fig. 10.1b).

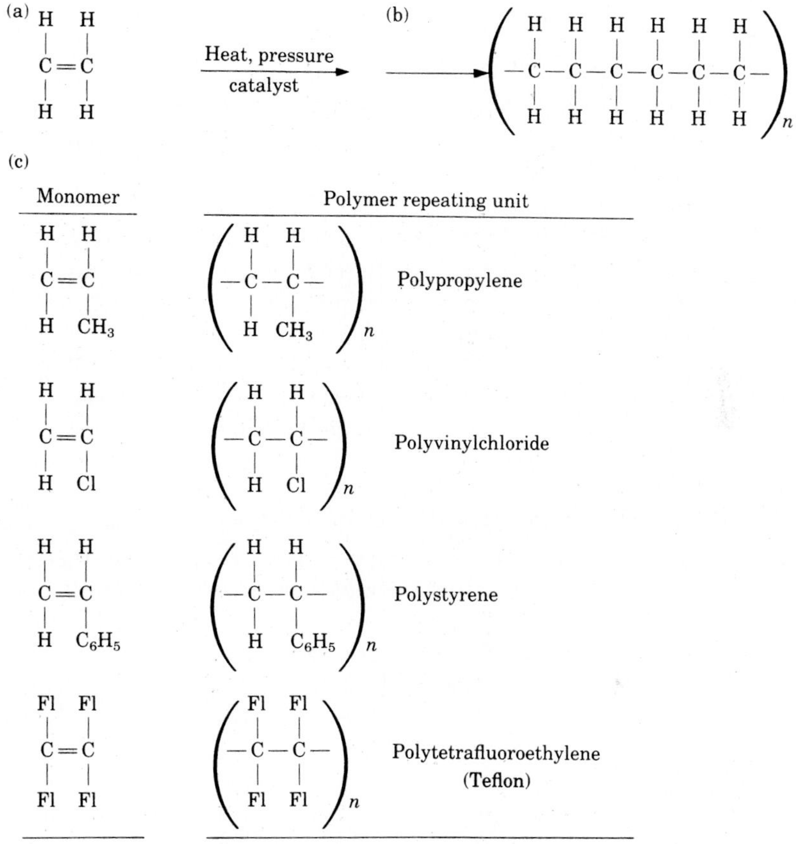

(a)

$$
\begin{array}{cc}
\text{H} & \text{H} \\
| & | \\
\text{C} & = \text{C} \\
| & | \\
\text{H} & \text{H}
\end{array}
\quad \xrightarrow[\text{catalyst}]{\text{Heat, pressure}} \quad
$$

(b)

$$
\left(
\begin{array}{cccccc}
\text{H} & \text{H} & \text{H} & \text{H} & \text{H} & \text{H} \\
| & | & | & | & | & | \\
-\text{C}-\text{C}-\text{C}-\text{C}-\text{C}-\text{C}- \\
| & | & | & | & | & | \\
\text{H} & \text{H} & \text{H} & \text{H} & \text{H} & \text{H}
\end{array}
\right)_n
$$

(c)

Monomer	Polymer repeating unit	
$\begin{array}{cc} \text{H} & \text{H} \\ | & | \\ \text{C} & =\text{C} \\ | & | \\ \text{H} & \text{CH}_3 \end{array}$	$\left(\begin{array}{cc} \text{H} & \text{H} \\ | & | \\ -\text{C}- & \text{C}- \\ | & | \\ \text{H} & \text{CH}_3 \end{array}\right)_n$	Polypropylene
$\begin{array}{cc} \text{H} & \text{H} \\ | & | \\ \text{C} & =\text{C} \\ | & | \\ \text{H} & \text{Cl} \end{array}$	$\left(\begin{array}{cc} \text{H} & \text{H} \\ | & | \\ -\text{C}- & \text{C}- \\ | & | \\ \text{H} & \text{Cl} \end{array}\right)_n$	Polyvinylchloride
$\begin{array}{cc} \text{H} & \text{H} \\ | & | \\ \text{C} & =\text{C} \\ | & | \\ \text{H} & \text{C}_6\text{H}_5 \end{array}$	$\left(\begin{array}{cc} \text{H} & \text{H} \\ | & | \\ -\text{C}- & \text{C}- \\ | & | \\ \text{H} & \text{C}_6\text{H}_5 \end{array}\right)_n$	Polystyrene
$\begin{array}{cc} \text{Fl} & \text{Fl} \\ | & | \\ \text{C} & =\text{C} \\ | & | \\ \text{Fl} & \text{Fl} \end{array}$	$\left(\begin{array}{cc} \text{Fl} & \text{Fl} \\ | & | \\ -\text{C}- & \text{C}- \\ | & | \\ \text{Fl} & \text{Fl} \end{array}\right)_n$	Polytetrafluoroethylene (Teflon)

FIGURE 10.1 Basic structure of polymer molecules. (a) Ethylene molecule.
(b) Polyethylene molecule. (c) Molecular structure of various polymers. These are
examples of the basic building blocks for plastics.

We now have a macromolecule, or a polymer, known as *polyethylene*. Other polymers
of a similar nature are shown in Fig. 10.1(c).

Polymers

The polymer shown in Fig. 10.1(b) is called a *linear polymer* because of its linear
structure. The number of molecules in this chain (length of chain) is known as the
molecular weight. A linear molecule does not mean that it is straight. These molecules
(chains) are of different lengths and their arrangement is *amorphous* (without any
long-range order). This arrangement is often described as a bowl of spaghetti, or
worms in a bucket, all intertwined with each other. Because of the differences in the
length of the chains, the molecular weight of a polymer is determined on a statistical
basis.

Bonding

Although within each long-chain molecule there is *covalent bonding* (primary bonds), no such intermolecular bonding exists between different chains. The bonds between different chains (intramolecular), and between the overlapping portions of the same chain, are known as *secondary bonds*. Secondary bonds are van der Waals bonds, hydrogen bonds, and ionic bonds. Secondary bonds are all much weaker than the covalent bonds within the chain.

The difference in strength between the two bonds is on the order of one to two orders of magnitude. This difference is important, because it is the weaker secondary bonds that determine the overall strength of the polymer. (Thus, for instance, an object is only as strong as its weakest part.)

If the repeating units in a chain are all of the same type, the macromolecule is known as a *homopolymer*. However, in order to obtain certain special properties, two or three different types of monomers can be combined in a polymer. These are known as *copolymers* and *terpolymers*, respectively.

Branching and Linking

The properties of a copolymer depend not only on the two monomers, but also on their arrangement in the molecular structure (Fig. 10.2). *Branching*, for instance, interferes with the relative movement of the molecules and affects the resistance to deformation. Another kind of chain is *cross-linking* (Fig. 10.2c), which is a spatial (three-dimensional) network structure with adjacent chains tied together. Cross-linking has great influence on the properties of the polymer, such as in the vulcanization of rubber (Section 10.10).

Plastics made of linear chain structures are called *thermoplastics*, and those made of cross-linked chain are called *thermosetting plastics*, or simply, *thermosets*.

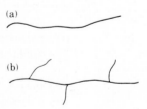

FIGURE 10.2   Schematic illustration of polymer chains. (a) Linear. (b) Branched. (c) Cross-linked. Thermoplastics, such as acrylics, nylons, polyethylene, and polyvinyl chloride, are linear-chain molecules. Thermosetting plastics, such as epoxies, phenolics, and silicones, have cross-linked chains. An example of cross-linking is the vulcanization of rubber for automobile tires, each tire being one giant molecule.

10.2.1 POLYMERIZATION

This is the process by which the monomers link together to make longer and larger molecules. (An automobile tire, for instance, is one giant molecule.) Polymerization requires that each small molecule have at least two reaction points for linking. Polymerization processes are complex; only a brief outline is given below.

Although there are many variations, two basic polymerization processes are: condensation and addition. In *condensation polymerization* the bonds are formed by the application of heat and pressure with a catalyst (initiater). The reaction by-products, such as water, are condensed. This process is also known as *step-growth* or *step-reaction* polymerization, because the molecule grows step by step. Linear, branched, or cross-linked structures can be obtained by various modifications of condensation polymerization.

In *addition polymerization*, also known as *chain-growth* or *chain-reaction* polymerization, bonding is carried out without reaction byproducts. It is called chain-reaction because of the high rate of long-molecule formation. The rate in this process is much higher than that of condensation polymerization.

The degree of polymerization (DP) is defined as the ratio of the molecular weight of the polymer to the molecular weight of the mer, or the number of mers per average molecule. This number can range from a few hundred to millions.

10.2.2 CRYSTALLINITY

In the foregoing section the arrangement of the long-chain molecules was described as amorphous and randomly intertwined, unlike the crystal (orderly) structure of metals (Chapter 3). However, it is possible to impart some *crystallinity* to polymers. This may be done either during solidification of the polymer or in subsequent processing by deformation. The crystalline regions in polymers are called *crystallites* (Fig. 10.3). These crystals are formed by the folding in an orderly manner of the long molecules over themselves.

By various techniques, it is possible to obtain different degrees of, although never complete, crystallinity. Crystallinity ranges from an almost complete crystal (95% by volume in the case of polyethylene) to slightly crystallized and mostly *amorphous* polymers. Common plastics that can have a high degree of crystallinity are polyethylene, polypropylene, polytetrafluoroethylene (i.e., Teflon), and nylon. Those that are almost completely amorphous are polyvinyl chloride (PVC), polybutadiene, and polystyrene. Examples of completely amorphous polymers are polycarbonate and acrylics. (See Section 10.8.)

Effects of Crystallinity

The mechanical and physical properties of polymers are greatly affected by the degree of crystallinity. As the degree of crystallinity increases, polymers become less rubbery (less ductile), stiffer, and harder. For example, the highly crystalline form of

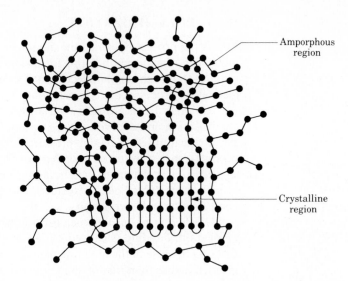

Amporphous
region

Crystalline
region

FIGURE 10.3 Amorphous and crystalline regions in a polymer. The crystalline region (crystallite) has an orderly arrangement of molecules, as in metals shown in Fig. 3.1. The higher the crystallinity, the harder, stiffer, and less ductile the polymer.

polyethylene (high-density polyethylene [HDPE]) has a specific gravity of 0.97 and is stronger, stiffer, tougher, and less ductile than low-density polyethylene (LDPE), which is about 60% crystalline with a specific gravity of 0.915. The increase in density with increasing crystallinity is due to crystallization shrinkage and is a result of more efficient packing of the molecules in the crystal lattice.

Optical qualities are also influenced by the degree of crystallinity. Because the index of refraction is proportional to density, the greater the density difference between the amorphous and crystalline phases, the greater the opacity of the polymer. Polymers that are completely amorphous can be transparent; examples are polycarbonate and acrylics.

10.2.3 GLASS-TRANSITION TEMPERATURE

Amorphous polymers undergo a distinct change in their mechanical behavior across a narrow temperature range. At lower temperatures they are described as hard, rigid, brittle, or glassy, and at higher temperatures as rubbery or leathery. The temperature at which this transition occurs is known as the *glass-transition temperature, T_g*. The term glass is included in this definition because glasses, which are amorphous solids (Section 11.8), behave in the same manner. Amorphous polymers do not have a specific melting temperature.

The glass-transition temperature varies with different polymers. For example, room temperature is above T_g for some polymers and below it for others. To determine T_g, the specific volume is plotted against temperature and the sharp change in the slope of the curve is observed (Fig. 10.4). In the case of highly cross-linked polymers, there is a gradual change in the slope of the curve near T_g; hence it is difficult to determine T_g.

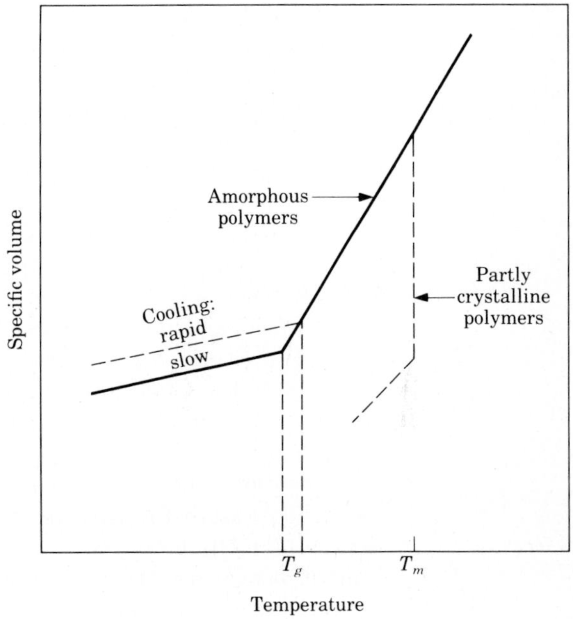

FIGURE 10.4 Specific volume of polymers as a function of temperature. Amorphous polymers (acrylics and polycarbonate) have a glass-transition temperature, T_g. They do not have a specific melting point, a behavior that is similar to glass. Partly crystalline polymers (polyethylene and nylons) undergo a sharp contraction at melting point, T_m, during cooling.

Typical glass-transition temperatures for some polymers and rubber are:

Polyethylene $-185°F$ $(-120°C)$

Polyvinylchloride $190°F$ $(87°C)$

Polypropylene $-15°F$ $(-10°C)$

Polystyrene $212°F$ $(100°C)$

Polymethylmethacrylate $220°F$ $(105°C)$

Rubber $-100°F$ $(-73°C)$

Partly crystalline polymers, unlike amorphous polymers, have a distinct melting point (T_m), as shown in Fig. 10.4. There is a sudden drop in the specific volume as the temperature is lowered.

10.3 THERMOPLASTICS AND THERMOSETTING PLASTICS

Thermoplastics

We noted earlier that, in the amorphous structure of a polymer, the bonds between adjacent long-chain molecules (secondary bonds) are much weaker than the covalent bonds (primary bonds) within each molecule. Hence, it is the strength of the secondary bonds that determines the overall strength of the polymer. Linear and branched polymers have weak secondary bonds.

If we now raise the temperature of this polymer above the glass-transition temperature, we find that it becomes softer and easier to form or to mold into a shape. The mobility of the long molecules (thermal vibrations) increases at T_g and above. If this polymer is now cooled, it returns to its original hardness and strength. In other words, the process is reversible.

Polymers that exhibit this behavior are known as *thermoplastics*. Typical examples are acrylics, nylon, polyethylene, and polyvinyl chloride. The diverse properties of thermoplastics are outlined in Section 10.8.

Thermosets

When the long-chain molecules in a polymer are cross-linked in a three-dimensional (spatial) network, the structure becomes in effect one giant molecule with strong covalent bonds. Cross-linking is done by the polymerization processes described in Section 10.2.1. Because of the nature of the bonds, the strength and hardness of such a polymer is not affected by temperature.

These polymers are known as *thermosets* because, during polymerization under heat and pressure, the network is completed and the shape of the part is permanently set (*curing*). Unlike in thermoplastics this reaction is irreversible and the polymer cannot be recycled. However, if the temperature is increased sufficiently the thermosetting polymer begins to decompose, char, and degrade. Thermosetting polymers do not have a sharply defined glass-transition temperature. Commonly, thermosetting resins become rubbery and compliant across a *narrow* temperature range.

The response of a thermosetting plastic to temperature can be likened to boiling an egg or baking a cake. Once the cake is baked and cooled, reheating it will not change its shape, and if the temperature is too high, it will burn. On the other hand, the response of a thermoplastic can be likened to ice cream. It can be softened, refrozen, and resoftened a number of times. It can be molded into shapes, frozen, and then softened again to be remolded into a different shape.

A typical example of a thermosetting plastic is phenolic, which is a product of the reaction between phenol and formaldehyde. Typical products of this polymer are the handles on cooking pots and pans and electrical components such as switches.

The polymerization process for thermosets generally takes place in two stages. The first one is at the chemical plant, where the molecules are partially polymerized into linear chains. The second stage is at the parts-producing plant, where the cross-linking is completed under heat and pressure during the molding of the part. (Processing methods for polymers are described in Section 10.12.)

Thermosetting plastics generally possess better mechanical, thermal, chemical, and electrical resistance and better dimensional stability than thermoplastics. (See Section 10.9.)

10.4 ADDITIVES

The earliest polymers (developed in the 1860s) were made of natural organic materials from animal and vegetable products, cellulose being the most significant example. With various chemical reactions, cellulose was modified into cellulose acetate. This plastic is used in making photographic films (celluloid), packaging sheets, textile fibers, and cellulose nitrate for use in plastics, explosives, rayon, and varnishes. The earliest synthetic polymer was a thermoset, Bakelite (a phenol-formaldehyde), named after L. H. Baekeland, was discovered in 1906.

The development of modern technology for plastics began in the 1920s with raw materials extracted from coal and petroleum products. In Section 10.2 ethylene was mentioned as the first example of the building block for polyethylene. Ethylene is the product of the reaction between acetylene and hydrogen, and acetylene is the product of the reaction between coke and methane. Other products of similar reactions are styrene and vinyl chloride. These materials are known as synthetic or manufactured organic polymers.

Although, in the example for polyethylene, only carbon and hydrogen were involved, other compounds can be obtained with chlorine, fluorine, sulfur, silicon, nitrogen, and oxygen. As a result, an extremely wide range of polymers with a wide range of properties is obtained.

10.4.1 ADDITIVES

In order to impart certain specific properties, polymers are generally compounded with additives. Additives modify and improve certain characteristics such as stiffness, strength, color, weatherability, flammability, arc resistance for electrical applications, and ease of subsequent processing. These additives are described below.

Fillers
These materials are generally wood flour, silica flour, various minerals, powdered mica, and short fibers of cellulose, glass, and asbestos. (See also Section 10.11.) Depending on their type, fillers improve the strength, hardness, toughness, abrasion resistance, and stiffness of plastics. These properties are maximized at various percentages of different types of polymer/filler combinations.

The effectiveness of a filler depends on the nature of the bond between the filler material and the polymer chains. Because of their lower cost, fillers also reduce the overall cost per unit weight of the polymer. Most thermoplastics and some thermosetting plastics contain fillers.

Plasticizers
These are added to give flexibility and softness to the polymer by lowering the glass-transition temperature. Plasticizers are low-molecular-weight solvents (with high boiling points, i.e., nonvolatile) that reduce the strength of the secondary bonds between the long-chain molecules of the polymer, thus making it soft and flexible. The most common application is in polyvinyl chloride (PCV), which remains flexible during its many uses.

Stabilizers
Most polymers are adversely affected by ultraviolet radiation and oxygen which weaken and break the primary bonds of the long-chain molecules. The polymer then

becomes rigid and brittle. This is known as *degradation*. A typical example of protection against ultraviolet radiation is the compounding of rubber with carbon black (soot). The carbon black absorbs a high percentage of the ultraviolet radiation. Protection against degradation by oxidation, particularly at elevated temperatures, is done by adding antioxidants to the polymer. Various coatings are another means of protection against degradation.

Colorants

The great variety of colors available in plastics is obtained by the addition of colorants. These are either organic (*dyes*) or inorganic (*pigments*). The selection of a colorant depends on service temperature and exposure to light. Pigments, which are dispersed particles, generally have greater resistance to temperature and light than dyes do.

Flame Retardants

If the temperature is sufficiently high a polymer will ignite. The *flammability* (the ability to support combustion) of polymers varies considerably, depending on their composition, such as the chlorine and fluorine content. Polymethylmethacrylate, for example, continues to burn when ignited, whereas polycarbonate extinguishes itself. The flammability of polymers can be reduced either by making them from less flammable raw materials or by adding flame retardants to the compound. Common flame retardants are chlorine, bromine, and phosphorus compounds.

Lubricants

To reduce friction during subsequent processing and use and to prevent sticking to the molds, lubricants are added to polymers. Lubrication is also important in preventing thin polymer films from sticking together.

Polyblends

To give them a rubbery behavior, polymers can be blended with small amounts of rubbery polymers. These are finely dispersed throughout the polymer and improve its impact strength. These polymers are known as rubber-modified.

10.5 BEHAVIOR OF THERMOPLASTICS

The behavior of thermoplastics depends on a large number of variables. Among the most important are temperature and strain rate. The behavior of thermoplastic polymers can be represented by the models shown in Fig. 10.5. Below the glass-transition temperature, T_g, the polymer is glassy (meaning brittle) and behaves like an elastic solid, i.e., the relationship between stress and strain is linear. Below its T_g, polymethylmethacrylate (PMMA) is glassy, whereas polycarbonate is not glassy below its T_g.

Glassy behavior can be represented by a spring whose stiffness is equivalent to the modulus of elasticity of the polymer (Fig. 10.5a). If the polymer is tested in torsion,

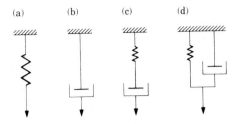

FIGURE 10.5 Spring-dashpot models for the stress, strain, and strain rate behavior of thermoplastics. (a) Elastic. (b) Viscous. (c) Viscoelastic, Maxwell model. (d) Viscoelastic, Kelvin or Voigt model.

we find the same linear relationship between shear stress and shear strain. These relationships are expressed as follows:

$$\sigma = E\epsilon, \tag{9.1}$$

$$\tau = G\gamma.$$

When the applied stress reaches the fracture stress, the polymer fractures, just as a piece of glass does at room temperature. If we now raise the temperature above the T_g, the polymer becomes a viscous fluid. Its viscosity decreases with increasing temperature. The relationship between the shear stress and shear strain rate is given by the expression for viscous flow,

$$\tau = \eta\dot{\gamma}, \tag{9.2}$$

where η is the viscosity and $\dot{\gamma}$ is the shear-strain rate. This behavior is known as *Newtonian*; that is, the shear stress is directly proportional to the shear-strain rate, assuming that η remains constant.

This behavior can be represented by a simple dashpot model shown in Fig. 10.5(b). We note that such a behavior is similar to the strain-rate sensitivity of metals described in Section 2.2.7 and represented by the equation,

$$\sigma = C\dot{\epsilon}^m, \tag{9.3}$$

where we note that for a Newtonian behavior, $m = 1$. Thermoplastics have high m values and, as we recall from the treatment in Section 2.2.7 on the effect of m on necking, this indicates that these polymers can undergo *large uniform elongation* in tension before fracture (Figs. 10.6a and b). This property enables the forming of thermoplastics into complex shapes involving large strains, such as in blow molding to make plastic bottles (Section 10.12.8).

If we raise the temperature of the thermoplastic from that of the glassy state to around T_g, we find that the polymer retains some of its elasticity while also becoming viscous. This is known as *viscoelastic* behavior and can be modeled as shown in Figs. 10.5(c) and (d), known as the Maxwell, and Kelvin or Voigt models, respectively. When a constant load is applied, the polymer first stretches at a high strain rate, and then continues to elongate over a period of time due to its viscous behavior. If the polymer is stretched at a high strain rate, it exhibits greater strength due to its viscosity, a behavior represented by Eq. (9.3).

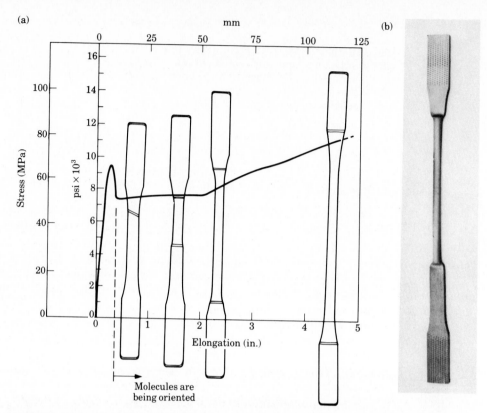

FIGURE 10.6 (a) Load-elongation curve for polycarbonate (a themoplastic). Note the extent of uniform elongation as compared to metals, shown in Fig. 2.2. (b) High-density polyethylene tensile-test specimen, showing uniform elongation (longest and narrowest region in the specimen). The second narrow region is that of the original cross section of the specimen. *Source:* (a) After R. P. Kambour and R. E. Robertson, General Electric Co. (b) Courtesy of A. Bell, Illinois Institute of Technology.

Note in these models that the elastic portion of the elongation is reversible (elastic recovery) but the viscous portion is not. It is also apparent that an understanding of these behaviors is essential for the proper use and processing of thermoplastic polymers.

When these polymers are deformed, say by stretching, the long-chain molecules align themselves along the general direction of elongation and, just like in metals, the polymer becomes anisotropic, i.e., the molecules become *oriented*. Thus, the specimen is stronger and stiffer in the elongated direction than in its transverse direction (see Section 10.6).

Thermosetting polymers do not exhibit the behavior of thermoplastics, i.e., their polymerization and cross-linking is irreversible. Neither temperature nor strain rate has any significant effect on the stress–strain curves for thermosets.

In addition to the deformation behavior described in this section, other factors, such as creep, stress relaxation, impact strength, fatigue, and various chemical, physical, and optical properties are also important. Treatment of these subjects in detail is outside the scope of this text and is best covered in texts on materials and design.

Crazing

Some thermoplastics, such as polystyrene and polymethylmethacrylate, when subjected to tensile stresses, develop localized, wedge-shaped narrow regions of highly deformed material. For isotropic specimens, these regions are at right angles to the direction of the principal tensile stress. (They do not occur under compressive stresses.) This phenomenon is known as *crazing*.

Crazing has been observed in transparent, glassy polymers as well as in others. Although they may appear to be like cracks, crazes are spongy material, typically containing about 50% voids (open-cell foam with interconnected voids). With increasing tensile load on the specimen, these voids eventually lead to fracture.

The environment and the presence of solvents and other liquids enhance the formation of crazes (*environmental stress cracking, solvent crazing*). Tensile residual stresses in the material also contribute to crazing and cracking.

A related phenomenon is *stress whitening*. When subjected to tensile stresses, such as by bending, the color of the material becomes lighter. This phenomenon is usually attributed to the formation of microvoids. As a result the material also becomes more opaque, as seen in Fig. 10.7.

FIGURE 10.7 Polyethylene tensile-test specimen. This photograph is taken with a back light, showing opaqueness of the stretched region. Note that the two unstretched ends of the specimen transmit more light (less opaque) than the center. *Source:* Courtesy of A. Bell.

10.6 MECHANICAL PROPERTIES

Extensive data are available on the mechanical and various other properties of polymers. Some of the typical properties are reviewed in this section.

Typical stress–strain curves for several thermoplastics are shown in Fig. 10.8. Note the wide range of strength, ductility, and toughness (area under the curve) in these materials. These curves can be described by the qualitative terms shown in Fig. 10.9.

The modulus of elasticity for several polymers are shown in Fig. 10.10. Note that polymers are about two orders of magnitude less stiff than metals. The ultimate tensile strength of several plastics and their strength relative to metals are shown in Fig. 10.11. The strength difference is about one order of magnitude. (See also Table 3.3.)

Effects of Temperature

The typical effects of temperature on the strength and elastic modulus of thermoplastics are shown in Figs. 10.12 and 10.13. With increasing temperature, the strength and modulus of elasticity decrease and toughness (area under the curve) increases. The effect of temperature on impact strength is shown in Fig. 10.14. Note the large difference in the behavior of various polymers.

Water Absorption

An important limitation of some polymers, such as nylons, is their ability to absorb water. The yield point, the elastic modulus, and the glass-transition temperature are all reduced with increasing moisture absorption (Figs. 10.15 and 10.16). Dimensional changes also take place due to water absorption or humid environment.

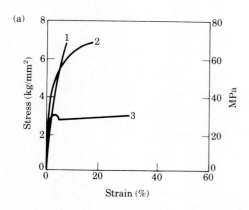

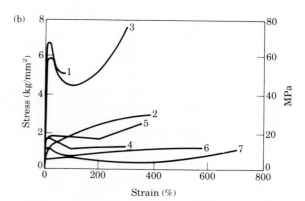

FIGURE 10.8   Stress-strain curves for various plastics. (a) 1. Polymethylmethacrylate. 2. Acetal resin. 3. Polystyrene. (b) 1. Polycarbonate. 2. Polyurethane rubber. 3. Polyamide. 4. Polyethylene. 5. Polytetrafluoroethylene. 6. Polyethylene. 7. Polyethylene. *Source:* After A. Kobayashi, *CIRP Annalen*, vol. 12, no. 2, p. 77.

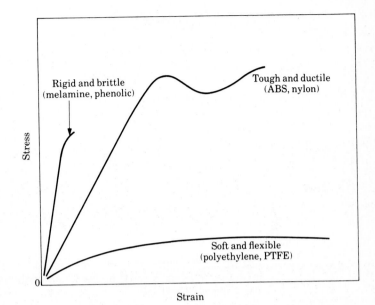

FIGURE 10.9 General terminology for describing typical stress–strain curves for three types of plastics. PTFE (polytetrafluoroethylene) is commercially available as Teflon. *Source:* After R. L. E. Brown.

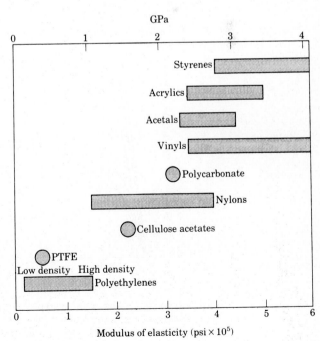

FIGURE 10.10 The range of modulus of elasticity for various plastics. Note that these values are about two orders of magnitude smaller than metals, as shown in Table 2.1.

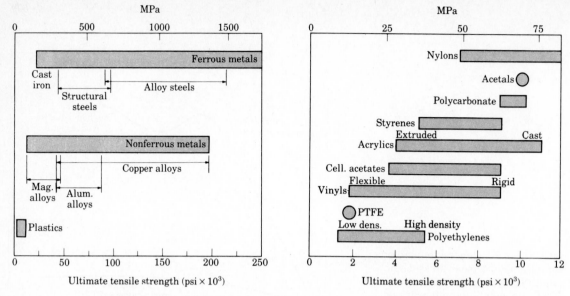

FIGURE 10.11 Ultimate tensile strength for various metals and plastics. Note that the strength of plastics is about one order of magnitude lower than that for metals. See also Tables 3.3 and 10.1.

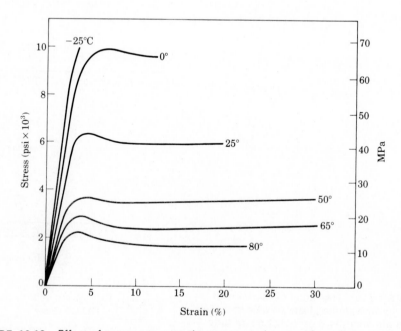

FIGURE 10.12 Effect of temperature on the stress–strain curve for cellulose acetate. Note the drop in strength and increase in ductility with a relatively small increase in temperature (below the boiling point of water). Compare with Fig. 2.10 for metals. *Source:* After T. S. Carswell and H. K. Nason.

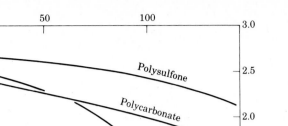

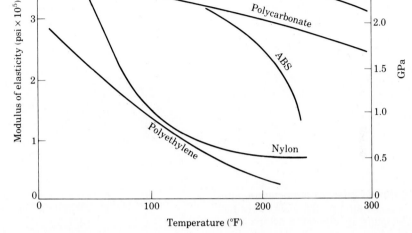

FIGURE 10.13 Effect of temperature on the modulus of elasticity for several plastics. Note that some plastics are more sensitive to temperature than others. Compare with Fig. 2.11 for metals.

FIGURE 10.14 Effect of temperature on the impact strength for various plastics. Note that small changes in temperature have a significant effect on the impact strength of some plastics. Compare with Fig. 3.46 for metals. *Source:* After P. C. Powell.

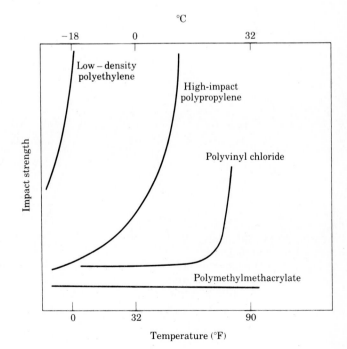

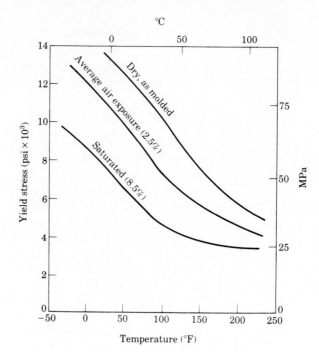

FIGURE 10.15 Effect of temperature and moisture absorption on the yield stress for nylon. Moisture absorption in nylons is an important factor in design with plastics.

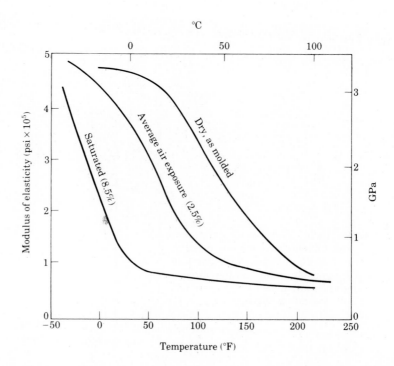

FIGURE 10.16 Effect of temperature and moisture absorption on the modulus of elasticity for nylon. Note that the modulus stabilizes after a certain temperature.

The environment can also have a significant effect on the strength of polymers. When subjected to static loading or due to tensile residual stresses, plastics can develop cracks in the presence of various solvents, lubricants, and water vapor. Radiation and visible light can also have detrimental effects on the strength of polymers.

Anisotropy

The deformation of polymers at room temperature not only increases their strength, but also results in anisotropy (orientation of the molecules), much like that observed in cold working of metals. Some typical results for polymers are shown in Figs. 10.17 through 10.20, where the strong anisotropy developed as a result of cold rolling is observed (see also Section 10.12.9). The increased strength and toughness of polymers, with cold working, makes these materials attractive for applications where improved mechanical properties are required.

Note in Fig. 10.17 that with greater cold-rolling reduction, the material becomes more and more strain hardening, i.e., the curve has an increasing positive slope. Thus, as described in Section 2.2.4, the uniform elongation increases with greater cold-rolling reduction, and hence the material can be stretched further during processing.

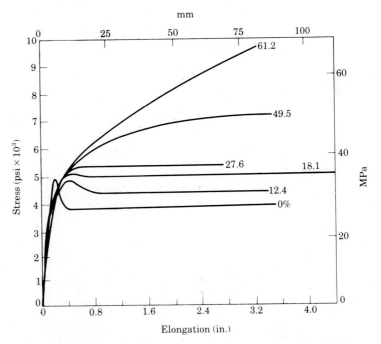

FIGURE 10.17 Stress-elongation curves for ABS (acrylonitrile-butadiene-styrene) as a function of thickness reduction by cold-rolling. Tested in the direction of rolling. ABS has many favorable mechanical, chemical, and electrical properties.
Source: L. J. Broutman and S. Kalpakjian, *SPE Jr.*, vol. 25, no. 10, 1969, pp. 46–52.

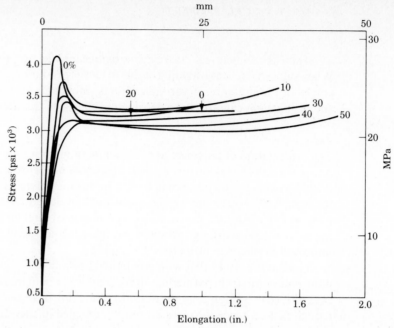

FIGURE 10.18 Stress-elongation curves for ABS as a function of thickness reduction by cold-rolling. Tested 90 degrees to the rolling direction. *Source:* After L. J. Broutman and S. Kalpakjian, *SPE Jr.*, vol. 25, no. 10, 1969, pp. 46–52.

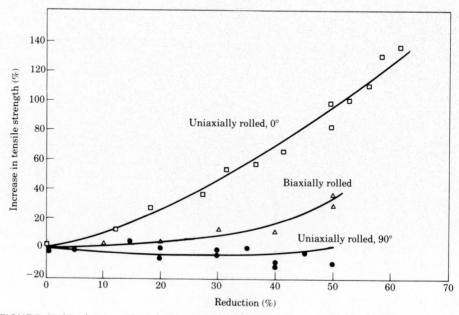

FIGURE 10.19 Increase in tensile strength for uniaxially and biaxially cold-rolled ABS as a function of reduction in thickness. Uniaxially rolled sheets were tested at 0 and 90 degrees to the rolling direction. *Source:* After L. J. Broutman and S. Kalpakjian, *SPE Jr.*, vol. 25, no. 10, 1969, pp. 46–52.

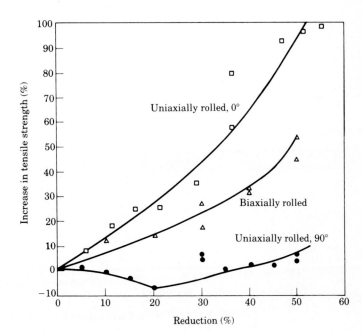

FIGURE 10.20 Same as Fig. 10.19, for polycarbonate. *Source:* After L. J. Broutman and S. Kalpakjian, *SPE Jr.,* vol. 25, no. 10, 1969, pp. 46–52.

10.7 PHYSICAL PROPERTIES

Plastics, in comparison with metals, are generally characterized by their low thermal (Fig. 10.21) and electrical conductivity, low specific gravity (ranging from 0.90 to 2.2), and by their relatively high coefficient of thermal expansion (about an order of magnitude higher than metals). The electrical conductivity of polymers increases with moisture absorption. The useful temperature range of polymers is much lower than that for metals as shown in Fig. 10.22.

General mechanical properties of major classes of polymers are given in Table 10.1.

10.7.1 FRICTION AND WEAR OF POLYMERS

Although their strength is low compared to metals, polymers generally possess low frictional and wear characteristics. This is why polymers are sometimes called *self-lubricating.* This property makes polymers attractive in many applications, such as sliding systems, bearings, gears, seals, prosthetic joints, and general friction-reducing applications. The factors involved in friction and wear of metals (Chapter 4) are also generally applicable to polymers.

In sliding, the ploughing component of friction (Section 4.6.1) is a significant factor because of the viscoelastic behavior of thermoplastic polymers. The adhesion component may or may not be as significant, depending on the polymer–metal pair.

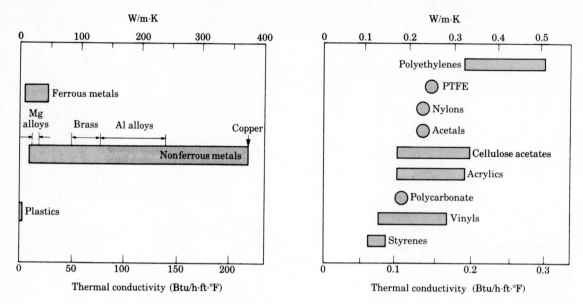

FIGURE 10.21 Thermal conductivity of metals and plastics. Plastics are poor thermal conductors. See also Table 3.3 and Fig. 11.15.

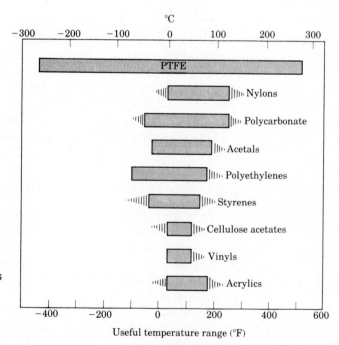

FIGURE 10.22 Useful temperature range for several plastics. The range for plastics is much smaller than for metals and ceramics. See also Figs. 3.52, 11.17 and 11.18.

TABLE 10.1
**RANGE OF MECHANICAL PROPERTIES FOR VARIOUS
ENGINEERING PLASTICS AT ROOM TEMPERATURE**

MATERIAL	UTS (psi, ×10³)	E (psi, ×10⁵)	ELONGATION (%)
ABS	4–8	2–4	5–75
ABS, reinforced	14.5	11	—
Acetal	8–10	4–5	25–75
Acetal, reinforced	19.5	14	—
Acrylic	6–11	2–5	5–50
Cellulosic	1.5–7	0.6–4	5–100
Epoxy	5–20	5–25	1–10
Epoxy, reinforced	10–200	30–75	2–4
Fluorocarbon	1–7	1–3	100–300
Nylon	8–12	2–4	60–200
Nylon, reinforced	10–30	3–15	1–10
Phenolic	4–10	4–30	0–2
Polycarbonate	8–10	3.5–4.5	10–125
Polycarbonate, reinforced	16	9	4–6
Polyester	8	2.8	5–300
Polyester, reinforced	16–23	12–17	1–3
Polyethylene	1–6	0.15–2	15–1000
Polypropylene	3–5	1–1.7	10–500
Polypropylene, reinforced	6–15	5–9	2–4
Polystyrene	2–12	2–6	1–60
Polyvinyl chloride	1–8	0.02–6	40–450

Temperature rise at the interface due to friction is an important factor. The combined effects of low thermal conductivity and low melting point are of major importance in friction and wear of polymers. Wear can take place by a variety of mechanisms, such as adhesion, abrasion, fatigue, and thermal degradation.

The frictional behavior of several polymers relative to metals is shown in Fig. 10.23. Note the well-known low coefficient of friction for PTFE (Teflon). It has been shown that an essential condition for the low friction of this polymer, when sliding over steel, is the formation of a transferred film of PTFE on the metal surface.

The relative abrasive wear resistance of a variety of polymers is given in Table 10.2. Abrasive wear behavior depends on the type of abrasion and the ability of the polymer to deform and recover elastically. It has been suggested that the parameter describing this behavior is the ratio of hardness to elastic modulus. The higher this ratio, the higher the abrasive wear resistance of the polymer. Resistance to abrasion can also be increased by special coatings on plastics.

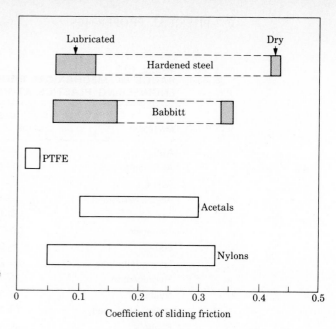

FIGURE 10.23 Coefficient of friction of hardened steel on metals and plastics. Note the low friction of PTFE (Teflon). The frictional behavior of polymers can be modified by the addition of lubricants to the polymer mix during processing.

TABLE 10.2
ABRASION RESISTANCE OF VARIOUS PLASTICS

High	Acetal, cellulosic, epoxy, nylon, phenolic, polycarbonate, polyester, polyimide, polyurethane, polyvinyl chloride
Medium	ABS, polypropylene
Low	Acrylic, fluorocarbon, polyethylene, polystyrene, silicone

10.8 GENERAL CHARACTERISTICS OF THERMOPLASTICS

In this and the following sections the general characteristics of plastics are outlined, as they relate to the manufacturing of plastic products. General recommendations for various applications are given in Table 10.3.

Acetals. Good mechanical properties, with strength, stiffness, creep and abrasion resistance. Processed by injection molding or extruding; also can be processed by other techniques depending on grade. Typical applications are mechanical parts and components where high performance is required over a long period, such as bearings, cams, gears, bushings, rollers, impellers, latches, wear surfaces, and housings.

Acrylics (polymethylmethacrylate, PMMA). Moderate strength, good optical quality and weather resistance, transparent but can be made opaque, generally resistant to chemicals. Can be processed by machining, molding, and thermoforming; molding compounds can be extruded or injection molded. Typical applications are optical use such as lenses, signs, displays, covers, window glazing, automotive lenses, lighting fixtures, and furniture.

TABLE 10.3
GENERAL RECOMMENDATIONS FOR PLASTIC PRODUCTS

DESIGN REQUIREMENT	APPLICATIONS	PLASTICS
Mechanical strength	Gears, cams, rollers, valves, fan blades, impellers, pistons	Acetal, nylon, phenolic, polycarbonate
Functional and decorative	Handles, knobs, camera and battery cases, trim moldings, pipe fittings	ABS, acrylic, cellulosic, phenolic, polyethylene, polypropylene, polystyrene, polyvinyl chloride
Housings and hollow shapes	Power tools, pumps, housings, sport helmets, telephone cases	ABS, cellulosic, phenolic, polycarbonate, polyethylene, polypropylene, polystyrene
Functional and transparent	Lenses, goggles, safety glazing, signs, food-processing equipment, laboratory hardware	Acrylic, polycarbonate, polystyrene, polysulfone
Wear resistance	Gears, wear strips and liners, bearings, bushings, roller-skate wheels	Acetal, nylon, phenolic, polyimide, polyurethane, ultra–high-molecular-weight polyethylene

Acrylonitrile-Butadiene-Styrene (ABS). Impact, abrasion and chemical resistance, good strength, dimensionally stable, rigid, and good low-temperature properties and electrical characteristics. Can be processed by all manufacturing methods for thermoplastics. Typical applications are pipes and fittings, tool handles, automotive components, communications equipment such as telephones, and appliances. Mechanical properties are further enhanced for glass-filled ABS.

Cellulosics. Wide range of mechanical properties depending on composition; can be made rigid and strong. Poor weathering, affected by chemicals. Can be processed by many conventional thermoplastic techniques. Typical applications are tool handles, pens, knobs, safety goggles, lighting fixtures, steering wheels, packaging, and toys.

Fluorocarbons. A family of polymers with good resistance to temperature and chemicals, and unique nonadhesive properties. Best-known example is polytetra-fluoroethylene (PTFE). Processing technique depends on the specific material. Can be molded and extruded. Typical applications are linings for chemical-processing equipment, coatings for home cookware, insulation for high-temperature wire and cable, gaskets, and low-friction surfaces.

Nylons (polyamide resins). Good mechanical properties, abrasion resistance, self-lubricating, and resistant to most chemicals. All nylons are hygroscopic (they absorb moisture); mechanical properties are reduced and part dimensions increase with moisture absorption. Can be processed by all thermoplastic techniques. Typical applications are industrial, electrical, consumer, and automotive uses such as gears, bearings, bushings, rollers, fasteners, zippers, electrical parts, and combs.

Polycarbonates.   Very versatile with good mechanical and electrical properties. Very high impact resistance; can be made resistant to chemicals. Processed by all thermoplastic techniques. Typical applications are safety helmets, safety shields, window glazing, bottles, windshields, load-bearing electrical components, electrical insulators, medical apparatus, business-machine components, and parts requiring dimensional stability.

Polyesters (thermoplastic, see also thermosetting polyesters).   Good mechanical, electrical, and chemical properties, good abrasion resistance, low friction. Can be processed by molding and extruding. Typical applications are gears, cams, and similar load-bearing members, and electrical components.

Polyethylenes.   A family of thermoplastics with good electrical and chemical properties. Mechanical properties depend on composition and structure. Two major classes are low-density (LDPE) and high-density polyethylenes (HDPE). Processed by common methods. Typical applications are housewares, bottles, garbage cans, luggage, toys, piping, and packaging.

Polypropylenes.   Good mechanical, electrical, and chemical properties. Easily processed by common methods. Typical applications are automotive components, medical devices, appliance parts, wire insulation, TV cabinets, piping, fittings, luggage, and ropes.

Polystyrenes.   Properties depend on composition, generally average properties. Processed by common techniques. Typical applications are disposable containers. Various grades have applications such as appliance, automotive, and radio and TV components, housewares, toys, and furniture parts.

Polysulfones.   Excellent resistance to heat, water, and steam; highly resistant to some chemicals but attacked by organic solvents. Processed by various techniques. Typical applications are steam irons, coffee makers, hot-water containers, medical equipment requiring sterilization, and power tool and appliance housings.

Polyvinyl Chloride (PVC).   A family of thermoplastics with a wide range of properties; may be made rigid or flexible. Rigid PVC is tough and hard; not suitable for applications requiring strength and heat resistance. Relatively easy processing by many methods. Typical applications are rigid PVC for the construction industry such as pipes, conduits, and various building components. Flexible PVC for wire and cable coatings, flexible tubing, footware, imitation leather, upholstery, records, and film and sheet for numerous uses.

10.9 GENERAL CHARACTERISTICS OF THERMOSETTING PLASTICS

Alkyds.   A family of compounds with good electrical insulating properties and low water absorption. Processed by molding. Typical applications are electrical and electronic components.

Aminos (urea and melamine). Properties depend on composition. Generally hard and rigid, resistant to abrasion, creep, and arcing, and flame retardant. Processed by molding. Typical applications are urea for electrical and electronic components; melamine for dinnerware. Also used for small appliance housings and handles.

Epoxies. A family of thermosetting resins with excellent mechanical and electrical properties, dimensional stability, adhesive property (see Section 12.12), and heat resistance. Processed by casting and molding. Typical applications are electrical components requiring mechanical strength and thermal insulation, and adhesives. Fiber-reinforced epoxies have excellent mechanical properties with applications such as pressure vessels, rocket-motor casings, tanks, and similar structural components.

Phenolics. Although brittle, phenolic resins have high resistance to heat, water, and chemicals. Processed by molding. Typical applications are knobs, handles, telephones, grinding-wheel bonds (see Section 9.3.1), and electrical components such as wiring devices, connectors, and insulators.

Polyesters (thermosetting). Generally reinforced with glass or other fibers. Good mechanical, chemical, and electrical properties. Typical applications are boats, luggage, chairs, automotive bodies, swimming pools, impregnant for cloth, paper, etc. Also available as casting resins.

Polyimides. Good mechanical, physical, and electrical properties at elevated temperatures, low friction and wear. Processed by molding. Typical applications are pump components (bearings, seals, valve seats, retainer rings, piston rings), electrical connectors for high temperature use, and aerospace parts.

Silicones. Properties depend on composition. Generally excellent electrical properties over a wide range of humidity and temperature, good weathering, and resistant to chemicals and heat. Processed by molding and other techniques. Typical applications are electrical components requiring strength at elevated temperatures, oven gaskets, and waterproof materials.

10.10 ELASTOMERS

Elastomers are a large family of compounds (amorphous polymers) with the characteristic ability to undergo very large elastic deformations without rupture. They are soft and have a low elastic modulus and a low glass-transition temperature (Section 10.2.3).

The term *elastomer* is derived from the words *elastic* and *mer*, and indicates a family of rubberlike materials. These polymers are highly kinked; they stretch but then return to their original shape. Cross-linking also takes place, the best example being the elevated temperature vulcanization of rubber with sulfur (discovered in 1839).

All rubbers are elastomers, but all elastomers are not rubbers. An elastomer has been defined as being capable of recovering substantially in shape and size after the load has been removed. Rubber has been defined as being capable of recovering quickly from large deformations.

Types of Elastomers

Elastomers may be classified into various categories. The major groups and their characteristics are outlined below.

Natural Rubber.   Has good resistance to abrasion and fatigue but low resistance to oil, heat, ozone, and sunlight; it has good frictional properties such as those required in tires.

Synthetic Rubbers (Neoprene, Buna N).   These rubbers are further developments with improved resistance to oil, heat, and gasoline, and higher useful temperature range.

Silicone.   Has the highest useful temperature range, up to 600°F (315°C), but other properties are generally inferior to other elastomers.

Polyurethane.   Has very good overall properties and high strength.

A characteristic of elastomers is their *hysteresis loss* in stretching (Fig. 10.24). The clockwise loop indicates energy loss, whereby mechanical energy is converted into heat. This property is desirable for absorbing vibrational energy.

Elastomers have extensive uses in industrial and consumer products, ranging from hoses to flexible dies in sheet-metal–forming operations (Section 7.10).

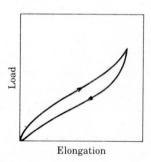

FIGURE 10.24   Typical load-extension curve for rubbers. The clockwise loop indicates hysteresis loss, hence the capacity of rubber for absorbing shock loading (as in automobile tires) by damping the vibrations.

10.11 REINFORCED PLASTICS

It was noted in Section 10.6 that, in spite of their many favorable characteristics, the mechanical properties of polymers are generally inferior to those for metals and alloys, such as their strength and elastic modulus. Materials such as carbon and glass are very strong and rigid when made into fibers with small cross-sectional area (size effect). Glass fibers, for instance, may have tensile strengths as high as 400,000 psi (2750 MPa), however, they are more brittle and sensitive to defects than polymers.

The strength and rigidity of a polymer can be improved by combining it with such fibers. This combination is called a *reinforced plastic* (sometimes also called *composite* materials). Table 10.1 shows the advantages of this combination with some common reinforcing fibers. Note the significant improvement in strength and elastic modulus of reinforced plastics, particularly relative to density.

The properties of a reinforced plastic depend on the type, shape, and orientation of the reinforcing material (Fig. 10.25) and its volume or weight fraction (Fig. 10.26).

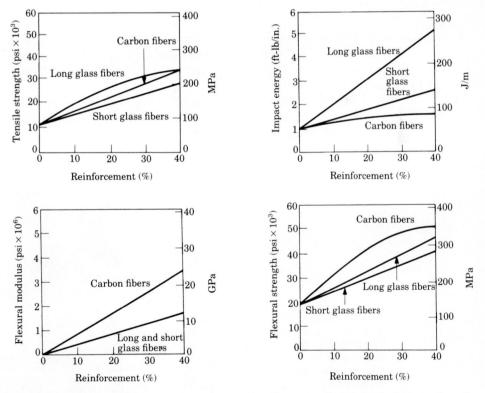

FIGURE 10.25 The effect of glass and graphite fibers on the mechanical properties of nylon 6/6. Note the significant improvement in properties with increasing amount of reinforcement. See also Table 10.1 and Fig. 10.29. *Source:* Fiberfil Division, Dart Industries Inc.

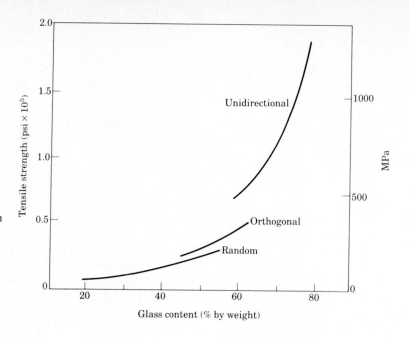

FIGURE 10.26 Tensile strength of polyester glass fiber composite as a function of testing direction and content of glass fiber. *Source:* After R. M. Ogorkiewicz.

Another significant factor is the strength of the bond between the fiber and the polymer matrix. This bond can be strengthened by special treatments to improve adhesion at the interface. Figures 10.27 and 10.28 show fracture surfaces of two reinforced plastics.

Generally, the highest stiffness and strength is obtained when the fibers are aligned in the direction of tensile stress, thus making the composite highly anisotropic. If the reinforced plastic part is to be subjected to forces in different directions, then the fibers are crisscrossed in the matrix in order to reduce anisotropy.

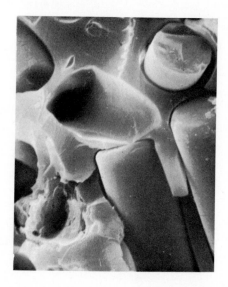

FIGURE 10.27 Fracture surface of glass-fiber reinforced epoxy (matrix). The fibers have random orientation and have a diameter of 10 micrometers. The bond between the glass fiber and the epoxy matrix can be improved by treating the surface of the fibers with a chemical compound, called silane. *Source:* Courtesy of L. J. Broutman, Illinois Institute of Technology.

FIGURE 10.28 Fracture surface of carbon-fiber reinforced epoxy (matrix). Note that the fibers, 9 to 11 micrometers in diameter, are in bundles and are unidirectional, unlike the random orientation of the glass fibers shown in Fig. 10.27. *Source:* L. J. Broutman.

Reinforced plastics may also be made with various other materials and shapes of the polymer matrix in order to impart specific characteristics such as ease of processing, lubrication, permeability, dimensional stability, and cost reduction. The cost of glass fibers is quite low; however, carbon fibers and especially boron fibers are expensive (Table 10.4).

The elements added to the polymer matrix may also be regarded as fillers (Section 10.4.1). In addition to discrete particles, reinforcements may also be in the form of woven cloth or mat (such as those used in reinforced grinding wheels, Section 9.3.1), and continuous fibers. Laminations, composed of sheets of different materials, are also used; the products are called *laminated plastics*.

A variety of reinforced plastics is available with a wide range of properties. Many thermoplastics can be reinforced, with glass contents ranging between 30 and 60%. Graphite fibers are also used extensively. The most commonly used thermosets as a matrix are polyesters and epoxies.

TABLE 10.4
PROPERTIES AND RELATIVE COST OF REINFORCING FIBERS

TYPE	TENSILE STRENGTH (psi, $\times 10^3$)	ELASTIC MODULUS (psi, $\times 10^6$)	RELATIVE COST
Asbestos	840	28	1
Boron	510	55	20,000
Carbon, high-strength	440	40	1,500
Carbon, high-modulus	290	60	1,500
Glass, E-type	250	10	20
Glass, S-type	380	12	50

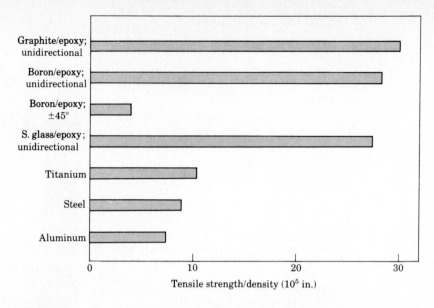

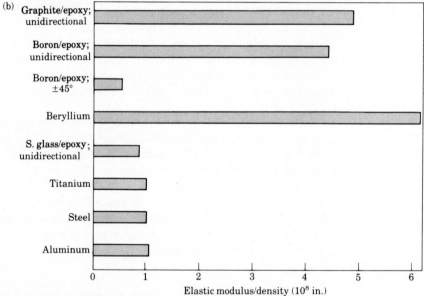

FIGURE 10.29 (a) Tensile strength/density ratio for several reinforced plastics and metals. (b) Elastic modulus/density ratio for several reinforced plastics and metals. Reinforced plastics have important applications in aerospace industries. See also last column in Table 3.3. *Source:* After M. J. Salkind.

Applications

Reinforced plastics are an important group of materials with many applications, such as aircraft and rocket components, helicopter blades, cabs for military vehicles, helmets, automobiles, leaf springs for trucks, boat hulls, and various other structures. Boron fibers are used for high-temperature applications.

Figure 10.29 shows the strength and elastic modulus of reinforced plastics relative to their density. The high ratios obtained with plastics, as compared to metals, make these materials attractive for applications where weight saving is important.

Careful inspection of reinforced plastics is essential for critical applications, in order to ensure that good bonding between the reinforcing fiber and the matrix has been obtained throughout the product (see Table 4.2). In some instances, the cost of inspection can be as high as one quarter of the total cost of the product.

10.12 PROCESSING OF PLASTICS

Plastics can be processed by a variety of methods, either in a molten state or as a solid. The basic processes are extrusion, molding, casting, and forming of sheet. Some of these processes can be used for both thermoplastics and thermosets as outlined below. (See also Tables 10.5 and 10.6.)

TABLE 10.5
COMMON SHAPING PROCESSES FOR THERMOPLASTICS.

	ACRYLICS	ABS	CELLULOSE ACETATE	NYLONS	POLYCARBONATE	POLYETHYLENES	POLYPROPYLENE	POLYSTYRENE	POLYSULFONE	POLYTETRAFLUOROETHYLENE	POLYURETHANE	POLYVINYL CHLORIDE
Compression molding			×									×
Transfer molding												×
Injection molding	×	×	×	×	×	×	×	×	×		×	×
Extrusion	×	×	×	×	×	×	×	×	×	×	×	×
Rotational molding		×		×		×	×	×			×	×
Blow molding	×	×		×	×	×	×	×	×			×
Thermoforming	×	×	×		×	×	×	×	×			×
Casting	×			×								
Forging		×		×		×	×					
Foam molding		×				×	×	×	×			×

Source: After R. L. E. Brown, *Design and Manufacture of Plastic Parts.* © 1980 by John Wiley & Sons, Inc. Reprinted by permission of John Wiley & Sons, Inc.

TABLE 10.6
COMMON SHAPING PROCESSES FOR THERMOSETS.

	ALKYDS	AMINOS (UREAS AND MELAMINE)	EPOXIES	PHENOLICS	POLYESTERS	POLYIMIDES	POLYURETHANES	SILICONES
Nonreinforced								
Compression molding	×	×	×	×	×	×	×	×
Transfer molding	×	×		×				×
Injection molding	×	×		×				
Rotational molding			×					
Casting	×		×	×	×		×	×
Foam molding		×		×	×		×	×
Fiber-reinforced								
Injection molding	×			×				×
Hand or spray-up			×		×	×		
Compression molding	×		×	×	×	×		×
Preform molding			×		×			
Cold-press molding			×		×			
Filament winding			×		×	×		
Pultrusion			×		×			

Source: After R. L. E. Brown, *Design and Manufacture of Plastic Parts.* © 1980 by John Wiley & Sons, Inc. Reprinted by permission of John Wiley & Sons, Inc.

10.12.1 EXTRUSION

This is one of the basic methods for processing thermoplastics. The raw materials, in the form of pellets, granules, or powder, are fed into a barrel through a hopper and are extruded with a screw-type conveyor or with a plunger or ram (Fig. 10.30). Most extruders are equipped with a single screw, although two-screw (or more) extruders are also available. The materials may be heated either through internal friction during extrusion, or by external means such as a heated barrel. The process blends, compounds, homogenizes, and extrudes the plastic simultaneously at temperatures between 275 and 700°F (135 to 370°C).

Because this is a continuous process, long products, with various cross-sections, such as solid rods, channels, and tubing with different profiles are extruded through dies (usually tool steel) of various geometries. Sheet a few thousands of an inch in thickness is also extruded. Plastic-coated wire, cable, or strips for electrical or other applications are also extruded by this process. The wire is fed into the die opening with the extruded plastic at a controlled rate.

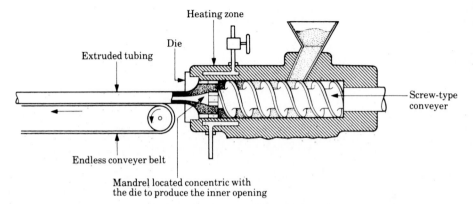

FIGURE 10.30 Schematic illustration of extrusion of plastic tubing. Plastic-coated electrical wiring is made by a similar extrusion process. Various solid cross-sections are also extruded by this process.

The extruded product is cooled by cold air, cold-water spray, or by running it through a water-filled channel. The rate and uniformity of cooling is important for dimensional control because of shrinkage and distortion. The extruded product is then coiled or cut off into desired lengths.

Pellets, which are used for other plastic-processing methods, described below, are also made by extrusion. Here the extruded product (small-diameter rod) is chopped continuously into short lengths.

10.12.2 INJECTION MOLDING

This is another commonly used and versatile process for both thermoplastics and thermosets. Injection molding is essentially similar to the hot-chamber die casting process described in Section 5.7.2. The pellets or granules are heated in a cylinder and the melt is then forced into a split-mold chamber (Fig. 10.31) either by a hydraulic plunger or by the screw system of an extruder at pressures that may exceed 20,000 psi (140 MPa). Viscosity of the melt is an important parameter because of its effect on material flow and forces.

The more recent equipment for injection molding is of the reciprocating screw type. As the pressure builds up at the die entrance due to the movement of the material, the rotating screw begins to move backward under this pressure to a predetermined distance (volume of material). The screw stops rotating and is then pushed forward hydraulically, forcing the molten plastic into the die cavity.

Although for thermoplastics the molds are relatively cool, thermosets are molded in heated molds, where polymerization and cross-linking take place. In either case, after the part is solidified (in thermoplastics), or set (cured, in thermosets), the molds

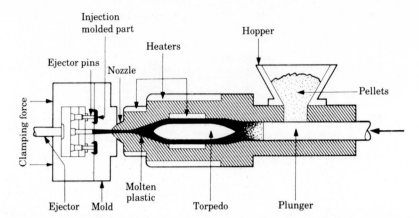

FIGURE 10.31 Schematic illustration of an injection molding machine. Injection molding of plastics is similar to the hot-chamber die-casting process, shown in Fig. 5.23.

are opened and the part is ejected. The molds are then closed and the process is repeated. Mold temperature is controlled by water flowing through channels in the mold block.

Injection molding is a high-rate production process with good dimensional control. The molds (generally made of tool steels or beryllium-copper) may have multiple cavities so that more than one part is made in one cycle of the machine. Because of the high cost of molds, production volume should be high for this process to be economical. (See also Table 6.4.)

Injection molding machines, which are generally horizontal, are rated according to the capacity of the mold and the clamping force on the molds, which in most machines is a few hundred tons. Typical injection-molded parts are cups and containers, various electrical and communication components, toys, and pipe fittings.

In *reaction-injection molding* (RIM) a mixture of two or more reactive fluids is forced under high pressure into the mold cavity. Chemical reactions take place in the mold rapidly and the polymer solidifies. Various fibers such as glass, graphite, or boron may also be used to reinforce the materials. *Structural foams* are also produced by a similar method (and by extrusion) using an inert gas (usually nitrogen) and resin mixture. The product consists of a rigid cellular (closed-cell) structure with a continuous solid skin as much as 0.080 in. (2 mm) in thickness. There are several methods for forming structural foams with density reductions as much as 40% from the solid structure. Because stiffness is proportional to the third power of the thickness of a part, for the same weight of material used, cellular structures are stiffer than solid plastics or metals.

Molds for RIM are generally made of polished metal, and pressures and temperatures involved are low. Major applications are automotive bumpers and fenders, thermal insulators for refrigerators and freezers, and stiffeners for structural components.

10.12.3 COMPRESSION MOLDING

In this process, a premeasured volume of powder, or a preformed part, is placed directly in a heated mold cavity and is formed by pressure with a punch or the other half of the mold (Fig. 10.32). Compression molding is thus similar to a forging operation with the same problem of flash formation (depending on the clearances) and the need for its removal (see Fig. 6.11a).

This process is used mainly for thermosets, with the original material in a partially polymerized state. Cross-linking is then completed in the heated mold under pressure and with curing times ranging from 0.5 to 5 minutes, depending on the material, part geometry and its thickness—the greater the thickness, the longer the curing time.

Because of their relative simplicity, mold costs in compression molding are generally lower than in injection molding. Three types of compression molds are available: *Flash-type* (for shallow or flat parts), *positive* (for high density), and *semi-positive* (for quality production), as shown in Fig. 10.32. Typical parts made by this process are knobs, handles, fittings, and housings. Fiber-reinforced materials (Section 10.11) may also be formed by compression molding.

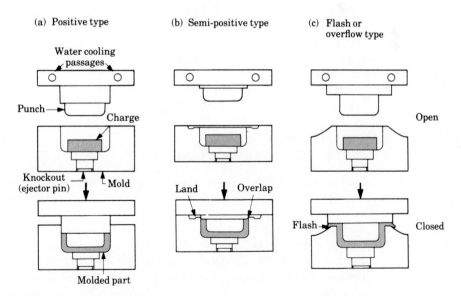

FIGURE 10.32 Types of compression molding. (a) Positive. (b) Semi-positive. (c) Flash. Compression molding is similar to closed-die forging of metals (Fig. 6.12). Note the flash in sketch (c), which has to be trimmed off, as it is with metals (Fig. 6.11a).

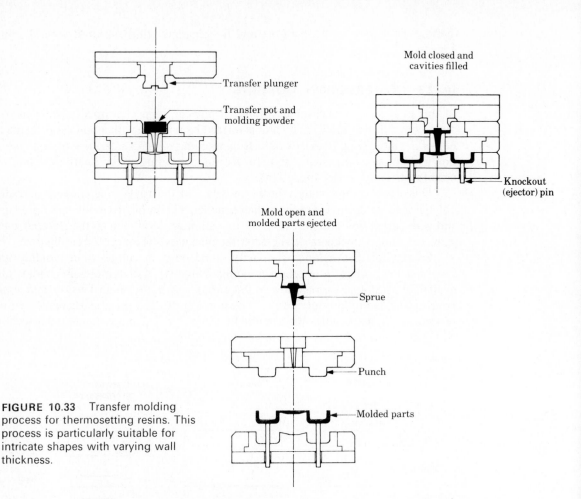

FIGURE 10.33 Transfer molding process for thermosetting resins. This process is particularly suitable for intricate shapes with varying wall thickness.

10.12.4 TRANSFER MOLDING

This process is a further development of compression molding. The preheated thermosetting material is placed in a transfer pot (Fig. 10.33), which is connected to the heated (about 325°F [160°C]) mold cavity through channels. Depending on the type of machine, a ram, plunger, or a screw feeder of an extruder forces the material (at pressures up to 12,000 psi [80 MPa]) to flow through the narrow channels into the mold cavity. This flow generates considerable heat, which raises the temperature and homogenizes the material. After filling the heated mold cavity, curing takes place by cross-linking.

Transfer molding is particularly suitable for parts with intricate shapes and varying wall thicknesses and has good dimensional control. Molds are usually made of tool steel. Typical parts are electrical and electronic components.

10.12.5 ROTATIONAL MOLDING

In this process (also called rotomolding) most thermoplastics and some thermosets can be formed into large hollow parts, such as tanks, drums, and buckets. The thin-walled metal mold (aluminum or steel) is made of two pieces and is designed to be rotated about two perpendicular axes (Fig. 10.34). A premeasured quantity of finely ground plastic material is placed inside the mold.

Liquid polymers, called *plastisols* (vinyl plastisols being the most common), can also be used where the process is known as *slush molding*. The mold is heated and rotated simultaneously. The particles of the material are forced against the inside walls of the heated mold by centrifugal force. Upon contact, the material melts and coats the walls of the mold. The part is cooled while still rotating and is removed by opening the mold.

This process can produce parts with complex hollow shapes with wall thicknesses as small as 1/64 in. (0.4 mm). Parts as large as 6 ft × 6 ft × 12 ft (1.8 m × 1.8 m × 3.6 m) have been formed. The outer surface finish of the part is a replica of the inside surface finish of the mold walls. Typical parts made by rotational molding are tanks of various sizes, boat hulls, buckets, drums, waste baskets, carrying cases, and plastic footballs. Various metallic or higher-melting-point plastic inserts may also be molded-in in parts made by this process.

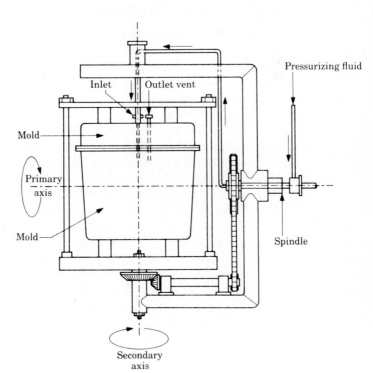

FIGURE 10.34 Rotational molding (rotomolding) process. Large tanks, buckets, and plastic footballs can be made by this process.

10.12.6 CASTING

A number of thermoplastics (nylons and acrylics) and thermosetting plastics (epoxies, phenolics, polyurethanes, and polyester) can be cast in rigid or flexible molds into a variety of shapes including sheet, rods, and tubing. In casting of thermoplastics, a mixture of monomer, catalyst, and various additives is heated and poured into the mold. The part is completed after polymerization takes place at ambient pressure. Intricate shapes can be formed with flexible molds, which are then peeled off. Typical parts are large gears, bearings, wheels, and components requiring abrasive wear resistance (see also Table 10.2).

Thermosets are cast in a similar manner. Degassing may be necessary for improved product integrity. Typical parts are similar to those in thermoplastic casting.

A variation of casting that is important to the electrical and electronic industry is *potting* and *encapsulation*. This involves casting the plastic around an electrical component, thus embedding it in plastic. Potting is done in a housing or a case that is an integral part of the product. In encapsulation, the component is covered with a layer of the solidified plastic. In both applications the plastic serves as a dielectric (nonconductor), and also as a structural member supporting the component.

10.12.7 THERMOFORMING

Thermoforming is a series of processes for forming thermoplastic film or sheet with the application of heat and air pressure, or a vacuum (Fig. 10.35). A variety of thermoforming processes have been developed; they are all based on the same basic principle. Because of the low strength of the thermoplastic materials, pressure difference due to vacuum is generally sufficient for forming, although air pressure is also applied in some processes.

Molds are usually made of cast aluminum or aluminum-filled epoxy since high strength is not a requirement and aluminum is a good thermal conductor. The holes in the molds (for vacuum) are generally less than 0.02 in. (0.5 mm) so no marks will be left on the formed sheets.

Because thermoforming is basically a stretching operation, the material should exhibit high uniform elongation, otherwise it will neck and fail. Thermoplastics have high values of m (strain-rate sensitivity [Section 2.2.7]) hence they can stretch considerably without failure (Fig. 10.6) and thus enable successful thermoforming operations into various shapes. These processes cannot form parts with openings or holes since the pressure difference cannot be maintained.

The sheets used in thermoforming are made by the *calendering* process (Fig. 10.36). A warm plastic mass is fed through a series of heated rolls (*masticated*) and is then stripped off in the form of a sheet.

Typical examples of thermoforming are advertising signs, liners for appliances, packaging, covers and housings, and panels for showers.

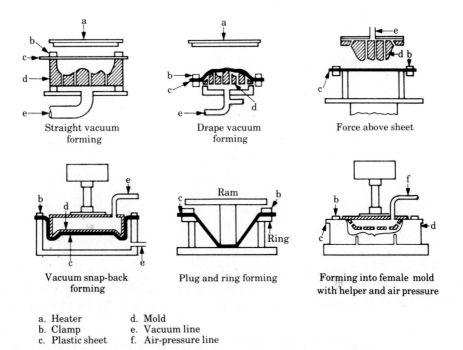

Straight vacuum
forming

Drape vacuum
forming

Force above sheet

Vacuum snap-back
forming

Plug and ring forming

Forming into female mold
with helper and air pressure

a. Heater
b. Clamp
c. Plastic sheet

d. Mold
e. Vacuum line
f. Air-pressure line

FIGURE 10.35 Various thermoforming processes for thermoplastics. These processes are commonly used in making advertising signs, packaging, and panels for showers.

FIGURE 10.36 Schematic illustration of calendering. The sheets produced by this method are subsequently used in thermoforming of plastics.

10.12.8 BLOW MOLDING

This is basically a bulging process. A tubular piece of plastic is heated and then pressurized internally and expanded into the cavity of a relatively cool split mold. Typical products are hollow, thin-walled containers or articles.

An example of an automated process is *extrusion blow molding*. Here a hollow tube of molten plastic (*parison*) is produced by an extruder (more recently, by an injection-molding machine) and is then pinched at one end. The parison is then placed

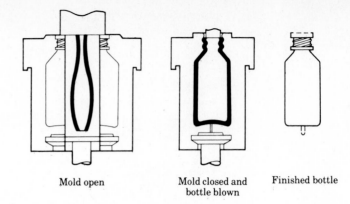

FIGURE 10.37 Schematic illustration of the blow-molding process. Plastic bottles for beverage and household products are made by this process. See also Fig. 11.12.

Mold open

Mold closed and
bottle blown

Finished bottle

into the open mold (Fig. 10.37), the mold is closed, and the parison is expanded (inflated) with an air nozzle (blowing pin) to a diameter that is usually three times the parison diameter. The air pressure is usually between 50 and 100 psi (0.35 to 0.7 MPa). The part cools off after contact with the die, which is then opened and the part is ejected. Molds are usually made of aluminum and are vented to prevent entrapment of air.

Typical products made by blow molding are containers for fluid and solid food, medicine, or other products, laboratory ware, toys, and storage tanks for liquids. Containers as large as 100 ft^3 (3 m^3) capacity have been blow molded.

10.12.9 COLD FORMING AND SOLID-PHASE FORMING

Processes that are used in cold working of metals (Chapters 6 and 7) can also be used to form thermoplastics at room temperature. Typical materials suitable for these operations are polypropylene, polycarbonate, ABS, and rigid PVC. The processes employed are rolling, deep drawing, extrusion, closed-die forging, coining, and rubber forming.

Important considerations are: the material must be sufficiently ductile at room temperature (hence polystyrenes, acrylics, and thermosets cannot be formed), and its deformation must be nonrecoverable, so that springback and subsequent time-dependent strains are minimized.

The advantages of *cold forming* of plastics are the following:

a. Increased strength and toughness, as shown in Figs. 10.17 through 10.20, and increased uniform elongation useful for subsequent processing,

b. Plastics with high molecular weight can be used to make parts with superior properties,

c. Forming speeds are not affected by part thickness since there is no heating or cooling involved, typical cycle times being shorter than molding processes.

Solid-phase forming is carried out at a temperature of the plastic that is 20 to 40°F (10 to 22°C) below its melt temperature (if a crystalline polymer), and the part is formed while still in a solid state. The advantage over cold forming is that the forming forces and springback are lower. Although not commonly used, solid-phase forming has been applied in forging, impact and direct extrusion, thread rolling, and rubber forming, with polypropylene as a common material.

10.12.10 PROCESSING OF REINFORCED PLASTICS

Reinforced thermoplastics and thermosets can usually be processed by many of the molding methods described above. The reinforcement may be in forms such as loose fibers, woven fabric or mat, roving (slightly twisted fiber), or continuous lengths of fiber. In order to have a good bond between the reinforcing fibers and the polymer matrix, it is necessary to first impregnate and coat the reinforcement with the polymer. This product is known as *prepreg* and should be stored at a sufficiently low temperature to delay curing.

Bulk molding compounds for compression or injection molding are available as premixes containing all the necessary ingredients and the reinforcements. Fiber lengths generally range from 1/8 to 2 in. (3 to 50 mm) although longer fibers (3 in. [75 mm]) may also be used.

Molding

Molding processes may be *compression*-type or *contact*-type. In the former, pressure is applied to the material (which is in bulk or sheet form), which is placed between two molds. Depending on the material, the part is formed either at room or at a higher temperature.

Contact-molding processes use single male or female molds. The materials are placed and formed in the mold by hand (*hand lay-up*) or they are sprayed (*spray-up*) (Fig. 10.38). These processes are relatively simple and inexpensive.

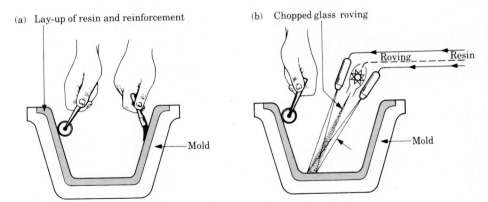

(a) Lay-up of resin and reinforcement   (b) Chopped glass roving

FIGURE 10.38 Methods of processing reinforced plastics. (a) Hand lay-up. (b) Spray-up. These processes are relatively simple and inexpensive. *Source:* After R. L. E. Brown.

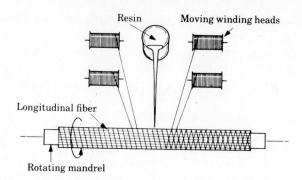

FIGURE 10.39   Filament winding process, used in making reinforced tubes and underground gasoline storage tanks for service stations.

Filament Winding

Filament winding (Fig. 10.39) is a process where axisymmetric parts, such as tubes and storage tanks, are generated on a rotating mandrel. The reinforcing filament, tape or roving, is wrapped continuously around the form. The reinforcements may be prepregs, or they may be impregnated by passing them through a polymer bath. The products made by filament winding are very strong and rigid because of their highly reinforced structure.

Pultrusion

Long extruded shapes with various constant profiles (extrusions) are made by the pultrusion process. This is a process where the continuous reinforcement (roving or fabric) is pulled through a thermosetting polymer bath and then through a long heated steel die. The product is cured during its travel through the die. The most common material for pultrusion is polyester with glass reinforcements.

10.12.11 MACHINING AND JOINING OF PLASTICS

General guidelines for machining plastics are given in Section 8.11.2. Joining of plastics are treated in Section 12.13.

10.13 DESIGN CONSIDERATIONS

We have seen that plastics have a wide range of properties and part shapes, with applications for numerous industrial and consumer products. The design procedures for plastics are somewhat similar to those in the processing of metals. However, the mechanical and physical properties of plastics should be carefully considered in design, material, and process selection.

It was noted that compared to metals plastics have lower strength and stiffness. This means that section sizes should be selected accordingly, with the view of maintaining a sufficiently high section modulus for improved stiffness. Fiber reinforcements are a highly desirable means of achieving this purpose (Table 10.1).

Physical properties, especially high coefficient of thermal expansion (and hence contraction), are important not only in mold design, but also in the compatibility of the plastic when used with metals. Plastics can easily be molded around metallic parts and inserts; compatibility is thus an important factor.

Selection of the appropriate material from an extensive list requires careful consideration of service requirements and possible long-range effects on material properties and behavior.

The overall part geometry determines the particular forming or molding process. Table 10.7 is a general guide in this selection. Even after a particular process is selected, the design of the part should be such that it will not cause problems concerning shape generation, dimensional control, and surface finish. Just as in casting of metals and alloys, material flow in the mold cavities should be controlled properly. Large variations in section sizes and abrupt changes in geometry should be avoided for improved product quality and increased mold life.

An essential aspect of all design decisions is cost, which includes the costs of equipment, tooling, and production (Table 10.8). As in all manufacturing processes, final selection depends greatly on the production volume (see Section 13.4). High equipment and tooling costs are acceptable if the production run is high.

The properties of the final product depend on the original material and its processing history. As noted, cold working improves strength and toughness. On the other hand, because of the nonuniformity of deformation (even in simple rolling of thin sheet) residual stresses develop in polymers, just as they do in metals (see Fig. 6.51). In fact, experimental evidence indicates that in cold rolling of polycarbonates, the increase in toughness, at low reductions, is due to the presence of residual stresses and not because of molecular orientation (Fig. 10.6a). The distribution of these stresses is also important. Only at higher reductions is the increase in toughness attributed to molecular orientation.

Residual stresses can also be generated by thermal cycling of the part. Also, the magnitude and direction of residual stresses (regardless of how they are produced) are also important factors, just as they are in metals. These stresses can relax over a period of time and cause part distortion.

Residual stresses can also lead to distortion if the part is processed subsequently, such as by removal of a section by machining (see Fig. 2.42). Tensile residual stresses can also be harmful in the fatigue life of the part and its resistance to chemicals and solvents. They can cause brittle behavior and rapid cracking of an otherwise ductile plastic.

TABLE 10.7
CHARACTERISTICS OF VARIOUS MOLDING AND FORMING PROCESSES FOR PLASTICS.

	SHAPE LIMITATIONS	INTRICATE, COMPLICATED SHAPES	CONTROLLED WALL THICKNESS	OPEN, HOLLOW SHAPES	ENCLOSED, HOLLOW SHAPES	LARGE ENCLOSED VOLUME	VERY SMALL ITEMS	PLAN AREA, 10 ft^2 (9 m^2)	FACTOR LIMITING MAXIMUM SIZE	INSERTS	MOLDED-IN HOLES	THREADS
Nonreinforced materials												
Compression molding	Moldable	Yes	Yes	Yes					Press	Yes	Yes	Yes
Transfer molding	Moldable	Yes	Yes	Yes				Yes	Press	Yes	Yes	Yes
Injection molding	Moldable	Yes	Yes	Yes				Yes	Press	Yes	Yes	Yes
Extrusion	Constant cross-section	Yes	Yes						Die	Die		
Rotational molding	Hollow			Yes	Yes	Yes		Yes	Available machine	Yes	Yes	Yes
Blow molding	Hollow, Thin-wall			Yes	Yes	Yes		Yes	Mold			Yes
Thermoforming	Thin-wall			Yes				Yes	Available machine			
Casting	Moldable	Yes	Yes						Mold	Yes	Yes	
Forging	Moldable	Yes	Yes						Die			
Foam molding	Moldable	Yes	Yes	Yes				Yes	Press		Yes	
Fiber-reinforced materials												
Injection molding	Moldable	Yes	Yes	Yes					Press	Yes	Yes	
Hand lay-up and spray-up	Large, thin-wall	Yes	Yes	Yes		Yes, by joining		Yes	Mold, or transport of parts	Yes	Yes	Yes
Compression-type molding	Moldable	Yes	Yes	Yes					Press	Yes	Yes	Yes
Preform molding	Moldable	Yes	Yes	Yes				Yes	Press			
Cold-press molding	Moldable	Yes	Yes	Yes				Yes	Press			
Filament winding	Surface of revolution		Yes						Available machine			
Pultrusion	Constant cross-section	Yes	Yes						Die			

Source: After R. L. E. Brown, Design and Manufacture of Plastic Parts. © 1980 by John Wiley & Sons, Inc. Reprinted by permission of John Wiley & Sons, Inc.

TABLE 10.8
COMPARATIVE COSTS AND PRODUCTION VOLUMES FOR PROCESSING OF PLASTICS.

	EQUIPMENT CAPITAL COST	PRO-DUCTION RATE	TOOLING COST	TYPICAL PRODUCTION VOLUME, NUMBER OF PARTS (10 10^2 10^3 10^4 10^5 10^6 10^7)
Nonreinforced materials				
Machining				10
Compression molding	High	Medium	High	10^4–10^7
Transfer molding	High	Medium	High	10^4–10^7
Injection molding	High	Fast	High	10^5–10^7
Extrusion	Medium		Low	
Rotational molding	Low	Slow	Low	10–10^3
Blow molding	Medium	Medium	Medium	10^4–10^7
Thermoforming	Low	Slow	Low	10–10^3
Casting	Low	Very slow	Low	10–10^2
Forging	High	Slow	Medium	10
Foam molding	High	Medium	Medium	10^2–10^7
Fiber-reinforced materials				
Injection molding	High	Fast	High	10^4–10^7
Hand lay-up		Slow	Low	10
Spray-up	Low	Slow	Low	10–10^3
Compression	High	Medium	High	10^3–10^6
Preform molding	High	Medium	Medium	10^3–10^5
Cold-press molding	High	Medium	Low	10^3–10^5
Filament winding	Low		Low	
Pultrusion	Low		Low	

Source: After R. L. E. Brown, *Design and Manufacture of Plastic Parts.* © 1980 by John Wiley & Sons, Inc. Reprinted by permission of John Wiley & Sons.

● **Illustrative Example 10.1**

The coin delivery cup of a vending machine was in one design made of an aluminum casting. This is the cup into which change is returned. This casting had to be trimmed of flash, then hand buffed to remove burrs and provide a smooth surface. Holes were then drilled and tapped for four machine screws. The part was then degreased for painting and loaded into a masking device to allow painting in the desired areas using a textured paint. To support the cup, a flange was made from sheet metal in a punch press. It was then degreased and chrome plated to be mounted on the cup.

Upon close examination of this part, it was determined that it could be made by injection molding using polyester, a thermoplastic. The production of this part by injection molding, Fig. P10.1, eliminated all the various operations required in the previous method. The part was molded in one piece with the desired surface texture and in the desired color (black in this case). Polyester has the necessary properties for this application, namely, mechanical strength, low friction, and good abrasion resistance.

FIGURE P10.1   *Source:* Courtesy of Plaspros, Inc.

Injection molding of this part has been found to be 35% more energy efficient than metal casting. Because of the elimination of many operations, handling, plating, etc., cost reductions as high as 75% can be obtained.   ●

SUMMARY

1. Plastics are an important class of materials because they possess a wide range of mechanical, physical, and chemical properties. Compared to metals, they are characterized by lower density, strength, elastic modulus, and thermal and electrical conductivity, and a higher coefficient of thermal expansion.

2. Plastics are composed of polymer molecules and various additives. The smallest repetitive unit in a polymer chain is called a mer. A monomer is the building block of polymers. Monomers are linked together by polymerization to form longer and larger molecules. By various techniques, it is possible to obtain some crystallinity in polymers. The degree of crystallinity greatly affects the properties of the polymers.

3. Two major classes of polymers are thermoplastics and thermosets. The former become soft and easy to form at elevated temperatures, and return to their original properties when cooled. Typical examples are ABS, acrylics, nylons, polyethylenes, and polyvinyl chloride. Thermosets, which are obtained by cross-linking polymer chains, do not to any significant extent become soft with increasing temperature. Typical examples are epoxies and phenolics.

4. Additives in polymers have various functions, such as improving strength, hardness, abrasion resistance, flame retardation, and lubrication. Other functions are to impart flexibility, softness, color, and stability against ultraviolet radiation and oxygen.

5. Another important class of materials is reinforced plastics. The reinforcements are generally graphite or glass fibers, and boron for high-temperature service. These are mixed in a polymer matrix with various proportions and orientations. The mechanical properties of the reinforced plastics are superior to nonreinforced plastics.

6. Whereas thermosets are molded or cast, thermoplastics are also formed by various processes such as extrusion and thermoforming. The high strain-rate-sensitivity index of thermoplastics allows extensive stretching in forming these materials. Thermoplastics can also be formed at room temperature by techniques similar to those used in metalworking.

7. The design of plastic parts should include consideration of stiffness. Thus, section sizes are important, as well as other factors involved in the service life of the component.

8. Elastomers are a large family of rubberlike amorphous polymers with a glass-transition temperature that is usually below the service temperature.

BIBLIOGRAPHY

Alfrey T, Gurnee EF. *Organic Polymers*. Englewood Cliffs, N.J.: Prentice-Hall, 1967.

Ash M, Ash I. *Encyclopedia of Plastics, Polymers and Resins* (3 vols.). New York: Chemical Publishing Co., 1980–81.

Beadle JD, ed. *Plastics Forming*. New York: Macmillan, 1971.

Beck RD. *Plastics Products Design*. New York: Van Nostrand Reinhold, 1970.

Bernhardt EC, ed. *Processing of Thermoplastic Materials*. New York: Van Nostrand Reinhold, 1959.

Billmeyer FW Jr. *Textbook of Polymer Science*, 2d ed. New York: Wiley, 1971.

Brandrup J, Immergut EH, eds. *Polymer Handbook*, 2d ed. New York: Wiley, 1975.

Broutman LJ, Krock RH, eds. *Modern Composite Materials*. Reading, Mass.: Addison-Wesley, 1967.

Brown RLE. *Design and Manufacture of Plastic Parts*. New York: Wiley, 1980.

Catherall JA. *Fibre Reinforcement*. London: Mills and Boon, 1973.

Crosby EC, Kochis SN. *Practical Guide to Plastics Applications*. Boston: Cahners, 1972.

Frados J, ed. *Plastics Engineering Handbook*, 4th ed. New York: Van Nostrand Reinhold, 1976.

Harper CA, ed. *Handbook of Plastics and Elastomers*. New York: McGraw-Hill, 1976.

Kaufman M. *Giant Molecules, The Technology of Plastics, Fibers and Rubber*. New York: Doubleday Science, 1968.

Kobayashi A. *Machining of Plastics*. New York: McGraw-Hill, 1967.

Lenk RS. *Plastics Rheology*, New York: Wiley, 1968.

Levy S, Davis JH. *Plastics Product Design Engineering Handbook*. New York: Van Nostrand Reinhold, 1976.

Middleman S. *Fundamentals of Polymer Processing*. New York: McGraw-Hill, 1977.

Milby RV. *Plastic Technology*. New York: McGraw-Hill, 1973.

Modern Plastics Encyclopedia. New York: McGraw-Hill, annual.

Nielsen LE. *Mechanical Properties of Polymers and Composites* (2 vols.). New York: Marcel Dekker, 1974.

Nutt MC. *Metallurgy and Plastics for Engineers*. New York: Pergamon, 1977.

O'Driscoll KF. *Nature and Chemistry of High Polymers*. New York: Van Nostrand Reinhold, 1964.

Ogorkiewicz RM. *The Engineering Properties of Plastics*. New York: Oxford, 1977.

Ogorkiewicz RM. *Thermoplastics*. London: CRC Press, 1969.

Parkyn B, ed. *Glass Reinforced Plastics*. London: Butterworths, 1970.

Parratt NJ. *Fibre-Reinforced Materials Technology*. New York: Van Nostrand Reinhold, 1972.

Powell PC. *The Selection and Use of Thermoplastics*. New York: Oxford, 1977.

Rodriguez F. *Principles of Polymer Systems*. New York: McGraw-Hill, 1970.

Roff WJ, et al. *Handbook of Common Polymers*. Cleveland, CRC Press, 1971.

Rosen SL. *Fundamental Principles of Polymeric Materials*, 2d ed. New York: Wiley, 1982.

Schultz J. *Polymer Materials Science*. Englewood Cliffs, N.J.: Prentice-Hall, 1974.

Tadmor Z, Gogos CG. *Principles of Polymer Processing*. New York: Wiley, 1979.

Throne JL. *Plastics Process Engineering*. New York: Marcel Dekker, 1979.

Tobolsky AV. *Properties and Structure of Polymers*. New York: Wiley, 1960.

Treloar LGG. *Introduction to Polymer Science*. London: Wykeham, 1974.

Wallace BM, ed. *Handbook of Thermoplastic Elastomers*. New York: Van Nostrand Reinhold, 1979.

Ward IJ. *Mechanical Properties of Solid Polymers*. New York: Wiley, 1971.

Williams DJ. *Polymer Science and Engineering*. Englewood Cliffs, N.J.: Prentice-Hall, 1971.

Williams HL. *Polymer Engineering*. Amsterdam: Elsevier, 1975.

Periodicals

Materials Engineering

Materials Science and Engineering

Modern Plastics

Plastics Engineering

Polymer Engineering and Science

Polymer Journal

Polymer–Plastics

PROBLEMS

10.1. Inspect Table 10.5 and explain why some plastic forming processes are more suitable for certain classes of plastics than for others.

10.2. Describe the advantages of applying traditional metalworking techniques to forming plastics.

10.3. Explain the advantages of reinforced plastics.

10.4. Based on the material covered in this chapter, explain why the abrasive wear resistance of plastics are as shown in Table 10.2 and what properties have an effect on wear resistance.

10.5. Inspect various plastic components in your car and attempt to identify the materials and the processes used in manufacturing those parts.

10.6. Assume that you are manufacturing a product in which all the gears are made of metal. A salesman visits you and asks you to consider replacing some of these metal gears with plastic ones. Make a list of questions that you would raise before you make a decision.

10.7. What design considerations should be taken into account in replacing a metal container for a beverage with one made of plastic?

10.8. Explain the basic differences between thermoplastics and thermosets.

10.9. Describe the mechanism by which most thermoplastics can be stretched to a far greater extent than metals.

10.10. Describe the differences between three types of compression molding.

10.11. List all possible applications for filament-wound plastics.

10.12. Plot the toughness (area under the stress–strain curve) for the material in Fig. 10.12 as a function of temperature.

10.13. Calculate the average increase in mechanical properties in Table 10.1 as a result of reinforcing.

10.14. Explain the difference between extrusion and pultrusion.

11 Processing of Powder Metals and Ceramics

11.1 INTRODUCTION

In all the processes described in the preceding chapters, the raw materials used for processing are either liquid or solid. The liquids are shaped by solidification techniques and the solid materials are shaped at room or elevated temperatures, basically either by forming or machining. In this chapter, processes that use materials as powder or in a slurry and the techniques involved in shaping them are described.

11.2 POWDER METALLURGY

Metal powders can be shaped into objects by compacting them in suitable dies and subsequently heating (sintering) them at elevated temperatures to obtain the required density and strength. The process is generally referred to as *powder metallurgy* (P/M) and was first developed in the 1840s to make parts and used commercially in the early 1900s to make the tungsten filaments for light bulbs. Using this process parts can be made from a variety of metals and alloys with good dimensional accuracy with sizes ranging from tiny balls (for ballpoint pens) to parts weighing 100 lb (50 kg).

Typical products made by powder-metallurgy techniques are gears, cams, levers, bushings, cutting-tool inserts, porous products such as filters and oil-impregnated bearings, a variety of relatively small components for office equipment, and electrical and various automotive and machine elements (Fig. 11.1). More recent developments are P/M tool steels for various cutting tools, and engine components.

Pure metals, alloys, or mixtures of metallic and nonmetallic materials can be formed by powder metallurgy techniques. The most commonly used metals are iron, copper, aluminum, tin, nickel, titanium, and refractory metals (tungsten, tantalum,

FIGURE 11.1 Typical parts manufactured by powder metallurgy techniques. Product sizes vary from tiny balls for ballpoint pens, to parts weighing 100 lb (50 kg). *Source:* Courtesy of Metal Powder Industries Federation.

molybdenum, and columbium). For alloys such as brass, bronze, steels, and stainless steels, prealloyed powders are used, where each powder particle itself is an alloy. Other metals and materials (used singly or in combination with different materials) are ceramics and various oxides, graphite, carbides, zirconium, and thorium.

The basic powder-metallurgy technique involves the following steps:

a. Powder production,

b. Blending,

c. Compaction,

d. Sintering.

For improved quality, or special applications, additional processing such as coining, sizing, forging, infiltration, and resintering may be carried out.

11.2.1 PRODUCTION OF METAL POWDERS

There are several methods for producing metal powder; the shape and properties of the powder particles depend on the method used (Fig. 11.2). The most common methods involve atomization of molten metal, reduction of metal oxides, electrolytic deposition, decomposition of metallic compounds, and mechanical comminution. These techniques are explained below. Other less commonly used methods are precipitation from a chemical solution, production of fine chips by machining, and vapor condensation. The source is generally bulk metals and alloys, ores, salts, and other compounds.

Atomization
Atomization involves forming a liquid metal stream by injecting molten metal through a small orifice. The stream is then broken up by jets of inert gas, air, or water, thus producing solid metal particles. Particle size depends on the temperature of the metal, rate of flow, nozzle size, and jet characteristics. In one variation of this method, a consumable electrode is rotated rapidly in a helium-filled chamber. The centrifugal force breaks up the molten tip of the electrode, thus producing metal particles.

Reduction
Reduction of metal oxides is carried out using gases, such as hydrogen and carbon monoxide, as reducing agents. Very fine metallic oxides are thus chemically reduced to the metallic state. This process produces metal powders of uniform size. The powders are spongy and porous and have spherical or angular shapes.

Electrolytic Deposition
Metal powders may also be produced by electrolytic deposition, either from aqueous solutions or from fused salts. These powders are among the purest. The size and shape of the particles depend on process parameters.

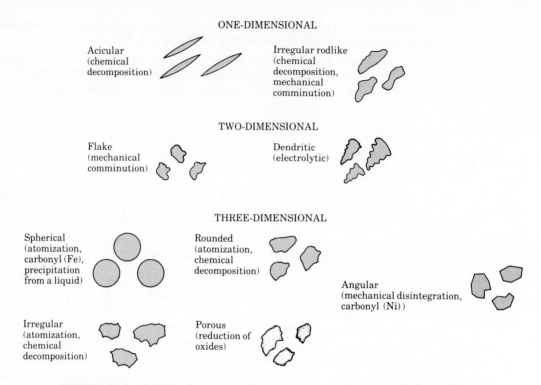

FIGURE 11.2   Particle shapes in metal powders and the processes by which they are produced. Iron powders are produced by many of these processes. See also Table 11.1. *Source:* After P. K. Johnson.

Thermal Decomposition of Carbonyls

Metal carbonyls, such as iron carbonyl, $Fe(CO)_5$, and nickel carbonyl, $Ni(CO)_4$, are formed by reacting iron or nickel with carbon monoxide. These products are then thermally decomposed to iron and nickel, producing small, dense, uniform spherical particles of high purity.

Comminution

Mechanical *comminution* (pulverization) involves crushing, milling, or grinding brittle metals or metals with low ductility into small particles. With brittle materials, the powder particles are angular in shape, whereas with ductile metals, they are flaky and not particularly suitable for powder-metallurgy applications.

Most metal powders can be produced by more than one method (Table 11.1); the choice depends on the requirements for the end product. Particle sizes range from 0.1 to 1000 μm. The shape (Fig. 11.2), size distribution, porosity, chemical purity, and bulk and surface characteristics of the particles depend on the process used. These properties are important because they have significant effects on permeability and flow characteristics during compaction and in subsequent sintering operations.

TABLE 11.1
METHODS OF PRODUCING METAL POWDERS

POWDER METAL	ATOM-IZATION	REDUC-TION	ELECTRO-LYTIC DEPOSI-TION	THERMAL DECOMPO-SITION	COMMI-NUTION	PRECIPI-TATION FROM LIQUID OR GAS
Aluminum	×					
Aluminum alloys	×					
Beryllium			×		×	
Cobalt		×				
Copper	×	×	×			×
Copper alloys	×					
Iron	×	×	×	×	×	
Iron alloys (low-alloy steel, stainless steel, tool steel)	×					
Molybdenum		×				
Nickel		×		×		×
Nickel alloys	×				×	
Silver	×		×			×
Tantalum		×	×			
Tin	×					
Titanium	×	×		×		
Tungsten		×				
Zirconium				×		

Iron powders are produced by many of the methods described above. Most powders, such as stainless steels, superalloys, and tool steels, are produced by atomization. Low oxygen content is important during atomization of many metals, especially for superalloy powders. Inert-gas atomization allows for oxygen contents of well below 1000 ppm.

Many metal powders with several grades of purity are commercially available. In addition to prealloyed powders, precoated powders are also available, where the metal particle is coated with another element.

11.2.2 BLENDING OF POWDERS

Blending (mixing) of powders is the second step in powder-metallurgy processing and is carried out for the following purposes:

a. Because the powders made by the various processes described above may have different sizes and shapes, it is necessary to mix them to obtain uniformity. The ideal mix is one in which particles of each material are distributed uniformly.

b. In order to impart special physical and mechanical properties and certain characteristics to the P/M product, powders of different metallic and other materials may be mixed.

c. In order to improve the flow characteristics of the powder metals, and to improve die life and reduce friction between the metal particles, lubricants or waxes may be mixed with the powders. Typical lubricants are stearic acid or zinc stearate in amounts of 0.25 to 5% by weight.

Mixing of powders must be carried out under controlled conditions to avoid contamination or deterioration of the powders. Excessive mixing may change the shape of the particles, and may also work-harden them, thus making the subsequent compacting operation more difficult. Powders can be mixed in air or other gaseous atmospheres. Liquids may also be used, depending on the metal powder. Liquids act as a lubricant, making the mix more uniform. There are several types of blending equipment available with various capacities and mixing characteristics.

Hazards

Some metal powders, because of their high surface-to-volume ratios, have explosive characteristics. Examples are aluminum, aluminum-magnesium and aluminum-silicon alloys, magnesium, titanium, zirconium, and thorium. Great care must be exercised during blending and during storage and handling. Some necessary precautions are: avoiding sparks, dust clouds, open flames, and chemical reactions; grounding of equipment; using nonsparking tools; and avoiding friction as a source of heat.

11.2.3 COMPACTION OF METAL POWDERS

This is the step where the blended metal powders are formed into shapes in dies using presses that are either hydraulically or mechanically actuated (Fig. 11.3). The purposes of compaction are to obtain the required shape, density, and particle-to-particle contact, and to impart sufficient strength to the part to enable handling for further processing.

Powder Flow and Pressing

Ease of powder flow is important for proper feeding of the powder into the die cavity. The pressed powder is known as *green compact*. Pressing is generally carried out at room temperature, although it can also be done at elevated temperatures. Because of friction between the metal particles in the powder and the friction between the punches and the die walls, there can be considerable variation in density within the part.

The density distribution for various die geometries and punch movements is shown in Fig. 11.4. For a cylindrical part, the ratio of length-to-diameter is important; the larger the ratio the less uniform the density distribution. (See Fig. 11.10.) Variation

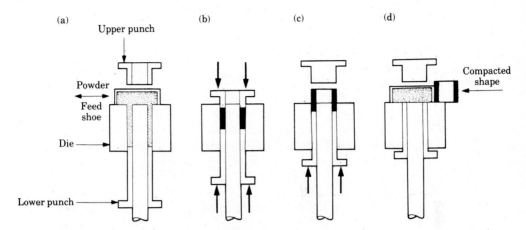

FIGURE 11.3 Schematic illustration of compaction of metal powders into shapes. The pressed powder part, known as green compact, is then sintered (heated) in a furnace to allow bonding of the particles.

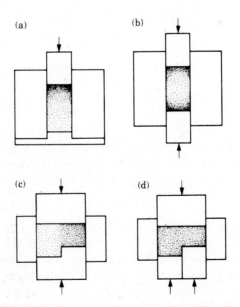

FIGURE 11.4 Density distribution in pressing metal powders. Note the greater uniformity of density in pressing with two-punches with separate movements. There are situations, however, where it is desirable to have a variation in density within the product. See also Fig. 11.10 and Illustrative Example 11.1.

in density can be minimized by proper design of punches and dies and by control of friction. Note, for instance, that it may be necessary to use multiple punches with separate movements to ensure that density is uniform throughout the part.

Equipment

Pressures required for pressing can range up to 120,000 psi (800 MPa). Press capacities are on the order of 200 to 300 tons (1800 to 2700 kN), although presses with much higher capacities are used for special applications. Most applications require less than 100 tons. For small tonnage, crank or eccentric-type mechanical presses are used; for higher capacities, toggle or knuckle-joint presses are employed (see Section 6.9).

Hydraulic presses, which are slower acting than mechanical types, can be used with capacities as high as 5000 tons (45 MN) for large parts. The choice of the press depends on parameters such as part size and configuration, density requirements, and production rate. An important consideration with regard to pressing speed is the entrapment of air in the die cavity at high speeds, as the presence of air will prevent proper compaction.

Compaction can also be carried out by a number of other processes, such as isostatic pressing, extrusion, forging, or rolling. These are described below.

Isostatic Pressing

In *cold isostatic pressing* (CIP), the metal powder is placed in a flexible rubber mold, made of neoprene rubber, urethane, polyvinyl chloride, or other elastomers. It is then pressurized hydrostatically in a chamber, commonly using water.

In *hot isostatic pressing* (HIP), the container is made of a high-melting-point sheet metal and the pressurizing medium is inert gas or vitreous fluid.

These processes have been used successfully for a variety of metal powders and product shapes. Parts with high length-to-diameter ratios have been produced with uniform densities and good surface details and properties. Typical applications of this technique are tool steels, carbide tools and dies, filters, and parts made of superalloys.

The main advantage of isostatic pressing is that, because of uniformity of pressure from all directions and the absence of die-wall friction, it produces compacts of practically uniform grain structure and density, irrespective of length-to-diameter ratio. Simple or complex shapes can be made with uniform strength and fracture toughness.

For CIP, the most common pressure is 60,000 psi (400 MPa) although pressures up to 150,000 psi (1000 MPa) have been used. Common conditions for HIP are 15,000 psi (100 MPa) at 2000°F (1100°C), although the trend is toward higher pressures and temperatures.

The main advantage of HIP is its ability to produce compacts with essentially 100% density, good metallurgical bonding, and very good mechanical properties. The process is relatively expensive and is used mainly for making superalloy components for the aerospace industry. It is also routinely used as a final densification step

TABLE 11.2
MECHANICAL PROPERTIES OF RENÉ 95* TURBINE WHEELS MADE BY VARIOUS PROCESSES. (After J. E. Coyne _et al_.)

PROPERTY	CAST AND FORGED	POWDER PREFORM FORGED	HOT ISOSTATIC PRESSED AND HEAT TREATED
ROOM TEMPERATURE			
Tensile strength, psi × 10^3	215	240	240
Yield strength, psi × 10^3	187	182	180
Elongation, %	7	15	20
Reduction in area, %	8	16	21
1200°F (650°C)			
Tensile strength, psi × 10^3	208	220	218
Yield strength, psi × 10^3	177	171	164
Elongation, %	7	14	13
Reduction in area, %	7	17	14

* An alloy of 14 Cr–61 Ni–8 Co and Mo, W, Nb and Ti.

for tungsten carbide cutting tools and P/M tool steels. More recently, HIP has been employed to close internal porosity in superalloy and titanium-alloy castings for the aerospace industry (Table 11.2).

Rolling and Other Processes
In _powder rolling_, or _roll compaction_, the powder is fed into the roll gap in a two-high rolling mill and is compacted into a continuous strip at speeds up to 100 ft/min (0.5 m/s). Sheet metals for various electrical and electronic components and for coins can be made by powder rolling. Compaction of powders can also be carried out by extrusion by encasing the powder in a container made of a metal with sufficient ductility (see also Section 6.16.1).

The density of the compacted part depends on the pressure applied (Fig. 11.5). As expected, the density of the compact approaches that of the theoretical density (bulk metal) as the pressure is increased. Another important factor in determining density is the size distribution of the particles of metal powder. If all particles are the same size, such as spherical particles, there will always be some porosity when they are packed together. However, if we now introduce smaller particles, these will fill the spaces in between the larger particles, resulting in a higher density.

Die and Punch Materials
The selection of die and punch materials for P/M processing depends on the abrasiveness of the powder metal and the production run, i.e., the number of parts to be made. Most common die materials are air- or oil-hardening tool steels, such as D2 or D3,

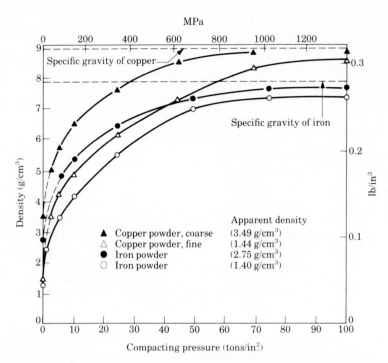

FIGURE 11.5   Density of iron and copper powders as a function of compacting pressure. Note that the increase in density stabilizes, approaching the density of the bulk metal, after a certain pressure is reached. Density has a major influence on the mechanical and physical properties of powder metallurgy products. See also Fig. 11.7. *Source:* After F. V. Lenel.

with a hardness of 60 to 64 HRC. Tungsten carbide dies are used for more severe applications. Punches are generally made of similar materials. (See Table 6.4.)

Close control of die and punch dimensions and tolerances is essential for proper compaction and die life. For instance, too large a clearance between the punch and the die will allow the metal powder to enter the gap and interfere with the operation and also will result in eccentricity. Diametral clearances are generally less than 0.001 in. (25 μm). Die and punch surfaces must be lapped or polished (in the direction of tool movements) for improved die life and overall performance.

11.2.4 SINTERING

Sintering is the process whereby the compacted metal powder is heated in a controlled-atmosphere furnace to a temperature just below its melting point, but sufficiently high to allow *bonding* of the individual particles. Prior to sintering, the compact is quite brittle and its strength (*green strength*) is low. In order to facilitate handling, compacts may be *presintered* by heating them to a temperature lower than the normal temperature for final sintering.

The nature and strength of the bond between the particles, and hence of the sintered compact, depend on the mechanisms of:

a. Diffusion,

b. Plastic flow,

c. Evaporation of volatile materials in the compact,

d. Recrystallization,

e. Grain growth,

f. Shrinkage.

The principal governing variables in sintering are temperature, time, and the atmosphere. Sintering temperatures are generally within 70 to 90% of the melting point, and sintering times at these temperatures range from a minimum of about 10 minutes for iron and copper alloys to as much as eight hours for tungsten and tantalum (Table 11.3). Continuous sintering furnaces are used for most production today. These furnaces have three chambers:

a. Burn-off chamber to volatilize the lubricants in the green compact in order to improve bond strength,

b. High-temperature chamber for sintering, and

c. Cooling chamber.

TABLE 11.3
SINTERING TEMPERATURE AND TIME FOR VARIOUS METALS

MATERIAL	TEMPERATURE		TIME, MIN.
	°F	°C	
Copper, brass, and bronze	1400–1650	760–900	10–45
Iron and iron-graphite	1850–2100	1000–1150	8–45
Nickel	1850–2100	1000–1150	30–45
Stainless steels	2000–2350	1100–1290	30–60
Alnico alloys (for permanent magnets)	2200–2375	1200–1300	120–150
Ferrites	2200–2700	1200–1500	10–600
Tungsten carbide	2600–2700	1430–1500	20–30
Molybdenum	3750	2050	120
Tungsten	4250	2350	480
Tantalum	4350	2400	480

Furnaces may be batch-type or continuous furnaces, with a variety of features, for high-production runs. The purposes of controlling the atmosphere during sintering are to:

a. Control the carburization and decarburization of iron and iron-base compacts, and

b. Reduce oxides or to prevent oxidation of compacts.

An oxygen-free atmosphere is thus essential for sintering. Although a vacuum is used mainly for refractory metal alloys and stainless steels, the gases most commonly used with a variety of other metals are hydrogen, dissociated or burned ammonia, and exothermic or endothermic-type atmosphere. Proper control of the atmosphere is essential for successful sintering and to obtain optimal properties.

Mechanisms of Sintering

Sintering mechanisms are complex and depend on the composition of metal particles, as well as processing parameters. As temperature increases, two adjacent particles begin to form a bond by diffusion (*solid-state bonding*) (Fig. 11.6a). As a result, the strength, ductility, and thermal and electrical conductivities of the compact increase, as well as its density. This mechanism leads to shrinkage of the compact.

If the two particles are of different metals, alloying can take place at the interface. It is also possible for one of the particles to be of a lower-melting-point metal than the other. In that case, the particle may melt and, because of surface tension, the liquid

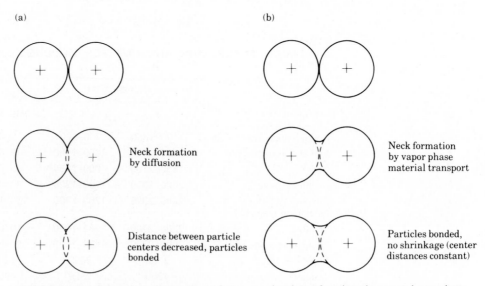

FIGURE 11.6 Schematic illustration of two mechanisms for sintering metal powders. (a) Solid-state material transport. (b) Liquid-phase material transport. Sintering is the process of bonding adjacent metal powders by heat. See also Table 11.3.

metal surrounds the higher–melting-point solid particle (Fig. 11.6b). This is known as *liquid-phase sintering*. An example is cobalt in tungsten carbide (see Section 8.10.4). In this way stronger and denser parts are obtained.

Depending on time, temperature, and processing history, different structures and porosities can be obtained in a sintered compact. This also depends on the extent of diffusion, recrystallization, and grain growth. Porosity cannot be completely eliminated because of the presence of voids during compaction and gases evolved during sintering. Porosity can be either a network of interconnected pores or cavities, or closed holes.

In addition to the commonly used furnace sintering, another method is *spark sintering*. In this process, which is still at an experimental stage, the loose metal powders are subjected to a high-energy discharge while in a graphite mold, heated by electrical current, and then compacted, all in one step. The rapid discharge strips any oxide coating (such as those on aluminum) or contaminants from the surfaces of the particles and thus encourages good bonding during compaction at elevated temperatures.

Typical examples of the effect of compacting pressure and density on the properties of sintered compacts are shown in Fig. 11.7. It can be seen that, as expected, strength, ductility, and electrical conductivity (because of the larger contact area between the particles) increase with increasing pressure and density. Such data are available in the literature to aid in designing P/M parts. (See also Section 11.9.1 on the effect of porosity on mechanical properties.)

11.2.5 FINISHING OPERATIONS

In order to further improve the properties of sintered powder-metallurgy products, or to give them special characteristics, several additional operations may be carried out. Among these are coining, sizing, forging, infiltration, and impregnation.

Coining, Sizing, and Forging
Coining and sizing are additional compacting operations carried out in presses. They are performed under high pressure and with very little deformation of the part. The purposes of these operations are to give final, precise dimensions to the sintered part, and to improve its surface finish and strength by additional densification.

An important development is the use of preformed, sintered alloy-powder compacts, which are subsequently cold or hot forged to the desired final shapes in closed dies. These products have good surface finish and dimensional tolerances, with uniform and fine grain size and distribution, and a microstructure relatively free of grain-boundary segregation and precipitates. The superior properties obtained make this technology particularly suitable for making automotive and jet-engine parts that are highly stressed.

Impregnation
The inherent porosity of powder-metallurgy components (those with an interconnected network of porosity) can be utilized to impregnate them, either with a fluid

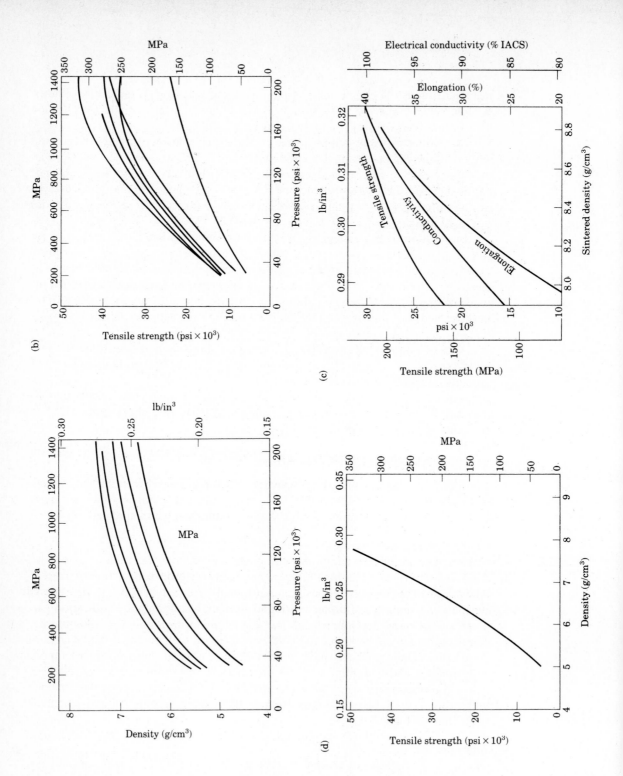

(b)

(c)

(d)

◀ **FIGURE 11.7** (a to c) Effect of compacting pressure on the density and strength for various iron powders. (d) Effect of density on tensile strength, elongation, and electrical conductivity of copper powder. IACS means International Annealed Copper Standard for electrical conductivity. *Source:* (a–c) After A. Squire. (d) After J. L. Everhart.

or with molten metal. A typical application of the former is to impregnate the sintered part with lubricating oil. This is done either by immersing it in heated oil or by vacuum. Bearings and bushings that are internally lubricated, with up to 30% oil by volume, are made by this method. There is thus a continuous supply of lubricant during the service life of the bearing. An example is bronze bearings in automobiles.

Infiltration
Infiltration is a process where a slug of lower-melting-point metal is placed against the sintered part and the assembly is then heated to a temperature high enough to melt the slug. By capillary action, the molten metal infiltrates the pores, resulting in a relatively pore-free part with good strength and density.

The most common application is the infiltration of iron-base compacts with copper. The advantages of copper-infiltrated P/M parts are that hardness and tensile strength are increased, and the pores are filled, thus preventing moisture penetration that would otherwise cause corrosion. There is, however, a lowering of the ductility of the part. Infiltration may also be done with lead, in which case the frictional character-istic of the product is lower than that for the uninfiltrated part (because of the low shear strength of lead).

In addition to the finishing processes described above, powder-metal parts may also be subjected to other operations, such as heat treating, machining, and grinding, resulting in improved properties and better dimensional control and surface finish.

11.3 DESIGN CONSIDERATIONS

Because of the unique properties of metal powders and their flow characteristics, and the brittle nature of powder compacts, there are general design guidelines that should be followed. The major considerations are outlined below:

a. The shape of the compact must be kept as simple and uniform as possible. Sharp changes in contour, thin sections, variations in thickness and high length-to-diameter ratios should be avoided (Fig. 11.8).

b. Provisions must be made in the design to enable the ejection of the green compact from the die; thus, holes or recesses in the part should be parallel to the axis of punch travel. Chamfers should also be provided to avoid breakage at the edges.

c. As with most other processes, P/M parts should be made with the widest tolerances consistent with their intended applications in order to increase tool life and reduce costs.

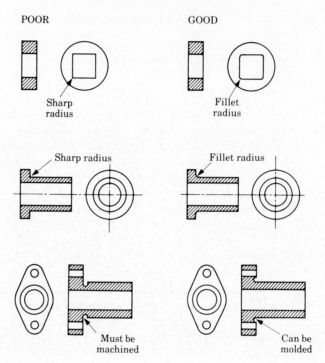

FIGURE 11.8   Examples of powder metallurgy products showing poor and good designs. As in all manufacturing processes, parts must be designed according to the characteristics of individual processing methods. Note that sharp radii and reentry corners should be avoided.

11.4 ADVANTAGES AND LIMITATIONS

As with any other manufacturing method, processing of powdered metals has its advantages and limitations. Although these should be apparent from the discussion thus far, they are outlined below.

Advantages

a. Availability of a wide range of compositions, with a combination of metals and nonmetals, capable of obtaining special physical and mechanical properties in a component (e.g., electrical, magnetic, stiffness, and damping characteristics, and hardness, density, and toughness, Tables 11.4, 11.5, and 11.6). Some of the highly alloyed new superalloys can be manufactured into parts only by powder-metallurgy processing.

TABLE 11.4
PROPERTIES AND TYPICAL APPLICATIONS OF COPPER-ALLOY POWDER METALLURGY PARTS

MATERIAL	MPIF TYPE	CONDITION	DENSITY, g/cm³	ULTIMATE TENSILE STRENGTH, psi × 10³	COMPRESSIVE YIELD STRENGTH, psi × 10³	HARDNESS, HRH	ELASTIC MODULUS, psi × 10⁶	IMPACT STRENGTH, ft-lb	ELONGATION, % (in 1 in.)	TYPICAL APPLICATIONS
P/M bronze 86.3–90.5 Cu, 9.5–10.5 Sn, 1.7 C max, 1 Fe max	N	Sintered and oil impregnated	5.8	8	7	—	—	—	1.0	Bearings or mechanical components resistant to atmospheric corrosion. Sleeve bearings, flange bearings, thrust washers, load-carrying bearing plates.
	R	Sintered and oil impregnated	6.6	14	11	—	—	—	1.0	
	S	Sintered and oil impregnated	7.0	18	18	—	—	—	2.5	
P/M brass 88.0–91.0 Cu, 83.0–12.0 Zn, 0–0.3 Fe	T	Sintered	7.4	20	9	57	—	—	13	Mechanical components requiring corrosion resistance and a pleasing appearance.
	U	Sintered	7.8	27	10	70	—	—	10	
P/M brass (leaded) 88.0–91.0 Cu, 1.0–2.0 Pb, bal Zn	T	Sintered	7.4	18	7	46	—	—	14	Same as above, free-machining quality.
	U	Sintered	7.8	26	8	60	—	—	20	
P/M brass (leaded) 77.0–80.0 Cu, 1.0–2.0 Pb, 0.3 Fe max, 0.1 Sn max, bal Zn	T	Sintered	7.4	24	12	55	12.0	10.0	13.0	Mechanical components resistant to atmospheric corrosion.
	U	Sintered	7.8	28	14	68	13.0	15.0	19.0	Ordnance components, builders' hardware, lock parts, housings, nuts, gears.
	W	Sintered	8.2	32	16	75	14.0	21.0	23.0	

(*continued*)

TABLE 11.4 (continued)

MATERIAL	MPIF TYPE	CONDITION	DENSITY, g/cm³	ULTIMATE TENSILE STRENGTH, psi ×10³	COMPRESSIVE YIELD STRENGTH, psi ×10³	HARDNESS, HRH	ELASTIC MODULUS, psi ×10⁶	IMPACT STRENGTH, ft-lb	ELONGA-TION, % (in 1 in.)	TYPICAL APPLICATIONS
P/M brass 68.5–71.5 Cu, 27.8–31.5 Zn, 0–0.3 Fe	T	Sintered	7.4	31	13	76	—	—	20	Mechanical components requiring corrosion resistance and a pleasing appearance.
	U	Sintered	7.8	37	15	85	—	—	26	
P/M brass (leaded) 68.5–71.5 Cu, 1.0–2.0 Pb, bal Zn	T	Sintered	7.4	28	11	65	—	—	22	Same as above, free-machining quality.
	U	Sintered	7.8	34	13	76	—	—	27	
P/M nickel silver 62.5–65.5 Cu, 16.5–19.5 Ni, bal Zn	U	Sintered	7.8	30	16	75	14.0	10.0	10.0	Mechanical components, corrosion resisting. Gears, levers, chuck jaws, electrical components, parts for marine exposure.
	W	Sintered	8.2	37	18	85	14.0	13.0	12.0	
P/M nickel silver (leaded) 62.5–65.5 Cu, 16.5–19.5 Ni, 1.0–1.8 Pb, bal Zn	U	Sintered	7.8	29	16	75	13.0	9.0	10.0	Same as above free-machining quality.
	W	Sintered	8.2	35	17	85	14.0	12.0	12.0	

Source: Metal Powder Industries Federation (MPIF).

MATERIAL	CONDITION	DENSITY g/cm³	ULTIMATE TENSILE STRENGTH psi × 10³	YIELD STRENGTH psi × 10³	HARDNESS	ELONGA-TION, % (in 1 in.)	TYPICAL APPLICATIONS
601AB (Alcoa) 0.25 Cu, 0.6 Si, 1.0 Mg, 1.5 lubricant, bal Al	Sintered	2.55	21	14	65-70 HRH	6.0	Similar to wrought 6061; strength, ductility, corrosion resistance.
	Heat treated	2.55	35	33	80-85 HRE	2.0	
201AB (Alcoa) 4.4 Cu, 0.8 Si, 0.5 Mg, 1.5 lubricant, bal Al	Sintered	2.64	30	26	70-75 HRE	3.0	Similar to wrought 2014 but without manganese. Good strength properties.
	Heat treated	2.64	48	48	70-75 HRE	2.0	
202AB (Alcoa) 4.0 Cu, 1.5 lubricant, bal Al	Sintered	2.56	23	11	55-60 HRH	10.0	Good ductility. Suitable for cold-formed parts.
	Heat treated	2.56	33	21	45-50 HRE	7.3	
602AB (Alcoa) 0.4 Si, 0.6 Mg, 1.5 lubricant, bal Al	Sintered	2.55	18	9	55-60 HRH	9.0	Good electrical conductivity (from 42.0 to 48.5% IACS, depending on treatment), ductility, and finishability.
	Heat treated	2.55	26	25	65-70 HRE	2.0	
22 (Alcan) 2.0 Cu, 1.0 Mg, 0.3 Si, bal Al	Sintered	2.53	24	16	83 HRH	6.0	Good mechanical properties in sintered or heat treated forms.
	Heat treated	2.53	38	29	74 HRE	3.0	
24 (Alcan) (2014) 4.4 Cu, 0.5 Mg, 0.9 Si, 0.4 Mn, bal Al	Sintered	2.54	24	14	80 HRH	5.0	Properties resemble wrought 2014. Good mechanical properties.
	Heat treated	2.54	35	28	72 HRE	3.0	

(continued)

TABLE 11.5 (*continued*)

MATERIAL	CONDITION	DENSITY g/cm³	ULTIMATE TENSILE STRENGTH psi × 10³	YIELD STRENGTH psi × 10³	HARDNESS	ELONGATION, % (in 1 in.)	TYPICAL APPLICATIONS
67 (Alcan) 0.5 Cu, bal Al	Sintered	2.52	15	8	60 HRH	12.0	High electrical conductivity (48% IACS) and ductility. Similar to wrought 1100.
68 (Alcan) 0.6 Mg, 0.4 Si, bal Al	Sintered	2.52	17	9	64 HRH	9.0	Good surface finish; high ductility and conductivity (42% IACS). Similar to wrought 6101.
69 (Alcan) (6061) 0.25 Cu, 1.0 Mg, 0.6 Si, 0.10 Cr	Sintered	2.50	18	10	66 HRH	10.0	Properties are similar to those of 6061. Good strength, corrosion resistance, ductility, and conductivity (40% IACS).
	Heat treated	2.50	30	28	71 HRE	2.0	
76 (Alcan) (7075) 1.6 Cu, 2.5 Mg, 0.20 Cr, 5.6 Zn	Sintered	2.51	30	22	90 HRH	3.0	Properties are similar to those of 7075. High-strength and hardness.
	Heat treated	2.51	45	40	80 HRE	2.0	
91 (Alcan) 26.3 Tribaloy	Sintered	3.05	14	—	—	2.0	Excellent wear resistance.
	Heat treated	3.05	15	—	—	1.0	

Source: Aluminum Company of America and Alcan Aluminum Corp.

(continued)

TABLE 11.6
PROPERTIES AND TYPICAL APPLICATIONS OF FERROUS POWDER METALLURGY PARTS

MATERIAL	MPIF TYPE	CONDITION	DENSITY, g/cm³	ULTIMATE TENSILE STRENGTH, psi ×10³	YIELD STRENGTH, psi ×10³	HARDNESS	ELASTIC MODULUS, psi ×10⁶	ELONGATION, % (in 1 in.)	IMPACT STRENGTH, ft-lb	FATIGUE STRENGTH, psi	TYPICAL APPLICATIONS
P/M iron 0.3 C max	N	Sintered	5.8	16	11	10 HRH	10.5	2.0	3.0	6,080	Structural (lightly loaded gears); magnetic (motor pole pieces); self-lubricating bearings; structural, wear resisting (small levers and cams) as carbonitrided.
	P	Sintered	6.2	19	14	70 HRH	13.0	2.5	4.5	7,220	
	R	Sintered	6.6	24	16	80 HRH	16.0	5	9.5	9,120	
	S	Sintered	7.0	30	21	15 HRB	19.0	9	15	11,400	
	T	Sintered	7.4	40	26	30 HRB	23.0	15	25	15,200	
P/M steel 0.3 to 0.6 C	N	Sintered	5.8	18	14	5 HRB	10.5	1.0	2.5	6,840	Structural (moderately loaded gears, levers, cams); structural
	P	Sintered	6.2	25	20	20 HRB	13.0	1.5	3.5	9,500	
	R	Sintered	6.6	32	23	45 HRB	16.0	2.5	5.0	12,160	
	R	Heat treated	6.6	60	57	100 HRB	16.0	0.5	—	22,800	
	S	Sintered	7.0	43	28	60 HRB	19.0	3.5	9.0	16,340	
	S	Heat treated	7.0	80	75	25 HRC	19.0	0.5	—	30,400	
P/M steel 0.6 to 1.0 C	N	Sintered	5.8	29	25	35 HRB	10.5	0.5	2.0	11,020	Structural (moderately loaded gears, levers, and cams requiring wear resistance) as heat treated.
	N	Heat treated	5.8	42	—	90 HRB	10.5	<0.5	—	15,960	
	P	Sintered	6.2	35	30	50 HRB	13.0	1.0	3.0	13,300	
	P	Heat treated	6.2	58	—	100 HRB	13.0	<0.5	—	22,040	
	R	Sintered	6.6	42	36	65 HRB	16.0	1.5	3.5	13,680	
	R	Heat treated	6.6	74	—	25 HRC	16.0	<0.5	—	28,120	
	S	Sintered	7.0	57	40	75 HRB	19.0	2.5	7.0	21,660	
	S	Heat treated	7.0	94	91	30 HRC	19.0	<0.5	—	35,720	
P/M copper iron 1.5 to 3.9 Cu, 0.3 C max	P	Sintered	6.2	23	17	80 HRH	13.0	2.5	5.5	8,740	Bearings or mechanical components
	R	Sintered	6.6	30	21	15 HRB	16.0	4.0	7.0	11,400	Mechanical components
	S	Sintered	7.0	37	23	30 HRB	19.0	7.0	17.0	14,060	Mechanical components
P/M copper steel 1.5 to 3.9 Cu, 0.3 to 0.6 C	P	Sintered	6.2	40	34	45 HRB	13.0	1.0	3.5	15,200	Bearings or mechanical components
	R	Sintered	6.6	50	38	70 HRB	16.0	1.5	5.5	19,000	Mechanical components
	R	Heat treated	6.6	85	81	30 HRC	16.0	<0.5	—	30,780	
	S	Sintered	7.0	62	45	80 HRB	19.0	3.0	9.5	23,560	
	S	Heat treated	7.0	100	95	35 HRC	19.0	<0.5	—	38,000	

TABLE 11.6   (*continued*)

MATERIAL	MPIF TYPE	CONDITION	DENSITY, g/cm³	ULTIMATE TENSILE STRENGTH, psi ×10³	YIELD STRENGTH, psi ×10³	HARDNESS	ELASTIC MODULUS, psi ×10⁶	ELONGATION, % (in 1 in.)	IMPACT STRENGTH, ft-lb	FATIGUE STRENGTH, psi	TYPICAL APPLICATIONS
P/M copper steel 1.5 to 3.9 Cu, 0.6 to 1.0 C	N	Sintered	5.8	33	30	45 HRB	10.5	<0.5	2.5	12,540	Bearings or mechanical components
	N	Heat treated	5.8	43	—	95 HRB	10.5	<0.5	—	16,340	
	P	Sintered	6.2	45	41	60 HRB	13.0	<0.5	3.0	17,100	
	P	Heat treated	6.2	55	—	25 HRC	13.0	<0.5	—	20,900	
	R	Sintered	6.6	60	48	70 HRB	16.0	1.0	5.0	22,800	Mechanical components
	R	Heat treated	6.6	80	—	35 HRC	16.0	<0.5	—	30,400	
	S	Sintered	7.0	80	57	80 HRB	19.0	1.5	8.0	30,400	
	S	Heat treated	7.0	100	95	40 HRC	19.0	<0.5	—	38,000	
P/M copper steel 4.0 to 6.0 Cu, 0.3 to 0.6 C	N	Sintered	5.8	35	30	50 HRB	10.5	0.5	3.0	13,300	Mechanical components
	N	Heat treated	5.8	—	—	90 HRB	—	—	—	—	
	P	Sintered	6.2	50	42	60 HRB	13.0	1.0	4.5	19,000	
	P	Heat treated	6.6	66	55	95 HRB	16.0	1.5	5.0	25,080	
	R	Sintered	6.6	—	—	75 HRB	—	—	—	—	
	R	Heat treated	6.6	—	—	25 HRC	—	—	—	—	
P/M copper steel 4.0 to 6.0 Cu, 0.6 to 1.0 C	N	Sintered	5.8	48	43	60 HRB	10.5	<0.5	3.0	18,240	Structural (medium loads including gears, cams, support brackets, levers, and ratchets). Materials can be heat treated to a high degree of wear resistance for any of these applications.
	N	Heat treated	5.8	—	—	95 HRB	—	—	—	—	
	P	Sintered	6.2	62	57	65 HRB	13.0	<1.0	3.5	23,560	
	P	Heat treated	6.2	—	—	30 HRC	—	—	—	—	
	R	Sintered	6.6	75	70	85 HRB	16.0	1.0	4.5	28,500	
	R	Heat treated	6.6	—	—	35 HRC	—	—	—	—	
P/M copper steel 6 to 11 Cu, 0.6 to 1.0 C	N	Sintered	5.8	36	—	55 HRB	—	0–0.5	—	—	Mechanical components
P/M copper iron 9.5 to 10.5 Cu, 0.3 C max	N	Sintered	5.8	30	—	70 HRF	—	0.5	—	—	Bearings or mechanical components

MATERIAL	MPIF TYPE	CONDITION	DENSITY, g/cm³	ULTIMATE TENSILE STRENGTH, psi ×10³	YIELD STRENGTH, psi ×10³	HARDNESS	ELASTIC MODULUS, psi ×10⁶	ELONGATION, % (in 1 in.)	IMPACT STRENGTH, ft-lb	FATIGUE STRENGTH, psi	TYPICAL APPLICATIONS
P/M infiltrated steel											
8 to 14.9 Cu, 0.3 to 0.6 C	T	Sintered	7.4	83	64	75 HRB	20.0	4.0	14.0	—	Mechanical components (special shapes)
	T	Heat treated	7.4	120	107	35 HRC	20.0	1.0	7.0	—	
8 to 14.9 Cu, 0.6 to 1.0 C	T	Sintered	7.4	90	75	80 HRB	20.0	2.5	12.0	—	
15 to 25 Cu	T	Heat treated	7.4	130	105	40 HRC	20.0	<0.5	7.0	—	
0.3 C max	T	Sintered	7.4	65	—	60 HRB	18.0	1.0	15.0	—	Mechanical components
15 to 25 Cu, 0.3 to 0.6 C	T	Sintered	7.4	75	50	75 HRB	18.0	1.5	9.5	—	
	T	Heat treated	7.4	115	95	30 HRC	18.0	<0.5	6.0	—	
15 to 25 Cu, 0.6 to 1.0 C	T	Sintered	7.4	85	75	80 HRB	18.0	1.0	10.0	—	
	T	Heat treated	7.4	125	107	42 HRC	18.0	<0.5	5.0	—	
P/M austenitic stainless steels											
303	P	Sintered	6.2	35	32	—	—	1.0	—	—	Type 303, mechanical components requiring secondary machining; type 316, structural, corrosion resisting, non-magnetic (small gears, levers, cams, and other parts for exposure to salt water and specific industrial acids); type 410, structural, corrosion resisting, (small gears, levers, cams, and other parts where applications require heat treating for wear resistance).
303	R	Sintered	6.6	52	47	—	—	2.0	—	—	
316	P	Sintered	6.2	38	32	—	—	2.0	—	—	
316	R	Sintered	6.6	54	40	—	—	4.0	—	—	
410	N	Sintered	5.8	42	41	—	—	<1.0	—	—	
410	P	Sintered	6.2	55	54	—	—	<1.0	—	—	

Source: Metal Powder Industries Federation (MPIF)

TABLE 11.7
EXPERIMENTAL P/M TITANIUM ALLOY AIRFRAME AND ENGINE PARTS AND WEIGHT COMPARISON. (After V. C. Petersen et al.)

	PART WEIGHT, lb		
PART	FORGING BILLET	HOT ISOSTATIC PRESSING	FINAL PART
Boeing 747 walking beam	55	30	21
General Dynamics F16 pivot shaft	148	53	32
General Electric TF34 compressor spool	147	65	15
McDonnell Aircraft F-15 drop-out link	115	56	14
McDonnell Aircraft F-15 keel splice	6	0.8	0.4
Northrop F-18 arrestor hook	181	55	28
Pratt & Whitney Aircraft F-100 fan disk	120	65	27
Williams International F-107 compressor rotor	32	6.2	3.6

b. A technique for making parts from high-melting-point refractory metals, which would be difficult to make by other methods.

c. High production rates for parts of relatively complex shape, with automated equipment requiring little labor.

d. Good dimensional control and, in many instances, elimination of machining, thus the elimination of scrap and waste (Table 11.7), hence energy saving.

e. Capability for impregnation and infiltration for special applications, such as in bearings and bushings.

Limitations

a. Limitations due to size and complexity of shape of parts, and also press capacity.

b. High cost of powder metals compared to other raw materials; high cost of equipment and tooling when production runs are low.

c. Mechanical properties, such as strength and ductility, that are generally lower than those obtained by other processing methods, such as forging, rolling, and extrusion. However, the properties of full-density P/M parts made by HIP or P/M forging can, in fact, be better than those made by other processes.

● **Illustrative Example 11.1**

Components such as gears, bushings, and some structural parts of garden tractors have been made by P/M techniques, replacing the traditional methods of making them by casting or forging. Gears have been manufactured competitively using high-quality powders with high compressibility and low compacting pressures. These parts range from medium to high density and are suitable for severe applications with high loads and for high-wear surfaces.

In one application, a reduction gear for a garden tractor was made from iron powder and infiltrated with copper. Although its strength was acceptable, the wear rate under high loads was very high. This resulted in loss of tooth profile, side loading on the bearings, and a high noise level. To improve wear resistance and strength, a new powder was selected containing 2.0% nickel, 0.5% graphite, 0.5% molybdenum, and the balance atomized iron powder. Because of the size of the part, the pressing loads were very high. The part was redesigned and the tooling was made with three lower punches to compact the part with three different densities. (See Fig. 11.4, which shows two lower punches.)

By this method high density was obtained in those sections of the part requiring high strength and wear resistance. The high density also permitted carburization for hardness improvement. The presintered part was re-pressed to a density of 7.3 to 7.5 g/cm^3 in the tooth area for improved strength, while the hub of the gear remained at its pressed density of 6.4 to 6.6 g/cm^3. The entire part required a compacting load of 450 tons (metric). The weight of the gear was 2.25 lb (1.02 kg). ●

11.5 CERAMICS

11.5.1 INTRODUCTION

Ceramics are compounds of metallic and nonmetallic elements. The term ceramics refers both to the material and also to the product; the Greek word *keramos* means potter's clay, and *keramikos*, clay products.

The earliest use of ceramics was in pottery and bricks. Because of the large number of possible combinations of metallic and nonmetallic elements, a variety of ceramics is now available for a number of different uses.

Typical examples of ceramic products are: construction materials such as brick and tile; refractory and insulating materials for melting of metals, furnaces, and as coatings for high-temperature applications; glasses and carbon products; fiberglass; earthenware such as china, stoneware, and ovenware; porcelain for ornamental objects, and other uses, electrical insulation, and dental work; various glasses for optical uses; cement and cement products; abrasives for grinding wheels; sandpaper; and cutting tools. Table 11.8 shows various categories and uses of ceramics.

TABLE 11.8
CATEGORIES AND USES OF CERAMICS

	TRADITIONAL CERAMICS
Abrasive products	Abrasive wheels, emery cloth and sand paper, nozzles for sandblasting, milling
Clay products	Brick, pottery, sewer pipe
Construction	Brick, concrete, tile, plaster, glass
Glass	Bottles, laboratory ware, glazing
Refractories	Brick, crucibles, molds, cement
Whitewares	Dishes, tiles, plumbing, enamels
	ENGINEERING CERAMICS
Automotive and aerospace	Turbine components, heat shields and exchangers, reentry components, seals
Electronics	Semiconductors, insulators, transducers, lasers, dielectrics, heating elements
High-temperature	Refractories, brazing fixtures, kilns
Manufacturing	Cutting tools, wear and corrosion resistant components, glass-ceramics, magnets, fiber optics
Medical	Laboratory ware, controls, prosthetics, dental
Nuclear	Fuels, controls

From these examples it is apparent that ceramics represent an important class of materials. The science of ceramics and their structure and properties are generally covered in materials science and engineering texts. In this chapter, we will deal only with general classes and properties of ceramics, their application, and methods of processing some of the engineering products listed above.

11.5.2 MATERIALS FOR CERAMIC PRODUCTS

One of the oldest raw materials used in ceramics is clay, which has a fine-grained sheetlike structure; the most common example is *kaolinite* $[Al_2Si_2O_5(OH)_4]$, with alternating layers of silicon and aluminum ions. When added to it, water attaches itself to the layers, makes them slippery, and gives clay softness and plastic properties that make it formable. This property is known as *hydroplasticity*.

Other raw materials available in nature are flint (rock of very-fine-grained silica), and feldspar (a group of crystalline minerals consisting of aluminum silicates and potassium, calcium, or sodium). However, these natural raw materials generally contain impurities of various kinds, which may have to be removed prior to further processing. The major classes of ceramic materials are described below.

Silica

Silica and silicates are among the most important classes of materials. *Silica* (SiO_2) is a polymorphic material. The cubic structure is found in refractory bricks for high-temperature–furnace applications. The most common form of silica is quartz, which is a hexagonal crystal. It is used extensively as oscillating crystals of fixed frequency for communications applications because it exhibits the piezoelectric effect. It is also hard, abrasive, and very resistant to water. Silica is the major constituent of glass, with most glasses containing more than 50% silica.

Silicates are the products of the reaction of silica with oxides of aluminum, magnesium, calcium, potassium, sodium, and iron. Examples are clay, asbestos, mica, and silicate glasses.

Alumina

Among nonsilicate materials for ceramics, alumina (or corundum, Al_2O_3) is perhaps the most widely used, either in pure form or as a raw material to be used with other oxides. The physical and mechanical properties of alumina are particularly suitable for applications such as electrical and thermal insulation, ceramic cutting tools (Section 8.10.5), and abrasives (Chapter 9). Structures containing various percentages of alumina and other oxides are known as *mullite* and *spinels* and are used as refractory materials.

Carbides

Carbides are another important class of materials. Typical examples are carbides of tungsten, titanium, and silicon, with applications in cutting tools, dies for metalworking, and abrasive products. (See Sections 8.10 and 9.2.1.)

Nitrides

Nitrides have also become important, particularly *cubic boron nitride* (CBN) and *titanium nitride*. The former has special applications in bonded abrasives (grinding wheels, Section 9.5.3) and the latter as a coating for cutting tools to improve tool life. *Silicon nitride* (Si_3N_4) is another relatively new ceramic. With its high resistance to creep at elevated temperatures, low thermal expansion, and high thermal conductivity (thus resisting thermal shock), it is suitable for structural applications such as in automotive and gas-turbine parts. A new development is cutting tool materials consisting of silicon nitride and aluminum oxide.

Carbon

The two forms of carbon of principal interest in manufacturing are *graphite* and *diamond*. Graphite is a crystalline form of carbon having a layered structure with basal planes or sheets of closely packed carbon atoms. As a consequence, it is weak when sheared in the same direction as the layers. This, in turn, has significance as far as its frictional properties are concerned, as discussed in Section 4.10.5. In addition to its lubricating characteristics, graphite is also very useful for electrical applications (high conductivity), as a seal (low friction), as a die material for casting (high-temperature resistance), and as reinforcing fibers (Section 10.11).

Diamond is the hardest substance known. This characteristic makes it very useful for cutting (either as a single crystal or in polycrystalline form) and grinding. Diamond is also used as a die material for drawing very thin wire.

Glass

Glass (Section 11.8) is an amorphous solid with the structure of a liquid; that is, it has been *supercooled*. It has no distinct melting or freezing point. Thus its behavior is similar to amorphous polymers, as shown in Fig. 10.4.

Although silica is the basic glass-forming oxide, the structure of glass can be modified by adding various other oxides to obtain a wide range of properties. Its use ranges from window glass, bottles, cooking ware, etc., to glasses with special optical, chemical, high-temperature characteristic, and as reinforcing fibers (see Section 10.11).

Glass ceramics (Pyroceram) have a high crystalline component in their microstructure, which is composed of large proportions of several oxides. They are processed as a glass.

These ceramics are first shaped and then heat treated wherein *devitrification*, or recrystallization, of the glass occurs. Glass ceramics have very low coefficients of thermal expansion and are strong, because of the absence of porosity found in conventional sintered ceramics.

11.6 PROCESSING OF CERAMICS

Processing of ceramics generally involves the following steps: crushing or grinding the raw materials into very fine particles, mixing them with additives to impart certain desirable characteristics, shaping by casting or molding, and drying and firing.

Crushing

Crushing is generally done in a ball mill. This may be done either dry or wet, the latter being more effective since it keeps the particles together and prevents the suspension of fine particles in air. The ground particles are then mixed with a variety of additives functioning as one or more of the following:

a. Binder

b. Lubricant for mold release, and to reduce internal friction between the particles

c. Wetting agent

d. Plasticizer to make the mix more plastic and formable

    e. Deflocculent (Na_2CO_3 and Na_2SiO_3, less than 1 %) to prepare the ceramic–water suspension. Water (about 25 %) is added to make the mixture more pourable and less viscous. Deflocculation changes the electrical charges on the particles of clay so that they repel each other, instead of attracting

    f. Agent for control of water, foaming, and sintering.

Shaping

The three basic shaping processes for ceramics are: casting, plastic forming, and pressing.

    Slip Casting.   The most common casting process is slip casting (also called *drain casting*) (Fig. 11.9). A *slip* is a suspension of ceramic particles in a liquid (generally water). In this process, a slip is poured into a porous plaster mold. The slip must have sufficient fluidity and low viscosity in order to flow easily into the mold.

    After the mold has absorbed some of the water from the outer layers of the suspension to the desired depth, the mold is inverted and the remaining suspension is poured out. The top of the part is then trimmed, the mold is opened, and the part is removed. (If a solid part is desired, the slip is supplied continuously to replenish the absorbed water; the suspension is not drained from the mold.) At this stage the part is a soft solid or is semirigid. The higher the concentration of solids in the slip, the less water has to be removed. The part is then fired.

    Plastic Forming.   Plastic forming, also called wet or hydroplastic forming, can be done by various methods such as extrusion, injection molding, or molding such as that done on a potter's wheel. Plastic deformation tends to orient the layered structure

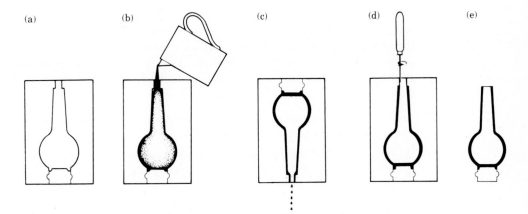

(a)   (b)   (c)   (d)   (e)

FIGURE 11.9 Schematic illustration of the slip casting process, also called drain casting. After slip casting, the part is dried and fired in an oven. *Source:* After F. H. Norton, *Elements of Ceramics*, Addison-Wesley, Reading, Mass., 1952.

of clays along the shearing direction (such as by the continuous rubbing of fingers on the clay mass on a potter's wheel), resulting in a preferred orientation. This leads to anisotropic behavior in subsequent processing of the ceramic product and also in its final properties.

Pressing.   The third method of forming ceramics is pressing, which is a compaction process. In dry pressing, the moisture content is generally below 4%. The process involves consideration of parameters and equipment characteristics similar to those in powder-metal compaction, with the same high production rates and close control of tolerances.

Lubricants, such as stearic acid or wax, are usually added to the powders. In pressing ceramics (as in P/M compaction) considerable pressure (and thus density) variations can occur due to internal friction of the particles and the friction at the die walls (Fig. 11.10). Wet pressing is generally used for compositions containing clay, and moisture content ranging between 10 and 15%. It is essentially a plastic-forming process, similar to a forging operation.

Hot Pressing.   In hot pressing (also called pressure sintering), the pressure and temperature are applied simultaneously. In this way porosity is reduced, making the part more dense, with a finer grain size. Because of the presence of both pressure and temperature, mold life may be low. Protective atmospheres are usually employed in hot pressing. Graphite is a commonly used punch and die material.

Depending on the shape and complexity of the product, other shaping processes, such as injection molding and those used in P/M, can also be used for ceramics.

Drying and Firing

After shaping the ceramic by any of the methods described above, the next step is to dry and fire it to give it the proper strength. Drying is a critical stage because of the tendency for the part to crack or warp from variations in thickness, moisture content, and the complexity of its shape. Control of the atmosphere (humidity and temperature) is important in order to reduce cracking and warping. This can be appreciated in view of the following:

a. Drying involves the movement of moisture from the interior to the surface of the part so that evaporation can take place.

b. The rate of diffusion of moisture throughout the part is temperature-dependent.

c. Loss of moisture results in shrinkage of the part by as much as 15 to 20% of its original size.

Furthermore, in a humid environment, the moisture gradient across the thickness of the part is lower than that in a dry atmosphere. This prevents the presence of a large gradient in shrinkage from the surface to the interior during drying. Hence a humid environment is desirable.

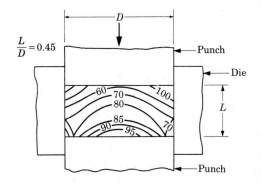

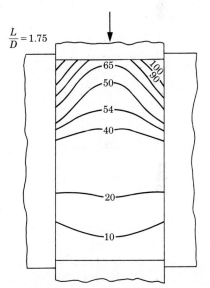

FIGURE 11.10 Density variation in pressed compacts. Note that the variation increases with increasing L/D ratio. See also Fig. 11.4. *Source:* Adapted from W. D. Kingery, et al., *Ceramic Fabrication Processes*, MIT Press, Cambridge, Mass., 1963.

Firing (also called *sintering*) involves densification by heating the part to an elevated temperature in a controlled environment. This process is similar to powder metallurgy, with the same complex interactions between adjacent particles. The strength and hardness of ceramics after firing is due to the glassy bond between the complex oxide particles in the ceramic. Because porosity is reduced, some shrinkage occurs during firing of ceramic products.

Finishing Operations
After firing, additional machining may be performed to remove flaws and to improve surface finish and tolerances. The processes that can be used are grinding, lapping, and ultrasonic, chemical, and electrodischarge machining. The choice of process is

important in view of the brittle nature of most ceramics and the additional costs involved in material removal. The effect of the machining process on surface properties of the product must also be considered, as described in Chapters 8 and 9.

11.7 COMPOSITE MATERIALS

A *composite* is a material with two or more different components combined to produce specific properties and characteristics that any one of the components cannot achieve singly. There are many examples of composite materials for different applications. Concrete with steel reinforcing bars is a well-known composite. (In fact, concrete itself can be considered a composite material.) Steel imparts the necessary tensile strength to the composite since concrete, being a brittle material, is generally assumed to have little or no tensile strength. An older example of a composite is the addition of straw to clay or brick for improved strength as a structural or construction material.

Another common example is reinforced plastics, the reinforcing elements being carbon, glass, or boron fibers (Section 10.11). In these examples, the reinforcing element is a fiber (as opposed to a particle) and, if aligned, imparts anisotropic properties to the product. Reinforcements will also significantly increase the stiffness of the part. Such reinforcement also has an important effect on the fracture toughness of the composite. The strength of the bond between the various components in a composite has a significant influence on its properties.

Another class of composites consists of coatings of various kinds on base metals or substrates. Examples are plating of aluminum, or other elements, on plastics for decorative purposes, and enamels (dating back to before 1000 B.C.) or similar vitreous coatings on metal surfaces for various purposes.

Composites can also involve particle reinforcement, the best examples being cemented carbides and bonded abrasives (grinding wheels). Carbides generally involve tungsten or titanium carbide, with cobalt or nickel as a binder (Section 8.10.4). Bonded abrasives are generally made of aluminum oxide, silicon carbide, diamond or cubic-boron-nitride grains with various organic, inorganic, or metallic binders (Section 9.3.1).

11.8 GLASSES

Most glasses are composed of more than 50% SiO_2 and other oxides; these are referred to as silicate glasses. Glasses are supercooled liquids—when the molten fluid is cooled rapidly, it does not have sufficient time to solidify into a crystalline structure. However, glass can crystallize and transform into a crystalline solid over a long period of time. This is known as *devitrified glass*, also called *glass ceramics*.

Glasses do not have clearly defined melting points. The term glass-transition temperature, T_g, is used to specify the temperature at which the slope of the specific volume–temperature curve undergoes a distinct change (see Fig. 10.4).

A wide variety of glasses is available with different mechanical, thermal, chemical, optical, and electrical properties (Table 11.9). The viscosity–temperature behavior of

TABLE 11.9
COMPOSITION AND PROPERTIES OF SOME GLASSES

TYPE	MAJOR COMPONENTS, %								CHARACTERISTICS AND USES
	SiO_2	Al_2O_3	CaO	Na_2O	B_2O_3	MgO	PbO	Other	
Fused silica	99.5+								Very low thermal expansion, high thermal shock resistance; high-frequency electrical insulation
96% silica (Vycor)	96.3	0.4		<0.2	2.9			K_2O <2	Very low thermal expansion; chemical ware, home appliances, sun lamps
Borosilicate (Pyrex)	81	2		4	13			K_2O 0.4	Low thermal expansion; oven ware, laboratory glasses, industrial glass piping, gauge glasses
Soda lime (Plate glass)	71–73	1	10–12	12–14		1–4			High durability, easily fabricated; windows, containers, ash trays, glass blocks, electric bulbs
Lamp stems	55	1		12			32		High electrical resistivity
Fiber (E-glass)	54	14	16–22	1.5	10	4			Fiber reinforcement (Table 10.4)
Thermometer	73	6		10	10				Dimensional stability
Lead glass	67			6			17	K_2O 10	High index of refraction; table ware, optical lenses, crystal glassware, neon tubes, capacitors
Optical flint	50			1			19	BaO 13 K_2O 8 ZnO 8	Specific index and dispersion values
Optical crown	70			8	10			BaO 2 K_2O 8	Specific index and dispersion values
Glass-ceramics	40–70	10–35				10–30		TiO_2 7–15	Easily fabricated, good mechanical properties

glasses can also be controlled over a wide range. (See also Sections 4.10.6 and 6.16.1 for glass as a lubricant in hot working of metals.)

Glass products (dating back to before 2000 B.C.) are numerous, ranging from window glass to fiberglass. Window glass, or flat-sheet glass, can be made by a combination of drawing and rolling processes from the molten state, or by a floating process whereby the molten glass floats on a bath of molten tin.

Techniques for Strengthening Glass

There are several techniques for strengthening glass. In one method, *tempered glass* (Fig. 11.11) is made by cooling the surfaces of the hot glass rapidly. Upon cooling, the surfaces develop compressive residual stresses (with tension on the inside), thus improving its strength. However, tempered glass shatters into a large number of pieces when broken, because of the rapid release of the energy stored as residual stresses.

Compressive residual stresses can also be generated on the surfaces of glass by *chemical tempering*. This is a process of ion exchange and is done by heating the glass in a bath of molten K_2SO_4.

Another method of strengthening is to laminate two pieces of flat glass with a thin sheet of plastic in between. When laminated glass breaks, its pieces are held together by the plastic sheet. A typical application is for automobile windshields.

Glass Bottles and Other Products

The steps involved in the production of a common glass bottle are shown in Fig. 11.12. Glass blowing dates back to about 250 B.C.

Glass for tubing is made by drawing it over a mandrel with air blowing through it. Processes such as casting, pressing, and blowing are also used in manufacturing glass bottles and various artifacts.

Glass Fibers

An important type of glass is *fiberglass*. Short fibers (used as insulating material for thermal or acoustical purposes) are made by a centrifugal spraying process. Continu-

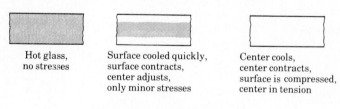

Hot glass,
no stresses

Surface cooled quickly,
surface contracts,
center adjusts,
only minor stresses

Center cools,
center contracts,
surface is compressed,
center in tension

FIGURE 11.11 Stages in inducing compressive residual stresses on the surfaces of tempered glass plate. This glass is strong, but shatters into many pieces when broken. Residual stresses may also be induced by chemical means. *Source:* After L. H. Van Vlack.

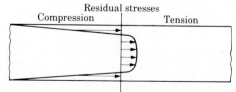

Residual stresses

Compression   Tension

ous fibers (for cloth, fiber optics for communication and optical use, and as reinforcements) are drawn through multiple orifices in heated platinum plates at speeds as high as 100,000 ft/min (500 m/s). Fibers as small as 2 μm in diameter can be produced. In order to protect their surfaces against damage, glass fibers are immediately coated with thermoplastics.

The tensile strength of fiberglass is on the order of 250,000 psi (1700 MPa), with a modulus of elasticity of 10^7 psi (70 GPa). However, because it is brittle, its elongation in tension is zero. (See also Section 10.11.)

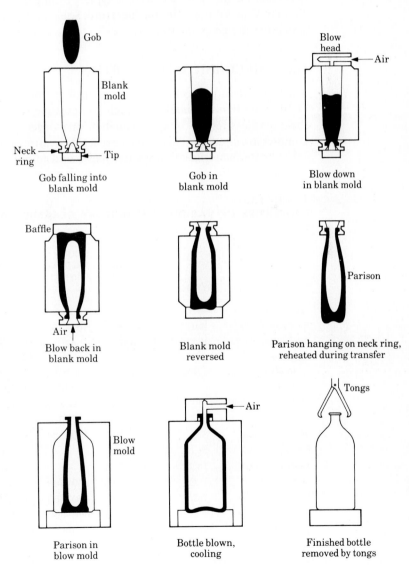

FIGURE 11.12 Schematic illustration of the stages in manufacturing a glass bottle. *Source:* After F. H. Norton, *Elements of Ceramics*, Addison-Wesley, Reading, Mass., 1952.

11.9 GENERAL PROPERTIES AND APPLICATIONS OF CERAMICS

In general, ceramics can be characterized as hard and brittle, with low thermal and electrical conductivities. However, because of the wide variety of available ceramic materials and their compositions, there are large variations in the mechanical and physical properties.

Furthermore, because of their sensitivity to internal and surface flaws and defects, and various levels of impurities, a range of values can be obtained for the same property and for the same material. A typical example is the variation in hardness and strength values reported in the literature. Some of the major mechanical and physical properties are given below.

11.9.1 MECHANICAL PROPERTIES

The mechanical properties for a number of engineering ceramics are given in Table 11.10. Note that ceramics have high compressive strength and hardness. However, their tensile strength is approximately one order of magnitude lower than their compressive strength. This is to be expected because of the sensitivity of ceramics and other brittle materials to porosity, flaws, and impurities. Such defects lead to the

TABLE 11.10
PROPERTIES OF VARIOUS REFRACTORY CERAMICS AT ROOM TEMPERATURE

MATERIAL	SYMBOL	TRANS-VERSE RUPTURE STRENGTH (psi × 10³)	COM-PRESSIVE STRENGTH (psi × 10³)	ELASTIC MODULUS (psi × 10⁶)	HARDNESS (HK)	POISSON'S RATIO (v)
Aluminum oxide	Al_2O_3	20–35	150–420	45–60	2000–3000	0.26
Beryllium oxide	BeO	12–18	100–300	40	750–1500	0.34
Boron carbide	B_4C	45	420	65	2800	0.21
Cubic boron nitride	CBN	105	1000	125	4000–5000	—
Diamond		200	1000	120–150	7000–8000	—
Magnesium oxide	MgO	5–15	120–200	15–50	500	0.36
Silica, fused	SiO_2	—	190	10	550	0.25
Silicon carbide	SiC	15–110	100–500	35–70	2100–3000	0.14
Silicon nitride	Si_3N_4	70	—	45	—	0.24
Tantalum carbide	TaC	30–35	—	40–75	800–2000	—
Thorium oxide	ThO_2	14	200–400	20–35	800	—
Titanium carbide	TiC	200–275	450–560	45–60	1800–3200	—
Tungsten carbide	WC	150–375	600–850	75–100	1800–2400	—
Vanadium carbide	VC	2–4	90	40	2100–2800	—
Zirconium oxide	ZrO_2	20	100–300	12–25	1000	—

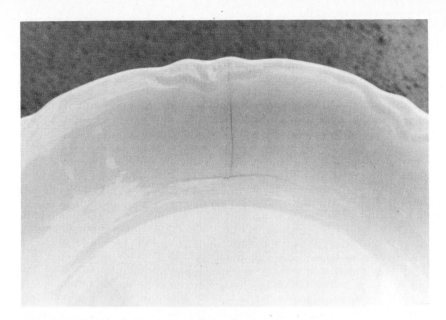

FIGURE 11.13 Cracks in a ceramic dish caused either by mechanical impact or thermal stresses. Such hairline cracks indicate perfectly brittle fracture of the material, i.e., no plastic deformation in the vicinity of the cracks. Whether or not a crack is new or old can be distinguished by its color. The dark color of the cracks in this photograph indicates that dirt has penetrated the cracks over a period of time. (See also Table 4.2 for detecting flaws by dye penetrants.)

initiation and propagation of cracks under tensile stresses, and severely reduce the tensile strength. (See also Section 3.7.1.)

Porosity has an adverse effect on mechanical properties of ceramics. Tensile strength is found to be empirically related to porosity as follows:

$$\text{UTS} \simeq \text{UTS}_0 e^{-nP}, \tag{11.1}$$

where P is the volume fraction of pores in the solid, UTS_0 is the tensile strength at zero porosity, and n ranges between 4 and 7.

The modulus of elasticity is likewise affected by porosity, as given by

$$E \simeq E_0(1 - 1.9P + 0.9P^2) \tag{11.2}$$

where E_0 is the modulus at zero porosity. Eq. (11.2) is valid up to 50% porosity. Common earthenware has a porosity ranging between 10 and 15%, whereas hard porcelain is about 3%.

Unlike most metals, ceramics lack impact resistance and toughness due to their inherent lack of ductility. Once initiated, a crack propagates rapidly (Fig. 11.13). Unlike ductile metals, it takes very little energy to sustain crack propagation in brittle materials. Toughness may be improved by selection of proper materials and control of structure, by reinforcements, and by design considerations as outlined in Section 11.10.

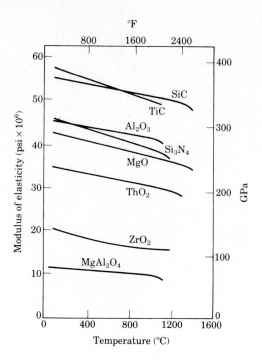

FIGURE 11.14 Effect of temperature on the modulus of elasticity for several ceramics. As a comparison, steel has a modulus of elasticity of about 30 million psi (200 GPa). See also Fig. 2.11 for metals and Fig. 10.13 for plastics. *Source:* After D. W. Richerson, *Modern Ceramic Engineering*, Marcel Dekker, New York, 1982.

Ceramics maintain their strength and rigidity at elevated temperatures, as seen in Fig. 11.14. This characteristic makes ceramics very attractive for high-temperature applications. (See also Section 11.9.3.)

Ceramics, especially glasses, exhibit a phenomenon called *static fatigue*. When subjected to a static tensile load over a period of time, glass may fail suddenly. This phenomenon occurs in environments where there is water vapor; it does not occur in a vacuum or in dry air. Static fatigue has been attributed to a mechanism similar to stress–corrosion cracking of metals (Section 3.7.1).

11.9.2 PHYSICAL PROPERTIES

Compared to most metals and alloys, ceramics have high melting or decomposition temperatures (Table 11.11). Their thermal conductivity has a wide range, by as much as three orders of magnitude (Fig. 11.15). The very low thermal conductivity of some ceramics makes them ideal for use as thermal insulators at elevated temperatures. Metals and alloys have, in comparison, much higher thermal conductivity.

The thermal conductivity k is related to porosity by,

$$k = k_0(1 - P),$$ (11.3)

where k_0 is the thermal conductivity at zero porosity.

TABLE 11.11
APPROXIMATE MELTING TEMPERATURES OF CERAMIC, METALLIC, AND ORGANIC MATERIALS.
(After D. W. Richerson.)

MATERIAL	°C	MATERIAL	°C
Polystyrene	65–75	Si_3N_4	~1750–1900
Polymethylmethacrylate	60–90	Mullite	1850
Polyethylene	120	Al_2O_3	2050
Nylon 6	135–150	Spinel	2135
Polyimides	260	B_4C	2425
PTFE	290	SiC	2300–2500
B_2O_3	460	BeO	2570
Aluminum	660	ZrO_2 (stabilized)	2500–2600
Nickel-base superalloy	1300	MgO	2620
Cobalt-base superalloy	1330–1410	WC	2775
Diamond	1400	VC	2820
Stainless steel (304)	1400–1450	TiC	3100
CBN	1540	Tungsten	3370
SiO_2 (fused)	~1650	TaC	3875

FIGURE 11.15 Effect of temperature on thermal conductivity for several ceramics and other materials. Note that conductivity varies by four orders of magnitude. See also Table 3.3 and Fig. 10.21. After various sources.

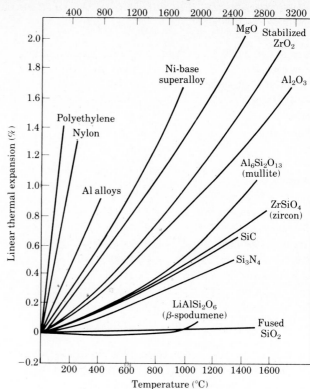

FIGURE 11.16 Effect of temperature on thermal expansion for several ceramics, metals and plastics. See also Table 11.12. After various sources.

The thermal expansion characteristics of ceramics are shown in Fig. 11.16. The behavior of silica (SiO_2) is particularly interesting because it has virtually no thermal expansion over a wide temperature range. (See also Section 3.11.3.)

As described in Section 3.11, the thermal properties of expansion and conductivity are important in inducing thermal stresses, which can lead to thermal shock or thermal fatigue. The tendency for thermal cracking (also called spalling, when a piece or a layer from the surface breaks off) is reduced with low expansion and high thermal conductivity.

Graphite has excellent resistance to thermal shock because of its high thermal conductivity (see Fig. 11.15); hence it is very suitable for die and mold materials at elevated temperatures. The virtually zero thermal expansion of fused silica (see Fig. 11.16) gives it high thermal shock resistance. A familiar example, illustrating the importance of low thermal-expansion properties, is the heat-resistant ceramics for cookware and stove tops. They can sustain very high thermal gradients, from hot to cold and vice versa. (See Illustrative Problem 11.1.)

An additional physical property of ceramics is the anisotropy of thermal expansion observed in noncubic crystals (Table 11.12). When a polycrystalline ceramic material is prepared with such crystals, thermal stresses develop and, if sufficiently high, they can lead to cracking.

TABLE 11.12
**COEFFICIENTS OF THERMAL EXPANSION FOR SOME
ANISOTROPIC CERAMICS**

| MATERIAL | COEFFICIENT OF THERMAL EXPANSION ($\times 10^6$/°C) | |
	NORMAL TO c-AXIS*	PARALLEL TO c-AXIS
Graphite	1	27
Al_2O_3 (alumina)	8.3	9
$3Al_2O_3$, $2SiO_2$ (mullite)	4.5	5.7
TiO_2	6.8	8.3
$ZrSiO_4$	3.7	6.2
SiO_2 (quartz)	14	9

* See Fig. 3.1.

● **Illustrative Problem 11.1**

Common household ovenware carries the following instruction (see Fig. P11.1):
For oven and microwave, no stovetop or broiler. Explain the reasons for this type of
instruction for this product.

SOLUTION.   This product is made of borosilicate (see Table 11.9) which has the
property of low thermal expansion, hence thermal stresses are low when subjected
to thermal gradients. Under extreme temperature conditions, however, thermal
stresses can be high enough to cause fracture of this material. The contact temperatures on a stovetop, or the temperatures due to radiation in a broiler, are very high
compared to those in an oven. Furthermore, this product is subjected to a rapid
temperature rise on its exposed surface on a stovetop or in a broiler, thus temperature
gradients are much more severe compared to a preheated oven where heat is transmitted mainly through convection at temperatures generally below 450°F (230°C).   ●

FIGURE P11.1

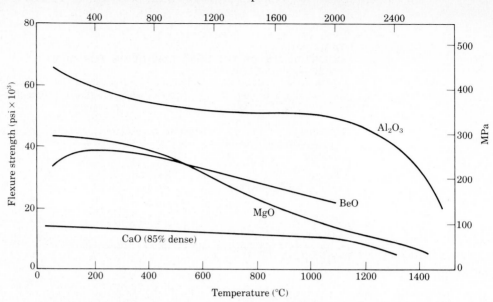

FIGURE 11.17   Effect of temperature on flexure strength for various ceramics. Note the high test temperatures. After various sources.

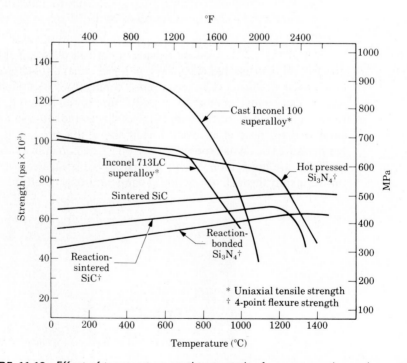

FIGURE 11.18   Effect of temperature on the strength of some ceramics and superalloys. See also Tables 3.21, 3.22, and 3.23 for high-temperature metals. After various sources.

11.9.3 APPLICATIONS

As seen in Table 11.8 ceramics have numerous applications. Some of these are traditional uses for various household and industrial needs. Several types of ceramics are available: enamels (which were fully developed by the Middle Ages) used for decorative purposes and porcelains (developed in the early 1700s) which are used in the electrical and electronics industry because of their high electrical resistivity and dielectric strength and magnetic properties. Certain ceramics also have good piezoelectric properties.

The mechanical properties of ceramics at elevated temperatures (Figs. 11.17 and 11.18) have led to their use in fuels and controls in nuclear applications, in aerospace components, such as heat shields (see Fig. 1.4), and in gas turbines, automotive components (Figs. 11.19 and 11.20), refractory furnace linings, heat exchangers, and seals. Various other applications are shown in Table 11.8.

(a)   (b)

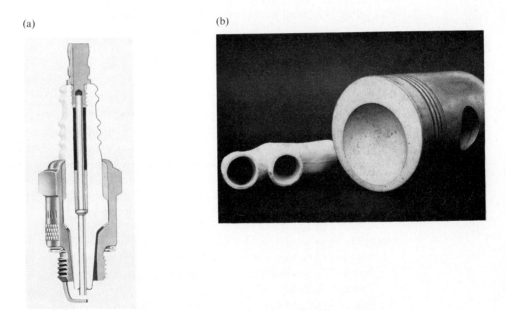

FIGURE 11.19 Examples of ceramic components for high-temperature applications. (a) Ceramic (aluminum oxide) insulator in an automotive spark plug. (b) Left: slip-cast fireclay ceramic exhaust port liner. Right: plasma-sprayed zirconia-coated piston for an adiabatic diesel engine. *Source:* (a) Courtesy of Champion Spark Plug Co. (b) Courtesy of R. F. Firestone, IIT Research Institute.

FIGURE 11.20   Precision cast silicon carbide stator for small gas turbine. This part is about 5 in. (127 mm) in diameter. *Source:* Courtesy of the Norton Company.

11.10 DESIGN CONSIDERATIONS

Depending on the particular product, the selection of ceramic materials, processing methods, and subsequent operations and assembly into other components require careful consideration, keeping in mind the various properties described above. It is essential to consider factors such as low tensile strength, sensitivity to defects, and the generally low impact toughness of ceramics. These limitations should be balanced against the many desirable characteristics of ceramics, such as their hardness, compressive strength, and various other physical properties.

Control of the quality of the materials used and the processing parameters is essential. As in all design decisions, there are priorities, limitations, and numerous other factors to be taken into consideration. There are situations where there is a lack of reliable data and sufficient past experience, particularly with newer ceramics.

Dimensional changes and possibilities of warping and cracking during the processing history are significant factors. Compatibility of a ceramic component, when it is a part of a larger assembly, is another consideration, particularly in regard to thermal expansion and type of loading. Reliability and consequence of part failure are always significant factors in the design process for ceramics. Techniques are being developed for predicting strength and toughness as a function of microstructure, for more reliable use of ceramic components.

● **Illustrative Example 11.2**

Ceramic materials are now being considered for applications in automotive engines, replacing traditional materials such as cast irons and steels. The following is a list of factors that should be taken into account in considering ceramics for these applications.

a. Necessity for developing new ceramics for this application with its special requirements.

b. Techniques to shape ceramics into desired products and the subsequent processing required.

c. Accuracies, tolerances, reproducibility, etc. that can be obtained and whether or not they are acceptable for this application.

d. Physical and mechanical properties of ceramics that are relevant for this application: frictional and wear properties, strength, toughness, resistance to thermal shock, notch sensitivity, thermal expansion, dimensional stability, resistance to fuel oils, change in properties over time, weight.

e. Effect of ceramic components on the dynamics of the engine.

f. The need for new techniques for cooling, assembly, repairs, maintenance, cleaning, lubrication, and seals.

g. Cost, new tooling and equipment required, training of service personnel, compliance with standards, codes, special methods of handling and shipping ceramic components.

h. Effects of catastrophic failure of ceramic components.

i. Public acceptance and consumer confidence.   ●

SUMMARY

1. A wide variety of products is made by the processes of powder metallurgy and ceramic and glass processing. They range from tools and dies (carbide, ceramic, or tool steels) to gears, tiles, glass bottles, automotive parts, and numerous other household and industrial products.

2. Powder metallurgy involves shaping metal powders under pressure and then heating them (sintering). Powders can be produced by various techniques, such as atomization, reduction from metal oxides, electrolytic deposition, and pulverizing. Most metal powders can be produced by more than one method,

3. The density, hence strength and ductility, of the compacted and sintered powder-metal products are less than those of the bulk metal processed by metal forming techniques. The inherent porosity can, however, be utilized to impregnate with a fluid, or infiltrate with molten metal, for specific applications such as in bearings.

4. Powder metallurgy has advantages such as a wide range of compositions, high production rate, and good dimensional control. However, part size and complexity can be limited, and costs can be high for low production runs.

5. Ceramics are compounds of metallic and nonmetallic materials. They are generally characterized by high hardness, compressive strength, high temperature resistance, and chemical inertness.

6. Ceramic products are first formed by casting, plastic forming, or pressing. They are then dried and fired to impart strength. Because of their inherent brittleness, ceramics are processed with due considerations to cracking and distortion.

7. Glasses are supercooled liquids, i.e., the rate of cooling is high and hence the time is insufficient to solidify into a crystalline structure. They do not have a clearly defined melting point. Glasses are available in a wide variety of forms, compositions, and mechanical, physical, and optical properties. The strength of glass can be improved by thermal or chemical treatments.

BIBLIOGRAPHY

Powder Metallurgy

Hausner HH, Mal MK. *Handbook of Powder Metallurgy*. New York: Chemical Publishing Co., 1982.

Hirschorn JS. *Introduction to Powder Metallurgy*. New York: American Powder Metallurgy Institute, 1969.

Jones WD. *Fundamental Principles of Powder Metallurgy*. London: Edward Arnold, 1960.

Lenel FV. *Powder Metallurgy: Principles and Applications*. New York: American Powder Metallurgy Institute, 1980.

Periodicals

International Journal of Powder Metallurgy and Powder Technology

Metal Powder Report

Powder Metallurgy

Ceramics

Doremus RH. *Glass Science*. New York: Wiley, 1973.

Engineering Properties of Selected Ceramic Materials. Columbus, Ohio: American Ceramic Society, 1966.

Henry EC. *Electronic Ceramics*. Garden City, New York: Doubleday, 1969.

Holloway DG. *The Physical Properties of Glass*. London: Wykeham, 1973.

Kingery WD, Bowen HK, Uhlmann DR. *Introduction to Ceramics*, 2d ed. New York: Wiley, 1976.

Lawrence WC. *Ceramic Science for the Potter*. Philadelphia: Chilton, 1972.

McMillan PW. *Glass-Ceramics*. London: Academic, 1964.

Norton FH. *Elements of Ceramics*. Reading, Mass.: Addison–Wesley, 1952.

Norton FH. *Fine Ceramics*. New York: McGraw-Hill, 1970.

Richerson DW. *Modern Ceramic Engineering*. New York: Marcel Dekker, 1982.

Samsonov CV, Vinitsku JM. *Handbook of Refractory Compounds*. New York: Plenum, 1980.

Shand EB. *Glass Engineering Handbook*, 2d ed. New York: McGraw-Hill, 1958.

Van Vlack LH. *Physical Ceramics for Engineers*. Reading, Mass.: Addison–Wesley, 1964.

Waye BE. *Introduction to Technical Ceramics*. London: Maclaren, 1967.

Periodicals

American Ceramic Society Bulletin

Brick and Clay Record

Ceramic Industry

Journal of the American Ceramic Society

Journal of Glass Technology

PROBLEMS

11.1. Explain why ceramics are weaker in tension than they are in compression. Is this also true for other brittle materials?

11.2. Make a list of all ceramic parts around your house and on your car. Explain why those parts are made of ceramics.

11.3. Describe the design considerations involved in making parts with powder metals.

11.4. Why is it that punches in powder-metal compaction are sometimes made of two or more pieces?

11.5. Explain the difference between impregnation and infiltration in powder metallurgy.

11.6. What are the advantages of cermets?

11.7. Explain the mechanisms involved in the sintering process.

11.8. Ceramics have generally been used in relatively small components. There is, however, a trend to use them as structural components also. Explain the difficulties that might be encountered in making larger ceramic components.

11.9. Describe the unique characteristics of glasses and make a list of the applications with which you are familiar.

11.10. Which of the types of metal-powder particles shown in Fig. 11.2 are likely to have high frictional resistance in a compacting operation?

11.11. How would you explain the dependence of electrical conductivity on density as shown in Fig. 11.7(d)? Would this also be true for thermal conductivity?

11.12. Explain why ceramics are effective cutting-tool materials. Would ceramics also be suitable for die materials for metal forming?

11.13. Why is it that in some P/M products density differences are introduced within the part?

11.14. Describe applications where a ceramic material with a zero coefficient of thermal expansion is desirable.

11.15. Add more specific items to Table 11.8.

12 Fastening and Joining Processes

12.1 INTRODUCTION

With few exceptions, all products, machines, or structures are an assemblage of parts that have been made as separate units. This is done either for ease of manufacturing the product, ease in assembling, convenience in transportation, or for economic reasons. Consequently, fastening and joining methods are an essential part of the overall design and manufacturing process. Generally, devices such as screws, bolts, rivets, and similar fasteners are used for fastening, and welding, brazing, soldering and adhesives are used for joining.

In addition to economic factors, there are important technical aspects in choosing the type of joint to be used. For example, the stiffness and dynamic characteristics of structures such as machine tools depend not only on the particular design and the materials used, but also on how the various components are assembled (Section 8.14.7). Bolted joints, for instance, have a higher damping capacity than solid or welded structures. Furthermore, in a particular component, different properties on different surfaces or regions may be desirable for functional purposes. Surfaces subjected to friction and wear, for instance, may require characteristics different from the bulk of the component. Examples are carbide inserts attached to the shank of a cutting tool by brazing (see Fig. 8.35b).

Joints may be permanent, such as welded bridge or machine components, or they may be semipermanent, such as bolts on an engine block. Various other fasteners are also of a semipermanent type.

12.2 MECHANICAL JOINING METHODS

When two or more components are to be joined so that they can be taken apart with relative ease, the most common method is by the use of threaded fasteners such as bolts, nuts, and screws. These generally require holes in the components through which the fasteners are inserted (Fig. 12.1). Depending on the type of fastener and the nature of the interfaces between different components, such joints can have both shear and tensile strengths, thus resisting forces in the shearing and tensile directions, respectively.

A wide variety of designs for fasteners is available, with standards specifying thread dimensions, tolerances, pitch, strength, and quality of the materials used. If the joint is to be subjected to vibration, a variety of specially designed nuts and lock-washers are available. Their main purpose is to increase the frictional resistance in the torsional direction, thus preventing the loosening of the fastener. Fasteners are discussed in detail in texts on machine elements and design.

Rivets

For permanent mechanical joints, the most common fastener is the rivet (Fig. 12.2), first used as far back as 3000–2000 B.C. Riveting requires the preparation of holes in the components to be joined. The installation of a rivet is done by deformation of its

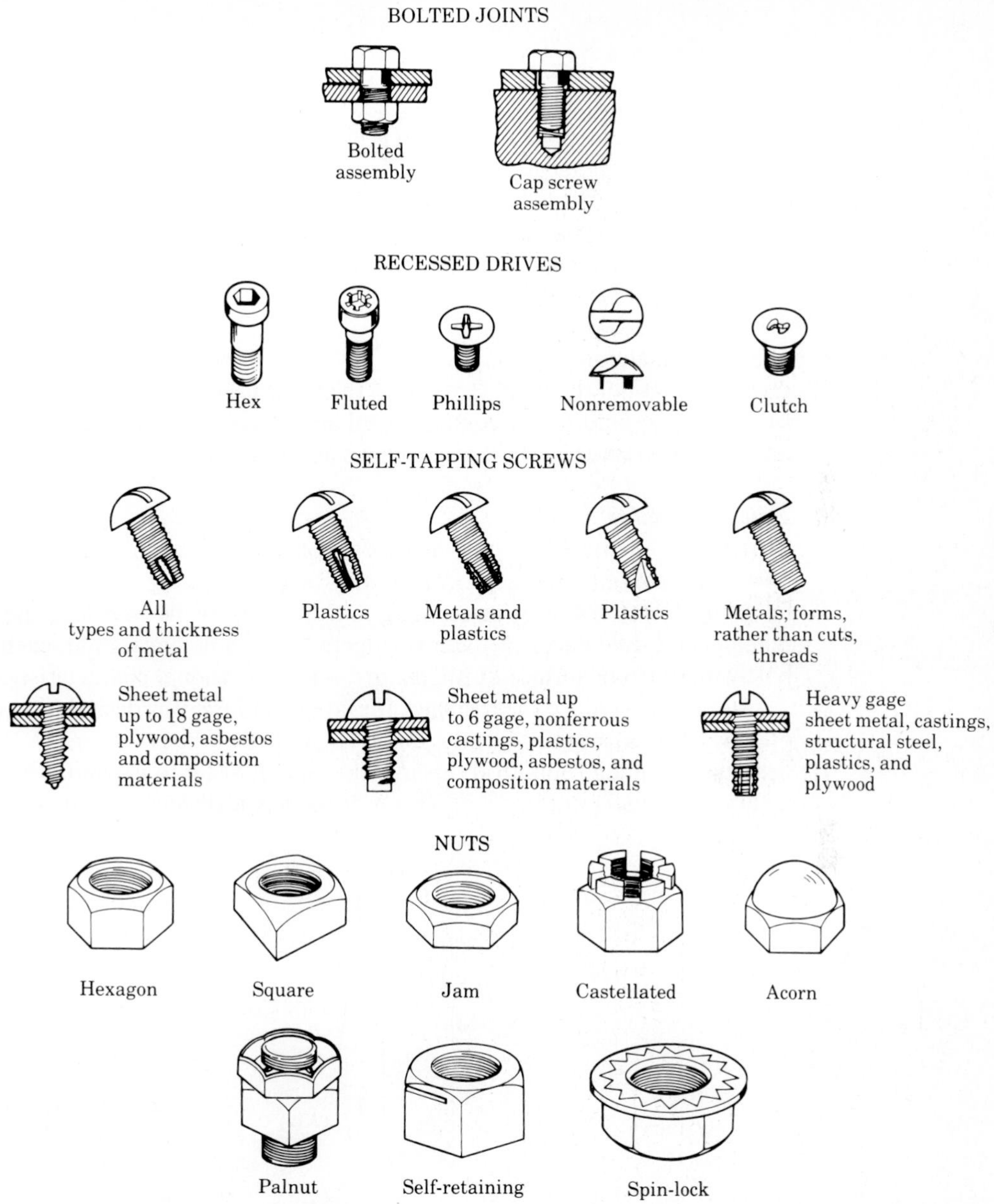

BOLTED JOINTS

Bolted
assembly

Cap screw
assembly

RECESSED DRIVES

Hex   Fluted   Phillips   Nonremovable   Clutch

SELF-TAPPING SCREWS

All
types and thickness
of metal

Plastics

Metals and
plastics

Plastics

Metals; forms,
rather than cuts,
threads

Sheet metal
up to 18 gage,
plywood, asbestos
and composition
materials

Sheet metal up
to 6 gage, nonferrous
castings, plastics,
plywood, asbestos, and
composition materials

Heavy gage
sheet metal, castings,
structural steel,
plastics, and
plywood

NUTS

Hexagon   Square   Jam   Castellated   Acorn

Palnut   Self-retaining   Spin-lock

FIGURE 12.1 Selected examples of threaded fasteners. There are many other types and sizes of fasteners available. *Source:* Adapted from *Tool and Manufacturing Engineers Handbook*, 3d ed., McGraw-Hill, 1976.

smaller end after being placed in the hole. This operation may be done either mechanically by expanding or upsetting, or through the use of explosives in the rivet cavity.

The design of riveted joints requires careful consideration of the type of loading to which the structure will be subjected, the size and spacing of holes, the compatibility

Semitubular   Tubular   Split   Compression

FIGURE 12.2 Selected examples of rivets. Rivets are generally permanent fasteners and are used extensively in steel construction and aircraft assembly. A typical aircraft may have hundreds of thousands of rivets.

of the rivet material with that of the components to be fastened, and the effect of holes on the strength of the joint. Because of the presence of a hole and its effect as a stress raiser, fatigue failure is a common problem in riveted joints, especially in aircraft and similar structures where vibrations cannot be avoided.

Hole Preparation

Because of the many fastening methods requiring holes, hole preparation is an important aspect of joint design. A hole may be generated by *drilling* (with subsequent finishing operations such as reaming and honing), or by *punching*. Because of the fundamental differences in these two methods of hole generation, each produces a hole with different surface finish, texture, and dimensional characteristics.

In order to reduce the tendency for fatigue failure, it is desirable to induce compressive residual stresses on the cylindrical surface of the hole (Fig. 12.3). This is best done by pushing a round rod through the hole in order to expand it by a very small amount (see insert in Fig. 12.3). This process plastically deforms the surface layers of

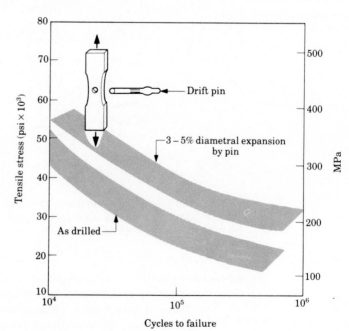

FIGURE 12.3 Effect of hole preparation on the fatigue life of 7075–T6 aluminum. The improvement in fatigue strength is obtained by imparting compressive residual stresses on the surface of the hole by the drift pin.

the hole in a manner similar to shot peening or roller burnishing (see Figs. 4.38 and 4.39) and induces circumferential compressive residual stresses in the hole.

The compatibility of the bolt or rivet material with that of the parts to be fastened is an important consideration with regard to *galvanic corrosion* (see Section 1.2.2). For example, in a system where an iron rivet is used to fasten copper sheets, the iron is anodic and the copper cathodic, thus leading to rapid corrosion. This is also the case with aluminum or zinc fasteners on copper. Numerous examples can be cited indicating the importance of proper choice of materials for fasteners.

Stapling

A third method of mechanical fastening is the use of staples. The process is much like an ordinary staple or stitching on a sewing machine. It requires no predrilled holes, it is fast, and is particularly suitable for joining thin metallic and nonmetallic materials.

Seaming

Mechanical fastening is also done without the use of bolts, rivets, or staples, and is based on the simple principle of folding two thin pieces of material together. This is much like joining two pieces of paper, for lack of a paper clip, by folding them at their corner.

Typical examples of this type of joining, called *seaming*, are shown in Fig. 12.4(a). The most common examples are the tops of beverage cans, food containers, or household products (Fig. 12.5a). In seaming, it is necessary that the materials be capable of

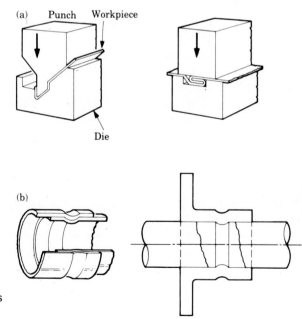

FIGURE 12.4 Examples of mechanical joining methods. (a) Method of producing lock seams. See also Figs. 7.28 and 7.31. (b) Two examples of crimping tubular products.

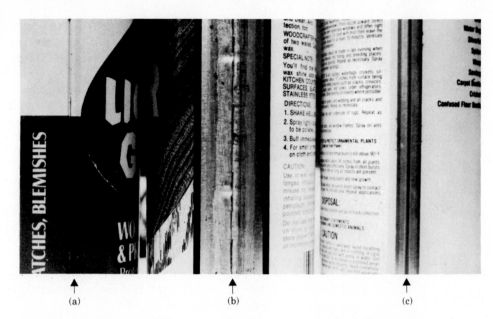

FIGURE 12.5 Three types of joints on the side of a can for household products. (a) Seam lock. See also Fig. 7.31. (b) Soldered joint. (c) Resistance seam welded joint. See also Fig. 12.23(b).

undergoing bending at very small radii, otherwise they will crack and the seams will not be watertight. Such seams may be improved with adhesives and other coatings at the interfaces to improve their quality and reliability.

Crimping

Another method of joining without using fasteners is crimping. This can be done with *beads* or *dimples*, as shown in Fig. 12.4(b). These processes can be used both on tubular or flat parts, provided that the materials are sufficiently thin and ductile to tolerate the large localized deformation.

Methods of joining by welding and adhesives are described throughout the rest of this chapter.

12.3 WELDING PROCESSES: ENERGY SOURCES AND CLASSIFICATION

Welding is the process of joining metallic or nonmetallic materials with heat and/or pressure. Whether it involves very small electronic components or large structures, welding is one of the most important aspects of manufacturing operations. There is a wide variety of welding processes, each group requiring a different source of energy.

The energy required for welding may be obtained from a variety of sources. The major sources generally fall into the following categories:

a. Mechanical,

b. Electrical,

c. Chemical,

d. Optical, and

e. Solid state (diffusion).

Because of the complex nature of the bond between two surfaces being joined, the form of energy may not always be directly identifiable. For example, processes such as friction welding, ultrasonic welding, and explosive welding, all use mechanical energy in the form of relative motion between the surfaces to be joined; yet the mechanism of bonding involves the solid-state process of diffusion. Likewise, processes using electrical energy, such as projection welding, also use mechanical forces and movement to complete the joint and give it full strength. Note that thermal energy is not included above. This is because the energy sources provide the heat necessary to produce the bond.

In spite of these difficulties, the sources of primary energy can be clearly identified. Processes typically using electrical energy are arc welding and resistance welding. Chemical sources of energy are used in gas welding and thermit welding, where the energy is converted into heat. Two major joining processes using optical energy are laser-beam welding and electron-beam welding. The high power density of energy in these processes is obtained by focusing the beams on a small area of the parts to be joined. Solid-state sources of energy are principally diffusion processes. The strength of the bond depends on the parameters of temperature, time, and pressure at the interface to be joined.

The categories described above can be further subdivided into a variety of welding processes involving different types of equipment and part geometries. Thus, welding processes can be tabulated in a variety of ways and may include different subdivisions.

The many categories and processes for joining are shown in Fig. 12.6. For the purposes of this textbook, only major categories are outlined and described below, including some subdivisions for industrially significant joining processes. The major categories of welding are solid-state and liquid-state joining. Brazing, soldering, and adhesive bonding are treated in other sections in this chapter.

12.4 SOLID-STATE JOINING

In describing the adhesion theory of friction in Section 4.6.1, it was noted that when two clean surfaces are brought into contact, the asperities bond together, forming microwelds. The strength of the bond depends, among other factors, on how clean the two surfaces are. Under ideal conditions and in the absence of any

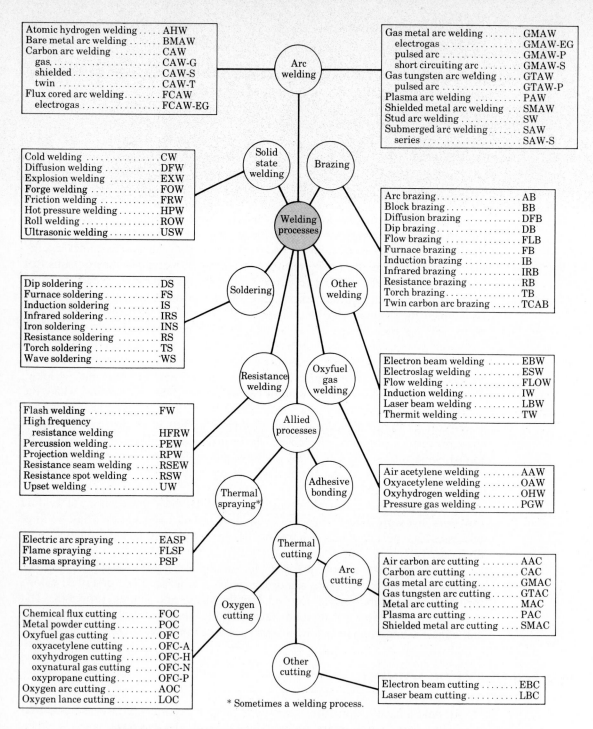

FIGURE 12.6 Master chart for welding and allied processes. *Source:* Courtesy of American Welding Society.

adsorbed layers of gas or liquid films, and other contaminants and oxide layers, the surfaces are close enough to each other to form interatomic bonds.

The important factors that determine the strength of these bonds are *pressure*, *time of contact*, and *temperature* of the interface. With an increase in one or more of these parameters, diffusion at the interface is accelerated, thus making the bond stronger. Depending on the degree of cleanliness and their roughness, a lateral movement of the mating surfaces under a normal load also improves adhesion by mechanically disturbing the surfaces, breaking up the oxide layers, and generating virgin surfaces.

It is apparent that this bonding mechanism can be utilized as a joining process for two similar or dissimilar materials. The process can be carried out at room temperature, or at elevated temperatures with the heat supplied either externally or through friction at the interface. In these bonds no liquid phase is present between the two mating surfaces, as it is in liquid-state joining processes described in Section 12.5.

Solid-state bonding may be classified into two general categories: those that involve some *plastic* deformation at the interface, and those that involve *diffusion* at the interface, which in turn requires high temperatures for a more effective bond. Bonding by plastic deformation can be subdivided into the individual joining processes of:

Cold welding or cold-pressure welding,

Explosive welding, and

Ultrasonic welding.

Bonding by diffusion and temperature can be subdivided into:

Forge welding,

Diffusion bonding, and

Friction welding.

12.4.1 COLD WELDING

In cold welding, pressure is applied to the workpieces either through dies or rolls. Because of the plastic deformation required, it is necessary that at least one, but preferably both, of the mating parts be ductile and that the interface be clean. It is also important to note that in bonding two dissimilar metals that are mutually soluble, brittle intermetallic compounds may form, thus resulting in a weak joint. An example is the bonding of aluminum and steel, where a brittle intermetallic compound is formed at the interface. The best bonding by cold welding is typically obtained with two similar materials.

When the pressure is applied through rolls, the process is called *roll bonding*. The technique presently used for manufacturing composite coins is roll bonding (Fig. 12.7). The starting material for the U.S. quarter is two outer layers of 75% copper–25% nickel, each 0.048 in. (1.2 mm) thick, with an inner layer of pure copper

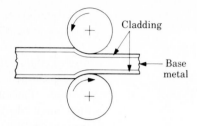

FIGURE 12.7   Schematic illustration of the roll bonding process. Materials for composite coins are made by this process. The U.S. quarter has two outer layers of 75% copper, 25% nickel alloy with an inside layer of pure copper.

0.20 in. (5.1 mm) thick. To obtain good bond strength, the surfaces are chemically cleaned and wire-brushed. The strips are then roll bonded to a thickness of 0.090 in. (2.29 mm). A second rolling operation reduces the final thickness to 0.0535 in. (1.36 mm). The strips thus undergo a total reduction in thickness of 82%.

12.4.2 EXPLOSIVE WELDING

In explosive welding (developed in the 1950s) the pressure is applied through an explosive layer (Fig. 12.8), producing very high pressure levels. The nature of the interface involves a wavy type of plastic deformation, resulting in mechanical inter-

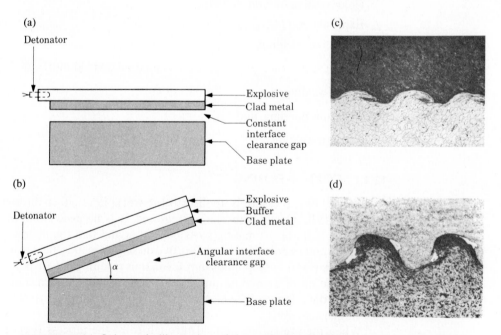

FIGURE 12.8   Schematic illustration of the explosive welding process. (a) Constant interface clearance gap. (b) Angular interface clearance gap. (c) Cross-section of explosively welded joint—grade 1 titanium on A516 steel. (d) Cross-section of explosively welded joint—Incoloy 800 (iron-nickel-base alloy) on A516 steel. *Source:* (c and d) Courtesy of E. I. DuPont de Nemours & Co.

locking of the surfaces (Fig. 12.8c and d). Cold-pressure welding also takes place in this process. Any oxide layer present at the interface is broken up by the high transient stresses at the interface during welding. The strength of the bond in explosive welding is very high and the process is suitable for joining a variety of materials. In addition to bonding flat plates by this technique, cladding can be done, as in composite coins.

Tubular products are often joined by explosive welding, where, for example, tube ends are expanded into the head plates of heat exchangers or boilers. This process is also used in the manufacture of equipment for the chemical industry.

12.4.3 ULTRASONIC WELDING

In ultrasonic welding, the interface is subjected to shearing stresses by an oscillating tip attached to a transducer (Fig. 12.9). The normal force is only nominal, but the shearing stresses cause a relative movement at the interface, thus breaking up oxide films or contaminants, allowing good contact, and producing a strong bond. The nature of the interface in ultrasonic welding is complex. Although the generation of heat is generally not significant, there are situations where interface temperatures can be high enough to cause metallurgical changes.

The process is versatile and can be used with a wide variety of metallic and nonmetallic materials. It is used extensively in the electronics industry, and in packaging involving metal foil.

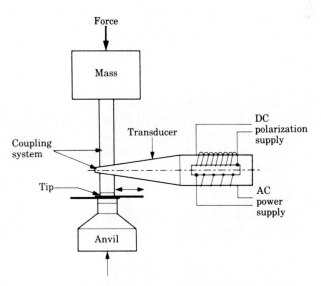

FIGURE 12.9 Ultrasonic welding. Bonding in this process is obtained by the lateral vibrations of the tool tip, which cause plastic deformation at the interface of the two pieces being joined.

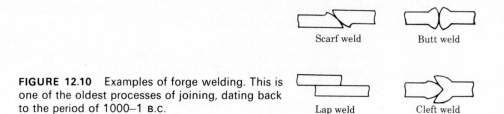

FIGURE 12.10 Examples of forge welding. This is one of the oldest processes of joining, dating back to the period of 1000–1 B.C.

12.4.4 FORGE WELDING

In the second category of solid-state bonding, elevated temperatures are required in order to obtain a strong bond between two bodies. The first category, *forge welding*, is one of the oldest joining processes, dating back to the period of 1000–1 B.C. It involves applying both heat and pressure at the interface.

The two components are heated to a high temperature and then pressed together with suitable hammers, dies, or rollers (Fig. 12.10). The plastic deformation at the interface breaks up the oxide layers and other contaminants. The interface is not particularly strong; consequently this process has been largely replaced by other joining processes.

12.4.5 DIFFUSION BONDING

In diffusion bonding (developed in the 1970s) the strength of the interface is primarily due to diffusion and, to a lesser extent, to the small plastic deformation of the surfaces. This process requires temperatures on the order of 0.5 T_m to achieve a sufficiently high diffusion rate. (See Fig. 3.17.)

The interface (Fig. 12.11) has essentially the same physical and mechanical properties as the base metal; its strength depends on pressure, temperature, time of contact, and cleanliness of the surfaces. While diffusion bonding is generally most suitable for dissimilar metal pairs, it is being used also for reactive metals such as titanium, beryllium, zirconium, and the refractory metal alloys.

An important development in diffusion bonding is combining it with superplastic forming (see Section 7.13) for components made of sheet metal. This technology is well advanced for titanium structures (typically, Ti–6Al–4V) for aerospace applications. Aluminum structures (7475–T6) are also being developed using this technique. It has been shown that design efficiency (by eliminating fasteners) and productivity is improved substantially by combining diffusion bonding with superplastic forming.

12.4.6 FRICTION WELDING

Whereas in forge welding and diffusion bonding the heat required is supplied externally, in friction welding (developed in the 1940s) the heat is generated through friction at the interface. A typical example is shown in Fig. 12.12. One of the members

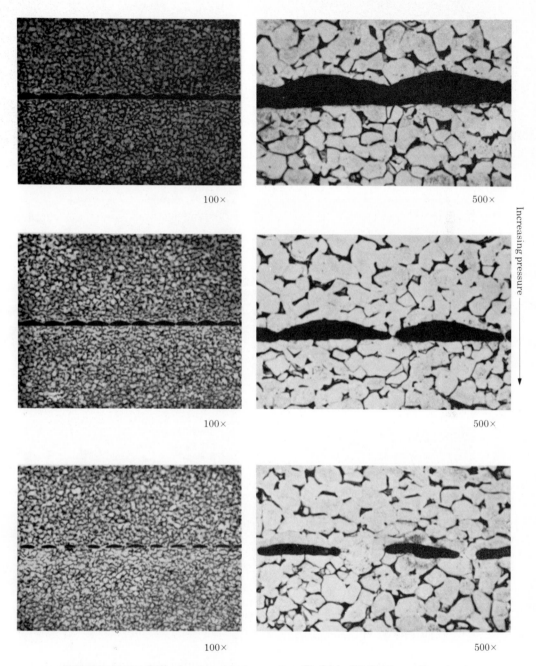

100×   500×

100×   500×

100×   500×

Increasing pressure →

FIGURE 12.11 Diffusion bond between two Ti–6Al–4V titanium alloy sheets. Temperature: 1700°F (925°C), time: 1 hour, pressure: 100–500 psi (0.7–3.5 MPa). Aerospace structures made of titanium or aluminum alloys can be manufactured economically by combining diffusion bonding with superplastic forming, described in Section 7.13. *Source:* Courtesy of Rockwell International Science Center.

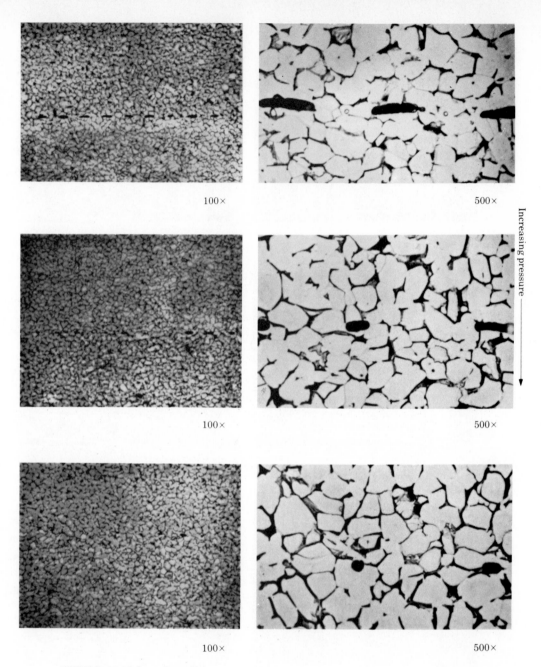

100×   500×

100×   500×

100×   500×

Increasing pressure →

FIGURE 12.11 (*continued*)

to be joined remains stationary, while the other is rotated at a high speed and then brought in contact, under pressure, with the stationary member.

The pressure at the interface and the resulting friction generate sufficient heat for welding to take place. The heated zone is generally confined to a narrow region. After sufficient contact is established, the rotating member is brought to a quick stop (so

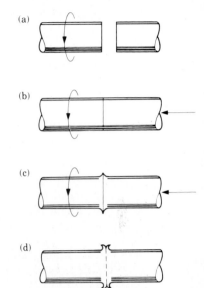

FIGURE 12.12 Schematic illustration of the friction welding process. The heat required for fusion is generated by the friction at the interface of the two bodies. See also Figs. 4.16, 8.18, and 8.19. Only one of the parts (left) has to have rotational symmetry in friction welding.

that the weld is not destroyed by shearing), while the pressure is increased. The process is thus somewhat similar to forge welding.

Because of the combination of heat and pressure, the interface in friction welding develops a flash by plastic deformation of the heated zone. This flash, if objectionable, can easily be removed by machining.

In a modification of this process, called *inertia friction welding*, the energy is supplied through a flywheel, which eventually comes to a stop as the weld is formed. A wide range of metallic materials can be joined by friction welding, provided that one of the components has some rotational symmetry, and both components are rigid enough to withstand the torque developed during the welding cycle.

● **Illustrative Example 12.1**

A square-head bolt shown in Fig. P12.1 was originally made by machining from a $2\frac{1}{2}$ in. square (64 mm) AISI 4140 steel bar. Because of the cost involved, inertia friction welding was considered as an alternative process. The shank was machined

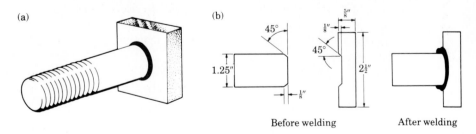

FIGURE P12.1 *Source:* Adapted from *Metals Handbook*, 8 ed., Vol. 6, 1971.

from a $1\frac{1}{4}$ in. (32 mm) diameter bar as shown in the figure, and the square head was recessed to a matching contour. The two parts were then inertia welded, the completed bolt was normalized, and the flash was removed by machining. Inspection of the friction welded joint indicated that the joint had a sound weld. The cost of the inertia-welded bolt was less than one half of the original method of manufacturing. ●

12.5 LIQUID-STATE WELDING PROCESSES

The vast majority of welding processes involve the melting of the interface between the two bodies to be joined. The energy required is supplied by electrical, chemical, or optical means, with or without the addition of filler metals to fill the gap at the interface. Fillers are materials added to the weld area in making a welded, brazed, or soldered joint.

Heating the joint to a high temperature involves important metallurgical changes. The rate of heat application and the physical properties of the metals to be joined are important, in that they control the level and distribution of temperature in a joint. As described in Section 3.5, the application of heat and the degree of prior cold work determine the grain structure in the joint area. Cooling, subsequent to the completion of the joint, results in recrystallized grains. Thus, the strength and toughness of the weld depends on all these factors.

The principles behind the major liquid-state welding processes, as well as the quality of welds produced and the equipment involved, are described in the next section.

12.5.1 CHEMICAL SOURCES OF ENERGY

Gas Welding

The most commonly used gas-welding process is *oxyacetylene welding*, developed in the early 1900s (also known as *oxyfuel gas welding*). It uses the heat generated by the combustion of acetylene gas (C_2H_2) in a mixture with oxygen. Gases such as hydrogen, propane, and natural gas are also used in this process. The technique involves melting the edges of a joint so that they are fused together. A filler wire or rod may also be used to supply molten metal to the interface.

Heat in gas welding is generated in accordance with the following chemical reactions. The primary combustion process, occurring in the inner core of the flame (Fig. 12.13), can be represented by the reaction,

$$C_2H_2 + O_2 \longrightarrow 2CO + H_2 + \text{heat}. \tag{12.1}$$

This is followed in the rest of the flame by the reactions,

$$2CO + O_2 \longrightarrow 2CO_2 + \text{heat}, \tag{12.2}$$

$$2H_2 + O_2 \longrightarrow 2H_2O + \text{heat}. \tag{12.3}$$

NEUTRAL FLAME

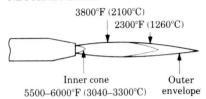

3800°F (2100°C)

2300°F (1260°C)

Inner cone
5500–6000°F (3040–3300°C)

Outer
envelope

OXIDIZING FLAME

Outer envelope
(small and narrow)

Inner cone
(pointed)

CARBURIZING FLAME

Acetylene feather

FIGURE 12.13 Three types of oxyacetylene flames. The gas mixture is basically equal volumes of oxygen and acetylene. These flames are used not only for welding, but also for cutting thick steel plates.

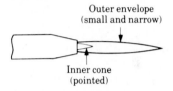

Bright luminous
inner cone

Blue envelope

The proportion of the two gases in the mixture is an important factor in gas welding. With no excess oxygen, i.e., at a ratio of about 1:1, the flame is considered neutral, whereas with higher oxygen supply the flame becomes oxidizing. The latter is harmful especially for steels because it oxidizes the steel. Only in copper and copper-base alloys is the oxidizing flame safe because a thin protective layer of slag forms over the molten metal.

When the supply of oxygen is lowered, the flame becomes reducing. The temperature of a reducing flame is lower, hence it is suitable for applications, such as brazing or soldering, that require low heat.

Filler Rods. Filler rods for welding steels are made of low-carbon, or low-carbon, low-alloy steel. These consumable rods may be bare, or they may be covered with flux. The purpose of the flux is to retard oxidation of the surface of the workpiece by generating a gaseous shield around the weld area. The flux also helps dissolve and remove oxides and other substances.

Gas-Welding Practice. Gas welding is essentially a manual process. Although it is relatively slow and requires a skillful operator, it has the advantages of being portable, versatile, and economical for low-quantity work. The equipment consists

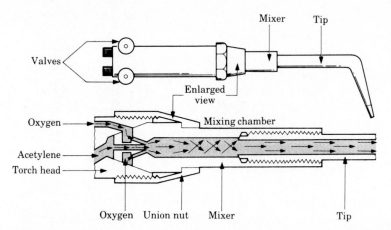

FIGURE 12.14   Cross-section of a torch used in oxyacetylene welding, showing the chamber where gases are mixed thoroughly for proper combustion. The acetylene valve is first opened, the gas lit with a match or other heat source, then the oxygen valve is opened and the flame adjusted. The tip is usually made of copper alloys.

of a welding torch (Fig. 12.14) connected by hoses to gas cylinders, and equipped with pressure gages and regulators, safety equipment such as goggles, and protective clothing. The process can be used with most ferrous and nonferrous metals.

Thermit Welding

Thermit welding (developed in the early 1900s) acquires its name from "thermite," which involves exothermic reactions between metal oxides and reducing agents. The heat generated is used in welding. The most common mixture is three parts of iron oxide (Fe_3O_4) to one part aluminum powder.

This nonexplosive mixture produces a maximum theoretical temperature of 5590°F (3090°C) within a period of one minute. The iron oxide and aluminum mixture may also contain other materials to impart special properties to the weld. The basic reaction is

$$3\,Fe_3O_4 + 8\,Al \longrightarrow 9\,Fe + 4\,Al_2O_3 + heat. \tag{12.4}$$

Other iron oxides and oxides of copper, nickel, chromium, and manganese can also be used in thermit welding, resulting in a temperature ranging up to 9025°F (5000°C). The reaction must be started with a special compound, known as the oxidizing agent.

Thermit welding involves aligning the parts to be joined with a gap, around which a sand or ceramic mold is built. The superheated products of the reaction flow into the gap, melting the ends of the parts to be welded. After the weld cools, the excess material is removed by machining, grinding, or other cutting methods. The thermit process is suitable for welding and repairing large forgings and castings, and also in joining pipes and railroad rails.

12.5.2 ELECTRICAL SOURCES OF ENERGY

In some welding processes (arc welding and resistance welding, developed in the early 1900s) electrical energy is converted into heat. In arc welding, this is done through either a consumable or nonconsumable electrode, wherein an arc is produced between the electrode and the workpiece. This arc produces high temperatures, ranging between 9000 and 54,000°F (5000 and 30,000°C), depending on the type of gas used. The temperature can be controlled through various electrical devices. It also produces radiation, which may be as much as 20% of the total energy. Resistance welding is a combination of pressure and heat produced by the electrical contact resistance of the pieces to be joined between two electrodes. These processes are described below.

Arc Welding Processes: Consumable Electrode
Major arc welding processes using a consumable electrode are described below.

Shielded-Metal Arc Welding (SMAW). This is one of the oldest, simplest, and most versatile processes in arc welding. Currently, about 50% of all industrial welding is performed by this process. The arc is generated by touching the tip of a coated electrode (developed in the 1920s) against the workpiece and then withdrawing it quickly to a distance sufficient to maintain the arc.

The heat generated melts

a. a portion of the tip of the electrode,

b. its coating, and

c. the base metal (workpieces) in the immediate area of the arc (Fig. 12.15).

A weld forms after the alloy (a mixture of the base metal, electrode metal, and substances from the coating) solidifies in the weld area.

The function of the coating (which is brittle) on the electrode is to:

a. Stabilize the arc,

b. Generate gases to act as a protective shield against the surrounding air,

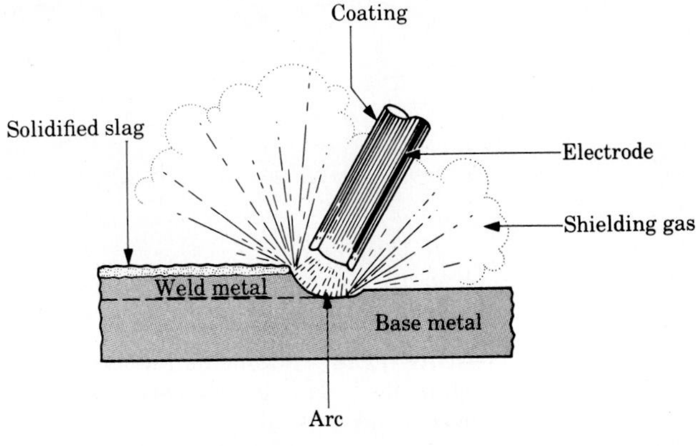

FIGURE 12.15 Shielded metal-arc welding process. About 50% of all large industrial welding operations are performed by this process, which is simple and versatile. The purpose of shielding is to keep air (oxygen) away from the weld zone.

c. Supply metal to alloy the molten base metal, and

d. Produce slags to protect the weld against formation of oxides, nitrides, and other inclusions.

The slag must be removed after each pass to ensure a good weld. Typical electrode dimensions are 9 to 18 in. long (230 to 460 mm) and 3/32 to 1/4 in. (2 to 6 mm) in diameter.

The electrical current in SMAW generally ranges between 50 and 300 A, depending on the type and cross-section of the electrode. The current may be alternating or direct current, and the polarity of the electrode may be positive (reverse polarity) or negative (straight polarity). The choice depends on factors such as the type of electrode, materials to be welded, arc atmosphere, and the depth of the heated zone.

The SMAW process has the advantage of being simple, requiring equipment consisting only of a power supply, power cables, and electrode holder. It is commonly used in construction, shipbuilding, and pipeline work, especially in remote locations, since this equipment is easily maintained. It is best suited for workpiece thicknesses between 1/8 and 3/4 in. (3 and 19 mm), although this range can be easily extended using special techniques.

Submerged Arc Welding (SAW). In this process (developed in the 1940s) shielding of the weld arc is provided by granular flux (lime, silica, manganese oxide, calcium fluoride, and other elements) fed into the weld zone through a nozzle by gravity flow (Fig. 12.16). The thick layer of flux completely covers the molten zone and

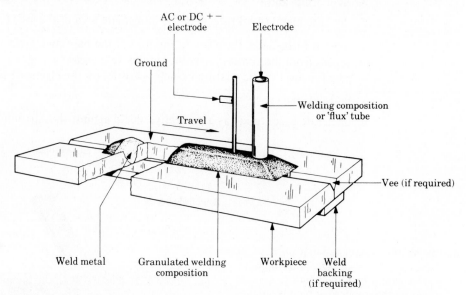

FIGURE 12.16 Submerged arc welding process. This technique produces welds of high quality and is widely used for a variety of steels for welding at relatively high travel speeds.

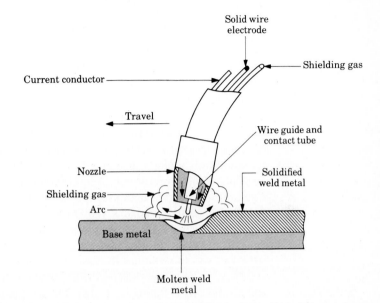

FIGURE 12.17 Gas metal-arc welding process, also known as MIG (for metal inert-gas). This process is used widely in the metal fabrication industry.

prevents spatter and sparks. The flux also acts as a thermal insulator, allowing deep penetration of heat.

The consumable electrode is a coil of bare round wire (1/16 to 3/8 in. [1.6 to 9.5 mm]) fed automatically through a tube (welding gun). High welding currents, ranging between 600 and 2000 A, from alternating- or direct-current power sources, may be used. Because the flux is fed by gravity the SAW process is limited to welds in a horizontal position. After the operation is completed, the flux can be recovered, treated, and reused.

This process can be either automatic or semiautomatic. It is widely used to weld a great variety of steels (sheet or plate) at relatively high speeds. The quality of the weld is high, with good ductility, toughness, and uniformity. The SAW process provides very high welding productivity, depositing from 4 to 10 times the amount of weld metal per hour as the SMAW process.

Gas Metal-Arc Welding (GMAW).   In this process, shielding is provided by an external source of gas, such as argon, helium, carbon dioxide, or various gas mixtures. The consumable wire is fed automatically through a nozzle into the weld arc (Fig. 12.17). The more common name for this process is *MIG* welding, for metal inert gas.

In addition to the use of inert gases, deoxidizers are usually present in the electrode itself, in order to prevent oxidation of the weld puddle. The welds are thus relatively free of any slag and hence multiple weld layers can be formed at the joint.

The GMAW process is suitable for a great variety of ferrous and nonferrous metals. This process (developed in the 1950s) is used extensively in the metal-fabrication industry. Because of the nature of the process, it is easy to train operators. It is also versatile and economical, as it increases welding productivity by a factor of 2 over the SMAW process.

With low currents and voltages, and small-diameter wire electrodes, the short-circuiting technique is used. The metal from the electrode is transferred to the workpiece in individual droplets when the electrode tip touches the molten weld arc and short circuits.

With higher currents, voltages, and larger-diameter wire electrodes, the mode of metal transfer changes and the amount of weld metal transferred across the arc increases rapidly. Depending on the type of shielding gas employed (i.e., argon, argon–oxygen, CO_2), the mode of metal transfer is altered to spray or globular. Using the larger-diameter wire electrodes with higher currents, heavier section sizes are commonly joined together.

The temperatures involved in GMAW are relatively low and, hence, the process is suitable for thin sheet and sections measuring less than 1/4 in. (6 mm). The GMAW process has virtually replaced the SMAW process in present-day welding applications in manufacturing plants. This process can also be easily automated and lends itself readily to flexible manufacturing methods, i.e., industrial robots (Section 13.8).

Fluxed-Cored Arc-Welding (FCAW).   In this process, the electrode is tubular and is filled with flux (Fig. 12.18). The flux is much more flexible than the brittle coating used on SMAW electrodes. The tubular electrode can therefore be in a coil form, thus the process combines the versatility of the SMAW process with the continuous and automatic electrode feeding (welding gun) feature of the GMAW process.

The FCAW process can also be referred to as self-shielded FCAW. This is because of the role of certain emissive fluxes in shielding the weld arc against the surrounding air. Normally, shielding gases are provided and nonemissive fluxes are employed. This process is thus very similar to gas metal-arc welding.

The cored electrodes are generally 1/16 to 3/32 in. (1.6 to 2.4 mm) in diameter. The process is versatile, easy to automate, economical, and is used for a wide variety of joints, mainly with ferrous metals. The higher weld-metal deposition rate of the FCAW process, over that of GMAW process, has led to its use primarily in joining heavy sections (i.e., 1 in. [25 mm] and thicker). The recent development of very-small-diameter tubular electrodes has extended the use of this process to the smaller section sizes.

A major advantage of FCAW is the ease with which specific weld-metal chemistries can be developed. By making alloy additions to the flux core of the wire, virtually any alloy composition can be manufactured. The process is also readily adapted to automated processes, especially the newer flexible automated or robotic systems (Chapter 13).

Electroslag Welding (ESW).   The ESW process (developed in the 1950s) is classified as a machine welding process because it requires special equipment (Fig. 12.19). The weld metal is deposited into a weld cavity between the two plates to be joined. The space is enclosed by two water-cooled copper dams or shoes to prevent the molten slag from running off. The filler wire is then fed through a guide tube and an arc is

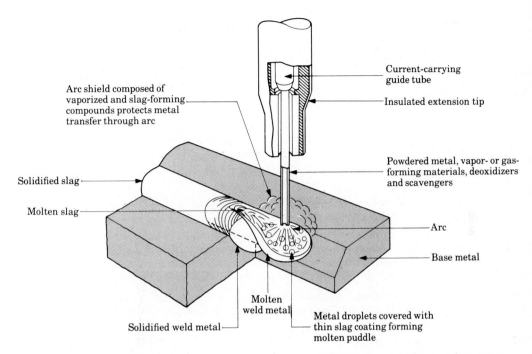

Current-carrying guide tube

Insulated extension tip

Arc shield composed of vaporized and slag-forming compounds protects metal transfer through arc

Powdered metal, vapor- or gas-forming materials, deoxidizers and scavengers

Solidified slag

Molten slag

Arc

Base metal

Molten weld metal

Solidified weld metal

Metal droplets covered with thin slag coating forming molten puddle

FIGURE 12.18 Flux-cored arc welding process. This process is similar to gas metal-arc welding, shown in Fig. 12.17, and can be readily adapted to automation.

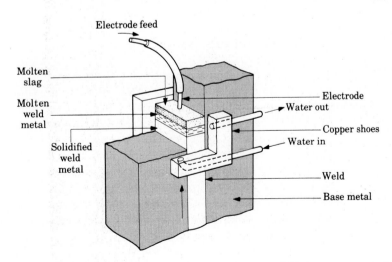

Electrode feed

Molten slag

Molten weld metal

Electrode

Water out

Copper shoes

Water in

Solidified weld metal

Weld

Base metal

FIGURE 12.19 Electroslag welding process, generally used for heavy structural steel sections.

used to initiate the melting process. After equilibrium is established, the arc is extinguished and the energy is supplied through the electrical resistance of the molten slag.

Because of the nature of the process, ESW is used primarily for welds in a vertical position. The upward movement of the shoes is obtained by mechanical drives. Single or multiple electrodes are used on plates with thicknesses ranging from 2 to 36 in. or more (50 to 900 mm) with a variety of joint designs, and the welding is done in one pass. The current is on the order of 600 A at 40 to 50 V, although higher currents are used for thicker plates. The welds are of good quality and the process is used in heavy structural steel sections, such as in machinery and nuclear-reactor vessels.

A variation of ESW is *electrogas welding* (EGW), where shielding is done by an inert gas, such as carbon dioxide, argon, or helium, and the heat is supplied by a continuous arc. The choice of gas depends on the type of material to be welded. Solid electrodes at up to 400 A, and flux-cored electrodes at up to 750 A, may be used in electrogas welding machines.

Arc Welding Processes: Nonconsumable Electrode

These processes, unlike consumable-electrode welding, typically use a tungsten electrode, as one pole of the arc, to generate the heat required. A shielding gas is supplied from an external source.

Gas Tungsten-Arc Welding (GTAW).   In gas tungsten-arc welding (also known as TIG welding, for tungsten inert gas) the filler material is supplied from a filler rod (Fig. 12.20). Alternately, joining may be accomplished without the use of any filler. The power supply may be alternating current (500 A) or direct current (200 A), depending on the metals to be welded. In order to improve electrical properties, the tungsten electrodes may contain thorium or zirconium.

With proper control of the electrical parameters, special arc characteristics that are preferred for certain metals can be obtained. The gas is usually argon or helium, or a mixture of the two. The GTAW process is used for a variety of metals. It is especially suitable for thin materials, producing welds of excellent quality and surface finish. The filler metals are similar in composition to the metals to be welded. The GTAW process is widely recognized as providing an extremely high-quality weld and is used in many critical applications.

Although industrial applications are limited, the *atomic hydrogen welding* (AHW) process uses an arc in a shielding atmosphere of hydrogen. Filler-metal rods may or may not be used. The arc is between two tungsten or carbon electrodes. Thus the workpiece is not a part of the electrical circuit as it is in GTAW. The hydrogen also serves to cool the electrodes.

Plasma-Arc Welding (PAW).   The second nonconsumable arc welding process is plasma-arc welding (Fig. 12.21), where shielding is obtained by an ionized hot gas (gas atoms with fewer electrons), which is an inert mixture of gases. A filler metal

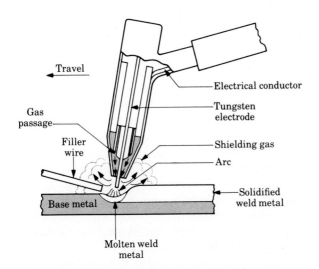

FIGURE 12.20 Gas tungsten-arc welding process, also known as TIG (for tungsten inert-gas). The welds are of very high quality and the process is used for many critical applications.

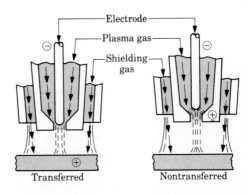

FIGURE 12.21 Plasma-arc welding process. Compared to other methods, deeper and narrower welds are made by this process and at higher welding speeds. See also Fig. 12.28.

may or may not be used. The electrode is surrounded by the ionized gas at the orifice and an extremely hot plasma jet is aimed at the weld area. Temperatures as high as 60,000°F (33,000°C) can be developed with a stable arc.

In PAW (developed in the 1960s) greater energy concentration, deeper and narrower welds, and higher welding speeds can be obtained compared with other arc-welding processes. With control of process parameters, other variations of PAW exist with different weld characteristics. Currents usually range between 1 and 100 A, and higher for special applications. (See also Section 9.9.)

Resistance Welding Processes (RW)
In these processes, the required heat is generated through the electrical resistance of the two bodies to be joined. The heat generated is given by,

$$J = I^2 Rt, \tag{12.5}$$

where

J = heat generated, in joules (watt-seconds);

I = current, in amperes;

R = resistance, in ohms; and

t = time of current flow, in seconds.

The actual temperature rise at the joint also depends on the specific heat and the thermal conductivity of the materials to be welded.

There are five methods of resistance welding: spot, seam, projection, flash, and upset welding. Lap joints are used in the first three processes (Section 12.10) and butt joints in the last two. The heat generated is followed by the application of pressure to ensure a good bond. Accurate control and timing of the electric current and application of pressure is an essential feature of resistance welding.

The strength of the bond also depends on the cleanliness of the mating surfaces, although the presence of oxide layers and contaminants is not as critical as in solid-state bonding (described in Section 12.4). Resistance welding processes require specialized machinery, but they have the major advantages of not requiring consumable electrodes, shielding gases, or flux.

Spot Welding (RSW).   In this process, the tips of two opposing electrodes contact the lap joint of two sheet metals and the resistance heating produces a spot weld (Fig. 12.22). In order to obtain a good bond, pressure is also applied until the current is turned off. The weld nugget formed (Fig. 12.22b) is generally 1/4 to 3/8 in. (6 to 9.5 mm) in diameter, with currents ranging from 3000 to 40,000 A, depending on the material and its thickness.

Welding may be performed by means of single or multiple electrodes, with the required pressure supplied through mechanical or pneumatic means. The power supply is generally alternating current, and the time of current application is typically measured in number of cycles of alternating current.

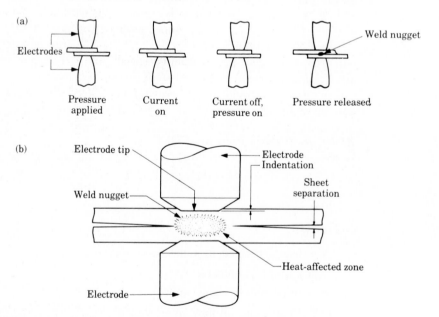

FIGURE 12.22   Resistance spot welding process. This is one of the most common processes used in metal fabrication and automotive body construction. It is readily automated and used widely with industrial robots (Section 13.8).

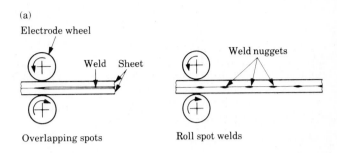

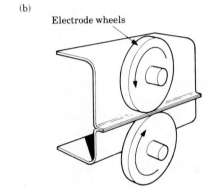

FIGURE 12.23 (a) Seam welding process. The rolls act as electrodes. The spacing of the welds can be controlled by the interrupted current application to the rolls. (b) An application of seam welding at an edge.

Seam Welding (RSEW). This process is a modification of spot welding, whereby the electrodes are replaced by rotating wheels or rollers (Fig. 12.23). With continuous alternating current power supply, the rollers make a spot weld whenever the current reaches a sufficiently high level in the alternating-current cycle. Overlapping spot welds can be made that produce a joint that is liquid- and gas-tight.

With intermittent application of current to the rollers, a series of spot welds can be made along the length of the seam. Welding speeds depend on the type and thickness of the materials. The RSEW process is used in making containers, such as in the longitudinal seam of beverage cans or household product containers (Fig. 12.5c).

Projection Welding (RPW). In this process, the electrical resistance at the joint is obtained by first embossing projections on one of the surfaces (Fig. 12.24) by the techniques described in Chapter 7. The electrodes, which are made of copper-base alloys and are water cooled, are flat. High localized temperatures are generated at the projections in contact with the flat mating part.

As in spot welding, weld nuggets are formed as the projections are compressed under the heat and pressure exerted by the electrodes. Although the embossing process is an added expense to the operation, RPW has the advantages of making a number of welds in one stroke of the welding machine, longer electrode life, and the ability to weld materials of different thicknesses.

FIGURE 12.24 Projection welding. The current is localized at the projections. A number of welds can be made with one stroke of this process, which is also suitable for welding two materials with different thicknesses, as shown in the figure.

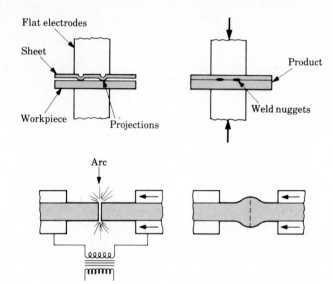

FIGURE 12.25 Flash butt welding. This process is suitable for end-to-end or edge-to-edge welding of parts.

Flash Butt Welding (Flash Welding) [*FW*]. In this process (Fig. 12.25) heat is generated from the arc as the two ends of the parts to be welded begin to make contact, thus developing an electrical resistance at the joint. Because of the presence of the arc, this process is also classified as arc welding. After reaching the proper temperature, pressure is applied at a controlled rate and a weld is formed by the plastic deformation of the joint.

Because impurities and contaminants are squeezed out during this operation, the quality of the weld is very good, although a significant amount of material may be burned off during the process. Later the joint may be machined for improved appearance.

Flash butt welding machines are usually automated, with a variety of power supplies, ranging from 10 to 1500 kVA. The process is suitable for end-to-end, or edge-to-edge, joining of similar or dissimilar metals.

Stud Welding. This process is similar to flash butt welding. The stud (a small part or a threaded rod or hanger) serves as one of the electrodes while it is being joined to another part, such as a flat plate (Fig. 12.26). In order to concentrate the heat, to prevent oxidation of the weld zone, and to retain the molten metal, a disposable ceramic ring (ferrule) is placed at the joint. The equipment for stud welding is automated with various controls for arcing and application of pressure.

Percussion Welding. Whereas the processes described above require a welding transformer for the necessary power requirements, the electrical energy for welding may be stored in a condenser. When the power is discharged in a very short time

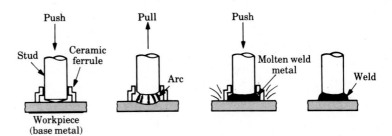

FIGURE 12.26   Stud welding. This process is similar to flash butt welding (Fig. 12.25) and is used for welding bars or threaded rods and fasteners on metal plates.

(usually less than 0.001s), high localized heat is generated at the interface. This process is known as percussion welding. It is suitable for joining dissimilar metals that are not weldable by flash butt welding, or when flash is not desirable at the weld joint.

12.5.3 OPTICAL SOURCES OF ENERGY

Electron-Beam Welding

In electron-beam welding (*EBW*), heat is generated by high-velocity electrons in a narrow beam. The kinetic energy of the electrons is converted into heat as they strike the workpiece. The process requires special equipment (Fig. 12.27) to focus the beam on the workpiece. This is generally done in a high vacuum although it can also be carried out at atmospheric pressure. Almost any metal can be welded by this process, with workpiece thicknesses ranging from foil to plate.

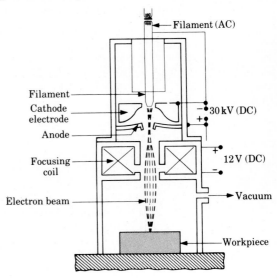

FIGURE 12.27   Schematic illustration of an electron beam gun. High-quality welds are made by this process, with deep and narrow welds.

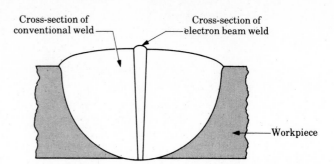

FIGURE 12.28 Weld beads in gas tungsten-arc and electron beam welding. Note the large difference in the size of beads.

The advantages of EBW (developed in the 1960s) are its capability to make excellent quality welds that are deep and narrow (Fig. 12.28) with small heat-affected zones (see Section 12.7). Very good control of parameters can be obtained, with welding speeds as high as 500 in./min (200 mm/s). Generally, no shielding gas or filler metal is required. Welds with depth:width ratios as high as 30:1 are possible with EBW.

Because of the nature of the process, welding can be done in otherwise inaccessible locations. Also, the low energy input produces welds with minimum shrinkage and distortion. However, the extremely high depth:width ratio of these welds can lead to centerline cracking (see Section 12.8). This can usually be eliminated by precise control of the welding parameters. (See also Section 9.9.)

Laser-Beam Welding

Laser-beam welding (LBW) uses a focused high-power coherent monochromatic light beam as the source of heat (Fig. 12.29). The beam has deep penetrating power and can be directed, shaped, and focused precisely. It is particularly suitable for narrow and deep joints. A major advantage of LBW is that, because the beam can be transmitted through the air, a vacuum is not required.

This process has been used successfully, with good weld quality, on a variety of materials with a wide range of thicknesses. It is ideally suited to automation, but

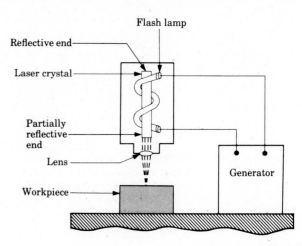

FIGURE 12.29 Schematic illustration of a laser beam gun. An advantage of this process is that a vacuum environment is not required, as it is in electron beam welding. Laser is an acronym for Light Amplification by Stimulated Emission of Radiation.

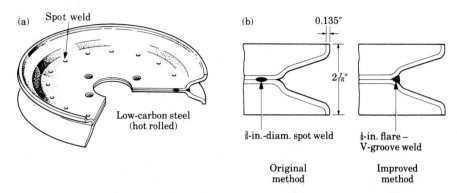

FIGURE P12.2   *Source:* Adapted from *Metals Handbook*, 8 ed., Vol. 6, 1971.

requires that the joint fit properly before welding. Depth:width ratios similar to those obtained with the EBW process are developed. The propensity for possible centerline cracking (Section 12.8) also exists in LBW. A major difference from EBW is that LBW does not generate x-rays. (See also Section 9.9.)

● **Illustrative Example 12.2**

A metal sheave (Fig. P12.2) consists of two matching pieces of hot-rolled low-carbon steel sheets 0.135 in. (3.4 mm) in thickness. The sheave is 20 in. (508 mm) in diameter and is used as a belt pulley. The original method of joining the two halves was by resistance spot welding, with 16 welds equally spaced around the periphery. Although the weld quality was satisfactory, the welding time per sheave was 1 minute. In order to increase production rate, an alternative process was chosen (gas metal-arc welding, GMAW) with a continuous weld around the periphery of the sheave as shown in the figure. With an automated welding process, the welding time per sheave was reduced to 40 seconds. ●

12.6 LIQUID–SOLID-STATE BONDING

In these processes, also known as *capillary joining*, the bond between the two surfaces to be joined is obtained by using a filler metal. The two basic processes are brazing and soldering.

12.6.1 BRAZE WELDING AND BRAZING

Brazing is a joining process where a filler metal is placed at the joint and the temperature is raised to melt the filler metal. Upon cooling a joint is obtained, the strength of which depends on the adhesion at the filler metal–workpiece interfaces. Filler metals that melt above 840°F (450°C) are used in brazing, whereas those that melt below this temperature are used in soldering (Section 12.6.2).

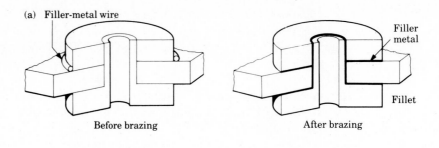

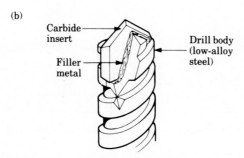

FIGURE 12.30 Examples of brazing, first used during the period of 3000–2000 B.C. (a) An assembly of parts where the filler metal wire has melted and, by surface tension, penetrated the interfaces between the two pieces. (b) Tungsten carbide insert attached to a drill by brazing. Masonry drills are made by this method. The hardness and abrasive resistance of the carbide tip is essential in drilling bricks, concrete, and similarly hard, brittle, and abrasive materials. *Source:* Adapted from *Metals Handbook*, 8th ed., vol. 6, 1971.

In brazing (first used as far back as 3000–2000 B.C.) the temperatures employed are below the melting point of the workpiece material (solidus temperature). This is unlike the liquid-state welding processes, where the workpieces in the weld area must melt for fusion to occur.

In *braze welding* (Fig. 12.30a), the joint is prepared as in liquid-state welding and the filler metal is deposited in the joint. After cooling, a solid joint is obtained. In *brazing*, the joint has little or no clearance and the filler metal, upon heating, fills the joint by capillary action. In both of these processes, to obtain maximum bond strength it is essential that the surfaces to be joined are clean and free of rust, oil, and other contaminants. In brazing, the clearances must fit within a very small tolerance range. The heat source in braze welding and brazing is generally an oxyacetylene torch.

Filler Metals

Filler metals used for brazing are shown in Table 12.1. Various other elements may also be added to impart special properties. The choice of the filler metal and other elements in it is important in order to avoid embrittlement of the joint by grain-boundary penetration of liquid metal (see Section 3.3.1). Brittle intermetallic compounds may also form, thus indicating the necessity of studying the relevant phase diagrams. Galvanic corrosion may also take place in braze joints. Filler metals are available in a variety of shapes, such as wire, sheet, rings, and filings.

Fluxes

The use of a flux is essential in brazing and braze welding in order to prevent oxidation, remove oxides, and reduce fuming. Brazing flux is generally made of borax, boric acid, borates, fluorides, or chlorides. Wetting agents may also be added to increase wetting characteristics of the molten filler metal and improve capillary action.

Applications

Although it can be used in various fabrication applications, the principal use of braze welding is in maintenance and repair of parts such as ferrous castings. However, brazing is primarily used as a joining process in manufacturing.

Examples of brazing are shown in Figs. 8.35(b) and 12.30. The filler metal fills the interface between two components by capillary action. The heating methods that are used identify the various brazing processes: torch brazing, furnace brazing, induction brazing, resistance brazing, dip brazing, infrared brazing, and diffusion brazing.

The torch in brazing is usually an oxyacetylene flame. Furnaces may be batch- or continuous-type for high production rates, especially for small parts. Furnace brazing has the advantage of enabling the brazing of complex shapes since the whole assembly is heated in the furnace. Induction and resistance heating have the high-frequency alternating current and electric-resistance features, respectively. Dip brazing is done in a molten salt or metal bath, where the salt acts as the flux. Infrared heating is done with a high-intensity quartz lamp; the process is particularly suitable for thin parts.

TABLE 12.1
TYPICAL FILLER METALS FOR BRAZING VARIOUS METALS AND ALLOYS

BASE METAL	FILLER METAL	BRAZING TEMPERATURE, °F
Aluminum and its alloys	Aluminum–silicon	1060–1150
Magnesium alloys	Magnesium–aluminum	1080–1160
Copper and its alloys	Copper–phosphorus	1300–1700
Ferrous and nonferrous (except aluminum and magnesium)	Silver, copper, copper–phosphorus	1150–2100
Iron-, nickel-, and cobalt-base alloys	Gold	1650–2000
Stainless steels, nickel- and cobalt-base alloys	Nickel	1700–2200

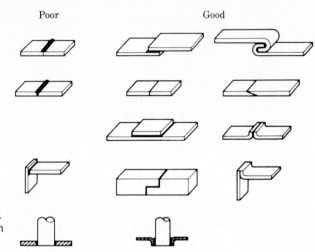

FIGURE 12.31 Examples of joints for brazing. Note that good design requires thin joints with large contact areas. (See also Fig. 12.46.)

For stronger lap or butt joints, and difficult operations, *diffusion brazing* is done in a furnace, where the filler metal diffuses into the workpiece with proper control of temperature and time (which may range from $\frac{1}{2}$ hour to 24 hours).

A number of brazing filler metals are available for different metals as shown in Table 12.1. Joint design is important. As seen in Fig. 12.31, strong joints require greater contact area for brazing (see also Fig. 12.46).

12.6.2 SOLDERING

In soldering, the filler metal (called solder) melts generally below 840°F (450°C). As in brazing, the solder fills the joint by capillary action, using similar techniques involving the same kind of heat sources.

The type of joint best suited for soldering is the lap joint, which exhibits some shear strength because of the larger contact area. For improved joint properties and special applications, other solder compositions that can be used are: tin–zinc, lead–silver, cadmium–silver, and zinc–aluminum alloys (Table 12.2). Soldering with copper–gold and tin–lead alloys was first practiced as far back as 4000–3000 B.C. Fluxes are generally inorganic or organic acids or organic resins. A common solution

TABLE 12.2
TYPICAL SOLDERS AND THEIR APPLICATION

Tin–lead	General purpose
Tin–zinc	Aluminum
Lead–silver	Strength at higher than room temperature
Cadmium–silver	Strength at high temperatures
Zinc–aluminum	Aluminum; corrosion resistance

is zinc ammonium chloride. Because of their chemical reactivity and corrosiveness, fluxes should be selected properly for each application. After soldering, fluxes should be removed thoroughly by washing with water to avoid corrosion. Organic resins should be used to avoid corrosion in electrical joints.

Soldering temperatures are generally in the range of 360 to 475°F (180 to 245°C). Hence, a soldered joint has limited usefulness for service at elevated temperatures. Also, because solders are usually tin–lead alloys (in proportions ranging from 5% Pb–95% Sn to 70% Pb–30% Sn), soldering is generally not used where structural strength is essential. The process is used extensively in the electronics industry, with automated equipment, and in manufacturing containers for liquid- or air-tight joints (Fig. 12.5b).

In *ultrasonic soldering*, the molten solder is subjected to ultrasonic cavitation by means of a transducer, which removes the oxide layer from the surfaces to be joined. The need for a flux is thus eliminated.

12.7 THE METALLURGY OF THE WELDED JOINT

A typical fusion weld joint is shown in Fig. 12.32, where three distinct zones can be identified:

a. Base metal (metal to be welded),

b. Heat-affected zone, and

c. Weld metal.

The metallurgy and properties of the second and third zones depend strongly on the materials used, the welding process, and the process variables. The metallurgical structure of the joint also depends on whether or not a filler metal is used in

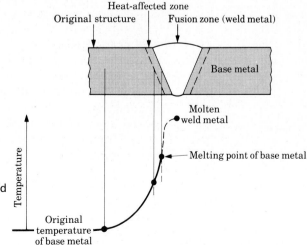

FIGURE 12.32 Characteristics of a typical weld zone in arc and gas welds. Note the three distinct zones: original structure of the base metal, heat-affected zone, and the weld metal.

welding. A joint obtained without a filler metal is called *autogenous*; the weld zone is composed of the molten and resolidified base metal.

A joint made with the common method of using a filler metal has a central zone called the *weld metal*, i.e., the region that has melted during welding. The weld metal is composed of a mixture of the base and weld metals. The *heat-affected zone* (*HAZ*) is in the base metal itself. It has a grain structure different from the base metal due to its being subjected to elevated temperatures for a period of time during welding. Because the base metal is far enough from the heat source, it does not undergo any changes as a result of the welding process.

The study of a weld joint thus requires an understanding of solidification of metals (Chapter 5) and phase diagrams described in textbooks on materials science. Because of the many variables involved, this is a complex area and only a brief, introductory description is given below.

Weld-Metal Solidification

After the application of heat and the introduction of a filler metal, the molten weld joint is allowed to cool. The solidification process begins with the formation of *columnar grains* (*dendritic*), which are relatively long and acquire a directional parallel to the heat flow. Thus, because metals are much better heat conductors than air, the grains are parallel to the plane of the two plates or sheets being welded (Fig. 12.33a). For a shallow weld, the grains are as shown in Figs. 12.33(b) and 12.34. The grain structure and size depend on the specific alloy, the welding process, the filler metal used, and the application (if any) of vibration during solidification to improve weld quality.

The weld metal is basically a cast structure and generally has a coarse grain. Usually this structure has low toughness, tensile strength, and ductility. However, with selection of proper weld-metal chemistry, or by post-welding heat treatments, the mechanical properties can be modified and enhanced. The results depend on the particular alloy, its composition, and thermal cycling. Cooling rates may also be reduced by preheating the weld area prior to welding.

Heat-Affected Zone (HAZ)

This is a zone between the weld metal and the base metal. It is a part of the base metal and, having been subjected to thermal cycling, its properties and structure depend on the rates of heat input and cooling, and the temperature to which this zone was raised. Thus, in addition to the metallurgical characteristics, the original grain size, orientation, and degree of cold work, physical properties such as specific heat and thermal conductivity are also important factors in determining the size and characteristics of the heat-affected zone.

FIGURE 12.33 Grain structure in (a) deep and (b) shallow welds. Note that the grains in the weld metal are perpendicular to the surface of the base metal. See also Figs. 5.4 and 12.34(a).

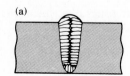
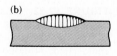

(a)

(b)

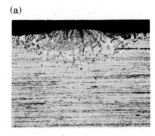

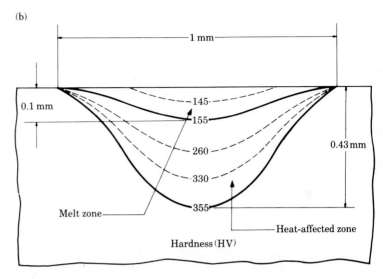

FIGURE 12.34 (a) Weld bead produced by a laser beam on cold-rolled nickel strip. (b) Microhardness profile of the weld bead shown in Fig. 12.34(a). Note the softer condition of the weld bead compared to the base metal. *Source:* Courtesy of S. Rajagopal, IIT Research Institute.

One of the most important considerations, as far as the strength of HAZ is concerned, is how the original strength and hardness of the particular alloy has been obtained. This may have been by cold working, solid-solution strengthening, precipitation hardening, or by thermal transformations.

Of these mechanisms, the simplest to analyze is the base metal that has been cold worked, such as by rolling. The heat of welding recrystallizes the thin, elongated grains of the cold-worked structure (see Figs. 3.13 and 3.18). The grains close to the bulk of the workpiece (base metal) will recrystallize into fine, equiaxed grains. However, those grains that are close to the weld metal, having been subjected to elevated temperatures for a longer period of time, will undergo grain growth, resulting in a softer and lower-strength region. Such a weld will thus be weakest in its heat-affected zone. The effects of the welding heat on HAZ of alloys subjected to other strengthening mechanisms prior to welding are more complex and are beyond the scope of this text.

12.8 WELD QUALITY: DEFECTS, RESIDUAL STRESSES, DISTORTION, AND TREATMENT OF WELDS

Because of its history of thermal cycling and attendant metallurgical changes (with severe gradients within the weld zone), a welded joint may develop imperfections and discontinuities. These are outlined below.

Porosity

Porosity is due to trapped air during welding or gases released either during solidi-fication, or from chemical reactions between various elements during welding. Most welded joints have some porosity, generally spherical in shape or in the form of elongated pockets. The distribution of porosity may be either random or concentrated in a certain region in the weld zone. Porosity can be controlled by deoxidizing elements, by having sufficient fluidity in the molten weld metal to allow the gases to escape, and by applying vibration during weld metal solidification. Porosity is generally harmless unless cracks are present, the material is brittle, or fatigue loading is involved. (See also Section 5.2.5 and Fig. 12.36.)

Slag Inclusions

These are materials, such as oxides, fluxes, and electrode-coating materials, that are trapped in the weld zone. If shielding gases are not effective, contamination from the environment also contributes to these inclusions. In multiple-layer welding opera-tions, it is essential that the weld bead surfaces be cleaned of all slag before the next layer is deposited. Welding conditions thus have an important role in slag entrapment.

Incomplete Fusion

Lack of fusion, or incomplete fusion (Fig. 12.35), may be due to a number of causes, such as insufficient temperature rise in the base metal, the presence of oxides, and improper cleaning of the weld area prior to welding. Inadequate penetration results when the welded joint is not deep enough. This is generally due to low heat input and high travel speed during welding.

Weld Profile

Weld profile is important not only because of its effect on the strength of the weld, but also because it can produce incomplete fusion or slag inclusions in multiple-layer welding operations.

Undercutting is a result of the melting away of the base metal and subsequent generation of a groove in the shape of a sharp recess or notch (Fig. 12.36). This recess acts as a stress raiser and reduces the fatigue life, and may lead to early failure of the welded joint.

Cracks

Cracks may be generated in many directions and locations in a weld area, as shown in Fig. 12.37(a). There are many causes for these cracks. Generally, they are due to a combination of temperature gradients, composition of the various regions in the weld area, and segregation of certain elements, such as sulfur, as the solid–liquid boundary moves during solidification from the base metal to the weld metal. Hydrogen can also cause embrittlement and cracking, especially in high-carbon and alloy steel welded joints. Common sources of hydrogen are welding electrodes and water vapor generated during welding.

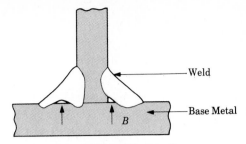

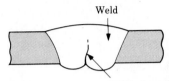

Incomplete fusion in fillet welds. B is often termed 'bridging'

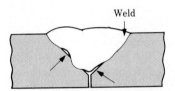

Incomplete fusion from oxide or dross at the center of a joint, especially in aluminum

FIGURE 12.35 Examples of incomplete fusion in welds. Defects reduce the strength of welds. Inspection of welds is therefore important, particularly for critical applications. Some regulations require that every inch of the weld be inspected by various techniques, outlined in Table 4.2. *Source:* Courtesy of the American Welding Society.

Incomplete fusion in a groove weld

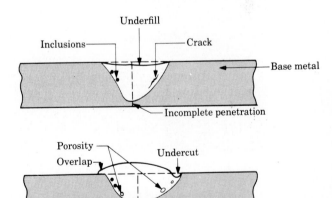

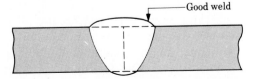

FIGURE 12.36 Examples of fusion-weld defects. Whereas external defects can be observed visually, detection of internal defects requires special techniques.

733

(a)

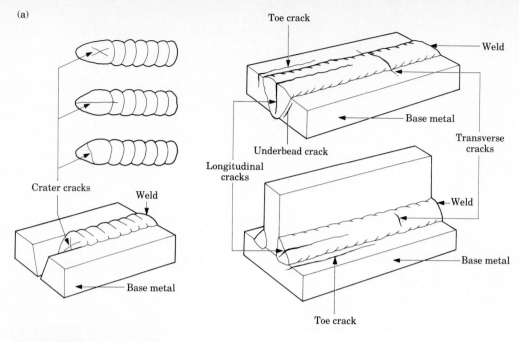

Toe crack

Weld

Base metal

Transverse cracks

Underbead crack

Longitudinal cracks

Weld

Base metal

Toe crack

Crater cracks

Weld

Base metal

(b)

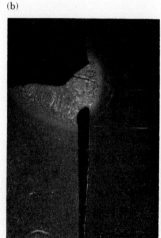

FIGURE 12.37 (a) Types of cracks observed in welded joints. These cracks are due to thermal stresses developed during solidification of the weld bead. (b) Cracks in a weld between two components of a structure. In this example, cracks developed because the weld bead between the two rigidly held surfaces could not shrink freely during cooling. See also Fig. 5.33.
Source: (a) Courtesy of the American Welding Society. (b) Courtesy of S. L. Meiley, Packer Engineering Associates Inc.

The inability of the weld metal to contract during cooling (thermal stresses) is also a cause for cracks (Fig. 12.37b). This situation is similar to hot tears developed in casting (see Fig. 5.33).

Residual Stresses

Because of the localized heating and cooling in welding operations, the uneven expansion and contraction of the weld area causes residual stresses.

The type and distribution of residual stresses in welds can best be described by referring to Fig. 12.38. When two plates are being welded, a long narrow, region *ab*

734

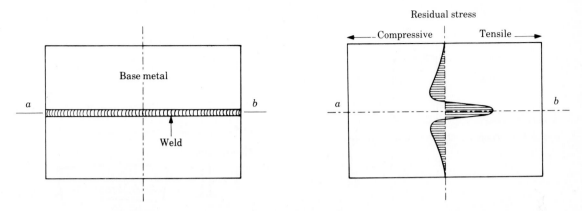

FIGURE 12.38 Longitudinal residual stresses developed in a straight weld between two plates.

is subjected to elevated temperatures, whereas the plates are essentially at ambient temperature. As the weld is completed and a certain time has elapsed, the heat from the weld area is dissipated to the plates while the weld area *ab* is beginning to cool. The plates thus begin to expand longitudinally while the weld length *ab* begins to contract due to cooling.

These two opposite effects cause thermal stresses distributed as shown in Fig. 12.38. The magnitude of the compressive residual stresses in the plates diminishes to zero at a point away from the weld area. Since there are no external forces acting on the welded plates, the tensile and compressive forces due to these stresses must balance each other. (See also Section 2.7.)

In complex welded structures, the residual-stress distributions are three-dimensional and difficult to analyze. The example described above involves two plates that are not restrained; in other words, the plates are not an integral part of a larger structure. If they are restrained, reaction stresses are generated because of the fact that the boundaries of these two plates are not free to expand or contract as they are in Fig. 12.38. This situation arises particularly in structures with high stiffness since they do not allow free movement of the members.

Residual stresses can cause:

a. Distortion, warping, and buckling of the welded parts

b. Stress–corrosion cracking (see Section 3.7.1)

c. Further distortion if a part of the welded structure is removed, such as by machining (see Fig. 2.42).

Examples of warping are shown in Fig. 12.39. Whether or not residual stresses have a significant effect on the fatigue life of a welded joint is somewhat controversial. However, it is to be expected that fatigue life is likely to decrease because of the presence of possible defects and discontinuities, as described above.

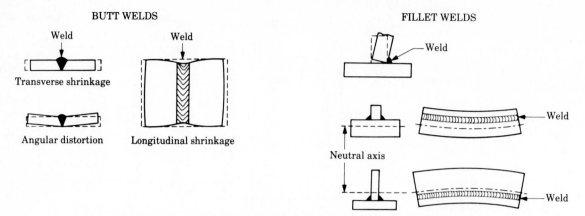

FIGURE 12.39 Distortion of components after welding, due to differential thermal expansion and contraction of different parts of the welded assembly. Warping can be reduced or eliminated by holding the pieces firmly with clamps or similar fixtures during welding and subsequent cooling to room temperature.

Stress Relieving

The problems caused by residual stresses can best be avoided by stress relieving the welded structure at an appropriate temperature and for a sufficient period of time. The temperature and time required depend on the type of material and magnitude of the residual stresses. They must be chosen properly in order to maintain the properties of the base metal so as not to reduce its stength or toughness.

For structures that are too large to be stress relieved in a furnace, the technique of in situ stress relieving may be employed. In this case, the structure such as a vessel is insulated and the entire structure is used as a furnace to bring it to the proper temperature. The technique of a partial stress relief may also be employed. In this case, only the areas immediately surrounding the welds are subjected to the stress relief cycle. (See also Section 3.5.)

Stress relieving may also be done by vibratory means, where the welded part is vibrated at one of its resonant frequencies. This is a relatively new technique and its results are controversial. The metallurgical structure and hardness of the weld area are not affected by the process. (This process should not be confused with the vibratory techniques used during welding to improve weld quality.)

Another method of stress relieving is to peen the weld bead area or to roll it (Section 4.13.2). This is a plastic deformation process and induces compressive residual stresses, thus reducing the overall tensile residual stresses in the weld. For multilayer welds, it is recommended that the first and last layers not be peened in order to protect these layers against any damage due to the peening action.

It was shown in Section 2.7.2 that residual stresses can be relieved, or reduced, by a small amount of plastic deformation of the part. This technique can be used in some welded structures, such as pressure vessels, by means of proof stressing them by pressurizing the vessels hydrostatically. To reduce the possibility of fracture under high internal pressures, it is important to make sure that the weld is properly made and is free of notches and discontinuities.

Residual stresses and distortion can also be avoided by preheating the base metal or the parts to be welded. This improves weldability by reducing the cooling rate and the level of thermal stresses (by reducing the elastic modulus), and also reduces shrinkage and cracking. Heating may be done either in a furnace, or by electrical or induction heating, and, for thin sections, by radiant lamps or hot-air blast. For optimal results, preheating temperatures and cooling rates must be controlled carefully in order to maintain acceptable strength and toughness of the welded structure.

Although the treatments described above pertain to fusion welding, control of residual stresses and warpage in various other welding processes, such as resistance welding, can be handled in similar ways.

12.9 WELDABILITY

In view of the procedures described thus far, it is apparent that the term *weldability* is difficult to define and specify. A general definition for the weldability of a metal is its capability of being welded into a sound structure with certain properties to meet specific service requirements. The same metal may thus be weldable by one process, yet not so by another. Table 12.3 is given as a general guideline for recommended welding methods for several metals and alloys.

The variables involved in weldability are numerous. For instance, they involve metallurgical considerations of the base metal, such as the alloying elements, impurities, structure, processing history, and presence of gases. Thus, a thorough knowledge of the phase diagrams and response of the metal to elevated temperatures over a period of time is essential. Mechanical and physical properties that have an influence on weldability are: hardness, strength, elastic modulus, toughness, ductility, notch sensitivity, specific heat, melting point, thermal expansion, surface tension characteristics at elevated temperatures, and corrosion.

Surface preparation is important, as are considerations of the nature and properties of surface oxide layers and adsorbed gases. The particular welding process has a significant effect with regard to temperatures involved and their distribution in the weld zone. Other operational factors are shielding, fluxes, and filler metals (including the moisture content of coatings on electrodes), speed of welding, position of welding, rate of cooling, preheating and postwelding techniques, and temperatures involved.

Because of these complex interrelations, it is beyond the scope of this text to treat the weldability of metals and alloys in detail. This information is readily available in various texts on welding as listed in the Bibliography at the end of this chapter.

Standards and codes have been developed for testing and inspection of welded joints. Among nondestructive techniques are visual, radiographic, magnetic-particle,

TABLE 12.3
A GENERAL GUIDE TO JOINING PROCESSES. (Courtesy of American Welding Society.)

JOINING PROCESS

Material	Thickness	SMAW	SAW	GMAW ST	GMAW B	GMAW P	GMAW S	FCAW	GTAW	PAW	ESW	EGW	RW	FW	OFW	DFW	FRW	EBW	LBW	B TB	B FB	B IB	B RB	B DB	B IRB	B DFB	S
Carbon steel	S	×	×			×	×		×				×	×	×		×	×	×	×	×	×	×	×	×	×	×
	I	×	×	×	×	×	×	×	×				×	×	×		×	×	×	×	×	×	×	×		×	×
	M	×	×	×	×	×	×		×				×	×	×		×	×	×	×	×	×				×	
	T	×	×	×	×	×	×		×		×	×	×	×	×					×						×	
Low-alloy steel	S	×	×			×	×		×				×	×	×	×	×	×	×	×	×	×	×	×	×		×
	I	×	×	×		×	×	×	×				×	×			×	×	×	×	×	×	×	×			×
	M	×	×	×			×		×						×		×	×	×	×	×					×	
	T	×	×	×			×		×			×		×	×		×	×	×		×					×	
Stainless steel	S	×	×			×	×		×	×			×	×	×		×	×	×	×	×	×	×	×	×		×
	I	×	×	×		×	×	×	×	×			×	×			×	×	×	×	×	×	×	×	×		×
	M	×	×	×		×		×	×				×		×		×	×	×	×	×	×				×	
	T	×	×	×		×			×	×			×		×		×	×	×		×					×	
Cast iron	I	×													×					×	×	×				×	×
	M	×	×	×					×						×					×	×	×				×	×
	T	×	×	×					×						×					×						×	
Nickel and alloys	S	×				×	×		×	×			×	×	×		×	×	×	×	×	×	×	×	×		×
	I	×	×	×		×	×		×	×			×	×			×	×	×	×	×	×	×			×	×
	M	×	×	×		×			×					×			×	×	×	×	×	×	×			×	
	T	×		×		×				×					×		×	×	×		×					×	
Aluminum and alloys	S			×		×			×	×			×	×	×		×	×	×	×	×	×	×	×	×		×
	I			×		×			×				×	×	×		×	×	×	×	×				×		×
	M			×					×					×	×	×		×	×	×	×	×			×		×
	T			×							×	×	×		×					×						×	
Titanium and alloys	S			×					×	×			×	×		×		×	×	×	×	×				×	×
	I			×		×			×				×			×	×	×	×	×	×						×
	M			×		×			×				×			×	×	×	×	×	×						×
	T			×		×					×	×		×		×		×			×						×
Copper and alloys	S			×					×	×			×				×	×	×	×	×	×	×			×	×
	I			×		×				×						×		×	×	×	×	×	×		×		×
	M			×		×										×		×	×	×	×	×				×	
	T			×		×										×		×	×	×						×	
Magnesium and alloys	S			×			×			×			×				×	×	×	×	×	×		×			×
	I			×		×			×	×			×			×	×	×	×	×	×	×		×			×
	M			×	×					×				×	×	×		×								×	
	T			×		×								×		×										×	
Refractory alloys	S			×					×	×			×	×			×	×	×	×	×	×	×			×	×
	I			×		×				×				×			×		×	×	×	×					×
	M													×													
	T																										

Code

SMAW–Shielded Metal Arc Welding
SAW–Submerged Arc Welding
GMAW–Gas Metal Arc Welding
ST–Spray Transfer
B–Buried Arc
P–Pulsed Arc
S–Short Circuiting Arc
FCAW–Flux Cored Arc Welding
GTAW–Gas Tungsten Arc Welding
PAW–Plasma Arc Welding

ESW–Electroslag Welding
EGW–Electrogas Welding
RW–Resistance Welding
FW–Flash Welding
OFW–Oxyfuel Gas Welding
DFW–Diffusion Welding
FRW–Friction Welding
EBW–Electron Beam Welding
LBW–Laser Beam Welding
B–Brazing
  TB–Torch Brazing
  FB–Furnace Brazing

IB–Induction Brazing
RB–Resistance Brazing
DB–Dip Brazing
IRB–Infrared Brazing
DFB–Diffusion Brazing
S–Soldering

Material Thickness

S–Sheet: up to 3 mm (1/8 in.)
I–Intermediate: 3 to 6 mm (1/8 to 1/4 in.)
M–Medium: 6 to 19 mm (1/4 to 3/4 in.)
T–Thick: 19 mm (3/4 in.) and up

liquid-penetrant, and ultrasonic testing methods (see Table 4.2). Each of these techniques has certain capabilities, sensitivities, limitations, and requirements for special equipment and operator skills.

Destructive testing involves making the welds and then testing them to failure in tension, bending, impact, fatigue, and tearing. Extensive standards and codes are available from *ASTM*, *AWS*, *ASME*, *ASCE*, *ANSI*, and various agencies of the federal government.

12.10 JOINT DESIGN

Five basic joint designs that are commonly used in welding are shown in Fig. 12.40. As can be seen from this figure, different types of welds can be made on the same type of joint. When selecting a joint, the following considerations must be kept in mind.

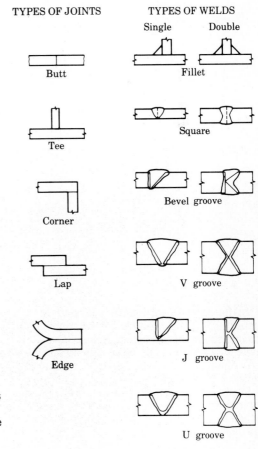

FIGURE 12.40 Standard terminology for different types of joints and welds commonly used in welding. The jagged double lines in the figures on the right depict the heat-affected zones in the weld area.

a. Configuration of the part or structure to be welded;

b. The methods and equipment used to provide component parts and service requirements for the welded structure, such as the type of loading;

c. Effects of warping;

d. Appearance;

e. Accessibility and ease of welding;

f. Costs involved, both in joint preparation and in welding.

As in all manufacturing processes, the optimal choice is that which meets all requirements at minimum cost (see Chapter 13). Some examples of welding are shown in Fig. 12.41, indicating the need for careful consideration of the factors outlined above. Detailed treatments of this topic are available in various specialized texts and reference books. The general characteristics of welding processes are given in Table 12.4 to serve as a guide in the selection process.

Since the type of weld and its characteristics must be specified in drawings, various welding symbols have been standardized, as shown in Fig. 12.42.

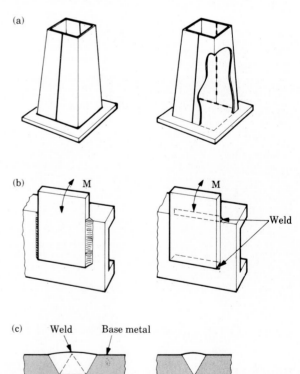

FIGURE 12.41 Examples of weld design selection. (a) Inside welds applied intermittently to this structural member improve appearance, reduce distortion, and conserve weld metal. (b) Although both designs require the same amount of weld material, the design on the right can carry three times the moment *M* of the one on the left. (c) The weld on the left requires about twice the amount of weld material than the design on the right.

TABLE 12.4
GENERAL CHARACTERISTICS OF JOINING PROCESSES

PROCESS	OPERATION	ADVANTAGE	SKILL LEVEL REQUIRED	WELDING POSITION	CURRENT TYPE	WELD DISTOR-TION*	COST OF EQUIP-MENT
SMAW	Manual	Portable and flexible	High	All	AC, DC	1 to 2	Low
SAW	Automatic	High deposition	Medium to low	Flat and horizontal	AC, DC	1 to 2	Medium
GMAW	Semiautomatic or automatic	Most metals	High to low	All	DC	2 to 3	Medium to high
GTAW	Manual or automatic	Most metals	High to low	All	AC, DC	2 to 3	Medium
FCAW	Semiautomatic or automatic	High deposition	High to low	Flat and horizontal	DC	1 to 3	Medium
OFW	Manual	Portable and flexible	High	All	—	2 to 4	Low
EBW	Semiautomatic or automatic	Most metals	Medium to high	All	—	3 to 5	High

* 1, highest; 5, lowest.

Basic arc- and gas-weld symbols							
Bead	Fillet	Plug or slot	Groove				
			Square	V	Bevel	U	J
⌒	◸	▽	‖	⋁	⋁	⋃	⋃

Basic resistance-weld symbols			
Spot	Projection	Seam	Flash or upset
✳	✕	✕✕✕	│

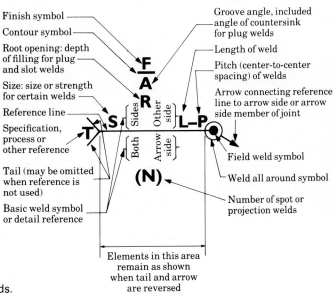

FIGURE 12.42 Standard identification and symbols for welds.

741

12.11 CUTTING OF METALS

The separation of a piece of metal into two or more pieces can be accomplished by removing a volume of material by nonmechanical means. The sources of heat described earlier in this chapter can be used for this purpose, namely, chemical, electrical, and optical.

Torch Cutting

The chemical sources of heat involve gaseous fuels, such as oxygen and acetylene or hydrogen in a torch. This is similar to oxyfuel welding, with the exception that the heat source is now used to melt a narrow path on a metal plate or sheet (Fig. 12.43).

Torch cutting is particularly suitable for steels, where the basic reactions are:

$$Fe + O \longrightarrow FeO + heat, \tag{12.6}$$

$$3Fe + 2O_2 \longrightarrow Fe_3O_4 + heat, \tag{12.7}$$

$$4Fe + 3O_2 \longrightarrow 2Fe_2O_3 + heat. \tag{12.8}$$

The highest amount of heat is generated from the second reaction above. This reaction requires a temperature of about 1600°F (870°C) to be initiated. Hence, the steel is preheated with oxyfuel, and oxygen is introduced later. The heat of reaction oxidizes and melts away the steel, generating a kerf. The higher the carbon content of the steel, the higher the preheating required. Torch cutting is most suitable for steels, although cast irons and steel castings can also be cut by this method.

The maximum thickness of the material that can be cut depends mainly on the gases used. With oxyacetylene, the maximum thickness is about 12 to 14 in. (300 to 350 mm) and with oxyhydrogen 24 in. (600 mm). Kerf widths range from about 1/16 to 3/8 in. (1.6 to 9.5 mm) with reasonably good control of tolerances. Torches may be guided along various paths either manually or mechanically with various program-

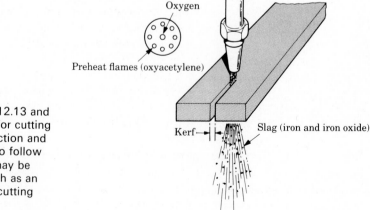

FIGURE 12.43 Flame cutting with oxyacetylene torch (see also Figs. 12.13 and 12.14). This is a common method for cutting thick steel plates for heavy construction and shipbuilding. It can be automated to follow cuts of complex contour. Cutting may be carried out by other techniques such as an arc, laser beam, a saw, or the wire cutting technique shown in Fig. 9.26.

Oxygen

Preheat flames (oxyacetylene)

Kerf

Slag (iron and iron oxide)

mable controllers. Underwater cutting is also done by this process using specially designed torches.

For materials that do not oxidize as readily or with high thermal conductivity, such as stainless steels and copper, iron powders or fluxes are introduced into the flame. These elements increase the temperature of cutting and help remove the oxides in the cutting zone, which otherwise would obstruct the heat transfer from the torch to the workpiece.

Arc Cutting

The electrical source of heat usually involves arc cutting using carbon electrodes, or plasma arc cutting with tungsten electrodes and various gases. A variety of materials can be cut at high speeds by this method.

Another development in this category is air carbon-arc cutting where the molten metal is blown away by a high-velocity air jet. Thus, oxidation of the metal being cut is not a requirement. This process is used especially for gouging and scarfing (i.e., removal of material from a surface).

Plasma arc cutting of steel plates under 3/4 in. (19 mm) thick yields considerably higher cutting productivity than conventional oxyfuel methods and is, therefore, the fastest-growing cutting process used in manufacturing for thinner sections.

The optical methods of cutting involve electron beams and lasers. For a brief description of these processes see Sections 9.9 and 12.5.3.

12.12 ADHESIVE BONDING

There are many metallic and nonmetallic components where joining and assembly can be done with the use of an adhesive. Adhesive bonding has been gaining increased acceptance since its first use on a large scale in assembling load-bearing components in military aircraft during World War II.

Adhesive bonding is used extensively in major industries such as aerospace, automotive, household appliances, and building products. It has been a standard method of joining and assembly in many other areas of manufacturing, such as packaging, home furnishings, and bookbinding.

An *adhesive* provides a bond at the interface either for structural strength (load-bearing), or for nonstructural applications, with properties such as sealing, insulating, preventing electrochemical corrosion between dissimilar metals, and reducing vibration through internal damping at the joints. Adhesive bonding has the additional advantage of distributing the load at an interface (because of the inherent plasticity of the adhesive), thus eliminating localized stresses that generally result from mechanical fastening or welding.

Moreover, structural integrity of the sections is maintained since no holes are required, and the appearance of the components is generally improved. Very thin sections can be bonded and, since the operation is generally carried out between room temperature and about 400°F (200°C), there is no significant distortion of the

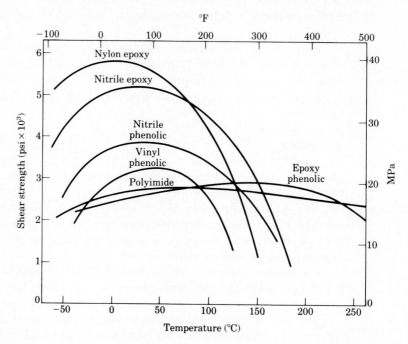

FIGURE 12.44 Shear strength for various adhesives as a function of temperature. (See Chapter 10 for characteristics of these materials.) *Source:* Aerospace Structural Adhesives, Pub. NMAB 300, National Academy of Sciences, Washington, D.C., July 1974.

parts or change in their original properties. A great variety of adhesives are available that provide adequate joint strength, including fatigue strength (Fig. 12.44 and Tables 12.5 and 12.6).

12.12.1 ADHESIVES

There are three basic types of adhesives:

a. Natural Products: starch, dextrin, soya flour, and animal products (commonly referred to as glues and first used around 1000–1 B.C.)

b. Inorganic Adhesives: sodium silicate and magnesium oxychloride

c. Synthetic Organic Adhesives: thermoplastics (used for nonstructural and some structural bonding) or thermosetting polymers (used primarily for structural bonding). Because of their strength, these adhesives are the most important for manufacturing processes.

Synthetic organic adhesives are formulated and classified into categories such as chemically reactive, pressure-sensitive, hot-melt, evaporative, and film adhesives.

TABLE 12.5

CHARACTERISTICS OF VARIOUS EVAPORATION-TYPE ADHESIVES. (From *Production Design Guide for Adhesive Bonding of Sheet Steel*, **AISI.)**

ADHESIVE	ADVANTAGES	LIMITATIONS	APPLICATIONS
Acrylics	Stable in sunlight; some flexible at low temperatures; good solvent resistance	Somewhat higher in cost	Bonding of cloth, plastics, and some metal foils
Butyl rubber	Resistance to water; low gas permeability	Low strength; tends to creep under load	Sealants; packaging applications
Natural rubber	Good tack, flexibility, adhesion, and water resistance	Oxidation results in loss of tack, flexibility, and adhesion; also attacked by solvents, oil, and grease	Rubber cements for porous and nonporous substrates, bonds dissimilar materials
Neoprene	Higher strength, better aging, and better heat resistance than natural rubber	—	Shoe adhesives; for structural bonding when combined with phenolic resin and curing agent
Nitrile-phenolic	Produces stronger bonds than nitrile	Heat and pressure required for strong bonds	Bonding abrasives, brake shoes; for laminating some metals
Nitrile rubber	Good resistance to oil, grease and water	—	Bonding vinyl, elastomers, and fabrics
Phenolics	High strength with modifications by elastomers or thermoplastic resins; good water resistance	Heat and pressure required for structural bonding	Bonding of plywood and particle board; used in structural adhesives
Phenoxy	Temperature resistant	Heat and pressure required for bonding	Bonding of metal, wood, and various nonmetallics
Polyimides (modified with nylon)	Good low- and high-temperature properties; good resistance to oils	Poor resistance to solvents	Bonding of metals
Reclaim rubber	Lower cost; usually blended with other rubbers	Lower elongation and tensile properties than natural rubber	Mastics and sealants
Silicone	Bonds wide variety of materials; adhesion at temperature extremes; good peel	Low cohesive strength unless cured with peroxide or amine catalyst; high cost	Bonding of dissimilar materials, plastics to metal and glass
Styrenebutadiene rubber (SBR)	Low cost	Less tack and film strength than natural rubber unless blended with reinforcing agents and resins	Bonding plastics, rubber, wood, and fabrics
Urethanes	Cure at room temperature in presence of moisture	—	Bonding of rubbers to fibers and metal; bonding of wood assemblies
Vinyls	Lower cost than acrylics	Unstable in sunlight	Bonding of many materials depending on base polymers. Bonding of plastics to metal, vinyl, and wood

TABLE 12.6
CHARACTERISTICS OF VARIOUS MODIFIED EPOXY ADHESIVES.
(From *Production Design Guide for Adhesive Bonding of Sheet Steel,* AISI.)

ADHESIVE	ADVANTAGES	LIMITATIONS
Epoxy-phenolics	Good heat resistance	Somewhat brittle; low peel strength
Nylon-epoxies	Strong; tough, high peel strength	May not be resistant to water
Nitrile-epoxies	Low-temperature curing; good resistance to water; good properties at low temperatures	Lower tensile, shear, and peel strength than nylon-epoxies

They are available in various forms, such as liquids, pastes, solutions, emulsions, powder, tape, and film.

The behavior of an adhesive during peeling may be brittle, or it may be ductile and tough, thus requiring higher forces to peel (Fig. 12.45). In addition to the type of adhesive used, joint design is also important in adhesive bonding (Fig. 12.46). In order to select an appropriate adhesive, all requirements of the component in regard to strength, service temperature, resistance to various fluids, chemical and environmental attack, and type of loading must be considered.

Surface Preparation and Application

Surface preparation is very important in adhesive bonding, because joint strength is affected by the presence of dirt, dust, oil, moisture, and various other contaminants. Contaminants also affect the wetting ability of the adhesive. Thick, weak oxide layers are detrimental, although a thin, strong, and adherent oxide layer may be desirable, particularly one with surface roughness to improve adhesion by mechanical interlocking.

Adhesive bonding can also be used in conjunction with other mechanical fastening methods to improve the strength of the bond (see Fig. 12.46). A variety of similar and dissimilar metallic and nonmetallic materials, with different thicknesses, can be bonded to each other by adhesives.

Adhesive bonding has certain limitations. Joint design and bonding methods require care and skill. Special equipment may be necessary such as fixtures, presses, and tooling, in addition to autoclaves and ovens for curing.

Curing cycles can range from 30 minutes to many days, depending on the adhesive; thus, production rates are low. Inspection of the quality and strength of the bond can be difficult (see Table 4.2). Service limitations can be important with regard to temperature, moisture, and other environmental conditions. Adhesive bonds for structural applications are rarely suitable for service above 500°F (260°C). The cost of the adhesive can be high; however, there are many applications where the overall economics makes adhesive bonding an attractive alternative to other fastening and joining processes.

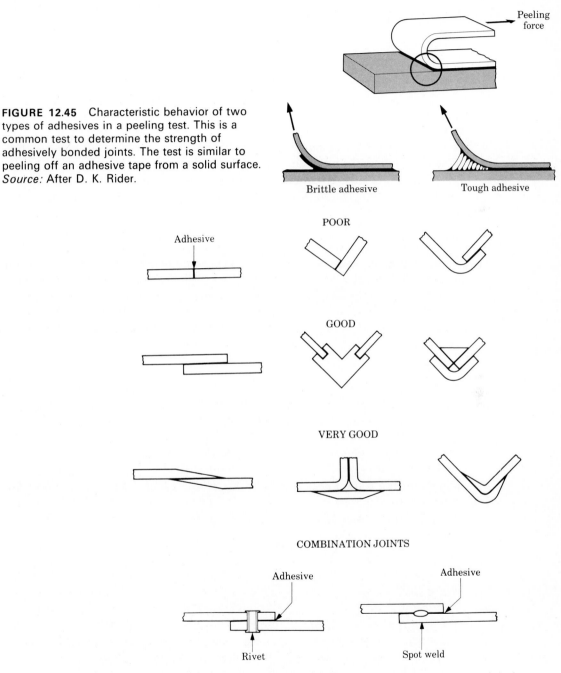

FIGURE 12.45 Characteristic behavior of two types of adhesives in a peeling test. This is a common test to determine the strength of adhesively bonded joints. The test is similar to peeling off an adhesive tape from a solid surface. *Source:* After D. K. Rider.

FIGURE 12.46 Various joint designs in adhesive bonding. Note that good design requires large contact areas between various members to be joined. Compare with Fig. 12.31. *Source:* After H. A. Perry.

12.13 JOINING OF PLASTICS

As described in Chapter 10, the two basic types of polymers are thermoplastic and thermosetting. Since thermoplastics soften and melt as temperature is increased, it is possible to join them by some of the techniques described earlier in this chapter.

The heat source can be hot air or other gases. The heat melts the joint and, with the application of pressure to ensure a good bond, allows fusion to take place at the interface. Plastics such as polyvinyl chloride, polyethylene, polypropylene, acrylics, and ABS (acrylonitrile-butadiene-styrene) can be joined in this manner. Filler materials of the same type may also be used.

Because of oxidation problems with such materials as polyethylene, an inert gas such as nitrogen may also be used. Also, because of the low thermal conductivity of polymers, the heat source may burn or char the surfaces, causing possible difficulty in obtaining sufficiently deep fusion. Other heating methods may also be employed. Ultrasonic- and friction-welding processes can also be used successfully. Thermoplastics can also be joined with the use of adhesives and solvents.

Because they do not soften or melt with increasing temperature, thermosetting plastics such as epoxy and phenolics are joined by adhesive bonding techniques. Generally, the process consists of roughening the surfaces with an abrasive and then wiping it with a solvent and pressing the surfaces together.

SUMMARY

1. Fastening and joining methods are important manufacturing steps because almost all products are an assemblage of parts made as separate units. These operations involve mechanical methods using fasteners, as well as a wide variety of welding and adhesive bonding processes.

2. The basic sources of energy for welding are mechanical, electrical, chemical, optical, and diffusion. From the earliest applications, such as forge welding, many joining processes have been developed for a wide variety of general or specialized applications.

3. Solid-state bonding utilizes pressure at the interface, with or without relative motion between the two bodies being joined. Examples are cold, explosive, ultrasonic, and friction welding, and diffusion bonding.

4. In liquid-state–welding processes the sources of energy are chemical (gas welding), electrical (arc and resistance welding), and optical (electron-beam and laser-beam welding). Liquid–solid-state bonding processes involve brazing and soldering. A welding torch or beam can also be used to cut materials.

5. The metallurgy of the welded joint is an important aspect in all welding processes as it determines the strength and toughness of the joint. The joint is composed of solidified weld metal and a heat-affected zone, with a wide variation in structure and properties.

6. Joint design, as well as surface preparation, protective atmospheres, and the quality of the weld, are important additional considerations. Because of thermal gradients in the weld area, distortion, cracking, and residual stresses can be a significant problem in welding.

7. Adhesive bonding has gained increased acceptance in major industries. In addition to good bond strength, adhesives have other characteristics, such as sealing, insulating, preventing electrochemical corrosion between dissimilar metals, and reducing vibration through internal damping within the bond. Surface preparation and joint design are important factors.

BIBLIOGRAPHY

Allen BM. *Soldering Handbook*. London: Iliffe, 1969.

Cary HB. *Modern Welding Technology*. Englewood Cliffs, N.J.: Prentice-Hall, 1979.

Davies AC. *The Science and Practice of Welding*, 7th ed. New York: Cambridge, 1977.

Fastener Standards, 5th ed. Cleveland: Industrial Fasteners Institute, 1970.

Giachino JW, Weeks W, Johnson GS. *Welding Technology*. Chicago: American Technical Society, 1968.

Gray TGF, Spence J. *Rational Welding Design*, 2d ed. Woburn, Mass.: Butterworths, 1982.

Gurney TR. *Fatigue of Welded Structures*. New York: Cambridge, 1968.

Harman RC, ed. *Handbook for Welding Design*. London: Pitman, 1967.

Houldcroft PT. *Welding Processes*. London: Cambridge, 1967.

Houldcroft PT. *Welding Process Technology*. New York: Cambridge, 1977.

Koenigsberger F, Adair JR. *Welding Technology*. New York: Hart, 1968.

Lancaster JF. *Metallurgy of Welding*, 3d ed. Allen and Unwin, 1980.

Laughner VH, Haugan AD. *Handbook of Fastening and Joining Metal Parts*, New York: McGraw-Hill, 1956.

Lindberg RA, Braton NR. *Welding and Other Joining Processes*. Boston: Allyn and Bacon, 1976.

Linnert CE. *Welding Metallurgy, Carbon and Alloy Steels*, 3d ed. New York: American Welding Society, 1965.

Manko HH. *Solders and Soldering*, 2d ed. New York: McGraw-Hill, 1979.

Masubushi K. *Analysis of Welded Structures—Residual Stresses and Distortion and Their Consequences*. New York: Pergamon, 1980.

Metals Handbook, 8th ed. Vol. 6, *Welding and Brazing*. Metals Park, Ohio: American Society for Metals, 1971.

Moss JB. *Science and Metallurgy for Students of Welding*. London: Butterworths, 1967.

Parmley RO. *Standard Handbook of Fastening and Joining*. New York: McGraw-Hill, 1977.

Patton WJ. *The Science and Practice of Welding*. Englewood Cliffs, N.J.: Prentice-Hall, 1967.

Phillips AL, ed. *Current Welding Processes*. New York: American Welding Society, 1964.

Phillips AL. *Modern Joining Processes*. Miami: American Welding Society, 1966.

Roberts PM. *Brazing*. New York: Oxford, 1975.

Romans D, Simons EN. *Welding Processes and Technology*. London: Pitman, 1974.

Soled J. *Fasteners Handbook*. New York: Reinhold, 1957.

Source Book on Brazing and Brazing Technology. Metals Park, Ohio: American Society for Metal, 1980.

Source Book on Electron Beam and Laser Beam Welding. Metals Park, Ohio: American Society for Metals, 1981.

Source Book on Innovative Welding Processes. Metals Park, Ohio: American Society for Metals, 1981.

Thwaites CJ. *Capillary Joining—Brazing and Soft-Soldering*. New York: Wiley, 1982.

Tylecote RF. *The Solid Phase Welding of Metals*. New York: St. Martin's, 1968.

Vill VI. *Friction Welding of Metals*. New York: American Welding Society, 1966.

Welding Handbook, 7th ed. (5 vols.). Miami: American Welding Society, 1976 on.

Adhesives

Adhesives in Modern Manufacturing. Dearborn, Mich.: Society of Manufacturing Engineers, 1970.

Aitken DF, ed. *Engineer's Handbook of Adhesives*. Brighton, England: Machinery Publishing Company, 1972.

Bikales NM. *Adhesion and Bonding*. New York: Wiley, 1971.

Bikerman JJ. *The Science of Adhesive Joints*. New York: Academic Press, 1968.

Cagle CV. *Adhesive Bonding Techniques and Applications*. New York: McGraw-Hill, 1968.

Cagle CV, ed. *Handbook of Adhesive Bonding*. New York: McGraw-Hill, 1973.

Delmonte J. *The Technology of Adhesives*. New York: Hafner Press, 1965.

Guttman WH. *Concise Guide to Structural Adhesives*. New York: Reinhold, 1961.

Houwink R, Salomon G, eds. *Adhesion and Adhesives* (2 vols.). New York: Elsevier, 1965 and 1967.

Katz I. *Adhesive Materials, Their Property and Usage*. Long Beach, Calif.: Foster Publishing Co., 1971.

Koehn CW, ed. *Industrial Adhesives*. Lancaster, Pa.: Armstrong Cork Co., 1959.

Parker RSR, Taylor P. *Adhesion and Adhesives*. New York: Pergamon, 1966.

Patrick RL, ed. *Treatise on Adhesion and Adhesives*. New York: Marcel Dekker, 1967.

Schneberger GL, ed. *Adhesives in Manufacturing*. New York: Marcel Dekker, 1983.

Shields J. *Adhesive Bonding*. New York: Oxford, 1974.

Skeist I, ed. *Handbook of Adhesives*, 2d ed. New York: Van Nostrand Reinhold, 1977.

Periodicals
Manufacturing Engineering

Metal Progress

Welding Journal

Welding Design and Fabrication

PROBLEMS

12.1. Would you classify rivets as permanent or semipermanent fasteners? Describe situations where rivets in a structure or assembly could be removed and riveted again.

12.2. Suggest methods of attaching a round bar (made of thermosetting plastic) perpendicular to a flat metal plate.

12.3. Explain the technique of inducing compressive residual stresses in a punched or drilled hole.

12.4. Make a list of products around your house that have been joined by (a) seaming, (b) seam welding, (c) spot welding, and (d) soldering.

12.5. How would you explain the large difference in the weld beads shown in Fig. 12.28?

12.6. Explain why the joint designs in Fig. 12.31 have been categorized as shown.

12.7. Explain why dendrites form in the particular directions shown in Fig. 12.33.

12.8. List the rules that must be followed to avoid cracking in welds.

12.9. Explain how residual stresses are developed in welded joints.

12.10. Explain how else you would weld the pieces shown in Fig. 12.41(a). If the part is about 1 m high, what other processes could you use to make this part? Make a list of the questions you would raise before making the final decision.

12.11. What is the effect of the thermal conductivity of the metal plate on the width of the kerf produced in Fig. 12.43? What is the effect of the speed of the torch?

12.12. Explain the reasons for the poor, good, and very good categories for the adhesive bonding joints shown in Fig. 12.46.

12.13. From Table 12.3, explain why some processes are not suitable for certain materials.

12.14. Describe various applications of the roll-bonding process shown in Fig. 12.8.

12.15. Explain why spot welding is so commonly used in automotive bodies and home appliances.

12.16. Describe the advantages of adhesive bonding.

13 Manufacturing Economical Products

13.1 INTRODUCTION

In the preceding chapters, the principles and applications of individual processing methods and equipment were described in detail. From this treatment, and from the review in Chapter 1, it can be seen that design, material selection, and manufacturing processes constitute a complex system.

Many factors are involved in a manufacturing system, including human and economic aspects, that are of major significance. The optimization of a manufacturing process, and the production of a part at the lowest overall cost, requires an understanding of the interrelation between many factors.

The design of a product, or a component of a product, and the selection of materials, depend first on service requirements, such as type of loading, temperature, and environment (see Chapter 1). Designs should be simple yet functional, using as many standard and commercially available parts as possible. Tolerances should be as wide as possible, and surface finishes as rough as possible, while meeting functional, appearance, and assembly requirements. The least expensive materials meeting all service requirements should be chosen.

13.2 SELECTION CRITERIA

The selection of the most economical material and manufacturing process depends on the following general factors:

a. Size, shape, service loads, and other functional requirements of the product,

b. Tolerances and surface-finish requirements,

c. Quantity required or lot size, and

d. Various other factors concerning the tooling and equipment, time schedule, and labor costs at the manufacturing facility.

A general review of the variety of processing methods available for all engineering metals and alloys is given in Fig. 13.1.

13.2.1 MATERIALS

The relation between materials and processing methods can be quite involved. Consider, as a simple example, a part that requires some machining and that can be made either from aluminum or steel (both meeting service requirements). The weight of the part is not a significant factor.

Since steel is less expensive than aluminum, the initial decision may well be to make the part from steel. However, the time required to machine steel is approximately three times that for aluminum. The manufacturing cost for steel will thus be higher than that for aluminum (Fig. 13.2). Hence the final decision depends on the total cost for each alternative.

Process	Irons	Steel (carbon, low alloy)	Heat and corr. res. alloys	Aluminum alloys	Copper alloys	Lead alloys	Magnesium alloys	Nickel alloys	Precious Metals	Refractory metals	Tin alloys	Titanium alloys	Zinc Alloys
Sand castings	■	■	■	■	■	□	■	■			□		□
Shell mold castings	■	□	□	■	■			□					
Full-mold castings	■	■	□	□	□	□		■					
Permanent-mold castings	■	□		■	□	□	■	□			□		□
Die castings				■	□	■	■				□		■
Plaster mold castings				■	■								
Ceramic mold castings	■	■	■	□	□		□	■					□
Investment castings		■	□	■	■		□	■	□				
Centrifugal castings	■	■	■	□	□			□					
Continous castings		□		■	■	□							
Open die forgings	□	■	■	□	□		□	□		□		□	
Closed die forgings Blocker type		■	■	□	□		□	□		□		□	
Conventional type		■	■	□	□		□	□		□		□	
Upset forgings		■	■	□	□		□	□		□		□	
Cold headed parts		■	□	■	■	□		□	□				
Stampings, drawn parts		■	□	■	■		□	■	□	□		□	□
Spinnings		■	□	■	■	□	□	■	□			□	□
Screw machine parts	□	■	□	■	■		□	■	□	□		□	□
Powder metallurgy parts	■	■	□	□	■			□	□	■		□	
Electroformed parts	□			□	■	□		■	□		□		□
Cut extrusions		□		■	■	□	■	□		□	□	□	
Sectioned tubing		■	■	■	■		■	■		□		■	
Photofabricated parts		■	□	■	■	□	■	■	■	■	□	■	■

□ Currently being used

■ Most frequently used

FIGURE 13.1 Manufacturing methods available for engineering metals and alloys. Note that aluminum and copper alloys can be processed by all the processes listed. Such lists may be revised periodically as new techniques and materials are developed. *Source:* After "Materials Selector," *Materials Engineering Magazine*, Penton/IPC.

Aluminum alloy
7075-T6

Plain carbon steel
1020, 111 HB

Ultra high strength steel
D6 ac, 212 HB

Stainless steel, martensitic
410, Ann, 163 HB

Low alloy steel
4340, Q and T, 332 HB

Stainless steel
310, 168 HB

PH stainless steel
17-7

Titanium alloy
Ti-6Al-4V

Titanium alloy
Ti-7Al-2Cb-1Ta, 285 HB

Low alloy steel
4340, Q and T, 52 HRC

Molybdenum alloy
Mo-5 Ti

Titanium alloy
Ti-3Al-13V-11Cr, 285 HB

Iron base high temp alloy
A-286, Aged, 320 HB

Ultra high strength steel
D6 ac, 47 HRC

Cobalt base high temp alloy
HS 25

Nickel base high temp alloy
Inconel-X

Tantalum alloy
90 Ta-10 W

Nickel base high temp alloy
Rene 41, Aged, 350 HB

Nickel base high temp alloy
Inconel 700, Aged, 400 HB

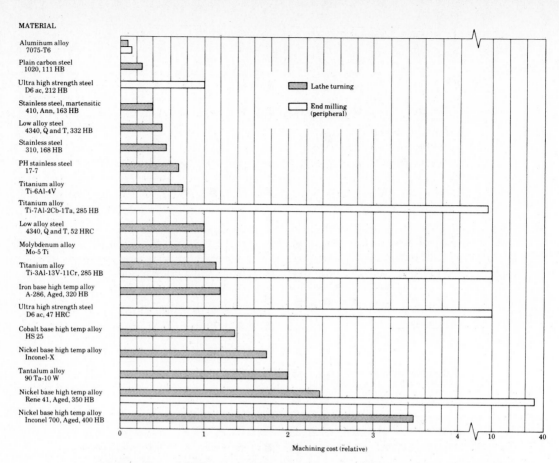

Lathe turning

End milling
(peripheral)

Machining cost (relative)

FIGURE 13.2 Relative machining (turning and milling) costs for several alloys used in aerospace applications. See Chapter 3 for properties of these materials. The manufacturing cost can be a deciding factor in material selection. *Source:* Adapted from Profile Milling for the Hard Metals, Report of the Ad Hoc Machine Tool Advisory Committee to the Department of the Air Force, May 1965.

The material selected for a product is a major factor in process selection. Some materials can be processed at room temperature, whereas others must be processed at elevated temperatures or by special techniques. Some materials are soft and ductile, others are hard, brittle, and abrasive. The tooling and equipment may require high strength, high wear resistance, or thermal fatigue life. As described throughout this text, the properties and surface characteristics of the manufactured product are affected by cooling rate in casting, deformation during processing, heat treatment, and many other factors.

13.2.2 SHAPE AND SIZE

The choice of a manufacturing process is often dictated by the complexity of the shape of the part. Castings, for example, are generally more complex than parts that can be made by other processes, such as forging or powder metallurgy. This is because the molten metal is capable of filling intricate cavities of the mold. This is also true in polymer processing.

On the other hand, forgings can generally be made into complex shapes by additional machining processes, such as by mechanical, electrical, or chemical means. In some designs, the shape may require that the part be fabricated from several sections and then joined together with the use of fasteners, or by brazing, welding, or bonding techniques. Even small parts may be manufactured in this way.

As described in various chapters in this text, all manufacturing processes have certain limitations. Among these are the minimum section size or thickness that can be obtained satisfactorily. This is an important aspect in weight reduction in manufactured products. For example, much thinner sections can be obtained by die casting or cold rolling than by sand casting or forging. The minimum sections usually produced by various processes are given in Fig. 13.3.

It is apparent that the design engineer must be receptive to changes in the original design to allow use of an economical manufacturing method.

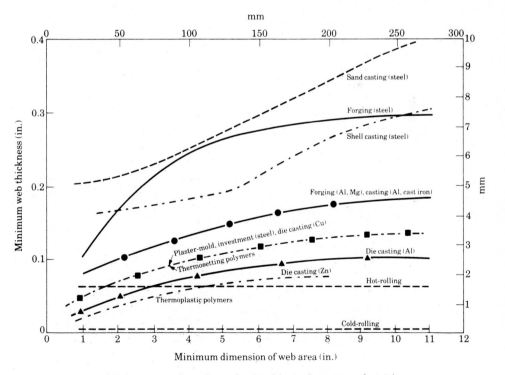

FIGURE 13.3 Minimum section sizes obtained in various manufacturing processes. Note that the cold-rolling process produces the thinnest sections; an example is aluminum foil used in packaging of cigarettes and candy. For a definition of web thickness and its area see Fig. 6.34. *Source:* Reproduced with permission from J. A. Schey, *Introduction to Manufacturing Processes*, McGraw-Hill, © 1977, p. 351.

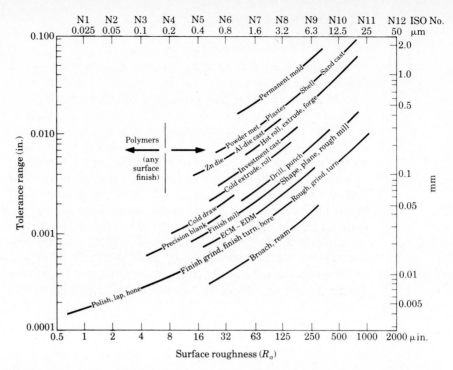

FIGURE 13.4 Tolerances and surface roughness obtained in various manufacturing processes. Note that tolerances vary by three orders of magnitude. These tolerances apply to a 1 in. dimension. See also Fig. 13.5. *Source:* Reproduced with permission from J. A. Schey, *Introduction to Manufacturing Processes,* McGraw-Hill, © 1977, p. 347.

13.2.3 SURFACE FINISH AND TOLERANCE

Surface roughness and its significance was described in detail in Chapter 4. The roughness obtained in various processes are shown in Fig. 13.4 and also in Figs. 4.8 and 9.12.

Tolerance is defined as the permissible variations in the dimensions of a part as specified on an engineering drawing. This subject, which is covered in detail in texts on machine design, is an important aspect of manufacturing, particularly in assembly and in the proper operation of all machinery, instrumentation, and equipment. Figures 13.4 and 13.5 show the range of tolerances usually obtained by various manufacturing processes. Note that this range is as much as three orders of magnitude, and that within each process there is a range of about one order of magnitude. Note also that tolerances increase with surface roughness and size of the component.

13.2.4 QUANTITY OF PRODUCTION

Depending on the type of product, the quantity of production can vary by many orders of magnitude. For example, paper clips, bolts and nuts, spark plugs, bearings, and ball-point pens are produced in very large numbers. On the other hand, jet engines for large commercial aircraft and propellers for ocean liners are manufactured in limited quantities. Because the quantity of production strongly affects manufacturing process selection, this subject is treated in Section 13.4.

13.2.5 LEAD TIME AND PRODUCTION RATE

The choice of a manufacturing process is also greatly influenced by the time required to start production. Processes such as forging and die casting involve extensive tooling, thus requiring long lead times before production can start. On the other hand, most machining processes involve tooling that can be adapted to a wide variety of situations in a relatively short time.

Another significant factor is the required production rate, that is, the number of pieces to be produced per unit time. From the treatment of various manufacturing processes in the preceding chapters, it can be seen that processes such as powder metallurgy, die casting, deep drawing, and roll forming are high-production-rate

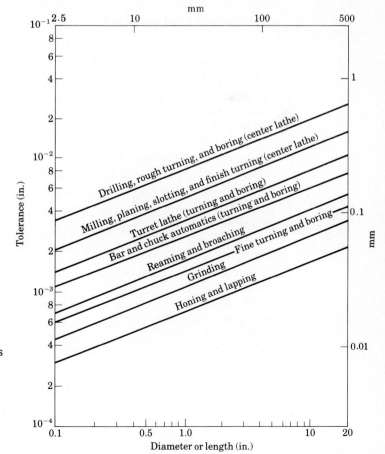

FIGURE 13.5 Relationship between part size and manufacturing tolerances in various machining operations. Note that tolerances increase with size of workpiece. *Source:* Adapted from *Manufacturing Planning and Estimating Handbook*, McGraw-Hill, 1963.

operations. On the other hand, sand casting, conventional and electrochemical machining, spinning, superplastic forming, diffusion bonding, and processing of reinforced plastics are relatively slow operations.

In the following sections, general aspects of manufacturing are discussed with regard to economics and methods for improved production.

13.3 MANUFACTURING ECONOMIC ANALYSIS

A product may be designed and manufactured according to all specifications and meeting all service requirements. However, in order for this product to be successful in the marketplace, its cost must be competitive with similar products.

There are different methods for subdividing the manufacturing cost of a product. Basically, however, the total cost is made up of the material cost, the tooling cost, the labor cost (direct and indirect), and the capital cost (plant and equipment).

13.3.1 MATERIAL COST

This is the cost of raw materials and depends not only on the type of material, but also on its shape. The raw material may be available in bulk or in the shape produced by casting, extrusion, forging, rolling or drawing, with various tolerances and surface finishes. The cost per unit weight depends on the form of the raw material. Table 13.1 shows the relative costs of wrought materials that are produced in different shapes. The cost of course increases with the number of operations involved to obtain the final product.

The cost of a particular material is subject to fluctuations, due to causes such as supply and demand and, if imported, geopolitical factors (see Fig. 1.6). If the cost becomes prohibitive or not competitive, alternative materials must be selected. For

TABLE 13.1
RELATIVE COST OF MATERIALS IN VARIOUS
PRODUCT SHAPES (After J. R. Miller.)

CONDITION	RELATIVE COST		
	STEEL	ALUMINUM	COPPER
Ore			
Raw	1	1	—
Refined	1.3	8	—
Pig iron	6	—	—
Raw metal			
Ingot, Billet or Bar	9	70	1
Sheet	14	113	2
Structural	14	106	2
Wire	23	90	1.5

example, when the price of copper increased substantially, electrical wiring for housing was, for a while, made from aluminum. Likewise, due to a shortage of copper in the 1940s, pennies were made from zinc-plated steel sheet.

If scrap is produced during manufacturing, the value of the scrap is deducted from the material cost. This value, which may be approximately one tenth of the original material cost, depends on the type of scrap and the demand for it.

13.3.2 TOOLING COST

These are the costs involved in making the tools and dies that are required in manufacturing the part. These costs are greatly influenced by the type of production process. For instance, if the part is a casting, the die cost for die casting is higher than the pattern cost for sand casting. An example of the wide range of tooling costs in polymer processing is shown in Fig. 13.6. These costs can be ascertained by reviewing the various equipment described in Section 10.12 on the processing of plastics.

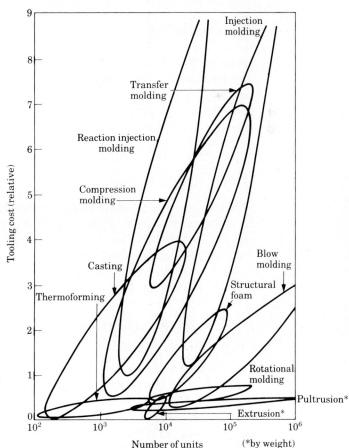

FIGURE 13.6 Typical tooling cost for processing of plastics. These processes are described in Section 10.12. Note the range of the cost and production quantity involved within each process. *Source:* Adapted from *Machine Design*.

In other examples, in a part manufactured by spinning (Section 7.11), the tooling cost for conventional spinning is much lower than that for power spinning. Likewise, in a machining operation, the choice of the cutting-tool material can be significant in that some tools, such as carbide tools, are more expensive than others, such as high-speed steel tools. (See, for instance, Table 13.2.)

13.3.3 LABOR COST

This cost is generally divided into direct and indirect costs.

Direct Labor Cost

The direct cost pertains to the labor involved in manufacturing the part (productive labor) from the time the raw material is first handled to the time the product is completed. This time is generally referred to as floor-to-floor time. For example, a machine operator picks up a round bar from a bin, machines it to the shape of a threaded rod, and returns it to another bin.

Indirect Labor Cost

Indirect labor costs relate to the costs involved in servicing the total manufacturing operation. The cost is accounted for by various personnel in supervision, maintenance, quality control, engineering, research, sales, and office staff. Indirect costs also include the costs of fuel, power, taxes on real estate, rent, and insurance. Because these items do not contribute directly to the production of finished parts, the costs are referred to as indirect costs, also known as overhead or burden rate, and the personnel are known as nonproductive labor.

13.3.4 CAPITAL COST

This item covers the costs of equipment and facilities. For example, a company may decide to manufacture a variety of valves. A new plant has to be built with all the necessary facilities. Melting furnaces, casting equipment, machine tools, quality-control equipment, and related equipment and machinery of various sizes must be purchased. If carried out in the plant, facilities must also be built for testing the valves prior to shipment to customers.

13.3.5 MACHINING COSTS: AN EXAMPLE OF MANUFACTURING COST ANALYSIS

The influence of various parameters involved in analyzing manufacturing costs may be illustrated by studying the effect on the various costs of changing the cutting speed in turning on a lathe.

The total production cost per piece, C, (unit cost) is composed of four items:

$$C = C_1 + C_2 + C_3 + C_4,$$

where C_1 is the nonproductive cost (labor, overhead, and machine-tool costs in setting up for machining, preparing the fixtures and the machine, loading and unloading parts, etc.); C_2 is the machining cost (labor, overhead, and machine-tool costs while the cutting operation is taking place); C_3 is the tool-change cost (labor, overhead, and machine-tool costs during tool change); C_4 is the cutting-tool cost.

The individual cost items above are all per piece. The dependence of each of these four items on cutting speed is shown in Fig. 13.7. As speed increases, the machining time (hence also cost per piece) decreases. However, as shown in Fig. 8.25 and described in Section 8.7, tool life decreases with speed. Hence the cutting tool cost increases, as does the tool-changing cost (since tools have to be changed more often). The nonproductive cost does not vary with cutting speed.

When the different cost items are added together, we find that there is an optimal speed for minimum cost per piece, and another optimal speed for minimum time per piece (high production rate) (Fig. 13.7). The range between the two optimal speeds is sometimes called the high-efficiency–machining range.

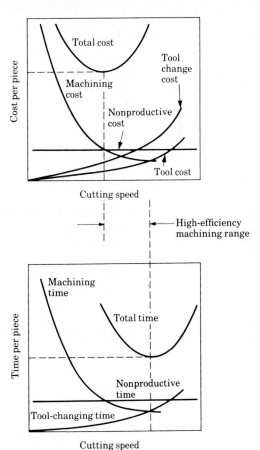

FIGURE 13.7 Schematic illustration of cost analysis in machining. Note that there is an optimum speed for minimum cost per piece and for minimum machining time per piece, respectively. See also Table 13.2.

TABLE 13.2
ECONOMIC ANALYSIS OF TOOL MATERIALS IN END MILLING OF ANNEALED 4340 STEEL (After V. A. Tipnis.)

CUTTER	RELATIVE PURCHASE COST	RELATIVE MATERIAL REMOVAL RATE	RELATIVE TOOL LIFE	RELATIVE MACHINING COST/UNIT VOLUME MACHINED
HSS	1	1	1	1.0
Cobalt HSS	1.25	1.5	2	0.6
Carbide	6	3	3	0.35
Brazed carbide	3	2.5	2.5	0.29

This analysis indicates the importance of identifying all relevant parameters in a manufacturing operation, determining various cost factors, obtaining relevant tool-life curves for the particular operation, and the proper measurement of the various time intervals involved in the overall operation. The importance of accuracy in obtaining the necessary data is evident from the observation in Fig. 13.7 that, depending on the shape of the total-cost curve, small changes in cutting speed can have a significant effect on the economics of the operation.

In another study in machining involving an end mill, the relative machining costs for four different types of cutting-tool materials are shown in Table 13.2. Note that each tool material has different cost and machining characteristics. In the final analysis brazed carbide has the lowest cost per unit volume of material machined.

13.4 MANUFACTURING COST AND VOLUME OF PRODUCTION

One of the most significant factors in manufacturing costs is the volume of production, that is, the number of parts of the same item required per unit time. Depending on the product, this number can vary by many orders of magnitude (Section 13.4.1).

Large production volumes require high production rates. This is accomplished by *mass-production* techniques, which involve special machinery and plant operations with two or three work shifts. On the other hand, small volumes of production involve proportionately larger direct labor. The qualitative relation between unit cost and production volume is given by various examples in Figs. 13.8 through 13.12.

In the example for drawing vs. spinning of the part shown in Fig. 13.12, note that deep drawing (see Fig. 7.40) requires detailed tooling, which is costly. On the other hand, spinning (conventional, in this example) requires simple tooling (see Fig. 7.61a). However, spinning a part takes much longer than deep drawing, hence labor cost is high. Consequently, if the number of parts required is high, it is more economical to produce this part by deep drawing.

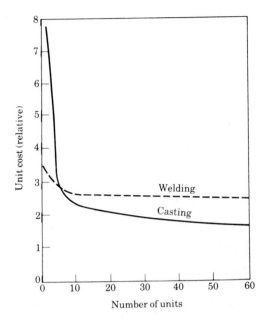

FIGURE 13.8   An example of unit cost (cost per part) for a machine tool base that can be manufactured either by welding or casting techniques. As the number of units to be manufactured increases, casting becomes the better alternative. *Source:* After L. E. Doyle.

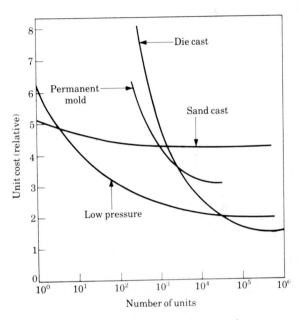

FIGURE 13.9   Unit costs in various casting processes for aluminum, as a function of number of units to be manufactured. See Chapter 5 for a description of these processes. *Source:* After R. J. Hickox, ASTME Paper SP 66–69.

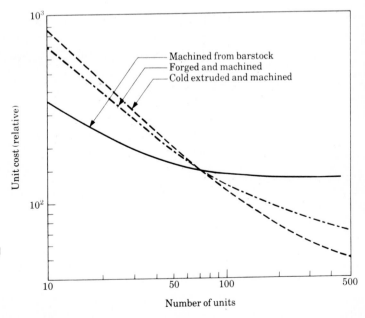

FIGURE 13.10   Unit costs for an aircraft part as a function of manufacturing methods. Note that, because machining involves considerable time, it is uneconomical for large-quantity production. *Source:* After L. E. Doyle

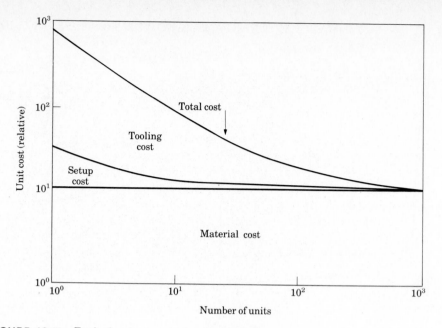

FIGURE 13.11 Typical unit costs in forging. Note that, because the same tooling is used, tooling and setup costs per part decrease as the number of units increase.

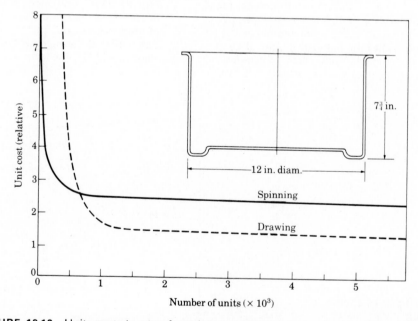

FIGURE 13.12 Units costs in manufacturing a round metal container by conventional spinning or deep drawing. Tooling for deep drawing is more expensive than for spinning. However, spinning requires more time per part, thus it is less economical for large-quantity production.

The downward trend of these curves for unit cost with increasing volume is an indication of the lesser direct labor involved. At low volumes, material costs are low as compared to direct labor, whereas at high volumes, direct labor costs are low as compared to material costs. Furthermore, once the tooling is made, the greater the number of parts produced, the lower the cost of tooling per part made.

In addition to meeting all design requirements, such as overall dimensions and tolerances, the product must also meet surface-roughness requirements. The roughness obtained depends on the particular finishing process used and can be a significant item in the overall cost. As surface roughness decreases the cost of manufacturing increases significantly (Fig. 13.13) because of the additional processing required.

For instance, in machining aircraft structural members made of titanium alloys it has been shown that as much as 60% of the cost of machining the part is expended

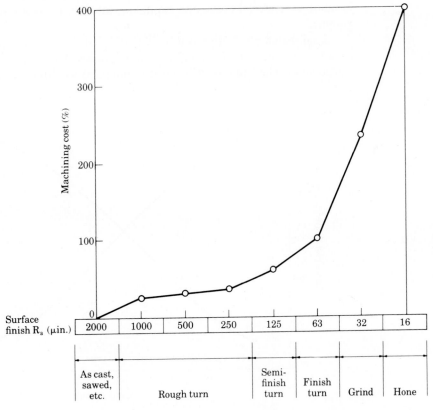

FIGURE 13.13 Increase in the cost of machining a part as a function of the surface finish required. To minimize manufacturing costs, surface roughness specifications and tolerances should be as coarse and as wide as acceptable for the service life of a part.

in the final machining pass in order to hold proper tolerances and surface finish. Thus, unless specifically required with proper technical justification (Chapter 4), a part should be made with as rough a surface and with as wide a tolerance as acceptable.

13.4.1 PRODUCTION QUANTITY

The terms generally used for the quantity of parts produced annually are as follows:

a. Mass Production: 10^5 units and higher
b. Large Batch or High Volume: 10^3 and 10^5
c. Small and Medium Batch: less than 10^2 to 10^3
d. Experimental or Prototype Production: less than 10 parts.

Since mass production involves special high-production machinery, the capital costs increase with increasing volume of production. Figure 13.14 shows the trends in labor vs. material costs, and capital vs. labor costs as a function of production volume.

Small-batch production is usually performed on general purpose machines, such as lathes, milling machines, and hydraulic presses. These machines are operated by skilled labor. The equipment is versatile and parts with different designs can be made

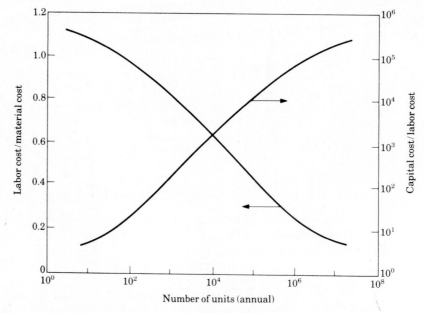

FIGURE 13.14 Relative costs of labor, materials, and capital as a function of annual production volume. Note the difference in the scales for the two ordinates in the figure. *Source:* After N. P. Suh.

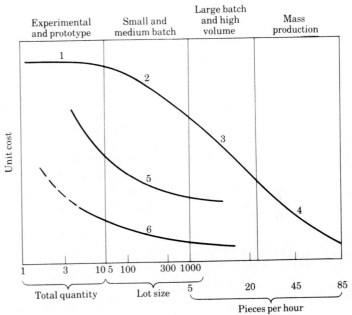

FIGURE 13.15 Unit cost as a function of method of production and production volume in machining. 1. Tool room machinery. 2. General-purpose machine tools. 3. Special-purpose machines. 4. Automatic transfer lines. 5. NC and CNC machine tools. 6. Computer-integrated manufacturing systems. *Source:* Cincinnati Milacron, Inc.

by appropriate changes in the tooling. Direct labor costs and unit costs are high (Fig. 13.15).

For larger quantities, such as in medium-batch production, the general-purpose machines are equipped with various jigs and fixtures, or they are numerically controlled (Section 13.6). To reduce the labor cost further, machining centers and flexible manufacturing systems have been developed. For quantities on the order of 10^5 and higher, the machines involved are designed for specific purposes (known as *dedicated machines*) and perform a sequence of different operations with very little human labor involved.

13.4.2 MATERIAL QUANTITY AND DESIGN

With high production rates and reduced human labor, the cost of the material becomes a significant factor in the overall cost of the product. Although this cost cannot be reduced below a certain level, efforts can be made to reduce the amount of material for items that are to be mass produced.

Since the shape of the part is usually optimized during design and prototype stages, further reductions in the amount of material needed can be obtained only by thickness reductions in the part. This process also involves selection of materials with high strength-to-weight or stiffness-to-weight ratios, the best example being aircraft and various other aerospace structures. (See Table 3.3 and Figs. 3.52 and 10.29.)

In engineering practice it has been observed that many designs are generally done intuitively and are usually based on empirical information. Although some parts in various products have been underdesigned, it has been noted that many parts are generally overdesigned (called *irrational design*), presumably with safety in mind.

As shown in various chapters in this text, however, thin cross-sections (in order to avoid overdesign) can present considerable difficulties in manufacturing parts, thus contributing to manufacturing costs. Typical examples where difficulties may arise are welding, casting, forging, and sheet-stretching operations on thin sections. In some operations, such as forging, thin parts may require excessive die forces. In welding, parts may distort. In castings, thin sections generally present difficulties in maintaining dimensional control and surface finish. In sheet-forming operations, parts may wrinkle and their formability may be reduced.

13.5 AUTOMATION

Until about three decades ago, most of the equipment and machinery used in manufacturing processes produced parts by batch operations. Generally this means that the processing methods were not efficient and that the cost of labor was substantial since each part had to be handled and processed individually. With changes in the structure of society and wage scales, competition, and with the advent on all fronts of technology, the need and possibilities for reducing the labor share of the cost of a product became apparent. This was particularly true during depressed economic periods and in the climate of increased competition from both industrialized and developing countries.

The obvious solution to this problem was automation of manufacturing operations. Basically, automation is a predetermined sequence of manufacturing operations in which human labor is replaced by a variety of mechanical and electronic devices, techniques, and equipment.

Automation can be applied to the manufacture of all types of goods—from raw materials to the finished product. It is applied to the following basic areas of activity:

a. Manufacturing processes,

b. Materials handling, before and after the product is made,

c. Inspection,

d. Assembly, and

e. Packaging.

The primary objectives of automation are as follows:

a. Reduce manufacturing costs through controlled production for improved accuracy and quality, thus reducing rejects;

b. Reduce labor cost, thus improve productivity;

c. Reduce damage to the workpiece in handling;

d. Higher level of safety for personnel, especially in hazardous working conditions;

e. Eliminate boredom in work requiring repetitive operations;

f. Economy of floor space in the plant.

The decision to automate a production facility requires the following additional considerations:

a. Type of product,

b. Phase of manufacturing operation to be automated,

c. High initial cost of equipment,

d. Reliability and maintenance of the automated system,

e. Overall economics.

13.5.1 EVOLUTION OF AUTOMATION

Although metalworking processes began to develop as early as 4000 B.C., it was not until the beginning of the Industrial Revolution in the 1750s that some measure of automation was introduced in the production of goods. As can be seen from Table 1.1, machine tools such as turret lathes and automatic screw machines, and automatic bottle-making equipment were developed in late 1890s and early 1900s. Mass-production techniques and transfer machines were developed in the 1920s. All these machines (except screw machines and turret lathes) had fixed automatic mechanisms and were designed to produce a specific product.

The major breakthrough in automation began with the first numerical-control machine tool developed at Massachusetts Institute of Technology in 1952. This was a large vertical milling machine (basically similar to Fig. 8.67a) performing end-milling and face-milling operations (such as those shown in Fig. 8.37) on an aluminum plate.

Beginning with this historic development, rapid progress has been made in various aspects of manufacturing, including the introduction of computers into automation, computerized numerical control, adaptive control systems, industrial robots, and computer-integrated manufacturing systems, which include computer-aided design and computer-aided manufacturing. These are described below.

13.6 NUMERICAL CONTROL

Numerical control (NC) is a method of controlling the movements of machine components by numbers (numerical). It is defined as a system in which actions are controlled by the direct insertion of numerical data at some point; the system must automatically interpret this data.

The data is stored on 1 in. (25 mm) wide punched paper or plastic tapes, or magnetic tapes. The basic concept of the punched tapes is that holes, whose positions

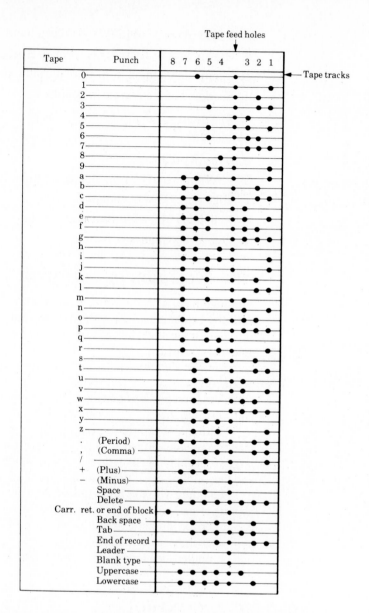

FIGURE 13.16 Standard punched tape used in numerical control, showing the EIC (Electrical Industries Association) code. A tape reader reads the characters (row of holes) successively, at rates from 30 to 300 characters per second, and gives commands to the machine tool for a particular operation to begin and to terminate. This information may also be stored on a magnetic tape.

represent specific information (Fig. 13.16), are read by sensing devices, which actuate relays and other devices. These devices then control various mechanical and electrical systems in the machine.

In this way, templates and other mechanical guides on machines are eliminated, and complex operations can be carried out, which would be almost impossible to perform manually with the same reproducibility from one production run to the next. A variety of programming systems have been developed for numerical control.

There are two basic types of control systems: *point-to-point* and *contouring (continuous-path)*.

In the point-to-point system, each axis of the machine is driven separately at different velocities, depending on the particular operation. The speed is as high as allowable, but is decreased as the tool reaches its final (and numerically defined) position. Thus, in an operation such as drilling or punching, the tool positioning is completed before the cutting begins. After the hole is drilled or punched, the tool is retracted, moved to another position, and the operation is repeated.

It is apparent that the stiffness of the machine tool and backlash in the gear drives and lead screws are important for accuracy. Backlash can be eliminated with special backlash take-up circuits, which make the approach to a particular position on the part always from the same direction. (This has been the usual practice in conventional hand-operated machine tools.)

In the contouring system, cutting takes place while the tool moves along a curved path (Fig. 13.17). The control of velocity is thus very important. The cutting path is in increments, each incremental path of the tool being linear, circular, or parabolic.

The contouring system is used on lathes, milling machines, grinders, welders, and on various other manufacturing equipment. Typical examples of operation are die sinking (i.e., producing a cavity in a die block) and welding along various paths. Machine tools are available with five axes (three linear and two angular) for milling complex contours.

13.6.1 COMPUTERIZED NUMERICAL CONTROL

The next step in the development of numerical control was in the 1970s when the control hardware in NC machines was converted to computer control with software. Two types of systems were developed: *computer numerical control (CNC)* and *direct numerical control (DNC)*.

CNC is a system in which a minicomputer or microcomputer is an integral part of the control panel of a machine or equipment. In DNC, several machines are directly controlled by a central computer. For such reasons as the declining cost and reduced

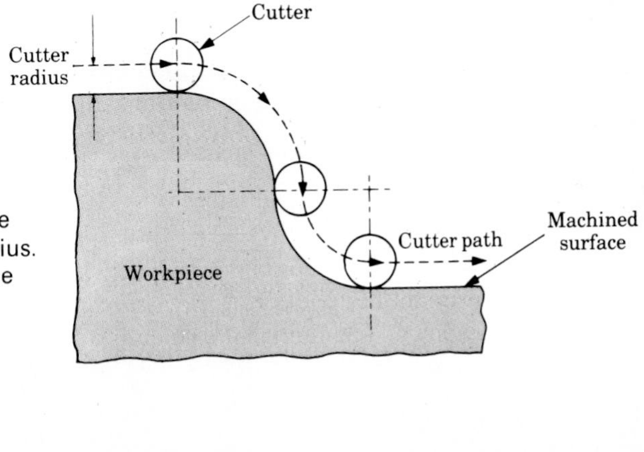

FIGURE 13.17 Path of a milling cutter performing a machining operation. Note that the cutter path is compensated for by the cutter radius. Cutter deflections, tool wear, and stiffness of the machine tool are important factors in the dimensional accuracy of parts produced by automated machine tools.

size of computers, lower investment, and flexibility, CNC systems are in much wider use than DNC in manufacturing operations today.

The advantages of CNC over conventional NC systems are the following:

a. Easier operation and simpler programming,

b. Lower maintenance costs due to the use of lesser control hardware,

c. Greater accuracy, and

d. Versatility, such as in editing and debugging programs, reprogramming, and in graphics to plot part shape.

The minicomputer in a CNC system is known as a *dedicated computer*; it is devoted exclusively to a single machine or application. (The term is also used in regard to machines, known as dedicated machines; Section 13.4.1.)

Machining Centers

A further development toward automation is the concept of machining centers. In general terms, a machining center is a machine that is equipped with automatic tool changers (Fig. 13.18) and designed to perform operations on different surfaces of a workpiece. Thus, the workpiece, after a particular operation, does not have to be removed and taken to another machine for further processing. In other words, the tools and the machine are brought to the work.

A machining center may be equipped with as many as 100 tools or more. It is capable of manufacturing large and complex parts efficiently and with high accuracy.

Advantages and Limitations

Numerical control of machine tools has the following advantages.

a. Flexibility in operation and ability to produce complicated contours with accuracy;

b. Short and consistent production time;

c. Experienced and skilled operators are not needed; the operator has more free time to attend to other matters in the work area;

d. Adjustments on machine are easy and require less time;

e. Planning of the manufacturing process is completed in the production office, rather than on the shop floor;

f. Cost and time estimates are accurate.

The major limitations of NC are:

a. Relatively high cost of the NC equipment;

b. The need for trained part programmers;

c. Special maintenance needed, requiring trained personnel.

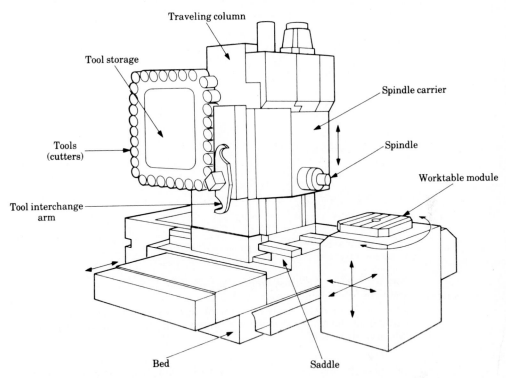

FIGURE 13.18 Schematic illustration of a machining center. A machining center is a machine tool equipped with as many as 100 tools or more and with an automatic tool changer. It is designed to perform different operations on various surfaces of a workpiece placed on the work modules. *Source:* Cincinnati Milacron, Inc.

13.7 ADAPTIVE CONTROL

Adaptive control (AC) is defined as automatic on-line adjustment of operating parameters in order to optimize production rate or to minimize cost in a manufacturing operation. Ideally, in an operation such as turning on a lathe or in milling, the adaptive control system senses some set of parameters, such as cutting forces, torque, temperature, tool wear rate, and surface finish. It then adjusts processing parameters accordingly (Fig. 13.19). Adaptive control is particularly important in situations where workpiece quality is not uniform, such as a poor casting or uneven heat treatment.

Adaptive control is done in real time, i.e., as the event is taking place. The system then converts this information into commands, which modify the operating parameters (such as speed or feed) to optimize the cutting operation. Optimization may involve maximizing material-removal rate between tool changes. For instance, if the forces increase excessively during machining (such as due to a hard spot in a casting),

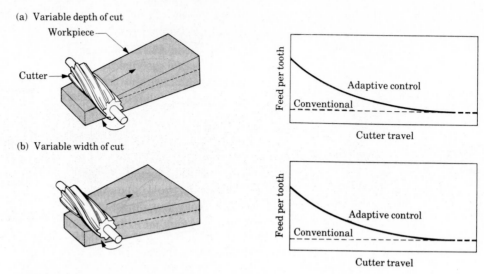

(a) Variable depth of cut

(b) Variable width of cut

FIGURE 13.19 An example of adaptive control. As the depth of cut or width of cut increases, the forces, torque, and power consumption increase. The system senses this increase and automatically reduces the feed to avoid tool breakage, or to maintain cutting efficiency. *Source:* After Y. Koren, *Computer Control of Manufacturing Systems*, McGraw-Hill, 1983.

AC changes the speed and/or feed to lower the forces to acceptable levels. Without AC (or operator interference) high cutting forces may cause tool chipping or breakage, excessive deflections, and hence loss of dimensional control and tolerances.

It can be recognized that human reactions to occurrences in everyday life already have adaptive control in themselves. If we drive a car on a rough road with potholes, we know how to steer to avoid potholes (visual sensing), or our body feels the movements and vibrations of the car on the rough road (tactile sensing). We then react by changing the speed or direction of the car.

It is apparent that, for AC to be effective in manufacturing operations, quantitative relations between the dependent and independent variables of that particular process must be known and stored in the computer memory. For instance, if tool wear rate in a machining operation is excessive, the computer must be able to know how much of a change (and whether to increase or decrease) in speed and/or feed is necessary to reduce the wear rate to an acceptable level. The system should also compensate for dimensional changes on the workpiece due to tool wear. If the operation is to bend a metal sheet in a V-die, the data must be stored in the computer as to how springback varies with punch travel and other process variables.

Compared to other parameters involved, forces, torque, and power have been found to be the easiest to monitor for AC. On-line monitoring of parameters, such as surface finish, tool wear rate, and residual stresses developed in a machined surface, present great difficulties and challenges. Adaptive control has been used in machining, grinding, and some sheet-forming and welding operations with reasonable success. Coupled with CNC, adaptive control is likely to be a powerful tool in manufacturing operations.

13.8 INDUSTRIAL ROBOTS

An *industrial robot* has been defined as a reprogrammable multifunctional manipulator designed to move material, parts, tools, or other specialized devices through various programmed motions to perform a variety of tasks. More generally, an industrial robot has also been defined as an all-purpose machine, equipped with a memory and appropriate mechanism to automatically perform motions replacing human labor.

The first industrial robots (1961) were used in hazardous operations, such as with toxic materials, or in loading and unloading hot workpieces from furnaces in metalworking plants and in foundries. From these early applications to protect the workers, industrial robots have been developed primarily for improving productivity in manufacturing operations.

Presently, the areas of application of industrial robots are:

a. Loading, unloading, and general handling of parts;

b. Assembly operations;

c. Inspection;

d. Welding operations (spot and arc);

e. Spray painting; and

f. Machining operations, especially deburring and grinding.

The basic components of an industrial robot are shown in Fig. 13.20. The robot has a main frame (which includes the *arm*) and a *wrist*. The motions of the arm and the wrist are similar to those of a human arm and hand. Each axis of motion is driven by an independent actuator, such as an electric or hydraulic motor.

As in NC machines, the type of controls in industrial robots are point-to-point and continuous-path. Depending on the particular task, the positioning accuracy required may be as small as 0.002 in. (0.050 mm), such as in assembly operations for electronic circuitry.

End-Effectors

The wrist in a robot is equipped with an *end-effector*. Depending on the type of operation, the end-effector may be equipped with:

Gripper, electromagnet, or suction cup for material handling;

Spray gun for painting;

Attachments for spot and arc welding;

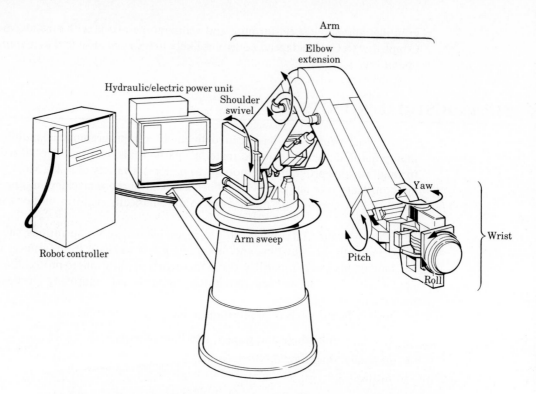

FIGURE 13.20   Components of an industrial robot. The end of the wrist is equipped with end-effectors that act like human fingers. Robots are used for a variety of operations such as welding, assembly, inspection, painting, materials handling, and some machining operations. *Source:* Cincinnati Milacron, Inc.

A power tool such as a burr, drill, or nut driver;

Measuring instruments such as a dial indicator.

End-effectors are generally custom-made to meet special handling requirements. Grippers are the most commonly used end-effectors. They are equipped either with two or three fingers, the latter being more versatile.

An industrial robot can be programmed to go through certain motions to reach a specific point and orientation in space through six degrees of freedom. However, a new generation of robots is being equipped with end-effectors that sense, by various means, the proximity to the object they are reaching. This sensing can be tactile. The force exerted by the gripper jaws on the object is fed back to the computer so that it can be controlled. For example, in the case of a glass bottle or a delicate object, if the movement of the jaws were not controlled, the gripping force could build up and break the object. (The human hand automatically adjusts this force.)

End-effectors are also being designed to sense the proximity to objects by visual (such as TV cameras) and other optical means. In this way, the end-effector will be able to pick up parts from a bin and grip a part in its proper orientation. (Picking parts from a bin has proven to be a difficult task for any automated system, because of the random orientation of the parts in close proximity to each other or to other similar parts in the bin.)

Robots with these type of sensing devices are known as intelligent robots because of their capabilities for perception and pattern recognition. Industrial robots are also being designed and developed to convert spoken words into operating commands.

13.9 COMPUTER-INTEGRATED MANUFACTURING

Computer-integrated manufacturing (*CIM*) can be defined as a system involving the coordinated participation of computers in all phases of manufacturing, from design to automatic testing and inspection of the final product. This is possible because of the development of sophisticated computers and various software and hardware components.

Computer-Aided Design

In the design process, *computer-aided design* (*CAD*) involves developing design data directly on a computer. In interactive computer graphics the user, in a graphics work-station attached to the computer, generates, manipulates, and modifies the design, which is graphically displayed. The final design is then drawn on a plotter interfaced to the computer.

CAD is used not only in mechanical design and drafting, but also in printed and integrated circuit design, and structural design and analysis. One very effective new tool is geometric modeling of solid objects as opposed to the usual edge and surface representations of the object.

Computer-Aided Manufacturing

Computer-aided manufacturing (*CAM*) is the use of computers to assist in all phases of the manufacturing of a part. For more efficient operation, CAM is usually interfaced with CAD, thus the term CAD/CAM.

CAD/CAM has had a major impact on manufacturing through standardization of product development, reduction of design effort and prototype work, improved productivity through better material use, reduced lead time, efficiency, speed and accuracy, and with fewer rejects. This is mainly accomplished through a single data-base to describe the design for function and tolerance analysis, stress analysis, planning for NC, technical illustrations, and other activities.

The development of CIM is generally regarded as revolutionary because of its major impact on all aspects of manufacturing operations. In addition to CAD/CAM, CIM aids in the management of the factory, labor, and inventory and integrates all of these activities under computer control. It has led to the development of the factory of the future, which is essentially unmanned. CIM has also made it possible to manufacture parts in small batches with the same degree of automation as in mass production.

The rapid development in CIM has involved various other systems, as described below.

13.9.1 MANUFACTURING CELL

The *manufacturing cell* is a system in which CNC machines and an industrial robot are used to manufacture a specific part, or several parts of similar geometry (group technology). A more complete definition is: A small unit within a manufacturing system, consisting of one to several work stations and having some degree of capability for:

Automatic loading and unloading of the work stations with workpieces,

Automatic tool changing at work stations,

Automatic transfer of workpieces (and sometimes tools) between work stations, and

Automatic scheduling and control of work-station loading.

All these operations are under computer control to produce a family of parts with similar features. Central to these activities in a manufacturing cell is an industrial robot. The work stations around the robot can be several machine tools, each performing a different operation on the part (Fig. 13.21).

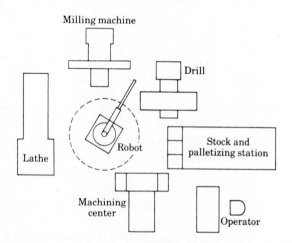

FIGURE 13.21   Schematic illustration of a manufacturing cell. A cell consists typically of two to five CNC machine tools, placed around an industrial robot that handles, loads, and unloads the machined parts. Each cell manufactures one part, or several parts with similar shape, This type of manufacturing is known as group technology.

13.9.2 FLEXIBLE MANUFACTURING SYSTEM

The *flexible manufacturing system* (*FMS*) is made up of manufacturing cells (each cell containing an industrial robot serving several CNC machines) and an automatic material-handling system interfaced with a central computer. This system is the highest level of sophistication that has been reached in manufacturing plants. As the name implies, FMS has great flexibility and can simultaneously handle more than one type of product.

Because of its capability to optimize each step of the total operation (from handling the raw material to the finished part), FMS can reach the highest possible level of efficiency in production (high productivity). By reducing the cost of direct labor and inventories, it is anticipated that FMS is capable of producing parts with cost savings as much as 80 to 90 % over conventional methods of production.

SUMMARY

1. We have seen that there are many different stages in manufacturing a product, involving not only materials, design, and processing, but also evaluation and reduction of costs. A product must be competitive in the marketplace.

2. A wide variety of processes and materials is available for manufacturing a particular part, as described in various chapters in this text. The final selection depends on factors such as service requirements, size and shape, tolerances and surface finish requirements, and the quantity and rate of production.

3. Manufacturing economics is an important subject that involves not only evaluating the actual but also the relative costs of competing selections of materials, design, tooling, labor, processing, and capital equipment. When added together to determine the final cost of a product, it is found that there is an optimal set of choices and decisions that minimizes cost. The quantity to be produced is a significant aspect of cost analysis.

4. Automation and the development of sophisticated computers has revolutionized manufacturing industries and the nature of future factories. Numerical-control machines, computer-aided design and manufacturing, adaptive control, industrial robots, and computer-integrated manufacturing systems, with their inherent flexibility and capabilities, have a major influence on productivity and cost reduction.

5. Decisions concerning the selection of any or all of these modern tools and equipment for a particular manufacturing plant requires careful consideration of many factors. The level of sophistication of these advanced systems requires skilled, knowledgeable, and trained personnel, otherwise major errors may be committed.

6. The high level of automation presently achieved, including unmanned factories, and the attendant savings in labor have far-reaching economic and social impact. Many discussions and predictions concerning automation can be found in the literature, dealing with its impact on unemployment, skill and training required, and the care needed for proper assessment of automation and its role in not only industrialized but also in developing countries.

BIBLIOGRAPHY

Amrine HT, Ritchey A, Hulley OS. *Manufacturing Organization and Management*, 3d ed. Englewood Cliffs, N.J.: Prentice-Hall, 1975.

Behrman JN, Wallender HW. *Transfers of Manufacturing Technology Within Multinational Enterprises*. Cambridge, Mass.: Ballinger, 1976.

Besant CB. *Computer Aided Design and Manufacture*, 2d ed. New York: Wiley, 1983.

Boothroyd G, Poli C, Murch LE. *Automatic Assembly*, New York: Marcel Dekker, 1982.

Childs JJ. *Principles of Numerical Control*. 3d ed., New York: Industrial Press, 1982.

Dieter GE. *Engineering Design*. New York: McGraw-Hill, 1983.

Eary DF, Johnson GE. *Process Engineering for Manufacturing*. Englewood Cliffs, N.J.: Prentice-Hall, 1962.

Enrick NL. *Quality Control and Reliability*, 7th ed., New York: Industrial Press, 1977.

Enrick NL, Mottley Jr HE. *Manufacturing Analysis for Productivity and Quality Cost Enhancement*, 2d ed. New York: Industrial Press, 1983.

Farago FT. *Handbook of Dimensional Measurement*, 2d ed., New York: Industrial Press, 1982.

Groover MP. *Automation, Production Systems and Computer-Aided Manufacturing*. Englewood Cliffs, N.J.: Prentice-Hall, 1980.

Gunn TG. *Computer Applications in Manufacturing*. New York: Industrial Press, 1981.

Halevi G. *The Role of Computers in Manufacturing Processes*. New York: Wiley, 1980.

Hitomi K. *Manufacturing Systems Engineering*, London: Taylor and Francis, 1978.

Holland FA, Watson FA, Wilkinson JK. *Introduction to Process Economics*. New York: Wiley, 1974.

Hunt VD. *Industrial Robotics Handbook*, New York: Industrial Press, 1983.

Koren Y. *Computer Control of Manufacturing Systems*. New York: McGraw-Hill, 1983.

Landau ID. *Adaptive Control*. New York: Academic Press, 1979.

Oliver S. *The Management of Production Technology*. Bury St. Edmunds, England: Mechanical Engineering Publications, 1978.

Ostwald PF, ed. *Manufacturing Cost Estimating*. Dearborn, Mich.: Society of Manufacturing Engineers, 1980.

Pressman RS, Williams JE. *Numerical Control and Computer-Aided Manufacturing*. New York: Wiley, 1977.

Puckle OS, Arrowsmith JR. *Introduction to Numerical Control of Machine Tools*. London: Chapman and Hall, 1964.

Rembold U. *Computers in Manufacturing*. New York: Marcel Dekker, 1977.

Smolik DP. *Material Requirements of Manufacturing*. New York: Van Nostrand Reinhold, 1983.

Vernon IR, ed. *Realistic Cost Estimating for Manufacturing*. Dearborn, Mich.: Society of Manufacturing Engineers, 1968.

Wage HW. *Manufacturing Engineering*. New York: McGraw-Hill, 1963.

Wilson FW. *Numerical Control in Manufacturing*. New York: McGraw-Hill, 1963.

Wilson FW, Harvey PD, eds. *Manufacturing Planning and Estimating Handbook*. New York: McGraw-Hill, 1963.

Periodicals

Computer and Industrial Engineering

Computers in Mechanical Engineering

Journal of Engineering for Industry

Journal of Manufacturing Systems

Machinery

Manufacturing Engineering

Robotics and Computer-Integrated Manufacturing

PROBLEMS

13.1. Explain the reasons for the trends in the curves shown in Figs. 13.8 through 13.13.

13.2. Inspect some products around your house or on your car and discuss ways and means by which you can simplify that product and manufacture it more economically.

13.3. Explain the reasons why the price of some products (such as calculators, computers, watches) has gone down over a period of time, whereas others (such as cars, houses, books) have gone up.

13.4. Inspect some of the manufacturing operations that you know of, or you have seen during visits to plants and factories, and suggest means of automating these operations.

13.5. Explain the reasons why some of the materials in Fig. 13.1 are not suitable for some of the processes listed in that figure.

13.6. Why is it that in Fig. 13.2 the same material costs more to machine by end milling than by turning?

13.7. Explain the reasons for the trend observed in Fig. 13.4, i.e., that tolerances increase with increasing surface roughness.

13.8. How would you design an end-effector for an industrial robot so that it can handle delicate objects without breaking them?

13.9. Why is it that in Fig. 13.3 the minimum web thickness obtained by hot or cold rolling does not depend on the web area as do other processes?

13.10. Would you have expected the upward trend in the curves shown in Fig. 13.5?

13.11. Explain why the cost for injection molding is higher than other processes shown in Fig. 13.6.

13.12. In Fig. 13.11 explain why the setup cost goes down with the number of units produced.

13.13. Explain the reasons for the trends in the two curves shown in Fig. 13.14.

13.14. Describe the techniques that you would use to compensate continuously for the cutter wear in the machining operation shown in Fig. 13.17.

13.15. Describe several applications of adaptive control for the processes covered in various chapters in this text.

13.16. Give examples of products that would be suitable for the manufacturing cell shown in Fig. 13.21.

13.17. Suggest various other arrangements for manufacturing equipment and machinery for a manufacturing cell.

13.18. Name types of products that are suitable for the four categories of annual production rates listed in Section 13.4.1.

Glossary

Glossary

AA. Arithmetical average. The Aluminum Association.

ABLATION. Removal of material from a solid surface by melting and vaporization.

ABRASION. Scratching or removal of material by a hard particle or protrusion.

ABRASIVE. A hard material capable of scratching or removing material from a surface.

ABRASIVE BELT. A flexible belt coated with abrasive particles, used for grinding and polishing.

ABRASIVE DISK. A grinding wheel or a disk-shaped coated abrasive.

ABS. Acrylonitrile-butadiene-styrene, a thermoplastic.

ACICULAR. Needlelike microstructure in cast iron containing martensite and other microconstituents.

ACS. American Ceramic Society.

ACTIVATION ENERGY. The energy required to initiate a reaction.

ACTIVE OIL. A lubricant containing chemically active ingredients, such as sulfur, chlorine, and phosphorus.

ADAPTIVE CONTROL. On-line adjustment of operating parameters to optimize production rate or minimize cost in a manufacturing operation; presently applied to machining and some forming and welding operations.

ADDITION POLYMERIZATION. Polymerization by sequential addition of monomers.

ADDITIVE. A material added for the purpose of imparting new properties or enhancing existing ones.

ADDRG. American Deep Drawing Research Group.

ADC. Analog-to-digital converter.

ADHESION. Molecular attraction between the surfaces of two bodies in contact.

ADMIRALTY BRASS. Brass containing copper, zinc, and tin.

ADSORPTION. Adhesion of thin layers of atoms or molecules, usually gas or liquid, on a solid surface by physical forces or chemical bonding.

AFS. American Foundrymen's Society.

AGING. A change in mechanical properties of metals and alloys occuring slowly at room temperature and more rapidly at elevated temperatures.

AIME. American Institute of Mining, Metallurgical and Petroleum Engineers.

AIRCRAFT QUALITY. High-quality metals and alloys that have been tested during manufacture and approved as suitable for aircraft parts.

AISI. American Iron and Steel Institute.

ALCLAD. Aluminum alloy sheet or plate coated with a very thin layer of high-purity aluminum.

ALKYDS. A family of thermoplastic or thermosetting resins made by a special process and generally used for protective coatings.

ALLIGATORING. Longitudinal splitting (along the thickness end parallel to the surface) of flat metal plates in rolling; also called fishmouthing.

ALLOTROPISM. Reversible phenomenon of metals existing in more than one crystal structure.

ALLOWANCE. The specified difference in dimensions between mating parts.

ALLOY. A metal composed of two or more chemical elements.

ALLOYING ELEMENT. An element added to a metal to impart certain specific properties.

ALLOY STEEL. Steel containing significant amounts of alloying elements such as nickel, chromium, molybdenum, and vanadium.

ALPHA IRON. Body-centered cubic form of pure iron.

ALUMEL. An alloy of nickel and aluminum, used in thermocouples.

ALUMINIZING. Coating of metal with aluminum by hot dipping, spraying, or diffusion.

AMALGAM. An alloy of mercury and one or more other elements.

AMORPHOUS. Noncrystalline, i.e., not having a crystal structure; without long-range order.

AMORPHOUS ALLOYS. Noncrystalline structures obtained by extremely rapid cooling (10^6 to 10^8 K/s) of a liquid alloy. *See* Splat cooling.

AMS. Aerospace Materials Specification.

ANION. A negatively charged ion.

ANISOTROPY. Exhibiting different properties when tested in different directions.

ANNEALING. The process of heating and cooling a material to impart changes in its properties, e.g., to reduce hardness and strength, to improve ductility, and change microstructure.

ANODE. The positive terminal of an electrolytic cell.

ANODIC CORROSION. Corrosion and dissolution of a metal acting as an anode.

ANODIZING. Forming of an inorganic oxide coating on a metal surface (usually aluminum) where the metal is the anode in an electrolytic cell.

ANSI. American National Standards Institute.

ANVIL. A block of steel on which metal is forged.

APMI. American Powder Metallurgy Institute.

APT. Automatically Programmed Tools.

ARBOR. Spindle on which a cutter or grinding wheel is mounted.

ARBOR PRESS. A hand press used for pressing arbors, shafts, etc., into holes.

ARC FURNACE. An electric furnace where the heat source is an arc between two electrodes or between an electrode and the metal to be melted.

ARC WELDING. Welding with an electric arc between an electrode and the metal to be welded.

ARCHITECTURAL BRONZE. An alloy of copper, zinc, and lead.

ARROWHEAD CRACKS. *See* Centerburst cracks.

ARTIFICIAL AGING. Aging above room temperature.

ARTIFICIAL INTELLIGENCE. The science of using computers to make intelligent choices in dealing with unanticipated situations, and to model aspects of human understanding.

ASCE. American Society of Civil Engineers.

ASLE. American Society of Lubrication Engineers.

ASM. American Society for Metals.

ASME. American Society of Mechanical Engineers.

ASNT. American Society for Nondestructive Testing.

ASPECT RATIO. Ratio of height to diameter.

ASTM. American Society for Testing and Materials.

ATACTIC. Random spatial arrangement of side groups along a long-chain polymer molecule.

ATOMIZATION. The dispersion of molten metal into particles by a stream of gas or liquid.

ATTRITIOUS WEAR. Wear and dulling of abrasive grains.

AUSFORMING. Thermal-mechanical processing of steels in which martensite is formed from an austenitic matrix which previously has been strengthened by plastic deformation.

AUSTENITE. A solid solution of one or more elements in face-centered cubic iron.

AUTOFRETTAGE. Prestressing a hollow metal cylinder, e.g., a gun barrel, by internal pressure.

AUTOGENOUS. Produced independently of external means.

AUTOMATION. Process of following a predetermined sequence of operations with little or no human labor.

AUTORADIOGRAPHY. A technique in which radiation emitted by a material is recorded photographically.

AWS. American Welding Society.

BABBITT. An alloy of tin, antimony, and copper, used for bearings.

BAINITE. A microconstituent of steel produced during the transformation of austenite, consisting of ferrite and carbide.

BAKELITE. A thermosetting resin.

BALL MILL. A mill for crushing materials in a rotating cylinder containing balls, usually made of steel.

BALL SIZING. Sizing and finishing the hole in a part by forcing a ball through it.

BANDED STRUCTURE. A segregated structure of nearly parallel bands aligned in the general direction of plastic deformation.

BARREL FINISHING. Improving the surface finish or deburring of parts by rotating them in a drum filled with specially shaped abrasives.

BARRELING. Curvature developed on the cylindrical surfaces of a specimen in compression.

BASAL PLANE. A plane perpendicular to the "c" axis of a close-packed hexagonal crystal.

BASE METAL. The metal in the highest proportion in an alloy. The metal to be welded, or that part of the metal that has not melted during welding.

BATCH PRODUCTION. Production where the lot size is usually less than 100,000 parts annually.

BAUSCHINGER EFFECT. The reduction of the yield stress of metal in one direction after being first subjected to plastic deformation in the opposite direction.

BAUXITE. The principle ore of aluminum.

BCC. Body-centered cubic.

BEADING. A forming operation in which an elongated projection or ridge is raised on sheet metal.

BED. The stationary part of a press that serves as a table or support for the bolster plate and/or die, or of a machine tool on which components such as the carriage, tailstock, etc., can travel.

BEILBY LAYER. A layer of metal disturbed by mechanical working.

BEACH MARKS. Progression marks on a fatigue fracture surface, also known as striations.

BEND ANGLE. The angle through which a part is bent.

BEND RADIUS. The inside radius of a bent section.

BEND TEST. A test for determining properties by bending the specimen between fixed points.

BENTONITE. A colloidal claylike substance.

BESSEMER PROCESS. Process for making steel by blowing air through molten pig iron to remove (by oxidation) most of the carbon, silicon, and manganese.

BHMA. British Hard Metal Association.

BHN. Brinell hardness number. (Obsolete. See HB).

BIAXIAL STRESS. A state of stress where stresses are applied along two principal axes only.

BILLET. A hot-worked, solid, semifinished, square or round product.

BIMETAL. A strip consiting of two layers of different metals.

BINARY ALLOY. An alloy containing two component elements.

BINDER. A material used to bond particles together, such as in sand molds or powder metallurgy.

BITE. One-half the difference between the original and final thicknesses of a plate or sheet in rolling.

BLACK OXIDE. A black finish on a metal produced by immersing it in hot oxidizing salts or salt solutions.

BLANK. A piece of sheet metal that is to be subjected to further forming operation. Powder metallurgy product requiring additional finishing operations.

BLANK DEVELOPMENT. The process of determining the optimal size and shape of a blank for a specific part.

BLANKHOLDER. Part of a die to hold sheet metal in place during forming; e.g., deep drawing.

BLAST FURNACE. A furnace in which solid fuel is burned with an air blast to smelt iron ore in a continuous operation to produce metallic iron, called pig iron.

BLASTING. Cleaning or finishing metal surfaces with abrasive particles at high speeds.

BLIND RISER. A riser that does not extend through the top mold (cope) in casting.

BLISTER. A raised area on a surface due to trapped gases or lack of adhesion.

BLOCKING. A preliminary forming operation in closed-die forging.

BLOOM. A semifinished hot-rolled product, rectangular in cross-section.

BLOOMING MILL. A rolling mill used to make blooms.

BLOTTER. A disk-shaped paperlike material used between a grinding wheel and its metal flanges to distribute contact stresses evenly.

BLOWHOLE. A hole in a casting or in a weld caused by entrapment of gas during solidification.

BLUE BRITTLENESS. Brittleness exhibited by some steels in a temperature range of 400 to 700°F (200 to 370°C).

BLUING (BLUEING). Formation of a thin, uniform bluish-black oxide film on polished steel to improve appearance and protect surface.

BOARD HAMMER. A forging hammer in which the upper die and ram are attached to boards that are raised by power driven rollers.

BOLSTER. A heavy plate or block to which dies are fastened.

BOND. The material that holds particles together, such as abrasives in a grinding wheel (bonded abrasives).

BONDED FILM LUBRICANT. Lubricating solids (graphite, molybdenum disulfide, etc.) in thermoplastic or thermosetting resins.

BORING. Removal of material from the internal cylindrical surfaces of a part.

BORT. Industrial diamond.

BOTTOMING. The forcing of a male punch against the die in bending.

BOUNDARY LUBRICATION. A condition of lubrication where the surfaces are covered with a layer of adherent lubricant or chemically formed coating, in which friction and wear are determined by the properties of the layer and not the viscosity of the fluid.

BOX ANNEALING. Annealing in a sealed container to minimize oxidation.

BRAKE FORMING. A forming operation on a press brake in which the principal mode of deformation is bending.

BRALE. A conical diamond indenter for Rockwell hardness testing.

BRANCHING. Bifurcation of a linear polymer chain.

BRASS. An alloy of copper and zinc.

BRAZING. Joining of two solid metal surfaces by capillary action of a molten nonferrous metal.

BREAKDOWN. A preliminary metalworking operation, usually on a casting that is reduced in shape by forging or rolling.

BRIDGE DIE. A die used in extrusion of hollow cross sections.

BRIGHT ANNEALING. Annealing in a controlled atmosphere to prevent oxidation and discoloration of a bright-surface.

BRINNELLING. Permanent indentation caused by repeated contact.

BRITTLE. Material behavior where fracture takes place with little or no plastic deformation.

BROACH. A straight cutting tool with multiple cutting teeth.

BRONZE. An alloy of copper and tin. Also, trade designation for some brasses.

B & S. Brown and Sharpe Gage.

BUE. Built-up edge, observed in metal cutting.

BUFFING. Process of imparting a lustrous surface finish by holding a workpiece against a rotating buffing wheel.

BUILT-UP EDGE. Workpiece material adhering to the tip of a cutting tool.

BULGING. Expanding the diameter of a tube or a shell.

BULL BLOCK. A power-driven revolving drum used in wire drawing.

BURNING. The discoloration or damaging of a surface due to high temperatures generated, such as in grinding or heating of steel.

BURNISHING. Smoothing of surface finish of workpieces (previously machined or ground) by frictional contact of fixed or rotating tools under high pressure.

BURR. A sharp edge on a workpiece caused by shearing or machining. A rotary cutting tool.

BUSTER. A pair of simple dies used in press forging.

BUTT JOINT. A joint between two abutting (adjoining without overlapping) plates or sheets.

BUTTON. That part of a weld that tears out in destructive testing of spot, seam, or projection welded specimens. Also, a part of a die.

B.W.G. Birmingham Wire Gage.

CAD. Computer-aided design. Computer-aided drafting.

CAM. Computer-aided manufacturing.

CAMBER. Roll geometry, in which the center of the roll has a larger diameter than the edges to compensate for deflection due to roll forces. *See* Crowning.

CANNING. Enclosing of a reactive metal in a sealed container for hot working. Also, a dished distortion in a flat sheet, sometimes referred to as oil canning.

CAPPED STEEL. A semikilled steel.

CARBIDE. A compound of carbon with one or more metallic elements.

CARBIDE TOOLS AND DIES. Tool materials generally made of tungsten carbide or titanium carbide in a matrix of cobalt and nickel, respectively.

CARBONITRIDING. Heat treatment in which carbon and nitrogen are introduced simultaneously into the surface of ferrous alloy parts.

CARBON STEEL. Steel containing carbon up to 2% and small (residual) quantities of other elements such as manganese, sulfur, silicon, and phosphorus; also called plain carbon steel.

CARBURIZING. Heat treatment in which carbon is introduced into the surface of ferrous alloy parts.

CARTRIDGE BRASS. Brass (70% Cu, 30% Zn).

CASE. Depth of surface hardened by heat treatment.

CASE HARDENING. Heat treatment in which the surface layers of a part (case) are hardened.

CASTING. Making objects by pouring molten metal into molds.

CAST IRON. Iron containing carbon (generally between 2 and 4%) in various forms as obtained in a casting, e.g., gray, white, malleable, and nodular.

CATALYST. A substance or agent that increases the rate of a reaction without itself undergoing any chemical change.

CATHODE. The negative terminal of an electrolytic cell.

CATION. Positively charged ion.

CAVITATION. Formation of partial vacuum in liquids, or formation of cavities.

CBN. Cubic boron nitride.

CDA. Copper Development Association.

CELLULOSE. A natural polymer.

CEMENTATION. The process of introducing elements into the outer layer of a metal object by means of high-temperature diffusion.

CEMENTED CARBIDE. Pressed and sintered carbide particles in a binder, e.g., carbide tools and dies.

CEMENTITE. A compound of iron and carbon, known as iron carbide (Fe_3C).

CENTERBURST (CENTRAL BURST) CRACKING. Rupture and void formation in the centerline of workpieces in plastic deformation, usually in drawing and extrusion.

CENTER DRILLING. Drilling of a shallow conical hole on a piece for mounting on a machine tool such as a lathe.

CENTERLESS GRINDING. Grinding of the outside or inside surfaces of a cylindrical workpiece supported between two grinding wheels and a work rest.

CENTRIFUGAL CASTING. Casting of tubular shapes in a rotating mold using centrifugal forces.

CERAMIC. A hard and brittle material consisting of compounds of metallic and nonmetallic elements.

CERAMIC GRINDING WHEEL. A grinding wheel with a vitrified bond, i.e., vitrified wheel.

CERAMIC TOOLS. Cutting tools that are generally made of sintered aluminum oxide.

CERMET. A cutting tool made of ceramic particles in a metal matrix by power metallurgy techniques.

CHAMFER. A beveled edge to remove sharpness.

CHAPLET. Metal support for holding cores in sand casting.

CHARPY TEST. An impact test in which a notched specimen supported at both ends is broken by the swing of a pendulum. The energy dissipated is a measure of impact strength or notch toughness.

CHATTER. Vibration of tool or workpiece in metalworking operations, originating from frictional or other instability.

CHECKING. *See* Heat checking.

CHEMICAL DEPOSITION. Replacement or precipitation of one metal from a solution of another metal or reagent.

CHEMICAL MACHINING. Removal of material from a surface by chemical etching.

CHEMISORPTION. Chemical bonding of an atom or molecule to the surface of a material.

CHEVRON CRACKS. *See* Centerburst cracks.

CHILL. A metal insert or piece, used in casting, to increase the rate of heat removal from the solidifying metal to avoid porosity.

CHIP BREAKER. Notch, groove, or a step on a cutting tool to break chips during machining.

CHROMEL. An alloy of chrome and nickel, used in thermocouples.

CHROMIZING. Forming of an alloy of chromium and iron by diffusion on the surface of steel.

CHUCK. A device for holding workpieces.

CIM. Computer-integrated manufacturing.

CLADDING. Deposition or bonding of metal on a surface by various means.

CLAY. A fine-grained natural material that can be shaped when moist, usually composed of alumina silicates and other materials; it becomes hard and brittle when fired.

CLEARANCE. The space between two mating parts.

CLEAVAGE. Fracture by splitting along a crystallographic plane.

CLIMB (DOWN) MILLING AND GRINDING. Cutting operation in which the tangential velocity of the tool is in the same direction as the workpiece movement.

CLOSED-DIE FORGING. Forging with dies that restrict the flow of the metal to the die cavities.

CLUSTER MILL. A rolling mill in which the smaller working rolls are supported by a number of larger diameter rolls.

CNC. Computer numerical control.

COALESCENCE. Joining or growing together.

COATED ABRASIVES. Flexible paper, cloth, or similar materials coated with abrasives, e.g., sandpaper, emery paper, and belt.

COGGING. Successive deformation of a bar along its length by open-die forging.

COHESION. Force of attraction between atoms or molecules.

COIL BREAKS. Creases or ridges on sheet or strip 90 degrees to rolling direction.

COINING. A cold squeezing operation in which all surfaces are confined.

COLD SHUT. A section of a forging or other process where the material is folded over itself, usually with an oxide layer in between.

COLD WELDING. Joining of metals by pressure at room temperature.

COLD WORKING. Plastic deformation at room temperature.

COLLET. A split sleeve to hold the workpiece during machining or grinding.

COLLOID. A dispersion of small particles (less than $0.5\ \mu m$) usually in a fluid.

COLUMNAR STRUCTURE. A coarse structure of parallel columns of grains with their axes perpendicular to the surface of the workpiece.

COMMERCIAL BRONZE. Brass (90 % Cu, 10 % Zn).

COMMINUTION. Pulverization by mechanical means.

COMPACT. A part made of compressed powder, metal, or other elements.

COMPACTED-GRAPHITE CAST IRON. Cast iron in which graphite is in the form of short thick flakes with rounded extremities and undulating surfaces.

COMPOSITE MATERIAL. Material that contains two or more different materials separated by a distinct interface.

COMPOUND DIE. A die in which more than one operation is performed with one stroke of the press.

COMPUTER-AIDED DESIGN. Design process in which data from design is fed directly into a computer, usually involving computer graphics but eliminating blueprints and punched tapes.

COMPUTER-AIDED MANUFACTURING. Manufacturing system in which computers are used to assist in the manufacture of a part, usually interfaced with CAD.

COMPUTER-INTEGRATED MANUFACTURING. Coordinated participation of computers in all phases of manufacturing, from design to automatic testing of the final product.

COMPUTER NUMERICAL CONTROL. A system in which a minicomputer is an integral part of the control panel of a metalworking equipment.

CONSTANTAN. An alloy of nickel and copper, usually used for thermocouples.

CONSTITUENT. A micrographically distinguishable part of an alloy or mixture.

CONTINUOUS PHASE. The phase that forms the matrix in which other phases are present as isolated units.

CONVERSION COATING. A compound of the surface of metal produced by chemical or electrochemical treatment usually for lubrication.

COOLANT. A metalworking fluid where the main function is to cool the tool (or die) and workpiece interface.

COPE. The upper half of a mold in casting.

COPOLYMER. A polymer with more than one type of mer.

COPPER–NICKEL. An alloy of copper, nickel, and iron.

CORE. A preformed sand aggregate inserted in a mold to shape the interior or that part of a casting that cannot be shaped by the pattern. Also, that portion of a heat-treated part that is not a part of the case.

CORING. Variable composition due to the solidification characteristics of an alloy. Also, a central cavity at the butt end of extrusions, also called extrusion pipe.

CORROSION. Surface alteration by chemical attack.

CORUNDUM. Natural aluminum oxide (an abrasive).

COST EFFECTIVE. Relative relation between cost and benefit in a process or operation.

COUNTERBLOW HAMMER. A forging hammer in which the ram and the anvil are simultaneously driven toward each other.

COUNTERBORING. Generating a flat bottomed hole by machining.

COUNTERSINKING. Generating a depression around the top of a hole by machining.

COUPON. A piece of metal or workpiece from which a test specimen is removed.

COVALENT BONDING. Bonding of atoms where outer electrons are mutually shared.

CR. Cold rolled.

CRANK PRESS. A mechanical press in which the ram is actuated by a crankshaft.

CRATER. A depression on the cutting face of the tool due to wear.

CRAZING. Development of narrow, cracklike, highly deformed and voided zones in a polymer subjected to tension; same as stress crazing.

CREEP. Elongation of a material under stress over a period of time, usually at elevated temperatures.

CRIMPING. A forming operation to close in a seam or to fold part of a sheet over or around another part.

CROPPING. Cutting of large bars and sections by shearing.

CROSS-LINKING. Three-dimensional linking of adjacent polymer chains, as in thermosets.

CROSS-ROLLING. Rolling of a sheet where the direction of rolling is at a 90-degree angle to the direction of previous rolling.

CROWN. A contour on a sheet (or roll) where the thickness (or diameter) at the center is greater than the edges. Same as camber.

CRT. Cathode-ray tube.

CRUCIBLE. A high-melting-point pot or vessel for melting metals.

CRYOGENIC. Relating to low temperatures.

CRYSTAL. A solid structure in which the atoms are arranged in a repetitive pattern.

CRYSTAL LATTICE. The spatial arrangement of atoms in a crystal.

CRYSTALLINITY. Volume fraction in a polymer that has a crystalline structure, the rest of the volume being amorphous.

CUPOLA. A vertical furnace for melting metal.

CUPPING. A forming operation in which a cup-shaped part is produced from a sheet-metal blank.

CUPRONICKEL. An alloy of copper and nickel.

CURING. Forming of polymer networks by heat and pressure, usually in thermosetting plastics.

CUTOFF WHEEL. A thin abrasive wheel for parting or slotting.

CUTTING FLUID. A metalworking fluid used in machining or grinding operations.

CYANIDING. Heat treatment in which carbon and nitrogen are introduced into the surface of ferrous alloy parts.

CYLINDRICAL GRINDING. Grinding of the outer surface of a rotating cylindrical workpiece.

DAC. Digital-to-analog converter.

DAMPING. Decreasing the amplitude or duration of vibration, usually associated with energy dissipation.

DAYLIGHT. The maximum distance between the pressing surfaces of a press.

DEAD SOFT. Fully annealed.

DEAD ZONE. Stationary material in a workpiece during metalworking, especially in extrusion.

DEBRIS. Particles that become detached in a wear process; also called detritus.

DEBURRING. Removal of burrs by various means.

DECARBURIZATION. Loss of carbon from the surface of a ferrous alloy as a result of heating in a medium, usually oxygen which reacts with carbon, or during hot working.

DEDICATED. Devoted exclusively to a single application, as in dedicated machines or dedicated computers.

DEEP DRAWING. Forming of a deep-recessed part with a punch and die.

DEGASSING. Removal of gases from molten metal, usually with inert gases.

DEGRADATION. Deterioration of the properties of a material, usually in polymers.

DEGREE OF POLYMERIZATION. The ratio of the average molecular weight of a polymer to that of the mer.

DELAMINATION. Removal of thin layers of material from a surface.

DENDRITE. A branched, treelike crystal structure, usually obtained during solidification of metals.

DEOXIDIZING. Removal of oxygen from molten metal.

DETRITUS. Wear debris.

DIE BLOCK. A block of die material to be shaped into a die.

DIE BUTTON. An insert in a die that matches the punch and is used for punching and piercing operations.

DIE CASTING. A casting process in which the molten metal is forced into the cavity of a metal mold.

DIE INSERT. A removable part or liner in a die.

DIE SINKING. Generating an impression in a die by material removal by various means.

DIELECTRIC. Electrically insulating.

DIFFUSION. Movement of atoms or molecules within a material or across a mating surface.

DILATATION. Change in dimensions or volume.

DIRECT NUMERICAL CONTROL. A system in which several machines are directly controlled by a central computer.

DIRECTIONALITY. Anisotropy.

DISLOCATION. A line defect in the regular structure of a crystal.

DNC. Direct numerical control.

D.O.M. Drawn over mandrel (for tubes).

DOPING. Alloying with small amounts of elements, usually in the semiconductor industry.

DOUBLE-ACTION PRESS. A press with two independent parallel movements, e.g., an inner slide to form the part and an outer slide for the blankholder.

DOWEL. A wooden or metal pin.

DOWN MILLING OR GRINDING. Same as climb milling or grinding.

DRAFT. Difference in the original and final thicknesses of a strip, as in rolling.

DRAFT ANGLE. The angle or taper on the surfaces of a die or mold pattern to facilitate removal of the workpiece.

DRAG. The lower half of a mold in casting.

DRAW BEAD. A bead or projection in dies for controlling metal flow in sheet forming.

DRAW BENCH. A horizontal machine for drawing heavy sections of rods and tubes.

DRAW RING. A ring-shaped part of the die in deep drawing.

DRAWABILITY. The ability of a sheet metal to be drawn into a cavity by deep drawing without failure.

DRAWING. Metalworking process such as wire or rod drawing, deep drawing, etc. Tempering.

DRAWING QUALITY STEEL. Flat rolled steel sheet which can undergo forming (especially deep drawing) without defects.

DRESSING. Sharpening of abrasive grains on a grinding wheel, usually with a diamond tool.

DROP HAMMER. A forging hammer where the energy is obtained from the release of potential energy of the ram.

DROSS. The scum that forms on the surface of molten metal.

DUCTILITY. The ability of a material to undergo plastic deformation without fracture.

DUMMY BLOCK. A thick disk placed between the ram and the billet in hot extrusion to prevent overheating of the ram.

EARING. The formation of wavy edges around the top edge of a deep-drawn part.

EBW. Electron-beam welding.

ECCENTRIC PRESS. A mechanical press in which the motion is obtained through an eccentric shaft.

ECG. Electrochemical grinding.

ECM. Electrochemical machining.

EDDY CURRENT. A current induced in the body of an electrically conducting material by a transient magnetic field.

EDGING. Gathering material toward the center in open die forging.

EDM. Electrical-discharge machining.

EIA. Electronic Industries Association.

EJECTOR. A device used to eject the formed part from the die.

ELASTICITY. Material property in which the

deformation caused by a load is removed after unloading.

ELASTOHYDRODYNAMIC LUBRICATION. A condition of lubrication in which the friction and film thickness between two bodies is determined by the elastic properties of the bodies, as well as the viscosity of the lubricant.

ELASTOMER. A polymer capable of undergoing large elastic strains.

ELECTROFORMING. Making parts by electrodeposition on a removable form.

ELECTROLESS PLATING. A plating process based on catalytic reduction of metal salts in the bath; also known as autocatalytic plating.

ELECTROLYTE. A fluid that conducts electricity.

ELECTROLYTIC COPPER. Copper that has been refined by electrolytic deposition.

ELECTROPLATING. Plating by electrodeposition in an electrolyte.

ELONGATION. Increase in length in tensile testing, usually expressed as a percentage of the original gage length.

EMBOSSING. Forming a design in relief in sheet materials.

EMBRITTLEMENT. Reduction in the ductility of a material.

EMERY. A natural abrasive composed of impure aluminum oxide.

EMULSIFIABLE OIL. An oil that forms a stable emulsion in water.

EMULSION. A suspension of one liquid phase in another, such as oil suspension in water (direct emulsion) or water suspension in oil (indirect emulsion).

ENAMEL. Vitreous material applied to the surface of metals, glass, and earthenware.

ENDURANCE LIMIT. Fatigue limit.

ENGINEERING STRAIN. *See* Strain (engineering).

ENGINEERING STRESS. *See* Stress (engineering).

EP. Extreme pressure, usually as additives in metalworking fluids.

EPOXY. A thermosetting resin, also used as an adhesive.

EQUIAXED GRAINS. Grains that have approximately the same dimensions in all directions.

EQUILIBRIUM DIAGRAM. A graphical representation of the temperature and composition limits of phase fields in an alloy system under conditions of equilibrium.

EROSION. Removal of material from a surface by abrasive particles carried by a gaseous or liquid medium.

ESW. Electroslag welding.

ETCHING. Subjecting the surface of a solid to chemical or electrolytic attack, usually to observe microstructural details.

ETP. Electrolytic tough pitch (copper).

EUTECTIC. The isothermal reversible reaction of a liquid that forms two different solid phases in a binary alloy system during cooling.

EXTENSOMETER. An instrument to measure changes in the length of a specimen caused by stress.

EXTREME-PRESSURE LUBRICANT. A lubricant that imparts increased load-carrying capacity to sliding surfaces under severe operating conditions.

EXTRUSION. Forcing of a material through a die to reduce and change its cross section.

EXTRUSION PIPE. A central oxide-lined discontinuity in the butt end of an extrusion, caused by the oxidized outer surface of the billet flowing around the end of the billet and into the center of the extruded product.

FACING. Generating a flat surface perpendicular to the axis of rotation by machining.

FATIGUE. Fracture under repetitive stresses.

FATIGUE LIMIT. The maximum stress below which a material can endure an infinite number of cycles of loading.

FATTY ACID. Saturated or unsaturated acids, such as oleic, stearic, and palmitic acids, derived from natural (vegetable or animal) fats, oils, and waxes.

FATTY OIL. Animal or vegetable oils such as palm oil, tallow, rapeseed oil, castor oil, lard oil, sperm oil.

FCAW. Fluxed-core arc welding.

FCC. Face-centered cubic.

FEED. The rate at which a tool travels along the length of the workpiece, such as in machining, usually expressed as in./revolution (mm/revolution) of the workpiece, or in./tooth (mm/tooth) for milling operations.

FERRITE. A solid solution of one or more elements in body-centered cubic iron.

FERRITE BANDING. Parallel bands of ferrite aligned in the direction of plastic working.

FERROGRAPH. An instrument used to determine the size distribution of wear particles in lubricating fluids.

FERROMAGNETIC. Exhibiting magnetic permeability, hysteresis, and saturation.

FIA. Forging Industries Association.

FIBER. Pattern of directional flow of metal crystals in plastic deformation. A finely drawn polymer or glass; also carbon fiber.

FIBER REINFORCED. Plastics or other materials mixed with fibers in one or random directions.

FILLER. Material used to modify the properties of a polymer or lubricant; also used in bonds for grinding wheels.

FILLER METAL. A metal that is added in making a welded, soldered, or brazed joint.

FILLET. A radius imparted to inside meeting surfaces in a part or a die.

FINISHING. Final machining or grinding operation.

FIRE CRACKING. Cracking due to rapid heating of a metal part with residual stresses.

FIRING. Sintering process for ceramics at elevated temperatures in a furnace.

FISHMOUTHING. *See* Alligatoring.

FISHTAIL. Overlapping back end of a rolled plate due to nonuniform deformation.

FIT. The amount of clearance or interference between mating parts.

FIXTURE. A device to hold the workpiece.

FLAME HARDENING. Heating the surface of an iron–base alloy with a flame and then quenching it.

FLAMMABILITY. Ability to support combustion; high capacity for combustion.

FLANK. Surface of a cutting tool adjacent to the workpiece surface.

FLASH. Excess metal between two dies.

FLASK. A metal or wood container or frame to make and hold a sand mold.

FLEXIBLE MANUFACTURING SYSTEM. A system made up of manufacturing cells (each cell containing a robot serving several computer numerical-control machines) and an automatic material-handling system interfaced with a central computer.

FLINT. A hard quartz.

FLOATING DIE. A die mounted in such a way that it can undergo some lateral or vertical motion in its holder during forming.

FLOW STRESS. The stress required to sustain plastic deformation at a particular strain.

FLUIDITY. The capacity of a molten metal to run into and fill the mold in casting.

FLUX. A material used to prevent formation of oxides and other contaminants, or to dissolve them and facilitate their removal, in joining operations.

FLY CUTTING. Cutting with a single-tooth milling cutter.

FLYING SHEAR. A shearing tool that travels with the strip during the cutting stroke.

FMS. Flexible manufacturing system.

FOIL. Very thin metal strip.

FORGE WELDING. Welding metal at elevated temperatures with a hammer.

FORGING. Plastic deformation of materials with compressive forces.

FORMABILITY. The capability of a material to be plastically deformed into a desired shape without fracture.

FOUNDRY. An establishment for casting metals; process of casting metals.

FRACTOGRAPHY. Study of fracture surfaces with photography.

FRACTURE. Failure, breakage, or fragmentation of a specimen or workpiece.

FRACTURE STRESS. Nominal stress at fracture.

FREE MACHINING. Machining characteristics of an alloy to which certain elements have been added to improve machinability.

FRETTING. Small lateral movements between mating surfaces under pressure, resulting in corrosion and wear (fretting corrosion).

FRIABILITY. Ability to be crumbled or pulverized.

FULLERING. Reducing the thickness at the center of a part in open die forging.

FUSION. A change of state from solid to liquid; melting.

GAGE (GAUGE). The thickness of sheet or diameter of wire. Also, an instrument used to measure thickness or length.

GAGE LENGTH. The original length of that portion of a specimen within which elongation and strains are measured.

GALLING. Localized metal transfer between mating surfaces; a severe form of scuffing associated with gross damage to surfaces.

GALVANIC CELL. A cell with two dissimilar metals and an electrolyte.

GALVANIC CORROSION. Corrosion at the interface of two dissimilar metals.

GALVANIZED STEEL. Steel coated with zinc to prevent or retard corrosion.

GARNET. A red silicate mineral, used as an abrasive

GIBS. Guides for the motion of slides in machinery, usually adjustable.

GLASS. An amorphous (noncrystalline) hard and brittle solid made of silicates, oxides, and other compounds.

GLASS-TRANSITION TEMPERATURE. Temperature at which the behavior and the specific volume of a polymer or glass undergoes a change.

GLAZING. Dulling of abrasive grains. Vitreous coating on ceramics.

GMAW. Gas metal-arc welding; same as metal inert-gas welding (MIG).

GRAIN. An individual crystal in a microstructure, such as in polycrystalline metal or alloy. Also, abrasive grain or grit.

GRAIN BOUNDARY. The interface between adjacent grains with different crystal orientations.

GRAIN GROWTH. An increase in the size of grains, usually during heating of the metal.

GRAPHITE. A layered (hexagonal) crystalline structure of carbon.

GRAY CAST IRON. Cast iron in which the graphite is present in flake form.

GREASE. A lubricant composed of an oil thickened with a soap or other thickeners.

GREEN. Unsintered, as in powder metallurgy. Damp or wet, as in sand molds.

GRINDING RATIO. The ratio of the volume of material removed to the volume of wheel wear.

GROUP TECHNOLOGY. A method of coding and grouping parts based on similarities in function, structure, or method of producing them.

GTAW. Gas tungsten arc welding; same as tungsten inert-gas welding (TIG).

GUN DRILL. A long drill with passages for coolant, used for deep holes.

GUNMETAL. Bronze, used for guns.

HADFIELD'S STEEL. A high-carbon, high-manganese (10 to 14%) steel with high strength, ductility, and wear resistance.

HARD FACING. Depositing of metal on a surface to improve wear resistance; also called hard surfacing.

HARDENABILITY. The property that determines the depth and distribution of hardness induced by quenching.

HARDNESS. Resistance to permanent indentation. Resistance to scratching.

HASTELLOY. An alloy of nickel, molybdenum, chromium, and iron.

HAZ. Heat-affected zone.

HB. Brinell hardness number.

HCP. Hexagonal close-packed.

HEADING. Upsetting the end of a bar to increase its cross section, as in a bolt head.

HEAT-AFFECTED ZONE (HAZ). The portion of the base metal that has not melted during joining, but whose properties have been affected by heat.

HEAT CHECKING. Cracks on the surface of a solid metal due to cyclic heating and cooling.

HEAT TREATMENT. Heating and cooling of a metal or alloy in order to impart certain properties.

HEMMING (FLATTENING). Bending sheet metal over itself to form a smooth edge.

HERF. High-energy-rate forming.

HETEROGENEOUS. Involving more than one phase.

HIGH-SPEED STEEL. Tool steels containing tungsten, molybdenum, vanadium, cobalt, and other elements, first used in machining at high speeds.

HIGH-TEMPERATURE ALLOY. Generally an alloy of nickel, cobalt, or iron used at temperatures above 1200°F (650°C) in applications such as jet engine blades and disks, and rockets.

HK. Knoop hardness number.

HOB. A rotary cutting tool with its teeth arranged along a helical path, used in generating gear teeth.

HOMOGENEOUS. Involving only one phase.

HOMOLOGOUS TEMPERATURE. The ratio of temperature to the melting point of a material, both measured on the absolute scale.

HOMOPOLYMER. A long-chain molecule composed of only one type of mer.

HOOP STRAIN AND STRESS. Circumferential strain and stress.

HOT DIPPING. Coatings of metals by immersion in molten metal baths of zinc, tin, or aluminum.

HOT HARDNESS. Hardness at elevated temperatures.

HOT MACHINING. Machining of metals at elevated temperatures to reduce cutting forces and improve machinability.

HOT PRESSING. Compacting at elevated temperature to induce sintering.

HOT SHORTNESS. Embrittlement of a metal at elevated temperatures caused by a low-melting constituent segregated at grain boundaries during solidification.

HOT TEAR. Fracture due to hindered contraction of metal during solidification.

HOT WORKING. Plastic deformation of metals above their recrystallization temperature.

HR. Rockwell hardness number, includes scales such as HRA, HRB, HRC, etc. Also, hot rolled.

HSLA. High-strength, low-alloy steels.

HUBBING. Producing a die cavity by pressing a male die on the surface of a die block.

HV. Vickers hardness number.

HYDRODYNAMIC LUBRICATION. Lubrication characterized by the pressure developed in the fluid film between two surfaces.

HYDROGEN EMBRITTLEMENT. Loss of ductility due to absorption of hydrogen.

HYDROPLASTIC. Formable when mixed with water, as in clay.

HYSTERESIS LOOP. Area bounded between the loading and unloading paths in a stress-strain curve, indicating energy dissipation or damping.

IACS. International Annealed Copper Standard (for electrical conductivity).

IDDRG. International Deep Drawing Research Group.

IMPACT ENERGY. The amount of energy required for fracture, usually measured by Charpy or Izod test.

IMPREGNATION. Filling of a porous matrix, such as powder metallurgy parts, with a liquid material.

IMPURITIES. Undesirable elements or compounds in a material.

INCLUSIONS. Nonmetallic elements or compounds (oxides, sulfides, silicates) in a metal that are either entrapped mechanically, or are formed during solidification or by subsequent reaction within the solid metal.

INCONEL. An alloy of nickel, chromium, and iron.

INDENTATION HARDNESS. Resistance to permanent identation.

INDEXABLE INSERTS. Cutting tools of various shapes and with multiple edges, usually made of ceramics or carbides.

INDUCTION HEATING. Heating with a coil through which high-frequency current is passed; used in melting or heat treating.

INDUSTRIAL ROBOT. A reprogrammable multifunctional manipulator designed to move material, parts, tools, or other specialized devices through various programmed motions; used in manufacturing operations.

INGOT. A casting, obtained through a primary ore smelting process, that is subsequently remelted or hot worked.

INGOT IRON. Iron of comparatively high purity.

INOCULATION. Addition of substances to molten metal to form nuclei for crystallization during solidification.

INSERT. A removable portion of a die or mold. A cutting tool, usually made of carbide or ceramic.

INTERFACE. A surface that forms the boundary between phases or systems.

INTERFERENCE. The difference in lateral dimensions between mating parts before assembly, such as a shaft and a hole.

INTERGRANULAR. At or along grain boundaries.

INTERMEDIATE ANNEALING. Annealing betweeen processing steps.

INTERMETALLIC COMPOUND. A hard and brittle phase formed between two or more metals.

INTERNAL FRICTION. Energy dissipated in a solid medium subjected to vibrations or fluctuating stresses, such as damping of vibrations in gray cast iron; usually measured by the logarithmic decrement.

INVAR. A nickel-iron alloy with a very low coefficient of thermal expansion.

ION. An atom that has gained or lost one or more outer electrons.

IONIC BOND. Bonding between ions of opposite charge.

IRONING. Thinning the walls of a deep-drawn cup.

ISO. International Standards Organization.

ISOMER. Molecules with the same composition but different structures.

ISOTACTIC. A regular arrangement of side groups in a long-chain polymer molecule.

ISOTHERMAL FORGING. Forging in which the workpiece and dies are heated to the same temperature.

ISOTROPIC. Having the same properties in all directions.

IZOD TEST. Similar to Charpy test with the exception that the specimen is supported at one end only.

JACKETING. Same as canning.

JIG. A device to hold a workpiece in place and guide the cutting tool.

JOGGLE. An offset in a flat plane consisting of two bends.

KERF. The space (width) that was occupied by the material prior to cutting, usually by a heat source.

KILLED STEEL. Deoxidized steel.

KIP. Kilopound (1000 lb).

KISSING. Touching of dies or mold halves.

KNURLING. Generating surface profiles with a shaped roller, such as on metal knobs or handles.

LABOR INTENSIVE. A manufacturing system or operation in which labor and labor costs play a dominant role.

LAMINATION. Superimposed layers of firmly bonded materials.

LAND. A straight section of a die or cutting tool at its exit.

LAP. A defect on a surface where metal has been folded over, usually appearing as a seam.

LAP JOINT. A joint formed with two overlapping members.

LAPPING. A surface-finishing operation with very fine abrasives.

LASER. Light Amplification by Stimulated Emission of Radiation.

LATTICE. Arrangement of atoms in a crystal structure.

LAY. Directionality in the surface finish of a manufactured part.

LBW. Laser-beam welding.

LEACHING. Extracting an element or compound from a solid piece by preferential dissolution.

LEVELING. Flattening of rolled sheet, plate, or strip.

LIGHT METAL. A low-density metal, such as aluminum, magnesium, titanium, and beryllium.

LINEAR POLYMER. A polymer in which the mers are joined in a line with no branches.

LIVE CENTER. A center that rotates with the workpiece, usually placed on the tailstock of a lathe.

LOADING. Filling of the spaces between the abrasive grains in a grinding wheel with chips from the workpiece.

LOGARITHMIC STRAIN. True strain.

LOT. A finite quantity of a manufactured product.

LOW STRESS GRINDING. Grinding under conditions that induce low residual stresses on the surface.

LUEDER'S BANDS. Stretcher strains.

MACHINABILITY. The capability of a material to be machined with relative ease with regard to tool life, surface finish, and power consumption.

MACHINING. Removal of material by mechanical, electrical, chemical, or other means.

MACHINING CENTER. A machine equipped with automatic tool changers and designed to perform operations on different surfaces of a workpiece.

MACROSTRUCTURE. Structure visible at magnifications from 1 to 10X.

MAGNETOSTRICTION. Change in dimensions due to magnetic fields.

MALLEABILITY. The cability of a material to undergo large plastic deformation under compression.

MALLEABLE IRON. Cast iron in which carbon is present as graphite clusters.

MANDREL. A bar on which the workpiece is mounted during forming or machining.

MANUFACTURING CELL. A system in which computer numerical control machines and a robot are used to manufacture a specific part or several parts of similar geometry.

MARAGING. Heat treatment of high-strength steels by aging of martensite.

MARAGING STEELS. A group of high-nickel martensitic steels with high strength and ductility.

MARTENSITE. A metastable phase of steel produced when quenching from the austenitic condition.

MASS PRODUCTION. Production where the lot size is usually more than 100,000 parts annually.

MATRIX. The principal phase in which another constituent is embedded.

MEAN FREE PATH. Average distance between discontinuities.

MECHANICAL ACTIVATION. The initiation or acceleration of a chemical reaction by mechanical exposure of a nascent (virgin) surface, as in metal cutting.

MECHANICAL FIBERING. Anisotropy due to alignment of impurities and inclusions in the workpiece during plastic deformation.

MECHANICAL PROPERTIES. Usually strength, ductility, elastic modulus, hardness, toughness, fatigue, and creep properties.

MECHANICAL WORKING. Plastic deformation by external forces.

MEMORY. Tendency to return to original shape upon heating.

MER. The smallest repetitive structural unit in a polymer chain.

MESH SIZE. The screen number of the finest screen through which almost all the particles of a powder sample will pass.

METALLIZING. Coating with atomized molten-metal spray or by vacuum deposition.

METALLOGRAPHY. The study of the constitution and structure of metals and alloys using microscopy and various other instrumentation.

METALLURGICAL BURN. Modification of the metallurgical structure near the surface due to excessive temperature rise, as in grinding.

METASTABLE. Tending to change to a different condition.

METROLOGY. The science and technology of measurements.

MICROHARDNESS TEST. An indentation test to measure the hardness of a microconstituent, usually under very light loads.

MICROSTRUCTURE. Structure (grains and phases) as revealed by magnifications greater than 10 X.

MIG. Metal inert-gas welding.

MILD STEEL. Carbon steel with a maximum of 0.25% carbon.

MILD WEAR. A form of wear characterized by removal of material in very small fragments, sometimes as oxide particles.

MILL SCALE. Scale formed on steel during hot working and subsequently cooled in air.

MILLING. Removing material with a rotary cutting tool.

MISCIBLE. Capable of being mixed, as in emulsions.

MISMATCH. Forging with dies that are not aligned properly.

MISRUN. A casting that is not fully formed.

MIST LUBRICATION. Lubrication in which the lubricant is propelled in a gas stream.

MIXED LUBRICATION. Combination of boundary and hydrodynamic lubrication.

MODELING. The description of a process and its behavior by mathematical relationships.

MODULUS OF ELASTICITY. The ratio of stress to strain in the elastic range in tension or compression.

MODULUS OF RESILIENCE. Area under the elastic portion of a stress-strain curve, representing the energy that can be stored without permanent deformation.

MODULUS OF RIGIDITY. The ratio of shear stress to shear strain in the elastic range.

MODULUS OF RUPTURE. The maximum tensile stress at failure in bending of a beam, usually in testing brittle materials.

MONEL. A nickel-copper alloy.

MONOCRYSTAL. Single crystal.

MONOMER. A mer that is not a part of a polymer chain.

MORPHOLOGY. Shape.

MOUNTED WHEEL. A small grinding wheel that is permanently mounted on a spindle.

MPIF. Metal Powder Industries Federation.

MULLING. Mixing of sand and clay.

MULLITE. Silicate of aluminum used as a refractory.

MUNTZ METAL. Brass (60% Cu, 40% Zn).

NATURAL AGING. Aging at room temperature.

NATURAL STRAIN. True strain.

NAVAL BRASS. An alloy of copper, zinc, and tin.

NC. Numerical control.

NDE. Nondestructive evaluation.

NDT. Nondestructive testing.

NEAR-NET SHAPE. Form of a part near the final desired dimensions, as in forging.

NEAT OIL. Hydrocarbon oil with or without additives, used undiluted.

NECKING. Localized reduction of the cross-sectional area of a specimen (or part) subjected to tensile stresses during plastic deformation.

NETWORK STRUCTURE. A structure in which the crystals of one constituent are partially or entirely surrounded by envelopes of another constituent.

NEUTRAL FLAME. A gas flame in which there is no excess oxygen or fuel.

NEWTONIAN FLUID. Fluid whose shear stress is proportional to the shear strain rate.

NIB. Die insert for wire drawing.

NICHROME. An oxidation-resistant alloy of nickel, iron, and chrome.

NIMONIC. A nickel-chrome-cobalt alloy.

NIOBIUM. Columbium.

NITAL. A solution of nitric acid in alcohol, used as an etching agent.

NITRIDING. Surface hardening of iron-base alloys by absorption of nitrogen.

NLGI. National Lubricating Grease Institute.

NMTBA. National Machine Tool Builders Association.

NOBLE METALS. Metals that resist oxidation and chemical reaction, e.g., gold, silver, and platinum.

NODULAR IRON. Ductile cast iron in which the graphite is in spheroid form.

NOMINAL STRAIN. Engineering strain.

NOMINAL STRESS. Engineering stress.

NORMAL ANISOTROPY. Anisotropy normal to the plane of a sheet or plate.

NORMAL WEAR. Wear within the design limits for a specific application.

NORMALIZING. Annealing process for ferrous alloys by cooling in still air to refine or homogenize the grain structure.

NOSING. Closing or reducing the diameter of the end of a tubular shape.

NOTCH SENSITIVITY. Reduction in mechanical properties due to the presence of notches, scratches, and other stress raisers.

NUGGET. The weld metal in spot, seam or projection welding.

NUMERICAL CONTROL. A system in which actions are controlled by the direct insertion of numerical data at some point in the system; the system automatically interprets at least some of the data.

OFHC. Oxygen-free, high conductivity, for copper.

OIL CANNING. Distortion in an area of a sheet-metal part where very little contour is present.

ORANGE PEEL. Rough surface appearance on metals due to coarse grain size.

ORIENTATION. Alignment of molecules in polymers by mechanical processing such as by stretching or rolling.

ORTHOGONAL. Mutually perpendicular.

OSHA. Occupational Safety and Health Administration.

OVERHANG. Unsupported area between a die and workpiece.

OVERHEAD. Expenses not chargeable to a particular product.

OXIDIZING FLAME. A gas flame with excess oxygen.

PACK ROLLING. Hot rolling of two or more sheets of metal.

PARISON. Hollow preform for blow molding of plastics.

PARTING AGENT. A material used to facilitate removal of the formed or cast part from the die or mold.

PARTING LINE. The line or plane separating the top and bottom dies or molds.

PATENTING. Heat treatment in wire drawing operations, for medium- or high-carbon steels, applied either prior to drawing or between drafts.

PAW. Plasma-arc welding.

PEARLITE. A lamellar aggregate of ferrite and cementite.

PEENING. Mechanical working of surfaces with shots.

PERFECTLY PLASTIC MATERIAL. A material with a constant flow stress, i.e., nonstrain-hardening.

PEWTER. An alloy, usually tin and lead, used for utensils.

PHASE. A physically distinct and homogeneous portion in a material.

PH. Precipitation hardening.

PHOSPHOR BRONZE. An alloy of copper, tin, and phosphorus.

PHYSICAL PROPERTIES. Usually density and electrical and thermal properties.

PICKLING. Removing surface oxides by chemical or electrochemical reaction.

PICKUP. Transfer of metal from workpiece to die surface.

PIEZOELECTRIC EFFECT. Reversible interaction between strain and an electric field.

PIG. A cast iron used in remelting.

PIG IRON. High-carbon iron from a blast furnace, or cast iron in the form of pigs.

PILGER MILL. A hot-working process with intermittent motions to reduce the diameter and thickness of tubes.

PIPE. Cavity formed due to contraction during solidification, or due to metal flow pattern in extrusion.

PITTING. Removal of material resulting in small sharp cavities on the surface of metal.

PLANAR ANISOTROPY. Anisotropy in the plane of a sheet or plate.

PLANE STRAIN. A state of strain where the strain in one of the principal directions is zero.

PLANE STRESS. A state of stress where the stress in one of the principal directions is zero.

PLANISHING. Producing a smooth surface on metals by mechanical working.

PLASMA. Ionized gas.

PLASTIC DEFORMATION. Permanent deformation.

PLASTICIZER. A chemical used to make polymers soft and flexible.

PLASTISOL. A polymer in liquid or paste form, used in casting or rotational molding.

PLASTOHYDRODYNAMIC LUBRICATION. A condition of lubrication in which friction and film thickness are determined by plastic deformation of the bodies in contact, as well as the viscosity of the lubricant.

PLUG. A mandrel, rod, or punch.

PLUNGE GRINDING. Grinding of cylindrical surfaces with feed in the radial direction only.

P/M. Powder metallurgy.

PMMA. Polymethylmethacrylate.

POINT DEFECT. Crystal imperfection involving one or a few atoms.

POINTING. Producing a tapered end by swaging.

POISSON'S RATIO. The absolute value of the ratio of the lateral to longitudinal strains.

POLYCRYSTALLINE. Structure with more than one crystal, and hence with grain boundaries.

POLYMER. A nonmetallic molecule made of a number of mers connected in a chain; a plastic.

POLYMORPHISM. Allotropism.

PORE. A small void or cavity in a solid body.

PORTHOLE DIE. A multiple-section extrusion die for extruding hollow shapes.

POWDER METALLURGY. The process of producing solid metal parts by compaction and sintering of metal powders.

PPM. Parts per million.

PRECIOUS METALS. Same as noble metals.

PRECIPITATION HARDENING. The process of hardening an alloy by heating it for the purpose of allowing a structural constituent to precipitate from a solid solution, as in stainless steels.

PREFERRED ORIENTATION. Orientation of crystallographic planes along the direction of maximum strain in plastic deformation.

PREFORMING. Preliminary forming or shaping operation.

PREPREG. Impregnation and coating of reinforcing fibers with polymers for subsequent use.

PRESINTERING. The heating of a powder metallurgy component to a temperature lower than the final sintering temperature, usually for ease of handling.

PRESS FIT. An interference fit or force fit made in a press.

PRESSWORKING. General sheet forming operations in presses.

PRINCIPAL STRESSES. Normal stresses on three mutually perpendicular planes on which there are no shear stresses.

PROCESS ANNEALING. Softening of a material by various treatments to improve workability during further processing.

PROGRESSIVE DIE. A die in which two or more sequential operations are performed with one stroke.

PROPORTIONAL LIMIT. The limit on a stress-strain curve where stress and strain are proportional.

PTFE. Polytetrafluoroethylene (Teflon).

PUCKERING. Wrinkling of a cup in deep drawing.

PUNCH PRESS. Mechanical press.

PVC. Polyvinyl chloride.

PYROMETER. A device to measure high temperatures.

QUARTZ. The mineral SiO_2, commonly colorless and transparent.

QUASI-HYDRODYNAMIC LUBRICATION. A regime of thin-film lubrication, usually observed in metalworking.

QUENCH CRACKING. Fracture of a metal piece during quenching from elevated temperatures.

QUENCHING. Rapid cooling in water or oil.

QUILL. A tubular shaft or spindle.

R_a. Arithmetical mean value.

R_q. Root-mean-square average.

RAM. A moving member to which a die or punch is attached.

RARE EARTH. Any group of metals with atomic numbers 58 through 72.

RARE GASES. Argon, helium, krypton, neon, radion, and xenon.

RD. Rolling direction.

REAL TIME. The ability of a computer to function and control a process as that process occurs.

RECARBURIZE. Carburizing in order to return to a metal surface the carbon lost during processing.

RECOVERY. Removal or reduction of the effects of cold working without motion of the grain boundaries.

RECRYSTALLIZATION. Formation of new, strain-free equiaxed grains from a cold-worked structure, usually by heating to or above the recrystallization temperature of the metal.

RED BRASS. Brass (85% Cu, 15% Zn).

RED SHORTNESS. Hot shortness.

REDUCING FLAME. A flame with excess fuel.

REDUCTION. Change in the thickness or cross section in processing. Removal of oxygen from an oxide.

REDUCTION OF AREA. The ratio of the change in the cross-sectional area to the original area in a tension test.

REDUNDANT WORK. Energy dissipated in nonuniform deformation during plastic working.

REELING. Straightening of round bars with concave rolls.

REFRACTORY. A material with a very high melting point, hence suitable for high-temperature applications, such as refractories for furnaces, or refractory metals and alloys for jet engines or rockets.

REFRACTORY METAL. A metal with a very high melting point (above 4000°F, 2200°C), usually tungsten, tantalum, molydbenum, or colombium (niobium).

REHBINDER EFFECT. Modification of the mechanical properties at or near the surface of a solid, attributable to interaction with a surfactant.

RELEASE AGENT. Parting agent.

RELIABILITY. The ability of a product to serve its intended function for a specified period of time.

RELIEF ANGLE. An angle on a tool or die to provide clearance for the workpiece at its exit.

RESIDUAL STRESS. Stress that remains in a body after removal of external forces (in plastic deformation), or due to thermal gradients (in casting or heat treatment).

RESILIENCE. Elastic energy released upon unloading.

RESIN. Unfilled and unreinforced natural or synthetic polymer.

RESINOID WHEEL. A grinding wheel in which the bond is a thermosetting resin, usually phenolic.

RESTRIKING. Obtaining final dimensions or correcting shapes with one or more additional blows, usually in forging or coining.

RHEOLOGY. The study of the deformation and/or flow of matter in terms of stress, strain, temperature, and time.

RIA. Robot Institute of America.

RIM. Reaction-injection molding.

RIMMED STEEL. Steel produced from an ingot with a case (rim) that is free of voids.

RINGING. Wear of wire drawing dies. Also, tapping of an abrasive wheel to determine the presence of cracks.

RISER. A reservoir in a mold providing additional molten metal to the casting.

RMS. Root mean square.

ROBOT. *See* Industrial Robot.

ROLLER LEVELING. Removal of warpage and distortion in a strip by passing it through staggered rollers.

ROUGHING. Machining or grinding without consideration of surface finish, usually followed by finish machining or grinding.

ROVING. A slightly twisted strand of fibers.

RUNNER. A channel through which molten metal flows in casting.

RUNNING-IN. The process by which machine parts improve in conformity, surface topography, and frictional compatibility during the initial stage of use.

RUNOUT. Total variation, in an axial or radial direction, of a surface of revolution from a reference surface.

RUST. Hydrated oxides of iron.

SAE. Society of Automotive Engineers.

SAFETY GLASS. Glass laminated with thin plastic sheet.

SALT BATH. A bath of molten salts used for heating steels during heat treatment.

SAMPE. Society for the Advancement of Material and Process Engineering.

SANDWICH ROLLING. Rolling two or more strips of metal together.

SAW. Submerged arc welding.

SCAB. A defect on the surface of a casting in the shape of a raised surface blemish.

SCALE. Thick layer of oxide formed at elevated temperatures.

SCALPING. Removing surface layers from ingots, billets, or slabs by mechanical means.

SCARFING. Cutting off surface projections, usually with a gas torch.

SCORING. Severe scratching. A wedge-type, longitudinal indentation on sheet metal to aid in subsequent tearing, as in flip-top beverage cans.

SCRAP. Discarded metal, usually to be remelted.

SCREW STOCK. Metal in the form of wire or rod used for making screw machine products.

SCUFFING. Localized damage caused by solid-phase welding between sliding surfaces.

SEAM. A cracklike unwelded fold or lap on the surface of metal, usually by rolling or forging.

SEASON CRACKING. Cracking over a period of time due to residual stresses and corrosion, usually in brass.

SEGMENTED DIE. A die made of many components that can be separated for removal of the workpiece.

SEGREGATION. Nonuniform distribution of alloying elements, impurities, and inclusions.

SEIZURE. Galling at an interface, hindering motion.

SELF-LUBRICATING. Showing low friction without application of a lubricant.

SEM. Scanning electron microscopy.

SEMICONDUCTOR. A material whose electrical conductivity can be controlled.

SEMIKILLED STEEL. Steel that is partially deoxidized.

SEVERE WEAR. Removal of material in relatively large fragments.

SFSA. Steel Founders' Society of America.

SHAKEOUT. Removing castings from a sand mold.

SHANK. That portion of a tool by which it is held, such as the shank of a drill held in a chuck.

SHAVING. Removal of a thin layer of material with a sharp cutter.

SHEAR DIE. Square die for extrusion, i.e., die angle is 90 degrees.

SHEAR FACTOR (m). Ratio of frictional shear stress to shear yield stress.

SHEAR LIP. A narrow, slanting ridge along the edge of a fracture surface; as in a tension test specimen.

SHEAR MODULUS. Modulus of rigidity.

SHEAR PLANE. A narrow zone along which shear takes place in metal cutting.

SHEAR THINNING. A decrease in viscosity of a lubricant with an increase in shear stress or time.

SHIM. A thin piece of material used to make adjustments in dies and in assembly of components.

SHOE. A metal block to support parts, or to retain slag and molten metal in electroslag welding.

SHORTNESS. Brittleness.

SHOT. Small spherical metal particles.

SHOT BLASTING. Shot peening, usually to remove scale or deposits.

SHOT PEENING. Cold working of the surface of a metal with a stream of shot at a high velocity, usually to improve the fatigue strength of the part by inducing compressive residual stresses on the surface.

SHRINKAGE CAVITY. Porosity in a casting due to solidification shrinkage.

SI. Système International d'Unites.

SILVER SOLDER. Silver-base brazing alloy containing silver, copper, and other elements.

SINKING. Drawing sheet metal or tube into a cavity with no inside tooling.

SINTERED CARBIDE. Cemented carbide.

SINTERING. Bonding of adjacent particles in a pressed powdered material or in a matrix by heating.

SIZE EFFECT. Dependence of properties on size of the part or specimen.

SIZING. A finishing operation to obtain final shape and dimensions of a part.

SKIMMING. Removing or holding back slag or dirt from the surface of molten metal before or during pouring.

SKIN. A thin surface layer that is different in structure and properties from the bulk.

SKIVING. Removal of material from a cylindrical body with an angled cutting tool.

SKULL. A thin cylindrical shell left in the container in extrusion, consisting of the oxidized surface of a billet.

SLAB. A block of metal with a rectangular cross-section shaped by a rolling operation; it may then be rolled into plates and sheets.

SLAG. Nonmetallic material produced during reduction of ores or welding processes by a combination of impurities in the molten metal and fluxes.

SLAG INCLUSIONS. Nonmetallic inclusions entrapped in solid metal.

SLIP. Plastic deformation by shear along a crystallographic plane. Also a suspension of ceramic particles in a liquid.

SLIP PLANE. Crystallographic plane along which slip takes place.

SLIP SYSTEM. A combination of a slip plane and its direction.

SLUG. A short piece of metal to be formed.

SLURRY. A suspension of particles in a liquid.

SMAW. Shielded metal arc welding.

SME. Society of Manufacturing Engineers.

SMELTING. Refining of metal from ores by separation in fused form from nonmetallic materials.

SOAKING. Holding of an ingot or part at a certain constant temperature for a long period of time to effect homogenization of structure or composition.

SOAP. Reaction product of fat or fatty acids with metal hydroxide (sodium or potassium), used for boundary or thick-film lubrication in metalworking.

SOLDERING. Joining solid metals with low-melting-point alloys.

SOLID-FILM LUBRICATION. Lubrication with application of solid lubricants such as graphite or molybdenum disulfide.

SOLID LUBRICANT. Any solid used as a powder or thin film on a surface to reduce friction and wear.

SOLID SOLUTION. A single homogeneous solid phase containing two or more chemical species.

SOLUBLE OIL. Oil that can be mixed with water as a dispesion (emulsion).

SOLUTE. The minor component of a solution.

SOLVENT. The major component of a solution.

SPALLING. Cracking or flaking of surface layers.

SPARK HARDENING. Formation of a thin film of carbide on a metal surface by vibrating a carbide rod against the surface while an electric arc is struck between them.

SPE. Society of Plastics Engineers.

SPECIFIC ENERGY. Energy per unit volume expended in material forming or removal.

SPEED CRACKING. Circumferential cracking of workpiece in hot extrusion.

SPHEROIDIZING. Heating and cooling of steel to produce carbide in spheroidal or globular form.

SPINDLE. Shaft to hold cutters or grinding wheels.

SPINELS. Group of hard crystalline minerals consisting of oxides of magnesium, aluminum, and other elements.

SPLAT COOLING. Extremely rapid cooling of a liquid alloy by propelling it at high speed against a heat-conducting surface, resulting in an amorphous alloy.

SPLIT DIE. A die that can be opened to remove the workpiece.

SPOT FACING. Machining a flat seat in a hole for bolts and nuts.

SPREAD. Lateral expansion of workpiece in plastic deformation, as in rolling of plates.

SPRINGBACK. Elastic recovery of a material after being processed, as in bending.

SPRUE. The channel connecting the pouring basin to the runner in a mold.

SQUEEZE CASTING. A process combining casting and forging.

STAMPING. A general term for pressworking operations on sheet metal.

STATIC FATIGUE. Delayed fracture due to stress corrosion, usually in glass and ceramics. Also, a form of hydrogen embrittlement of metals.

STECKEL ROLLING. Rolling of metal by pulling it through idling rolls.

STEEL. An iron-base alloy containing a maximum of 1.7% carbon.

STELLITE. A cobalt-base superalloy.

STERLING SILVER. An alloy of 92.5% Ag, 7.5% Cu.

STICKING. The phenomenon at an interface where the frictional stress is equal to the shear yield stress of the softer metal.

STICK-SLIP. A jerky motion between sliding members due to the formation and destruction of junctions.

STORED ENERGY. That portion of the energy of plastic deformation that remains in the metal in the form of lattice defects.

STRAIN (ENGINEERING). The ratio of the change in dimension to the original dimension.

STRAIN (TRUE). The natural logarithm of the ratio of the final to original dimensions.

STRAIN AGING. Aging induced by plastic deformation at room temperature.

STRAIN ENERGY. Energy expended in plastic deformation.

STRAIN HARDENING. Increase in strength and hardness as a result of plastic deformation, same as work hardening.

STRESS (ENGINEERING). The ratio of load to the original cross-sectional area.

STRESS (TRUE). The ratio of load to the instantaneous cross-sectional area.

STRESS-CORROSION CRACKING. Season cracking.

STRESS CRAZING. Formation of cracklike regions in plastics when subjected to tensile stresses.

STRESS RAISERS. Changes in the contour of a structure or discontinuities that cause local increases in stress.

STRESS RELIEVING. Heating to a certain temperature to reduce or eliminate residual stresses.

STRESS WHITENING. Change in color of plastics when subjected to tensile stresses.

STRETCH LEVELING. Removal of warpage and distortion by stretching a formed part.

STRETCHER STRAINS. Elongated markings on sheet metal due to localized yielding, usually in mild steel exhibiting yield-point elongation; also known as Lueder's lines or bands.

STRIATIONS. Progression marks on a fatigue fracture surface, also called beach marks.

STRINGER. Elongated impurities or inclusions (oxides and sulfides) in wrought metals in the direction of plastic deformation.

STRIP. Narrow sheet metal.

STRIPPING. The removal of a formed part from dies.

STRUCTURAL DAMAGE. Defects produced within a material during plastic deformation.

STRUCTURAL STEEL. Steel used for structural purposes.

STUFFING BOX. A box filled with powdered soap used to coat wire prior to wire drawing.

SUBGRAIN. A portion of a grain that is oriented differently from the neighboring portions of the same grain.

SUBSTRATE. Layer of material below a coating.

SULFO-CHLORINATED OIL. Lubricant containing sulfur and chlorine as additives.

SUPERALLOY. An alloy for high-temperature service.

SUPERCOOLING. Cooling below the temperature at which an equilibrium phase transformation can take place, without actually obtaining transformation.

SUPERFINISHING. An abrasive finishing process to remove surface irregularities such as grinding and chatter marks.

SUPERPLASTICITY. Capability of a material to undergo large uniform strains before necking and failure.

SURFACE TEXTURE. Geometric and physical characteristics of a surface.

SURFACING. Deposition of metal on surfaces to impart special properties.

SURFACTANT. A chemical substance characterized by a strong tendency to form adsorbed interfacial films.

SWEATING. A soldering technique in which the mating parts are precoated with solder and then joined together with additional solder.

S.W.G. Steel-wire gage.

SYNTHETIC FLUID. Lubricant that does not contain any mineral oil.

TANDEM. Two or more pieces of metalworking equipment arranged in a straight line for continuous operation, as in tandem rolling or drawing.

TAPPING. Removing molten metal from a furnace.

TEM. Transmission electron microscopy.

TEMPER. Condition of an alloy, such as annealed, cold-worked, heat-treated.

TEMPERED GLASS. Glass with residual compressive stresses on its surfaces.

TEMPERING. Thermal treatment of quenched steel to relieve residual stresses, reduce hardness, improve toughness and ductility.

TENSILE STRENGTH. Ultimate tensile strength.

TEXTURE. The state of crystal orientation; same as preferred orientation.

THERMAL FATIGUE. Fracture resulting from cyclic thermal stresses and temperature gradients.

THERMOCOUPLE. A device with two wires to measure temperature.

THERMOPLASTIC. A polymer of long-chain molecules not interconnected by strong bonds that softens when heated and hardens when cooled.

THERMOSETTING. A polymer of long-chain molecules interconnected by strong chemical bonds that does not soften to any significant extent by heating.

THICK-FILM LUBRICATION. Lubrication in which the friction and wear between the surfaces are determined by the viscosity of the lubricant. *See* Hydrodynamic lubrication.

THIN-FILM LUBRICATION. A regime between thick-film and mixed lubrication where friction and wear are determined by the viscosity of the lubricant and partly by the properties of the surfaces.

THIXOTROPY. The property of recovering consistency after a decrease due to shearing.

TIG. Tungsten inert-gas welding.

TINPLATE. Low-carbon steel coated with commercially pure tin by hot dipping or electrolysis.

TIR. Total indicator reading.

TMS. The Metallurgical Society of AIME.

TOLERANCE. Permissible variation in the dimensions of a part.

TOOL LIFE. The time that it takes for a cutting tool to reach a certain wear land, usually in minutes.

TOOL STEEL. Any high-carbon or alloy steel used to make tools and dies for metalworking; usually includes alloying elements such as chromium, tungsten, and molybdenum.

TOUGH-PITCH COPPER. Refined copper containing from 0.02 to 0.05% oxygen.

TOUGHNESS. Energy expended in plastic deformation before fracture, usually the area under the true stress-true strain curve.

TRACE. Extremely small quantity of an element in a material.

TRAMP OIL. Oil from the hydraulic or lubricating system of machines which leaks into the metalworking fluid.

TRANSFER. Moving of parts from one station to another.

TRANSGRANULAR. Across the grains.

TRANSITION TEMPERATURE. Temperature at which the ductility of a material undergoes a rapid change.

TRANSVERSE DUCTILITY. Ductility in a direction 90 degrees from the principal working direction, as in rolling.

TREPANNING. An annular cut through the thickness of a material, producing a solid cylindrical slug.

TRIBOLOGY. The science and technology of interacting surfaces in relative motion, i.e., friction, wear, and lubrication.

TRIMMING. Removal of excess material from a formed part.

TRIPLE-ACTION PRESS. A press in which three operations are performed on the same part in one stroke.

TRUE STRAIN. *See* Strain (true).

TRUE STRESS. *See* Stress (true).

TRUING. Removal of material from the circumference of a grinding wheel to restore its grinding face.

TURK'S HEAD. An arrangement of four idle rollers in forming square or rectangular sections.

TURNING. Machining on a lathe with a single-point cutting tool.

TWINNING. Crystal lattices of two parts that are related to each other in orientation as mirror images.

TYPE METAL. An alloy of lead, antimony, and tin, used in making printing type.

UL. Underwriters Laboratory.

ULTIMATE TENSILE STRENGTH. The maximum value of engineering stress in a tension test.

ULTRASONIC. Sound waves inaudible to the human ear.

UNIAXIAL STRESS. Stress applied in one direction only.

UNIFORM STRAIN. Strain prior to necking.

UNIT CELL. The smallest repetitive unit in a crystal structure.

UNS. Unified Numbering System (ASTM-SAE).

UP MILLING OR GRINDING. Opposite of climb (down) milling or grinding.

UPSETTING. Reducing the height of a piece by compression, similar to heading.

U.S.S.W.G. United States Steel Wire Gage.

UTS. Ultimate tensile strength.

VACANCY. An unoccupied position for an atom in a crystal structure.

VALUE ENGINEERING. A technique that analyzes the value of each component of a product or a system with respect to the overall operation.

VAPOR DEPOSITION. Coating formed by condensation of metal vapor or thermal decomposition.

VENT. A small opening in a mold to allow the escape of gases.

VISCOELASTIC. Exhibiting both viscous and elastic behavior.

VISCOSITY. The bulk property of a fluid, semifluid, or semisolid substance that causes it to resist flow.

VISCOSITY INDEX. A measure of a fluid's change of viscosity with temperature; the higher the index, the smaller the relative change in viscosity.

VITRIFIED WHEEL. A grinding wheel with a glassy bond, also called ceramic wheel.

VITREOUS. Glassy or glasslike.

V/O. Percent by volume.

VULCANIZATION. Treatment of rubber with sulfur to cross-link the elastomer chains and impart certain properties to the rubber.

WARM WORKING. Forming at a temperature above the ambient but below the recrystallization temperature of a metal.

WAX. A substance of plant or animal origin that is harder, more brittle, and less greasy than fats.

WEAR. Loss of material from a surface.

WELD METAL. The portion of a metal that has melted during welding.

WELDING. Joining of materials with heat and/or pressure.

WET DRAWING. Drawing of rods and wire in which the die is immersed in a lubricant.

WETTING. Spreading of a fluid or molten metal on a surface.

WETTING AGENT. An additive in fluids to reduce surface tension and facilitate spreading of the fluid over a surface.

WHISKER. A needlelike single crystal.

WHITE CAST IRON. Cast iron with large amounts of iron carbide.

WHITE METAL. A light-colored tin- or lead-base bearing alloy.

WIRE ROD. A hot-rolled coiled stock to be cold drawn into wire.

W/O. Percent by weight.

WOOD'S METAL. An alloy of bismuth, lead, tin, and cadmium.

WORKABILITY. Formability.

WORK HARDENING. Strain hardening.

WORKPIECE. A piece of material to be processed.

WROUGHT. Condition of metal that is shaped by hot or cold plastic deformation such as rolling and forging, usually from a cast state.

WROUGHT IRON. Iron that contains very little carbon and is easily workable.

YIELD STRESS. Stress at which the material yields and begins to deform plastically.

YOUNG'S MODULUS. Modulus of elasticity.

ZIRCALOY. Alloy of zinc, tin, and columbium.

Index

Index